W9-DJM-777

NARROW ROADS OF GENE LAND

The Collected Papers of
W. D. HAMILTON

NARROW ROADS
OF
GENE LAND

—

The Collected Papers of
W. D. HAMILTON

—

VOLUME 2

The Evolution of Sex

—

OXFORD
UNIVERSITY PRESS

OXFORD

UNIVERSITY PRESS

Great Clarendon Street, Oxford OX2 6DP

Oxford University Press is a department of the University of Oxford
It furthers the University's objective of excellence in research, scholarship,
and education by publishing worldwide in

Oxford New York

Athens Auckland Bangkok Bogotá Buenos Aires Cape Town
Chennai Dar es Salaam Delhi Florence Hong Kong Istanbul Karachi
Kolkata Kuala Lumpur Madrid Melbourne Mexico City Mumbai Nairobi
Paris São Paulo Shanghai Singapore Taipei Tokyo Toronto Warsaw

with associated companies in Berlin Ibadan

Oxford is a registered trade mark of Oxford University Press
in the UK and in certain other countries

Published in the United States
by Oxford University Press Inc., New York

© W. D. Hamilton, 2001

The moral rights of the author have been asserted
Database right Oxford University Press (maker)

First Published 2001

All rights reserved. No part of this publication may be reproduced,
stored in a retrieval system, or transmitted, in any form or by any means,
without the prior permission in writing of Oxford University Press,
or as expressly permitted by law, or under terms agreed with the appropriate
reprographics rights organization. Enquiries concerning reproduction
outside those terms and in other countries should be sent to the Rights Department,
Oxford University Press, at the address above

You must not circulate this book in any other binding or cover
and you must impose this same condition on any acquirer

A catalogue record for this book is available from the British Library

Library of Congress Cataloging in Publication Data
Data available

ISBN 0 19 850336 9 (Pbk)

Typeset by Footnote Graphics, Warminster, Wilts

Printed in Great Britain on acid-free paper by
Biddles Ltd, Guildford & King's Lynn

Contents

—

15
Time Like a Dripping Tap

16
The Three Queens

17
Uccello/Othello

18
Health and Horsemen

APPENDICES

Our Paper Then and Now

Further Evidence

Publisher's Note

=

The first volume of *Narrow Roads of Gene Land*, on 'Evolution of Social Behaviour', was published in 1996 and contains all of Hamilton's publications prior to 1981—a set especially relevant to social behaviour, kinship theory, sociobiology, and the notion of 'selfish genes'.

The format of this second volume, devoted to the middle third of Hamilton's life's work, on sex and sexual selection, follows that of the first: each paper is restyled and republished, with corrections made for the book by Hamilton, and is preceded by an autobiographical introduction. The papers follow mostly chronologically from the 15 in Volume 1 and include every major publication of Hamilton's up to 1990 and also one from 1991.

Some of the introductions were much revised by Hamilton during 1999, and at the time of his death in March 2000 he had left the book in an almost-finished state. Sarah Bunney, who edited Volume 1, was responsible for the editing of Volume 2, and she and Hamilton collaborated during 1999, their work continuing until just before Hamilton's departure for the Congo in late December. During that time he was able to check most of the editing. He had earlier read the first proofs of all but one of the reset papers. After Hamilton's death, Sarah Bunney was greatly assisted in the resolution of the remaining textual and reference queries by Mark Ridley. Help from Ruth Hamilton was also invaluable at this stage.

Volume 1 contains a frontispiece illustration at the beginning of each of the 15 chapters. Although Hamilton had prepared some notes on possible frontispieces for Volume 2, it was decided not to embark on the considerable search for the images that would have been necessary to do justice to his wishes.

The proofs for Volume 2 were checked by Sarah Bunney and Mark Ridley, and the two indexes were prepared by Ann Parry. Jeremy John, who prepared the index for Volume 1, is at present engaged in the creation of a W. D. Hamilton archive at the British Library.

It is planned in due course to publish a third and final volume of *Narrow Roads of Gene Land* to contain all of Hamilton's remaining papers but without, sadly, his introductions.

William Donald Hamilton, FRS (1936–2000).
(James King-Holmes/Science Photo Library)

Foreword

The following eulogy was delivered by
RICHARD DAWKINS
at the Memorial Service held for
William Donald Hamilton FRS (1936–2000)
in the Chapel of New College, Oxford, on 1 July 2000

———

THOSE of us who wish we had met Charles Darwin can console ourselves: we may have met the nearest equivalent that the late twentieth century had to offer. Yet so quiet, so absurdly modest was he that I dare say some members of this college were somewhat bemused to read his obituaries—and discover quite what it was they had harboured amongst them all this time. The obituaries were astonishingly unanimous. I'm going to read a sentence or two from them, and I would add that this is not a biased sample of obituaries. I am going to quote from 100 per cent of the obituaries that have so far come to my notice (my italics throughout).

> Bill Hamilton, who has died aged 63 after weeks in intensive care following a biological expedition to the Congo, was *the primary theoretical innovator in modern Darwinian biology, responsible for the shape of the subject today*
> (Alan Grafen in *The Guardian*).

> … *the most influential evolutionary biologist of his generation*
> (Matt Ridley in *The Daily Telegraph*).

> … *one of the towering figures of modern biology*
> (Natalie Angier in *The New York Times*).

> … *one of the greatest evolutionary theorists since Darwin*. Certainly, where social theory based on natural selection is concerned, he was easily our deepest and most original thinker
> (Robert Trivers in *Nature*).

> … *one of the foremost evolutionary theorists of the twentieth century*
> (David Haig, Naomi Pierce, and E. O. Wilson in *Science*).

> A *good candidate for the title of most distinguished Darwinian* since Darwin
> (my offering, in *The Independent*, reprinted in *Oxford Today*).

... one of the leaders of what has been called 'the second Darwinian revolution'

> (John Maynard Smith in *The Times*; Maynard Smith had earlier called him, in language too informal to be repeated in *The Times* obituary, 'The only bloody *genius* we've got.')

Finally:

> All his life, Bill Hamilton played with dynamite. As a boy, he nearly died when a bomb he was building exploded too soon, removing the tips of several fingers and lodging shrapnel in his lung. As an adult, his dynamite was more judiciously placed. He blew up established notions, and erected in their stead an edifice of ideas stranger, *more original and more profound than that of any other biologist since Darwin*
>
> (Olivia Judson in *The Economist*).

Admittedly, the largest gap in the theory left by Darwin had already been plugged by R. A. Fisher and the other 'neo-Darwinian' masters of the 1930s and 40s. But their 'Modern Synthesis' left a number of important problems unsolved—in many cases even unrecognized—and most of these were not cleared up until after 1960. It is certainly fair to say that Hamilton was the dominant thinker of this second wave of neo-Darwinism, although to describe him as a solver of problems somehow doesn't do justice to his positively creative imagination.

Bill frequently would bury, in throwaway lines, ideas that lesser theorists would have given their eye teeth to have originated. He and I were once talking termites at coffee time in the Department of Zoology. We were especially wondering what evolutionary pressure had driven the termites to become so extremely social, and Hamilton started praising 'Stephen Bartz's Theory'. 'But Bill,' I protested, 'that isn't Bartz's theory: it's your theory. You published it seven years earlier.' Gloomily, he denied it. So I ran to the library, found the relevant volume of the *Annual Review of Ecology and Systematics*, and shoved under his nose his own buried paragraph. He read it, then conceded, in his most Eeyorish voice, that, Yes, it did appear to be his own theory after all. 'But Bartz expressed it better.' As a final footnote to this story, among the people whom Bartz acknowledged in his paper, 'for helpful advice and criticism' was—W. D. Hamilton!

Similarly, Bill published his theory of the sex ratio of honeybees, not in a Letter to *Nature* devoted to the topic, as a normally ambitious scientist would have done, but buried in a review of somebody else's book. This book review, by the way, carried the unmistakably Hamiltonian title, 'Gamblers since life began: barnacles, aphids, elms'.

If his famous theory of kinship occupied the early part of Bill's career, the second half was dominated by his obsession with parasites, and his theory of how they might answer the greatest outstanding riddle of Darwinism—the problem of the existence of sex. And, as a spinoff, his theory of sexual selection, proposed and tested in collaboration with Marlene Zuk—who, by the way, has flown all the way from Los Angeles to be here today. Bill would have been very touched.

The two towering achievements for which Hamilton is best known were the genetic theory of kinship, and the parasite theory of sex. But, alongside these two major obsessions, he also found time to answer, or play a major part in the co-operative answering of, a whole set of other important questions left over from the neo-Darwinian synthesis. These questions include:

- Why do we grow old, and die of old age?[1]

- Why do population sex ratios sometimes depart from the normally expected 50/50? In the course of this short paper[2] he was one of the first to introduce the Theory of Games to evolutionary biology—a development that was, of course, to prove so endlessly fruitful in John Maynard Smith's hands.

- Can active spite, as opposed to ordinary selfishness, be favoured by natural selection?[3]

- Why do so many animals flock, school, or herd together when at risk from predators? This paper had another very characteristic title: 'Geometry for the selfish herd'.[4]

- Why do animals and plants go to such lengths to disperse their progeny far and wide, even when the places they are dispersing to are inferior to the place where they already live? This work[5] was done jointly with Robert May, the Chief Scientific Advisor to the Government,[6] who is here today.

- In a fundamentally selfish Darwinian world, how can co-operation evolve between unrelated individuals? This work was done jointly with the social scientist Robert Axelrod.[7]

- Why do autumn leaves turn so conspicuously red or brown? In a typically audacious—yet compelling—piece of theorizing, Hamilton suspected that the bright colour is a warning given by the tree, a warning to insects not to lay their eggs on this tree, a warning backed up by toxins just as a wasp's yellow and black stripes are backed up by a sting.[8]

This last, extraordinary idea is typical of that youthful inventiveness that seemed, if anything, to increase as Bill grew older. It was really quite recently that he proposed a proper theory for how the hitherto rather ridiculed theory of 'Gaia' could actually be made workable in a true Darwinian model. At his burial on the edge of Wytham Wood near Oxford,[9] his devoted companion of his last 6 years Luisa Bozzi spoke some beautiful words over the open grave, in which she made allusion to the astonishing central idea of this paper—that clouds are actually adaptations, made by microorganisms for their own dispersal.[10] She quoted Bill's remarkable article 'No stone unturned: a bug-hunter's life and death' (*Times Literary Supplement*, 11 September 1992), in which he expressed a wish, when he died, to be laid out on the forest floor in the Amazon jungle and interred by burying beetles as food for their larvae:

> Later, in their children, reared with care by the horned parents out of fist-sized balls moulded from my flesh, I will escape. No worm for me, or sordid fly: rearranged and multiple, I will at last buzz from the soil like bees out of a nest—indeed, buzz louder than bees, almost like a swarm of motor bikes. I shall be borne, beetle by flying beetle, out into the Brazilian wilderness beneath the stars.

Luisa read this, then added her own elegy, inspired by his cloud theory:

> Bill, now your body is lying in the Wytham woods, but from here you will reach again your beloved forests. You will live not only in a beetle, but in billions of spores of fungi and algae. Brought by the wind higher up into the troposphere, all of you will form the clouds, and wandering across the oceans, will fall down and fly up again and again; till eventually a drop of rain will join you to the water of the flooded forest of the Amazon.[11]

Hamilton was garlanded eventually with honours, but in a way this only underlined how slow the world was to recognize him. He won many prizes, including the Crafoord Prize and the Kyoto Prize. Yet the first volume of his disturbingly candid autobiography[12] reveals a *young* man tormented by self-doubt and loneliness. Not only did he doubt himself. He was led to doubt even whether the *questions* that obsessively drove him were of any interest to anybody else at all. Not surprisingly, this even occasionally led him to doubt his sanity.

The experience gave him a lifelong sympathy for underdogs, which may have motivated his recent championing of an unfashionable, not to say reviled, theory of the origin of human AIDS. As you may know, it was this that was to take him on his fateful journey to Africa earlier this year.[13]

Unlike other major prize winners, Bill really needed the money. He was the despair of his financial advisers. He was interested in money only for the good that it could do, usually to others. He was hopeless at accruing the stuff, and he gave away much of what he had. It was entirely characteristic of his financial astuteness that he left a will that was generous but—unwitnessed. Equally characteristic that he bought a house in Michigan at the top of the market, and later sold it at the bottom of the market. Not only did Bill's investment fail to keep up with inflation, he actually made a substantial loss, and could not afford to buy a house in Oxford on his return from Michigan. Fortunately the university had a nice house in its gift in Wytham village, and, with Dick Southwood, as ever, quietly taking care of things behind the scenes, Bill and his wife Christine and their family found a place where they could thrive.

Every day he cycled into Oxford from Wytham, at enormous speed. So unbecoming was this speed to his great shock of grey hair, it may have accounted for his numerous cycle accidents. Motorists didn't believe that a man of his apparent age could possibly cycle so fast, and they miscalculated, with unfortunate results. I have been unable to document the widely repeated story that on one occasion he shot through the windscreen of a car, landed on the back seat and said, 'Please drive me to the hospital'. But I have found reliable confirmation of the story that his start-up grant from the Royal Society, a cheque for £15,000, blew out of his bicycle basket at high speed.

I first met Bill Hamilton when he visited Oxford from London, in about 1969 to give a lecture to the Biomathematics Group, and I went along to get my first glimpse of this man, who was already my intellectual hero. I won't say it was a let-down, but he was not, to say the least, a charismatic speaker. There was a blackboard that completely covered one wall. And Bill made the most of it. By the end of the seminar, there wasn't a square inch of wall that was not smothered in equations. Since the blackboard went all the way down to the floor, he had to get on his hands and knees in order to write down there, and this made his murmuring voice even more inaudible. Finally he stood up and surveyed his handiwork with a slight smile. After a long pause, he pointed to a particular equation (aficionados may like to know that it was a version of the now-famous 'Price Equation') and said: 'I really like that one.'

I think all his friends have their own stories to illustrate his shy and idiosyncratic charm, and these will doubtless grow into legends over time. Here's one that I have vouched for as I was the witness myself. He appeared for lunch in New College one day, wearing a large paperclip attached to his glasses. This seemed eccentric, even for Bill, so I asked him: 'Bill, why are you wearing a paperclip on your glasses?' He looked solemnly at me. 'Do you really want to know?', he said in his most mournful tone, though I could see his mouth twitching with the effort of suppressing a smile. 'Yes,' I said enthusiastically, 'I really really want to know.' 'Well,' he said, 'I find that my glasses sit heavily on my nose when I am reading. So I use the clip to fasten them to a lock of my hair which takes some of the weight.' Then as I laughed, he laughed too, and I can still see that wonderful smile as his face lit up with laughing at himself.

On another occasion, he came to a dinner party at our house. Most of the guests were standing around drinking before dinner, but Bill had disappeared into the next room and was investigating my bookshelves. We gradually became aware of a sort of low murmuring sound coming from the next room. 'Help.' 'Er, Help . . . I think. Er, yes, Help! Help.' We finally realized that, in his own uniquely understated way, Bill was murmering the equivalent of 'HEEEELLLLP!!!!!!' So we rushed in, to find him, like Inspector Clouseau with the billiard cues, struggling desperately to balance books that were falling all around him as the shelves collapsed in his arms.

Any other scientist of his distinction would expect to be offered a first-class air fare and a generous honorarium before agreeing to go and give a lecture abroad. Bill was invited to a conference in Russia. Characteristically he forgot to notice that they weren't offering any air fare at all, let alone an honorarium, and he ended up not only paying for his own ticket but obliged to bribe his own way out of the country. Worse, his taxi didn't have enough petrol in its tank to get him to Moscow Airport, so Bill had to help the taxi driver as he siphoned petrol out of his cousin's car. As for the conference itself, it turned out when Bill got there that there was no venue for it. Instead, the delegates went for walks in the woods. From time to time, they would reach a clearing and would stop for somebody to present a lecture. Then they'd move on and look for another clearing. Bill wondered if this was an automatic precaution to avoid bugging by the KGB. He had brought slides for his lecture, so they had to go for a *night*-time ramble, lugging a projector along. They eventually found an old barn and projected his slides on its whitewashed wall. Somehow I cannot imagine any other Crafoord Prizewinner getting himself into this situation.

His absentmindedness was legendary, but was completely unaffected. As Olivia Judson wrote in *The Economist*, his duties at Oxford required him to give only one undergraduate lecture a year, and he usually forgot to give that. Martin Birch reports that he met Bill one day in the Department of Zoology, and apologized for forgetting to go to Bill's research seminar the day before. 'That's all right', said Bill. 'As a matter of fact, I forgot it myself.'

I made it a habit, whenever there was a good seminar or research lecture on in the Department, to go to Bill's room five minutes before it started, to tell him about it and encourage him to go. He would look up courteously from whatever he was absorbed in, listen to what I had to say, then rise enthusiastically and accompany me to the seminar. It was no use reminding him *more* than five minutes ahead of time, or sending him written memos. He would simply become reabsorbed in whatever was his current obsession, and forget everything else.

For he was an obsessive. This is surely a large contributor to his success. There were other important ingredients. I love Robert Trivers's musical analogy: 'While the rest of us speak and think in single notes, he thought in chords.' That is exactly right. He was also a wonderful naturalist—he almost seemed to prefer the company of naturalists to that of theorists. Yet he was a much better mathematician than most biologists, and he had the mathematician's way of *visualizing* the abstract and pared-down essence of a situation before he went on to model it. Although many of his papers were mathematical, Bill was also a splendidly individual prose stylist. Here's how, in the first volume of his auto-anthology, *Narrow Roads of Gene Land* (1996), he introduces the reprinting of his 1966 paper on 'The moulding of senescence by natural selection'.[14]

He first transcribes for us a marginal note that he wrote on his own copy of this paper:

> Thus ageing animal should climb *down* his evolutionary tree: young man's youthful features in trends which made *old* gorilla.

This leads his older self into a magnificently Hamiltonian set-piece:

Therefore, one last confession. I, too, am probably coward enough to
give funds for 'elixir' gerontology if anyone could persuade me that
there is hope: at the same time I want there to be none so that I will
not be tempted. Elixirs seem to me an anti-eugenical aspiration of the
worst kind and to be no way to create a world our descendants can
enjoy. Thus thinking, I grimace, rub two unrequestedly bushy eyebrows
with the ball of a happily still-opposable thumb, snort through nostrils
that each day more resemble the horse-hair bursts of an old Edwardian
sofa, and, with my knuckles not yet touching the ground, though
nearly, galumph onwards to my next paper.

His poetic imagination is constantly surfacing in little asides, even in his
most difficult papers. And, as you would expect, he was a great lover of
poets, and carried much poetry in his head, especially A. E. Housman. Perhaps he identified his young self with the melancholy protagonist of *A
Shropshire Lad*. In his review of my own first book (and can you imagine my
joy at receiving a review from such a quarter?), he quoted these lines:

> From far, from eve and morning
> And yon twelve-winded sky,
> The stuff of life to knit me
> Blew hither: here am I.
>
> Now—for a breath I tarry
> Nor yet disperse apart—
> Take my hand quick and tell me,
> What have you in your heart.
>
> Speak now, and I will answer;
> How shall I help you, say;
> Ere to the wind's twelve quarters
> I take my endless way.[15]

He ended the same review by quoting Wordsworth's well-known lines on
the statue of Newton in the Antechapel of Trinity College, Cambridge. Bill
didn't mean it this way, of course, but the last words of the poem fit *him* as
well as they fit Newton, and I want to leave you with them.

> . . . a mind forever
> Voyaging through strange seas of thought, alone.

1. 'The moulding of senescence by natural selection', *Journal of Theoretical Biology* **12**, 12–45 (1966).

2. 'Extraordinary sex ratios', *Science* **156**, 477–88 (1967).

3. 'Selfish and spiteful behaviour in an evolutionary model', *Nature* **228**, 1218–20 (1970).

4. *Journal of Theoretical Biology* **31**, 295–311 (1971).

5. 'Dispersal in stable habitats', *Nature* **269**, 578–81 (1977).

6. Now President of the Royal Society and no longer the Government's Chief Scientific Advisor.

7. 'The evolution of co-operation', *Science* **211**, 1390–6 (1981).

8. W. D. Hamilton and S. P. Brown, 'Autumn tree colours as a handicap signal, *Proceedings of the Royal Society B* **268**, 1489–93 (2001).

9. 18 March 2000.

10. W. D. Hamilton and T. M. Lenton, 'Spora and Gaia: how microbes fly with their clouds, *Ethology, Ecology and Evolution* **10**, 1–16 (1998).

11. At the Memorial Service, Luisa read both these passages herself.

12. *Narrow Roads of Gene Land*, Volume 1.

13. Bill had gone to a remote and hazardous part of the Congo, on a stubborn mission to test an unfashionable theory of the origin of human AIDS. While there he contracted severe malaria, was rushed by his two brave companions back to England, seemed to recover, then suddenly succumbed to a massive haemorrhage (which may have been independent of the malaria). He was unconscious in hospital in London for five weeks until he finally died, on 7 March 2000, leaving this book, and so much else, unfinished.

14. Chapter 3, 'Live now, pay later', pp. 83–93.

15. Read at the Memorial Service by Ruth Hamilton.

Happy is the man that findeth wisdom

This anthem, composed especially for the funeral in Westminster Abbey of Charles Darwin, was sung at the memorial service for W. D. Hamilton in the Chapel of New College, Oxford on 1 July 2000.

Proverbs iii. 13, 15–17.

J. Frederick Bridge, Mus. Doc., Oxon.

Preface

═

THIS book is the second volume of the history of an obsession, and the obsession is evolution. Ever since I first understood Darwin's idea I have applied it everywhere, attempting by it to understand all aspects of life, including, of course, human. The plan for conveying the results remains as in the first volume of *Narrow Roads of Gene Land*. Following on mostly chronologically from the 15 papers that were in that volume, every major paper I have published up to 1991 is republished and every paper provided with an introduction about what I was doing at the time and why I wrote it: I also discuss how I see the work affecting life, including our lives, more generally than I do in the paper; I tell how the work has been received, and what may have followed that bears on each idea. Writing the introductions has been harder for this volume than it was for the first because many of the ideas are still controversial. I often feel a need to go into detail and cite many publications to support a point of view: in the last volume most of the ideas had achieved a degree of stability.

The main topic of the book is why sexual ways of reproduction are so abundant, when, for natural selection to work and for progressive evolution to occur, much simpler and more efficient ways would seem to suffice. Those very efficient and non-sexual ways do, in fact, occur quite commonly. 'Greenfly' on your rose bushes (properly they are aphids and more properly still, regarding the usual culprit, *Macrosiphum rosae*) show you asexual reproduction every summer, and I think no gardener will argue as to the efficiency of their method. A live-fish-food producer also won't argue but looks on asexual ways more benignly. They are partly why he turns to the species he does to form his stocks: tubifex, whiteworm, daphnia, brineshrimp—the favourites of the trade; all these are normally asexual. If you live in the tropics and especially if on a Pacific island then vertebrates no less may also be showing the mode. The small lizards on your walls—geckos—and especially those species of habitations and human-disturbed places are quite commonly asexual. Dandelions in your lawn, blackberries draped on your fence are, habitat-wise, similar examples on the plant side, and I could name many more.

Very clearly it is not the case that simpler reproduction ways *cannot* be re-invented time and time over and, in fact, for a great many species the asexual mode seems to stand, as it were, 'near by' in the sense that closely related species or strains within a particular species may be doing it in some hole or corner (though actually in the latter case it is more usually the sexual strain that ends up in the small part of the range).

As befits a way that is simpler, all this parthenogenesis—reproduction without males—looks to be attainable from the sexual state through just a few changes in the genetic code, with perhaps a few more needed to perfect it. Geckos are vertebrates, our distant cousins—so why not us? Why doesn't asex take over, and what made the baroque peculiarity of its counterpart, sex, ever start up in the first place? Why is it that most of the organisms I have mentioned are small and there are not even any asexual large lizards, leave alone sharks, dogs, cats, or (leaving aside intense human intervention) cows?

The astonishing thing, and the scandal to biology, is that, almost 200 years after Lamarck focused our minds on change and origin in living things, and almost 150 years since Darwin showed us a broad swathe of mechanism, still we have no mechanism to give us satisfying and universally accepted answers to these questions, and there are many others. We are near to a mechanism as I believe and this field of the study of sexuality is an exciting one to be in. It is as if we are all in a wood—a wood on a hillcrest. We scramble around seeking the gap that will let us look out on the landscape of the other side, which we glimpse tantalizingly through the trees. A few rocks are seen, more misty trees—and, hurray, there's a stream; but is it all a rocky, mountainous landscape? Other glimpses are suggesting a riverine plain. Could there be two landscapes? . . . Eventually as we move on the ridge, a gap will come, and we will see through clearly. Many of us say, let's open a gap in the canopy for ourselves; some say let it be here, some there; and several groups are already busy at the work. No axe-man myself but not bad at scrambling (well, metaphorically), I move around and try to compose the picture by integrating all the glimpses I can find and I think that this way works too. In this book I will be explaining the main theories that might be out there and you will soon know what the landscape is that I am fairly sure is going to be revealed.

Those who have read my first volume will find that the characteristics of the author apparent before—notably a trait approaching to autism about what most regard as the higher attributes of our own species—have not gone away. In spite of a re-orientation that may have seemed occurring (and was in its small way) in those feelings of change in the late 1970s that I described in the Epilogue to the last volume, readers will soon realize they have still to deal with a person who, for example, believes he understands the human species in many ways better than anyone and yet who manifestly doesn't understand in any practical way how the human world works— neither how he himself fits in and nor, it seems, the conventions limiting what he is allowed to discuss. He proudly claims to centralize a concept of genetic kinship that he applies not only in biology but in anthropology and sociology: yet if you travel with this man—that is, with me—to the Amazon, you find in the wilds an irresponsible child, almost idiot savant, much more excited to be noticing a cousinship between white *Corynostylis* flowers in a spray on a high vine in the forest and a low sweet violet flower he knows from the chalk hills of Kent, than by meeting that one-in-ten-thousand man of his own country, culture, and race whom he chances upon in a riverside bar. And in that bar, how much more eagerly I talk to a local man who knows the *Corynostylis* and who knows of and can describe to me four different *qualidades* of strychnine vines tangling the branches over the local bay and who can tell me how one has large edible fruits whereas berries of all the rest are lethal and useful only for arrow poison.

At the same time, friendly as I may seem to the local naturalist, I still for weeks forget to ask him any of the things he, like everyone, wishes to be asked most—the matters more important to him even than his knowledge of the *qualidades* of *Strychnos* and all the rest that he tells me, which he feels to be a slightly trivial hobby, almost a weakness. For months I may fail to ask such a person who helps me whether he is married and has children and what he does for a living. Meanwhile the tales hinted to me by *Strychnos* and *Corynostylis* (names that I remember far more easily than that of my human informant), concerning their travels as evolving species through geological ages and across partings of continents, interest me far more than those told me, even when with fantastic yet believable detail, by the wandering Briton in the bar. Thus, where Hamlet holds up a skull and muses on the character of its inhabitant, whom he knew, I instead hold a brood-skull of a babassú palm—that is, hold its massive stone-hard but decaying fruit. I am musing on that, found by kicking apart a clump of seedling palms to see if they are genuinely separate; under the stems (which proved indeed separate) is the nut. What struggles, what silent self-sacrifice (perhaps to bruchid beetles, perhaps to agoutis, perhaps simply to the pressure of stronger companions), I wonder, may have occurred as these six brethren germinated together in a situation where the mother has ordained (through her slow evolution of the massive and hard mesocarp) that one only can possibly take forward her gift of life.

I will explain all this more in Chapter 12. The point here is that matters even of such plant sibling competition you will find me persistently believing to be highly relevant to ethical issues of, for example, human abortion. And later, how will the babassú's story illuminate that of Jacob and his brothers in the Bible . . . , or that of me and my own brothers . . . ? And so on. Fights between human brothers—for a kingship say, as so often in the history books—are common enough and they are interesting, but I have to see all these matters as culture-infected and therefore vastly more complicated than what I understand. At best I see their dramas played out in a thicket, all of it on uneven terrain way off somewhere to the side from the 'Narrow Roads' of these books—a thicket rife with stepfathers, with morals, with the doings and worries exactly of the likes of Hamlets, those parts of human stories I didn't much want to hear about in the Amazon bar. It's a thicket through which I have never seriously thought-fought or learned my way. Yet it is mostly those rule-bound half-fights that are the real business of human living, the issues that 'catch conscience', snare the interest when at random we switch on the TV. It is with such and with understanding particular human decisions that you will find me naive.

After these admissions you may feel, what is the point?—and you may right here close the book. Can this man really be as narrow as he says, you ask yourself; if so, with my own interests as they are, why read any further? Perhaps you have already scouted a bit ahead and seen where I am clearly writing *something*, indeed a lot (and all out of mostly ignorance as I admit!), about humans. But perhaps two stories will give you pause before closing, or at least give you sympathy. Possibly they may also persuade you that, crude as my views may be of those fields where I seldom go—sociology and politics, and the like—the ideas do appear, within my lifetime, to have had some predictive value.

It seems to me that in the concept of Culture we should recognize a spirit akin to some clever and determined braggart; if true, then be cautious about what he says. Of course it's not that this guy is actually a nobody or unimportant, rather that always he likes to pretend himself more powerful than he is. And part of his bragging is to diminish, where he can, status of his senior colleagues—that is, he diminishes 'instinct' or 'nature' as variously called.

I see it as Culture's very nature to put out a continual propaganda for its predominance. It is known now how autists, for all that they cannot do in the way of forging human relationships, detect better out of confusing minimal sketches on paper the true, physical 3-D objects an artist worked from, than do ordinary un-handicapped socialites; in short, autists often develop as gifted visualizers. Like these, then, and like also those colour-blind, who, as it is likewise now known, crack visual crypses and camouflage of some kinds all the better for having their 'handicap', so may some kinds of autists, unaffected by all the propaganda they have failed to hear, see further into the true shapes that underlie social phenomena.

My first story. When I was a junior lecturer at Imperial College in London in the mid-1960s, still trying to understand the evolution of altruism and sex ratio, the Royal Society of London set up a Population Study Group to 'become thoroughly well informed about the scientific aspects of important population problems' and to report to the society these causes and implications. In Britain, products of the 'baby boom' that had happened just after the Second World War were approaching adulthood and their numbers were demanding government attention; in most parts of the world a population explosion based on the spread of Western scientific technology was in full swing. To alleviate anxiety as we listened to predictions made of human doublings in extraordinarily short periods, there was much talk of the 'demographic transition', that pattern of preference and practice making for small families, begun in Western Europe and thence spreading eastwards and also to the whole of North America—in short, spreading in a mere 50 years through all of what were then the world's most affluent nations. At the time, however, and in most areas of the world, the future arrival of such a transition had to be taken on faith. In spite of encouraging far-off examples occurring in, for example, post-war Japan, it was not clear then (and I would say still isn't in, for example, Africa) that all the world was going to follow the 'Western' pattern of abundant use of birth control that made the transition possible.

It is a part of the constitutional intention of the Royal Society to be ready to give advice to the British government on anything within the scientific ambit. At the same time plain curiosity makes members of the society always keen to advise themselves. So interested fellows, perhaps in this case spear-headed by their president (of whom more in a moment), had brought together a study group of experts and arranged for them to have space and to meet once a month in the society's rooms. The RS rooms were then in Burlington House, that ornate building in Piccadilly, London, which houses also the Linnean and Geological societies and, as the best known to the public, the Royal Academy (visual arts). The group was to discuss what was happening to population, what was likely to happen, and, although any pronouncement on desirable policy was very far from its main objective, the group was to try to tease out causative threads that might be useful. Demographers, economists, public health people, and reproductive physiologists were among those invited. The chairman of the group was Lord Florey, co-Nobel laureate (with Ernst Chain) for developing penicillin. In the same year as the group was started he was made a British life peer, and simultaneously was the President of the Royal Society itself. Sir Alan Parkes, a reproductive physiologist (I had attended a few lectures from him while I was an undergraduate at Cambridge), and Dr G. A. Harrison, human geneticist, were the group's secretaries and the main organizers. My supervisor at the London School of Economics (LSE), John Hajnal, a demographer whose work was very much within the field to be covered (he had published, for example, a wide-ranging study of the demographic transition as it had occurred in Eastern Europe—indeed, I think the very idea of the demographic transition stemmed partly from this pioneer paper), was a starting member of the group.

After a few meetings John suggested my name as one who could profitably attend. Whether he meant profitably from the group's point of view or from mine I have never been quite sure. I went in the spirit that being invited was an honour and that it was indeed a subject in which I was interested, and one on which I had been accumulating facts and ideas relating to 'fitness' and reproductive altruism; and, finally, I went thinking that possibly something out of what I was thinking for myself or collecting in libraries might be useful to them. I expected from the first to be learning far more than contributing; but most who attended also spoke at least in the discussion time and I didn't suppose that I was expected to be just listener.

I remember vividly the dim-level theatre in Burlington House with its dark upholstered chairs and the dais at the front uplifting a rather grand and heavily leathered throne for the President (a reddish colour comes to my mind); oddly enough I don't remember very much of who the speakers were or what they talked about. Nor do I remember much of the subsequent discussions nor any drastic new insight coming. But was it there, perhaps, that I learned from one speaker that the humble potato brought from the Andean civilizations of America soon after Columbus might count as the prime cause of the 'spawning' populations in Western Europe, and particularly in Britain, during the industrial revolution—this introduced new crop doing more for the mass well-being that sooner or later yields babies than did the revolution itself, and more also than the advent of hygiene and scientific medicine that shortly followed?

It may have been so, and in that case there was at least one important idea. But what I certainly learned fastest from listening, was that population growth and all thoughts associated with it were obviously subjects of intense and acrimonious differences of opinion amongst everyone, whether experts or not (the 'nots' I knew about already from ordinary conversations, it was the 'experts' that were a surprise to me). Some present thought the world-wide situation very serious; others equally definitely thought it was not; some thought growth was always good for economies and for standards of living—all peoples deserved their growth spells and the 'West' couldn't tell others not to do that which, with great profit, it had done for itself; others argued passionately that this was shortsighted and the cultures would regret their present explosions and should be encouraged to slow down; some believed modern birth control was a heaven-sent answer, others seemed vaguely opposed to it; as to its methods, some appeared to believe in the one range while intensely disapproving others; and so on.

As I listened in silence I drew one general conclusion: for us to be so passionate about a topic we must be close indeed here to that centre of my actual and hoped-for expertise—biological fitness. It must be because of such a proximity to the deepest evolved roots of our psyche that no one seemed able to address the subjects of reproduction and population in a dispassionate way (I could tell from my own feelings as I listened to some of the points that ready-made passions and lack of objectivity were present in myself). Well, wasn't this all just as I should expect; wasn't it indeed a topic in which I should expect our deepest urges to be concealed almost from our very selves only in order that, in our everyday commerce with others, we would avoid being forced to expose ultimate objectives in 'everyday' discussion—not expose, that is, personal, family, class, or racial ultimate biases, rather to put on view an agreeable and softened version, a general hypocrisy, something to the effect that it doesn't matter who reproduces, that we treat all people and groups with equal favour? That we all hold, whatever our specific denomination, a pan-religious view to the effect that 'all men are brothers' when actually we know very well, deep down, it isn't true?

After sitting through several meetings in silence, I began to think it was possibly unsocial of me to remain so completely non-contributory; I felt almost as if I was a spy. Surely I ought to be able to say at least something out of my studies—all those long hours in the LSE library, my laborious extraction and differencing of figures from the heavy United Nations paperbacks on population growth, even if it had to be merely some off-the-cuff points about hiddenness of motive and potential hypocrisy such as I have just written. Those points, however, were difficult to put over and they weren't yet very clear in my own mind: for example, I foresaw justifiably irate rejoinders from psychologists challenging this or that about my naive version of the subconscious (a basically Freudian scenario of the subconscious was still very ascendant in those days; the more rational and data-supported evolutionary version still waited to begin with Bob Trivers, and his first paper was some 6 or so years in the future). Therefore I took my cue from something else.

At the beginning of one of the sessions there was a discussion about the general course of the group's programme for the future—who we should invite to talk and what about. I suggested it might be useful for us to discuss the *psychology* of population situations and to give special attention to those where closely placed or intermixed distinct groups had strikingly different rates of increase. In particular, it might be useful to consider what this might do to competitive birth rates and aggressive instincts connected with population perceptions—in fact, also with the inception of wars. There was silence as I stopped. I'd wanted to explain my thought as far as I could in words that didn't bring in my pet and as yet little accepted views about the importance of genetical kinship for human altruism and aggression. It had seemed to me that my case for the interest of this topic could be made for present purposes without that and based on known historical instances by themselves.

The silence that came surprised and unsettled me, so I added something about every one having pride in his or her family and, perhaps not wanting to see descendants lost in a sea of strangers; while, in anything like a democracy, people would be not liking to imagine their own preferences and way of life being over-ridden by decisions deriving from ways of life either—for example, not caring about the countryside, urbanizing as far as possible, and so on ('and so ons', together with a good sprinkling of 'ums' and 'ers', which I omit, were inevitable in my speech of those days, as I remember it). But I think it wasn't any of these traits that caused the silence that again followed. Then someone said something like: 'This suggestion goes a bit beyond our brief, doesn't it?' What, couldn't they see that these ideas *did* affect population issues?

In an effort to be more explicit and to be taken more seriously, I then exposed some corner of my actual work, saying something about how we were all expected, as a result of population genetical processes—natural selection in fact—to have psychological biases that wouldn't necessarily be easily visible on the surface but whose reality would come to the fore in situations where these rapid changes in a population's composition were imminent. There was a matter of within- and between-group variances involved here, this applying to the very genes that made us. It wasn't necessary to such ideas, I added, that shortages of land or whatever would be apparent right when divisive psychology took effect; it would be in this nature of the group psychology to anticipate what might be *about to happen*. (Somewhere around this point Lord Florey came to his feet.) If we really wanted to understand why population is a difficult issue to discuss and to do anything about it in the world, I continued (rather desperately and speaking into a still intense silence), it is very essential that we understand the evolutionary forces that have moulded reproductive and territorial psychology in humans—the features must be old, of course, started doubtless mainly in our Old Stone Age past. If we wanted to recommend policies to affect population trends in any direction today, we perhaps needed to discuss first the underlying motivations that all people *had* to possess—that must be there from the very fact that they themselves came from successful parentage and successful families of the past . . .

I'm sure I wasn't quite as coherent in this as I make myself appear now, but in essence that was my suggestion of a new topic. My description had come out a bit longer than I had intended because as I started I sensed the opposition the idea seemed to generate. I have put in the word 'policies' above because without it Lord Florey's response to me would be even more incomprehensible than it was: I don't recall that I used it, don't see why I should, but I might have. I cannot reproduce Florey's exact words as he interrupted me but remember something like this: 'I really must interject here and make a clarification. My companions in the Royal Society decided it would be worthwhile to set up this group so that we could seriously inform ourselves about all factors bearing on human population, a topic which seemed to be becoming a vexed one in the mind of the public. But there have to be limits, a division between facts and fantasies. If our young friend here thinks that I, as their representative, am able to persuade them to march with banners in the streets on any particular issue concerning population, he is quite mistaken. I suggest (he turned to the rest) we discard this suggestion and move on to something more serious.'

Of course, I lapsed back into my habitual silence, but I felt extremely hurt at such a response to my first attempt to say anything and I wondered yet again why anyone had thought to invite me. Florey was very obviously agitated and angry and I couldn't see why: why didn't this group even *discuss* what they thought I had said? What *had* I said? I sensed from looks a rather similar haughty annoyance in several others in the audience. Yet where had I even hinted anything to suggest banners and marching? It was true, of course, that racism was indeed an underlying theme in what I had mentioned and racists probably did sometimes march with banners; but what had that to do with the investigative spirit of my suggestion? Might it be actually I had said then and there—had he perhaps read my paper about kin selection, imagined racism somehow in that, and formed his comment accordingly? I felt, actually, that my study had indeed enabled me to understand racism better than most, and, well, if that was implied, why not discuss it—didn't a population study group need to understand so widespread a phenomenon? Wasn't it true that racism had had indeed a very big effect on population composition in Europe itself even within the past 25 years? As it was, Florey was not done with me even in silence. In summing up the decisions of that session at its end he came back again to the talk of banners and marching and his colleagues in the society being not of that type, not subject to fanciful exhortations, and how he would wish this to be kept in mind during all future meetings.

After Florey had said this the second time and as the group began to disperse I wondered whether I should go to him at once to try to find out what he imagined I had said. I waited in the hall as it emptied but saw him immediately enter conversation with two or so others and they were still talking as they passed me on their way out. As he passed me I caught a glance that might have been somewhere between curiosity and guilt, as if he knew he had been unfair; but with it there still was a stony and lofty air that promised no further discussion—also I thought later he had the look of a man who was tired. I had seen the same expression many times in the past on my professor's, Lionel Penrose's, face many times when he passed me in silence in the corridors of the Galton Laboratory, and my returned expression probably wasn't too dissimilar, perhaps betraying my thought that I, too, can do without his discussion, but won't go away. But in the case of the Royal Society Population Study Group I did soon go away. I attended for perhaps three more meetings; then I decided I had absorbed enough of all the leather smell and panelling and the (to me) superficial and quarrelsome opinions: I made an extra shadow in the dark seats no more. Later I thought how my attendance had been like my presence in my school chapel: there, too, it was awesome and dim and had become boring by repetition, and there was a silent boy. In that case, having been always totally unable to sing any sort of correct note or harmony, I would, when pressed by a prefect, open and shut my mouth silently like a fish, mouthing hymns as required, but mostly thumbing through my book and trying to find for my own inward recitation the few hymns that were passable as poetry (mighty few). In Burlington House it was as if a prefect had been watching me once again: but well, wasn't I a step further? At least this one hadn't ordered me to mouth out the beliefs I didn't hold.

That perhaps two-minute speech suggesting a topic was my only attempt to contribute anything to the Population Study Group meetings. It was the only time I opened my mouth apart from occasional hushed comments to John Hajnal, who sometimes sat beside me. I think the group itself didn't last very much longer. Shortly I learned that Florey himself was seriously ill with cancer. I felt more understanding towards him then and guessed why he had seemed tired. In all of his handling of the meetings there had been something of the same disappointing impatience as with me, although it was less intense. It seemed to me the same peevish spirit as I had noticed in Sir Ronald Fisher the last (and only second) time I had seen him at his Storey's Way edition of the Genetics Department of Cambridge; he, too, had been due to die of cancer in about two years. Fisher had been talking to his assembled tea-time group about the evidence for the link between smok-

ing and cancer—Austin Bradford Hill and Richard Doll's results, all deep
and rational but slightly high-pitched in tone. But not all elderly, sick, and
eminent men seem to have to be like that: I didn't see the same in J. B. S.
Haldane, for example, who was also pale with cancer and dying when I had
seen him lecture in University College in about 1962; perhaps I did see a
trace in Allan Wilson in Kyoto in 1989, as will come in Chapter 18. Florey
died in 1968.

The Royal Society's next president, Lord Blackett, oversaw what Florey
had begun—the society's move to its present, more magnificent and less-
shared building in Carlton Terrace, the big cream-coloured block near the
Duke of York's statue and overlooking St James's Park. I was to begin a
vastly more satisfying and respected relation with the Royal Society; in fact,
in that very year 1968 when they first financed me to go to Brazil to study
wasps and at the same time selected me to join the big Royal Society/Royal
Geographical Society expedition in Mato Grosso; 10 years later still I myself
became one of those august, never marching (and, I hope, seldom un-listen-
ing) Fellows of the society.

My second story. In the autumn of 1998, perhaps because of having now
several international prizes behind me, I found myself invited to join
another study group and this time specifically to address it. It was to be a brief
meeting in the Vatican; again there was some kind of 'society' behind the
meeting that wanted (it said) to become more informed by science. The subject was definitely not population although that topic was far from banned:
we were being encouraged to cover the whole of the philosophical ambit of
science. The Pontifical Academy of Sciences of the Vatican had called a
meeting of scientists, historians, and philosophers to discuss 'Nature': what
did this 'Nature', human or otherwise, mean and how had the idea of it
changed down the ages? How would Nature be seen in the early years of the
next millennium? Quantum physics, cosmology, genetics, evolution,
humanity, and culture were all to be included.

I thought it an excellent idea. Were the conference proceedings to be
published? This was an important point for me from the start. It wasn't
clear. The preliminary guiding documents that I and some dozen or so outsider invitees received indicated that written versions for our lectures were
indeed part of the deal—the free trip, etc.—but no promise of publication
seemed to be made (or threat as I more usually see it, because that prospect
with its thousand eyes inspecting for all time involves me in vastly more

work). I was listed in the draft programme to talk about developmental biology as affecting this topic of the nature of Nature. It wasn't my field at all and there had been some mistake. In my reply to the inviting letter I therefore pointed out my complete inexpertise on development and in general I was a bit discouraging about being able to find the time even if given a task in my real field, if a serious written contribution was expected. Father Pittau, SJ, the then organizer, wrote that they would like to have a talk from me anyway; I could suggest my own title in the general area of evolution.

I then asked myself what was the real objective of the Pontifical Academy (which to my mind, for all the claim that it stood completely apart from any ecclesiastical control, must mean the Vatican and all its hierarchy) in holding the conference. What did they expect from the face-to-face discussion? If Roman Catholics really wished to inform themselves on the state of any branch of science, it was all out there in textbooks and journals anyway; indeed, in my field of evolution I knew they already had excellent scientists within the Pontifical Academy (Luca Cavalli-Sforza and Peter Raven were examples). Why couldn't they simply consult them—why call in outsiders like me? Was it just that they wanted the *loan of names* of scientists like me that they could attach to a conference they would announce to have been held within the Vatican, and use this to reinforce amongst their own as well as outside intellectuals an impression that the Pontifical Academy, and Catholic thinking generally, was seriously in touch with science and was deemed by the outsiders to be a worthwhile scientific institution?

If that was all in the present case, I was not keen to lend. At the same time, I was aware that the Pope had seemed recently to be trying to achieve some rapprochement of his religion with science. After 300 years (!) of silence he had at last admitted to a mistake made by the Catholic Church in bringing Galileo to trial and forcing him to recant over his support for Copernicus; he had likewise regretted the burning of Bruno; not least in his last encyclical he had explicitly admitted that evolution, at least within limitations to the non-human (and especially non-mental world), certainly had occurred by natural selection just as Darwin had said. There seemed, in short, to be a genuine spirit and apology, rectification and updating abroad in the Catholic Church—a clearing of old rubbish, one may say, from its decks before its grand ship went steaming across its own (in a sense) self-made line into the new millennium.

If this second spirit, or something like it, was the real one, I thought, might the Catholic Church be receptive to further evidence that various of their present doctrines were, in fact, travesties of any 'accords with Nature'; indeed, were such frontal assaults upon the principles of evolution and ecology, including as these affected humanity, that in the long run (at least here on earth) they could not possibly persist? And could they be persuaded to see that, pending the collapse that must come, these doctrines were making the world into a much worse and less happy place for humanity to inhabit? If such openmindedness was the spirit of the day I would be glad to help. Any change in Roman Catholic doctrines that are contributing to over-population, for example, would be a big step and I would much like to con-tribute to that.

There was also the issue of human health. I knew well that it was not deep in the nature of any branch of Christian religion (or perhaps of any religion whatever) to care much about this. Because according to many religions all our physical troubles are put directly or indirectly in our way by an omnipotent Being in order to test us; and because faith and miracles supposedly can cure them; and because, on the more pragmatic plane, beliefs of such a kind held by the afflicted faithful keep their minds and donations oriented towards churches, it is easy to see that theological and pragmatic viewpoints accord quite neatly: in short, there is no good reason for religions and their ministrants to want to see better average health. The extra interest the chronically sick have in religion is evident everywhere but is especially prominent at particular centres such as Lourdes and Loreto. But not even Catholicism could want all people to be so physically incompetent that civilization, and, with it, all the splendour and effectiveness of their earthly world would have to break down. It wasn't just the ballrooms that would fall silent: the very engine rooms themselves of the grand liner Catholicism, on which decks they were effecting minor tidyings for a com-ing 'Crossing of the Line', would eventually go silent, too, and the ship split apart. To my mind this actually happening is not such a remote possibility.

Thinking in this way about what I could say about average human health as its basis was changing under present civilized and sentimental policies, with these commonly abetted by having their sentiment recast as religious dogma, it occurred to me that here was a useful thing I could possibly do for the meeting and this perhaps could be without too much time committed. One of the 'introductions' I had already drafted for this book seemed almost as if ready drafted for a paper on this topic. I could rewrite and re-orient it a bit to suit the symposium and, with an effort, have ready a publishable manuscript in good time. I was thinking of the introduction that I had already completed as a first draft for Chapter 12 of this volume. I therefore replied to Father Pittau that I could come if he would accept a talk from me on the following theme: 'Changing and evolving nature: health and human paths in the next millennium'. Father Pittau replied that the title was fine and he was pleased that I had accepted. As often happens my first choice of title soon seemed to me clumsy and by the time of the meeting I had changed it to 'Changing and evolving nature: the human and the rest'. But the idea was the same.

I intended to cover on the one hand how drastically we were indeed, in the short term, changing our external and environmental 'Nature' by the combination of our technology and our overpopulation and on the other how, in the long term, we were changing (and in this case micro-evolving for the worse) our own internal 'Nature'—that is, our own genome. To a substantial extent the latter trend was coming through our recent and unnatural ethic that every conceptus, no matter how mutated, was deserving of every technical effort we knew to make it survive. I would try to convince the conference that these two trends in the present situation of sentiment tied to ever-advancing technology had to be on a collision course from which our only escape, other than the collapse of civilization and thus collapse of the infrastructure to continue the policy, would be the advent of truly superorganismic, superhuman integration. But this, if it came, would also bring in the end attitudes towards frail human bodies that neither the Catholic Church nor irreligious people of the present day would at all like to contemplate, an opposite to that valuation of individual life intrinsic to the utopia inspiring our initial efforts. Like Alice in the looking-glass garden we did not yet realize that as we headed childlike for where we wanted to be, we actually found ourselves carried the opposite way.

You will find this argument in more detail in the introduction to Chapter 12. The manuscript I submitted to the Pontifical Academy organizers can be imagined from this introduction fairly accurately by just assuming absent from the essay all the parts that refer to Gustavus Adolphus College and the symposium held there (that was where the paper the introduction is introducing was delivered—sorry, if this is all a bit complicated) and also by imagining stripped away all the parts referring directly to the 'introduced' paper itself. The whole, it has to be admitted, has ended a rather rambling and long essay, and I also admit that the version that went to the Vatican was similar; however, when you have glanced through it I ask readers to judge for themselves whether the events that followed with my contribution at the Vatican symposium, which I now describe, were justified.

The setting for the symposium in October 1998 was delightful, even awe-some—at once intimate and grand. The Pontifical Academy had long ago been given as its headquarters a villa built in the mid-sixteenth century for the accommodation for Pope Pius IV in the gardens behind St Peter's cathedral. These gardens covered a section of hillside that sloped down towards the foundations of the cathedral. They were lovely in themselves and for the faithful visiting St Peter's (visitors who, unlike us, would only excep-tionally be allowed to wander there but who might from various points dis-tantly see them) their palms and cycads, their lawns, and their wonderfully constructed rockeries and fountains might seem like a foretaste of paradise itself. Over the gardens and our building on one side loomed the bulk of the cathedral, the world's largest church, topped by Michelangelo's soaring and perfect dome; on a second side the gardens were shut in from a view towards the city of Rome by long ranks of the windows of another almost equally massive building, the Vatican Museum. The past papal villa, the conference building, was a museum piece itself, a treasure (especially within) of antique architecture and decoration. In its central hall where we talked (and where Pope Pius IV perhaps had given audiences) now tiers of high-backed wooden seating arrayed three of the sides while provision for a screen was set against the fourth. Every seat was provided with a microphone and when in session all the seats of the room were filled. The conference was maximally attended. Perhaps there were about a hundred of us in all and of those pre-sent perhaps one quarter were showing in their clothing at least some token of ecclesiastical calling, but my impression from some types of questions and comments was that a somewhat higher proportion might have had some connection with theology.

The day after my manuscript had been handed in, it turned out that the programme for the meeting had changed. I found myself scheduled to give my talk as the first in the morning of the third day instead of on the second. I arrived that following day at the appointed 9 o'clock ready to speak but found a blackboard announcing still further rescheduling; among the new changes my talk was now put back to be second last of the morning, of what was to be the last half day of the meeting. I began to feel a bit worried and suspicious that I might have wasted my long efforts with talk and manuscript.

Next day at 9 o'clock the morning and last session was launched; and, rather as I anticipated, the earlier talks ran on in quite leisurely fashion. First, unfinished matters had to be wrapped up from the previous day. The conference had gained time from the fact that Stephen Jay Gould, scheduled in the programme as we received it on arrival, had not come. But other things running late had already eaten up his time. Now this morning, however, the organizers tried hard to fill in for Gould. Every speech was supposed to be 'commented' upon in a counter speech given by an appointee from the Pontifical Academy who had been set to study each submitted paper. Although Gould had not spoken the previous day, and no one knew what he would have said, the counter speech by his commentator seemed still to be necessary. This speech was given and after it an elderly philosopher in the audience (who, in another *ex tempore* address on a previous day had informed us he had published 49 books) gave a little speech of his own in the question time, stating his view that, as regards this question of evolution, all 'ultra-Darwinism' was now proven to be nonsense. As a valid contrast to 'ultra-Darwinism', he enlarged on the philosophic importance of the verb 'to be'—'*essere*' as it is in the Italian, he explained. When he had finished, the chairman of the morning session then added some 15 minutes of his own comments on evolution theory. Seemingly this was a further fill-in for Gould and he included mention with no detail of recent theory that he believed in (and I imagine Gould does not) 'evolution genes'—that is genes differing in function from all other classes of genes and whose specific purpose was to speed up evolution progress when such speeding was needed. At the end of this, the time for me had arrived, but just before I took the floor the chairman warned me that I now had only 20 minutes for my talk and that this would have to include the questions (the original allocation had been 45 minutes): we were desperately tight for time, he told me, and I must remember there was another speaker to come after me and before lunch and thus before the end of the presentations.

It hardly needs saying that under continual contraction of my time relative to what I had prepared for, my presentation was hurried and truncated. There was time for just one question and two comments; both the latter were hostile and dismissive of my points and there was virtually no time for me to respond to them. My appointed commentator then took the floor and told us very amiably how he so disagreed with my paper he could not try to address it; instead he would talk about historical issues in the history of science and art that had interested him; these had been raised in another talk.

We descended through the ornate passages and stairs to lunch, which, as usual, was very good. After lunch those at the meeting who were members of the Pontifical Academy entered a closed session on academy affairs and the rest of us drifted away. I left the meeting with a sensation of how almost nothing had changed in the 40 years since I had been similarly 'put down' by the President of the Royal Society when I tried to suggest something at the Population Study Group meeting as I described above. Some subjects are simply taboo in our society and they have remained so in both London and Rome for at least 50 years. In the London case I imagine I had been invited to attend the meetings because John Hajnal had billed me to the group as a 'serious young man'; now, known as a Crafoord and a Kyoto Laureate, I was often sought after for conferences like the present one—but, when the conference topic was like that of the present one, it seemed I was still only sought by people who didn't know me, and who had not been warned! Just as in London, when it became apparent what sorts of things I had to say, the organizers and chairmen, who tend to be, even among scientists, the more political types, would avoid, almost at any cost or rudeness, allowing me to be heard. Will the reader who has reached to this paragraph be different from this? I must wait to know.

Meanwhile, I believe, the warnings I try to give at those kinds of meetings are one by one coming true. The streets of Belfast in 1966 seemed a long way from the dignity of Burlington House but at the same time perhaps too close and too much of a political issue, and if I remember there was actually not much fuss in Belfast in 1966. Likewise the rounded and over-cultivated, lived-on hills in the then infant state of Rwanda were also far away and no one was worrying—and nor were they about the hills of Kosovo. Yet, now, how can it be that people can still shut their eyes to those *so obvious* causes of group strife that I had said we should discuss? Even in the past few years a book has been written (E. Sober and D. S. Wilson, *Unto Others: The Evolution of Altruism*, as cited in the introduction to Chapter 4, note 18), claiming for group selection a power almost equal to individual selection; all of the rosy side of this picture—the group amity, the advantage within the group of honest communication, and so on—is fully detailed in the book; the other side of group selection, exactly the group strife that in 1966 I was noting and saying we should consider (in 1971 I called this the 'other' or 'warfare' picture, on Achilles' shield—see the frontispiece and p. 217 in Chapter 6, Volume 1 of *Narrow Roads of Gene Land*) is given the briefest of attention and is treated as if it was largely avoidable through acculturation of disadvantaged groups into the positively selected and dominating ones.

As I write and as 'ethnic cleansing' continues in the supposedly unified nation that was Yugoslavia, I ask myself, on lines of such acculturation being a conceivable answer, is it possible for a young male Kosovar to avoid being 'cleansed' out of his life or, if luckier, simply out of his country, by his saying he wishes to convert to 'being a Serb'? Can an Arab in *de facto* Israel keep his time-honoured land on the West Bank by saying he wants to be a Jew—can he even retain it if he adds that he will scrupulously observe observances and turn up each week at the synagogue? Back again in Yugoslavia, if a woman Kosovar states that she accepts racial justice in her Serbian rapist's act, acknowledging his great genetic superiority to her husband, and that she will try henceforth to rear a good 'Serb' out of the man's (or men's) issue, is she then spared and supported?

To all this, the answer is obviously not. Far better would it be then, if the prevention of what is actually happening in these places had been started at a much more basic and philosophical level and even with scientific, dispassionate discussion in such places as Burlington House, and thus started out with evolutionary understanding of what we are. With or without believing in my particular line of theorizing, wouldn't that have been better than sending out the present mix of bombs and aid teams to Yugoslavia? Wouldn't it have been better for the United Nations, 40 years ago, to have at least been given the grounds to understand that all areas in the world where two races (or indeed any two forms of cultural identity) show strikingly different and above-replacement reproductive rates are likely soon to turn flash-points for war and genocide—and for the UN (or UNESCO) to have tried at least to warn the occupants of such regions where they are heading? I even see a role for the World Bank in the sorts of preventive inducements that could be applied: money should only be lent to the degree that receiving cultures show they are serious in reducing the inflammatory growth rates they currently support (or/and in reducing those rates likely to cause outward conflagration between their own culture and another). A

simple pragmatic ground for such refusal of aid would be that there is no
point to try to better a country economically by making it loans if all gains
are soon to be burnt up very shortly in war and massacre. But there are also
reasons that are much more ethical than this: perhaps it would be possible
for all gypsies, for instance, to disappear suddenly and 'cleanly' from a country with little economic consequence; but that doesn't make such a disappearance any less wicked and undesirable. Persuasion about foreseeable
consequences, and misery to come, is surely the route of reason and it is this
route that establishment interests seem still to attempt to deny to me.

I don't mean that I could not have got around these setbacks, and, in the
Vatican, I wasn't hauled off to be shown, as gentle persuasions for me to
stop rocking the boat of virtue, the instruments of torture in the cellars as
may have been shown to Galileo and Bruno. There were just polite good-
byes. But what, for example, should I have done after that earlier parallel
interaction with Britain's most august scientist—laureate, lord, and PRS—
in Burlington House? Should I indeed have dived back into geography, into
history, back into the library at LSE, ferreting there still further through the
UN tomes on population growth plus mining all the older books on historical demography—and, out of all this, proved my point in print, a matter of
history, no theory involved? I am sure I could have done it. But really the
point at the time seemed to me to be already too obvious to be very interesting. Besides this I was engaged in another rather similar but more intricate
endeavour that was taking up all my time. I was trying to prove myself right
against all those, including my former professor, Lionel Penrose, who had
told me that there was no such thing as genetic altruism or genetic selfishness or aggression, and that, here again, I was on a wild-goose chase.

I had already published my first big paper but it seemed to generate little
reaction. Hard to catch were the geese indeed, but they existed and were
still for me the exciting hunt. All that is described in Volume 1, including
another theme I suppose I might have been ripe at that time to offer in a
more serious and extended write up than that I provided—a sketch in a letter
to New Scientist (see note 19 of the introduction to Chapter 6 in Vol. 1, pp.
196–7). The idea was of a system of universal birth rights solving the population problem. I still think this the justest and most painless way forward
out of the population crisis as well as being a death blow to all the cruelty
that natural selection normally inflicts on our species. I could have expanded
on that. But, again, science was too interesting and I didn't feel myself cut

out for such ethicizing and politics as would be necessary for the course of convincingly explaining those ideas: please judge on my decision to leave it to others by what I write here.

Now, what should I do to reply to the event in Rome? I suppose once more the most direct answer is to gather evidence documenting the abundance, the insidiousness, and the inevitability of human deleterious mutation and write a book about it, convince people that something we don't like will indeed happen and we can either forestall it or not. But, again, the matter seems to me so obvious, so almost like offering a tedious re-explanation of the second law of thermodynamics, that I don't want to waste my time: what I do here instead is simply to re-polish my Chapter 12 essay and write this preface to enable you to picture what I had meant and what happened.

I predict that in two generations the damage being done to the human genome by the ante- and postnatal life-saving efforts of modern medicine will be obvious to all and be a big talking point of science and politics. It has taken about one generation for the trend I spoke of in that former suggestion—the matter of the strife-promoting effects of differential and above-replacement birthrates—to reach the point of obviousness that it has now. Because even that species of obviousness is not yet translated much into journalism or everyday political consciousness, it seems we might have to wait another 40 years before the world has come to the point where I feel I stood in Burlington House in 1966. Meanwhile on this other issue it seems it is going to be about the same. In 40 years civilized countries will have become uncomfortably aware of, for example, the increasing load of the intrinsically unhealthy on health services, the increasing dominance of both sport and everyday health-demanding activities by nations that have enjoyed the worst records of the practice of scientific medicine. The burden imposed by heavily 'mutationally challenged' people on the less challenged and the able-bodied will already be more obvious. But probably in 40 years most will still be saying it doesn't matter, we can cope, and that the answer to everything—now fixing up the germ-line—is just around the corner (another implied prediction here is that because of a growing need to have at least some such hope while continuing to take the technological health route, the present restriction on experimentation with human germ-line and/or stem cells and with few-cell embryos will have to be lifted very soon). Anyway, step forward by another 40 years and I believe the validity of the themes I was trying to raise in my Vatican lecture will be easily apparent to every thinking person: it will be seen that *Homo sapiens*, at least in the so-called 'first world', is indeed entering a phase of medical metastability.

Metastability—the possibility of non-self-limiting breakdown—will hold with respect to potential epidemic disease agents, to potential physical disruptions (sudden climate change, asteroid impact, and the like), and with respect to sheer physical inabilities in daily life. The servicing of an essentially machine- and electronics-based culture will be ever more liable to an escalating breakdown. After 80 years I predict people will have begun to think seriously about these points. By then, I admit, there will still be enough time to back out, although it will only be to an accompaniment of considerable suffering. One of the ways in which I think backing plus curbing of the hypocrisies of individualism will come about will be through a greater measure of *family* responsibility that political parties will see it as necessary to impose: as example, if a family wants to keep a particular vegetable baby alive, the family must pay for it. Similarly, if a church objects to the alternative—letting the baby die—extra taxes to pay for the baby's special care will be required from specifically that church's coffers, backing its beliefs. None will be demanded from another church that agrees the baby should die. In general along such lines, it will be a great step in the equitable running of modern society if a sincerity tax comes to be imposed on all propaganda—what you say you believe in you must show you believe in through hard cash and sacrifice; as example again, there should be no option but that your own child attends the idealistic comprehensive school you say you believe in.

But why not start to discuss and make decisions and try experiences now? Just as I believe Kosovars should be allowed to practise Islam in Kosovo, in the heart of Christian Europe, if they want to (including even practise the purdah and female circumcision if they want that) but *not* including any right of unlimited increase (or at least providing very strong disincentive, in contrast to the recent supposedly liberal and yet embittering spirit of toleration), so I believe that other self-defined yet quite different groups of idealists should also be allowed to practise a religion that includes parent-decided selective infanticide, provided this is done under good safeguards against cruelty. Through free choice in idealism this will become real, effective group selection on the cultural plane, hopefully no longer contaminated from its warlike predecessor out of biology.

Obviously always it will be better to terminate defective zygotes pre-emptively. It should be as soon as a precondition can be detected, before birth where possible and, of course, the earlier the better. Fortunately it is already happening and the popular and commonsense will is sweeping aside religious objections. In the background to this popular will, a huge range of fellow beings of the planet—in short, all that is now known of natural history of charming, admired, and loved animals and plants—is teaching us that there is nothing intrinsically wrong or inhumane in the course I am suggesting and that the taking of such early life has actually potential to enable us to be far more 'humane' to far more sentient and, above all, to more fearful older people—that is, children and adults—in the future. No religion, I am glad to say, has tried to teach me that a coconut palm is wicked because it forms three embryos and gives the reward of life to only one or that God slipped up with his intentions when he created this wonderful plant; nor has a religion tried to teach me that a doe rabbit is hell bound because she eats the runt of her litter. Let us allow the human trials of a like *natural* philosophy, where if anything there is so much more joy to be gained by it. The idea is emphatically not cruel in any ordinary sense of the word.

Watching, let the present probable majority who don't want such steps in their lives, record what happens. If it all turns out a disaster, if all the people in the cult are unhappy or they revert to barbarity in other ways, or if the dying embryos and babies can be shown by ingenious physiological measures to be unhappy, too, in some sort of anticipation of fate (I can't think how), and if (as is more possible) the parents are simply unhappy about what they have done, let this be published for all to judge and be warned.

On the other hand, if we refuse even to talk about such new or corrective ideas (and if, in parallel, we refuse to talk plainly about the role of increase in human 'group selection'), let us at least forgo all the present hand-wringing, the alases, and the 'what-might-we-have-done-earliers' when the physically incompetent are suddenly found to be dying in full states of human sentience, as is happening in events like those of Kosovo and Rwanda. Under a policy of disaster-acceptance, if that is how it has to be, there is, of course, no obligation for the competent person to stand back, to not help—indeed it will be hard not to help; but in such disasters she or he who does help shouldn't lament, shouldn't be stunned at what they are seeing. In my introduction to Chapter 2 in my Volume 1 (p. 18), I suggest that just possibly the seemingly almost universal sentimental perception is the right one and my own wrong: my very different view may be just one of the ways I am socially colour-blind. Just possibly the long-term kindest course in human eugenics, *unavoidably necessary* to keep the human genome competent (leaving aside the 'devil we know' or, if you like, the other deep-blue sea of a return to savagery, to bans on all effective doctors and health care), is indeed just this *laissez-faire* policy—plain acceptance that catastrophes will come and we will suffer them meekly; after each (speaking for those who survive) we will simply wait to see what the principle of immediate kindness demands that we do next. According to such a course, we are refusing— explicitly, culturally, and most certainly not naturally—to think much about disasters at all, while, for present expedience/obedience, we deploy all possible kindnesses and 'humanity' we can. Expedience may even include, presumably, doling heroin sweeties to crying children if we find that that is what makes them happiest fastest . . . but surely you can see that this gets very absurd, and it is exactly why I break away.

Throughout these two volumes I try to express how strongly I don't believe in the above *laissez-faire* and short-term view of our future—and to give reasons. To be more serious than about crying children (though I'm serious there, too, within its time frame—I think it's a case where most of us have already learned through some bitter lessons, either in our own present generation or by natural or cultural selection in the past, what the limits are), let me say that for a start I think the disaster-acceptance course may lull us into not noticing many other truly deadly dangers for our species. Indeed I think we are already lulled. In the bliss of the always easiest, quickest benefits, we may be forgetting to look much to our distant future at all. Deep in responding to ever-increasing clamour to be 'humane', to concentrate on the individual and immediate (and increasing) human needs— issues of Viagra for the elderly, for example—we may miss noticing in time, say, the coming asteroid, then find too late we have already forgone the chance to deflect it. Thus we may come to our extinction even without the help of our own asteroidal-scale wars that are so likely if we let population increase continue un-reprimanded and uncontrolled as we do.

As another example (and also to point a grim message of hope, because it may actually be an answer: epidemics hardly can succeed to kill everybody), transplants of pig organs to humans may indeed soon be endowing more years of life to millions—and all this still in the little-questioned Hippocratic tradition of patient primacy. At the same time some shareholders will make a killing. But in making these 'ethical' gifts we will be setting aside those extremely well-founded concerns about pig viruses jumping to humans from living pig tissues. Those wonderful millions of immuno-suppressed human bodies are the prepared feather-beds for potentially vaster billions of virus bodies to lie in and try in, cosy as kicking babes, and a few of them hatching, via mutations and recombinations, great hopes for the future. Evolution is relentless, undirectional, caring not whom it slays: these viruses, too, in a few years may be acquiring capacity almost to end our species. In the year almost ending it may be only the Virus Empire's Nobel Prize Committee that sits to award prizes and that for medicine will go for the human work that was done, brilliant and serendipitous, on pig trans-plants, virus culture, and species jumping . . . Never mind, there is just a chance the better prizes will be awarded in heaven: for a few years, on the long-taught principles of 'humaneness', 'immediacy', and 'primacy of the individual' (and with just a little subtracted perhaps about selfish 'market forces'), there will have been put to our credit all those millions of extra moderately good years of life that were lived. And to that credit also all the joys of the shareholders in transplant companies, seeing their portfolios shoot up, those who not only financed all the benefit but made for them-selves a 'killing', or, as the other saying of the market sometimes is, a 'bomb'.

Still, even when all this has been said, it has to be admitted that the course of 'catastrophe acceptance' as our strange way of keeping our car of civilization running is at least consistent and rational; if you also believe in God and eternal heaven, it may be, for you, the course that genuinely makes most sense. Tracing the consequences of such a spirit, we should help the casualties (the short-termism of looking benignly on them and their large families still applies, presumably) but definitely we shouldn't sigh when we find the futures of the likes of Cambodia, Timor, Rwanda, Yugoslavia, and so on, breaking again into factions, tearing and thinning themselves, and re-establishing thereby not only tolerable densities but better standards of health. Quite obviously, unless you strongly believe in rewards in an afterlife and/or believe in an imminent end to the world, advising such peoples to go on doing *all* of the following—increasing, loving one another, fixing up all defects, including of foetuses, and rejoicing to have indefinitely increasing density because of more souls (which set is what at present the Roman Catholic Church explicitly advises)—makes no sense at all. But so also advise, by somewhat more confused implication and with some of the above items deleted, most of our democracies, though not, of course, China, which isn't one.

Sadly I think few will read even this far. But for those who do read to here, and perhaps further, what more needs to be said in overall introduction to the series of essays and papers that follow? I have said little yet about the volume's main theme, the evolutionary reasons for sex, even though these reasons as I will explain them are very much in the background of the above issues and, as you will see in the introduction to Chapter 8, they even provide the best evolutionary argument against the racism that so often comes to underpin the conflicts I have just mentioned (see the introduction to Chapter 12). I think for the whole of the book I just have to plead sincerity. I didn't go to the Population Study Group meetings wanting to make a fuss, instead wanting to learn and, if I could, to help. Likewise I didn't go to the Vatican meeting wanting to make a fuss. Rather, again, it was that they had asked for a contribution to an important theme and I wanted to give it. I am aware that vast numbers of admirable and idealistic people, including some members of that true superspecies of *Homo*, the true human altruists, are enrolled in RC ranks. Usually I seem to see them among minor clergy; they are those out among the people and in most cases if too sincere seem not greatly liked by their own superiors—or at least not until after they are dead and their reputations perhaps begin to demand what the people think. I would like to help all such sincere workers to be a more effective force than they are, to help them to play less self-defeating parts in the supra-animal future that is indeed opening before us. Basically, because this course is going to be above biology even while it still must take its models from biology, I even admire Catholicism in the way I admire any other vital and unusual organism—like admiring a beetle for its mysterious but beautiful horns or a honeybee hive for its workforce. Especially in the mix that this lot have created I admire the attempted celibacy of the RC priests—though I concur with others that they have created for themselves a crisis about keeping in touch. It is in such a spirit of sincerity, accepted, as I hope, by both religious people and agnostics, that I would like the book to be read.

We are indeed hurtling into a very different time. By odd combination of a 200-million year reverence for five-fingered limbs and a 2000-year reverence for a man from Nazareth, we characterize our immediate tran-

sition as 'entering a Third Millennium', but really what makes our condition special is that we come to it with all the 10 fingers loaded and 10 toes
staggering beneath incredible and massive gifts, not actually from Jesus but
from science. They are gifts that, even a century ago, we could not have
dreamed of and that still leave us gasping: gadgets without manuals;
machines that look drivable. But dare we? With some that looked so drivable, like nuclear power, I think we are already aware that we were too
young and we have gone off the road. Uncle was too generous . . . On the
philosophy side the present state looks better, and, as I repeatedly express in
this book, I have the hope that this side may bring us a proper caution with
the other gifts in time to prevent disaster. Here, indeed, we fly forward with
new and still-changing conceptions of nature—that's one notion I fully
accept out of the intentions of that rather disappointing and skewed Vatican conference—and I am proud to think I personally may have had some
part in providing this new view, the new glimpse through the canopy.
Unexpectedly I had found also a part in influencing the 'human' side of the
philosophy, too, a part seemingly as great as that which I more intended and
aimed for—the 'rest', as I had put it in my title. In the two volumes I have
now completed I would like the reader to find a change to be—well, what
exactly? Inspired with me by the new nature that we see, the view over the
hill? Well, at least to be forewarned.

So far I have written of some themes in the chapters that follow but only
a minority. The matter of what health is in large living creatures and how
health when we have understood it bears on what sex itself is (well, I agree,
essere actually *is* central, he of the 49 books is right; but then surely
Descartes and many others knew it too) has been so far mentioned only in
one paragraph of the opening page. This is the major theme of the book, but
the aspect of the effect of sex most emphasized above—beating back bad
mutation—while inevitable and undoubtedly playing a part in sex's maintenance, you will soon find that I am not treating as the primary one. The
mutation themes are addressed again only in a divided two-and-a-half-chap-
ters (especially it's Chapters 12 and 16, and the Appendix to Chapter 16,
but there will be many other brief passages). This is to say that, in so far as I
treat the subthemes above, I have devoted most space to my *second* preferred
theory of sex—that which somewhere (Chapter 16) I name the 'Mutation
Black Queen' hypothesis. My own favourite theory, however—the 'Parasite
Red Queen' (PRQ) hypothesis—is also very relevant to the question of the

degeneration of the human genome in the absence of selection; that is, the phase we are entering. Still more it is relevant to the degeneration of the environment, provided you recognize the new diseases to emerge as a part. PRQ has colour and bravura built into its very name and nature—peacocks, stags, narwhals, bowerbirds as the older themes, lesbian gulls, dwarf and family males, canids in labour pains and acting anxious husbands, recreational sex in beetles as newer ones . . ., all are there in plenty in the pages to come. If you possess my Volume 1 you may also see what I wrote there in a preliminary way, on what sex is and how it spices and colours the world, in the 'paper' of Chapter 10. I believe, therefore, that this Queen can be trusted to speak out for herself as the pages turn and I need say no more here.

To have *some* idea about sex, in the opening years of the new and 'sex-war' century—I try not to say millennium—is, I suggest, essential for every well-groomed thinking person, and this whether you are the one who cares most about human nature or the one caring most about 'the rest', the natural history. If you hadn't realized that you needed to be groomed with a theory and hadn't decided—you thought love was simply love, perhaps, as I once did—then this book as a whole is a comb for a top pocket, a haircut, an idea that I think it can't be denied is interesting, by good luck humane, and also very likely to be true. To my mind the varied surround of the theory's evidence as this stands augmented and diversified since my early suggestive papers in the early 1980s (Chapters 2 and 5), and all of which surround I shall try to present in the papers and commentaries—may already give you confidence that parasites, themselves undoubtedly as old as life (barring perhaps the submicrobial generations of initial replicators), do indeed lurk somewhere down there near to the root of this other ancient and weird property that we call sex. Adopting my opinions on it all, I think you may find many biologists still who disagree, but they will not, today, be calling you a fool.

People to be thanked for their help in creating this volume can make a shorter list than I allotted in the last because quite largely it is matter of the same people over again—if you want the full detail please see (or, better, buy) my Volume 1. But some are so important they need to be repeated. Colleagues in the Museum of Zoology at the University of Michigan, together with all of that university's vast facilities that were provided to me in the early years of my work on sex, must be acknowledged. This includes that university's contribution to my life's experience of inspiring and beautiful librarians and its scented subenvironment, in the museum, of books, birds, and parasites plus enough of dismissal there of the ideas to get me irritated. Over here, colleagues in the Royal Society shortly afterwards continued my support through a Royal Society Research Professorship (thereby, incidentally, making my point posthumously to Lord Florey, because his excellent idea it had been as PRS to create the professorships: no doubt of all the kinds of scientific talent he would have imagined his idea supporting, mine, linked with its mysterious tramp of marching fellows' feet, would have been last). My appointment has been through two of its five-yearly crises of renewal. This has certainly helped make possible both this book and the last one and together I admit they have ended taking out far more time than I had thought necessary and thus proportionately they have reduced my output of real science: I simply have to hope that enough colleagues will like what all that time has produced and make apology for not being a better and faster writer. Actually I doubt that most RS Fellows *will* care for the key ideas any better than they (or others) seemed to care for them in Burlington House in 1966, but whatever—undoubtedly overall, and in British tradition, great patience and democracy have been shown.

Similarly I thank Oxford and Zoology at Oxford and, still more, New College at Oxford for patience with their sociality-studying but socially-blind professor. I thank Michael Rodgers again for his continued interest as my publisher and Sarah Bunney, again my copy editor, for her continued efforts to disentangle prose, and make my hail of references and notes fall and melt into sensible channels. I thank all my co-authors, of course—Rick Michod, Marlene Zuk, Jon Seger, Ilan Eshel, Bob Axelrod, and Reiko Tanese—for their parts in papers that wouldn't otherwise have been written; likewise my great thanks to my various heroes in the fields of evidence, with this not only for their studies themselves but also keeping me up to date with published results. Here I mention Dieter Ebert, Stephen Emlen, Steve Frank, Curt Lively, Mark Dybdahl, Bobbi Low, Anders Møller, Eibi Nevo, Wayne Potts, Paul Schmid-Hempel, Paul Sherman, and Claus Wedekind. In a more personal way I must thank Carlos Fonseca, Paul Ewald, Peter Henderson, Nigella Hillgarth, Jeremy John, Olivia Judson, and Servio Ribeiro for their general faith in the worth of the ideas as it all went on and for patience with my slowness of writing—in many cases cutting from my time with them and this through difficult periods. Lastly and most of all I thank Luisa Bozzi, bee-eater sharing a northern wader's stone, for her fount of cheerfulness and all the places and kinds of support, too numerous to mention, that she has poured over me since my work on this volume began.

Oxford
December 1999

GHOSTS OF THE MUSEUM

Fluctuation of Environment and Coevolved Antagonist Polymorphism as Factors in the Maintenance of Sex

That I could think there trembled through
His happy good-night air
Some blessed Hope of which he knew
And I was unaware.

THOMAS HARDY[1]

SEX is an intellectual problem for some of us but a personal one for us all. The intellectual problem has affected me much only since, in the mid-1960s, I first understood kin selection and sex ratio, but from then on it became my main scientific preoccupation, as will be seen in this volume. The personal one, of course, had been with me for very much longer, since puberty or perhaps ever since there arose those early Oedipal feelings about my parents—sexual love for my mother, jealousy and resentment against my father. These feelings I certainly had; it was one of the few things Freud seemed to me to have had right although even there I felt he made too much fuss. But that personal problem had been far from reflecting or even suggesting the existence of sex's other and scientific aspect. Via introspection, personal sex did shed on scientific sex just a few extra spots of light but, far more, it shed lies. The matter came to me at first all entangled with the ideas of my upbringing—what I *ought* to think about love, marriage, duty, honesty . . . this because having babies and rearing them is often almost if not absolutely the most important thing we do in our lives. So the two sides are inextricably entangled. In particular, all that I learned or failed to learn about 'love' rattled me from the first. Love was written about in every novel I picked up. But what was it—how defined? Haldane once aptly described, for the benefit of John Maynard Smith, another tricky biological concept, 'fitness', as simply 'a bugger'. Well, love was another bugger, something we biologists couldn't very exactly define. Or else was it, perhaps, like Law about whose resemblance to Love W. H. Auden once wrote a longish poem both to lament and explain?[2] Or otherwise was it like Justice to which George Eliot devoted the opening (and near to closing) paragraph of a long novel?[3] Must one have, instead of science, such accepting and poetic viewpoints in order to understand love? Maybe I am—slowly—coming to understand better, but I could write a half dozen of comedies and still more tragedies about mistakes I have made along the way.

Perhaps, indeed, it needs a post-sexopausal man to look at the sexual side of the concept of love soberly, and what man alive is quite that? Old women should do better, although in parallel case, just because they are so much more certain who their offspring are compared with men; they lose objectivity about motherly sides of love. And so, perhaps, do men lose in respect of the loves for partners, and for comrades in war. But I should mention that even in treating the scientific aspect, in which for the most part the dread word needs no mention, progress has not been always emotionally straightforward for me. Sex has been a little easier.

Once I had seen how easily and completely I could justify my small animals in their excess production of females, the painful left boot that sex ratio had been jumped suddenly to my other foot and there started even worse blisters. The problem became why a species would ever put a half of its effort into producing males. The waste—the gratuitous births allowed to create such as myself—agonized me increasingly as I saw more and more examples where a possibility of having nothing but females among one's offspring could be only a species-split away from a given sexual race. The gap could be less, it could be merely the difference between one population and another; cases were rumoured even of that ultimate triumph of feminine disdain in which an individual female might choose personally whether to invite a male to contribute to offspring or instead to keep the whole damned love affair to herself. Such situations naturally set me wondering how males and Mendelizing genetics ever came into life at all. It must be somewhere very far back and very unicellular where lay (or floated perhaps in 'soup') the events setting out ground plans for all the sexuality of higher organisms, making the seeds even of all my own social and sexual errors. If God in 1 BC, correcting his mistake of a few billion years earlier (that his 'sixth day of creation' when he made Adam before Eve), had made his eventual human offspring a female, what trouble he might have saved in the long run! Had he made a Daughter to continue that trait he already endowed to her mother, all humankind could by now be one parthenospecies—a single clone. Human generations of 2000 years are enough for such a replacement. Everyone on the planet now would be like that good Mary whom he originally chose; everyone would be happy; he himself would have nothing more to complain of. But how boring the observation of human life would become for him! And how would he escape the outcome that others of his creation—certainly in his eyes deserving space in his universe, because he put them there—would evolve to shadow with sickness this endlessly expanding nimbus of the chosen and then perfected Marys? That such shadowing, short of his intervention every moment, would be inevitable is also going to be a major theme of this volume. And what about those organisms that depend on people having sex—the *Trepanema pallida* of syphilis, and the human pubic louse? In this alternative course of history, these creatures, in which word I again stress the 'create', would have faced rapid extinction.

The paper of this chapter was my first wholly devoted to the scientific problem. Previously I had published only an essay review of two relevant books (see Chapter 10 of Volume 1 of these reprinted papers). The present paper was thought out in the mid- to late 1970s and completed in 1980. Two graduate students I had taught at different times joined me in the work, one early and one later. Peter Henderson was a student from during my long lectureship at Imperial College London. I had known him and we had begun the work well before my move to the USA in 1977. We discussed the main computer simulation that is in the paper and he began programming it in about 1974. Nancy Moran, already represented in these re-publications (see Chapter 14 of Volume 1), joined us in the literature review and writing stage after a meeting on evolution that the three of us attended at the University of Michigan, Ann Arbor in late 1978 and that I describe below. By the time of the meeting I was in the stage of gathering library material for the discussion of the results of Peter's simulation and of others similar I had done myself. Nancy, likewise, was busy with library work, preparing to define her thesis topic, which overlapped with this one.

The writing of this paper seems, in fact, a tangle as I look back on it, a mess further complicated by a very slow editorial progress resulting in publication coming well after two others that actually followed in my research sequence (Chapter 14 of Volume 1 and Chapter 2 of this volume). But I remember vividly one incentive for the growth of the whole topic—a winter male cardinal. In the poem of the verse I set at the head of this introduction, the thrush sang to Hardy in a winter evening of 1900. It set him thinking of Hope and of the frustrations of humanity. It is a beautiful poem and, reading it, how well I seemed to know Hardy's hedge, his rain, the very English cold of the particular day, even that 'winter' mood in which he had walked. When, rather similarly, an unexpected cardinal puffed his breast on a tree at dawn and sang out over the snow as I skied to my work at the University of Michigan in the late 1970s, it was winter, too, but outwardly very different—a swish of snow rising from my skis, sometimes accompanied by a crackle of dry weed stems that they broke, but above all a cold much deeper than ever in Dorset, England. But I, too, thought about Hope—the idea of hope and some of my own. Particularly I thought about evolutionary hopes that cannot have troubled Hardy. Ann Arbor in midwinter was tough, and what was the bird's brilliance of song and colour doing here in the Nichols Arboretum—at 5 Celsius below? What did this male expect from his song? Most other birds of Ann Arbor,[4] as well as a considerable part of the town's humanity, had gone south long ago. Even the pair of belted kingfishers were no longer fishing the nearby ice-free rapid of the Huron River; they were in a state or two to the south. But this one cardinal, red as the misty sun that lit the twigs from which he sang and, above those, lit the forehead and sceptical eyebrows of a great university hospital towards which (in many senses, as you will see) I was gliding, this one tiny crazy, feathered being, an evolutionary cipher who ought not, by most reasoning I could so far muster to his support, to exist at all leave alone to sing, was seeing the Ann Arbor winter through. And he wanted the world to know it.

Racoons and students likewise saw those winters through in Ann Arbor; they, too, mostly weren't hibernating. The former I'd see occasionally at dawn or dusk climbing into or out of trash barrels as I passed through the arboretum. Students also could be there in the 'Arb', gliding on paths like mine among the trees, but on the whole student watching was better done on the central quadrangle of the campus where the type was more abundant

—20 000 or so in the whole university. Here along the paths and sidewalks their movements pulsed with the hours of lectures like corpuscles propelled by a giant heart. Nowhere were any idling outdoors as they might be in summer. They were muffled, unterritorial, crackling re-frozen slush underfoot, and generally swiftly in passage between (dare an Englishman say this?) their overheated buildings. Even so, quite commonly in winter you saw them paired and quite commonly, like the cardinal, they were beautiful.

All of such interesting activities (I might classify them as the indirect manifestations of sociality and sexuality, as these become entwined) became much more so in other seasons, but the Michigan non-winter always seemed too short for what had to be done and seen in it. Winter was mainly for thinking and not field work, and perhaps it was because of this that the one winter cardinal seemed to mean so much.

With my family I was living in one of three houses grouped far from others by the Huron River and I had many possible routes to walk (or, in winter, to ski or skate) to work. Usually I crossed by the bridge in the Gallup Park and then, traversing the railway by one of the ragged holes in its fence, I would walk up via the arboretum just as on the day when I watched the cardinal. Another route pierced the woods by Huron High School towards North Campus where I could catch the university bus across to the main campus. Yet another used the bike path on the same side as far as the Veterans' Hospital and then zig-zagged over the river by the railway bridge, there again entering the arboretum and so up. I was amazed at the freedom pedestrians had to cross where they please or even to walk along railways in America. Perhaps a right to walk on railroads is in some amendment of the US Constitution; if so, it would be nice to think a freedom for yucca seeds to ride on trains is constitutional there too—there was one yucca plant in a grassy area I used to pass on that route and it was quite out of its natural range. Its parent seed was perhaps dropped by some truck of the 2-mile-long trains that converged to us from the fan of steel roads beyond Chicago, where somewhere the plant is surely common.

Now I turn back to sex, and also to where, not far from that lone yucca as it happens, on a spring evening and on one of the less frequent of my daily routes over the railway and Huron on the one bridge, I heard and faintly saw, high over a patch of marshy woodland by the river, a lek of male woodcocks. So high, clustered as they were and yet playing lonely in his air—dusk skylarks almost one might call them—why would these mud-

probers of American woodland be so different from ours, the lone, 'roding'
British woodcock males that flew low on their display flights, though
likewise at dusk, over the woods of my childhood in Kent? Yet both still
cried their hopes by feather rather than vocal sounds, as also in early
summer did the night hawks (really a kind of nightjar) that made crazy
dives to the flat roofs of our central campus, where I had to imagine their
trim females pretending disinterest while attracting the bravado and
waiting to nest. And why would dusk species choose feather sounds—could
there be a meaning, what had happened?

By the week of the woodcock displays, sex was everywhere around
Ann Arbor. It was not long after the winter cardinal that black stoneflies
were out on the melting snow by the railway bridge, and often copulating;
willows were bursting their yellow and silky blooms; and, in the woods,
skunk cabbage flowers, by spring metabolic heat alone, melting their way
up through the snow. About this time each year would come that special
day when from far ahead beside the railway I would hear the first redwings
creaking among the cattail reeds, and shortly would see them settling aslant
the high stalks of the reeds, flashing their brilliant epaulettes, and chasing
each other and working out who was to be top in the marsh this year.
Later, it would be yellowthroats among the railway-side willows, now in
leaf, while out among the weeds and bushes of the arboretum's lower field,
amorous pairs, triangles, and even quadrangles of the still more brilliant
goldfinches would be fluttering, and perhaps an indigo bunting—potential
polygynist too—would be singing from a high twig or wire (never seeming
as brilliant at a distance as he really is).[5] Later, like a sudden blast from the
rain forest, came summer-seeding rain-forest families—sumachs, sassafras,
hop tree, the prickly heat of the hot spells joining the prickly ash and rain-
forest swarms of mosquitoes in the woods. I watched the sumachs
intermingling the edges of their hunched clones along the roadsides,
though whether for fighting or amorous embracing (or for both) I still wait
to know, while all over them amidst their palm-like tufts of leaves there lit
up the russet and crimson candles of their flowers.

Of course not all the summer surprises were tropical. Rhubarb
crossed to a sunflower; yellow prairie dock was in the redwing marsh; cone
flower daisies were on the island opposite our house. But in waste places
Dave Queller's milkweeds broke out their balls and domes and orange
plates of flowers,[6] complexly tropical, catching at insects' feet. Yet ever

what strange mixtures! There with the last in the waste lots stood black raspberry, delicious fruit of summer, and just as its cousin bramble on Badgers Mount in Kent had called down to their white flowers (and sometimes to my net) the great British 'fritillaries', so these called the amazingly differing sex forms of the diana butterfly, another of the same tribe. Tropical again beside our lake and seeming to pour their wet heat into our air, weirder still in pollination than any yet mentioned, pickerel weed and loosestrife flowered almost tri-sexed; while in season I saw these same weeds splashed by temperate sex-crazed carp that sometimes almost climbed from the water in their efforts to escape attentions of what I supposed to be their less-favoured suitors—or, in the case of the suitors themselves, they were battling to keep up. So it went, until, towards the summer's end, the blue-black oil beetles were courting and joining obesely on the flowering goldenrods, while even overlapping the first sharp frosts that so distinguish a Michigan 'Fall' from British 'Autumn', crickets would still trill into the dusk advertisements of seniority and survival,[7] last fading echoes of a vast Midwest cacophony of summer nights . . .

My mystery in all of this was simply WHY? What 'blessed Hope' was it that made life pour so much of itself away so wastefully and so ecstatically? Or, seeing the matter in another light, why were so extremely few (as I might here instance crow garlic interspersing purple bulbils among pink flowers plus a rare viviparous grass that I found on the bike path),why were only these few masturbating into existence near-perfect and clonal duplicates (barring mutations), alone and hidden?[8]

Parallel to such inspiring if baffling natural performances, there were also darker sides to my walk, although 'dark' here cannot really be quite the right word. It is hard to trace now the gradual course of how two particular sights began to trouble me in a way that was to continue even after each had proven, in its separate way, to have offered me a part of an answer to my most burning question. The Nichols Arboretum is on an old moraine high enough almost to qualify as a hill (at least for the Midwest) but this one became almost twice as high again when the university decided to perch its huge new hospital and medical school upon its top. So it came about that from various points of my walk, but especially when I was on the Veteran's Hospital and Huron Towers route, the great block building peered at me, Big Brother-like as I said, over the arboretum trees. Even before it was so rebuilt during the later part of my time in Ann Arbor to become the vast complex it now is, the hospital had already been the biggest building of the university except for the stadium. There came to be something symbolic in the growth of this particular pair and by the time I left Ann Arbor I thought I understood a little better the tension of dread and longing stretched between the late-born colossi of the great seat of learning. As with Rome's own Coliseum, they seemed products of a kind of 'Empire' period; and their atmosphere had come to pervade all the campus between.

Often on my walk I would watch helicopters fly in with the seriously sick (heart attack victims mostly I supposed) to a pad on the hospital roof. On Saturday afternoons I would see farther off a light aircraft trailing its advertisement banners over Ann Arbor's other pole, the stadium, where a hundred thousand would be watching a celebration of Michigan's peaks of health and wholeness. Well, what could be distressing or 'dark' for me in all of that, you may wonder. For the moment I can only ask for your patience until the later chapters of this volume. By the end of my stay the whole of a brand-new floor, overlooking the river that I daily crossed and re-crossed, was devoted to the Medical School's library—it may have been Big Brother's left cheek approximately, I am not sure. More and more my quest for sound reasons for having sex drew me to this library while I visited correspondingly less the very efficient but much less richly endowed library of Biological Sciences—a rabbit warren of packed books by comparison.

What treasure and what excess that medical library had! Here, for example, I found that old, mainly veterinary book of Hutt on the disease resistance of domestic animals[9] and read his claim that virtually every domestic species could be rapidly improved in their resistance to any disease by breeding. Hell, then, what were the animal breeders at that this hadn't been done long ago—why were diseases still a problem? And for that matter, couldn't the same unsubtle (and purely natural) breeding principle—working, one would have supposed, automatically in humans and their progenitors over the past millions of years—have already made every Michigan inhabitant not only healthy but into a potential 'Blue' for the stadium? Was our extreme diversity in health, so contradictory to such an idea, simply an ill chance; or was it something at once more inevitable and more understandable? Here for a third time I signal the intention of this volume: answering these two last questions is going to be its most major theme.

The other 'dark' side of my walk is similar but it was a shadow that only a naturalist would have noticed and then probably only one straight out of southern England like I happened to be. First it was just a shrub that I recognized: *Rhamnus catharticus* or purging buckthorn. This had been in my earliest flower book and also was the food plant of my favourite yellow brimstone butterflies of Kent. On my first walks in Ann Arbor I had found this buckthorn, immediately recognizable, extremely abundant all through the woods near the town, including those of the arboretum. There I saw it press back and made rare by comparison such true Michigan undershrubs as the wych hazel, prickly ash, and ironwood, which, because of their novelty, I was looking for with far more interest and attention. Then amazingly in the woods across my path on my other main walk, the one to North Campus, I soon found Britain's only other buckthorn, *Frangula alnus*. It likewise was abundant. Near my childhood home discovering that species in a wet valley had once been an exciting find for me, having come at a time when I thought I already knew all my local native trees; now in the Huron High School woods I found it, like its cousin across the valley, existing in an unnatural abundance, filling the undergrowth and crowding the natives locally to extinction. Only the lianes—wild grape and the celastrus bittersweet—could still thrive above it.

Later I realized that both woodlands had also huge stocks of two shrubby honeysuckles that likewise weren't native Americans. We foreign professors had to struggle for years for our 'green cards' permitting us to work in the USA and thus to replace our irritatingly short-lived visas; but these buckthorns and honeysuckles had them already by forgery by the thousand; their green cards were simply their very perfect, vigorous, and large leaves unmarked by insects. Ignoring 'Immigration' in Detroit, honeysuckles and buckthorns, wetbacks without shame, had worked here already for years and they grew like fury. Later these particular bushy and small-tree honeysuckles admitted to me what I had from the first suspected: they were Eurasians, though this time not from Britain. All four species I have mentioned flowered and fruited abundantly.[10]

And I suppose that about here I need to warn both myself and my readers that it is likely to be only my fanaticism for the 'disease–sex connection', which the present paper begins to bring into the open, that made me imagine that, though perfect and unbitten like their leaves, their green or creamy flowers seemed to me to be somehow lacking the true enthusiasm for the thing they were built for—not like that I was seeing in, say, the native milkweeds and goldenrods of the same area. It was as if these newcomers would as happily have set their prolific fruit though self-pollination or without any sex at all, like the green-whiskered grass by the bike path, even though in Britain (as Darwin himself, as it happens, had reached exactly among the hills of my childhood) they are outbreeders and are half way even to having separated the two sexes, as have hollies and humans. The caterpillars that shredded buckthorn leaves on those hills where later I idly chased the brimstone adults of the caterpillars for my collections (and besides these I cannot resist to mention the green hairstreaks and dark umbers that likewise fed on the tree and which I also hunted, and the buckthorn-potato aphid and all those other potential nibblers and suckers that I came to know later but whose damage doubtless was always evident in the relative stunting of the bushes compared with the transatlantic cousins I was now seeing)—all of these enemies had been left behind in Kent or wherever the American colonists came from that brought the shrubs.[11] Anyway, be these speculations as they may, it is certainly true that by the end of my time in Ann Arbor various plants—those I have mentioned and many others—were continually bringing to my mind a particular question: with no enemies evolving against a species, should it continue to have sex?

What, then, exactly is this problem about sex? Readers of Volume 1 may know already roughly what I mean from my essay entitled 'Gamblers since life began . . .' (Chapter 10), but I will explain again briefly here. It is a much deeper problem than that which is so often immediately troubling for humans—of there being enough of it in one's life or of the right kind (the old right-partner problem or whatever). The problem central to this book is why sex exists at all. To humans it is so intuitive that sex is needed for babies that few of us question the inefficiency of the strange way in which we, together with most other large living things, continue our species. Yet any one who grows roses and who has watched them catch a greenfly infestation (and who is 'green' enough not to blast them immediately with insecticide?) should know the efficiency I am referring to.

If you really want to see this efficiency at work, it is best to do what Harvard entomologist Edward O. Wilson suggested to the lady who wrote to him about ants in her sitting room—buy a magnifying glass: in short, to treat your colony as a pet and return to it a few days later. Or perhaps, better still, just read a little and realize that the first generation of those green blebs, so easily to be seen extruding from the bulbous mothers, especially if you use the lens, can already contain within themselves their *own* embryos, which, in turn, in a further few days will be extruded and be walking and plugged-in grandchildren—and these, of course, already have *their* grandchildren forming within. . . . This model of the progress and overpopulation on the roses is surely enough to tell you how ridiculously inefficient is our own way of proceeding. For a start just think of that most anguished part, the searching for the mate in humans; then think of the still more pointless production of the males who must search or be searched for. . . . Throughout summer among the aphids there is none of that— never a male nor an act of sex in sight. Isn't Darwinian natural selection supposed to love efficiency? If so, how was it ever efficient to create males, a sex commonly so 'lazy' in family matters (as, in fact, are all aphid males when, rarely, they occur) as to give no help whatever to females in garnering the resources that will make the next generation?

With this chapter and the next both devoted to the topic, and with various comments and the mentioned essay review re-published, there is no need to elaborate on the matter here; but in case any doubt still lingers concerning the efficiency to be gained by the elimination of males, those who know my first book could look back to Fig. 4.2 (p. 146) of Volume 1, Chapter 4. The upper perfectly straight line of dots in the diagram represents the changes of frequency under selection when fertility of any variety is doubled: it was drawn to illustrate the increase of a fully driving Y' chromosome as described in the text of that chapter; but this line could apply equally to the increase of a parthenogenetic biotype of insect competing with a sexual one. Clearly if you are a gene with some say in what the genome does, then to allow your bearer to waste half her descendant biomass in each generation on a seemingly pointless production of maleness certainly isn't in itself a good idea; in fact you lose out and disappear extremely fast compared with a gene preventing maleness. But if being around with males, however idle they may be, or having your daughters be around with them, mysteriously brings into your stock some splendid and at least twofold vigour, even when this is through qualities that, due to the mixing, you cannot call wholly your own, the case might be very different.

What are the qualities and what can such vigour be? A powerful genie must lurk in this ancient lamp if it is to bring forth the palace of

Mendelizing genetics and sex. And the genie's magic must be ubiquitous—
it must work in Alaska and in the deep ocean trenches. The name of the
problem for which we ask assistance surely has to be, in the first place,
change: equally the magic talons of achievement must be *recombination*. In
short, the environment of most organisms has to be changing very rapidly
and the changes must demand new *combinations* of cut-out sections of the
existing genetic 'blueprint' rather than designs or material that is truly
novel in itself. All this follows because we know from classic Mendelism
that new combinations are really all that sex can produce which asexual
reproduction can't:[12] mutation, the generator of more fundamental novelty,
works equally for both kinds of reproduction. The idea that genomic
juggling with the tricks of life already invented, tricks held within the
populations and species that had perfected them, could provide all the time
new or repetitive recombinations to bewilder living and evolving enemies,
is for me the essence of what has come to be called the 'Red Queen' theory
of sexuality.[13] Therefore I am pleased to note, on reading the present paper
again, how clearly (if briefly) we had specified our expectation along these
lines as early as 1980.[14]

It was about 1975 that Peter Henderson and I at Imperial College
began to try a combination of maths and computer simulation that might
decide whether, in such a context, it would be enough to have rapid and
potentially large *random* changes in the environment—changes not yet
imagined to be biotic. We weren't particularly hopeful but felt we should
try this as the most obvious possibility. Two or three years later and now in
Ann Arbor, and with Nancy Moran joined, our conclusions in the work
remained not as clear as we would have liked but they still came as our first
'modelled' defeat of the twofold inefficiency and they sketch ahead those
broad but narrowing roads of Gene Land that I, at least, was to wander on,
back and forth, for the next 12 years.

As already mentioned I am placing the paper in this book not quite
in its proper sequence of publication. Because of the seemingly unavoidable
slowness of multi-author edited volumes the paper came out after that
which is going to be Chapter 2 even though it was written earlier.[15] It
seemed better here to keep to the order of the work so as to show the train
of thought: for example, the paper of Chapter 2 will have various models
and thoughts that hadn't occurred to us when I was working on this one.
On the whole I think that later the models improve our attack on sex quite

dramatically; however, the introductory sections of the present paper set out fairly clearly what we saw and still see to be the main problems. They are thus a good introduction to the theme of this volume while at the same time they show how the new line connects with sociality, which was the main theme of my Volume 1.

One thing the work I started with Peter Henderson quickly made clear to us was how very hard the twofold cost is to beat. I don't consider we ever do beat it realistically in this paper. We established that randomly changing patterns of advantage dependent on genotype could be devised to preserve variability and even that the changes could be technically such as to justify full Drones Club membership for males—that is, it would preserve them even when totally idle from the utilitarian point of view and when capable parthenogenetic females were at all times trying to invade and oust them. Thus far the work seemed encouraging; the damning bad, however, was that all cases of such preservation necessitated patterns of fitness so extreme and peculiar that we couldn't imagine them existing in the real world. Even sympathetic reviews of the ecological and selection effects of fluctuating physical environment since then have not changed our opinion.[16] Secretly I think we were both a little relieved because we were already nurturing ideas about other possibilities; and as it turned out some of these did prove much more robust.

Soon after we had begun to get our results Peter and I had to separate. I left for Michigan as described in Chapter 14 of Volume 1, while Peter, after completing his PhD, went to his first job. Not long previously another student, Victoria Taylor, had left me to change from study of a half-millimetre parthenogenetic beetle at Silwood Park (Imperial College's Field Station) *Ptinella* (see Volume 1, Chapter 12), to working on elephants in Africa. Likewise from his parthenogenetic ostracods, which were smaller even than *Ptinella*, Peter proceeded to the autecology of oil-fired power stations. In keeping with my rule above about big organisms, these, too, as it happens are far from parthenogenetic. If sceptical of this think of how, for example, when crossed with a certain weapon-making technique, the old 'fossil-carbon' species spawned the nuclear kind, which is still extant. As you will read again and again in this book, my sex theory explains that while its process generates more failures than successes a few of the latter can be so good as to heavily outweigh the failures in the long run. At present nuclear power stations may seem to illustrate the success side of such a process, but my hope is that eventually it will become recognized that they were actually the failures—a form of techno-life too unstable and having excreta too toxic for the frail life form that builds them. But, ending this digression (which I don't know if Peter would agree with anyway), I don't think it had been boredom with parthenogenesis or even with smallness that made either Vicky or him choose sexual and large organisms as their next topics; rather it was simply a need for a change and probably also for some cash.

The immediate result for Peter and me was that for a considerable time we couldn't easily exchange ideas. IBM card readers at South Kensington and at Silwood ceased to clatter with our decks. But as we parted we had two things on our minds. The first was that some responsiveness on the part of environment to a population's current genetic state, absent in the case of our random 'physical' changes, seemed urgently needed if variability (and, dependent on this, maleness) were to be kept alive. A connected point was that in the real world, as in our simulations, a high degree of involvement with inanimate food (that is, detritus— material that might vary a lot but would not exactly bite back) might be among the circumstances that favoured parthenogenesis. The latter point had been first suggested by Peter's increasingly detailed knowledge of his tiny clam-like crustaceans, the ostracods; however, the idea was soon reinforced by our reading about other groups. In general, parthenogenesis seemed most apt to appear in species living in simple food webs and with their numbers determined by physical forces rather than by living ones.

In several of the introductions of Volume 1 and in the book review that is its Chapter 10, I commented that we weren't by any means alone in seeing the looming crisis about sexuality in biological thought. As usual George Williams had been in the lead in seeing this and he had already published his book;[17] shortly afterwards John Maynard Smith brought out another book.[18] But these books seemed to me aimed mainly to clarify the problem and to suggest limited solutions and avenues for investigation: neither author even claimed to have found an embracing solution. Yet the uniformity of diploid genetics and of sexual epiphenomena always seemed to me very impressive, and, as I have already mentioned, I was sure that a cause as monolithic as the product would eventually be found. In the year when I brought to Michigan the half-digested results of my collaboration with Peter, which for me amounted effectively to a rejection of the idea of sex as completely ruled by physical changes (a tidying of the workshop, as it were, and putting away of the broom ready for other endeavours), I discovered the important review paper by Robert Glesener and David Tilman,[19] supporting, through known facts of ecology, the alternative path that I was already favouring—to wit that changes in the *living* environment of a species were likely to be most important in maintaining its sexuality.

Later, while I was working on my cycling model (see Chapter 2 of this volume) and was trying to obtain sex-support out of a simple two-species interaction, graduate students of my class drew my attention to yet another paper that had been published the same year (1978) by John Jaenike.[20] Building on work by Bryan Clarke, Jaenike likewise brought biotic ideas, in his case of host–parasite coevolution and cycling, into the realm of sex. From the perspective of my own model Jaenike's idea was timely and exactly in accord with what I was thinking, except that he envisioned longer cycles (which were to come to me, though with a vengeance, only later). Consequently it has always been a puzzle to me with all four authors that I have mentioned, though especially so with Jaenike,[21] that none followed through their ideas with much vigour on the issue of sex. Was the problem yet again that I was the crank, persisting through failure to see some flaw that had become plain to them? Or am I more 'dinosaur' or 'bird' as against 'reptile'—more inclined, that is, to warm my egg/ideas and nurture my nestling/hypotheses to their independence, while these other authors instead, in more reptilian way, left theirs to fend for themselves? Anyway, more support shortly came. In the year before the present paper, Hans Bremermann, a mathematician at Berkeley, published an argument that parasitism could be the likely genie of sex.[22] Although I was pleased to see this further convergence, Bremermann's argument wasn't of the kind I was looking for. Having essentially based his approach on a group-benefit and on long-term advantages, I saw him as failing to confront the difficulty of the very high demographic inefficiency, the point I considered to be crucial. Hans needed a model: I had one even if crude and a partial failure in not realistically conserving its variation; but also, it seems, I had a bigger stock of fanaticism.

Beating that twofold 'cost', or inefficiency of sex as it might equally be called, is the biggest challenge. People who haven't faced this either through simulation or a mathematical formulation usually don't realize how awesome a problem it is. It is that run in my old diagram of 1967 (Fig. 4.2

in Volume 1), carrying as it does a frequency from 0.1 to 99.9 per cent in 20 generations, or from one in a million to a whole million less one in 40 generations, both of them periods that are mere blinks of an evolutionary eye. On top of this, George Williams's book[17] had thrown out yet another challenge: asex must be shown beatable even for small families if we are fully to explain the mammals and various other slow reproducers. This isn't quite asking just for a defence of our dearly respected *Homo sapiens* on the grounds that he (or really she) happens to fall near the least-fecund end of the spectrum of the animals. Women need men for a lot of aid besides making them fertile, much more in fact than do the general run of female mammals. That early diagram of mine had seemingly implied that a virgin about in the time of Jesus giving rise to two daughters could have led by now to the extermination of all males in Europe, with probably all those of the Middle and Near East thrown into the competitive holocaust as well.

Probably, as I mentioned earlier, with a little doubt simply because I'm not sure of a realistic human generation time over the 2000 years in question, there has even been sufficient time already for a feminization of the whole world; and all could have happened by way of a perfectly peaceful competition even if some quite drastic Malthusian unhappiness had to occur among both sexes in the male-producing remainders (they would see they were dying out—but events of this kind are eternal in life anyway). Male help means that this scenario, quite apart from no authentic human parthenogenesis being known, doesn't quite work for our species.[23] Indeed the advantage of the parthenogenetic women could be fully expressed even at the beginning only if the new females behaved in some way similar to prostitutes and obtained their child support through an income, or else were crypto-asexual and just fooling their diligent husbands as to biological fatherhood. That support and that diligence, of course, would have to dwindle as the parthenogenetic type spread because the type's very success kills off the source. In short, unless technology could step in to a degree not yet possible, an unfailing supply of kindly and foolish 'Josephs' is needed if such parthenogens are even to persist in, let alone take over, our world; and, in fact, even the predicted early burgeoning of a clone is likely to slow as soon as the new lines are recognized for what they are.[24]

You will notice that my parthenogenetic subjects are now no longer being described as a cloud of morally ideal Marys and are being given a much more sinister aspect; indeed even *H. sapiens* microsp. *virago* begins to seem not an exaggerated trinomial for the new type. Yet even if we could imagine selection being checked during the spread of the microspecies, there remain in the world plenty of species in which males are supplying far less paternal aid than present-day men do, so the problem of sex (and within it the subproblem of sex accompanied by low family size raised by Williams) is still acute. In animal groups where the male does contribute substantially, as happens in butterflies and moths through a food package that comes with the sperm,[25] parthenogenesis is particularly rare.

My striving to find models that could cope with low fecundity as well as with twofold cost is apparent in all my sex models in this volume and, as far as I have noticed, I am the only person to have solved the puzzle for an infecund species by a model demonstration. As answers, the models of the present chapter, and others only slightly less simplified and artificial that come in the second half of this book, are suggestive only: I didn't find a model about which I could feel really happy until, along with Robert Axelrod and Reiko Tanese, I had worked for a considerable time on the one I describe in Chapter 16. Our success came only after a progression of models that I constructed during the intervening years.

The models I describe in the following paper were first presented by Peter Henderson and myself at a symposium on social evolution, which Don Tinkle, the Museum of Zoology's director, and Richard Alexander, whom I might describe as its guerrilla commander and general bomb thrower, had organized in Ann Arbor in the year after I started work there. Peter was given leave to attend by the Central Electricity Generating Board of Great Britain, with which he also had recently begun work. I remember the meeting as coming a little after the tail end of an 'Indian Summer' with the last fall leaves of Michigan tumbling and sliding in dry companies on the smooth and still thin black ice of ponds and lakes. Best remembered by me out of the whole meeting is a party that we held after hours in the public wing of the museum. This was the first of several museum parties I was to attend there. The big hall was always a splendid setting as well as being, for any British naturalist, quite a cultural surprise in the small town. On this night the hall was charged further by a special atmosphere. We staff and graduate students of the museum were being joined by evolutionists from all corners of North America and even by some (like Peter Henderson) from abroad. How unlike it all was to any party I had ever been to in Britain!

There was a social chatter at once deep, reflexive, and ironic, and it covered even such topics as why we humans should ever chatter at all. No mere Detroit car-builders or mechanic analysts in the field of biology were we despite our location; instead we met as the high priests of the Ultimate Explanation. Flattered by the praise coming for my previous work, made perhaps arrogant even above the rest by the special atmosphere, even so, throughout the meeting I felt myself a little apart and even sad. I had a new theme; I kept pulling it from my bag and no one around me wanted to look at it much. I even felt myself being a slight disappointment to the others; they expected more enthusiasm on the 'kinship' themes, such as I had once really had. Was it at this meeting, or one of those later ones in the same hall, that I saw Venus of Botticelli stand in that giant clam shell of the hall of the exhibit wing of the museum, and thought how her main message for us mortals, ever ancients, had mythed her into being and Botticelli of quattrocento Florence had painted her, had stayed pent up within her calm lips—until perhaps now when she might open them for us and begin to speak?

Mostly, I admit, my memories of the meeting were more positive. I remember how good it was to stroll beyond that shell of the goddess with an informed colleague, to face some lovingly constructed diorama of the ancient scenes of life—say a diorama of the Cambrian sea floor—and to discuss with whoever was with me the sociability or the sex that might have been there in that sea. What of the trilobites, for example? They had indeed bizarrely ornamented males in some species—already, so long ago; that much was known! But what did the trilobites do wrong that their cousins the crustaceans and the insects did right, enabling those other two groups to survive? What might we now be doing wrong, and did it matter if we were? Were we becoming too successful, too common? Whatever, any doom we were inviting we faced at the meeting laughingly and for now stood as the priestly bearers of a vast new optimism, an understanding not only of all the 'proletariat' of life that had so long been our professional concern as zoologists, but increasingly we were understanding ourselves as well.[26]

Why are 'mother's brothers' so popular in anthropology? This is just one example of topics that were prominent at that meeting. The 'avunculate', as it is called, is an old and favourite chestnut in the discipline of anthropology and has been played with in many ways. It illustrates how deeply different human cultures can be for some have the avuncular favouritism and some don't. But we had a new angle, started by Richard Alexander. He had seen how in all cultures, indeed in all *Homo*, you can be sure you are related to your mother's brother whereas your certainty of how you are related to your supposed father may vary very greatly. We were asking what exactly did happen that time 9 months before your zeroth birthday?[27] Your father's cultural relatedness might be fatherhood; his genetic relatedness zero. We were proud that it should be we zoologists who were pointing out these possibilities while it was the anthropologists (some were even on the floors over our heads during the meeting) who were being forced to huff and puff, like the wolf confronted with the cheeky brick-built house of the junior pig. But—and this was such a huge 'but' for me—why was there even a problem of deciding from whom such support should come, uncle or any other, why did such ultimate ciphers as fathers and mother's brothers ever start to exist? To what extent were they indeed ciphers? What did the anthropologists' accounts say— how much did the men help with village matters, with digging the fields? As for the latter, mostly no, actually; but such facts were for the anthropologists to decide, if we could trust them.[28]

Problems like that of the 'mother's brother' still seem to me a very long jump ahead even from the differing sexes of an avian cardinal, leave alone from explaining the two mating types of an alga, which was roughly where I was at the time. All the same, at whatever level, the debate could remain scientific. I admit to being a pretty free sort of 'jumper ahead' myself. But if we were to be so deeply concerned, as we were at the meeting, with the likes of *human* cardinals—the meanings of robes of office, ultimate bases for cultural celibate castes to evolve, and so on— wasn't it premature to forget the huge gaps we had left behind us: why my *avian* cardinal had been there in the snow; why one bird, one species, should so advertise? Certainly over the years at Ann Arbor I became increasingly troubled by my failure to persuade others, especially to persuade my colleagues and students, to see my point that this was a supreme problem still waiting (perhaps the most profound problem in all biology) to be solved.

Ever since I first saw it, the facade of the Museum of Zoology reminded me a little of Waterloo Station in London. The building's design even had the same widening as if built for spreading railways behind that never came. Likewise later, working inside the museum, there came a return of that lofty indifference of the world to my concerns that I had felt in the early 1960s, as if I was again thinking about kin selection in the midst of end-of-leave soldiers in the real Waterloo Station. A smoking battlefield evoked by the station's name and then remembering shrapnel pits of that equally 'damned close thing', the Battle of Britain, that later had marked the Portland stone of the real station in my own lifetime—all indeed of King and Country, Mother Britannia, for which we fought, and the tired and ironic soldiers who had later shown me, during my national service and after it, how we had won—these had been my ghosts of the background in the great station hall in London. Here, beside the clam shell at the museum parties it was the challenge of this calm-lipped Venus who stood there. Not that I could discuss exactly this in my classes, nor admit to hallucinations, but rather that we might discuss what made the mighty *Tridacna*, for example, to be the most genetically polymorphic as well as the largest of all bivalve molluscs[29] while in passing joking of the myths of Venus and Adonis; but that was only conversation and to draw smiles and I had the lasting impression that none of my students was ever going to be as desperately distracted as I was by this problem of sex. Perhaps in that, however, I was a bit overpessimistic, for later during my time there came Marlene Zuk and Steven Frank—certainly better and more independent allies in the quest than I was to find, bar one, after my return to Britain.

Maybe my failure has something to do with a weakness in explaining my insistent though vague senses of 'problem' that affect me in various fields of knowledge, feelings that what most people are currently believing just cannot be right. It is an uneasiness I keep even when I have little idea what could be true in place of the theme I believe to be suspect—a crank's sense of a problem existing even when everyone around me is showing by extensive arguments that there can't be one. By the time I am ready with a clear explanation of what my problem is then at least for me it has been more or less solved and then, of course, I feel much less reason to talk about it. So it had been when I would think again and again during the early 1960s about writing to J. B. S. Haldane in India (R. A. Fisher was the superior alternative to my mind, but he was already dead) about my problem of altruism and never did once write to him. By the time I *could* write to ask him an unmuddled question that I need not be ashamed of, my problem was gone (and, as it happens, so, too, had Haldane). Perhaps my difficulty in the Ann Arbor period had to do also with poor personal ability for the kind of rapid verbal interchange so loved in America—that 'thinking on one's feet', implying all that 'Muhammad Ali' quick intellectual counterpunch and footwork, or, differently imaged, withstanding and returning a Hollywood gunfire of witty and down-putting answers. Some Americans seem to check your ability in both these ways before believing you worthy even to be listened to: surviving—and not being cowed by—a test is rather like the joining rites of a club.

Still, looking beyond my class rooms and around the world, as if those railways had been built out of the back of the museum building, I certainly wasn't this time really so alone as I had been in 1960. From virtually every nation that is at all near to the front line of advancing Darwinian theory, voices were being raised over a current crisis about sexuality. My own favoured idea for an answer—the pressures from a coevolving world of parasites—was sometimes mentioned although never by anyone (it is my impression) as persistently as by me. Hence, in spite of the local disappointments that probably stemmed more from my character than from the actual attitudes of peers and students, either there in Ann Arbor or in Britain, I had much less doubt about a problem being real than I had had when engaged with the other 'social' issues of my earlier research years. About what other favoured ideas for an answer were at the time you will hear more in the following chapters, although never so much as about my personal favourite theory.

References and notes

1. T. Hardy, *The Poetical Works of Thomas Hardy in Two Volumes. Vol. 1. Collected Poems: Lyrical, Narratory, and Reflective* (Macmillan, London, 1928).
2. W. H. Auden, *Collected Poems* (Random House, New York, 1976).
3. G. Eliot, *Romola*, World's Classics (Oxford University Press, Oxford, 1971 (1863)).
4. R. B. Payne, A distributional checklist of the birds of Michigan, *Miscellaneous Publications of the Museum of Zoology, University of Michigan* **164**, 1–71 (1983).
5. R. B. Payne and D. F. Westneat, A genetic and behavioural analysis of mate choice and song neighbourhoods in indigo buntings, *Evolution* **42**, 935–47 (1988).
6. D. C. Queller, Proximate and ultimate causes of low fruit production in

Asclepias exaltata Oikos **44**, 373–89 (1985); D. C. Queller, Sexual selection in flowering plants, in J. W. Bradbury and M. Andersson (ed.) *Sexual Selection: Testing the Alternatives*, pp. 165–79 (Wiley, Chichester, 1987).

7. M. Zuk, Variability and attractiveness of male field crickets (Orthoptera: Gryllidae) to females, *Animal Behaviour* **35**, 1240–8 (1987).

8. T. Elmqvist and P. A. Cox, The evolution of vivipary in flowering plants, *Oikos* **77**, 3–9 (1996).

9. F. B. Hutt, *Genetic Resistance to Disease in Domestic Animals* (Comstock Publishing Associates, Ithaca, NY, 1958).

10. All four species have berries and the American woods have many birds that eat these and spread their seeds. The black berries of purging buckthorn seemed among the least preferred in the woods and only towards the end of winter when others were gone would I see flocks of cedar waxwings feeding on them. The purgative effect of berries of this species seemed very great and doubtless were evolved to prevent digestion of seeds. Against the waxwing the effect seemed almost excessive because the snow under the trees often was purple stained by the liquid droppings inspired by the fruits: thus most seeds seemed to be deposited in the thicket where they were feeding. As shown by the plant's success and dispersion since introduction, however, the dispersal clearly succeeds.

11. The buckthorn-potato aphid, *Aphis nasturtii*, has reached North America but probably arrived quite independently of *Rhamnus*.

12. Sex can and does, of course, play a part in DNA repair and a whole book attests to the possibility that this is even a sufficient cause for sex to exist—see R. E. Michod, *Eros and Evolution: A Natural Philosophy of Sex*, Addison-Wesley, Reading, MA (1995). It seems to me, however, that there are various objections. If repair were the main incentive one would expect (a) there would not be the emphasis on *outbred* sexuality nor would exist all its ecological variation (inbred sex or else polyteny with no sex would be adequate and would occur rather uniformly) and (b) a mechanism of excision and insertion creating no overall recombination appears more efficient than the existing mechanism that brings about recombination of long stretches of DNA. For Rick Michod's latest defences on points of these kinds, I refer the reader to his recent paper: Origin of sex for error repair. III. Selfish sex, *Theoretical Population Biology* **53**, 60–74 (1998).

13. M. Ridley, *The Red Queen: Sex and the Evolution of Human Nature* (Viking, London, 1993).

14. It should not be overlooked, of course, that 'new' as used has a wide scope because if there are numerous polymorphic loci the number of possibilities for

their combination is astronomical and in this sense 'new' could be interpreted as 'real novelty'. In order to produce an ever more complex organism, however, something more than rearrangement is necessary: truly new biochemical structure must be produced. A common road to this involves a duplication of a locus and then further mutations that give the genes in the duplicated site new functions, these becoming eventually remote from the original functions.

15. A similar situation arose with the first two chapters of Volume 1, but in that case because the first paper was really a kind of abstract of the long publication that followed, and because it contained no independent material, I chose to leave the publication sequence unchanged. The same policy is applied with Chapters 15–17 of this volume, where the problem came again.

16. E. Nevo, Adaptive speciation at the molecular and organismic levels and its bearing on Amazonian biodiversity, *Evolución Biolóica* **7**, 207–49 (1993); J. M. C. Hutchinson, Evolution in fluctuating environments: a game with kin, *Trends in Ecology and Evolution* **11**, 230–1 (1966); A. Sasaki and S. Ellver, Quantitative genetic variation maintained by fluctuating selection with overlapping generations: variance components and covariances, *Evolution* **51**, 682–96 (1997); B.-E. Saether, Environmental stochasticity and population dynamics of large herbivores: a search for mechanisms, *Trends in Ecology and Evolution* **12**, 143–9 (1997).

17. G. C. Williams, *Sex and Evolution* (Princeton University Press, Princeton, 1975).

18. J. Maynard Smith, *The Evolution of Sex* (Cambridge University Press, Cambridge, 1978).

19. R. R. Glesener and D. Tilman, Sexuality and the components of environmental uncertainty: clues from geographical parthenogenesis in terrestrial animals, *American Naturalist* **112**, 659–73 (1978).

20. J. Jaenike, An hypothesis to account for the maintenance of sex within populations, *Evolutionary Theory* **3**, 191–4 (1978).

21. A recent informant tells me that Jaenike's paper arose out of a drinking bet in graduate student days, a bet no doubt about what would or wouldn't yield cycles of selection and thus a potential for sex preservation. Perhaps the author wasn't entirely serious and this was how he did more towards convincing me than himself.

22. H. J. Bremermann, Sex and polymorphism as strategies in host parasite interactions, *Journal of Theoretical Biology* **87**, 671–702 (1980).

23. The general need for two parents to be happy about an offspring is another reason not to worry too much about human experiments following lines of that which created the sheep Dolly. Basically I believe that (a) people who try

for these offspring and succeed will find the outcome dull and frustrating compared with the interactions (sibling and parent–child) they witness around them in normal families, and (b) over generations, populations will come to notice my message: clones decline in health.

24. Mild sex-ratio catastrophes somewhat similar to that suggested are known in a few other animals. Most notably humanoid are cases in some gull populations where the male deficiencies may possibly be due to increase of sex-hormone-like chemicals in the environment (D. Cadbury, *The Feminisation of Nature*, Hamish Hamilton, London, 1997). In these gull populations 'sapphic' pairs form with the two females taking male and female 'roles'. The birds court, pseudo-copulate, lay eggs (often noticeably too many eggs due to both partners laying) and they share the incubation and feeding of young in a deceptively normal way. Some eggs hatch, but the lesbian pairs produce in the end much fewer fledglings than normal pairs. The hatching of eggs is evidently due to extra-pair copulations with the true males of the population (K. M. Kovacs and J. P. Ryder, Reproductive performance of female–female pairs and polygynous trios of ring-billed gulls, *Auk* **100**, 658–69 (1983); G. L. Hunt, Jr, A. L. Newman, M. H. Warner, J. C. Wingfield, and J. Kaiwi, Comparative behavior of male–female and female–female pairs among Western gulls prior to egg-laying, *Condor* **86**, 157–62 (1984)).

That such a repertoire of behaviour exists and allows some young to be raised suggests to me that anthropogenic chemicals in the environment may not be the whole basis of these strange situations, or at least it suggests that there have been other causes for similar sex-ratio biases in the past; otherwise such seemingly adaptive 'last-ditch' resorts would not have appeared so smoothly in several gull species.

25. B. Karlsson, O. Leimar, and C. Wiklund, Unpredictable environments, nuptial gifts and the evolution of sexual size dimorphism in insects: an experiment, *Proceedings of the Royal Society of London B* **264**, 475–9 (1997); N. Weddell and P. A. Cook, Determinants of paternity in a butterfly, *Proceedings of the Royal Society of London B* **265**, 625–30 (1998).

26. R. D. Alexander, *The Biology of Moral Systems* (Aldine de Gruyter, New York, 1987).

27. R. R. Baker, *Sperm Wars* (Fourth Estate, London, 1996).

28. D. Freeman, *The Fateful Hoaxing of Margaret Mead: A Historical Analysis of her Samoan Research* (Westview, Boulder, CO, 1999).

29. J. A. Ayala, D. Hedgecock, and G. S. Zumwalt, Genetic variation in *Tridacna maxima*, an ecological analog of some unsuccessful evolutionary lineages, *Evolution* **27**, 177–91 (1973).

FLUCTUATION OF ENVIRONMENT AND COEVOLVED ANTAGONIST POLYMORPHISM AS FACTORS IN THE MAINTENANCE OF SEX[†]

W. D. HAMILTON, PETER A. HENDERSON, and NANCY A. MORAN

———

At this symposium there seem to be no papers on the social behaviour of prokaryotes. The large long-lived species that tend to attract study by behaviourists and ethologists—those species, no doubt, which are considered to have social behaviour worth discussing—are almost always eukaryotes and *sexual*. This fact, intrinsically curious, poses a dilemma for genetical kinship theory. Our outline of the dilemma can be starting point and justification for the somewhat non-social theme of natural selection that follows.

The theory has it that the strongest positive social interactions will occur between the closest relatives: in the most straightforward view, then, clonal relatives ought to be the most co-operative. This prediction is fulfilled admirably in the very diverse and integrated behaviour shown by the cells of the metazoan body; but it is fulfilled rather poorly in a diversity of other cases—for example, in the behaviour of aphids, rotifers, armadillos, and human monozygotic twins. An excuse has commonly been offered that these clonal situations are too short-lived in evolutionary time or too rare for much progress with social development to be made before they end. But if this is true, why? At once, in so far as it is acceptable, the excuse confronts us with the problem of sex.

Data bearing on the excuse are of mixed import. It is satisfactory, for example, that aphids have now yielded at least a few cases of a sacrificial and sterile soldier caste;[1,2] aphids are a group where the lease of survival can be considered renewed each year in the annual generation of sex. Rotifers, on the other hand, contribute a negative impression: the order Bdelloidea has undergone considerable adaptive radiation without, apparently, any sex at

---

[†]In R. D. Alexander and D. W. Tinkle (ed.), *Natural Selection and Social Behavior*, pp. 363–81 (Chiron, New York, 1981).

all, and such radiation surely implies quite enough time for the formation of social colonies in suitable situations of non-panmixia if relatedness were the main controlling factor.

Even with such dialectical items aside, acceptance of the excuse merely transfers us to another kind of difficulty about relatedness. If sex is so important then our reliance on coefficients of relatedness in genetical kinship theory is in doubt: the coefficients of relatedness currently used fail to assess special advantages possessed by sexual progeny through which, so the excuse claims, they eventually outlast the progeny of clones. In explaining why identical twins co-operate as little as they do we are forced the more to wonder why husband and wife co-operate so much, or, to put the main issue more precisely, to wonder what ultimate value it is that females get from the act of mating and from the production of unprovenly useful sons. This second horn of the dilemma sharpens a little as further facts come to light concerning how males, beyond mere idleness, sometimes indifferently trample juveniles to death in the course of their fighting and mating (elephant seals) or even deliberately kill them (langurs).

Evidently the question of sex has a bearing on many fine details of social behaviour. Issues of mate choice, for example, might be understood better if we understood first, from an evolutionary point of view, why mating occurs at all.

PREVIOUS VIEWS

Interest in the evolutionary problem of sex took renewed impetus in the mid-1960s. It came to be realized that as regards any constant selection pressure asexuals have less disadvantage in their speed of adaptive response than had previously been supposed.[3–6] In fact, in some circumstances they have an advantage.[7,8]

In general the rate of response is much more dependent on the rate of occurrence of new mutations of the right kind than on how sexual mechanism or the lack of it allows the mutations to be assembled. About the same time in the 1960s genetical kinship theory bought into focus the fact that a sexual offspring is only a 'half' offspring to its parent, so raising the spectre of a widespread and often powerful selection incentive for females to produce 'full' offspring instead—by parthenogenesis. Obviously the incentive exists only to the extent that parental investment by males is absent,[9] for if the sexes invest equally the half-as-many 'full' offspring that a female could produce working alone would give no advantage over the full number of 'half' offspring she could have through co-operation with a mate. While blatantly uninvolved fathers are sufficiently common in the living world for the problem of sex to remain very acute, this point about male investment cautions against magnifying the problem too much. Unobvious forms of investment by the male might be more common than is realized. For example, it is tempting to connect the generally very low rate of incidence of parthenogenesis in moths (compared to other insects) with the long-mooted suspicion that males may contribute importantly to the food reserves of eggs through materials transferred in copulation. Gilbert[10] has demonstrated such transfer in a heliconiine butterfly. (If such transfer is important generally in Lepidoptera, a reported case of parthenogenesis in moths[11] strengthens the point made here. The case was of gynogenesis; thus, even though the females seemed not to have the option of not mating, they could be considered cheating the males in an evolutionary sense if males contribute substantially to the eggs.)

Returning from these asides to the main problem, from the renewed doubts and from the rather limited success of models referring to unequivocally 'good' and 'bad' mutations, which, when present at different loci, could be either in 'positive' or 'negative' linkage disequilibrium (LD), it is a fairly obvious step to argue that if constant evolution in one direction cannot give advantage to sex then the advantage is likely to come from some sort of varying selection pressure that continually favours different combinations among a set of alleles already present in the population. Actually this type of 'reassortment theory' of sex began long before the 1960s, indeed can be dated back to Weismann in the last century (see ref. 12). Yet in most versions of the reassortment view, environmental patchiness is considered to determine the need for continual reassortment rather than any population-wide and intergenerational changes.[13,14] Where such changes are admitted, they are thought to pose 'qualitatively new challenges' rather than simply to rehash old ones.[15]

It hardly need be said that our own model will differ in these various respects. It will have no unequivocally 'good mutations', its only patches will be temporal and coextensive with the whole population; and they will be endlessly rehashed, giving by themselves no progressive evolution at all—no more, basically, than amounts to a permit to the species to stay alive. The view of sex that the model implies approaches that of Treisman[16] and, most of all, Jaenike.[17] It also approaches the view of Thompson,[12] who emphasized a function of sex in slowing down adaptive responses to environmental changes. However, whereas Thompson's discussion was openly group selectionist and disclaimed attempt on the problem of short-term advantage against mutant asexual strains, our models can show such advantage on a short time scale—not, admittedly, the scale of a single generation but with definite advantage to sex achieved in a few.

When environmental autocorrelation is incorporated (although this development is not reported here), the general model of this paper will bear some resemblance to another of Maynard Smith's models[14,18]—one which requires that correlations between different environmental features change signs between generations, generating a negative heritability of fitness. He considered this requirement unrealistic; however, he was emphasizing abiotic factors of selection that, admittedly, are expected to show correlation, which at best is zero, and otherwise is positive, over generations. The stronger reactivity that is possible in the biotic environment in contrast provides a more plausible basis for negative heritability.

Returning to other recent versions of the reassortment approach, it has to be admitted that models with even shorter-term advantages to sex have already been obtained.[13,14,19] What were these models and why do we still seek others? The models in question postulated patchy environments, possibly but not necessarily, also fluctuating over generations and also had an essential element of sibling competition. While they had outstanding success in rendering sex immune to inroads of parthenogenesis in the short term, such success, unfortunately, could be claimed only for a rather special class of lifestyles: competition had to be very intense and consequently minimum fecundities had to be very high. Since in nature large long-lived species—which are, to repeat, the favourites of symposia on animal behaviour—are often of low fecundity and yet monotonously fail to abandon their male sex (whether industriously paternal, polygynous, or promiscuous), these existing successful models fail rather awkwardly just where, in a sense, they are most needed. Moreover, it is unappealing to have to believe that the pervasive uniformity of meiosis and fertilization is propped up by various different causes in different parts of the eukaryote kingdom.

TEMPORAL FLUCTUATION

The idea that 'varying selection pressure' or 'environmental heterogeneity' might support sex is really a compound idea; variation, heterogeneity, can be either in place or in time. The preceding successful models emphasize the spatial variation. Because it is known that temporal fluctuation of selection can have quite different results from spatial variation (see review in ref. 20), we decided to try to extend some of the models of temporally fluctuating selection beyond the question of polymorphism, which has been their usual focus in the past, to the problem of sex. Two principal considerations encouraged this approach.

One arose out of a model on the problem of sex proposed by Treisman.[16] The underlying idea here is very simple and can be explained without reference to details of the model—indeed, it had been roughly stated in connection with the disadvantageous inflexibility of asexual strains many times before.[12] If for the range of variability manifested in a species, every genotype sooner or later encounters a state of environment where its fitness is zero, then every asexual clone that a sexual species is capable of producing must correspondingly die out; on the other hand, with suitable adjustment of the 'sooner or later' (so that the sexual species is not itself forced to extinction during the period while clones are flourishing) the sexual species may well survive all of the crises that are deadly for individual genotypes.

If clones continually die out like this and the sexual species survives, there is a slow selection of genes that preclude the option of parthenogenesis more and more thoroughly—a process somewhat analogous to Darwin's evolutionary loss of wings in insects of windy islands. In fact, even death of clones is unnecessary for this selection (instead a clone might become, for example, almost permanent in some part of the original niche); it is merely necessary that clones have no possibility to recontribute to the sexual gene pool. Thus, except insofar as facultative parthenogenesis occurs, a certain metastability of sex is to be expected. This may help to explain why the disappearances of hopeful clones are not as common as the model suggests they should be. Actually we do not know of a single clear case of the re-establishment of a sexual strain in an area previously occupied by an apomict strain. And apomictic strains often seem too demographically stable and geographically uniform relative to their nearest sexual congeners to satisfy this view although admittedly the evidence is equivocal,[15] and many of the most uniform are somewhat unnatural outbreaks into new continents arising from human activity, so that it seems possible that more natural situations in which various clones spring up, flourish, and wither in succession as offshoots from a persistent 'stolon' of sexuality may still await recognition. Perhaps some cases of claimed facultative parthenogenesis may prove to be of this kind, being actually cases of the sexual parent species in process of reoccupying the subniche formerly occupied by a derived parthenomorph.

The second consideration that directed us to models with temporal fluctuation of environment derives from recent reviews of the ecological circumstances of known cases of parthenogenesis. Evidence accumulates to suggest factors of the biotic environment as being particularly supportive of sex:[17,21—24] it seems that a species is more likely to go asexual in parts of its range where the variety of its important interactant species is reduced. This suggests the possibility of certain kinds of strong reactivity in the environmental fluctuations. Whether the biotic interactions involved present themselves as prey evasions, food qualities, or tactics of predators or parasites, all are subject to frequency- and density-dependent jostling of the kind that theoretical community ecology has sought to analyse. Even the possibility of 'chaotic' fluctuations involving two or more species implies at least some degree of autocorrelation, positive or negative, of the environment of each species involved, and, of course, much more regular patterns fairly easily occur in simple models.

Limit cycles in the interaction of polymorphic host and parasite have recently been emphasized by Clarke.[25] Clarke's paper also well reviews the literature on parasite selection, and frequency dependence more generally, as sources of polymorphism. He concludes that proper appreciation of this kind of selection will largely exorcise the difficulties that seem to favour the 'neutralist' view on the abundance of protein polymorphism. Of most interest in the present context, however, is the demonstrated robustness of limit cycles as contrasted to conditions of final stasis (i.e. balanced polymorphism or allele fixation) in this kind of selection situation.

As well as a possibility of regular and quasi-regular types of limit cycle, the so-called 'chaotic' interaction[26] has been found in theoretical models where one host species is exposed to a variety of pathogens[27] and also where a prey's (or host's) genotype array was matched as regards evasion and susceptibility by a corresponding array of variation in a predator (or parasitoid).[28] The last-mentioned study took as a possible real counterpart a recorded case of irregular fluctuation in an insect host and parasitoid (see ref. 29 and fig. 1 in ref. 28): it was suggested that the variation might concern the depth to which the parasitoid had to probe for host larvae in the medium. Actually the postulated variation in the parasites as to the preferred depth of probing for prey cannot apply to the Hassell–Huffaker experiment[29] because the parasite they used is thelytokous. Nevertheless, in general, a guess about the depth of prey as a crucial variable in a host–parasitoid would not be unreasonable because polymorphisms in ovipositor length, very possibly arising out of agonistic coevolution of the kind postulated, are known in other parasitoids.[30,31] Auslander and his coworkers[28] did not relate their findings to the problem of sex. However, a similar two-species system (with a true predator, not parasitoid) was discussed independently by Hubbell and Norton[32] in connection with the very unusual occurrence of parthenogenesis in a cave cricket.

Coincidental to the variation that Auslander et al. hypothesized,[28] with the cave cricket the depth of burial in a substrate is known to be a factor in escaping the predator (a carabid beetle) although here the ovipositor in the story belongs to the prey and appears to be the device through which the prey's eggs partially escape predation. The ovipositor has proved to be of significantly greater mean length in caves where the predator is present. Data are not given on whether it is also relatively more variable in such caves. Hubbell and Norton[32] draw attention to the absence of the predator from caves inhabited by the parthenogenetic strain in contrast to its usual presence in the caves occupied by the sexual counterpart, and they tentatively link this absence to the thesis of a connection of sex with biotic diversity.[21] Obviously any linkage with the mixed strategies and cyclical and strange chaotic attractors of Auslander et al. must as yet be even more tentative. It will be shown later there is reason to suspect that one predator alone would not easily provide sufficient complexity of selection to ensure the retention of sex.

The biotic environment of the parthenogenetic cricket may well be simple not only as regards its predators but also in one sense as regards its own food. Although it is a very omnivorous scavenger, it hardly takes any food that is alive. A survey of feeding habits in relevant animal groups suggests that a diet of dead organisms or detritus is disproportionately common in the known cases of parthenogenesis. Of course, dead food and detritus (apart from associated fungi and microorganisms, which might sometimes gain from at least contact with a feeder for the sake of dispersal) have no interest in trying to avoid being eaten and hence do not coevolve in a manner likely to select for genetic diversity in whatever feeds on them. Live foods on the contrary not only are expected to flee from or oppose, in an evolutionary sense, any important feeder[33] but also are expected to diversify their evasive behaviour, toxins, etc. within the species.[25] To pursue this kind of strategy they have to retain sex themselves and in retaining it may impose a similar retention by the predator. Some hint of the degree to which living foods tend to demand specialist feeders (implying, within the more generalist species, specialist genotypes maintained in polymorphism) is perhaps illustrated by the contrast between the insect faunas of living and dead trees.[34,35]

As regards living foods, the idea of the constant and generally hostile coevolution going on between these species and those that feed on them leading to the diversification of species, and within species, to adaptive polymorphisms, has been well reviewed by Levin.[22] Although his main concern was with why recombination is kept so open (i.e. linkage so loose), Levin included parthenogenesis as the extreme beyond inbreeding of a trend that he identified as towards genetic conservatism in biotically simpler habitats. The problem of sex versus non-sex was not clearly focused, however, perhaps because of inclination to a group-oriented view of natural selection. Before Levin, other authors had already documented enough cases of situations involving so-called 'gene-for-gene' resistance and susceptibility (really a particular genotype-for-genotype pattern) in parasites and hosts to suggest that, if searched for with the thoroughness so far devoted only to agricultural and disease situations, such interacting polymorphic systems will be found to be abundant almost everywhere (refs 36 and 37; see also refs 25, 38–42). We must emphasize the 'almost everywhere' here; this paper, of course, endorses cited previous suggestions that such interactions will be found to be less abundant at the margins of a species' range and wherever else adoption of extreme sex ratios and parthenogenesis is apt to occur. The initial theoretical viewpoint hardly differs from that explained recently by Jaenike;[17] but where some statements of Jaenike were unquantified, and could be considered theoretically doubtful, we here attempt to provide confirmatory models. These will now be described. In a brief preview it may be stated that their outcomes somewhat complicate the ideas so far outlined by showing that at least two factors of fluctuating selection may be needed to stabilize sex on a reasonable basis. Looking back, for example, the Auslander model as it stands, based on single-factor differences, seems almost certain to be insufficient. Real cave crickets likewise may need a second genetic hassle with a predator—over, it might be, distastefulness of eggs as well as that already indicated concerning depth of burial—if the one predator is to provide a sufficient basis for retaining sex. In general, our models suggest better hopes of a given species for stable sexuality if it faces two important antagonist species rather than one.

THE MODELS

We confine attention to models based on two simplifying assumptions. First, we assume that asexual strains arise easily from the sexual species, so that for every genotype of the sexual species there exists a corresponding asexual clone. Admittedly, this assumption out-does reality in favour of the clones, but it gives impartial symmetry to the analysis and, in the light of various large-scale tests of apparently non-parthenogenetic invertebrates revealing occasional offspring arisen from unfertilized eggs, a viewpoint of this kind seems reasonably acceptable.[43]

Preparing for our second assumption, consider statistical descriptors required for clone and species growth, 'mean fitness', and natural selection in the fluctuating environment. Consider the multiplication of any one of the asexual clones over a series of generations: let the sequence of fitnesses in the varying successive environments be $A_1, A_2, A_3, \ldots$. At the end of the nth generation the population of this clone will have grown from its initial value by a factor equal to the product

$$A_1 A_2 A_3 \ldots A_n.$$

Thus the fitness which, if expressed constantly, would have given the same total product is

$$(A_1 A_2 A_3 \ldots A_n)^{1/n}.$$

The ideal 'mean fitness' that assesses the growth or decline of a clone would use this expression evaluated for the limit $n \to \infty$. Similar expressions, which will be referred to as LGMFs (long-term geometric mean fitnesses), can be found for the other clones; and, if they differ, comparison would show which clone will win.

The second simplifying assumption of our models, then, is that the LGMFs of all our clones are equal; fitness parameters and frequencies for the various states of environment are to be applied such that this occurs. This second assumption is obviously even more artificial than the first. Its aim is to de-emphasize competition between clones and bring to focus the one other LGMF that is crucial to assessing the role of sex, the LGMF of the sexual population as a whole. For this sexual *geometric* mean, obtained over generations in the same way as the last, the A_t in the formula will themselves be means over the various genotypes within the tth generation, these being the more familiar *arithmetic* means of genotype fitnesses (e.g. $w = p^2 a + 2pqb + q^2 c$ under random mating). The separate applications of geometric and arithmetic means for between- and within-generation contexts of selection are well known.[20,44] Our present account combines such applications and its limited achievement with respect to sexuality depends implicitly on a fundamental inequality, $AM > GM$, which holds whenever the means are composed from positive numbers.

For sex to be 'stable' we assume that the LGMF of the sexual species must be higher than the highest LGMF among the clones. Suppose that the highest such LGMF is M, and that this is a number within, or at least not far outside, the range 1 to 2. M is to be related to an LGMF of the corresponding sexual genotype that is set at 1, so that the excess of M above 1 represents the cost of meiosis in the species. In other words, when the asexual female is producing M offspring it is supposed that the corresponding sexual female produces two—of which, on average, only one is female. Obviously M will sink towards 1 insofar as male parental care is important (the unhelped parthenogenetic female producing fewer offspring), and insofar as the apomictic process is imperfect; but M could also conceivably rise above 2, as when, for example, mate-finding is a particularly limiting difficulty for the sexual species. It is arranged that, in all the models that follow, all clones have an LGMF of M.

One-locus models

We show first a simple model demonstrating a theoretical possibility for a sexual species to outcompete its clones in spite of (1) low maximum fecundity (in contrast to models of Williams[13,19] and of Maynard Smith[14]) and (2) no fitnesses that are zero (in contrast to Treisman's model[16]). It is supposed that the environment has two states conferring fitnesses on the clones and sexual genotypes as follows:

	Sexual species			Clones		
	gg	gG	GG	$\overline{gg}$	$\overline{gG}$	$\overline{GG}$
Environment 'A'	r	r^{-1}	r	Mr	Mr^{-1}	Mr
Environment 'B'	r^{-1}	r	r^{-1}	Mr^{-1}	Mr	Mr^{-1}

In accord with our assumption the LGMF of all the clones is M. Thus the analytical task is to determine the demographic behaviour of the sexual population under a specified pattern of successive environments: does the LGMF of the sexual population exceed M? Figure 1.1 shows at once how the chosen fitness sets holds out a hope for sexuality in the fact that at any given gene frequency other than 0 or 1 the LGMF of the sexual species is greater than unity. The heavy-line GM mean curve is shown for the assumption that the two environments occur with equal frequency irrespective of gene frequency. Obviously at any non-terminal gene frequency LGMF >2 can be

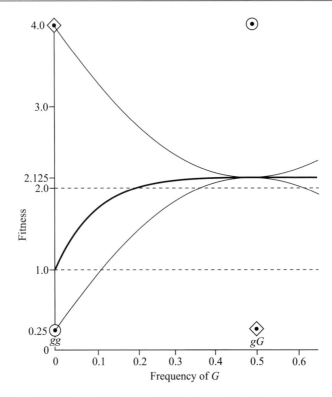

Figure 1.1 A model giving evolutionary stability of sex at low fecundity. The environment has two states occurring with equal frequency. One state, ◇, gives fitnesses 4, $\frac{1}{4}$, and 4 to genotypes *gg*, *gG*, and *GG*; the other state gives fitnesses $\frac{1}{4}$, 4, and $\frac{1}{4}$. The diagram shows various relevant fitnesses over half the gene-frequency range, the other being the mirror image of that shown. The symbols show genotype fitnesses; the thin curves show population mean fitnesses as a function of the gene frequency for each environment; and the thick or centred curve is the geometric mean between the thin curves as appropriate to showing the expected contribution to the multiplicative growth rate of the population at given gene frequencies. The dashed lines show the probable range of mean fitnesses of clones derivable from the three sexual genotypes. It can be seen that over a broad range of gene frequencies (~0.2 to 8) the growth rate of the sexual population exceeds the likely fitness maximum of 2 for the clones; and under random or regular alternation of the environments described the gene frequency moves asymptotically towards 0.5, assuring eventual renewed superiority of the sexual population after disturbance. Such final concentration at 0.5 gene frequency only holds if the two genotype fitness patterns are symmetrical; departures from symmetry result instead only in centralized distributions with a consequent chance of fixation, although this may be very small. If while maintaining symmetry the *gG* fitness value is dropped from 0.25 to zero, the LGMF of the model still exceeds 2 and the state of equal frequencies is still globally approached. However, asymmetry of homozgote fitnesses now has a more radical effect. It causes both fixation extremes to attract over nearby ranges of frequencies: each homozygous population is uninvadable by the missing allele.

obtained by a sufficient increase of r. However, gene frequency may change—and perhaps by large steps if r is high—because of the selection process, so our focus quickly transfers to the problem of the stochastic behaviour of gene frequencies over a long series of generations. Is there a steady-state distribution and, if so, is this sufficiently concentrated centrally to allow stability of sex at values of r that are biologically feasible?

It is easily seen from symmetry that a gene frequency at $q = 0.5$ is an equilibrium and it turns out that this equilibrium, although not 'protected' in the sense of Felsenstein,[20] is stable both for a regular alternation of the two environments and for random sequences based on equal frequency. Thus, the final state is concentrated ideally for giving advantage to sex, and the LGMF is simply $\frac{1}{2}(r + r^{-1})$. When $r = 4$, as in Fig. 1.1, this is 2.125, so sex is stable even against a full cost of meiosis ($M = 2$) for a fecundity of 8, and this is about as low as any known for a sexual species.

The problem with this model is, of course, that there is neither evidence nor *a priori* plausibility to the postulated pattern of fitnesses: even with $r = 4$ the degree of over- and underdominance is extreme compared with known natural cases, and nothing like the supposed switching from over- to under-dominance has ever been recorded. Worse than this, such artificiality seems ineradicable. It appears to be impossible in a one-locus model to modify fit-ness patterns or sequence of environments so as to retain the desired high LGMF in a more realistic setting. Beyond the problem of retaining a stable and central-peaked distribution, which can be overcome if necessary by bringing in frequency dependence in the occurrence of the environmental states, difficulties loom in the fact that the most hopeful patterns of symmetry in fitness coefficients force conditions where stability implies that alleles must have LGMFs of unity.

Consider, for example, the regular sequence of fitness patterns:

$$(r^{-2}, r, r) \rightarrow (r^{-2}, r, r) \rightarrow (r, r^{-2}, r^{-2}).$$

This model has a more plausible pattern of dominance and the sequence gives a stable cycle of gene frequencies. The stable cycle has been achieved through reduction of the LGMF of one homozygote. The main point, how-ever, is that even with such reduction allowed, the model remains hopeless for sex, for it is easily seen that allele G is experiencing always the same sequence of fitness, r, r, r^{-2}, etc., irrespective of tenure in heterozygote and homozygote: hence its branching process cannot grow. Next consider an even simpler and more attractive pair of patterns:

$$(r^{-1}, 1\ r) \quad \rightleftharpoons \quad (r, 1\ r^{-1});$$

and suppose here that environments occur in some regular or random pattern that ensures a non-terminal stable distribution. Almost the same drawback operates: an allele must, while in the homozygous state, meet the two types of environment, in the long run, with equal frequency; on occasions when it escapes from this state it gets only the standard unit; so overall its LGMF is 1.

In the light of such examples it appears unlikely that a viable one-locus model for the maintenance of sex based on the fluctuation of the environment can be devised. This at least gives some cool comfort with regard to traditional views of sex. If a one-locus model had proved plausibly 'sufficient' an equally important problem would be left outstanding—that of explaining the near universality of crossing-over. Even more important, such a model based on special values of the three diploid genotypes would have failed to explain the persistence of sex in the many eukaryote groups that spend almost their entire lives as haploids. In these organisms (e.g. green algae, bryophytes, and fungi), the advantages of sex must depend on the effects of recombination between loci; from this point of view a generally applicable model needs to treat at least two loci. In the many primarily haploid organisms that are isogamous, the cost of meiosis is eliminated, reducing the magnitude of the advantage needed to maintain sex.[45] Still, many forms exist that combine anisogamy and a predominance of haploidy (e.g. the genera *Pleodorina* and *Volvox* of the green algae). It is also noteworthy here that, as with aphids, rotifers, and many plants, most largely haploid species possess mechanisms for asexual reproduction; thus, the argument that sex is a maladaptive vestige, maintained only because wholly asexual lines fail to arise, is particularly weak. In addition, the fact that sex probably originated in haploids makes it especially desirable that a theory with pretensions to generality should cover life cycles that are still primarily haploid.

Two-locus models

Success in the first described one-locus model, such as it was, was caused by the possibility under sex that an allele over a series of generations can slip from homozygosity to heterozygosity and back in ways that take some advantage of the changing favourabilities of the diploid associations. As presented, each environment occasioned special production of the genotype or genotypes that will be most successful in the next environment. With two segregating loci the number of avenues open to an allele for 'changing alliance' is greatly expanded. As already indicated cyclical patterns of environmental change, ranging from regular to chaotic, strongly suggest themselves as providing advantage to continual change of alliance. It is intuitively plausible that the modes of such regrouping would themselves range correspondingly from regular to 'chaotic', so that the randomness of segregation and crossing-over could partly reflect a prevalence of chaotic cycling in nature. These issues remain vague in the present account because work in progress on two-locus models has not yet covered the promising role foreseen for cyclicity in changes of environment. The results to be presented here aim to clear the ground respecting the necessity for frequency dependence or cyclicity in gaining stability for sex.

The main question so far addressed to our model is whether a *purely random* sequence of environments is capable of stabilizing sex on a reasonable basis. The concept of null gene pair at a different locus along with our first one-locus model shows that stability certainly can be obtained: but can it occur more realistically? The answer, obtained first from simulation studies and later backed by analysis (see the Appendix on p. 44), has been a limited 'yes'. The new best-model state described as follows is somewhat less incredible; yet on present evidence is hardly to be described as realistic. The result once more suggests that while the trick can just be done with the materials allowed, much more impressive and satisfying effects await when auto-correlations in the environmental sequence can be brought in.

The model has two alleles at each of two loci. Because of the usual analytical difficulties we have used mainly Monte Carlo simulation of the genetical-demographic process. So far we have investigated only random sequences involving four states of environment supposed to favour the four types of double homozygote, *AABB*, *AAbb*, *aaBB*, and *aabb*: we shall refer to these as the 'AB', 'Ab', 'aB', and 'ab' environments.

Consider the construction of a symmetrical 3×3 fitness matrix that generalizes the property chosen for the one-locus case; namely, that the LGMF of every genotype taken individually should be unity. Let there be symmetry with respect to the different environments: each environment produces the same fitness pattern but favours different alleles. Such symmetry at once excludes forms analogous to that which gave the only success in the one-locus case.

The symmetries required in the construction of the matrix are most easily seen if the logarithms of fitness are shown: set out in such a form that the most general matrix, given as for an 'AB' environment, can be presented as follows:

$$
\begin{array}{cccc}
 & bb & bB & BB \\
aa & (-1 - y + z)\delta & (-\tfrac{1}{2} + y)\delta & (0 - z)\delta \\
aA & (-\tfrac{1}{2} + y)\delta & 0 & (\tfrac{1}{2} - y)\delta \\
AA & (0 - z)\delta & (\tfrac{1}{2} - y)\delta & (1 + y + z)\delta \\
\end{array}
$$

Adequacy of three parameters is set by the degrees of freedom, but in choosing three to apply as shown here we are guided by the intention of the e^δ can serve like the multiplicative fitness factor r in the simpler no-dominance model, so setting a general 'slope' to the present log fitness matrix, or, equivalently, a general 'degree of concavity' to the matrix of fitness itself. But, as the log fitness expression for $AABB$ shows, e^δ does not any longer necessarily set the highest fitness attained: indeed, when either or both of y and z are negative $AABB$ might not be, even in its best environment, the genotype with the highest fitness. In searching the now extensive parameter space for realistic models relevant to sex, however, it seemed reasonable to constrain y and z against patterns of co-epistasis that seem too bizarre, and we therefore mainly confine attention to cases where a specific rule of 'no over- or under-dominance' applies. That is, apart from a few trial excursions, we treat only cases where fitness of the favoured double homozygote is at least as high as any other. This rule confines (y, z) to a kite-shaped region around the zero point, with corners at $(\frac{1}{2}, 0)$, $(0, \frac{1}{2})$, $(-\frac{1}{4}, 0)$, $(0, -\frac{1}{2})$. The zero point $(y = z = 0)$ has fitness interactions multiplicative both within and between loci. It is well known that this pattern of selection makes the behaviour of the two loci effectively independent: any initial linkage disequilibrium dies away. Thus, the possibility of success of sex under any pattern of environmental change where the selection on one locus is uncorrelated with selection on the other is excluded by the argument already given for the one-locus case. Likewise, by an argument used already for the one-locus examples, the possibility of success is excluded at the kite's extreme corner at $(\frac{1}{2}, 0)$, where the fitness pattern has r^{-1} and r at opposite corners of the matrix and units everywhere else.

For our trial study of the behaviour of the model by computation, we assumed non-overlapping generations and an infinite random-mating population. Fitness was assumed to affect mortality although, given certain assumptions about breeding system, the parameters could be considered to include effects on gamete output as well. A parameter was provided for the recombination fraction between the two loci. With these assumptions and any state of the 3×3 fitness matrix and any four initial frequencies of chromosome types ab, aB, Ab, and AB, the chromosome frequencies in the next generation are easily calculated. Using a pseudo-random number generator to give sequences of the four states of the environment we followed movements of gene frequency and linkage disequilibrium over long series of generations. So far the effects of autocorrelation or gene-frequency dependence in the sequence of environments have not been studied. During the runs logarithms of population mean fitness were accumulated with a view to forming at the end, as antilog of the total, a geometric mean fitness for the whole run.

Even if gene frequencies terminalize, the expected LGMF falls only to 1. Thus values close to 1 (or, in occasional runs, just below—these caused by, presumably, accidentally hostile sample sequences) were taken to indicate failure to establish polymorphism. Interest focused on parameter states giving high GMFs, especially when these could be shown to be repeatable and steady for longer runs.

It was found that compared with the one-locus successful model fairly high best fitnesses (i.e. $\delta > 5$ or $e^{\delta} > 45$) were necessary to get any GMFs above 2; however, the really novel feature was that certain patterns having highest fitnesses of this order could predictably produce LGMF > 2 under pure stochasticity of environment; in other words, such LGMFs > 2 are possible *at stable equilibria*. The first result can be understood if thought is given to the maximum value that the GMF curve or surface attains as more loci enter: although important, this point will not be expanded here. The second result was more surprising in view of statements in the theoretical literature that seemed to give hardly any hope for continuing polymorphism when there is no overdominance in geometric mean fitness for any heterozygous geno-type.[20] Before giving further attention to this, it is convenient to mention that the linkage parameter played no important part in specifying the region where polymorphism is possible. Starting from a mid-range linkage value, we found that tighter linkage lowers any otherwise promising GMF, while less linkage, on the other hand, makes hardly any difference—such change as was shown in the Monte Carlo simulations was once more downwards although very slight. Such findings are not very surprising in view of the disadvantage that tight linkage of alleles is expected to confer in models with varying combination optima like ours.[46] Therefore in most runs we used $c = 0.25$, and this value is also assumed in the discussion that follows.

Within our 'kite' of (y, z)—this being, to repeat, the region that excludes all overdominance, and corners at $(\frac{1}{2}, 0)$, $(0, \frac{1}{2})$, $(-\frac{1}{4}, 0)$, and $(0, -\frac{1}{2})$—we found almost no suggestion of possible polymorphism when $0 \leqslant y \leqslant \frac{1}{2}$. Nor was any favourable suggestion obtained in trials of larger values of y (i.e. outside the kite). Increase of z along $y = 0$ gave even less positive indication than increase of y along $z = 0$; increase of negative z along $y = 0$, on the other hand, was bet-ter but still suggestive that slow terminalization was everywhere continuing.

In the blunt corner of the kite, with the effect sharply accentuating around the extremity at $(-\frac{1}{4}, 0)$, a region was found where protected polymorphisms can occur: here, consequently, if δ is high enough, sex is stable. In other words, following the guaranteed survival of its various alleles, the sexual strain itself can be protected from extinction from pressure of its asexual mutants. In our runs, $\delta = 5$ appeared to be just sufficient to give LGMF > 2.

The point $y = -\frac{1}{4}$, $z = 0$ corresponds to what might be called a 'flat-top, flat-bottom' fitness pattern for the loci, an example being:

$$\begin{bmatrix} 1/r & 1/r & 1 \\ 1/r & 1 & r \\ 1 & r & r \end{bmatrix},$$

where r as before is some fitness value >1. Since $\delta = 5$ corresponds to $r = e^{\delta} = 42.52$, the lowest fecundity capable of stabilizing sex in this random environment model is about 85.

Can the model really achieve what, at first reading of the literature of selection in fluctuating environments (Gillespie[47]; see ref. 20 for a review), seems to have been shown impossible? The seeming contradiction depends on the fact that Gillespie summarized a sufficient condition for polymorphism but not a necessary one (see ref. 20). The polymorphism generated by this system is 'unprotected' in the sense that it has a vanishing tendency to preserve itself at extreme gene frequencies, and this property was the matter of concern in Gillespie's proof. However, small as it may become, an overall centralizing tendency exists as long as the gene frequency is not 0 or 1. When with the present model at $y = -\frac{1}{4}$, $z = 0$, we temporarily focus on one locus and treat the other as if merely contributing to 'variability' in fitness within each of the three genotypes, it turns out that overdominance in heterozygote geometric mean fitness actually holds for every combination of gene frequencies (varying both q and Q, between 0 and 1) and of linkage disequilibrium. That this is true in the framework of assumptions reasonable for our model (random mating, equal frequency of the four environments, neither locus already fixed and at least four of the nine genotypes present) is proved in the Appendix (p. 44).

A less rigorous but perhaps useful intuitive approach to the working of this model may also be outlined as follows. The fitness matrix oriented as before, at the same time that it tends greatly to increase the frequency of the favoured double homozygote (bottom right), tends also to create linkage disequilibrium in favour of the homozygotes on the non-principal diagonal (top right and bottom right). While by chance one environment is repeating itself, the increasingly common homozygote is also the most highly fit, so that very high mean fitnesses are attained. But obviously this cannot continue if polymorphism is to be preserved, and if the environment completely reverses (so as to favour the opposite-corner genotype), the common genotypes are then extremely unfit, and previous gains to the LGMF product are offset. But the direction of the linkage disequilibrium means that if, instead of reversing, the environment turns (i.e. favours top right or bottom left after a run on top left or bottom right), there tend to be substantial numbers of homozygotes and heterozygotes able to express moderate or high fitness. The amount of 'turning' in a random sequence is evidently enough for this effect to achieve results of the kind needed: the same line of thought, however, obviously extends greater hopes as regards what may be found with models where cyclical tendencies are built in and cause turning at a suitable rate. One simple case already tried, for example, had the factor that switches selection for A versus a subject to random change only every fifth generation while the factor switching for B versus b was subject to random change every fourth: it was then found that $r = 9$ was enough to give advantage to sex. This would correspond to a fecundity of 18. This schedule does not give a cycle, of course, but it does provide for many more turns than reversals and also for many short runs of the same environment. Special kinds of cycles can probably lower the requisite fecundity still further; and, as already indicated, biotic interreactions in nature may supply cyclical pressures rather freely—Baltensweiler,[48] for example, documents one example.

Turning back from this preview to limitations of the present model, the reason why other authors attached importance to there being more than negligible centripetal tendency from selection at extreme gene frequencies is that a polymorphism is likely to be lost by drift if extreme frequencies are reached, and here it has to be admitted that the model described with $r = 42.5$ and random equiprobable environments gives a very concave distribution (Fig. 1.2) and quite frequently visits gene frequencies so extreme (e.g. 10^{-7}) that in any reasonable population an allele would be lost. In the case just mentioned with $r = 9$ similar extremes were visited. Considerations that oppose this tendency for fixation, however, are that (1) extremes for an allele tend to occur when frequencies at the other locus are in the mid-range and here the tendency to recover is stronger, and (2) weak density-dependent effects and also recurrent mutation could, in nature, provide for potential buffering or recovery at extremes. Altogether the need for polymorphism protection at the extremes seems overemphasized in the literature and may have caused neglect of otherwise plausible models.

How plausible is the specific fitness pattern suggested? So far variants of the general matrix have not been much explored although it has been con-

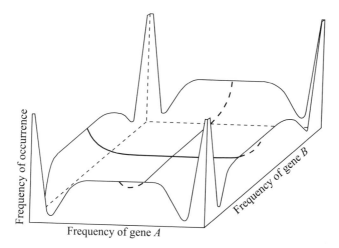

Figure 1.2 Roughly sketched form of the ultimate distribution of gene frequencies at two unlinked loci when each serves as epistatic modifier of multiplicative interaction at the other, so as to produce the 'flat-top, flat-bottom' pattern described in the text. The environment is fluctuating at random (i.e. no gene-frequency dependence) among four states, these each favouring a particular corner of the diagram (i.e. favouring an '*ab*', '*aB*', '*Ab*', or '*AB*' allele combination).

firmed that the seemingly even more artificial condition $y < -0.25$ on $z = 0$ permits stability of sex at much lower values of δ.

Jointly the 'flat top' and 'flat bottom' condition as it stands does not suggest any easy genetical or ecological interpretation. Equality of the three higher fitnesses could be subject to a claim similar to one sometimes brought in for the common occurrence of dominance: if having three out of four alleles 'right' for a current environment permits full fecundity (say), then having all four right can gain no additional advantage. But this cannot be applied to equality of the three low fitnesses. If we let the low-corner fitness sink to $1/r^2$ (or completely to zero) it is obvious that all four doubly homozygous asexual strains are at once rendered inviable compared with the rest. Perhaps this is not unreasonable: apomictic strains are often claimed to be highly heterozygous. It will be of interest to see how the distribution and the conditions for sex change in these circumstances, and also perhaps in various others that will leave the doubly heterozygous asexual strains as the only possible contender against sex. However, a hint that with or without such modification the 'flat-top' condition in our model at least could correspond to a relevant feature of the real world comes from various studies of the so-called 'gene-for-gene' interactions that often hold between the genetic polymorphisms of parasites and of their hosts.[36,40] It is very commonly reported that resistance to a given strain of parasite is dominant. Because the genetic tests were often made by first selecting for the trait that would be advantageous to the host (in the given context) and then backcrossing, and because reference to modifiers or an element of polygenicity often occurs in the reports as well, it seems possible that, seen in terms of our model, the experimental procedure is unintentionally selecting only the high side of the fitness matrix for examination; along this side, of course, our matrix shows dominance as found. But if the 'polygene' or 'modifier' effects are indeed accountable to another locus of equal importance, as suggested here, we have to confess to complete ignorance as to the realities of selection with which that other locus might be concerned.

A weakness in the above tentative view is that for variants of the parasite it is more often the apparently less advantageous avirulent type that appears to be dominant. But perhaps neither artificial selection in the maintenance of parasite cultures nor natural selection of wild variants is easily able to press virulence to its extreme, this being caused, in each case, by the difficulty of 'maintaining the culture'.

Involvement of these gene-for-gene systems would, as already argued, greatly improve the situation for sex because they must bring at least some element of gene-frequency dependence into environmental changes. Once this or another source of autocorrelation of environment is brought in, a wide variety of other forms of the matrix, including, it is to be hoped, forms that are much more realistic, may prove successful.

What has been shown is that at least for one special pattern of epistasis such dependence or autocorrelation is not strictly necessary: the sequence of environments can be random. This suggests that climatic variability, for example, could after all be contributing substantially towards the maintenance of sex, even though on general evidence it seems very unlikely to be the sole agent. Because one recent study found very high correlations between prevalence of enzyme polymorphisms and climatic variability, leaving apparently not too much polymorphism in need of explanation by anything else,[49] extension of the model in a way that would help climate to play a part might seem an almost necessary step to secure its survival. However, while climate may be very important generally for polymorphism, Bryant's analysis did not exclude the possibility of other agencies, possibly biotic ones, being important for at least a few polymorphisms. While principal factors in his factor analysis of polymorphisms generally regressed on climatic variability very well, the same regression for second factors was usually weak and there were always some loci that loaded very heavily on second factors while light on the first. Perhaps these second factors were concerned with the kind of hostile biotic forces that have been the main focus of the present discussion. Of course, complete independence of climate would not be expected; if the biotic emphasis in the problem of sex is right, climatic fluctuation is still certain to play some part in the detail of how an array of co-antagonist species keep constantly on the move.

CONCLUSIONS

Compared to parthenogenetic offspring, sexual offspring transmit genes of a mother diluted by one half. Yet the record of the living world shows that in most circumstances females 'prefer' sexual offspring. In some cases it is known also that females prefer some mates to others; and that they do so on some sort of eugenic grounds is plausible. In spite of these indications and the ramification of sex-dependent phenomena throughout biology (e.g. diploid genetics) we still do not know what sex is for.

In the field of social behaviour such lack of understanding leaves large gaps in genetical kinship theory. For example, benefits of outbreeding versus inbreeding underpin breeding structure, and breeding structure itself determines, through kinship considerations, major trends of social behaviour. Again, if benefits of sex are immediate and substantial, there is a missing factor based on outbreeding that ought to be included in assessments of reproductive value.

Circumstantial evidence accumulates to suggest that complexity of the biotic environment is particularly supportive to sex. This paper has furthered such evidence by examples and study of possible mechanisms. Mechanisms with cyclical interactions among species are highly promising. In our attempt to clear the ground of possible non-cyclical, although fluctuating, selection, we find simple models where even random environmental sequences give sex short-term stability against parthenoform competition, and that succeed in this (1) without high fecundities, (2) without sibling competition, (3) without involvement of more than one locus, and (4) without overdominance in the fitness pattern manifested in any single generation. However, in these models (3) and (4) could not be achieved together, and both the one-locus and two-locus successful cases had certain grave and seemingly ineradicable artificial features. Hence, former theoretical emphasis on recombination as intrinsic to the function of sex was upheld by the results: at least two loci and two at least partially independent agencies of selection appear to be necessary for a reasonable model. As a whole the results suggest that cycles of selection and/or selection responsive to the dynamics of the species themselves (as in interspecies interaction) may be crucial, especially in the cases hitherto treated as the most puzzling (low fecundity organisms with no paternal care). It is suggested that pure random fluctuating selection may play a supportive part that is secondary to the effects of cycling selection. Further work in progress (not reported in this paper) shows that indeed conditions (1), (2), and (4) can be retained under much more natural patterns of dominance and epistasis when forms of environmental cyclicity are added.

It seems justified to conclude the seemingly rather non-social theme of theoretical biology reported here by the suggestion that to understand social behaviour at a deeper level evolutionary behaviourists may soon need to pay closer attention to transgenerational patterns in the major forces of selection on their organisms: in such patterns eventually may be found reasons why one species tolerates or prefers incest, why one has lower fecundity, more parental care, and why one has males that are more showy, less industrious, and yet more preferred.

APPENDIX

We assume two loci each with two alleles (a, A and b, B), and an environment that has four equally probable states labelled 'ab', 'aB', 'Ab', and 'AB'. The labels designate the type of double homozygote favoured in each state and indicate the orientation of the fixed fitness matrix. This matrix is shown below for the 'aB' state (the form for the 'AB' state is shown in the text):

$$
\begin{array}{c c c c}
 & bb & bB & BB \\
aa & 1 & r & r \\
aA & r^{-1} & 1 & r \\
AA & r^{-1} & r^{-1} & 1
\end{array}
\tag{1}
$$

Given (1) random mating, (2) loci A and B not completely linked, and (3) $r > 1$, it is required to prove that occurrence of the four environments in random sequence establishes polymorphism for both loci.

From symmetry of the environment and of the matrix it is obviously sufficient to prove this for one of the two loci. We proceed, therefore, to show
that the LGMFs for the three genotypes at locus A fulfil a condition of overdominance in geometric mean fitness, treating locus B as a modifier.

Let p, q, P, and Q denote the frequencies of alleles a, A, b, and B ($p + q = 1$
and $P + Q = 1$) and let D denote 'linkage disequilibrium', implying that the
frequencies of the four chromosome types ab, aB, Ab, and AB are $pP + D$,
$pQ - D$, $qP - D$, and $qQ + D$. Under random mating, the state of the population is then completely determined by the three parameters q, Q, and D. If
polymorphism and a steady-state distribution over q, Q, and D is attained,
then any small region in the neighbourhood of a specified (q, Q, D) will be
visited by the population repeatedly, albeit with very varying frequency
according to the location. With one quarter of these visits an 'ab' environment will follow, with one quarter an 'aB' environment, and so on: hence
for the very long-term assessment of the geometric mean performances
of aa, aA, and AA we can group visits in fours that cover instances of all
four environments, and hence see that for each genotype at the A locus the
quartic root of the product of the four different mean fitness performances
in each environment will evaluate the long-term contribution of that genotype.[47]

Let $\tilde{W}_0$, $\tilde{W}_1$, and $\tilde{W}_2$ denote LGMFs of aa, aA, and AA in this sense (the
tilde is used for a GM over the four environments; the period in the subscript
is used to denote AMs within each environment and within the A-locus
genotype taken over three genotype conditions at the B locus). Then our
argument states that $\tilde{W}_{0.} < \tilde{W}_{1.} > \tilde{W}_{2.}$ for all (q, Q, D) will be a sufficient condition for polymorphism and for the existence of an asymptotic distribution
as assumed.

We proceed to prove that this condition holds for q and Q in the closed
interval 0 to 1 with D at any feasible value in the closed interval $-\frac{1}{4}$ to $\frac{1}{4}$.

From the frequencies of the four chromosome types and random mating
the distribution prior to the selection of B genotypes within each A genotype
is easily determined to be as follows:

	bb	bB	BB		
aa	P_0^2	$2P_0Q_0$	Q_0^2	/1	
aA	P_0P_2	$P_0Q_2 + Q_0P_2$	Q_0Q_2	/1	(2)
AA	P_2^2	$2P_2Q_2$	Q_2^2	/1	

where $P_0 = P + D/p$, $Q_0 = Q - D/p$, $P_2 = P - D/q$ and $Q_2 = P + D/q$.
When D is such that one of these four frequencies is zero while the rest are
positive, then D is at maximum or minimum of its range for the (q, Q) in
question (however, at the one point $q = Q = 0.5$ either both P_0 and Q_2 or P_2
and Q_0 reach zero together at the extreme of D). We may take as typifying

the form of (2) at an extreme of D the case where $Q \leqslant p$ and D is at its minimum, $D = -qQ$, which gives $Q_2 = 0$, $P_2 = 1$:

$$
\begin{bmatrix}
P_0^2 & 2P_0Q_0 & Q_0^2 \\
P_0 & Q_0 & 0 \\
1 & 0 & 0
\end{bmatrix}. \tag{3}
$$

Other extreme forms (i.e. for other (q, Q) and/or D_{max} rather than D_{min}) always have (a) 'convergence' to a unit in some corner of the matrix, (b) either P_0 and Q_0 or P_2 and Q_2 placed so as to be either right- or left-adjusted in the middle row, and (c) the remaining row filled by a distribution that can be considered as that 'Hardy–Weinberg' that derives from frequencies recorded in the middle row. When $D = 0$, $Q_0 = Q_2 = Q$ and all rows are 'Hardy–Weinberg' and identical.

Turning to fitnesses, imagining (1) rotated to its four positions, it is seen that for the middle row (aA) only two equiprobable patterns occur. Hence, the LGMF of Aa, W_1, is particularly simple. For $D = -qQ$, for example, enabling (3) to be used, producing frequencies with fitnesses and adding to obtain AMs, then producing again and taking square root to obtain the GM, we have

$$
\tilde{W}_{1.} (D = -qQ) = \sqrt{(P_0 r^{-1} + Q_0) (P_0 r + Q_0)}. \tag{4}
$$

By like procedure for $D = 0$ we have

$$
\tilde{W}_{1.} (D = 0) = \sqrt{(P_0^2 r^{-1} + 2P_0Q_0 + Q_0^2 r) (P_0^2 r + 2P_0Q_0 + Q_0^2 r^{-1})}.
$$

Multiplying the first factor by r^{-1} and the second by r, so as to form perfect squares, we find

$$
\begin{aligned}
\tilde{W}_{1.} (D = 0) &= (P_0 r^{-1} + Q_0) (P_0 r + Q_0) = \{\tilde{W}_{1.} (D = -qQ)\}^2 \\
&= P_0^2 + P_0Q_0 (r + r^{-1}) + Q_0^2. \tag{5}
\end{aligned}
$$

For $r > 1$ we have $r + r^{-1} > 2$ and hence $\tilde{W}_1(D = 0) > \tilde{W}_{1.} (D = -qQ) > 1$, except that when $Q_0 = 0$ or 1 both LGMFs are unity. With the general form of the mid-row as in (2) and treating Q_0 as a constant and Q_2 as a variable reflecting D, it is easily shown that $\tilde{W}_1(D = 0)$ is a maximum of a wholly upward-convex function. Hence (4), which applies both for D_{min} and D_{max} (with Q_0 fixed), gives a minimum for the LGMF of the heterozygote. If it can be shown that even this minimum exceeds $\tilde{W}_{0.} (Q_0)$, for all Q_0 then $\tilde{W}_{1.} > \tilde{W}_{0.}$ will be proven for all Q_0 and Q_2.

The distribution of B genotypes within the genotype aa may be combined with fitnesses of B genotypes in all four environments to give the following formula for the LGMF of aa analogous to (4):

$$\tilde{W}_{0.} = \left\{ \begin{array}{l} (P_0^2 r^{-1} + 2P_0Q_0 r^{-1} + Q_0^2) \\ \times (P_0^2 \quad + 2P_0Q_0 r \quad + Q_0^2 r) \\ \times (P_0^2 r \quad + 2P_0Q_0 r \quad + Q_0^2) \\ \times (P_0^2 \quad + 2P_0Q_0 r^{-1} + Q_0^2 r^{-1}) \end{array} \right\}^{\frac{1}{4}}.$$

With some effort this can be reduced to the form:

$$\tilde{W}_{0.}^4 = 1 + 2(1 - P_0Q_0)P_0Q_0 Z + (2 + P_0Q_0)P_0^3 Q_0^3 Z^2, \tag{6}$$

where $Z = r + r^{-1} - 2$.

From (5) we find in similar notation:

$$\{\tilde{W}_{1.} = (D = -qQ)\}^4 = (1 + P_0Q_0 Z)^2$$
$$= 1 + 2P_0Q_0 Z + P_0^2 Q_0^2 Z^2. \tag{7}$$

From $r > 1$ we have $r + r^{-1} > 2$, so that $Z > 0$; and we also have $0 < (1 - P_0Q_0) \leqslant 1$ and $0 \leqslant (2 + P_0Q_0)P_0Q_0 < 1$. Hence Z and Z^2 are positive and for each the coefficient in (6) is smaller than in (7), so that $\{\tilde{W}_{1.}(D = -qQ)\}^4 > \tilde{W}_{0.}^4$. Quartic roots of these are unequal the same way. As already explained it follows from this generally that $\tilde{W}_{1.} > \tilde{W}_{0.}$.

The inequality $\tilde{W}_{1.} > \tilde{W}_{2.}$ can be obtained by exactly the same argument except that at the start Q_2 is fixed while D is varied such that Q_0 passes from zero to one: this argument, in effect, explores the entire feasible space of (q, Q, D) a second time. Finally, of course, the whole argument can be repeated to demonstrate overdominance in the LGMFs and consequent polymorphism at the B locus.

We have not yet tried very hard to extend the foregoing proof to the more general matrix given (in log form) in the text; but preliminary points about work in this direction seem worth making. First, there is a large region within our kite where polymorphism will not be obtained. When the function $\tilde{W}_{1.}(D = 0)$ lies wholly below the function $\tilde{W}_{0.}(Q_0)$, then the required over-dominance cannot occur: it appears that this may apply to the whole region $y > 0$. Second, our proof as given identifies merely a sufficient condition for polymorphism, not a necessary one, and it is obvious that there must exist a variety of related fitness matrices that give polymorphism, but have $\tilde{W}_{1.}(D = -qQ) < \tilde{W}_{0.}$. In spite of such relation, it is easy to image that $\tilde{W}_{1.} > \tilde{W}_{0.}$ could hold overall if, for example, the more extreme values of D are rarely visited. For such cases other criteria, perhaps based on behaviours at low gene frequencies, need to be developed.

References

1. S. Aoki, *Colophina clematis* (Homoptera, Pemphigidae) an aphid species with 'soldiers', *Kontyu* **45** (2), 276–82 (1977).
2. S. Aoki, S. Yamane, and M. Kiuchi, On the biters of *Astegopteryx styracicola* (Homoptera, Aphidoidea), *Kontyu* **45** (4), 563–70 (1977).

3. J. F. Crow and M. Kimura, Evolution in sexual and asexual populations, *American Naturalist* **99**, 439–50 (1965).

4. J. F. Crow and M. Kimura, Evolution in sexual and asexual populations, *American Naturalist* **193**, 89–91 (1969).

5. J. Maynard Smith, Evolution in sexual and asexual population, *American Naturalist* **102**, 469–73 (1968).

6. J. Felsenstein, The evolutionary advantage of recombination, *Genetics* **78**, 737–56 (1974).

7. S. Karlin, Sex and infinity: a mathematical analysis of the advantages and disadvantages of recombination, in M. S. Bartlett and R. W. Hiorns (ed.), *The Mathematical Theory of the Dynamics of Biological Populations*, pp. 155–94, (Academic Press, London, 1973).

8. I. Eshel and M. W. Feldman, On the evolutionary effect of recombination, *Theoretical Population Biology* **1**, 88–100 (1970).

9. R. L. Trivers, Sexual selection and resource-accruing abilities in *Anolis garmani*, *Evolution* **30**, 253–69 (1976).

10. L. E. Gilbert, Postmating female odor in *Heliconius* butterflies: a male-contributed antiaphrodisiac, *Science* **193**, 419–20 (1976).

11. C. Mitter and D. Futuyma, Parthenogenesis in the fall cankerworm, *Alsophila pometoria* (Lepidoptera, Geometridae), *Entomologia Experimentalis et Applicata* **21**, 192–8 (1977).

12. V. Thompson, Does sex accelerate evolution? *Evolutionary Theory* **1,** 131–56 (1976).

13. G. C. Williams, *Sex and Evolution* (Princeton University Press, Princeton, NJ, 1975).

14. J. Maynard Smith, A short-term advantage for sex and recombination, *Journal of Theoretical Biology* **63**, 245–58 (1976).

15. J. Maynard Smith, *The Evolution of Sex* (Cambridge University Press, Cambridge, 1978).

16. M. Treisman, The evolution of sexual reproduction: a model which assumes individual selection, *Journal of Theoretical Biology* **60**, 421–31 (1976).

17. J. Jaenike, An hypothesis to account for the maintenance of sex within populations, *Evolutionary Theory* **3,** 191–4 (1978).

18. J. Maynard Smith, What use is sex?, *Journal of Theoretical Biology* **30,** 319–35 (1971).

19. G. C. Williams and J. B. Mitton, Why reproduce sexually?, *Journal of Theoretical Biology* **39,** 545–54 (1973).

20. J. Felsenstein, The theoretical population genetics of variable selection and migration, *Annual Review of Genetics* **10,** 253–80 (1976).

21. R. R. Glesener and D. Tilman, Sexuality and the components of environmental uncertainty: clues from geographical parthenogenesis in terrestrial animals, *American Naturalist* **112,** 659–73 (1978).

22. D. A. Levin, Pest pressure and recombination systems in plants, *American Naturalist* **109,** 437–51 (1975).

23. M. T. Ghiselin, *The Economy of Nature and the Evolution of Sex* (University of California Press, Berkeley, 1975).

24. O. Cuellar, Animal parthenogenesis, *Science* **197,** 837–43 (1977).

25. B. Clarke, The ecological genetics of host–parasite relationships, in A. E. R. Taylor and R. Muller (ed.), *Genetic Aspects of Host–Parasite Relationships*, pp. 87–103 (Blackwell, London, 1976).

26. R. M. May, Biological populations with nonoverlapping generations: stable points, stable cycles and chaos, *Science* **186,** 645–7 (1974).

27. I. Eshel, On the founder effect and the evolution of altruistic traits: an ecogenetical approach, *Theoretical Population Biology* **11,** 410–24 (1977).

28. D. Auslander, J. Guckenheimer, and G. Oster, Random evolutionarily stable strategies, *Theoretical Population Biology* **13,** 276–93 (1978).

29. M. Hassell and C. Huffaker, Regulatory processes and population cyclicity in laboratory populations of *Anagasta kühniella* (Zeller) (Lepidoptera: Phycitidae). III. The development of population models, *Research in Population Ecology* **11,** 186–210 (1969).

30. R. R. Askew, *Parasitic Insects* (Heinemann, London, 1971).

31. W. D. Hamilton, Wingless and fighting males in fig wasps and other insects, in M. S. Blum and N. A. Blum (ed.), *Reproductive Competition, Mate Choice and Sexual Selection in Insects*, pp. 167–220 (Academic Press, New York, 1979) [reprinted in *Narrow Roads of Gene Land*, Vol. 1, pp. 435–82].

32. T. H. Hubbell and R. M. Norton, The systematics and biology of the cave-crickets of the North American tribe Hadenoecini (Orthoptera Saltatoria: Ensifera: Rhaphidophoridae: Dolichopodinae), *Miscellaneous Publications of the Museum of Zoology, University of Michigan* **156,** 1–124 (1978).

33. D. C. Olson and D. Pimental, Evolution of resistance in a host population to an attacking parasite, *Environmental Entomology* **3,** 621–4 (1974).

34. C. S. Elton, *The Pattern of Animal Communities* (Methuen, London, 1966).

35. W. D. Hamilton, Evolution and diversity under bark, in L. A. Mound and N. Waloff (ed.), *Diversity of Insect Faunas*, Symposia of the Royal Entomological Society of London, No. 9 (Blackwell Scientific, Oxford, 1978) [reprinted in *Narrow Roads of Gene Land*, Vol. 1, pp. 394–20].

36. P. R. Day, *Genetics of Host–Parasite Interaction* (Freeman, San Francisco, 1974).

37. P. M. Dolinger, P. R. Ehrlich, W. L. Fitch, and D. E. Breedlove, Alkaloid and predation patterns in Colorado lupine populations, *Oecologia* **13,** 191–204 (1973).

38. G. F. Edmunds, Jr and D. N. Alstad, Coevolution in insect herbivores and conifers, *Science* **199,** 941–5 (1978).

39. J. A. Dunn and D. P. H. Kempton, Resistance to attack by *Brevicoryne brassicae* among plants of Brussels sprouts, *Annals of Applied Biology* **72,** 1–11 (1972).

40. A. R. Barr, Evidence for the genetical control of invertebrate immunity and its field significance, in K. Maramorosch and R. E. Shope (ed.), *Invertebrate Immunity Mechanisms of Invertebrate Vector–Parasite Relations*, pp. 129–35 (Academic Press, London, 1975).

41. C. S. Richards, Genetics of the host–parasite relationship between *Biomphalaria*

glabrata and *Schistosoma mansoni*, in A. E. R. Taylor and R. Muller (ed.), *Genetic Aspects of Host–Parasite Relationships*, pp. 45–54 (Blackwell, London, 1975).

42. J. M. Rutter, M. R. Burrows, R. Sellwood, and R. A. Gibbons, A genetic basis for resistance to enteric disease caused by *E. coli*, *Nature* **257,** 135–6 (1975).

43. M. J. D. White, *Animal Cytology and Evolution*, 3rd edn, pp. 701–3 (Cambridge University Press, Cambridge, 1973).

44. J. F. Crow and M. Kimura, *Introduction to Population Genetics Theory* (Harper and Row, New York, 1970).

45. Williams 1975 (ref. 13), p. 113.

46. B. Charlesworth, Recombination modification in a fluctuating environment, *Genetics* **83,** 181–95 (1976).

47. J. Gillespie, Polymorphism in random environments, *Theoretical Population Biology* **4,** 193–5 (1973).

48. W. Baltensweiler, The relevance of changes in the composition of larch bud moth populations for the dynamics of its numbers, *Proceedings of the Advanced Study Institute on Dynamics of Population Numbers*, pp. 208–19 (Oosterbeck, 1970).

49. E. H. Bryant, On the adaptive significance of enzyme polymorphisms in relation to environmental variability, *American Naturalist* **108,** 1–16 (1974).

MESSING THE PLOTTER

Sex Versus Non-sex Versus Parasite

It was said even at the time of Drelincourt that no less than two hundred and sixty-two groundless theories of sex had been suggested; and it may be added that since that time there has been no falling off of interest in the sex question if the number of new theories proposed is a criterion. The latest theory which I venture to formulate, is sure, therefore, to meet with scepticism.

G. A. REID[1]

LONG before I had finished my share of the writing of the paper of the last chapter I was already working on other models. Many were tried before I came to those selected for the publication of this chapter. I mathematized them as far as I could and then used simulation on the Michigan main computer, accessing it either from a huge cave-like room called NUBS in the basement of the herbarium (our nearest-neighbour building) or, later, from a smaller terminal room in my own floor of the Museum of Zoology. NUBS had firebrats plus a spectrum of dazed, earnest, and sometimes frighteningly expert freshmen. Freshmen are first-year university students; firebrats, not arsonists but primitive insects. Smartly striped like football players, the latter dashed swiftly about on the floor under the piles of unwanted output paper, especially favouring that mounded against walls. I think their name's origin lies in their being commonly found near bakery ovens. The nature of their food there is obvious; but what it was under the paper in NUBS is hard to imagine unless perhaps there were mummified students, dead of their sorrow at their unco-operating programs.

My programs were fickle and certainly caused me rage, sorrow, and humility as well as that unparalleled blank feeling that the programmer suffers where logic itself seems to have abandoned the universe—the program has to have been right and the computer wrong. But eventually my programs did co-operate. Their aim was to simulate a coevolutionary process in which a population of hosts was being affected by a population of parasites and the parasites by the hosts. At least in principle the hosts always and the parasites sometimes were arranged to be active in the models in two alternative versions: in one they reproduced sexually while in the other they were parthenogenetic. I was trying to see what environments and what interactions would make life easy for the sexual variants and hard for the parthenogenetic. Once my lines of work were established, model grew from model under a kind of secondary Darwinian selection with ideas effective for success of sex continued and explored and ones that didn't work dropped. High ability to overcome costly sex, low fecundity, and, above all, my hunches about credible fitness patterns were the conditions of my 'niche'. An example of a significant 'mutation' or even perhaps 'macropreadaptation' (if that is the word I need to create) will be shown shortly: as will be seen, the finding in this case—paths to slower cycles and implications of these—caused a burgeoning of my interest in sexual selection (Chapter 6).

Almost all large organisms are diploid. In other words, every cell in the body reads a programme on how to develop out of two sets of chromosomes, one originating from one sexual parent and one from the other. If soon after such a double set has formed by a successful fertilization, the new offspring revert to the gametic chromosome number by a 'meiotic' or 'reduction' division of the assembled chromosomes, with the main series of body-building divisions following this reduction, then a so-called haploid multicellular organism results. It is very important to note that the reduction here is not a return to having all of the one parent's chromosomes or all of the other's but to having a mixture from them, and a mixture even within individual chromosomes. At least for higher organisms the occurrence of recombination is the hallmark of sexual reproduction. Recombination of genes by sexuality occurs whenever there is a 'reduction' or 'meiotic' division. Thus recombination is just as characteristic of haploid organisms as it is of diploids.

An example of a haploid organism is a moss plant. Provided a moss performs its sexual cycle, it fully participates in whatever the benefits may be of genetic recombination.[2] Because of the simplicity of the genetics and frequency changes under selection in haploids I decided that it would be better both for my programming and my interpretations if I focused on haploid recombination as in a moss. Haploids are by no means genetically impoverished in their variation,[3] a fact which by itself eliminates some whole theories for the maintenance of polymorphism. Nor are the tissues made of haploid cells insignificant in the affairs of the world. Most of the moss plant one notices is haploid but parts of it sometimes present may not be. The whiskery spore shakers that grow above the plant are diploid; they are a brief stage growing parasitically on each haploid mother. The whiskers can be conspicuous and sometimes quite relatively large when present, I have to admit: they make, for example, that silvery or golden hair standing over the tuft on the old wall or your roof tiles or that which adds a reddish fuzz to the green carpet spread on the black soil where the old bonfire was. Still, they stay a very small fraction of a typical moss. In other places, through the peat they diligently pack away, waterlogged and unrotting, beneath them and built out of the billions of their haploid remains, some mosses, notably some *Sphagnum* species, have sequestered for the general benefit or otherwise of our planet, probably greater deposits of non-gaseous carbon than have all diploid 'higher' plants of the taiga and temperate and tropical forests put together.[4] If you doubt my claim here, and point out that also included in the peaty remains would be the diploid spore-making parts just mentioned, I have to counter that the peat mosses have, as it happens, very small diploid sporophores—they are like the buttons on that loose-knit pastel pullover compared to the pullover itself. 'Woolly' *Rhacomitrium*, an abundant and characteristic moss on our mountain tops and in other British abodes of bleakness and rain, is another very successful haploid. Growing where no other moss nor leafy plant can, often it carpets the ground rocks to a depth of tens of centimetres. The carpet it spreads, it is admitted, is patchy and threadbare, but how it still clings and grows where nothing else can! And again its capsules, its only diploid stage, tend to be small and few.[5]

Male bees, wasps, and ants are also haploid and they thus join the mosses in giving us this clear hint that diploidy of bodies cannot have a leading role in the drama of sex. Sex of the kind that was interesting me most—that with the neat symmetrical chromosome performances of fertilization and meiosis, as mentioned above—seemed to me clearly an evolutionary unity with all its procedures and genetic results much the same whether one looked at it in plants, animals, or unicells. This unity and the implied age for sexual phenomena—back to the earliest single-celled organisms of the Precambrian era—made for me an awesome phenomenon, a spectacle I viewed as I might a cathedral. It was hard to believe, as well as unpleasing to imagine, diploidy as being necessary in the engineering of, say, the nave of this cathedral while for its transepts some quite different principles were being applied. Would a church historian see one ogival door arch in an otherwise Gothic cathedral to imply inroads of Islamic influence in the whole? I think not: to assume so would ignore an obvious unity. Moreover in discussing haploidy, diploidy, and sex we are certainly not discussing just ornaments but rather how sex functions in life's strategic engineering. A door very explicitly Islamic, or else decorative motifs and tesselations that are superficially upon the 'spandrels', or on whatever[6]—these might be another matter, indeed be admired variants that were requested from some convert craftsman from the East. Yet still they would remain superficial, could not be the key to the whole. As will be seen later in this volume (Chapter 6), it could be that even the seeming mere ornaments of the structure of sex might really, endorsing Daniel Dennett[7] and me and Marlene Zuk, be conveying an underlying functional message or purpose.

Male honeybees—drones—have sometimes been treated as useless ornaments of their species and they happen to be haploid. Maurice Maeterlinck viewed them this way but was in serious error.[8] That the drone is no dwindling appendix or toenail to his species could be emphasized from two points of view, that of population and of individual. First, if we may judge by a remarkable but very local and seemingly inefficient South African race of honeybees, which happens almost to lack males and to reproduce mainly by parthenogenesis,[9] the near loss of sperm service and consequent loss of recombination have not been a good thing for it, because, had it been, the system of avoiding sex having once been invented would have spread all over Africa instead of existing in just a small corner. In line with my theme throughout this volume haploid males, then, are probably necessary for honeybee health. This conclusion is backed by Swiss findings of how honeybee cousins, bumble bees, suffer in colony health when the extent of their sex-generated diversity in the nest is reduced.[10]

Second, the drones' facilitation of recombination seems good for their own health too. There is nothing weak or unhealthy about the individual drone.[11] It's true he can't sting and doesn't work except in his own interest.[12] But when engaged in what does interest him (which is, of course, mating) he is a superb athlete, soaring in airborne assemblies, racing for queens, and aching as he does so for a glorious death. When he wins an embrace, hardly a Semtex-wrapped terrorist annihilates himself more instantly and completely than the drone honeybee as he explodes his genitalia together with his life's supply of sperm into the queen. In drone honeybees, 'extroversion' carries a new meaning and attains a new level. Shortly afterwards, his dangling body is bitten away by the mated queen and it falls as a spent shell. Certainly his extroversion is no more a design fault than is the worker's rather similar autotomy and suicide when she, too, gives up her life, in this case along with her sting, in defence of her colony—both features are very clearly designed in by natural selection (see Volume 1). I sometimes think God placed the honeybee in the path of human attention, not at all as eighteenth-century theologians would have it, for our appreciation of the benefit of honey and of his beneficence, but simply as one of his greatest puzzles and his best jokes—another 'riddle of the Sphinx', if you like, from which he hoped we would learn; and for me at least merely pondering the absurdity of the reproductive system of male-haploidy has made me at times break into loud guffaws and also to cry, so infinitely teasing are its problems. One thing I am sure about in all this is that both the above acts of suicide must betoken for their performers the very opposite of despair.

So also, though to lesser degree, must mating betoken for the queen: each copulation and each drone she annihilates as she engages with some dozen of them in very quick succession, all of this up in the air as described and in a single flight, must seem pleasurable to her. The pleasure is expected the lesser than that of the drone, however, because the acts are so far from being her own finale, indeed quite the opposite. Here, of course, is yet another way the honeybee seems expressly created to boggle and box in our minds through the extremity of the case which it presents. Where the drones, after the act, may have little more than seconds or minutes to live, the queen has years—years of potential fertility. This is a situation that is extremely unusual for an insect but at the same time a beautiful illustration of the evolutionary theory of senescence.[13]

Through his death the male who has reached the queen has provided her with the best of well-tested genes. As to his physique, it is true you sometimes see 'runt' drones (much smaller than usual) in hives but they are perfectly formed and the explanation of their size is that they have been reared, due to various accidents, in worker brood cells that are too small for normal development. In short, the cause of the drone's normal social limitation lies quite elsewhere than in bad development or in bad genes that have become exposed due to his haploidy.[14] Certainly it isn't in any immediate manifestation of mutation or the haploid's inevitable lack of heterozygosity.[11,15] A similar assessment follows if we compare haploid males of bees, wasps, ants, and sawflies with the diploid males of the comparable huge order of flies: haploid males, then, are just like all insect males—superbly engineered machines built all for the usual double purpose of courting and mating.

If yet more is needed beyond the mosses and the hymenopteran males to emphasize the irrelevance of diploidy to sex, still more discomfort for the theory exists in various seaweeds in which it seems a matter of indifference to the plant body whether it is haploid or diploid, the two states sometimes alternating in the life cycle and being outwardly hardly distinguishable.[16]

With all these facts in mind, models stabilizing sex on the basis of fitness patterns at only one locus—that is, by models that are completely dependent on diploidy—seem to me like playing scales on the piano or, worse, like the playing of two-finger exercises. Even two-locus models, whether haploid or diploid and including the example I give the main attention to in the paper of this chapter, are absurdly limited compared with the real multi-locus stabilization that occurs in nature. Even the best played of scales aren't sonatas and I have therefore watched with surprise how ideas about pure diploid recombination have sometimes been published in top journals as though they might by themselves be serious theories addressing the problem of sex[17] (see also note 18). The best one can say is that the ideas suggest minor tributary benefits. Just as I reckon of my own diploid models of the present paper that they may be useful as parables, their shortcomings as regards sex as a whole seem overwhelming. Walk out (up to your knees in the water usually) on a haploid quaking bog and tell the bog these diploid stories and I promise you will feel it quake with its laughter. My own 'two-locus haploid' model in this chapter I also see as parable but it is a bit more open-ended and generally useful because of the possibility of an unlimited increase in the number of loci that it treats. In spite of this, because of its very simplistic nature, I would not have thought of sending such a model to, say, the journal *Nature*.

Maybe, then, the others come into *Nature* more as indices of fashion or of mutual support systems. For me the papers were nowhere near to showing practical ways to pay for the costliness of sex. More on these papers, and what I think to be better models, will come in the introductions to Chapters 15 and 16. When the present paper was published, I had perhaps been too modest because before long I was surprised to find my own model taken as if it, too, had been serious in suggesting that sex might be stabilized in nature by just two parasite-resisting loci; almost needless to say the discussion of it against such a background was dismissive.[19]

Most of my 1980 *Oikos* paper just extended the idea I had started with Peter Henderson previously (Chapter 1) and it made in the spirit of my opening quote, I suppose, somewhere around Theory 300. Its main novelty lay in my way of generating and handling the fluctuations the parasites are supposed to create. The procedure wasn't realistic but simple; it could be analysed. The previous effort with Peter and Nancy Moran had tidied away—for my mind at least—the attempt to give a major role to purely random physical changes. Even cyclical physical factors, such as conditions upon us in the sunspot cycles, Milankovitch cycles, and so on, were proving too relentless as destroyers of variation unless some extra responsiveness of environment to the state of the population was brought in. As I learned later, others were focusing on the difference of behaviour of 'abiotic' and 'biotic' resources as they affected more purely ecological theoretical problems and came up with rather the same conclusions.[20] Pure 'random walkers' may seem the most hopeless ditherers if you graph their progress but eventually they always stride disastrously too far, which means in our case they step over the edge to extinction. Nothing warns that the edge is coming and a gene, precious variant perhaps of a chromosome's long evolution, goes extinct. Generally this loss of variation under buffeting by physical factors seemed inescapable.

When I turned to more hopeful and 'lively' alternatives, I found myself soon discontented with the predator possibility, even though predators surely would have to be responsive to some degree. Kestrels dropping on mice this year must have had their parentage and their present survival chances moulded at least slightly by the abundance and the genotypes of the mice of the recent past years. But if the kestrels are the common predators of mice just now, this may be due to last year's abundance of voles, not mice. Vole genotypes as well as mouse genotypes may be moulding the kestrels and so may the beetles, bird chicks, and etceteras that they also feed on. The point is that predators take many species so that, except in the very simplified ecosystems of extreme habitats (snowy owls, say, pouncing on hares in the Arctic), each predator species looms—or 'hovers' in the case of kestrels—over a particular prey species with no more persistent and reactive menace than loom (or hover) the snow clouds of a hard winter. The mammal may adapt to the snow, but (at least with the remote possibilities of Gaian circuits put to one side) the snow has no reason or mechanism to adapt to the mammal. This must be the case with predators too.

Thus for parasites to be the interactive hostile agency was left to me almost by exclusion and the more I thought of them the more enthusiastic I became. They certainly did change their attack modes, and fast. They are unkind to their hosts; and yet they cannot fail to be responsive to host genotype frequencies. Even before AIDS recently drove these messages home to us the facts were clear enough. In the models of my paper, by using initially parasite fitnesses made instantaneously responsive to host frequencies, I tried taking the implications of all this 'frequency dependence' of response almost to its limit.

As I will describe further in the introduction to Chapter 3, I was at the time finding myself invited to campuses all around North America. Usually invitations came with a request for a talk about kin selection or 'within the field of sociobiology'. Increasingly, however, I would reply that I wanted to talk about sex. Sometimes, needing not to sound too stark and to forestall student counsellors being circulated with warnings about me, I would offer a few words of explanation, say how I felt all kin-selection theory to lie under a still-unremoved shadow, that of sex, and why all Mendelian genetics should exist. If in the lecture I mentioned AIDS, as I increasingly did for those were days when everyone knew its great power to wake people up and focus the mind, especially a youthful one, and as I came to cite it as a good illustration of the ubiquitous and unexpected nature of disease challenges that come at a species, there was, of course, an immediate difficulty I had to meet. I had always to explain that, of course, it made little sense to suggest that sex owed its existence only to the defeat of kinds of pathogens that were spread through having sex . . . but most pathogens didn't spread this way. Above all, I pointed out, AIDS was the case where we had the most recent incentive and, under pressure of urgency, were applying the best skills that humans had ever had available to see what was happening. All severe parasites likewise must change fast when invading a new, abundant, and highly polymorphic host like *Homo sapiens*. We were a challenge to HIV-1 just as HIV-1 was to us.

But this is jumping ahead a little. Even in the year (1980) the paper came out, leave alone when it was first drafted, as I now recall, 'AIDS' was just puzzled doctors phoning each other across the USA about unusual symptoms. The talk that I based on the models in the paper was given at a meeting on 'Theories on Population and Community Ecology' near Uppsala, Sweden, and this was in February 1980, at which time the first medical-journal recognition of a new epidemic in progress was still a year away. I must in the paragraph above have been thinking of how I did quickly come to use AIDS to illustrate the ubiquity and changeability of disease threats in the years when the model of this chapter became the main thread of my most common talk. As to Sweden, I remember little about the meeting except for deep dusk, deep cold (though no worse than in Michigan), and deep snow. I remember snow covering all but the tips of reeds bordering a marsh outside the double-glazed windows through which I peered eagerly, for the first time, into the home countryside of Linnaeus. Perhaps I was seeing here what gave that great man time and the courage . . . no point to go outside, sit, and write, because few creatures plant or animal would be stirring. That perhaps made the task seem finite, and, above all, the winters gave him the time. Besides those tips of reeds I remember only the huge alder trees in dim screens behind. Michigan hardly had alders; those of Britain, though the same species, were smaller than these. Hereditarily so? But that was Lamarck and Darwin . . .

The snow, the alders, and the cold, misty dimness couldn't hide a generally lumpy and ice-scraped landscape not unlike some of what I knew near to Ann Arbor, and it was easy to see why so many Scandinavian immigrants had come to the Great Lakes region. Another memory of the trip is the surprise I had at the excellence of the English of the Swedes; indeed, this was so for all the Scandinavians I met. To judge by those I mixed with, the dialect of Midwest America would not have bothered the settlers unless it was on the side of a problem of *un*-Anglicizing and—dare I say—de-grammaticizing so as not to be laughed at. I recall Staffan Ulfstrand, our meeting's organizer, speaking an Oxford English so perfect it made me ashamed of my own. I have scattered here almost my only separable memories of the meeting; I have since been to Sweden several more times.

Later in the trip I visited and gave a talk in Oslo where I recall a misty dimness exceeding even that of Uppsala and, replacing the alders, a backdrop of a steep-piled, almost mountainous city. Under street lights that I cannot remember being ever extinguished, business men with smart brief cases instead of ski sticks glided down paths connecting their homes in high suburbs to, presumably, their work places in the city centre. I had seen Oslo before but that had been summer and I a student. This present visit was a brief stay; one day more and all the Cimmerian visions vanished, the nuclear winter was over and I was back with my class in Michigan. There I skied to and from work amid deep snow and cold but did so in decently regulated daylight of the latitudes of Spain.

I had come home feeling burdened a little by my commitment to write up what I had reported at Uppsala within about a month for *Oikos*, the Scandinavian journal specializing in ecology and evolution. I worked at it hard and subsequently so did the editor, Pehr Enckell, for the paper came out in the same year. In retrospect the writing, though hasty, seems to have been not bad, though I notice that my copy editor for the present volume has had to eliminate a few out-of-date forms of expression used in the paper, which even to me, normally quite insensitive to all the fuss about hidden gender assumptions, now seem odd. Under her slight modification you will read, for example, how 'even for *humans* an expressed birth fecundity of 14 would not be extreme': I had had 'for man' in the original. Worse, a little below, I see myself in the old copy speaking of 'fetal wastage' again in 'man'; and this again she has tidied. Feeble men, it seems to me now, who couldn't, given the chance, do better than 14, and very ambiguous men, too, if sometimes losing their embryos. Was my wording always sounding so silly, I wonder, or has it just become so in the present climate?

As a first step in testing my expectation that a very open-ended support for sex would come through recombining genes scattered through the whole genome (as contrasted to recombining them within the diploid pairs that are formed in each generation at each locus—the theme mentioned above and in note 2), I chose in my own model to start with the simplest two-locus' situation possible. If support could be found in this then surely I could get more support, possibly very much more, when numerous interlocus interactions were brought in. Eventually I'd try out the whole necklace of all the chromosomes at once, but, for now, my displayed jewels were to be few: just two alternative gene forms (alleles) in one position and two more in another. This gave me four possible combinations in the genotypes I was considering: *AB, Ab, aB, ab*. The two locations could be on the same chromosome or on different ones. For the case where the two 'loci' were to act as one I incorporated a variable (a 'parameter' as I will often call such a number as this when it is to affect a whole model) by which I could pre-assign how tightly my 'loci' were to be 'linked'. In effect, I put into the model for each run a value saying how often from the diploid stage (supposed very brief in this model, as I explained) you would get a gamete with a changed association (or a 'recombination') of two variant genes that were previously on one strand of the DNA necklace. In other words my linkage parameter said how often a new combination that neither parent had carried would be formed.

The mechanism for this is a kind of reciprocal 'cut and paste' of homologous parts of DNA molecules occurring when the two chromosomes are closely juxtaposed. But it is not the mechanism *per se* that we are concerned with here; more, it is the consequences. Because 'recombination' and 'linkage disequilibrium' are key ideas for almost all theories of sex and these terms recur throughout this book, it is worth pausing to explain these particular consequences at the population level a little further. If an African marries a Scandinavian, how often will a single egg or sperm in one of their children carry both mother's (say) gene for blue eyes and father's for dark hair? For sure the characters of different races are always brought together in one sense, in being present from the start in the offspring of such a cross, but now we are asking a different and deeper question about possible long-term consequences: how often are they, thereafter, in the same haploid genotype, perhaps on the very same chromosome? In other words, how often do they emerge in grandchildren transposed into new alliances?

This is where the linkage parameter is needed. If the genes are close together on one chromosome (I don't know what is the truth for eye and hair colour, probably they aren't on one chromosome, but for my illustration this doesn't matter) then it will be rare that a grandchild will receive other than the old grandparental combinations from, let us say again, their hybrid mother: blue eyes will mostly come out with fair hair and brown eyes with dark hair as in the formative races. Given that crossover events are about equally common all along the chromosomes, a close spacing of genes on the physical double helix ensures that only rarely will there be a 'crossover' in the relevant short region of the inherited giant molecule during that chromosome pairing and halving (not replicative) division that gives rise to the egg. If, on the other hand, the gene loci are far apart (and most of all, of course, if they are on different chromosomes) then the grandchild is very likely to have the new, and unusual, combination. But even for genes very distinct on chromosomes, or on separate ones, the chance of carrying recombined rather than 'grandparental' gene combinations is never greater than 50 per cent.

You may call this production of novelty cautious on the part of Mother Nature or you may call it simply even-handed; but remember that if MN errs this way for so-called 'free recombination', she is vastly more cautious about breaking up her 'tested' combinations when it comes to parthenogenesis—too cautious, as it turns out, for long-term survival of a stock, as evidence has already shown. With parthenogenesis it is as if a knob of the model—the recombination parameter[21]—had been tuned right to its 'off' position: no new combinations are allowed to occur at all except such as may arise through gene mutation. Linkage is complete for all loci under parthenogenesis.

Attacking again this awkward to understand but important little quantity, 'linkage disequilibrium' (LD) (justly called 'little' because confined to the range $-1/4$ to $+1/4$), one way to explain it is to point out that for the human population of the world as whole, the fact that, when hair colours and eye colours are coded for at different chromosomal loci, as they are, we find in Scandinavia a type with blue eyes and fair hair while in Africa one with mainly brown eyes and dark hair. This is a good example of linkage disequilibrium except for one point that isn't typical—that is, that this case is geographic. Because the facts are on an ethnic or geographic scale and because they are easily reasoned to be connected with adaptation to the lack of sunshine in Scandinavia and to the excess of it in Africa, the example doesn't show LD of the sort typically in the mind of a population geneticist when discussing it. Far more interesting examples from our point of view of gene associations along chromosomes (or holding across sets of chromosomes) are those found within single endogamous populations—unusually frequent two-locus (or many-locus) genotypes found, for example, within one village in Africa or within one village in Scandinavia. These are much more interesting than the racial/geographic example because, in a freely interbreeding sexual population, if there isn't something specially fit about them and about their being different from the rest, they wouldn't be there. If in fully intermixing population units we still find genes distant on chromosomes commonly connected in particular combinations instead of associated at random, geneticists have to blink and wonder how it has happened. It could be because there has been a recent hybridization in the history of the population—like if all recently arrived Saxons had taken Celtic wives. This in effect brings us back to the geographic example, but the event has to be really very recent because random mating plus recombination works the association quickly back towards zero—that is to proportional association, unless the linkage is very tight. If there is no reason to expect such hybridization the explanation for the association almost has to be that differential survival or reproduction of gene combinations has been occurring.[22]

To avoid at this point sending you to your bookseller asking for your money back, rather in the way that my undergraduates used to picket the professor at Imperial College to ask for theirs in respect of my lectures—or else my removal or, best of all, for a lenient upgrading of all marks given on my exam paper—when I tried to teach them the matters I am covering here (including perhaps my mention of Saxons and Celts, and Swedes and Africans, which was an even worse thing to do in the 1960s than it is now), I have, first, just above put the essential quantitative idea into a note (22) and, second, will now try yet again to give as gently as possible the gist of what the decay of associations brought about by sexuality really means. I start by pointing out what the note explains in more detail, that having genes close together on chromosomes (and thus linked by a short molecular chain, and thus unlikely to be separated by an event of crossing-over) obviously opposes the decay of associations. If natural selection sets up a new gene association—that is, if two particular alleles at different loci synergize survival in some way (they create extra babies bearing the association)—how can the decay of association over the generations ever be a good thing? A good question, and one that has been asked frequently by the best population geneticists since the Neodarwinian synthesis in the

1920s but never with any universal resolution. For a start let us recognize that saying one association, *AB* say, is good just has to imply that some other associations are relatively poor. Suppose another, *aB*, is becoming not just relatively but downright and absolutely bad—in fact, it's the worst genotype to have—and yet it currently happens to exist in large numbers. Evading how such a situation came about, we can see immediately that in this case recombination is likely to be currently a good thing: it is busy creating extra instances of what selection is saying, likewise, to be best. It is bursting apart *aB* and creating out of that the relatively good genotypes *AB* and *ab*. More of these good ones are being formed than would appear if the population were asexual. Obviously for sex to be beneficial there have to be many changes going on.

Observing the states and changes of LD along with the states and changes in the patterns of selection is going to be, eventually, a bright, clear torch to shine on the central conundrum of this volume, the purpose of sex. Unfortunately it is not a torch easy to grab and to use at a moment's notice; in fact, you probably agree at this point that the concepts, though precise, are a bit slippery. Obviously, too, one needs to observe many generations to see the effects of recombination processes occurring even if the changes of selection are indeed quite fast. Except for very small and short-lived organisms—exactly the ones difficult to keep track of in the most relevant setting for them, the great outdoors—research grants are not normally given nor are experimenters sufficiently patient, to cover the periods needed. Moreover, small creatures that do breed fast enough for experiments over many generations are, alas, exactly those that are least typical of the problem of sex. Small organisms are the ones most often parthenogenetic or highly inbred (similarly parthenogenesis inbreeding beyond a very few generations foregoes the effects of recombination—see Chapters 4 and 12 of Volume 1).

Even without experiments and tests, however, the general answer concerning LD is obvious. The decay of a combination that became statistically common by selection can be good only if the selective advantage that set it up has changed. As to how the pattern of selection across the loci can be changing fast enough to justify the presence of an expensive mechanism for causing associations to alter, my answer, based on a mass of circumstantial observations, has become very simple: coadapting parasites. Why these in particular? Because unlike anything else we know that their own evolution very swiftly enables them to exploit that which has become common, such as, exactly, the gene combinations showing linkage disequilibrium in their frequencies that I have just described. But if elimination of something that is in continual obsolescence is sex's function, isn't this process, with its need of many generations to destroy old combinations, too slow—this even when linkage on the chromosome is loose, leave alone when it is tight? Yes, it is slow, but in the long run that, too, can turn out an advantage. In a world full of cycles perhaps the host may soon need again the very combination its recombination is currently destroying, so it can be best not to destroy it too fast and too completely. There is much more to say about all this but in essence the above questions and answers should provide you with a thumbnail guide to aid your thought as you follow how the two- and more-locus models of this paper and of this volume are achieving their results; if you want to follow more deeply the more quantitative aspects of the story it will be best to consult a text of population genetics, such as that of Crow and Kimura.[23] A recent review article should also help.[24]

Even two-locus population genetics under constant selection is difficult mathematically if one demands complete algebraic accounts. Three-locus genetics is worse; here even excellent mathematicians soon admit to much *terra incognita*. If selection is not constant everything is worse yet again for both cases and becomes worse still if selection is not only inconstant but strong. Against such background it came as a surprise and delight to me to find that for the *Oikos* paper I could trace a fairly clear algebraic path through my model's forest, provided I used one simplifying assumption. This was, as already mentioned, that the countermoves of the parasites were to be so maximal and so predictable that I could in effect omit the genetics of the parasites altogether. Of course, I was not omitting the effects of them, I was simply putting the effects straight into the fitness formulas I was using for the hosts. This brought me the neatest and most suggestively open-ended model for sex I had yet devised; perhaps it was even the neatest I had yet seen—certainly it was far better than any of my single-locus models, either those that come in the first part of this paper or those of my previous notes (at that time no one else, so far as I know, had tried single-locus ones; the rather silly vogue for them, seemingly so seriously taken, that I have already mentioned, came later). Because of the literal 'liveliness;' in its response to parasites and for the other reasons given, my model was also better than the diploid two-locus model of Chapter 1. Having found my way to the analytical solution, my initial hesitance about contributing a written paper to the special *Oikos* issue that Staffan Ulfstrand and Pehr Enckell were planning changed at about the time of the meeting to a positive enthusiasm to contribute; I saw it as a fairly painless way to get into print what I had come to believe a neat parable for the more general problem.

The messages of the parable to me were principally two. First, under sufficiently strong selection a frequency-dependent response will stabilize sex against the twofold cost and can do this even when fecundity is extremely low. Second, even with strong selection and even when two loci only are varying, the process needn't involve any dramatic changes in gene frequency; statistical changes of gene associations (LD) are enough. Thus very simple conditions might set in motion a process that was highly beneficial to sex and yet at once give absolutely no danger of extinction for any allele and partly conceal what was going on in the sense that records of gene frequencies with no attention paid to genotype frequencies (implying no LD) might reveal almost nothing. Changes of the tightness of particular associations have to happen if two-locus haploid models are to have any hope against a cost of sex; otherwise there are no 'obsolescent gene structures' that recombination can work to undo nor new ones needing to be made. Because gene associations (and, more rarely, proven changes in them over time) are continually being discovered as populations are more and more sampled genetically, and gene frequencies in polymorphisms turn out less extreme than one might expect of them on a classic Neodarwinian view, I felt the hints that had emerged so readily from my simple model might be far reaching and that more realistic multilocus versions of the same type might show these virtues strengthening. Possibly they would even be able to explain that non-extinction of ancient gene variants that has become an ever greater puzzle to geneticists as more and more examples are found (see Chapter 16). Linkage disequilibrium certainly can't be fluctuating as widely or as abruptly in nature as it is in my model, still less can it fluctuate with such complete regularity; but I guessed that adding more loci might give me the same achievements under a less-stereotypical performance. The hunch later proved correct; the virtues of the present case did continue and strengthen as I extended to more loci and more realistic conditions (Chapter 16).

For the present, the extremity of frequency response that I found and the corresponding magnitude of the fitness differential needed to start the needed patterns of change soon brought the paper to the critical attention of more realism-loving theoreticians, as I have already mentioned. Robert May and Roy Anderson demonstrated that it was extremely unlikely that the conditions my model needed to 'oscillate' in the way I had found, and by these oscillations to protect sex, could be achieved on any quantitative assumptions that they could justify with the known parameters of epidemiology.[19] As I have said, I felt this was taking my model too literally; it was like complaining about the parable of the Good Samaritan on the grounds that wayside assaults were too rare. A Good Man today could not hope to prove himself in the way the Bible described (actually probably not a good example: muggings may well be more common per capita in London now, say, than they were in biblical Judaea—but I am sure you see my point). On the whole, however, the critique pleased me. Having one's work noticed over-seriously is much better than not having it noticed at all, which is by far the more usual fate of theoretical papers (nor just for theoretical ones either). I was also pleased that May and Anderson at least admitted the correctness of my analysis, which some other critics (mistakenly) had denied. Perhaps their most painful point for me (suggestively though not provenly confirmed in another of May's genetical re-analyses[19]) was a claim that the bold single 'pitchfork' I present in Fig. 2.1 of the paper wasn't an isolated singularity like I showed it. Separately I was assured by Bob May that I had overlooked an endless cascade of similar pitchforks somewhere off to the right in the diagram, these equally being latent in my assumptions. In short, an infinite series of the forks exists basically similar to the cascade May had discovered in his pioneering work on demographic chaos: my case was the genetical parallel to this. As an amateur enthusiast for modern ideas of chaos and fractal structure, following the published spectaculars of these topics from afar, I had suspected as soon as I saw my pitchfork appearing and so similar to that in a paper by May and Oster[25] that the further bifurcations would occur in my model too; but when I looked for them, even with what seemed to me very extreme states of selection, I seemed to get only the single two-point cycle that I illustrate. Presumably I have not gone far enough to the right; but to go farther looking for these more complex frequency oscillations and eventually for chaos would have carried me farther still from real biology than what I describe. May, too, was generally shrugging at the biological significance of such extra oscillations. Personally I still need to be fully convinced that they exist, wondering if this might be a case where sex smoothes and simplifies wild judderings that could otherwise occur.[26]

Apart from the above criticisms, what did this model lead to—what in myself for example? At a fairly early stage after the two-locus model had completed its primary checks I wrote a program to give me '3-D' graphical output via the Michigan Terminal System (really, of course, one speaks of 2-D shadows of 3-D objects). This way I could see records of every step that my population was taking, and I always like to have visible images. In this case the plotting exercise was fun and instructive but proved aggravating to the computer technicians running the colour plotter in the Computer Center. Their machine of that time moved an inked stylus across paper and when my model took up any regular cycle the stylus usually, as if with a generous intention and excitement of its own at my discovery, would proceed to use its fine point and wet ink to cut the cycle right out of the paper. This, of course, made a mess and resulted in some huffy phone calls and some later sour confrontations when I went to the Computer Center to pick up my output on my way home. Apart from a modern historical library close by the Computer Center on North Campus was the closest university building to my home and the direct way to it was the pleasant walk I have mentioned in Chapter 1, the one that cut through the Huron High School woods in which in passing I could re-inspect the plagues of the alder buckthorn or marvel at a new wonder I had found there, the horned *Bolitotherus* beetles conducting their rhinoceros-like but upside-down amours on the white spore pastures of shelf fungi of dying elms. The males look like small triceratops dinosaurs but their fights can have no thunderous rhinocerine charges: even for beetles such must be virtually impossible to execute when clinging by one's toenails to a white powdery ceiling. It was rather similar in my tussles with the technicians; circumspection and delicate passive resistance, keeping good toe-holds in the ceiling, were the theme. I would mutter apologetically, speak of 'taking steps' and slink to a quiet desk or terminal to try to see, firstly and excitedly, what my new run had found, and, secondly, in a rather dull spirit of procrastinating if I could, to think what I needed to change so as to not cut holes in the plotter paper quite so quickly next time. Almost for sure the cure lay in ornamenting my main generation loop for the program with some of those execrated but so serviceable Fortran 'go to' statements: these certainly stopped the problem of the plotter eventually but just as I fixed it I lost interest in the program, having seen something else.

My first sight of the trajectory of my two-locus model moving into the vertical oscillation while parthenogenetic fitness at the same time was decisively beaten was very exciting for me; I guess I felt like a schoolboy who has walked into a dark archway of a cave and discovered a veritable Lascaux of graffiti—and, one might say, very sexy graffiti at that! How could such a model, with such vertical cycles (in terms of LD), fail to be on the right track. In contrast, how flat, how impotent, one might put it, were other oscillations the plotter at other times drew for me under the parabolic arch of maximal LDs I had put in to give a sense of scale. I found that the model would go to these 'flat' cycles sometimes from very nearly the same starting point as I had in a run that led to a vertical performance . . . In these others the dynamic varied gene frequency but made no change to the degree of gene association (LD) so that, for reasons already explained, sex lost all chance at once even though, again, no alleles were being eliminated. But this weak joke with myself concerning the erect cycles, making those 'sexy' graffiti I just referred to (the reader of Volume 1 has probably noticed how I like such feeble parallels!), didn't last long because soon I had seen better ideas looming very large behind those that first struck me: it was as if I had found outlines of a mammoth sketched there as if from life—paintings incomparably more ancient as well as better drawn, and underlying the crudely outlined schoolboy dreams on the damp wall that had first caught my attention.

My attempts to find different and interesting dynamical behaviours in my model that would not mess up plotting machines by drawing cycles, nor mess my programs with too many 'go tos', had brought me to some runs where instead of going into either of the two-point cycles, the population point just dawdled along for 40 or so generations at a time through the turns of a very rough and only vaguely repeating cycle. The motions it made tore no paper and brought me no more irritated phone calls; but at the same time they unfortunately revealed themselves less capable and prompt supporters of sex than my vertical two-pointer had been, so I was not at first greatly thrilled. The most sure way to produce the slow cycles was to introduce a 'lag'. This meant that instead of having the selection determined just by the last generation's genotype frequencies I would make it respond to what they had been several generations back. In nature this might happen if it had taken time for a parasite or predator to complete a dormancy or if it had enjoyed a few free-living generations eating fungi and dead leaves in the soil or had spent time with another host before returning to its parasitism of the host under consideration. According to my books about larger parasites such 'indirect' life-history cycles are common. Much longer delays could also be speculated. Spores of a deadly fungus really do wait regularly in the soil 17 years for the re-emergence of adult 17-year cicadas. What human epidemiology would we expect if spores from the diseases of Tutankhamun could actually survive in the tomb for 3000 years? What is the genetic epidemiology of anthrax or of other spore-borne diseases in dry places?

With minor lags of fixed period added it was easy to have my model enter permanent and slow, if usually erratic, cycling; but the motion normally would not be forceful enough for asex to be defeated and the swings tended to be so wide I was having again to worry about genes being lost. In effect I had put a sleepy drunken driver—'Lag' let us call her—at the wheel of my program's car and it gave me a gambler's kind of fun to watch this driver careering about the road. But I knew well that it was disaster for my whole enterprise if my settings sent her right off it. Along with Lag in the car rode Variation and, along with her, her sister Sex—and she, so varied and so lovely, it was my main concern to save.

That is to say, I would so muse like this until I saw in a diagram suddenly a more serious thing. I remember one case about which I was feeling particularly disappointed because not even the latest adjustments had quite yielded a long-term mean fitness able to beat the twofold asex advantage. That average figure, which came onto the plotter sheet under the caption 'LGMF' (long-term geometric mean fitness), was always the first thing I looked at. Present, too, as an output figure on the plot was the length of the cycle. It was depressing enough in this case and bringing in again that old daunting spectre of the possible 20-generation take-over in a population of 1000: even if my threshold LGMF > 2 was achieved there were these various Romeo-and-Juliet-like disasters by which I could still be defeated. Gloomily I traced the back and forth of this neatly drawn curve—there were no cuts in the paper this time, all was neat and coloured—noticing where the line in true 3-D space must pierce and

pierce again the LD = 0 plane (that plane which implies zero association of
the genes, or that having them paired as if at random[22]). If I were one
animal in the population that was 'averaged' by the moving point, could I
know whereabouts in the cycle we—I, my fellows, and our parasites—
were, I wondered idly? Could I detect our local-state LD? Where in the
cycle would I feel at my best? Obviously the answer to the last would
depend on my genotype. Suddenly I stared at my coloured lines snaking the
page with a renewed interest. Feeling good, being fit: what did this imply to
be happening to biological fitness of the various genotypes as this cycle goes
around? Of course I could tell if I was fit and others could tell too. So, apart
from how I might feel about myself, which would be the best genotype at
any point for me to be with? And who would I most envy? Suddenly a lot
of further questions came fast. When, if ever, would fitnesses of parent and
offspring be negatively heritable—like I was now realizing they had to have
been all the time in my two-point cycle—and when instead would they be
positive? This last plainly held right here in the diagram and it wasn't a
failure after all.

Shortly the run positively shone for me with its further possibilities:
I found more realistic sex in it, and that of a far more sensible kind than all
that had come from the two-point cycles I summarize in the paper of this
chapter, and I was now sure I was moving beyond parable. Much of it
implied stuff I could suggest right away to my winter cardinal in Ann
Arbor's arboretum, and to my upside-down 'triceratops' on the shelf fungi
in the Huron High School woods. Perhaps all I was seeing now had long
been obvious to others but, as for me, I felt I had chanced upon a minor
grail (or a dragon as it might turn out) in my understanding of population
genetics. I had found a rationale for a permanent heritability of fitness.
Suddenly it seemed that it was the old vertical two-pointer that was victim
fit for the Ann Arbor Veteran's Hospital, no longer the hero of my 'virility'
parable as he had been at first. On grounds that the model shows that the
genotype that is unfit this generation is bound to be very fit in the next,
those pure 'vibrations' I had been watching—why, obviously, these hint at
a selection to mate with the sickliest, the most parasitized, of all possible
partners! An absurd hero surely, was this, even for a parable. No one's
darling could be, I had to suppose, at least in the biological world, quite so
sickly as to be unable to stand up and . . . well, do it. But short of that . . .
anyway, I continued, the simplistic model was all of it a definite absurdity; I

wasn't going to need my future Oxford colleagues, in a year or so, to point this much out to me.[19] Those drone honeybees throwing their lives away: at least they proved to her that they were no disguised polygynists. But they weren't weaklings either, they had raced for her and won. No, that couldn't work, the cycle in honeybees couldn't be period 2 any more than in the others; and in wasps, where there was sometimes a period 2 cycle, the males didn't self-sacrifice, or not quite anyway . . .

Finish, I thought, what *Oikos* is kindly publishing with so little question; after that, go for this new idea.

Soon that half hour of revelation from my chart (probably, if I am to be accurate, it was more like several weeks of a deepening conviction) plunged me once more into the libraries. Mainly now it was to the bird library of the museum that I went. My first target was, of course, my cardinal, this most local example and so abundant in Ann Arbor. I became obsessed with cardinals—their colours, their blood, their shit, their worms, their pairing, their care on the nest . . . Cardinals sang, puffed brilliant feathers for me on snowy trees; ruby-quilled waxwings jetted their spore- and egg-laden diarrhoea deep in my mind just as I had seen them, in the reality, in late winter jet it to soak purple into the old snow under the foreign berry-laden purging buckthorn trees. Books, bones, and birds of many kinds swayed and swooped around me. Hoopoes it was first and then blue jays and Canadian jays ('whisky jacks') that soon eclipsed the cardinal and the waxwing as my focal birds. My soon sole supporter in my new addiction, Marlene Zuk, trod close on my heels. Fortunately Robert Payne and Robert Storer, curators of the Bird Division which we had started to haunt, were men not easily surprised: if someone's new enthusiasm was about birds, for them it must be at worst a worthy lunacy and I found myself very welcome in the library. As with the crows' parliaments reported to the curators by old ladies writing from the country, there might somewhere lie concealed a grain of truth in what I and Marlene thought we were doing. But concerning what the model of the present chapter was to lead on to under Darwin's heading of 'sexual selection', this bare sketch of the new idea has to be enough and now I must back off; the rest of the consequences of those exciting moments with the coloured chart will come in Chapter 6. Before that several other topics intervene.

References and notes

1. G. A. Reid, *The Laws of Heredity* (Methuen, London, 1910).

2. As with diploid plants some strains of moss never get around to the necessary structures and acts but again, as with diploids, such asexual states seem to be terminal twigs on the bryophyte evolutionary tree. As usual, geographical patterns apply with asexuality being much more common in the more extreme environments—for example, in high mountain and polar regions (P. Convey and R. I. L. Smith, Investment in sexual reproduction by Antarctic mosses, *Oikos* **68**, 293–302 (1993)).

3. Y. Yamazaki, The amount of polymorphism and genetic differentiation in natural populations of the haploid liver wort *Conocephalum conicum, Japanese Journal of Genetics* **59**, 133–9 (1984); D. Mishler, Reproductive ecology of bryophytes, in J. Lovett Doust and L. Lovett Doust (ed.), *Plant Reproductive Ecology: Patterns and Strategies*, pp. 285–306 (Oxford University Press, Oxford, 1988); R. Wyatt, A Stoneburner, and I. J. Odrzykoski, Bryophyte isozymes: systematic and evolutionary implications, in D. E. Soltis and P. S. Soltis (ed.), *Isozymes in Plant Biology*, pp. 221–40 (Chapman and Hall, London, 1989); R. Wyatt, I. J. Odrzykoski, and A. Stoneburner, High levels of genetic variability in the haploid moss *Plagiomnium ciliare, Evolution* **43**, 1085–96 (1989); H. N. Kim, K. Harada, and T. Yamazaki, Isozyme polymorphism and genetic structure of a liverwort *Conocephalum conicum* in natural populations of Japan, *Genes & Genetic Systems* **71**, 225–35 (1996).

4. L. G. Franzen, D. L. Chen, and L. F. Klinger, Principles for a climate regulation mechanism during the late phanerozoic era, based on carbon fixation in peat-forming wetland, *AMBIO* **25**, 435–42 (1996).

5. J. H. Tallis, Studies in the biology and ecology of *Rhacomitrium lanuginosum* Brid. II. Growth, reproduction and physiology, *Journal of Ecology* **47**, 325–50 (1958).

6. I allude here to a well-known paper by Gould and Lewontin (S. J. Gould and R. C. Lewontin, The spandrel of San Marco and the Panglossian paradigm: a critique of the adaptationist programme, *Proceedings of the Royal Society of London B* **205**, 581–98 (1979)), which, however, seems to have misinterpreted the name and function of the architectural device referred to—see D. C. Dennett (*Darwin's Dangerous Idea: Evolution and the Meanings of Life* (Simon and Schuster, New York, 1996)), and also A. I. Houston (Are the spandrels of San Margo really panglossian pendentives?, *Trends in Ecology and Evolution* **12**, 125 (only) (1997)). But whether the features they visualized in San Marco in Venice were pendentives, squinches, or yet something else, the general idea of

the analogy to architecture used by all the above writers seems a good one and I follow it here.

7. Dennett (1996) in note 6.

8. M. P. M. B. Maeterlinck, *The Life of the Bee* (George Allen, London, 1901).

9. J. M. Greeff, Effects of thelytokous worker reproduction on kin-selection and conflict in the Cape Honeybee, *Apis mellifera capensis, Philosophical Transactions of the Royal Society of London* **351**, 617–25 (1996); M. H. Allsopp and H. R. Hepburn, Swarming, supersedure and the mating system of a natural population of honey bees (*Apis mellifera capensis*), *Journal of Apicultural Research* **36**, 41–8 (1997).

10. This point is strongly supported for bumble bees (J. A. Shykoff and P. Schmid-Hempel, Parasites and the advantage of genetic-variability within social insect colonies, *Proceedings of the Royal Society of London* **243**, 55–8 (1991)); B. Baer and P. Schmid-Hempel, Experimental variation in polyandry reduces parasite load and increases fitness in a social insect, *Nature* **397**, 151–4 (1999), but recently there has been controversy over whether it is also true for honeybees—see the comment (B. Kraus and R. E. Page, Jr, Parasites, pathogens and polyandry in social insects, *American Naturalist* **151**, 383–91 (1998) and counter comment (P. W. Sherman, T. D. Seeley, and H. K. Reeve, parasites, pathogens and polyandry in honeybees, *American Naturalist* **151**, 392–6 (1998)) on the 10-year previous paper of P. W. Sherman, T. D. Seeley, and H. K. Reeve (Parasites, pathogens and polyandry in social Hymenoptera, *American Naturalist* **131**, 602–10 (1988)), and also see Chapter 10 and the general survey by Schmid-Hempel (*Parasites in Social Insects* (Princeton University Press, Princeton, NJ (1998)).

11. G. M. Clarke, The genetic basis of developmental stability. III. Haplo-diploidy: are males more unstable than females?, *Evolution* **51**, 2021–8 (1997).

12. Maeterlinck's romantic account of social affairs of male honeybees hive were probably largely based on sounder first-hand accounts by J. H. Fabre and to this extent his 'massacre of the drones' concept was largely correct. The fact that males are sometime hustled by the workers and towards the end of the season are thrown out of the hive or/and killed is certainly true. Very recently the maltreatment of drones by workers has found an amusingly neat parallel in the behaviour of paper wasps. Adult males often stay on the small open nests of *Polistes dominulus* and try to obtain food from returning foragers. Two naturalists at Cornell University have observed that when a load of food is brought back to a nest, workers may attack males and seemingly force them to take refuge, head inward, in empty cells where they remain for several minutes long enough for the arrived load of food to be distributed without the males

participating: the two authors call this 'stuffing' of the males (P. T. Starks and E. S. Poe, 'Male-stuffing' in wasp societies, *Nature* **389**, 450 (1997)). The authors point out that this performance accords perfectly with kin-selection theory. On a nest founded by a singly inseminated queen (the usual state), males are less related to workers than are either other workers or the coming new queens. Therefore males should be the disfavoured targets for the workers' altruism.

13. See Volume 1 (Chapter 3) of *Narrow Roads of Gene Land* and also L. Keller, Indiscriminate altruism: unduly nice parents and siblings, *Trends in Ecology and Evolution* **12**, 99–103 (1997).

14. See Volume 1 (Chapters 2 and 8) of *Narrow Roads of Gene Land* and also Mishler (1988) in note 3.

15. W. E. Kerr, Sex determination in honey bees (Apinae and Meliponinae) and its consequences, *Brasilian Journal of Genetics* **20**, 601–11 (1997).

16. C. Destombe, M. Valero, P. Vernet, and D. Couvet, What controls the haploid–diploid ratio in the red alga, *Gracillaria verrucosa?*, *Journal of Evolutionary Biology* **2**, 317–38 (1989).

17. M. Kirkpatrick and C. D. Jenkins, Genetic segregation and the maintenance of sexual reproduction, *Nature* **339**, 300–1 (1989).

18. D. Weinshall, Why is a two-environment system not rich enough to explain the evolution of sex?, *American Naturalist* **128**, 736–50 (1986).

19. R. M. May and R. M. Anderson, Epidemiology and genetics in the correlation of parasites and hosts, *Proceedings of the Royal Society of London B* **219**, 281–313 (1983).

20. R. A. Armstrong and R. McGehee, Competitive exclusion, *American Naturalist* **115**, 151–70 (1980).

21. It will be useful to define a parameter for recombination and another for its complement, linkage. If we define recombination r as the frequency per meiosis of a crossover of alleles at two loci in a double heterozygote (frequency of an a–B chromosome arrangement, say, appearing in a gamete from a diploid made up of A–B paired with a–b), then the chance of no change, or the degree of linkage, can be defined as $s = 1 - r$.

22. An example of a state-of-linkage disequilibrium might be useful here. Suppose that in a population of 100 the numbers of the four genotypes

ab	*aB*	are as	5	20
Ab	*AB*		35	40.

    We note that B is much more commonly associated with *a* than *b* is (about 4 times as frequently). If we work out how the genotypes should be such that the ratios would be equal in the rows without changing the numbers of any of the alleles, we find they should be as

10   15   or in fractions or frequencies   0.1   0.15
30   45   0.3   0.45.

Note that the numbers within columns also now show proportionality: B is as commonly partnered with a as it is with A. This arrangement is said to show zero-linkage disequilibrium, as can be tested by checking that the diagonal products of the frequencies are equal ($0.1 \times 0.45 = 0.3 \times 0.15$). If we perform the same cross-products for the set of frequencies first given we find

$$D = 0.05 \times 0.4 - 0.35 \times 0.2 = -0.068.$$

D so calculated is often used as a quantitative measure of LD. Note that the sign of D is arbitrary: had we changed the arrangement of the four genotypes at the start by exchanging the two columns we would have obtained instead

$$D = +0.068.$$

But if we stick to one definition throughout a particular discussion, say we define

$$D = \text{Freq}(ab)\text{Freq}(AB) - \text{Freq}(Ab)\text{Freq}(aB),$$

then no problem arises. As far as advantages to sex are concerned, any change $+$ to $-$ is as good as any change $-$ to $+$, while to visualize what is going on when a change happens it is helpful to think of a 'frequency-excess' diagonal in tables like the above, either the Northwest–Southeast diagonal in excess for $D > 0$, or else SW–NE for $D < 0$.

Random mating plus recombination in any situation of no selection is continually rearranging genotype frequencies such that $D = 0$ is being asymptotically approached; in fact $D' = sD$ for each successive generation where s is the linkage fraction.

23. J. Crow and M. Kimura, *An Introduction to Population Genetics Theory* (Harper and Row, New York, 1970).

24. S. P. Otto and Y. Michalakis, The evolution of recombination in changing environments, *Trends in Ecology and Evolution* **13**, 145–51 (1998).

25. R. M. May and G. F. Oster, Bifurcations and dynamical complexity in simple ecological models, *American Naturalist* **110**, 573–99 (1976).

26. M. Doebeli and J. C. Koella, Sex and population dynamics, *Proceedings of the Royal Society of London B* **257**, 17–23 (1994); recently Akira Sasaki told me that he has proved that further bifurcations do not occur in my frequency-dependent genetic case.

SEX VERSUS NON-SEX
VERSUS PARASITE[†]

W. D. HAMILTON

———

Pressure of parasites that are short-lived and rapid-evolving compared to the hosts they attack could be an evolutionary factor sufficiently general to account for sex wherever it exists. To be such a factor, parasites must show virulences specific to differing genotypes. Models are set up on this basis (one-locus diploid-selection and two-locus haploid-selection) in which the rapid demographic reactivity of parasite strains to abundance of susceptible hosts becomes represented in a single frequency-dependent fitness function which applies to every host genotype. It is shown that with frequency dependence sufficiently intense such models generate cycles, and that in certain states of cycling sexual species easily obtain higher long-term geometric mean fitness than any competing monotypic asexual species or mixture of such. In the successful cycle of the two-locus model, both population size and gene frequencies can be steady while only oscillating linkage disequilibrium reflects the intense selection. High levels of recombination work best. Fecundity in the models can be low and no incidence of competition of siblings or other relatives is required.

Given acellular simple organisms in the early history of life it is not difficult to imagine selection that would favour multicellularity. Provided cell aggregates were clonal, so that they would co-operate well, differentiation of somatic cells could give several advantages. There are parallels with the trends to eusociality currently occurring in some insects. Here defence against predators and parasites—including conspecific parasites—and increasing ability of a colony to buffer local changes of the physical environment seem to be very important. Similar factors most probably applied to simple multicellular organisms. Achievements with regard to the second factor permit occupation of new habitats.

Specialized somatic cells can increase the sophistication of defence but also introduce new vulnerability. One new weakness is in the inevitable slowing of the intrinsic growth rate due to physico-chemical logistics. Thus all parasites that remain much smaller than a host (these parasites might be called, in a broad sense, pathogens) have an advantage in a rate of evolution that will help them to keep abreast.

†*Oikos* **35**, 282–90 (1980).

A second weakness, brought in by multicellularity and crucial to the theme of this essay, is that body-building in itself requires that cells adhere to cells: the difficulty then may be to recognize as non-self, and to deny attachment, all other cells that present themselves at a cell surface. Non-self cells are potential parasites. Attached without having alerted any defence system in the host, an alien parasite cell is strongly placed with regard to further exploitation. Somewhat similar considerations can apply rather more weakly to unicellular organisms as hosts under attack by smaller and still shorter-lived parasites, notably viruses: here the argument would be cast in terms of chemical systems and organelles.

Thus an ongoing antagonistic coevolution is to be expected over the matter of recognition. The parasite evolves towards a presentation which is either so bland that to the victim it seems like being touched by an inert body or by nothing at all or else involves some positive mimicry of attributes of host cells. The host on its side would evolve ever keener methods for discriminating what are truly the cells of its own clone. Provided each individual host is able to know its own idiosyncracy, mimicry by the pathogen is an incentive to variation by the host, for a host that has a new mutation in a recognition substance is able to react to the existing mimetic race of pathogen. But the pathogens by virtue of short generation time can be expected to evolve an appropriate mimetic presentation soon after the new type's advantage has made it common. Obviously a frequency-dependent polymorphic equilibrium is likely. Thus in somewhat vague outline we can imagine this coevolution becoming more complex and diversified, with arbitrary password-like identification substances (histocompatibility antigens?) and facultative responses (current pathogen strains matched by clone proliferation of specific defence cells, as in the immune system?) being brought in by the host, and special difficulties for the host (such as mimicking common small particles like pollen grains or self-enwrapping in the membranes of the previous host) being invented by the parasite.

To elaborate on this coevolution in terms of immunology and microbiology is beyond my present competence and beyond the scope of this paper. Instead I intend to consider very simple model systems involving the general kind of frequency dependence just outlined and consider what 'steps' the host might take (to adopt for the moment a loose teleology) to make a password system as effective as possible—effective, I will assume, not through facultative adjustment, as with an immune system, but through continual random recreation of 'passwords' by sex and recombination.

Historically my theme makes up a thread concerning the role of parasitism in evolution recently followed by Clarke[1] in pursuit of reason for the abundance of protein polymorphism. Before Clarke, the thread started perhaps (as so often) with Haldane;[2] after Clarke, it was traced in my present direction—decisively towards sex although not very far—by Jaenike.[3] Besides the stimulus of Jaenike's paper, a similar suggestion from William Irons (personal communication), emphasizing pathogen mimicry of host self-recognition antigens, contributes direction to the present paper.

Randomness *per se* in recreation of passwords will not be a particular focus in this account. However, it seems worth noting here in passing that randomness of that kind first identified in the rules of Mendel and subsequently proven so universal in sexual processes, already suggests a theme of escape from enemies. Such randomness, in other words, might be a parallel to that entering the solution of the game of 'matching pennies' and others of similar conflictual coevolutionary slant.[4] Thus randomness in cellular events, which it is as easy to imagine following fixed courses as random ones, seems a further hint that in this area of antagonistic coevolution may lie the answer to why sex arose and how it is maintained.[5]

Specifically, in the models that follow, a cell is supposed able to make antibodies to a certain class of substances and actually to make them to those variants within the class that it itself possesses. Detection, by use of these antibodies, of self-antigens in another cell results in a pacification of the defence system and facilitation of attachment. Any exterior living object in which the self-antigens are not detected is not allowed attachment and is treated as potentially dangerous.

ONE-LOCUS MODEL

In the case of a diploid there are imagined to be two quasi-independent genomes checking for their own passwords: hostility results if either genome fails to find its password. A pattern of genetical partial resistance that is interpretable roughly on these lines is that shown by mice to three strains of a leukaemogenic virus.[6] This particular virus, however, seems unlikely to be a major selection factor in the wild.

In general, the heterozygote might be expected to have an advantage over homozygotes because it imposes a harder test on a mimetic parasite cell and will be harder to fool. However, where the parasites are abundant and short-lived, heterozygote-mimetic strains can be expected to appear all the same. This is especially true because in the balanced polymorphism earlier outlined heterozygotes tend to be the commonest genotype. Once the heterozygote mimic is present and itself common the tendency of the heterozygote to be common is to its disadvantage, so that it is likely, over a series of generations, to become the least-fit class. But because of the frequency dependence neither homozygote will fix; we see the possibility of an interesting interaction of stabilizing and destabilizing influences.

Before proceeding to examine this interaction in a specific case it may be noted, by way of introduction to the later model of this paper, that for a host to make its password depend on two syllables coded at independently segregating loci increases the range of passwords. So also, of course, does adding new alleles at a single locus. But eventually adding alleles distributed between two loci makes the number of types rise as a fourth power instead of as a square. For this among other reasons multi-locus password systems certainly merit attention.

Here it also seems appropriate to note that while the story has been told so far in terms that imply microbial parasites, it is by no means only these that practise highly specific wiles against their hosts, or that are short-lived enough to show the very reactive frequency dependence that the models will be found to require. Special resistance in particular genotypes of host, special virulence in particular genotypes of parasite, and matched polymorphic systems for attributes of these kinds, are being found ever more widely. In the case of wheat and hessian fly,[7] for example, it is seen that the parasite in such a system does not have to be a microbe. The models that follow are much too simple to fit any such known cases but hopefully their very simplicity may serve to bring out basic principles that might apply in a wide set of more complex realistic interactions.

Consider a population with three mutant genotypes of an asexual host species (A, B, C). To each genotype a parasite species has produced a virulent pathotype. The presence of these pathotypes affects the fitness of each host genotype in the same frequency-dependent fashion: let $w_i(f_i)$ be the fitness where i is the genotype symbol, f_i is the frequency of type i in the population, and w is a fitness function that monotonically declines with increasing f and is such that always $w\left(\frac{1}{3}\right) = 1$. Thus the system has a fixed point at $\left(\frac{1}{3}, \frac{1}{3}, \frac{1}{3}\right)$ in the frequency space, and the decline of w with frequency means that at least for some functions this point represents a stable equilibrium. If, however, the decline of w is steep in the neighbourhood of the fixed point it can happen that the equilibrium becomes unstable. Then any disturbance gives rise to an oscillating departure.

For example, suppose the fitness function is

$$w_i = r^{1-3f_i}.$$

If $r = e^g$ and $f_i = \frac{1}{3} + d_i$ then this may also be written

$$w_i = \exp(-3gd_i).$$

In the neighbourhood of the fixed point ($gd = 0$), this is approximately

$$w_i = 1 - 3gd_i.$$

Multiplying by the frequency, $\frac{1}{3} + d_i$, summing for all three types, normalizing (actually unnecessary here because mean fitness approximates to 1 near the fixed point), and neglecting terms in d^2, the approximate frequency in the next generation is $\frac{1}{3} + d_i(1 - g)$. Hence the recurrence relations near the fixed point are of the simple form

$$d_i' = d_i(1 - g).$$

This shows that for $0 < g < 1$ there is a monotonic approach to the fixed point; for $1 < g < 2$ there is an oscillatory approach; and for $g > 2$ there is an oscillatory departure from which a permanent oscillatory state can be predicted to result.

A more general analysis using a Taylor theorem expansion of w gives the recurrence equation

$$d' = d \{1 + \tfrac{1}{3} \dot{w} (\tfrac{1}{3})\},$$

where $\dot{w}$ is the first derived function of w with respect to f.

Now consider a sexual host population with two alleles at a single locus such that antigenically, relative to the asexuals just considered, we have equivalences $AA \equiv A$, $AB \equiv C$, and $BB \equiv B$. Suppose the fitness function is the same. This sexual system in the presence of the parasite species does not have a fixed point for the three phenotypes at $(\tfrac{1}{3}, \tfrac{1}{3}, \tfrac{1}{3})$; for, if A and B are equally frequent, genotype AB occurs with frequency $\tfrac{1}{3}$ and hence has lower fitness than the homozygotes.

It is clear, however, that equal frequencies of alleles correspond to a fixed point. Stability at this point can be examined. It is now convenient to cast the argument in terms of gene frequencies. Thus d will now be used differently, as the measure of the departure of a gene frequency from $\tfrac{1}{2}$.

Using the Hardy–Weinberg ratio and with objective as before we obtain

$$d' = 2 \frac{w(\tfrac{1}{4}) + \tfrac{1}{4}\dot{w}(\tfrac{1}{4})}{w(\tfrac{1}{4}) + w(\tfrac{1}{2})} d,$$

which can be abbreviated to $d' = \lambda_s d$.

With the fitness function as before, $w(f) = r^{1-3f} = \exp\{g(1-3f)\}$, it is found that

$$\dot{w}(f) = -3g.\exp\{g(1 - 3f)\}.$$

So

$$w(\tfrac{1}{2}) = \exp(-\tfrac{1}{2}g), \; w(\tfrac{1}{4}) = \exp(\tfrac{1}{4}g), \text{ and } \dot{w}(\tfrac{1}{4}) = -3g.\exp(\tfrac{1}{4}g).$$

Hence

$$\lambda_s = 2 \frac{\exp(\tfrac{1}{4}g) - \tfrac{3}{4}g.\exp(\tfrac{1}{4}g)}{\exp(\tfrac{1}{4}g) + \exp(-\tfrac{1}{2}g)}$$

$$= 2 \frac{1 - \tfrac{3}{4}g}{1 + \exp(-\tfrac{3}{4}g)}.$$

When $g = 2$ we find $\lambda_s = -0.82$, but for $g = 2\tfrac{1}{6}$, $\lambda_s = -1.04$, so as g increases stability gives place to oscillations at a value a little higher than was found for the asexual population.

Location of the breakpoints where permanent oscillation appear as frequency dependence becomes more sensitive is not, however, the main point of the model: it is enough to have shown that when once the fitness function enters a certain range of steepness oscillations of some kind are almost certain to occur in any mixed population. It is assumed for a mixed population that fitness depends on the joint frequency of the sexual genotype and its asexual counterpart. Regarding a possible advantage to the asexual population in a mixture, interest centres on the separate *long-term geometric mean fitness* (LGMF) of the sexual strain versus the best such mean achieved by an asexual when all asexuals are given a twofold fitness advantage, corresponding to the worst case of wastage due to unproductiveness of males.[8, 9] I have not attempted formal analysis of the dynamics of mixtures but instead have studied the situation by computer simulation. Results indicate that analysis would certainly have to be complex: various oscillatory patterns were observed including six-point cycles and also seemingly chaotic fluctuations. Often both sexuals and asexuals persisted together indefinitely. Sometimes just one asexual strain went extinct, sometimes two did so while one persisted. But in general the advantage which the sexual strain gets from having, each generation, the arithmetic mean of a concave fitness set (see Chapter 1) ensures that if g is made large enough all asexual strains go extinct, or are at least kept at such low frequencies that, in nature, extinction would be probable. For example, if $g = 4.67$ (fitness function illustrated in Fig. 2.1(a)), total frequency of the persisting asexuals was kept below 0.0001. Polymorphism here was apparently chaotic. With this value of g the maximum fitness possible in the model (given by $r = e^g$) is 106.3 and the minimum is 0.00009. The maximum fitness proves less important for predominance of the sexual strain than the form of the lower part of the curve: it was found that with g slightly higher than the minimum needed to give predominance of the sexuals a 'plateau fitness' could be imposed which was much less than the natural maximum and yet still high enough to enable the sexual strain to drive out the asexual. For example, this happens when g is 5.33 (giving $r = 207.1$) and a plateau fitness is imposed at $w = 7$ (Fig. 2.1(b)). In this case a steady six-point cycle occurred. Using a different fitness function, $w_i = \{\frac{3}{2}(1 - f_i)\}^g$, success with an even lower plateau was shown: with $g = 10$ asexuals could be kept below one per thousand when maximum fitness is only 4.

Such low maximum fitnesses are of interest because of the difficulty previous authors have noted of finding any models where the two-fold effective fecundity advantage of an asexual can be overcome unless the maximum fecundity is set rather high, roughly in hundreds or thousands.[8,9] However, the 'plateau' models just mentioned have not really evaded this difficulty as regards the origin of sex, although they may possibly reveal an aspect of the strength of sex to defend itself once it is well established. This is because in the plateau models the asexuals on their own cannot even constitute a viable population. In the last-mentioned model, if the sexual strain is absent the asexuals fluctuate wildly and go effectively to extinction in a few generations.

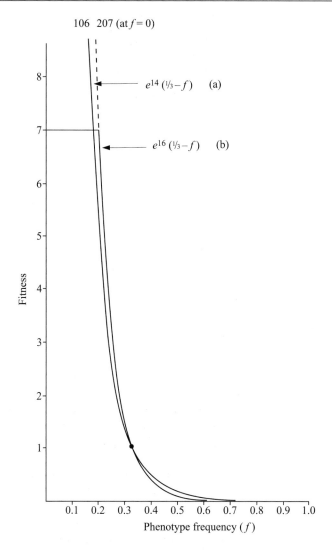

Figure 2.1 Two examples of frequency-dependent fitness functions that permit a sexual species to preponderate against its set of asexual emergent strains despite effectively doubled fecundity in the latter. In case (a) the limiting fitness of a very rare phenotype is 106. This case gives rise to a chaotically changing polymorphism in which asexual strains go to very low frequencies (below 10^{-20} for homozygous strains; below 10^{-3} for heterozygous strains). The maximum and minimum fitnesses observed are about 3.0 and 1.7, and the overall LGMF is 2.6. In case (b) a maximum fitness of 7 is imposed on a function otherwise going to 207 at $f = 0$. This case gives rise to a six-point cycle with highest mean fitness at 2.03, lowest at 0.88, and LGMF at 1.41. Here, asexual genotypes reduce to very low frequencies and may be going to zero: in the simulation cycle preceding generation 200 none is above 10^{-5} and the heterozygotes are below 10^{-33}.

In the preceding exponential model, the three strains enter a stable three-point cycle. When without plateau, the fitness function dictates that the LGMF of each strain must be at two (twofold fecundity advantage of asexuality being already entered); but imposing the plateau causes this mean fitness to drop. In the case where $g = 5.33$ and the plateau is at 7, the LGMF is 0.228. Obviously such a population is inviable. But the model does show that once sex is established it can evolve into regions of stable coevolved fluctuation with its parasites that are uninhabitable by any mixtures of parthenogenetic lines that the sexual could give rise to. The sexual species in the last case is viable, with LGMF at 1.407. In summary, artificial as it is this simple one-locus model may be claimed to be a little more realistic and relevant to sex than that model which appears to be its only one-locus predecessor in the literature so far (see Chapter 1).

TWO-LOCUS HAPLOID SELECTION MODEL

Close to a model already outlined, without formal analysis, by Jaenike,[3] I next consider a system with two loci each with two alleles. For simplicity, selection will be supposed to operate only in the haploid phase. This is an unwished-for assumption from the present point of view because elaborate multicelled sexual organisms that we aim to explain are mostly diploid. But if viable haploid-selection models can be found, it is to be expected that diploid-selection versions can likewise be devised (see Chapter 1) although so far, it has to be admitted, it has been easier to make haploid ones realistic.

There are now four haplotypes, AB, Ab, aB, ab, and each is supposed to confront a corresponding pathotype in the parasite species. We assume that asexual variants of each haplotype exists, each as before of twice the effective fecundity of its sexual counterpart.

The following picture of the life cycle can be suggested. After gamete fusion diploids persist only briefly and undergo no selection in this stage (they might be resting eggs, say). Then after meiosis haploid unicells are formed which multiply by division and then initiate the multicellular bodies to whose parasitism and selective elimination our previous story applies. Once past a certain stage of development parasite infection no longer kills them but the infected survivors can still harbour and multiply parasites. Thus each haplotype breeds up the clone to which it is susceptible. From among these clones, that corresponding to the most numerous genotype among current adults decimates the young of the same genotype next season. So a high fitness one season tends to be followed by low fitness next season, and so on. Alternatively, with probably little difference to the behaviour of the model, it could be supposed that the young haploid hosts support infectious epidemics such that each host genotype suffers as a function of its density, which is directly dependent on frequency. However, strictly it is only the first of the above alternatives that has been investigated so far.

Fitness in the model is assumed to be frequency dependent as before but a frequency of $\frac{1}{4}$ instead of $\frac{1}{3}$ now plays the crucial role. Corresponding as closely as possible to the last model although more general (see also Chapter 1 for the rationale) the following fitness function is chosen:

$$w_{11} = \exp\left[-\delta\left\{(1-z)P_{11} + zP_{12} + zP_{21} + (-1-z)P_{22}\right\}\right]$$
$$= \exp\left[-\delta\left\{v_{11} - v_{22} - 2z(v_{11} + v_{22})\right\}\right],$$

where P_{ij} are frequencies of haplotypes, and $v_{ij} = P_{ij} - \frac{1}{4}$. Likewise $w_{12} = \exp\left[-\delta\left\{v_{12} - v_{21} - 2z(v_{12} + v_{21})\right\}\right]$, and likewise for the others.

These functions have the property that when all four are present and persist and a sexual strain is absent the LGMF of every strain is one (or two after twofold fecundity advantage of asexuals has been applied). However, if a sexual strain is present with asexuals or is present alone then the situation is more complicated. Whatever the case a haplotype, whether sexual or asexual, is to have its fitness formulated from the above expressions using the total frequency of the haplotype (i.e. sexual-type frequency + asexual-type frequency). We consider a pure sexual population; analysis of this case covers that of a pure mixture of asexuals if the parameter for linkage, c, is set to zero.

Let P_{11}, P_{12}, P_{21}, P_{22} be the frequencies of adult haplotypes just after a round of selection. Then after meiosis and fertilization corresponding frequencies among offspring are $P_{11} - cD$, $P_{12} + cD$, $P_{21} + cD$, $P_{22} - cD$, where c is linkage and D is the initial linkage disequilibrium, $D = P_{11}P_{22} - P_{12}P_{22}$.

The transformation of the system from one generation of adults to the next is given by a system of four equations of which the following is typical:

$$P'_{11} = (P_{11} - cD)w_{11}/\bar{w},$$

where $\bar{w}$, the mean fitness, is the sum of the four products of frequency and fitness like the product shown.

When all the v are small, all w approach to 1. Thus $\bar{w}$ can be ignored and, using the linear approximation to the exponential function, the above equation becomes approximately

$$\tfrac{1}{4} + v'_{11} = (\tfrac{1}{4} + v_{11} - cD) \times [1 - \delta\{v_{11} - v_{22} - 2z(v_{11} + v_{22})\}].$$

Again, ignoring terms in v^2, this is

$$v'_{11} = v_{11} - cD - \tfrac{1}{4}\delta\{v_{11} - v_{22} - 2z(v_{11} + v_{22})\}.$$

D can be approximated by $\frac{1}{2}(v_{11} + v_{22})$; hence we obtain the system of four linear transition equations of which the above is the first. This linear system can be analysed for stability, and its eigenvalues prove to be $1 - \frac{1}{2}\delta$ and $1 - c + \delta z$.

We are interested only in the positive range of the parameter δ, which roughly specifies the severity of selection. The first eigenvalue shows that $\delta>4$ is a sufficient condition for instability at the fixed point, tending to specify an oscillatory departure. Independence of this eigenvalue of c indicates that both pure sexual and pure asexual populations will show cycles in this range of δ. However, instability due to the second eigenvalue can occur at lower values: assuming z to be in the range $-\frac{1}{2}, \frac{1}{2}$ instability can start as low as $\delta>3$ if $c = \frac{1}{2}$ and $z = -\frac{1}{2}$ (oscillatory departure), and occurs at every positive value of δ if $c = 0$ and $z > 0$ (non-oscillatory departure).

A non-oscillatory departure does not imply extinction or that, once far from the fixed point, cyclical patterns cannot occur. In fact, $z > 0$ can give cycles and these in turn can lead to stability of sex if δ is high enough. But much easier conditions for stability of sex are obtained in the cases where $z < 0$. Since in this range $z = -\frac{1}{2}$ (assuming this to be the lower limit) is both best for sex and gives the simplest equations and closest similarity to the previous one-locus model (giving $w_{11} = \exp\{2\delta(\frac{1}{4} - P_{11})\}$, etc.), I will now confine attention to this case.

Numerical simulation of the system in this case for $3<\delta<4$ shows that a stable two-point cycle develops.

In this cycle allele frequency does not vary at all at either locus: instead each locus has frequencies constant at $\frac{1}{2}$. As to absolute values, all v's are also identical and constant but cycle in the following way: in one generation v_{11} and v_{22} are positive while v_{12} and v_{21} are negative, then v_{11} and v_{22} are negative while v_{12} and v_{21} are positive, and so on.

Knowing that this cycle existed, it was possible to locate the equilibrium absolute value of D analytically. In the briefest form I could obtain, the equilibrium equation is

$$\cosh\{\ln(4\sqrt{bD})\} - \cosh(\ln\sqrt{b})\coth 2\delta D = 0,$$

where $b = 1 - c$.

Treating the LHS expression as $f(D)$, we obtain, in preparation for a Newton method approximation to the solution for D,

$$f'(D) = 1/D \sinh\{\ln(4\sqrt{bD})\} + 2\delta \cosh(\ln\sqrt{b}) \operatorname{cosech}^2(2\delta D).$$

Using this I obtained the results graphed in Fig. 2.2. As can be seen, the two-point cycle described above potentially occurs at all $\delta>3$. But when δ rises through 4 a further pair of stable cycles is potentiated. These are actually homologous to the case already implied in the figure where $c = 0$: if loci are completely linked the system becomes equivalent to one of four independent asexual strains. The figure shows that there is a stable cycle where $c = 0$ and D alternates. D is $P_{11}P_{22} - P_{12}P_{21}$: thus, considering the other ways of pairing among four independent asexual strains, there must also exist stable cycles for $D' = P_{11}P_{12} - P_{22}P_{21}$ and for $D'' = P_{11}P_{21} - P_{12}P_{22}$. As is implied in the graph and easily confirmed by simulation experiments, cycles

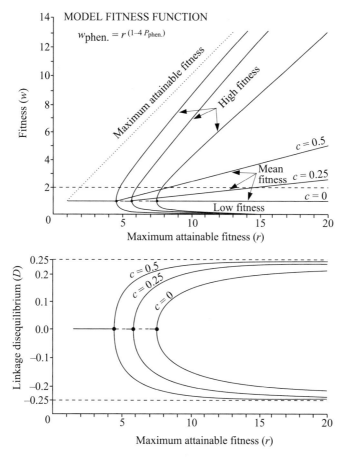

Figure 2.2 Linkage disequilibrium (*D*) and fitness performance (*w*) displayed as functions of maximum attainable fitness (*r*) for a model of haploid selection on two alleles at each of two loci. At a bifurcation point depending on the linkage parameter (*c*) increasing *r* renders a stable equilibrium for four constant equi-frequent genotypes unstable and a two-point cycle supervenes. Unless there is no recombination (*c* = 0), almost linearly rising geometric mean fitness in the cycles indicates that from some *r* upwards the sexual species will out-compete the set of its emergent asexual strains. The critical *r* occurs where the mean fitness rises through *w* = 2 (dashed line; 2 represents effectively doubled fecundity of the asexual strains). High and low fitness of a particular genotype in the cycle are also shown.

for D' and D'' exist as attractors for all $\delta > 4$, whether or not $c = 0$. A little thought based on what has been said about the cycle already analysed shows that these two new cycles, respectively, must arrange that $P_{11} = P_{12}$ and $P_{21} = P_{22}$, or else that $P_{11} = P_{21}$, and $P_{12} = P_{22}$; in other words, either the gene frequency of A goes to 0.5 while that of B alternatives, or vice versa. As apparent from the $c = 0$ case in Fig. 2.1 (and as designed for in construction of the fitness function) cycles of this kind give no advantage to sexual reproduction for any magnitude of δ. Thus when $c > 0$ and there is some hope for sex from the cycle and circumstances mentioned previously, the possibility that the system enters the domain of the alternative type of cycle must be considered a danger for sex, for as soon as the alternative cycle occurs the system is likely to be invadable by asexual mutants. Unfortunately, there is little that can be reported here on the sizes of the domains of 'good' and 'bad' cycles. When $c = 0$ and $δ > 4$ it is obvious from symmetry that all three cycles have equal domains; but then all three are equally bad. Since the inception of the cycle favourable to sex 'grows' back to $δ = 3$ as c increases to 0.5, it seems conjecturable that this cycle will have the largest domain at all $δ > 3$ for all $c > 0$. But one finding suggesting this may not be true is that in one set of simulation runs a certain fixed starting point was observed to pass out of the attractance of the good cycle into that of the bad as δ increased from 6 to 7.

Having given a cautionary outline concerning the existence of cycles unsatisfactory for sex, I now resume consideration of the hopeful two-point cycle.

For the purpose of displaying results Fig. 2.2 does not use δ itself as abscissa but instead that highest fitness which a given δ makes possible. This highest fitness occurs when the frequency of a type is zero and is given by $r = \exp(\frac{1}{2}δ)$. Here r is equivalent to what Williams[8] called the ZZI ('zygote-to-zygote increase') of a species. Its interest for the theory of sexuality, arising from the difficulty of devising successful models when r is low, has already been mentioned.

In the case of the present model values of r capable of making sex succeed against parthenogenesis are satisfyingly low. It can be seen from the figure that the changes accompanying the onset of instability as r increases are quite dramatic. Most relevant here, in best case, $c = \frac{1}{2}$, the mean fitness of the sexual begins a very nearly linear ascent at the bifurcation and passes 2 when $r \simeq 8.6$. The fecundity of a sexual species with a 1:1 sex ratio corresponding to this is 17.2. The actual highest fitness expressed in the cycle at this point is about 6.8, corresponding to fecundity 13.6.

There are hardly any sexual species—if any at all—whose normal potential fecundity is less than this. Even for humans, for example, an expressed birth fecundity of 14 would not be extreme. Of course, a severe caution has to accompany such an example, for even fetal selection in humans is on diploids: for humans and most other sexual species only wastage of gametes could be strictly relevant to the model. A previous model (Chapter 1) however, gave some ground for expectation that similar models with diploid selection can be produced.

Another cause for caution in applying the above results lies in the fact that the model studied so far covers a sexual species alone, or the set of four parthenogenetic strains alone (the case $c = 0$) but does not cover the behaviour of a mixed sexual–asexual population. Thus it is not clear that as soon as the LGMF of a pure sexual species rises above 2, the same species in a mixed population will be able to shed its asexual competitors. However, it is clear that the sexual population with LGMF>2 is an evolutionarily stable strategy (ESS) with regard to low-frequency invasion by asexuals, and it will be surprising if results suggested in Fig. 2.2 are not also robust against high-frequency invasions.

As Fig. 2.2 shows, and as all the foregoing discussion has led to expect, the best result for sex comes when c is maximal (i.e. there is independent assortment of loci). For the benefit of sex as in these models, different syllables of a password are best coded in different chromosomes. While I am not aware that contributions to self-identifying antigens can come from unlinked loci, there is evidence, again for mice, that such loci may contribute to the competence of antibodies against complex artificial antigens.[10] This and other recent evidence that recognition of foreign substances is not wholly dependent on a single short region of chromosome,[11] perhaps adds plausibility to the recombinant password idea.

Again we have not strictly shown that a high-recombination population beats a low-recombination one, still less that a linkage-modifying locus is selected to cause high values of c, but these outcomes seem probable. The demand for high-recombination values in models like these which assume fluctuating environments is in contrast to the requirements on linkage found in other broad classes of models concerned with sex, where usually the puzzle is to see why selection for closer linkage doesn't make the genotype 'congeal'.[9, 12]

Attempting to resolve this puzzle, Charlesworth[13] showed that alleles causing higher rates of recombination can be selected when linkage disequilibrium is made to fluctuate; and Graham Bell (personal communication) has shown this also in a model where the changes in linkage disequilibrium result from spontaneous cycling. Bell's model is frequency dependent and coevolutionary like the present one but uses different fitness functions. His analysis is focused on the length of time lag rather than on selection intensity in the induction of cycling, and on the selection for degree of linkage rather than ability of sexuals to beat a doubled effective fecundity of asexuals.

In the present model, lowest possible fitnesses are very low [$\exp(-\frac{3}{2}\delta)$]. Low fitnesses in the stable cycles are less low—in fact, merely reciprocals of the high fitnesses—but their contrast to the high fitnesses still implies an intensity of selection that may seem very implausible. However, bearing in mind another feature, that all this selective elimination goes on with both mean fitness and gene frequencies remaining constant, it seems that the occurrence of such intense selection might easily escape notice. Ideally, to reveal the kind of process envisioned, mortality statistics are needed classified for both genotype and major biotic causes. Lacking such data the most hopeful simple sign would be linkage disequilibrium observed to fluctuate radically from one generation to another. Hitherto there have been too few genetic surveys of natural populations that both cover enough loci to make likely the detection of polymorphism loci that are at least close to 'password syllable' loci, and at the same time cover enough generations. Data for the same natural population over many generations are particularly lacking.

CONCLUSIONS

The above model is, of course, bound to be an extreme oversimplification of any real situation. However, this model, plus some simple modifications and extensions which cannot be detailed here, plus also the even more artificial one-locus system described earlier, can be summarized as having made certain possibly useful points about what might be expected in nature if biotic interactions,[14] and especially interactions with short-lived parasites, are responsible for maintenance of sexuality. The points emerging from the model are:

1. Frequency-dependent selection acting hardest against the most common genotype easily sets up cyclical processes.[3]

2. In such processes population size may remain relatively constant and so fail to suggest intensity of selection.

3. When more than one locus is involved gene frequencies also may remain relatively constant; but intense fluctuating selection then remains reflected by fluctuating linkage disequilibria.

4. In a two-locus model cyclical or fluctuating processes tend to onset at lower intensities of selection for sexuals than for asexuals.

5. As selection intensity increases there comes a point, usually achievable at moderate levels of fecundity, where a sexual species has an advantage over any asexual strain even when the latter are given a twofold advantage in effective fecundity. Asexuals then die out or are maintained only at very low frequency.

6. High levels of recombination facilitate such exclusion of asexuals by sexuals.

7. Pressures on a host species by a set of varieties, or of species, of parasites seems particularly likely to engender the cycling discussed. In particular, shortness of life cycle of parasites relative to hosts gives their populations the kind of overreactive frequency dependence that is particularly favourable to sex.

8. Success of sex due to the frequency-dependent selection processes described does not require competition between sibs or other relatives.

References

1. B. Clarke, The ecological genetics of host–parasite relationships, in A. E. R. Taylor and R. Muller (ed.), *Genetic Aspects of Host–Parasite Relationships*, pp. 87–103 (Blackwell, London, 1976).

2. J. B. S. Haldane, Disease and evolution, in *Symposium sui Fattori Ecologici e Genetici della Speciazione negli Animali, Supplemento a La Ricerca Scientifica* Anno 19°, pp. 68–76 (1949).

3. J. Jaenike, An hypothesis to account for the maintenance of sex within populations, *Evolutionary Theory* **3**, 191–4 (1978).

4. R. A. Fisher, Randomisation and an old enigma of card play, *Mathematical Gazette* **18**, 294–7 (1934).

5. W. D. Hamilton, Gamblers since life began: barnacles, aphids, elms, *Quarterly Review of Biology* **50**, 175–80 (1975) [reprinted in *Narrow Roads of Gene Land*, Vol. 1, pp. 357–67].

6. F. Lilly, in H. O. McDevitt and M. Landy (ed.), *Genetic Control of Immune Responsiveness: Relationship to Disease Susceptibility*, pp. 279–88 (Academic Press, New York, 1972).

7. J. H. Hatchett and R. L. Gallun, Genetics of the ability of the Hessian fly, *Mayetiola destructor*, to survive on wheats having different genes for resistance, *Annals of the Entomological Society of America* **63**, 1400–7 (1970).

8. G. C. Williams, *Sex and Evolution* (Princeton University Press, Princeton, NJ, 1975).

9. J. Maynard Smith, *The Evolution of Sex* (Cambridge University Press, Cambridge, 1978).

10. J. L. Caldwell, Genetic regulation of immune responses, in H. H. Fudenberg *et al.* (ed.), *Basic and Clinical Immunology*, pp. 130–9 (Lange, Los Altos, 1976); here p. 135.

11. D. L. Rosenstreich, Genetics of resistance to infection, *Nature* **285**, 436–7 (1980).

12. J. R. G. Turner, Why does the genotype not congeal?, *Evolution* **21**, 645–56 (1967).

13. B. Charlesworth, Recombination modification in a fluctuating environment, *Genetics* **83**, 181–95 (1976).

14. R. R. Glesener and D. Tilman, Sexuality and the components of environmental uncertainty: clues from geographical parthenogenesis in terrestrial animals, *American Naturalist* **112**, 659–73 (1978).

CHAPTER 3

UNEQUAL COUSINS

Coefficients of Relatedness in Sociobiology

I am his Highness' Dog at Kew;
Pray tell me, Sir, whose Dog are you?
ALEXANDER POPE[1]

IN MY first years at Ann Arbor I was invited to many campuses and more occasionally was invited abroad. Almost always I wanted to accept but not for the honour or the discussion of ideas, instead more for the prospect of seeing new landscapes and new natural history. It was also interesting, I admit, even within America, to see new local ways and new styles of people but these differences did not seem to me great. I noticed them but they were far less exciting for me than the differences in natural history. As for the towns I wasn't thrilled by those either; species of shade trees in the streets were usually the best novelties from my point of view. Otherwise it was mostly the same high blocks, the same intersecting planes, and the same high human pretensions. Steel, concrete, polished granite, aluminium, glass: then the same statues tempting you not even to read the name.

Even stores and restaurants often had the same names. A main road into Albuquerque, for example (which town is host to the University of New Mexico), with its signs uplifted on huge steel tubes, its fast-food places, its carpet stores, its motels and supermarkets, was not so different to me from any of several roads into Ann Arbor—Washtenaw Avenue, say—unless you were to glance above the neons and see ragged cliffs and sunset mountains or if, crossing a parking lot, you were to notice yellow flowers on a bush resolve as you came nearer to be long yellow fruits on a cholla cactus. I am not at all saying that the giant malls and chain stores of such a road are bad; in fact they are obviously a part of all that tremendous

American efficiency I greatly admire. And some special and perhaps unexpected advantages that the big chains have will be discussed rather seriously in Chapter 4 (in which I will again instance, more specifically, Albuquerque). But the shopping malls, even those that are run down and are risking typical American lawsuits by letting cholla cacti lurk in their parking lots, are not what you choose to write home about. So, summarizing cities without their special plants, let me state that I feel at one with that folk singer who wrote: 'I've seen your towns—they're all the same/The only difference—is in the name.' She of the song found a lover in one town, lost him in another, and ended disillusioned; instead the trees and the weeds and the moth larvae I have found in foreign towns, with these never even pretending to care for me, have stayed with me all my life.

But the name of a town—that's different from all the steel and concrete! Cincinnati, for example, Kalamazoo . . . America has many wonderful names and this brings me to the case of Tucson. I had first heard of this southwestern city when an undergraduate at the University of Cambridge. Although I gather Tucson is often in 'Western' films, I didn't know of it from these because I never saw any until my travels to America began when I started to see some on the transatlantic jets—in silent versions because I never bought the headsets. Instead I knew about Tucson for a reason starting soon after I visited Waterbeach gravel pits near Cambridge and started to grow the shepherd's-purse plants I found there, trying to elucidate the strange and heraldic variation of their leaves. Waterbeach gravel pits were simply where I had first noticed these variations and where my best examples grew, with their rosettes becoming more and more dramatically different as autumn approached and they headed for diapause—saving in their tap root and intending to flower early next spring. Shepherd's-purse (*Capsella bursa-pastoris*) was abundant everywhere but on good soils the variation, though present, is not well expressed. Soon after starting my experiment, I had found in the library a paper by the American botanist George H. Shull about this polymorphism. He told that the sharp-ended leaflets versus the polygonal were controlled by a single gene, but he also claimed that another gene locus contributed as well and therefore could get 9:3:3:1 ratios. Of this other locus I could never convince myself—it seemed to me more a range of inclination to be a faster grower. Shull had collected the plants widely in the USA and seemed to know a lot about their genetics, and it was he who had mentioned particular seed lots from this town with a strange name, Tucson, seemingly out in some desert.

From that time on I had always wanted to see these places and also, of course, to see their shepherd's-purse plants. Shull had collected his most distinctive variety ever, 'var *heegeri*', in Tucson. This one wasn't distinctive for a leaf character; seemingly once, and only in Tucson, the European weed had mutated to a form that loses the neat heart-shaped 'purse' of a capsule that provides the popular name and acquires instead a more compact globular one—an atavism (reversion to a primitive type) one suspects. It was on account of this memory of a long past interest and of the name of the place, not at all as the place near where Cochise and his Apache Indians roamed, nor even because of its desert of giant cacti (plants whose presence there I didn't yet know of), that I felt a quiver of eagerness when I read the letter from Richard Michod inviting me to talk and to discuss with him some technical questions we had been corresponding about concerning coefficients under inbreeding.

Mostly, as I wrote before, I was invited to places because of kin selection and the social and behavioural issues I had worked on—the topics of my early papers of Volume 1. Kin selection was very popular in the late 1970s, especially at graduate levels of teaching. On the other hand a lot of the discussion coming out in journals about my work was quite critical of the idea. The validity of my emphasis on relatedness in evolution seemed by now admitted; but was it all working quite the way I had claimed? Ever more acute (and dare I say fussy?) brains were prowling the theoretical journals; ponderous mathematical cortices skimmed my pages like flying saucers and back at their base did not always pronounce favourably on what they saw. Inclusive fitness wasn't 'well defined', it was said; maybe there wasn't after all going to be a useful predictive province for all this new stuff in social evolution. Of course it made me tense to see my eldest child challenged but usually I became relaxed soon again after noticing some misunderstanding or that my critic was studying cases I had not claimed to cover and, as biological reality, didn't believe in. Usually they were the cases with strong selection and/or unconditional gene expression. Sometimes knowing some major theoretical figure at a university that was inviting me (a person quite separate from the enthusiastic junior faculty member who, like Rick Michod in the present case, had sent me the letter), I guess I felt a bit like Giordano Bruno receiving a friendly note from a cardinal inviting him to Rome to discuss philosophy—plus, possibly, for a chat concerning some of those recent and entertaining notions that Copernicus had brought up concerning the sun. Nothing resembling a stake or a faggot adorned the Vatican seal on the letter at that stage I'm sure.

On one of my trips a fairly senior member of a biology department and a keen follower of Robert MacArthur (the famed North American theoretical ecologist who had died young, while I was a graduate student I think) invited himself to the room where I was revising my notes a half hour before my advertised talk. He told me he had noticed I was billed for a talk on sex ratio: was it to be the ideas of my recent paper? Yes, I replied, wondering if this really was the first moment he had noticed all the posters, and detecting as I thought something artificial in his manner. Well, he went on, a while ago he had spotted a fundamental flaw in the argument behind all my sex-ratio conclusions and felt he should tell me . . . Judging by his continued manner he had been very sure he would soon convince me that all my papers had been rubbish, and his timing made his intention to effect a complete sabotage of my talk very obvious. Fortunately I, too, had been busy with my homework, and attending to exactly the points he was bringing up. There turned out to be sufficient time not only to point out my visitor's own error but to finish my note reading. But even though Rick Michod and I had derived slightly different formulas for what seemed to be the same thing, I am glad to say that nothing like the earlier event threatened me at Tucson either from Rick or anyone else. Rick and I had a very profitable interaction of which the paper that follows was the outcome. Quite contrary to the other case, already at an earlier brief meeting Rick had shown himself to be a friendly and gentle person to be at odds with—if any disagreement developed between us other than the usual merely semantic ones. Whether we did argue and, if so, who was right, I confess I do not remember. I recall only our decision that we might usefully update and extend recent theory of the necessary coefficients by writing the paper.

While sometimes accepting to talk about kin selection, mostly I was trying to offer topics that were currently more exciting for me, such as sex. And I was often also struggling to resist the temptation to accept the invitation at all. As I see from filed letters, evidently this had been my first response when I answered Rick in February of 1979. I said I would 'love to come some time but had made a resolution not to travel to give seminars this year' and I explained that 'I spent so much time travelling last year to the detriment of my research—including, for instance, keeping abreast of the present multiplication of coefficients of relatedness!' The last comment was an oblique reference to the manuscript Rick had sent me showing some formulas he had derived that were slightly different from mine; but there were others too.

But could I hold out against a name like Tucson and an atavistic seed capsule for long? No. My copies of correspondence show me being coy with Tucson only until late summer. Then with the excuse that a trip to India had recently flatted (this was true, I am afraid; an Indian temptress, I confess, for a month or two, had beaten the seed capsule and the South West) I wrote to Rick telling him I was free and would like to come after all.

Therefore on the morning of 19 November 1979 that old kilometre-squared and varicoloured chessboard of the Great Plains went flowing once again below my Boeing's window. After a brief landing and plane-change for a new direction at Denver, and then with the Flat Irons on my right and later even the far snow-tipped higher Rockies slipping behind, a redness crept into the cloud-like and still flowing patterns below me. They were new patterns, new colours I had never seen. Long roads snaked into dark and townless distances studded with faint buttes or mesas, while the rare fields near to the roads changed their squares for circles (the irrigated fields) and shouted up at the plane in green or black set against a red, yellow, or a yellow-brown background that I knew must be incipient desert. More scarce even than towns, the rivers mostly had to be deduced from the courses of their spiky and devious canyons—buttes, as it were, that had been switched to their negatives and then brought together into sinuous ranks. This was the South West and all new to me. Even with my neck cricked by hours of the sideways and downward staring still I found I had no will to attend to the rough sheets in my hand that bore such titles as 'Sociobiology of bizarre males', 'Pricean levels in haplodiploids', and 'Relatedness in multi-queen S.I.s under inbreeding'. Idle in my bag stayed also the notes I had prepared for the defence of my formulas of relatedness against Rick—that is, my struggles yet again through all that logic about how knowing that either he or that other are inbred should change, for example, how Henry IV felt about his cousin, Richard II, or how Henry's son, soon to be Henry V, felt about his cousin (second cousin actually in this case), Edmund Mortimer, and how Edmund Mortimer about him.

Usually when explaining to non-biologists how kin are to be valued in the evolutionary calculations, I start with a deliberate vagueness about what might be meant by such a phrase as 'the proportion of genes in common' between two relatives and I concentrate on the obvious importance of the probabilities of gene survival. What 'genes in common' means it takes too long to explain and perhaps even at the end I don't completely understand the matter myself . . . Suppose Harry Plantagenet, Prince of Wales, is inbred and hence has two copies, *aa* say, of some altruism-determining gene, while his second cousin Edmund Mortimer (who, as it happens, was another very plausible contender for the throne) happens to have one only—his genotype is *ab* let us say. What proportion at this locus do we say the two have in common? Obviously the question hardly makes sense or at least doesn't unless one says a lot more first. Had they been haploid like two moss plants who were cousins all would be easy; either plant X would have the altruism gene possessed by plant Y or it wouldn't: one could give a probability and then simply average such chances for all the gene loci. But notwithstanding the thoughts given in the introduction to Chapter 2, most of the socially most interesting organisms are diploid. As I have pointed out, diploids have two copies of every kind of chromosome, one from their mother and one from their father. And in fact this problem about 'genes in common' was an old chestnut for me and I had pricked fingers on it from the very earliest days of my 'altruism' obsession. I had dealt in outline with questions of relatedness when one or both of two relatives could be inbred in some of my recent papers (see Chapters 5 and 8 of Volume 1) but I was very much aware of various issues still unresolved or remaining unpublished.

About the time of my first invitation from Rick Michod, Nathan Flesness, for example, had published in *Nature* a paper entitled 'Kinship asymmetry in diploids'. He showed clearly that if one relative was inbred and another not or less so, as in my example of the first two pretenders to the English throne above, the coefficient of relatedness that the inbred should use when deciding how to behave towards the outbred relative (a cousin once again, let us say) was not the same as the coefficient that the outbred cousin should apply if making the reciprocal decision. The theory says that the outbred person in fact, to a slight degree, should tolerate the unfairness in the interest of his genes, or, to put it more properly, his genes should even see to such an unfairness coming about: he should behave more generously towards the inbred than he would expect the inbred to reciprocate. The key to my understanding this had come when I realized, a year or two after my paper of 1964, that Sewall Wright's correlation concept of relatedness hadn't been quite what I needed for my own calculus of relatedness. For a Machiavellian world of pure gene-controlled behaviour I saw that what Henry IV and Henry V need to do to decide their cousinly status with regard to their respective copretenders to the throne is to predict the genotypes of Richard II and cousin Mortimer with reference to their own, and to do so by the best possible statistical reasoning. Just as a statistically minded agronomist uses a regression coefficient based on his knowledge of responses to fertilizer to predict as best he can the yield of crop, so must the Henrys use regression coefficients of genes to best predict how genes in common may have descended to the cousin. This, of course, is not an adequate explanation but it gives the idea. The upshot was that I, too, needed the regression coefficient, not the correlation one of Wright.[2]

Wright had practically provided both of them but the one he himself had used had been the correlation. It is well known that if a bivariate scatter of values is not a straight line, the regression of Y on X is not the same as the regression of X on Y. The simplest 'regression' relatedness formula is $b_{XY} = 2r_{XY}/(1 + F_X)$ where F_X is a measure of individual X's inbreeding: this formula had been given in my papers of 1970 and 1971. The consequence for asymmetry of altruism was left implicit in the formula at that time—increase F_X and the value of the formula is obviously reduced.[3] On seeing Flesness's paper I was a little chagrined to realize that I had often mentioned such asymmetry at meetings and to students, usually via some half-joking reference to a British (and also, it seemed, American) inexplicable generosity and tolerance towards European inbred aristocracy, a deference accorded often even without any real evidence that recipients either were what they pretended or were providing any return benefit.[4] In real life false aristocracy, as shown in Mark Twain's fictional tricksters encountered by Huckleberry Finn on the Mississippi, of course never reciprocated at all; they simply moved on down the river. Real aristocracy, usually stuck in one place, usually did seem to reciprocate something to the somewhere but it was often only a slightly elusive currency, which another author, Evelyn Waugh, chose to characterize as 'charm'.

Enthusiasm for aristocracy and acceptance of the sufficiency of charm as repayment for huge public gifts and indulgence seemed to me well illustrated, much later, in the case of Princess Diana. Meanwhile Flesness's paper reminded me I had neither explicitly put the 'lopsided' implication of inbred relatedness into print nor had even taken it into a bar where beer and *bonhomie* plus, at that time, plenty of cigarette smoke, might fix it for ever as 'my discovery' in younger minds . . . But, to be more serious, the point that Flesness made was, as I said, just the signal of those times and the beer-stained envelopes bearing the lost calculations of Haldane's 'two brothers or eight cousins', my own anecdotes of the two kings or the two princes, of how inbred Darwin outranked outbred Galton, were all being flung out as part of a mood to tidy and make everything exact, with no one seeming to care that the differences were quite trivial compared with the forces implied in the fact that cousinships—that is, sexuality—existed at all, or indeed that the issue of passive smoking had never yet faced up to even its first case of 'monozygotic twins reared apart' (and reared also, of course, in smoking and non-smoking households). For me the exciting issues were always the biggest I could see, the farthest yet highest of blue mountain tops, which means I was always getting a little tired of my old ones.

There were relaxations, of course, and among these were names, and, with names, the mystique of coincidence—or was it more? Should it ever be proven that an '-ov' or '-ove' patronymic termination marks an ex-Russian Y chromosome coding for a special interest in relatedness and ancestry I would be better able to understand the high frequency of names of this termination that were busy on the 'relatedness' front I am talking about—Orlove, Charnov, and Abugov are ones I remember, but I recall also Chekhov's Russian story *Three Years*, in which a brother goes mad over his noble ancestry. No one else in the story is of noble origin—all the story's dull yet picturesque business people are descended from serfs. But then '-ov' itself is only the Russian equivalent of Scottish 'Mac-' or English '-son': it tracks back but a single generation. Then again, perhaps that's enough to show interest and could be the first step for an induction . . . In contrast, plenty more names followed in the rush that were Teutonic, Jewish, or more generally European: Eshel, Cohen, Feldmann,

Cavalli-Sforza, Scudo, and Ghiselin are examples trailing a little behind the '-ovs' into my memory on the theme of relatedness. Names that were British originally also are there—Charlesworth, Benford, Norman, Williams, and Darlington—but, sadly for my pride both of self and of race, these bearers tended to be both more critical and, towards the end of the list (though please note that the 'Williams' referred to isn't 'George C.'), they are also more wrong, which last facts perhaps help to explain why my place of work at this time was America.

Controversy is attention to one's work so that apart from addressing middling mistakes, both mine and those of critics, which I tend to think the most serious for immediate reputation because they can be found plausible, I tend to ignore most criticism, expecting ultimately all errors to self-refute and that surely the most transparent ones will be those to go soonest.

Often, however, this proves wrong: a demagogic 'science' of public sentiment and myth, fuelled onward by populist rhetoric, to which kinds of attack the field of evolution is especially susceptible, has allowed many absurdities to persist. I have ignored truly ridiculous errors and misunderstandings put out by P. J. Darlington[5] as also, up to Volume 1 of these memoirs, absurdities of social evolution that are believed, seemingly, by Lynn Margulis, and, even up to this volume, also those of Amotz Zahavi about kin selection.[6] Generally I have ignored also the errors that I see as necessary to religious Marxism.[7] Recognizing a fellow revolutionary in Darwin, Marx asked to dedicate his great work to him and yet never really understood the evolutionist's thought or even the principles on which science worked. Like wrong science, wrong political ideas are also self-correcting but they are so much more slowly and on a scale of greater pain to humanity. This is because the whole aim of political ideas is to be put quickly into effect and once they find an initial apparent success, careers are committed, and the changes they cause gain great inertia. Change thereafter is slow. In contrast, in all matters that are of much importance in science, change occurs quickly: wrong ideas fade simply because all attempts to build on them fall down. The provocation and recognition of collapse, sometimes begun by the very scientist who started a given false line or sometimes induced by his peers, is the very core of the process. Its pain, unlike that begun by the politician, is only that of the initiator and those who tried to further a wrong idea. The method has set a pace of progressive change far above that of all other kinds of public thought. At times the pace may be too great for our frail frames and emotions evolved under ice-age timescales; but this is a problem of things working too well, our getting too much of what we ask for, and not one of beating one's head and those of others against a brick wall, as with the political ideas.

As I remember, at our meeting in Tucson we talked as much about the philosophical issues of evolution theory and about science in general as we talked about the refinements of kinship. Trained as a mathematician Rick Michod is a philosophical person as his publication list well shows. Most of the algebra in the paper is his; I supplied a critique of some points and my notion of where we were heading. Our paper was succeeded in the following year by a further review by Rick of 'kin selection' as a whole and this is probably the better historical platform from which to look back on the views of the time than is the paper of this chapter.[8] Even Rick's review, however, was by no means the end of the story; as I pointed out in Volume 1 the development of more and more inclusive coefficients that would use and interpret inclusive fitness theory was to continue for some time. Designed to be omnivorous from the very first by the name I chose, the concept has even attempted some true predations, as when it attempted to gobble up, not terribly successfully in the event though its approach was sound, 'reciprocation' and 'interspecies interactions' as extensions of its normal fare.[9]

I have missed out how my sidelong and downward staring at the red West in the plane ended—that is, my landing in Tucson and my first impressions. Did I, in fact, land directly in Tucson or was it perhaps at Phoenix? Perhaps my guilty sense of breaking my resolution makes me not remember. Perhaps I did land at Phoenix. That was another good name, but after one or two more visits Phoenix became remarkable to me mainly for the proportion of grey heads one could see there—heads that, if not daring to hope for actual phoenix of renovation in the fire of the desert, at least obviously hoped to end their lives warm, like Sam McGee in Robert Service's poem. I am now fairly sure that I did land at Phoenix and that it was probably Rick who drove me to Tucson through, as I remember from another trip, a rather spoiled strip of desert although it still had stretches of the wonderful creosote bush that needs no 'phoenix' name because it seems immortal in very nature; 9000 years for the oldest clones, so I saw it written!

Of Tucson itself I recall a desert brightness and (if this makes any sense) an extreme dryness in the sunshine itself; also the campus date palms, and those 'washington' other palms in the streets with their tall stems either woolly with dead leaves like sheep, or shorn, according to suburban taste. Out of a purple haze small mountains resembling stony turds overlooked the town on several sides. The haze was probably a poisonous smog rather than the distilled poisonous terpenes of desert bushes like from the creosote ones that I would prefer to have breathed, but, whatever, its brown-purple seemed pretty to me and part of this South West atmosphere because unfamiliar: all of it was my foreign land and my substitute for India. For the rest that I wanted to see, my stay was all too brief. Guilt about my resolution drove me, I expect, and it determined the short time I allowed. Fortunately I have had occasions to go back. From that first one I remember a trip by car with students over (or rather between) two of the minor guardian 'turds' and out into the desert where saguaro cacti grow like a sparse plantation of green-painted telegraph poles, receding into a more purplish and less brown haze than that over the town—this to the south and towards the Mexican frontier, perhaps evidence that some of the smog was terpenes after all. A few minutes of walking into this stony and at first hostile wilderness, violent with its thorns, a new wonder at my every careful stride, was all I had time for before I had to be taken to the airport.

Thus in its setting Tucson was as good as its name for me, a real real desert town. My first visit there counts as a statistical outlier in all my experience. But before being too fulsome about Tucson and its desert, I had better be careful because of another that comes much earlier in my experience although not in my conscious memory—I mean the desert around my birth place in Cairo, Egypt. Here my mother often took me by pram out from the officers' married quarters a little way into the desert, or, occasionally it seems, my father drove us out farther from the town. In photos I sit sepia in a flat sand and seem to be hitting a faintly sepia desert, in this case with a horse-hair fly whisk—a fat baby absorbed in what it is doing. Neither I nor the desert nor the ruler-drawn horizon behind me, undented in the shot my father made by any eminence including pyramid (still less by sacred Baboquivari Mountain), look in the least inspiring. Still, I seem to be enjoying myself and doubtless was acquiring a taste for that sand and gritty air that was perhaps to rejuvenate me in deserts far in my future, as on the Serra Geral de Goiás in Brazil in 1964, or during this visit to Tucson, or in Israel as I will describe in Chapter 8. From my mother's milk at the time, and perhaps earlier, via our mutually negotiated but probably large placenta,[10] I believe I acquired more literal tastes from her, unconscious and for long unused. My mother admitted she had been very fond of fruit during my pregnancy: figs, dates, and mangoes were certainly among them and all three are now close to being my favourites. Inspired perhaps by just those flavours I may have fought hard to have a big placenta for I came out a big baby. As to more general American deserts and those of Tucson as I was to revisit them, I drove later around Arizona with my family and saw all from close up—the reds and yellows that I had watched the first time I flew in. But surely to a naturalist who has once been in it, the Sonoran desert that begins by Tucson has to be the most wonderful in the world. What other has such cacti, for example, indeed how can desert vegetation ever seem right without cacti afterwards? Where else, to start another tack, has so much 'virgin birth' been going on and opened for study? The prickly pears and the brine shrimps, lizards, and fish; possibly even some honeybees of Arizona have become parthenogenetic, as I once heard though I have not yet seen it confirmed in print. During my first visit, however, I think I was not aware of any unusual frequency of this state around me and any doubts concerning the worth of maleness that I brushed with were not intrinsic to the harsh environments, rather to the air-conditioned halls where I lectured and that occasional fire from feminists that I grew used to after my talks in universities—sociobiology as the new fascist and male-dominated doctrine, and the like.

Rick and I ran on over the years from that meeting on parallel tracks to some extent. I think both of us had seen that the relatedness we were modelling in this paper was just one corner of a land unknown that had far more to it than the niceties of relatedness. Later, we have taken up somewhat different interpretations of the nature of some of that territory. Rick already has out a full book giving his views about sex,[11] while, as you see here, I am still writing mine. Darwin made natural selection a part of a science so strong and novel as to be now able to invade, for example, industrial technique (see Chapter 4) as well as even questions about the adaptedness of universes.[12] But, as I stated earlier, science is also itself, in part, a natural selection of ideas. So for the present I just bow to Rick's truths of DNA repair as undoubtedly a part of what will be our full story.

But let us wait to see whose 'truths', in the end, are more primary and have the better survival. I never saw var *heegeri* in Tucson but didn't look all that hard; I suspect that even when George Shull collected it, it was a lingering mutation, a toe-hold reversion that had been given its chance but even then was half way to its final demise. But, then, it is commonplace for me that when I go to look for a rare plant I don't find it but find others instead.

References and notes

1. A. Pope, Epigram engraved on the collar of a dog given to His Royal Highness, Frederick, Prince of Wales, in *The Poems of Alexander Pope. Volume VI. Minor poems*, ed. N. Ault (Methuen, London, 1954).

2. We are here very close in the required relatedness measures to the literal 'regression' concept of Francis Galton whose human statistics had shown him that on average the offspring of a person Y with a certain quantitative character—arm length, say—'regressed' halfway to the population mean. Y was supposed to have married a randomly picked person, or at least one of no particular similarity, so Galton's result is exactly what we would expect under most theories of inheritance that assume that mother and father contribute equally. But that 'most theories' includes unfortunately the invalid 'blending theory' of inheritance. The superficial seductiveness of this gave Darwin some trouble when he realized its consequences, so hard to reconcile with the maintenance of variation, and, it is said, forced him into unnecessary compromise with Lamarckian views.

   Mendel meanwhile had the answer for everyone but neither Darwin nor anyone of his time saw it, nor did anyone for a while even after Mendel's work had been rediscovered. Only after a brilliant paper by R. A. Fisher in 1918 were the studies of the two cousins, Darwin and Galton, synthesized through simultaneous and brilliant illumination by Mendelism. In this paper more than in any other, emerged the current still-triumphant paradigm of Neodarwinism. Fisher's paper had been characteristically refused by the two mutually antagonized 'Galtonian' and 'Mendelian' referees (Karl Pearson and William Bateson) to whom it had been sent after submission to the Royal Society of London—exactly the two men its contents should have reconciled. In consequence it came to be published by the Royal Society of Edinburgh.

3. I thought initially that Henry V, as the more aristocratic, would be more inbred than Mortimer, but this turns out to be incorrect. At least as far as my *Encylopedia Brittanica* (*EB*) information goes (1967 edition, ed. W. E. Preece, article on 'Plantagenets'), both had mothers unrelated to their husbands. Both were fighters in high health.

For a more interesting comparison I consequently transferred attention to Henry V's father, Henry IV, and his cousin, Richard II, both grandchildren of Edward III. Using these cousins to address the question hinted, we find that they were indeed both complexly but weakly inbred. Remarkably, none of their four parents was inbred, at least according to my source. Their two mothers, however, were linked (through their fathers) into the web of European aristocracy of their time. I counted for each seven independent paths by which a gene in a mother could be identical to one in her husband— this within the limited one-page pedigree provided in EB. Amongst such paths, Richard II had the two shortest (via his ancestors Edward I and Philip III of France: six- and eight-ancestor chains, respectively; I will indicate such chains by 6 and 8). Henry IV had Henry III (father of Edward I) as his most recent ancestor, implying a 9-chain, and, more distantly, has to go back to Blanche, daughter of Alfonso VII of Castile, to find his next-best 13-chain. As to the longer chains, 5 for each, all were quite different and none shorter than 13. Altogether, with his coefficient at $F = 164/8192$ against Henry IV's $F = 19/8192$, Richard II is the more inbred.

Correspondingly to our theory Richard II usurps the throne; however, obviously such small Fs, through contribution to the $1 + F$ divisor in the regression formula, are going to make, actually, a fairly negligible difference to the regression coefficients. Consequently (in the spirit of my argument and showing the method) we need not be surprised that neither pretender to the throne deferred to the other and they fought out their claims.

Turning to another 'pretender' in this picture, Edmund Mortimer, already mentioned, by my data he has $F = 0$. In this light it fits again that (a) he joined the rebellious faction of Hotspur and Glendower as its 'pretender' only by their persuasion; (b) after his capture, by Henry IV, he stayed tamely (and was tolerated) for the rest of his life as a semi-prisoner at Windsor; and (c) my EB source (Vol. 15, pp. 867–8) tells us that 'Edmund seems to have rewarded Henry V with persistent loyalty', including informing him of a plot by others to depose the king and put him (Edmund) on the throne. All this has the expected slant of the theory. The man to whom he gave loyalty, as noted, was seemingly no more inbred than he was; however, his pattern of behaviour was probably set during the reign of his relative's inbred and doubtless (to the youth) dominating and awesome father.

But, the method now shown, it has to be said that it is those two last adjectives that probably convey the real factor; whether their degree, and the young man's response, have any epigenetic input in this case stays inscrutable.

4. M. Twain, *The Adventures of Huckleberry Finn* (New English Library (Signet Classic), London, 1884 (1962)).

5. P. J. Darlington, Jr., Non-mathematical models for evolution of altruism, and for group selection, *Proceedings of the National Academy of Sciences, USA* **69**, 293–7 (1972); P. J. Darlington, Jr, Genes, individuals, and kin selection, *Proceedings of the National Academy of Sciences, USA* **78**, 4440–3 (1981).

6. A. Zahavi, *The Handicap Principle* (Oxford University Press, Oxford, 1997).

7. For example, M. Sahlins, *The Use and Abuse of Biology* (University of Michigan Press, Ann Arbor, 1976); see also *Narrow Roads of Gene Land* Vol. 1, pp. 192, 267–8, and 331.

8. R. E. Michod, The theory of kin selection, *Annual Review of Ecology and Systematics* **13**, 23–55 (1982).

9. See *Narrow Roads of Gene Land* Vol. 1, pp. 262 and 268; also S. A. Frank, *Principles of Social Evolution* (Princeton University Press, Princeton, NJ, 1998).

10. D. Haig, Genetic conflicts in human pregnancy, *Quarterly Review of Biology* **68**, 495–532 (1993).

11. R. E. Michod, *Eros and Evolution: A Natural Philosophy of Sex* (Addison-Wesley, Reading, MA, 1995).

12. L. Smolin, *The Life of the Cosmos* (Weidenfeld & Nicolson, London, 1997).

COEFFICIENTS OF RELATEDNESS IN SOCIOBIOLOGY[†]

RICHARD E. MICHOD and W. D. HAMILTON

———

A much-discussed, quantitative criterion for the spread of an altruistic gene is Hamilton's rule[1,2]

$$\frac{c}{b} < R, \tag{1}$$

where c and b are additive decrement and increment to fitness of altruist and recipient, respectively, and R is a measure of genetic relatedness between the two individuals. When rearranged as $-c.1 + b.R > 0$, the rule can be interpreted as requiring that the gene-caused action increase the 'inclusive fitness' of the actor. Since its introduction, Hamilton's rule and the attendant concept of inclusive fitness have gained increasing acceptance and use among biologists and have become integral in the field now named sociobiology. However, the essentially heuristic reasoning used in deriving these concepts, along with the lack of a complete specification even in the original outbred model,[2] have led to many investigations into the population genetical underpinnings of Hamilton's rule.[3–18] On the basis of these considerations, several reports[3,10,12,13,18] have proposed new formulae for R. These formulae have no obvious relation to each other or to the coefficients originally suggested by Hamilton. This proliferation of coefficients is undoubtedly confusing to many and the net effect may be to generate distrust both of the rule and of the notion of inclusive fitness. Our purpose here is to show that these various formulae for R (refs 3, 10, 12, 13, and 18), although independently derived, are actually the same.

What is R in equation (1)? On the basis of a particular outbred model, Hamilton[1,2] claimed that Wright's[19] 'coefficient of relationship' was the required R. However, later, giving a heuristic development but no further model, Hamilton[20,21] modified his identification of R with Wright's coefficient by claiming that equation (1) requires, in principle, a regression coefficient of genotype of recipient on genotype of altruist, whereas Wright's coefficient is

[†]*Nature* 288, 694–7 (1980).

the corresponding correlation coefficient. Such a correlation coefficient will often be the same as the regression coefficient but differs when the inter-actants are inbred to different extents.[18] These discussions[1,2,20,21] implied that R was independent of gene frequency and selection, and that equation (1) held for inbred populations (with the regression coefficient as R).

However, in recent models,[3–17] R is taken to be the threshold value of c/b, below which the cost–benefit ratio must be for the gene to increase. The R formulated in this operational way need not necessarily correspond to any simple measure of genetic relationship. Some of these models[3–11,14,15,17] specify the whole mating process with respect to interactions within families. Other models[12,13,16] maintain the generality of Hamilton's[2] original approach, and apply to interactants of arbitrary relationship, but fail to specify the population and mating system processes which give rise to R. If Hamilton's rule is to be useful, it must turn out that R, as formulated by these various approaches, must not vary widely as gene frequency or the parameters of selection change, except as implied by the simple proportionality to c/b. In many cases of interest, R is constant,[4,7–9,15] but in others[3,4,7,10–13,16] dependencies on gene frequency and dominance enter into calculations of R.

Our analysis here will not consist of any derivations of the coefficients from the gene-frequency dynamics, as that is done from different points of view in the various works to be considered.[3,10,12,13] Instead, we will simply convert these various coefficients into a common symbolism and show that they are, in fact, identical. However, before doing so, we should acknowledge an important aspect of the derivations: either selection must be weak so that standard identity coefficients can be used as measures of genetic relationship, or rather special conditions must be imposed on the models so that there is no selection at internal points of the pedigree patterns studied.

The genetic relationship between two diploid individuals, X and Y, at a single locus can be summarized by the following nine 'condensed' identity coefficients $\Delta_1, \Delta_2, \ldots, \Delta_9$, where Δ_i is the probability of genetic identity event i obtaining (Fig. 3.1 upper left-hand corner).[22] These coefficients consider identity states among all four alleles present in X and Y. The traditional two-allele coefficients, interpreted as probabilities, are

$$f_X = \Delta_1 + \Delta_2 + \Delta_3 + \Delta_4 \qquad f_Y = \Delta_1 + \Delta_2 + \Delta_5 + \Delta_6 \qquad (2a)$$

$$f_{XY} = \Delta_1 + \tfrac{1}{2}(\Delta_3 + \Delta_5 + \Delta_7) + \tfrac{1}{4}\Delta_8, \qquad (2b)$$

where f_X and f_Y are the inbreeding coefficients of X and Y, respectively, and f_{XY} is the 'coefficient of consanguinity' between X and Y, or the probability that a random allele from X is identical by descent with a random allele from Y at the locus of interest. Letting X denote the altruist and Y the recipient, the coefficient which we will use to relate the others may now be defined as

$$R = \frac{f_{XY}\alpha + (\Delta_1 + \tfrac{1}{2}\Delta_3)\,(1 - \alpha)}{\tfrac{1}{2}\alpha + f_X\,(1 - \tfrac{1}{2}\alpha)}, \qquad (3)$$

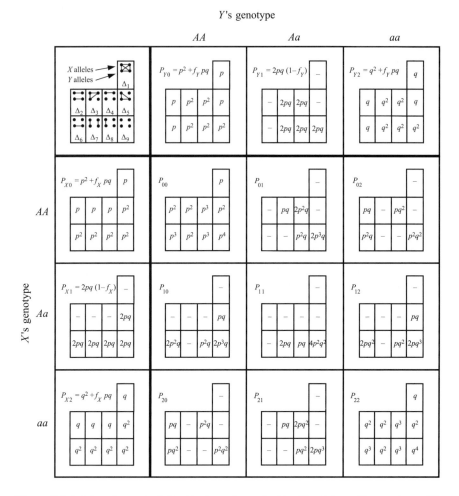

Figure 3.1 Joint and marginal distribution of genetic identity states with genotypes. The nine 'condensed' identity states are given in the upper left-hand corner with probabilities Δ_i $(i = 1, 2, \ldots, 9)$, $\sum_i \Delta_i = 1$. The alleles of X are on top and the alleles of Y on the bottom. A line connecting alleles indicates identity by descent. Given the occurrence of each identity state, the distribution of the genotypes of X and Y are given within the large side and top three boxes, respectively. For example, if identity state 5 obtains, then X is AA, Aa, aa with probabilities p^2, $2pq$, q^2, respectively; the like probabilities for Y are p, 0, q, respectively. Given the distribution of identity states, the marginal distribution of genotypes is given in the top left corner of each of these six large boxes. The joint distributions of genotypes are given in the nine large boxes towards the bottom right of the figure. For example, if identity state 3 obtains, then

$$P_{00} = p^2, P_{01} = pq, P_{03} = P_{10} = P_{11} = P_{12} = P_{20} = 0, P_{21} = pq, \text{ and } P_{22} = q^2.$$

where $\alpha = 2q + 2h - 4qh$, p is the frequency of the non-altruist allele A with $q = (1 - p)$ the frequency of the altruistic allele a, and h is the probability that a heterozygote performs an altruistic act. If we express the phenotypes as the probabilities of behaving altruistically (0, h, 1 corresponding to AA, Aa, aa), and the genotypes as the fractions of altruistic alleles (0, $\frac{1}{2}$, 1 corresponding to AA, Aa, aa), then α is simply the regression of phenotype on genotype in a randomly mating population. This convention regarding the genotypes and phenotypes will be used throughout. With a change in notation, R as given by equation (3) is identical to the coefficient derived by Flesness and Holtzman[13] (equation (10), where $P(H) = \Delta_1 + \frac{1}{2}\Delta_3$ and their $h = \frac{1}{2}\alpha$). It is important to note that if the gene effects are additive, if there is no inbreeding, or if $p = q = 0.5$, then equation (3) reduces to the coefficient proposed by Hamilton:[20,21]

$$R = \frac{2f_{XY}}{1 + f_X}. \tag{4}$$

Consequently, the regression coefficient (equation (4)) will be a good estimate of R in many cases of interest. We now show that the coefficients derived elsewhere in the literature[3,10,12,18] are identical to equation (3).

Below, we index genotypes with a single subscript i, with $i = 0, 1, 2$ corresponding to genotypes AA, Aa, aa, respectively. Let P_{ij} be the joint distribution of genotypes i and j in the population with i being the genotype of X and j the genotype of Y (Fig. 3.1, Table 3.1). Let X_g, Y_p and Y_g, Y_p denote the genotype and phenotype of X and Y, respectively.

Orlove[3] and Orlove and Wood[10] derive $R = \text{Cov}(X_g, Y_p)/\text{Cov}(X_g, Y_p)$ in an outbred model of kin selection in haplodiploid families. Their coefficient[3,10] equals

$$R = \frac{\text{Cov}(Y_g, X_p)}{\text{Cov}(X_g, X_p)} \tag{5}$$

Table 3.1 Joint and marginal distributions of X_g and Y_g and X_p and Y_p

X \ Y		Y_g	0	$\frac{1}{2}$	1	
		Y_p	0	h	1	
X_g	X_p					
0	0		P_{00}	P_{01}	P_{02}	$P_{X0} = p^2 + f_X pq$
$\frac{1}{2}$	h		P_{10}	P_{11}	P_{12}	$P_{X1} = 2pq(1 - f_X)$
1	1		P_{20}	P_{21}	P_{22}	$P_{X2} = q^2 + f_X pq$
			$P_{Y0} = p^2 + f_Y pq$	$P_{Y1} = 2pq(1 - f_Y)$	$P_{Y2} = q^2 + f_Y pq$	1

P_{ij} is the joint distribution of genotypes i and j and P_{Xi}, P_{Yi} are the marginal distributions of the genotypes of X and Y, respectively ($i, j = 0, 1, 2$). From the table, the following expressions can be calculated for use in calculations concerning equations (5): $E(X_p) = hP_{X1} + P_{X2}$, $E(Y_p) = hP_{Y1} + P_{Y2}$, $E(Y_g) = \frac{1}{2}P_{Y1} + P_{Y2} = q$, $E(X_g) = \frac{1}{2}P_{X1} + P_{X2} = q$, $E(Y_g X_p) = \frac{1}{2}hP_{11} + hP_{12} + \frac{1}{2}hP_{21} + hP_{22}$, $E(X_p X_g) = \frac{1}{2}hP_{X1} + P_{X2}$, $E(X_g Y_g)P_{11} + \frac{1}{2}P_{12} + \frac{1}{2}P_{21} + P_{22}$, $E(X_g^2) = \frac{1}{4}P_{X1} + P_{X2}$.

if there is no inbreeding. However, as we now show, equation (5) is more generally correct, as well as being the most concise and intuitive representation of equations (3) or (17).

To relate equation (5) to equation (3), we must first calculate the required covariances. The information needed to do so is given in Table 3.1, in which the genotypes and phenotypes of X and Y are given along with the joint and marginal probabilities. Using this information to evaluate the covariances in equation (5), we obtain directly

$$\frac{\text{Cov}\,(X_g, Y_p)}{\text{Cov}\,(X_g, X_p)} = \frac{\frac{1}{2}P_{11}h + hP_{12} + \frac{1}{2}P_{21} + P_{22} - (\frac{1}{2}P_{Y1} + P_{Y2})\,(hP_{X1} + P_{X2})}{\frac{1}{2}hP_{X1} + P_{X2} - (hP_{X1} + P_{X2})\,(\frac{1}{2}P_{X1} + P_{X2})}. \tag{6}$$

From Table 3.1, we can also obtain for future use

$$\text{Cov}\,(Y_g, X_g) = \tfrac{1}{4}P_{11} + \tfrac{1}{2}P_{12} + \tfrac{1}{2}P_{21} + P_{22} - (\tfrac{1}{2}P_{X1} + P_{X2})(\tfrac{1}{2}P_{Y1} + P_{Y2}) \tag{7}$$

$$\text{Var}\,(Y_g) = \tfrac{1}{2}qp(1 + f_Y) \tag{8a}$$

and

$$\text{Var}\,(X_g) = \tfrac{1}{2}qp(1 + f_X). \tag{8b}$$

Substituting equation (7) into the numerator of equation (6)

$$\text{Cov}\,(X_g, Y_p) = \text{Cov}\,(Y_g, X_g) + (h - \tfrac{1}{2})\,(\tfrac{1}{2}P_{11} + P_{12} - qP_{X1}). \tag{9}$$

The denominator of equation (6) can be simplified directly to

$$\text{Cov}\,(X_g, X_p) = pq[\tfrac{1}{2}\alpha + f_X(1 - \tfrac{1}{2}\alpha)]. \tag{10}$$

As shown in Fig. 3.1 (see also refs 12, 22, and 23), it is possible to express the joint probabilities of the interactions, P_{ij}, in terms of the condensed identity coefficients and the gene frequency. In particular,

$$\begin{aligned} P_{11} &= 2pq(\Delta_7 + \tfrac{1}{2}\Delta_8 + 2pq\Delta_9) \\ P_{12} &= pq(\Delta_5 + 2q\Delta_6 + q\Delta_8 + 2q^2\Delta_9). \end{aligned} \tag{11}$$

These formulas (equation (11) and Fig. 3.1) are only strictly correct for neutral genes; however, it is hoped that they are approximately correct if selection is weak. Substituting equation (11) into equation (9) and recalling from Table 3.1

$$P_{X1} = (1 - f_X)2pq,$$

we obtain after some algebra

$$\text{Cov}\,(Y_g, X_p) = \text{Cov}\,(Y_g, X_g) + \tfrac{1}{2}pq\,(\alpha - 1)[\Delta_5 + \Delta_7 + \tfrac{1}{2}\Delta_8]. \tag{12}$$

Letting r_{XY} be the correlation between the genotypes of X and Y, we have

$$\text{Cov}\,(Y_g, X_g) = r_{XY}\,[\text{Var}\,(X_g)\,\text{Var}\,(Y_g)]^{1/2}. \tag{13}$$

However,[19]

$$r_{XY} = \frac{2f_{XY}}{[(1 + f_X)(1 + f_Y)]^{1/2}}. \tag{14}$$

On substituting equations (8) and (14) into equation (13), we obtain

$$\text{Cov}(Y_g, X_g) = f_{XY}pq,$$

and so equation (12) becomes

$$\text{Cov}(Y_g, X_p) = f_{XY}pq + \tfrac{1}{2}pq(\alpha - 1)[\Delta_5 + \Delta_7 + \tfrac{1}{2}\Delta_8]. \tag{15}$$

After rearranging equation (15) and using equation (2b)

$$\text{Cov}(Y_g, X_p) = pq[f_{XY}\alpha + (1 - \alpha)[\Delta_1 + \tfrac{1}{2}\Delta_3]]. \tag{16}$$

On substituting equations (16) and (10) into equation (5), we obtain equation (3) as was to be shown.

Michod[12] (equation (3)) derived the following coefficient for use in Hamilton's rule

$$R = \frac{2xp\rho + (1 - x)(1 - 2q)\rho'}{1 + x - 2q} \tag{17}$$

$$\rho = \frac{(\Delta_1 + \tfrac{1}{2}\Delta_3)}{q + pf_X} + \frac{q[\Delta_5 + \Delta_7 + \tfrac{1}{2}\Delta_8]}{q + pf_X}, \tag{18a}$$

and

$$\rho' = \frac{(\Delta_5 + \Delta_7 + \tfrac{1}{2}\Delta_8)}{1 - f_X}, \tag{18b}$$

which are Jacquard's conditional genic structures.[12,22,23] These conditional coefficients in equation (18) give the probability with which X can predict, on the basis of pedigree ties, the allelic distribution in gametes of Y. This probability depends on whether X is homozygous (equation (18a)) or heterozygous (equation (18b)). In addition, the variable x appearing in equation (17) is the frequency of altruistic homozygotes among the total class of altruists and is

$$x = \frac{q^2(1 - f_X) + f_X q}{q^2(1 - f_X) + qf_X + (1 - f_X)2pqh}. \tag{19}$$

Equation (19) corrects a mistake in table 1 of Michod[12] where it was thought that the average inbreeding coefficient (averaged over the frequencies of X and Y in the population) should be used in equation (19). This is incorrect due to the conditional nature of the altruism. As the gene is only expressed in X, when X is in a certain relationship to Y (the relationship being summarized by the Δ values), it is only the frequency of homozygous altruists among all X individuals which matter in equation (19). This fact alone accounts for the incorrect impression, given in fig. 2b of that paper,[12] that dominant genes could never increase in frequency (fig. 2a of ref. 12 is unaffected by these considerations). After substituting equations (19) and (18) into equation (17) and collecting the Δ values in terms of f_{XY}, equation (3) is obtained after some rearrangement.

In conclusion, the various coefficients derived in the literature[3,10,12,13,18] for use as R in Hamilton's rule are equivalent. These coefficients were originally derived from very different points of view. Given the results here, these various approaches support each other and suggest that the one pleomorphic coefficient (equations (3), (5), or (17)) is, indeed, correct.

Acknowledgements

This work was supported in part by NSF grant DEB 79–10191 to R.E.M. We thank Bob Abugov for discussion and comment.

References

1. W. D. Hamilton, *American Naturalist* **97**, 354–6 (1963) [reprinted in *Narrow Roads of Gene Land*, Vol. 1, pp. 6–8].

2. W. D. Hamilton, *Journal of Theoretical Biology* **7**, 1–52 (1964) [reprinted in *Narrow Roads of Gene Land*, Vol. 1, pp. 31–82].

3. M. J. Orlove, *Journal of Theoretical Biology* **49**, 289–310 (1975).

4. P. R. Levitt, *Proceedings of the National Academy of Sciences, USA* **72**, 4531–5 (1975).

5. F. M. Scudo and M. T. Ghiselin, *Journal of Genetics* **62**, 1–31 (1975).

6. E. L. Charnov, *Journal of Theoretical Biology* **66**, 541–50 (1977).

7. L. L. Cavalli-Sforza and M. W. Feldman, *Theoretical Population Biology* **15**, 268–81 (1978).

8. B. Charlesworth, *Journal of Theoretical Biology* **72**, 297–319 (1978).

9. M. J. Wade, *Proceedings of the National Academy of Sciences, USA* **75**, 6154–8 (1978); *American Naturalist* **113**, 399–417 (1979).

10. M. J. Orlove and C. I. Wood, *Journal of Theoretical Biology* **73**, 679–86 (1978).

11. R. Craig, *Evolution* **33**, 319–34 (1979).

12. R. E. Michod, *Journal of Theoretical Biology* **81**, 223–33 (1979).

13. N. R. Flesness, *Nature* **276**, 495–6 (1978).

14. R. F. Michod, *Genetics* **96**, 275–96 (1981).

15. R. F. Michod and R. Abugov, *Science* **210**, 667–9 (1980).

16. B. Charlesworth, in H. Markl (ed.), *Dahlem Workshop on Social Behaviour: Hypothesis and Empirical Tests*, pp. 11–26 (Verlag Chemie, Weinheim, 1980).

17. R. Abugov and R. E. Michod, *Journal of Theoretical Biology* **88**, 743–54 (1981).

18. R. E. Michod and W. W. Anderson, *American Naturalist* **114**, 637–47 (1979).

19. S. Wright, *American Naturalist* **56**, 330–8 (1922).

20. W. D. Hamilton, in J. F. Eisenberg and W. S. Dillon (ed.), *Man and Beast: Comparative Social Behavior*, pp. 57–91 (Smithsonian Press, Washington, DC, 1971) [reprinted in *Narrow Roads of Gene Land*, Vol. 1, pp. 198–227].

21. W. D. Hamilton, *Annual Review of Ecology and Systematics* **3**, 193–232 (1972) [reprinted in *Narrow Roads of Gene Land*, Vol. 1, pp. 270–313].

22. A. Jacquard, *The Genetic Structure of Populations* (Springer, New York, 1974).

23. R. C. Elston and K. Lange, *Annals of Human Genetics* **39**, 493–6 (1976).

BEST AND WORST HOTELS

The Evolution of Co-operation

Eye for eye, tooth for tooth, hand for hand, foot for foot.
EXODUS 21:24

There is a destiny that makes us brothers.
None goes his way alone;
Everything we send into the lives of others
Comes back into our own.
E. E. MARKHAM[1]

I MENTIONED in the last introduction one of my passings through Denver. There are many ways to turn there besides towards the deserts, indeed all the West opens, but it is also a good place simply to alight. A few miles out from the airport, in a countryside where prairie dogs still dig their towns and shady cottonwood trees cluster along the creeks that fan into the plain, one can feel to be on the foreshore of a new continent. The awesome wall of the Rocky Mountains faces you a few miles distant and all the Great Plains are behind you like a sea. I stopped at a small motel in this zone—I think this was when I was visiting the University of Colorado—and from the motel I collected a surprising beach pebble, foreign perhaps for the huge cross-roads of Denver Airport I had just come from, but appropriate enough, it seemed to me, for the land of valley settlements just beginning to the west.

A grey-haired man at the checkout gave me a leaflet about his motel. He intended no doubt for me to pass it on to others or for it to bring me back. With no explanation the leaflet had, along with a sketch of the motel, the above beautiful verse—that is, the second and less bitter one.

No authorship was on the leaflet, which I soon lost, but the verse stuck in my mind and much later with help at New College, Oxford, where most of such problems can be solved, I found out the name of the American poet who wrote it. In a way the poem then became a disappointment because I couldn't any longer think of a grey-haired motelier deserving a prize for poetry, although actually the dated style had already made this idea unlikely; moreover, the other poetry of this poet I found I didn't like—not even the weak and rather detached second verse that follows the one given. Nevertheless, together, the two 'verses' I have set above well express how humans, at least since the time when any one could call them humans and probably for very much longer, have acted, philosophized, and, above all, have *felt* about what is to be the central idea of this chapter, although to begin to be accurate to its stricter theme perhaps one or two more lines should be added to my 'Denver' verse. Enlarging on what we send into the lives of others and what they send back, and unfortunately spoiling the aura, the added lines need to go something like:

> When bad it comes not few,
> But much—and PDQ.

One day in the Museum of Zoology at Ann Arbor there came a phone call from a stranger asking what I knew about evolutionarily stable strategies and for some guidance to relevant literature. I told him that the so-called 'ESS' was getting to be a big field—and, yes, I had had some part in it in some studies of 10 years ago, and it would help if I knew his particular topic. With each word that followed my caller seemed to come leaping out of the vastness of the USA nearer and nearer. He was on my own campus of U of M, he said, in a unit I hadn't at that time heard of, the Institute for Public Policy Studies (now renamed the School of Public Policy). He told me the street and it was one I connected less with university sites than with occasional Burger King lunches and a clothing store where once, acting in ignorance just after first arriving in AA, I had bought some unnecessarily expensive pyjamas. But why not be there, I wondered as I listened, if you studied public policies. I thought how they could at least see in the shopping street below their windows more typical effects of the policies, whatever these might be, than would be revealed through windows that looked out through the rarefied air of the campus—

like me in the arboretum watching birds and flowers, for which again their behaviour on the campus, if they came there, would be less natural.

As often when I have not yet fully locked on to what someone is saying, such semi-loony thoughts run a sort of parallel race in my mind and I heard Robert Axelrod tell me how he specialized in international relations, and used game theory a bit (out of a political science, not a maths, background) but recently had been working on decisions about co-operation with models that he could apply to individuals or to groups. Specifically he'd looked at some versions of Prisoner's Dilemma. Richard Dawkins, whom he had phoned, had told him that I had discussed Prisoner's Dilemma in one of my papers—an application to sex ratio, right?—and that I had also discussed a general idea of evolutionary stability. The situations that he was simulating by computer were getting to be a bit biological, too, though he thought of this as more ecological than evolutionary. Still, it seemed some outcomes were quite like some of the things in Dawkins's *The Selfish Gene* . . .

Fast food and clothing, squirrels and pigeons faded as this went on; a name, a game, an old and a renewed puzzle took their place. It was no surprise to me that social scientists thought about co-operation and about the decisions made by individuals and by groups. But I had had an impression (naive as Bob Alexrod would soon show me) that not many gave attention to game theory or approved it. In reviews I had read in London dating from the origin of the theory onward I seemed to see mainly cold water and invective being poured on it by various economists and other commentators. At the time when I was reading the theory and these reviews (mainly in Senate House Library as I remember) I had been too green to realize that this was an almost sure sign that the topic, as shown by the dislike it was arousing in some, has made people nervous and angry because it is important and is sure to grow. Personally from my own first readings, while I had at first disliked the seeming cynicism of the ideas, I had simultaneously appreciated a much greater realism than what I had been used to in this field—it was like the revelations of Adam Smith and Thomas Malthus coming to me over again. I had also been very keen about potential applications the theory might have to biology. The whole approach had that retrospective obviousness that dubs an idea as important. What is more it seemed all applicable over an immense scale— levels of a bacteriocin, a sex ratio, a human act of betrayal: all these could come in.

Problems in the theory related to human acts of betrayal of trust I soon found to be enshrined rather explicitly in the Prisoner's Dilemma game. Its maddening simple story and resolution, seeming to challenge every moral that I believed in, soon wrestled its way to the frontmost lobes of my brain. The Prisoner's Dilemma in fact came to me like second and third heads of a hydra after I had cut off the first and thought that thereby the whole animal was killed—and the new heads were even uglier and they had no convenient hair of genetical kinship such as I had used to seize on and wrestle with in the kinship story. I say two heads here because I saw a distinction between the original problem for human and rational action in the dilemma (as with the two separate prisoners deciding whether to turn state evidence against each other), and, contrasted to this, evolution's analogous problem about what natural selection would 'decide', given many generations and many mutations to fit together in the case of unthinking organisms. Eventually I felt I had got some way with this last problem through my findings about sex ratio in semi-subdivided populations and because I believed that at least the incentives of human rationality had to be derived largely from the evolutionary struggle at this preceding level, there seemed at least an appendage—an ugly ear maybe—that would let me wrestle with the other new head as well. But for the most part such ideas just rattled with me for years in confusion, together with Bob Trivers's vague but hopeful insights. This had been the best I had been able to put together.

Now on the phone to me was someone out of political science who seemed to have just the sort of idea I needed. A live games theorist was here on my own campus! Nervously, and rather the way a naturalist might hope to see his first mountain lion in the woods, I had long yearned for and dreaded an encounter with a games theorist. How did they think? What were their dens full of? (The gnawed bones of dead ideals it often seemed to me.) And, as People, were they as cynical as the notions they treated? For example, had my hero, John Von Neumann, once really put Kyoto at first in front of Hiroshima and Nagasaki on the list of cities to be destroyed with the new bomb he was creating?[2] Axelrod on the phone sounded nice and, very surprising to me, he was more than a bit biological in his manner of thinking. I sensed at once a possibility that the real games theorists might be going to turn out to be a kind of kindred to us; perhaps political science had its demagogues too, its face-lift masseurs, academics whose whole scholarly approach was an endeavour to make themselves seem much nicer than you were, as well as claiming to be more effective. We evolutionists had such competitors; perhaps games theorists, too, must confront the smirks of those cosmeticians in their corridors—feel forced to scowl back even as more bitter Jeremiahs and Cassandras of their discipline than they were, be for their public the honest but dour Machiavellis at last telling them truly what dishonesty was.

The voice went on; maybe I had dropped down to one-ear listening again. But presently both came up sharply as I heard something about 'the same answer a second time' and also about 'conditionality' and 'tit for tat'. It was beginning to sound really odd. 'But surely it's proven, isn't it, that you must always defect in that game, repeated or not?', I broke in. 'If it's repeated you think about the last move and there it's just ordinary Prisoner's Dilemma, so you defect; then the next move back—now that's the last—and so it unravels right to the start: so you defect then too.' 'Ah, but I just told you there's no last move, not one you can know of,' the voice said patiently, 'in my model I just have a *probability* . . .' (Bob has a special raised voice and a drawl for this word, which I wish I could convey in print).

Evidently here was something very new and surprising; also a romantic 'shadow of the future' came into it that I liked the sound of. On my desk something levitated, spun, and sped for the window of my office; I couldn't listen and at the same time hold this thing down. It was a resolution I had formed in the year past and which was supposed to affect the whole of the present one: that I would concentrate on the problem of sex and would put to one side all the old ESS stuff that bombarded me still and was now so fashionable. The voice was telling me how simple programmes beat complex ones . . . there were 'ecological' runs . . . contestant programmes allowed to reproduce . . . the simplest rule ending as the fastest growing . . . Dawkins had been phoned, and why hadn't he (Axelrod) contacted me, who was right there—and therefore he now had. Finally, we agreed he would send me some manuscripts he had submitted and, if I found them interesting, we would have lunch together some time and talk some more.

It was to be many months until I retrieved my resolution from where it had lodged like a torn, blown-away kite in the branches of the honey locust below my window. That was the thornless variety of the species (which, by the way, I never liked as much as that fierce, natural mastodon-defying thorny and wild kind of the honey locust that some gardens in Ann Arbor also had—this was the native, the real thing) and thus my paper-thin resolution had been not much damaged and I resumed it. Thereafter a lot of people who wrote to me about sex-ratio papers as well as the new one about co-operation came to be spared my opinions. But by then also a lot else had happened and many of a new set of letters I received I did answer. After the phone call I sent Bob Axelrod some ESS references and he sent me some manuscripts. Soon after the first contact we met for lunch and it was the first of many. Soon after the lunch again I proposed that the work seemed so interesting biologically we might try writing it up for a joint paper in *Science*; his contribution would be the basic ideas plus the description of his tournaments, and mine to add a natural scientist's style and some biological illustrations.

Far from being the ruthless or cynical practitioner some of the critics of game theory had suggested to me, Robert Axelrod was a friendly, undominating man and an intuitive understanding between us was immediate. Both of us always liked to be always understanding new things and to be listening more than talking; both of us had little inclination for the social manoeuvring, all the 'who-should-bow-lowest' stuff, which so often wastes time and adrenalin as new social intercourse starts. Bob is the more logical, but beyond this what we certainly share strongly is a sense for a hard-to-define aesthetic grace that may lurk in a proposition, that which makes one want to believe it before any proof and in the midst a confusion and even antagonism of details. Such grace in an idea seems often to mean that it is right. Rather as I have a quasi-professional artist as my maternal grandmother, Bob has one closer to him—his father. Such forebears perhaps give to both of us the streak that judges claims not in isolation but rather by the shapes that may come to be formed from their interlock, rather as brush strokes in a painting, shapeless or even misplaced considered individually, are overlooked as they join to create a whole. (In science it is all actually better than this because hopelessly inharmonious brush strokes are wiped away by the work of others fairly soon.)

A common aesthetic sense for the distant goals is perhaps what makes Bob a less dismissive critic than most of my Swiss-cheese logic. More than once I recall him defending me against colleagues over some of my more tenuous early arguments about sex. That brilliant cartoonist of the

journal *American Scientist*, Sidney Harris, has a picture where a
mathematician covers the blackboard with an outpouring of his formal
demonstration. It's a big spillage of the alphabet soup, as I would have said
in Volume 1, and it starts top left on the blackboard and ends bottom right
with a triumphant 'QED'. Halfway down, though, one sees a gap in the
stream where is written in plain English: 'Then a miracle occurs', after
which the mathematical argument goes on. Chalk still in his hand, the
author of this *quod est demonstrandum* now stands back and watches with a
cold dislike an elderly mathematician who peers at the words in the gap
and says: 'But I think you need to be a bit more explicit—here in step two.'
I easily imagine myself to be that enthusiast with the chalk and I also think
of many castings for the elderly critic. Yet how easy it is to imagine a third
figure—Bob—in the background of this picture, saying cheerfully: 'But
maybe he has something all the same, maybe that piece can be fixed up.
What if . . .'

　　It was following some such defence that much later we started to
work together on an idea about sex. Just as this chapter is our first 'CC'—
that is, a mutual co-operation (see our notation in the paper), that later
paper was to be our second and it makes Chapter 16 in this volume. It is
the paper that I regard as containing the second most important of all my
contributions to evolution theory. From that work onwards a huge problem
for me was solved and I no longer needed to worry about it, or at least I
need worry not nearly as much—all I had to think more about were the
details of how it worked and the extensions. So the sense for what is worth
doing that Bob and I seem to have shared has been, for me at least, not
trivial. That second joint paper of 1990 (actually mainly written some
three or so years before) was to be the first model where sex proved itself
able to beat any asex competitor immediately and under very widely
plausible assumptions. In that paper I was as much the leading spirit—the
first 'C' as it were in the Prisoner's Dilemma—as Bob was leader and first in
the paper that follows. In our first collaboration I have to say that I didn't
provide much. There were amendments and metaphors in the text,
abbreviations for a few lines of the proof, and then the biological examples.
And, sad to say, none of the last has quite stood the test of time unscathed,
although other examples have come in their place.[3]

　　Anyway, between us and with surprisingly little difficulty we pushed
our paper into *Science*. Once published it drew so much interest that it won

us the Newcomb–Cleveland Prize as *Science*'s supposed best paper for its year, 1981. Bob was very pleased to have it in *Science*, telling me this was a rare achievement for a social scientist. Probably I had indeed helped in getting it there because a certain style and the type of referencing as well as demonstration of relevance to non-humans (to the supposed hapless and unteachable of lower life because otherwise it had to be purely an encroachment on sociology and should be published in their journal) were almost essential.

Of course both of us were very pleased about the prize. Although I find it hard to believe in the validity of the American Association for the Advancement of Science deciding on the worth of a paper so soon after publication, in this case it was true that our explanation of the special property of TIT FOR TAT began an immediate landslide of further studies, which continues to this day.[4] It became a slight embarrassment to me that, because of the joint paper being in such a prominent journal, the discovery of the special strength of TIT FOR TAT and, through this, of the potential simplicity of the evolution of reciprocation often appears ascribed by modern writers as if Bob and I had contributed equally, which, as I have explained, was far from true. Most citers are probably aware of this: the big factor is probably that a citation to *Science* is more easily grabbed in a rush for a reference list than are citations to the more special and long-named journals where Bob had also (and earlier) published. For non-specialists, however, possibly a false impression is still perpetuated.

In early experimentation with iterated Prisoner's Dilemma Bob had noticed the general robustness of the simple policy of TIT FOR TAT. I think, however, it came as a surprise to him, as to the rest of us, when it turned out in his tournament that nothing better seemed findable than this briefest of all the submitted strategies. When I received Bob's manuscripts after our contact by phone I was pleased to find that it had been Anatol Rapoport who had twice submitted the surprise winner. Rapoport was the writer, who, through his articles in *Scientific American* in the early 1960s, had first brought game theory to my notice. His articles had led me to Von Neumann, who in turn had opened up the new world. Games were suddenly not merely fun but a deep theme of life, and, in strict parallel to my own endeavours with kinship, there now came this whole other swathe, perhaps, at first sight even more vague, more tangled, more emotion-loaded than the kinship one had been, of social matters about which I saw that one could again reason—and, at stretch, even reason quantitatively. Because Rapoport had long been interested in the Prisoner's Dilemma himself, even to the point of running studies on real people playing the game,[5] unsurprisingly he had had a hunch about a simple scheme of play that would be hard to beat. His submitted strategy, which Bob named TIT FOR TAT (or, as I'll abbreviate it henceforth, TFT), was to co-operate on the first move and thereafter simply to do whatever the other player did on the last move. Thus if he cheated you last move, you cheat him next; if he co-operated, then you co-operate with him next. Obviously, for humans, the idea of this as a sound rule of response is ancient. As my opening quote shows, one of the oldest books of the Old Testament has it: return good for good, evil for evil. The latter was made even into a principle of justice. But that anything could be proved mathematically about such a simple theme seems to have occurred to no one before Bob.

What tortuous paths my life has taken through dark and frightening forests of reason, and paths of what different kinds! Woodland lanes I used to wander barefoot as a boy on Badgers Mount, Kent, in shorts, with an insect net in my hand, a killing bottle carried like a false pregnancy in my shirt, seem in retrospect almost like a model for all I have done later. Already perhaps in those lanes I am little past the pure animism of a hunter and collector and was wanting to understand. In this line I was deciding to work 'upwards', as I saw it, out of my primary inspirations, like the butterflies were flying, towards a goal of a theory of pure pattern. My aim would be to understand patterns generally, but I would start with the patterns laid out on my butterflies' wings, and follow these perhaps with the winged waifs' patterns in time too—like that curl and glide of my speckled wood as it flew from its sunspot on the leaf for a brief survey of its wood-edge realm, ever returning, or that still more superb glide, in the same way but wider and perhaps taking in a bramble flower tuckshop on the way, of the white admiral!

How easy in another context I found the speckled wood (*Pararge aegeria*) to catch; how different and more difficult the clouded yellow (*Colias*), which I used to chase in the fields of red clover or lucerne. Especially it was difficult when a missed first swipe had 'turned on' your prey. It was no help that you could run much faster in a straight line than the butterfly could; straight lines were gone, fractals the rage amidst the alfalfa 50 years before they became so in academia. It seemed as if literally some 'worse-than-random' generator of dodges had switched on the working of the butterfly's wings—a pre-programmed switchable option in how it was built to fly. Therefore later I knew I needed to understand this particularly *un*graceful flight, this optimal dodging, and it, too, needed to be considered as a part of 'pattern'. I expect those readers who know the theory of games and its earliest interesting theorems see here at once one reason why some of those earliest results were exciting for me.

At the time when I was that path-wanderer with the net again and going back from these erratic 'wing-beat' patterns to the orderly ones on the wings, more specifically I saw my way lying through melanin, carotenoid and flavone pigments, and stuffs of those kinds that I had read about in E. B. Ford's 'New Naturalist' books. I imagined myself to grow competent. I would use insights from genetics (again got from Ford,[6] for he was my most important, indeed almost my only, scientist I cared to read at this time) and also from how the pigments were produced and dispersed;[7] in this way I'd synthesize new, wonderful wings for butterflies, new designs Rhopalocera had never seen. Through such a path, I believed, I might come eventually to my theory of pattern.

Much later I learned with astonishment that the English mathematician and code-breaker Alan Turing, starting with much more maturity and insight—and indeed genius—than I had, had tried an almost identical beginning out of butterfly and moth wings when he was looking for a general theory of the spatial patterns of development. With far more sense, he had come at it using the maths and physics of diffusion rather than through chemistry. My own idea was quickly abandoned and indeed I came to pay less and less attention to chemistry—God forbid I should be too reductionist, as is explained in my comments on others' attitude to this in Volume 1 of *Narrow Roads of Gene Land* (see note 11 on p. 267). I realized that what I had been thinking of was like starting to lay stones for a cathedral without having first given thought to the architecture. But I would have thought it a leg-pull at the time if someone had told me of a future when I would find it more rewarding to talk 'patterns' to political scientists rather than to fellow biologists; or told that I would shun the journals publishing on the more chemical side of evolution; or, most improbable of all as it would then have seemed, that I would tell serious men—and, in fact, political scientists and computer scientists, classes of thinkers either I didn't know about or at the time didn't actually exist—

how a speckled wood butterfly on a sunlit bramble spray, just by being there and not wandering, was displaying a 'bourgeois' strategy.[8] And I would find these serious men listening with interest and making pertinent replies! All science is finding patterns, and those on which I have chanced to make most progress have concerned behaviour. Even with these I have ended looking at matters many levels more 'emergent' (if that is the modern word) than the level of the curl and glide, which still is in my mind an awesome and almost pure note of graceful motion, even now so beautiful to me. Instead I discuss the butterfly's apparent intentions, examine that apex of a pyramid of metaphors by which we can say what the butterfly hopes to achieve in a space (which may be actually just the bramble leaf and its surrounds) that it has tried to reserve. The geometry of the spatial patterns on the wing I now leave to the developmentalists and the glide to the applied mathematicians.

Perhaps already at school when I was reading in *New Biology* (the now long extinct Penguin-paperback style of a semi-popular journal) about Alan Turing's approach to butterfly and moth patterns, I was seeing my ideas suddenly overtaken (even though far from solved) by a stronger mind, and at the same time was experiencing that disillusion that has come so often—that this wasn't quite what I had wanted anyway. Clouds dissolving and reforming, my 'patterns' were moved on. Possibly for Turing they moved too, for, so far as I know, he did nothing more. Had Turing once been a butterfly collector? Sewall Wright, who by my time at Ann Arbor had become my favourite evolutionist, had been one as a boy; Charles Michener the bee biologist, whom I was to admire greatly later, had been too.

As you can see, as a result of the meeting with Bob Axelrod, I was myself spending time with a new set of people. At Bob's invitation I joined a small, interdisciplinary, lunch-time discussion group of which he was a member; a group of four at the time who met once or twice a month. Before my arrival the group was called 'BACH' after the initials of their surnames; my addition made it, a little more awkwardly, 'HBACH'; and now for 14 years it has been 'BACH' again. Common to us all was interest in the computer both for itself and for the new views of the world, and even of philosophy and mathematics, that it was opening. No actual computer was in the big room where we took our lunches but it was, I suppose, a brazen idol standing near, hidden by a screen. If so our senior

priest was Arthur Burks from Philosophy (also with an appointment in
Computer Science), who had worked with valve-based machines in the
earliest days and—a great distinction for me—had personally known Von
Neumann. The other members were Michael Cohen, like Bob from the
Institute of Public Policy Studies, and John Holland, physicist by origin but
with his current affiliation hovering as undecided, seemingly, as a quantum
state somewhere between Electrical Engineering and Computer Science.
John was rapidly acquiring fame as the originator of an ingenious
'biological' approach to the solution of complex problems. Rick Riolo, a
master of programming, was also often present, and, later and more seldom,
Mike Savageau came from microbiology.

In spite of having pride in the practical applications, I think that for
John the appeal of his 'live' software problem-solvers was mainly
philosophical: what set limits to what artificial processes using natural
selection could achieve? Perhaps nothing—there were no limits except to
how big you could make an artificial universe in your computer. This seems
an obviously important question for us now, but it didn't then. Called by
him 'The Genetic Algorithm', his method (actually only the first of a
hierarchy) involved programming for a computer a certain handling of
populations of trial attempts on the problem. The attempts, which were
specified in a linear genetic code somewhat like the real one, were set to
reproduce, mutate, recombine, compete, and so forth as if they were
organisms. How much each reproduced was determined by its success with
the problem.

This seemingly small and crude set of properties was sufficient to
create a true evolutionary system, and with astonishing vigour and
ingenuity for the simple beings they were, the 'GA' robots—'motives
without minds', as someone said of viruses—went at their job of sniffing
out, over the generations, ever improved and even sometimes provably
perfect, solutions to difficult problems. 'GAs' are a big field today,[9] not only
in academia but increasingly in industry. Later during the 1990s I played
with the method myself but mostly was just trying to twist the artificial life
idea back on itself to throw light on its parent regime, real life. Since a
claim of the GA has been the achievement of a kind of 'hang-gliding'
transition, a floating forth and waiting upon chance updraughts, crossing
valleys, spiralling from peak to higher peak, as it were, across a potentially
endless mountainscape of attempted optimization, I hoped to use the

process to gain better insight into how the really radical changes—the macroevolutionary transitions—occur in organic evolution. For this, too, involved the problem of peak transitions. For example, how does a clam evolve to be a snail (two shells to one) or else vice versa, or how does a coat-of-mail shell (*Chiton*) get to be either? But that study cannot be covered in this volume.

I think of the (H)BACH group as groping for a new discipline, perhaps a bit like that which now sometimes takes the name 'complexity theory'. Our day-to-day aims, however, were not so ambitious; enough that we could interest each other, or each produce once in a while an introductory talk arising out of our work or out of a recent trip to a meeting or a recent paper read, which could start a discussion.

By the time I joined that group Public Policy Studies had jumped like a courting crane out of worldly Liberty Street and the company of the Burger King restaurant, where I first found it, and, soaring over the Harlan Hatcher Graduate Library (which, as is usual with the biggest library on any campus, I seldom used) as well as over the crowns of many lofty elms of the main campus quadrangle, had alighted finally in a rather old but handsome building on the other side called Lorch Hall. I imagine now that high-ceilinged rather chapel-like room in this building where we met and with my vision comes cream cheese crammed into a bagel and a crackling brown bag on the long table, this still containing my Granny Smith apple and cold coke, all of which I would have bought minutes before in the 'Village Corner' store. Many of the discussions of that group proved a bit too economical or computerly-scientific for me to follow but surprisingly frequently our discussion drifted towards some manifestation of Life. All guises of life and natural selection were up for our grabs and when these topics came then, of course, I could contribute and myself gain new ideas.

Because relevant to a vast range of living phenomena, I was particularly interested in Bob's discoveries concerning the Prisoner's Dilemma, but I was equally enthralled by John's simple and audacious notion that natural selection, implemented purely artificially but aided by additional life-inspired features such as mutation and sexual recombination, could generate discoveries that were sometimes beyond the range of the unaided human mind. At the very least his 'GA' was providing very handy rocks to throw at anti-evolutonists.

Which metaphor conveniently reminds me of a scene out of one of the fundamentalists' own pamphlets: it shows a man trying to mend his car from a distance of 5 yards by hurling rocks at the engine. The cartoon, of course, is supposed to highlight the impossibility of using random mutation to do anything constructive for an organism. Yet think of this: if such simple Neodarwinian assumptions—rocks to be thrown in truly vast numbers, and in harmless, program-controlled and progressive ways within the virtual worlds of a computer (as happens in a GA)—could indeed 'evolve' new and better ways not only to fix a car but to build, say, a better aircraft, and if, as will probably be particularly impressive to Americans, you could shortly be seen to be making or saving money through having these evolving GA slaves working for you, who was going to deny any more that 'blind' mutation and selection could create things? Thus if the conditions of life were the same (which they certainly ought to be because they provided the inspiration for the GAs that John Holland constructed), the equivalent of a rock-throwing procedure in the real world (where the rocks are cosmic particles, free radicals liberated from reactions, and so on) could hardly not be creating new structures in nature too. Well, of course, it is guaranteed that the fundamentalists will still find ways to deny it even though I don't know how. Under pressure especially from my colleague,[10] their rock-throwing mechanic, he who has been their steady assistant for so long, may possibly have to move on. Nevertheless, we may be sure that something or someone else will be sure to replace him in supposedly proving to us how evolution cannot work.

Another interest for me from the (H)BACH meetings was to realize that such matters as Bob's discoveries and Oskar Morgenstern's assistance to Von Neumann were just items out of a long chain of the unintended contributions economists and their like have been making to evolution theory for a long time. Out of the recent past there had been, for example, the 'Nash equilibrium'. Virtually this was the biologist's 'ESS' prefigured under a more 'rationalist' hat, and it had come from an economist, John Nash. And long before him, the essence of the same idea had been formulated even early in the nineteenth century by Augustin Cournot, a French economist, who is sometimes treated as ancestor to the subfield microeconomics.[11] Back even from Cournot, of course, loomed the great Scottish sociologist/economist writer Adam Smith whose realism about human economy, along with that of Malthus, I have mentioned already and who undoubtedly gave important inspiration to Darwin. Adam Smith's work I had read when I was at Cambridge; but the news about Cournot came to me, if I rightly recall, from Bob Axelrod. Thorstein Veblen, American economist and philosopher, to be mentioned under the topic of sexual signalling in Chapter 17, was yet another in this stream.

How does the co-operation Axelrod and I discuss in our paper relate to kin selection? Here there is really little that I need to add to what I wrote in Volume 1 (Chapter 8, pp. 262–4) about Robert Trivers with his prescient stress on the long continuance of interactions; in a sense our paper is merely mathematizing and quantifying an idea that Trivers had displayed heuristically. In a letter I sent to him along with a copy when our paper came out, I told him that the theory I had helped with was to be considered my revenge—my 'tit' for a 'tat' of his from long ago. In that he had shown how a 'patents-pending' sex-ratio idea of mine put together with a clearly 'patents-granted' kin-selection idea could yield, greatly to my chagrin, something completely new and unexpected about the social insects—about, in fact, various 'patents-uncertain' as well as erroneous examples I had used to illustrate kin selection.[12] Just as I'd kicked myself for not having noticed the 'Trivers and Hare' applications in that case, so the present paper showed something I thought Bob T would be sorry to have missed—a way to quantify in effect the poetic 'long shadow' he had plainly, in its essentials, recognized.

Convincing though our proof was, how final was our own account in this paper for the real issues of the world? Of course, it wasn't at all. Within our stated limits we made no mistake, but TFT, it turned out, hadn't been put on so secure a plinth as we made it seem in the paper. Certainly we had toppled the old, ugly ALWAYS DEFECT (or 'ALL D') from its primacy in the theory of games but I guess since then that ALL D has had many a grim giggle from where it has lain, watching as TFT has suffered challenges and setbacks itself. Other plinths, in brief, have since been set up. But, and this may illustrate what I meant about Bob Axelrod reaching for larger designs that turn out to work better than he had good reason to expect,[13] TFT seems to stand increasingly firm again at the present, or at least it does so after a few minor extra conditionalities have been added to its decisions.[14] As can be guessed from points we make in our paper, the major trouble with TFT lies in it being merely neutrally stable against strategies that are equally 'nice' but less cautious and less provokable than it is. Any strategy whatever that has this 'nice' property—that is, of being never the first to defect—can drift into the picture if it occurs as a mutant in a TFT population or becomes otherwise infused there. A mutation knocking out the response to a prompt for vengefulness in a situation in which revenge is never called for, is not a difficult one to imagine occurring. Unfortunately the foolishly nice strategies that arise this way, lacking all policy to cope with anything bad, are in danger eventually to hold out the stirrup to others that are much worse than themselves; ALL D, for example, can rise out of the ivy where it had fallen, and come galloping back. In short, as a chapter in the history of a theory of Good versus Evil, our paper records a battle won but certainly not an end to the war.[15]

The most dramatic later contender in the line of 'success via simplicity' that we started to advertise in our paper was another equally simple strategy, which, while less able to score a high average success in very mixed situations such as of Bob's tournaments, is more resistant than TFT is to reinvasion once established. This strategy was named 'PAVLOV' by Nowak and Sigismund,[16] because, like a dog of the famous Russian behaviourist, it responds not socially at all but simply according to whether it is fed or not (though I am sure Pavlov's actual poor animals were not really so socially oblivious). 'Don't play with your mind' an ill-tempered hotelier in Copenhagen once said to me when I tried to suggest a slightly different from normal arrangement for paying for my room; the PAVLOV strategy obeys this hotelier. It forms no image of the other player at all, instead it plays an asocial, almost autistic strategy: if its last move obtained a good outcome it uses the move again; if it receives a worse outcome it tries the other play. PAVLOV can't be assumed quite so simple at the very start of a run of interactions, however, and one might wonder how this can be. Innately PAVLOV seems to categorize two of the four possible outcomes (S and P) as 'bad' and the other two (T and R) as 'good' (see the paper for the definition of these symbols). Thus if it starts off again a defector, it seemingly somehow knows its payoff is bad and it changes immediately from co-operation to defection; then if it gets a better result it stays defecting—in effect (and unlike TFT), it stays exploiting the other player's attempts to co-operate. But if instead it received a 'bad' result (as when the other has also defected) it tries a change and goes back to co-operation. In short, it is restless until its state is 'good' and then steadies, all of which admittedly sounds very biological. But, without previous experience, how did it know a first result of getting a bad payoff (S) was worse than it might have been? Would an autist know this? Why should it not equally guess this payoff to be good and continuing to accept it from an ALL-D for ever? Surely to be realistic there should be a trying of this and that and a building on experience before it begins the PAVLOV schedule. On the other hand it seems to me more reasonable that TFT, explicitly assumed socially aware and able to recognize individuals, knows 'intelligently' whether its partner was mean to it or not on the last move. The most recent results coming from both simulation in models and observations on humans suggest that behaviour may and should hover somewhere between PAVLOV and TFT with the 'niches' appropriate to each depending on how far back an interaction can be remembered and on the degree to which moves are simultaneous.[17]

Our arguments in the paper about the possibility of initiating co-operation in an asocial environment apply almost equally to TFT and to PAVLOV. Both are 'nice' strategies and so both do well if there is likelihood for them to be playing with their own kind. Both are helped by minor modifications to their 'simplicity' if their environments are 'noisy'—that is, if there are occasional unintended moves and misinterpretations. A patchy environment that puts the players into clusters while also confining their reproduction locally, creates inevitably local concentrations of 'niceness', which, by their success, impel 'nice' migrants to other clusters. If the localities that receive them happen to be demographically weak through their inhabitants being generally not nice, there is a good chance that the immigrants will take them over. This is especially likely, of course, when several reciprocators come to a locality together.

Bob and I, at somewhere about this point in our writing of the paper, had a slight disagreement concerning how to present the initiation of the spread of co-operation. Bob insisted that there was no necessity at the start to refer to 'kin' and that even the early stages didn't have to be a version of a takeover through kin selection. I contended that while there might be no absolute necessity, kinship would, in fact, be the usual reason for the similarity seen within groups in nature. When this was the case, kin selection would be the appropriate term for the preliminary phase. Moreover such a starting scenario could be described in terms of 'inclusive fitness' if one chose, although an account of this kind might not be neat. The similar individuals found together in clusters normally are each others' relatives even if their joint presence arises only from the chance inequalities of reproduction. The only way to avoid such kin-based clustering is to allow phases of population where the clustering ends and all organisms mix. If they afterwards become more clustered again than is expected by chance, they must have sorted themselves by their properties—properties which, one had to admit, need not have been social. If humans, for example, they could sort themselves by their loves or hatreds of vintage cars, or of living in hilly country. Then, in addition, such non-social preferences might just happen to correlate with social traits; plains men might be all PAVLOVs, hill lovers all ALL DEFECTs (to be fair I am suggesting here for the moment the opposite to what I did at the start for the actual Rocky Mountains and the Great Plains). Bob's line was: well, why not? And mine: but how unlikely and where was even a single plausible example? This as it happens had been also exactly my reply to David Sloan Wilson on the issue of non-kin group selection, which he had raised as possible in a paper of 1975 (for a recent summary of his line, which remains little changed, still without convincing example, see note 18).

Let me play advocate for Bob's and David's points of view for a moment here. Better than hills, where our arguments might unnecessarily acquire a political tinge, consider whether the connecting non-social character could be a love of the sea. The opening of the novel *Moby Dick* beautifully depicts how utterly unrelated and different-natured people do, in fact, meet in the crews of ships: all they have in common, unless they are press-ganged and have not even that, is their seamanship—they decided to be seamen. Crews surely are much more likely to survive the dangers of the sea if they are co-operative. Yet immediately, against the rosy dawn of such a potential, think of how fishing trawlers are all caught in their own nets right now—even sunk by them one might say, with all of them losers—in a kind of gigantic World Cup series of the Tragedies of the Commons game[19] about overfishing. It is a dilemma in which all sailors in all crews seem to be defecting. Admittedly, however, they could all still be altruistic within the crews and even, as I suggested in more than one place in Volume 1, perhaps the less co-operation there is at the one level the more is likely to exist at another. All the same, it is just not plausible that all sea goers are altruistic any more than that all red-heads are. In contrast investors in a mutual provident society may have more chance to be at least somewhat similar—it is clear that they are all thinkers to the future, all expect the society's promises to be kept, and so on. And yet still I do not hear that co-investors often pair off in personal friendships just because of such similarity, and in summary cannot think what the non-social character that brings about the assortation is supposed to be.

Our argument, I may add, didn't finish completely but because the possibility of pure non-kin assortation was clear, even where its practicality wasn't, we were both glad to include its mention in the paper. Far more important overall were that (1) once a threshold of frequency has been crossed, tit-for tatters could finish their invasion without further help from clustering, and (2) neither clustering nor any threshold effect is helpful to defectors—they cannot not re-invade. I may add here that not only are origins still matters for debate but the evidence concerning what humans actually do, rationally or otherwise, in PDs and Tragedies of the Commons is still unclear, as will be seen in Chapter 14. Personally I am satisfied that, whatever the details are, the three main bands of reasoning that Bob and I were involved with—the hierarchical analysis of selection processes, kin selection, and reciprocation theory including its interspecies aspects (mainly this chapter)—are an adequate foundation for all biological social phenomena.[20] The degree to which humans in the long term can escape the bounds of these theories is less clear to me, but most modern findings concerning the best of animal examples and the average of human ones (including those on PD-like situations that will be outlined in Chapter 14) make me pessimistic. One possibility that we can expect is that, just as in the past we have been taken over and 'ordered about' into acts that are both better and worse than is natural for us by religions and like moral philosophies, so in the future we will be increasingly taken over and ordered about (as well as drug-altered and physiologically tinkered) by technological superorganisms that will come to control us (see Volume 1 and Chapter 12 of this volume). This may happen and we may thus escape the bounds—but will it be 'we' that escape?

Returning to the origins that I argued with Bob, to the acute reader it may seem that my side of this argument was admitted to be lost through some of the examples I used in the paper: the grouper fish at the coral reef, for example, clearly isn't related to the cleaner fish that cleans him. But the situation here is actually more subtle. The customer who ate his barber at the end of the haircut would certainly be 'defecting' on their deal but in a clustered world he would also be defecting on his brother and cousin, if they live anywhere near, through destroying, very likely, a mutualist on which all their haircuts depend. If in contrast the grouper fish doesn't eat the cleaner, and fellow groupers on his reef likewise never do so, how is this correlation likely to have come about? We are back with the same most likely answer that I have already hinted: the grouper fish of one reef are probably related and this is how their system with the cleaners, which eventually does not need relatedness, was started; and it is the same if there are biters and non-biters among the cleaner fishes—the former class is perhaps the equivalent of a whole guild of Sweeney Todd barbers. But in so far as sorting by phenotypes occurs, the start from kin selection certainly greatly broadens and secures its base as the non-kin additional process kicks in.[21]

Humans have so many ways of sorting themselves that examples of non-kin sorting among us may seem at first particularly likely. But against this must be set the equally high human aptitude for dissimulation. Humans can certainly show many parallels to being false cleaner fish. We easily simulate a kindly phenotype without actually being one and this is likely to set back progress in the situations that are not initially underpinned by kinship. Both the positive and the negative opportunities envisioned here are less likely for non-human life, but a successful maxim of trusting kin most and of particularly distrusting all highly mobile interactants, whether they are of your own species or another, can hold for both us and for the rest. The emphasis in the paper on mobility as a factor in the evolution of virulence, foreshadows (in parallel to a theme initiated earlier by Ilan Eshel[22]), a lot of recent further discussion within the theme of the evolution of virulence—that is, whether diseases generally evolve to be more severe or less after their first establishment in a new host species, and where does their equilibrium level of virulence settle?[23] To a degree this issue is the same as the question of whether, over generations or within a lifetime, old neighbours become 'dear neighbours'[24] even when they are rivals and parasites. Although the strong involvement of mobility in these issues is based on abundant evidence and theory,[25] it is still much overlooked.[26]

Hotels along the arteries of a big country like the USA and the equivalents of our English 'B&Bs',[27] where situated in tourist hotspots world-wide, are far from being coral reefs as regards reliable trust in receiving a return benefit: in such places the false cleaner-fish side of human nature seems always to be winning and on the part of both 'hosts' and 'guests'. 'Take a bite and flee' seems to be the maxim. Such hotels have given me some of the worst value-for-money experiences I have ever had during travel, and yet the businesses look as though they might have originated as those 'nice' family concerns, which, on the Margulis view of life, one might be expecting to be eagerly waiting to do me real service and thus to awaken my better nature. Friendly and fair small motels, of course, do exist. The one near Denver was such and its poem might well have reflected its genuine idealism. Is it the hotel or the customers that take the first step on the route to a DD (Defect–Defect interaction—both act mean)? That motel was on the road to the great and cluster-enforcing mountains. Grimly in contrast to it I recall a motel on the outskirts of Albuquerque, inevitably near an intersection of giant interstate highways. In the morning after a hot and noisy night, I left its dirty bathroom, its cheesecloth sheets in a bedroom that had everything chained to something else, I stood at the checkout counter contemplating a bill filled with unforewarned additions—and, of course, the proprietor still had my passport and I was in a hurry . . . Alas I forget the name or I would tell it here to shame them and bring them less custom—that is, less T (Temptation to Defect, as the Prisoner's Dilemma notation would have it).

Strongly competing, however, for Greasy Sheet Award of my lifetime has to be a B&B on the outskirts of Galway town in Ireland, featuring sheets not just thin but used and soon to be further trampled by the many feet of bed bugs who visited me in the night. There was also the landlady's son, who likewise visited, or rather came back into, what was obviously his room, this after he or she supposed wrongly that I had gone to sleep, and slept on the floor. Finally in the morning there was the second 'B' that never came—no breakfast, the sign had been a fraud. No food in the house, the landlady explained to me, was the trouble, 'the shops not having been open and won't be until ten'; also for sure she would look for my passport where she had put it for the gardai (police) but would I be paying her for the night, first of all? Even so, hiking as I was with only a sleeping bag during a week of wind and rain (i.e. Irish weather), perhaps I was better off in that dreary house found at a wet fork when entering the town than I would have been sleeping out in a country side, where, even on the fine days, only the tops of the wide stone walls were free from the universal rain-sodden sponge of the ground.

It has been seen in Chapter 2 and will again in Chapter 16 how much I love haploid mosses. I love their modesty and economy in size and their delicate and singly sufficient chromosomes. Among them I especially love the sphagna. I revere these plants in the first place for their additions to the sheer bulk of northern and western countries, like Ireland, over which they lie like an immense blanket thrown on a restless sleeper, trapping carbon for him and ameliorating not just for these fringe lands but for all of us our greenhouse influences. But I also revere sphagna for what they have taught me by their very success. They teach about haploidy, sex, and coexistence in communities. But as a bed when soft rain has saturated their every billionth hollow leaf with water, their soft enticements can seem, amongst the offered sleeping places of the West, enduringly worse even than those of a 'D' landlady in a sluttish house or than even the stony top of a wall. For one night my sleeping bag could soak and sink in the sweet (and not cold) Irish moss water—I have been very intimate with sphagna at times—but the bag's down is a sponge, too, and next day I have to carry the water. Moreover, in such weather how will it ever dry?

The more curious thing that I started to learn from both these experiences with hotels, and later had explained to me by colleagues in the (H)BACH group, is that you can be much more certain of a fair deal if you go to the big hotel chains. It is indeed a point intrinsic to the paper that follows and is worth emphasizing. A customer of chain organization, be he meeting the links of the chain in cities far distant from each other, still visits the same organism. He remembers the organism and, if he pokes the organism in nasty-enough ways, the organism remembers him. Thus the repeatedness of potential encounters enters. Reciprocation arises just as our paper shows. It does not pay a chain to defect on customers—they'll desert the chain—and nor does it pay customers to steal the towels—such customers will be blacklisted and eventually find themselves only accepted by hotels that bite back, like the one that lacerated me at Albuquerque.

All this, of course, is simply life as we learn it and in the end we are not so much moved. It is different and more difficult with more personal betrayals, however, perhaps because life often isn't long enough for us to learn from these more serious mistakes and change our policy. Shakespeare, as always, has this side well understood, as below, with his song from *As You Like It*, in which his sardonic Jacques is trying to make light entertainment out of events he has deeply felt. I didn't sleep on the walls in the west of Ireland for the rest of my trip because just one landlady had cheated me in one B&B but I might have done so if it had been a person I thought a friend who had cheated me. Perhaps this is because we as a species learned personal trust for a million and more years in much smaller and more intrarelated communities than those of the world we now inhabit. In the present world our mobility is such that betrayals have come to look ever better as a pragmatist's option. Indeed it shapes up in most ways as a cheater's paradise and perhaps it is only giant banks watching us with unsleeping eyes through their slots in walls the world over, and keeping track of everyone's credit rating, that we save ourselves from anarchy. But, alas, cheats can work through banks and even be banks too.

> Freeze, freeze, thou bitter sky,
> That dost not bite so nigh
>   As benefits forgot:
> Though thou the waters warp,
> Thy sting is not so sharp
>   As friend remembered not.
>   SHAKESPEARE, *As You Like It*

References and notes

1. E. Markham, *Poems of Edwin Markham,*1st edn (Harper, New York, c.1950); the first verse of his poem *A Creed*. The second verse, I was sorry to discover, is less good and strangely disconnected with the first. My thanks to Craig Raine for finding poet and poem.

2. W. Poundstone, *Prisoner's Dilemma* (Oxford University Press, Oxford, 1993).

3. R. Axelrod and D. Dion, The further evolution of cooperation, *Science* **242**, 1385–90 (1988).

4. R. Axelrod, *The Complexity of Cooperation* (Princeton University Press, Princeton, NJ, 1997); J. Bendor and P. Swistak, The evolutionary stability of cooperation, *American Political Science Review* **91**, 290–307 (1997).

5. A. Rapoport and A. W. Chammah, *Prisoner's Dilemma* (University of Michigan Press, Ann Arbor, 1965).

6. E. B. Ford, *Butterflies* (Collins, London, 1945).

7. W. D. Hamilton, On first looking into a British treasure, *Times Literary Supplement* 13–14 (12 August 1994); P. Marren, *The New Naturalists* (HarperCollins, London, 1995).

8. M. Mesterton-Gibbons, Ecotypic variation in the asymmetric hawk-dove game: when is Bourgeois an evolutionary stable strategy?, *Evolutionary*

Ecology **6**, 198–222 (1992); I. C. W. Hardy, Butterfly battles: on conventional contests and hot property, *Trends in Ecology and Evolution* **13**, 385–6 (1998).

9. See, for example, C. Davidson, in *New Scientist* (11 April 1998) pp. 36–9. A search on the terms 'genetic algorithm' and 'genetic programming' in titles, abstracts, or keywords of scientific journals indexed by Science Citation Index showed me in Autumn 1998 about 1700 references over the past 4 years.

10. R. Dawkins, *The Blind Watchmaker* (Longman, Harlow, 1986).

11. J. M. Henderson and R. E. Quandt, *Microeconomic Theory: A Mathematical Approach* (McGraw-Hill, New York, 1971).

12. R. L. Trivers and H. Hare, Haplodiploidy and the evolution of social insects, *Science* **191**, 249–63 (1976).

13. Bendor and Swistak (1997) in note 4.

14. J. Wu and R. Axelrod, How to cope with noise in the Iterated Prisoner's Dilemma, *Journal of Conflict Resolution* **39**, 183–9 (1995).

15. R. Boyd and J. Lorberbaum, No pure strategy is evolutionarily stable in the repeated Prisoner's dilemma game, *Nature* **327**, 58–9 (1987).

16. M. Nowak and K. Sigismund, A strategy of win-stay lose-shift that out performs tit-for-tat in the prisoner's dilemma game, *Nature* **364**, 56–8 (1993).

17. C. Wedekind and M. Milinski, Human cooperation in the simultaneous and the alternating prisoners-dilemma—Pavlov versus Generous Tit-For-Tat, *Proceedings of the National Academy of Sciences, USA* **93**, 2686–9 (1996).

18. E. Sober and D. S. Wilson, *Unto Others: The Evolution of Altruism* (Harvard University Press, Cambridge, MA, 1997).

19. The 'Tragedy of the Commons' game is a many-person version of the 'Prisoner's Dilemma', named after a paper by Garett Hardin (*Science* **162**, 1243–8 (1968)), in which he pointed out how failure to agree on how to apportion grazing (and other personal use) of the common land possessed by a community usually led to its over-use and often to its permanent ruin. It is not necessary to specify the detailed rules for this 'game' to see that situations of this general kind, in which when everyone 'defects' almost everything is lost but when everyone 'co-operates' everything is productive and fair EXCEPT THAT—and alas that—a lone defector (thief, trickster, traitor, parasite, whatever it may be appropriate to call him) can do better than anybody, represent a very common problem for social animals and that there is a huge variety of human examples fanning out beyond the 'common' of Hardin's example.

20. For the hierarchical analysis, see Vol. 1 of *Narrow Roads of Gene Land*, especially Chapters 5 and 9, and also recently the comprehensive account by

S. A. Frank in *Principles of Social Evolution* (Princeton University Press, Princeton, NJ, 1998). For kin selection, see the whole of my Vol. 1 and references therein. For reciprocation theory, see this chapter and its references, also noting M. Doebeli and N. Knowlton, The evolution of interspecific mutualisms, *Proceedings of the National Academy of Sciences, USA* **95**, 8676–80 (1998) and G. Roberts and T. N. Sherratt, Development of cooperative relationships through increasing investment, *Nature* **394**, 175–9 (1998).

21. D. S. Wilson and L. A. Dugatkin, Group selection and assortative interactions, *American Naturalist* **149**, 336–51 (1997).

22. I. Eshel, On the founder effect and the evolution of altruistic traits: an ecogenetical approach, **11**, 410–24 (1977).

23. Tangentially the basic idea had also been touched on by me in 1964 in my mention that kinship theory indicated that 'uninhibited competition should characterize the species with the most freely mixing populations' (see Vol. 1, p. 40).

24. J. B. Falls, J. R. Krebs, and P. K. McGregor, Song matching in the Great Tit (*Parus major*)—the effect of similarity and familiarity, *Animal behaviour* **30**, 997–1009 (1982).

25. Doebeli and Knowlton (1998), Frank (1998), and Roberts and Sherratt (1998) in note 20.

26. L. Margulis and D. Sagan, *Origins of Sex* (Yale University Press, New Haven, CT, 1986); L. Margulis and R. Fester (ed.), *Symbiosis as a Source of Evolutionary Innovation, Speciation and Morphogenesis* (MIT Press, Cambridge, MA, 1991); L. Margulis and D. Sagan, *What is Sex?* (Simon & Schuster, New York, 1997).

27. Common throughout the British Isles, these are private homes offering overnight accommodation plus breakfast usually at a rate considerably less than that of hotels.

THE EVOLUTION OF CO-OPERATION[†]

ROBERT ALEXROD and W. D. HAMILTON

———

Co-operation in organisms, whether bacteria or primates, has been a difficulty for evolutionary theory since Darwin. On the assumption that interactions between pairs of individuals occur on a probabilistic basis, a model is developed based on the concept of an evolutionarily stable strategy in the context of the Prisoner's Dilemma game. Deductions from the model, and the results of a computer tournament, show how co-operation based on reciprocity can get started in an asocial world, can thrive while interacting with a wide range of other strategies, and can resist invasion once fully established. Potential applications include specific aspects of territoriality, mating, and disease.

The theory of evolution is based on the struggle for life and the survival of the fittest. Yet co-operation is common between members of the same species and even between members of different species. Before about 1960, accounts of the evolutionary process largely dismissed co-operative phenomena as not requiring special attention. This position followed from a misreading of theory that assigned most adaptation to selection at the level of populations or whole species. As a result of such misreading, co-operation was always considered adaptive. Recent reviews of the evolutionary process, however, have shown no sound basis for a pervasive group-benefit view of selection; at the level of a species or a population, the processes of selection are weak. The original individualistic emphasis of Darwin's theory is more valid.[1–3]

To account for the manifest existence of co-operation and related group behaviour, such as altruism and restraint in competition, evolutionary theory has recently acquired two kinds of extension. These extensions are, broadly, genetical kinship theory[4] and reciprocation theory[5] (for additions to the theory of biological co-operation see refs 6–8). Most of the recent activity, both in field work and in further developments of theory, has been on the side of kinship. Formal approaches have varied, but kinship theory has increasingly taken a gene's-eye view of natural selection.[9] A gene, in effect,

[†]*Science* **211**, 1390–6 (1981).

looks beyond its mortal bearer to interests of the potentially immortal set of its replicas existing in other related individuals. If interactants are sufficiently closely related, altruism can benefit reproduction of the set, despite losses to the individual altruist. In accord with this theory's predictions, apart from the human species, almost all clear cases of altruism, and most observed co-operation, occur in contexts of high relatedness, usually between immediate family members. The evolution of the suicidal barbed sting of the honeybee worker could be taken as paradigm for this line of theory.[10]

Conspicuous examples of co-operation (although almost never of ultimate self-sacrifice) also occur where relatedness is low or absent. Mutualistic symbioses offer striking examples such as these: the fungus and alga that compose a lichen; the ants and ant-acacias, where the trees house and feed the ants, which, in turn, protect the trees;[11] and the fig wasps and fig tree, where wasps, which are obligate parasites of fig flowers, serve as the tree's sole means of pollination and seed set.[12,13] Usually the course of co-operation in such symbioses is smooth, but sometimes the partners show signs of antagonism, either spontaneous or elicited by particular treatments.[14] Although kinship may be involved, as will be discussed later, symbioses mainly illustrate the other recent extension of evolutionary theory, the theory of reciprocation.

Co-operation *per se* has received comparatively little attention from biologists since the pioneer account of Trivers;[5] but an associated issue, concerning restraint in conflict situations, has been developed theoretically. In this connection, a new concept, that of an evolutionarily stable strategy (ESS), has been formally developed.[9,15—17] Co-operation in the more normal sense has remained clouded by certain difficulties, particularly those concerning initiation of co-operation from a previously asocial state[18] and its stable maintenance once established. A formal theory of co-operation is increasingly needed. The renewed emphasis on individualism has focused on the frequent ease of cheating in reciprocatory arrangements. This makes the stability of even mutualistic symbioses appear more questionable than under the old view of adaptation for species' benefit. At the same time other cases that once appeared firmly in the domain of kinship theory now begin to reveal relatednesses of interactants that are too low for much nepotistic altruism to be expected. This applies both to co-operative breeding in birds[19,20] and to co-operative acts more generally in primate groups.[21—23] Here either the appearances of co-operation are deceptive—they are cases of part kin altruism and part cheating—or a larger part of the behaviour is attributable to stable reciprocity. Previous accounts that already invoke reciprocity, however, underemphasize the stringency of its conditions.[24]

Our contribution in this area is new in three ways.

1. In a biological context, our model is novel in its probabilistic treatment of the possibility that two individuals may interact again. This allows us to

shed new light on certain specific biological processes such as ageing and territoriality.

2. Our analysis of the evolution of co-operation considers not just the final stability of a given strategy, but also the initial viability of a strategy in an environment dominated by non-co-operating individuals, as well as the robustness of a strategy in a variegated environment composed of other individuals using a variety of more or less sophisticated strategies. This allows a richer understanding of the full chronology of the evolution of co-operation than has previously been possible.

3. Our applications include behavioural interaction at the microbial level. This leads us to some speculative suggestions of rationales able to account for the existence of both chronic and acute phases in many diseases, and for a certain class of chromosomal non-disjunction, exemplified by Down's syndrome.

STRATEGIES IN THE PRISONER'S DILEMMA

Many of the benefits sought by living things are disproportionally available to co-operating groups. While there are considerable differences in what is meant by the terms 'benefits' and 'sought', this statement, insofar as it is true, lays down a fundamental basis for all social life. The problem is that while an individual can benefit from mutual co-operation, each one can also do even better by exploiting the co-operative efforts of others. Over a period of time, the same individuals may interact again, allowing for complex patterns of strategic interactions. Game theory in general, and the Prisoner's Dilemma game in particular, allow a formalization of the strategic possibilities inherent in such situations.

The Prisoner's Dilemma game is an elegant embodiment of the problem of achieving mutual co-operation,[25] and therefore provides the basis for our analysis. To keep the analysis tractable, we focus on the two-player version of the game, which describes situations that involve interactions between pairs of individuals. In the Prisoner's Dilemma game, two individuals can each either co-operate or defect. The payoff to a player is in terms of the effect on its fitness (survival and fecundity). No matter what the other does, the selfish choice of defection yields a higher payoff than co-operation. But if both defect, both do worse than if both had co-operated.

Figure 4.1 shows the payoff matrix of the Prisoner's Dilemma. If the other player co-operates, there is a choice between co-operation which yields R (the reward for mutual co-operation) or defection which yields T (the temptation to defect). By assumption, $T > R$, so that it pays to defect if the other player co-operates. On the other hand, if the other player defects, there is a choice between co-operation which yields S (the sucker's payoff) or defection

Player B

	C Co-operation	D Defection
Player A		
C Co-operation	$R = 3$ Reward for mutual co-operation	$R = 0$ Sucker's payoff
D Defection	$T = 5$ Temptation to defect	$P = 1$ Punishment for mutual defection

Figure 4.1 The Prisoner's Dilemma game. The payoff to player A is shown with illustrative numerical values. The game is defined by $T > R > P > S$ and $R > (S + T)/2$.

which yields P (the punishment for mutual defection). By assumption $P > S$, so it pays to defect if the player defects. Thus, no matter what the other player does, it pays to defect. But, if both defect, both get P rather than the larger value of R that they both could have got had both co-operated. Hence the dilemma.[26]

With two individuals destined never to meet again, the only strategy that can be called a solution to the game is to defect, always despite the seemingly paradoxical outcome that both do worse than they could have had they co-operated.

Apart from being the solution in game theory, defection is also the solution in biological evolution.[27] It is the outcome of inevitable evolutionary trends through mutation and natural selection: if the payoffs are in terms of fitness, and the interactions between pairs of individuals are random and not re-peated, then any population with a mixture of heritable strategies evolves to a state where all individuals are defectors. Moreover, no single differing mutant strategy can do better than others when the population is using this strategy. In these respects the strategy of defection is stable.

This concept of stability is essential to the discussion of what follows and it is useful to state it more formally. A strategy is evolutionarily stable if a population of individuals using that strategy cannot be invaded by a rare mutant adopting a different strategy.[15–17] In the case of the Prisoner's Dilemma played only once, no strategy can invade the strategy of pure de-fection. This is because no other strategy can do better with the defecting individuals than the P achieved by the defecting players who interact with each other. So in the single-shot Prisoner's Dilemma, to defect always is an evolutionarily stable strategy.

In many biological settings, the same two individuals may meet more than once. If an individual can recognize a previous interactant and remember some aspects of the prior outcomes, then the strategic situation becomes an iterated Prisoner's Dilemma with a much richer set of possibilities. A strategy would take the form of a decision rule that determined the probability of co-operation or defection as a function of the history of the interaction so far. But if there is a known number of interactions between a pair of individuals, to defect always is still evolutionarily stable and is still the only strategy which is. The reason is that defection on the last interaction would be optimal for both sides, and consequently so would defection on the next-to-last inter-action, and so on back to the first interaction.

Our model is based on the more realistic assumption that the number of interactions is not fixed in advance. Instead, there is some probability, w, that after the current interaction the same two individuals will meet again. Factors that affect the magnitude of this probability of meeting again include the average lifespan, relative mobility, and health of the individuals. For any value of w, the strategy of unconditional defection (ALL D) is evolutionarily stable; if everyone is using this strategy, no mutant strategy can invade the population. But other strategies may be evolutionarily stable as well. In fact, when w is sufficiently great, there is no single best strategy regardless of the behaviour of the others in the population (for a formal proof, see ref. 28; for related results on the potential stability of co-operative behaviour, see refs 29–31). Just because there is no single best strategy, it does not follow that analysis is hopeless. On the contrary, we demonstrate not only the stability of a given strategy, but also its robustness and initial viability.

Before turning to the development of the theory, let us consider the range of biological reality that is encompassed by the game theoretic approach. To start with, an organism does not need a brain to employ a strategy. Bacteria, for example, have a basic capacity to play games in that (1) bacteria are highly responsive to selected aspects of their environment, especially their chemical environment; (2) this implies that they can respond differentially to what other organisms around them are doing; (3) these conditional stra-tegies of behaviour can certainly be inherited; and (4) the behaviour of a bacterium can affect the fitness of other organisms around it, just as the behaviour of other organisms can affect the fitness of a bacterium.

While the strategies can easily include differential responsiveness to recent changes in the environment or to cumulative averages over time, in other ways their range of responsiveness is limited. Bacteria cannot 'remember' or 'interpret' a complex past sequence of changes, and they probably cannot dis-tinguish alternative origins of adverse or beneficial changes. Some bacteria, for example, produce their own antibiotics, bacteriocins; those are harmless to bacteria of the producing strain, but destructive to others. A bacterium might easily have production of its own bacteriocin dependent on the per-ceived presence of like hostile products in its environment, but it could not aim the toxin produced towards an offending initiator. From existing evidence, so far from an individual level, discrimination seems to be by species rather even than variety. For example, a *Rhizobium* strain may occur in nodules which it causes on the roots of many species of leguminous plants, but it may fix nitrogen for the benefit of the plant in only a few of these species.[32] Thus, in many legumes the *Rhizobium* seems to be a pure parasite. In the light of theory to follow, it would be interesting to know whether these parasitized legumes are perhaps less beneficial to free-living *Rhizobium* in the surrounding soil than are those in which the full symbiosis is established. But the main point of concern here is that such discrimination by a *Rhizobium* seems not to be known even at the level of varieties within a species.

As one moves up the evolutionary ladder in neural complexity, game-playing behaviour becomes richer. The intelligence of primates, including humans, allows a number of relevant improvements: a more complex memory, more complex processing of information to determine the next action as a function of the interaction so far, a better estimate of the probability of future interaction with the same individual, and a better ability to distinguish between different individuals. The discrimination of others may be among the most important of abilities because it allows one to handle interactions with many individuals without having to treat them all the same, thus making possible the rewarding of co-operation from one individual and the punishing of defection from another.

The model of the iterated Prisoner's Dilemma is much less restricted than it may at first appear. Not only can it apply to interactions between two bacteria or interactions between two primates, but it can also apply to the interactions between a colony of bacteria and, say, a primate serving as a host. There is no assumption of commensurability of payoffs between the two sides. Provided that the payoffs to each side satisfy the inequalities that define the Prisoner's Dilemma (Fig. 4.1), the results of the analysis will be applicable.

The model does assume that the choices are made simultaneously and with discrete time intervals. For most analytical purposes, this is equivalent to a continuous interaction over time, with the time period of the model corresponding to the minimum time between a change in behaviour by one side and a response by the other. And while the model treats the choices as simultaneous, it would make little difference if they were treated as sequential.[33]

Turning to the development of the theory, the evolution of co-operation can be conceptualized in terms of three separate questions:

1. *Robustness*. What type of strategy can thrive in a variegated environment composed of others using a wide variety of more or less sophisticated strategies?

2. *Stability*. Under what conditions can such a strategy, once fully established, resist invasion by mutant strategies?

3. *Initial viability*. Even if a strategy is robust and stable, how can it ever get a foothold in an environment which is predominantly non-co-operative?

ROBUSTNESS

To see what type of strategy can thrive in a variegated environment of more or less sophisticated strategies, one of us (R.A.) conducted a computer tournament for the Prisoner's Dilemma. The strategies were submitted by game theorists in economics, sociology, political science, and mathematics.[34] The rules implied the payoff matrix shown in Fig. 4.1 and a game length of 200 moves. The 14 entries and a totally random strategy were paired with each other in a round-robin tournament. Some of the strategies were quite intricate. An example is one which on each move models the behaviour of the other player as a Markov process, and then uses Bayesian inference to select what seems the best choice for the long run. However, the result of the tournament was that the highest average score was attained by the simplest of all strategies submitted: TIT FOR TAT. This strategy is simply one of co-operating on the first move and then doing whatever the other player did on the preceding move. Thus TIT FOR TAT is a strategy of co-operation based on reciprocity.

The results of the first round were then circulated and entries for a second round were solicited. This time there were 62 entries from six countries.[35] Most of the contestants were computer hobbyists, but there were also professors of evolutionary biology, physics, and computer science, as well as the five disciplines represented in the first round. TIT FOR TAT was again submitted by the winner of the first round, Professor Anatol Rapoport of the Institute for Advanced Study (Vienna). It won again. An analysis of the 3 million choices which were made in the second round identified the impressive robustness of TIT FOR TAT as dependent on three features: it was never the first to defect, it was provocable into retaliation by a defection of the other, and it was forgiving after just one act of retaliation.[36]

The robustness of TIT FOR TAT was also manifest in an ecological analysis of a whole series of future tournaments. The ecological approach takes as given the varieties which are present and investigates how they do over time when interacting with each other. This analysis was based on what would happen if each of the strategies in the second round were submitted to a hypothetical next round in proportion to its success in the previous round. The process was then repeated to generate the time path of the distribution of strategies. The results showed that, as the less-successful rules were displaced, TIT FOR TAT continued to do well with the rules which initially scored near the top. In the long run, TIT FOR TAT displaced all the other rules and went to fixation.[36] This provides further evidence that TIT FOR TAT's co-operation based on reciprocity is a robust strategy that can thrive in a variegated environment.

STABILITY

Once a strategy has gone to fixation, the question of evolutionary stability deals with whether it can resist invasion by a mutant strategy. In fact, we will now show that once TIT FOR TAT is established, it can resist invasion by any possible mutant strategy provided that the individuals who interact have a sufficiently large probability, w, of meeting again. The proof is described in the next two paragraphs.

As a first step in the proof we note that since TIT FOR TAT 'remembers' only one move back, one C by the other player in any round is sufficient to reset the situation as it was at the beginning of the game. Likewise, one D sets the situation to what it was at the second round after a D was played in the first. Because there is a fixed chance, w, of the interaction not ending at any given move, a strategy cannot be maximal in playing with TIT FOR TAT unless it does the same thing both at the first occurrence of a given state and at each resetting to that state. Thus, if a rule is maximal and begins with C, the second round has the same state as the first, and thus a maximal rule will continue with C and hence always co-operate with TIT FOR TAT. But such a rule will not do better than TIT FOR TAT does with another TIT FOR TAT, and hence it cannot invade. If, on the other hand, a rule begins with D, then this first D induces a switch in the state of TIT FOR TAT and there are two possibilities for continuation that could be maximal. If D follows the first D, then this being maximal at the start implies that it is everywhere maximal to follow D with D, making the strategy equivalent to ALL D. If C follows the initial D, the game is then reset as for the first move; so it must be maximal to repeat the sequence of DC indefinitely. These points show that the task of searching a seemingly infinite array of rules of behaviour for one potentially capable of invading TIT FOR TAT is really easier than it seemed: if neither ALL D nor alternation of D and C can invade TIT FOR TAT, then no strategy can.

To see when these strategies can invade, we note that the probability that the nth interaction actually occurs is w^{n-1}. Therefore, the expression for the total payoff is easily found by applying the weights $1, w, w^2 \ldots$ to the payoff sequence and summing the resultant series. When TIT FOR TAT plays another TIT FOR TAT, it gets a payoff of R each move for a total of $R + wR + w^2R \ldots$, which is $R/(1 - w)$. ALL D playing with TIT FOR TAT gets T on the first move and P thereafter, so it cannot invade TIT FOR TAT if

$$R/(1 - w) \geq T + wP/(1 - w).$$

Similarly, when alternation of D and C plays TIT FOR TAT, it gets a payoff of

$$T = wS + w^2T + s^3S \ldots$$
$$= (T + wS)/(1 - w^2).$$

Alternation of D and C thus cannot invade TIT FOR TAT if

$$R/(1 - w) \geq (T + wS)/(1 - w^2).$$

Hence, with reference to the magnitude of w, we find that neither of these two strategies (and hence no strategy at all) can invade TIT FOR TAT if and only if both

$$w \geq (T - R)/(T - P) \text{ and}$$
$$w \geq (T - R)/(R - S). \tag{1}$$

This demonstrates that TIT FOR TAT is evolutionarily stable if and only if the interactions between the individuals have a sufficiently large probability of continuing.[28-31]

INITIAL VIABILITY

TIT FOR TAT is not the only strategy that can be evolutionarily stable. In fact, ALL D is evolutionarily stable no matter what is the probability of interaction continuing. This raises the problem of how an evolutionary trend to co-operative behaviour could ever have started in the first place.

Genetic kinship theory suggests a plausible escape from the equilibrium of ALL D. Close relatedness of interactants permits true altruism—sacrifice of fitness by one individual for the benefit of another. True altruism can evolve when the conditions of cost, benefit, and relatedness yield net gains for the altruism-causing genes that are resident in the related individuals.[37-39] Not defecting in a single-move Prisoner's Dilemma is altruism of a kind (the individual is foregoing proceeds that might have been taken) and so can evolve if the two interactants are sufficiently related.[27] In effect, recalculation of the payoff matrix in such a way that an individual has a part interest in the partner's gain (i.e. reckoning payoffs in terms of inclusive fitness) can often eliminate the inequalities $T > R$ and $P > S$, in which case co-operation becomes unconditionally favoured.[27,40] Thus it is possible to imagine that the benefits of co-operation in Prisoner's Dilemma-like situations can begin to be harvested by groups of closely related individuals. Obviously, as regards pairs, a parent and its offspring or a pair of siblings would be especially promising, and in fact many examples of co-operation or restraint of selfishness in such pairs are known.

Once the genes for co-operation exist, selection will promote strategies that base co-operative behaviour on cues in the environment.[5] Such factors as promiscuous fatherhood[41] and events at ill-defined group margins will always lead to uncertain relatedness among potential interactants. The recognition of any improved correlates of relatedness and use of these cues to determine co-operative behaviour will always permit advance in inclusive fitness.[5] When a co-operative choice has been made, one cue to relatedness is simply the fact of reciprocation of the co-operation. Thus modifiers for more selfish behaviour after a negative response from the other are advantageous whenever the degree of relatedness is low or in doubt. As such, conditionality is acquired, and co-operation can spread into circumstances of less and less relatedness. Finally, when the probability of two individuals meeting each other again is sufficiently high, co-operation based on reciprocity can thrive and be evolutionarily stable in a population with no relatedness at all.

A case of co-operation that fits this scenario, at least on first evidence, has been discovered in the spawning relationships in a sea bass.[42,43] The fish, which are hermaphroditic, form pairs and roughly may be said to take turns at being the high investment partner (laying eggs) and low investment partner (providing sperm to fertilize eggs). Up to 10 spawnings occur in a day and only a few eggs are provided each time. Pairs tend to break up if sex roles are not divided evenly. The system appears to allow the evolution of much economy in the size of testes, but Fischer[42] has suggested that the testis condition may have evolved when the species was more sparse and inclined to inbreed. Inbreeding would imply relatedness in the pairs and this initially may have transferred the system to attractance of tit-for-tat co-operation—that is, to co-operation unneedful of relatedness.

Another mechanism that can get co-operation started when virtually everyone is using ALL D is clustering. Suppose that a small group of individuals is using a strategy such as TIT FOR TAT and that a certain proportion, *p*, of the interactions of members of this cluster are with other members of the cluster. Then the average score attained by the members of the cluster in playing the TIT FOR TAT strategy is

$$p[R/(1 - w)] + (1 - p)[S + wP/(1 - w)].$$

If the members of the cluster provide a negligible proportion of the interactions for the other individuals, then the score attained by those using ALL D is still $P/(1 - w)$. When *p* and *w* are large enough, a cluster of TIT FOR TAT individuals can then become initially viable in an environment composed overwhelmingly of ALL D.[28-31]

Clustering is often associated with kinship, and the two mechanisms can reinforce each other in promoting the initial viability of reciprocal co-operation. However, it is possible for clustering to be effective without kinship.[4]

We have seen that TIT FOR TAT can intrude in a cluster on a population of ALL D, even though ALL D is evolutionarily stable. This is possible because a cluster of TIT FOR TAT's gives each member a non-trivial probability of meeting another individual who will reciprocate the co-operation. While this suggests a mechanism for the initiation of co-operation, it also raises the question about whether the reverse could happen once a strategy like TIT FOR TAT became established itself. Actually, there is an interesting asymmetry here. Let us define a nice strategy as one, such as TIT FOR TAT, which will never be the first to defect. Obviously, when two nice strategies interact, they both receive R each move, which is the highest average score an individual can get when interacting with another individual using the same strategy. Therefore, if a strategy is nice and is evolutionarily stable, it cannot be intruded upon by a cluster. This is because the score achieved by the strategy that comes in a cluster is a weighted average of how it does with others of its kind and with the predominant strategy. Each of these components is less than or equal to the score achieved by the predominant, nice, evolutionarily stable strategy, and therefore the strategy arriving in a cluster cannot intrude on the nice, evolutionarily stable strategy.[28–31] This means that when w is large enough to make TIT FOR TAT an evolutionarily stable strategy it can resist intrusion by any cluster of any other strategy. The gear wheels of social evolution have a ratchet.

The chronological story that emerges from this analysis is the following. ALL D is the primeval state and is evolutionarily stable. This means that it can resist the invasion of any strategy that has virtually all of its interactions with ALL D. But co-operation based on reciprocity can gain a foothold through two different mechanisms. First, there can be kinship between mutant strategies, giving the genes of the mutants some stake in each other's success, thereby altering the effective payoff matrix of the interaction when viewed from the perspective of the gene rather than the individual. A second mechanism to overcome total defection is for the mutant strategies to arrive in a cluster so that they provide a non-trivial proportion of the interactions each has, even if they are so few as to provide a negligible proportion of the interactions which the ALL D individuals have. Then the tournament approach demonstrates that once a variety of strategies is present, TIT FOR TAT is an extremely robust one. It does well in a wide range of circumstances and gradually displaces all other strategies in a simulation of a great variety of more or less sophisticated decision rules. And if the probability that interaction between two individuals will continue is great enough, then TIT FOR TAT is itself evolutionarily stable. Moreover, its stability is especially secure because it can resist the intrusion of whole clusters of mutant strategies. Thus co-operation based on reciprocity can get started in a predominantly non-co-operative world, can thrive in a variegated environment, and can defend itself once fully established.

APPLICATIONS

A variety of specific biological applications of our approach follows from two of the requirements for the evolution of co-operation. The basic idea is that an individual must not be able to get away with defecting without the other individuals being able to retaliate effectively (for economic theory on this point, see refs 44–46). The response requires that the defecting individual not be lost in an anonymous sea of others. Higher organisms avoid this problem by their well-developed ability to recognize many different individuals of their species, but lower organisms must rely on mechanisms that drastically limit the number of different individuals or colonies with which they can interact effectively. The other important requirement to make retaliation effective is that the probability, w, of the same two individuals' meeting again must be sufficiently high.

When an organism is not able to recognize the individual with which it had a prior interaction, a substitute mechanism is to make sure that all of one's interactions are with the same interactant. This can be done by maintaining continuous contact with the other. This method is applied in most inter-species mutualism, whether a hermit crab and his sea-anemone partner, a cicada and the varied microorganismic colonies housed in its body, or a tree and its mycorrhizal fungi.

The ability of such partners to respond specifically to defection is not known but seems possible. A host insect that carries symbionts often carries several kinds (e.g. yeasts and bacteria). Differences in the roles of these are almost wholly obscure.[47] Perhaps roles are actually the same, and being host to more than one increases the security of retaliation against a particular exploitative colony. Where host and colony are not permanently paired, a method for immediate drastic retaliation is sometimes apparent instead. This is so with fig wasps. By nature of their remarkable role in pollination, female fig wasps serve the fig tree as a motile aerial male gamete. Through the extreme protogyny and simultaneity in flowering, fig wasps cannot remain with a single tree. It turns out in many cases that if a fig wasp entering a young fig does not pollinate enough flowers for seeds and instead lays eggs in almost all, the tree cuts off the developing fig at an early stage. All progeny of the wasp then perish.

Another mechanism to avoid the need for recognition is to guarantee the uniqueness of the pairing of interactants by employing a fixed place of meeting. Consider, for example, cleaner mutualisms in which a small fish or a crustacean removes and eats ectoparasites from the body (or even from the inside of the mouth) of a larger fish which is its potential predator. These aquatic cleaner mutualisms occur in coastal and reef situations where animals live in fixed home ranges or territories.[5–8] They seem to be unknown in the free-mixing circumstances of the open sea.

Other mutualisms are also characteristic of situations where continued association is likely, and normally they involve quasi-permanent pairing of individuals or of endogamous or asexual stocks, or of individuals with such stocks.[10,48] Conversely, conditions of free-mixing and transitory pairing conditions where recognition is impossible are much more likely to result in exploitation—parasitism, disease, and the like. Thus, whereas ant colonies participate in many symbioses and are sometimes largely dependent on them, honeybee colonies, which are much less permanent in place of abode, have no known symbionts but many parasites.[49,50] The small freshwater animal *Chlorohydra viridissima* has a permanent stable association with green algae that are always naturally found in its tissues and are very difficult to remove. In this species the alga is transmitted to new generations by way of the egg. *Hydra vulgaris* and *H. attenuata* also associate with algae but do not have egg transmission. In these species it is said that 'infection is preceded by enfeeblement of the animals and is accompanied by pathological symptoms indicating a definite parasitism by the plant'.[51] Again, it is seen that impermanence of association tends to destabilize symbiosis.

In species with a limited ability to discriminate between other members of the same species, reciprocal co-operation can be stable with the aid of a mechanism that reduces the amount of discrimination necessary. Philopatry in general and territoriality in particular can serve this purpose. The phrase stable territories means that there are two quite different kinds of interaction: those in neighbouring territories where the probability of interaction is high, and strangers whose probability of future interaction is low. In the case of male territorial birds, songs are used to allow neighbours to recognize each other. Consistent with our theory, such male territorial birds show much more aggressive reactions when the song of an unfamiliar male rather than a neighbour is reproduced nearby.[52]

Reciprocal co-operation can be stable with a larger range of individuals if discrimination can cover a wide variety of others with less reliance on supplementary cues such as location. In humans this ability is well developed, and is largely based on the recognition of faces. The extent to which this function has become specialized is revealed by a brain disorder called prosopagnosia. A normal person can name someone from facial features alone, even if the features have changed substantially over the years. People with prosopagnosia are not able to make this association, but have few other neurological symptoms other than a loss of some part of the visual field. The lesions responsible for prosopagnosia occur in an identifiable part of the brain: the underside of both occipital lobes, extending forwards to the inner surface of the temporal lobes. This localization of cause, and specificity of effect, indicates that the recognition of individual faces has been an important-enough task for a significant portion of the brain's resources to be devoted to it.[53]

Just as the ability to recognize the other interactant is invaluable in extending the range of stable co-operation, the ability to monitor cues for the likelihood of continued interaction is helpful as an indication of when reciprocal co-operation is or is not stable. In particular, when the value of w falls below the threshold for stability given in condition (1), it will no longer pay to reciprocate the other's co-operation. Illness in one partner leading to reduced viability would be one detectable sign of declining w. Both animals in a partnership would then be expected to become less co-operative. Ageing of a partner would be very like disease in this respect, resulting in an incentive to defect so as to take a onetime gain when the probability of future interaction becomes small enough.

These mechanisms could operate even at the microbial level. Any symbiont that still has a transmission 'horizontally' (i.e. infective) as well as vertically (i.e. transovarial, or more rarely through sperm, or both) would be expected to shift from mutualism to parasitism when the probability of continued interaction with the host lessened. In the more parasitic phase it could exploit the host more severely by producing more infective propagules. This phase would be expected when the host is severely injured, contracted some other wholly parasitic infection that threatened death, or when it manifested signs of age. In fact, bacteria that are normal and seemingly harmless or even beneficial in the gut can be found contributing to sepsis in the body when the gut is perforated (implying a severe wound).[54] And normal inhabitants of the body surface (e.g. *Candida albicans*) can become invasive and dangerous in either sick or elderly persons.

It is possible also that this argument has some bearing on the aetiology of cancer, insofar as it turns out to be due to viruses potentially latent in the genome.[55,56] Cancers do tend to have their onset at ages when the chances of vertical transmission are rapidly declining.[57] One oncogenic virus, that of Burkitt's lymphoma, does not have vertical transmission but may have alternatives of slow or fast production of infectious propagules. The slow form appears as a chronic mononucleosis, the fast as an acute mononucleosis or as a lymphoma.[58] The point of interest is that, as some evidence suggests, lymphoma can be triggered by the host's contracting malaria. The lymphoma grows extremely fast and so can probably compete with malaria for transmission (possibly by mosquitoes) before death results. Considering other cases of simultaneous infection by two or more species of pathogen, or by two strains of the same one, our theory may have relevance more generally to whether a disease will follow a slow, joint-optimal exploitation course ('chronic' for the host) or a rapid severe exploitation ('acute' for the host). With single infection the slow course would be expected. With double infection, crash exploitation might, as dictated by implied payoff functions, begin immediately, or have onset later at an appropriate stage of senescence.[59]

Our model (with symmetry of the two parties) could also be tentatively applied to the increase with maternal age of chromosomal nondisjunction during ovum formation (oogenesis).[60] This effect leads to various conditions of severely handicapped offspring, Down's syndrome (caused by an extra copy of chromosome 21) being the most familiar example. It depends almost entirely on failure of the normal separation of the paired chromosomes in the mother, and this suggests the possible connection with our story. Cell divisions of oogenesis, but not usually of spermatogenesis, are characteristically unsymmetrical, with rejection (as a so-called polar body) of chromosomes that go to the unlucky pole of the cell. It seems possible that, while homologous chromosomes generally stand to gain by steadily co-operating in a diploid organism, the situation in oogenesis is a Prisoner's Dilemma: a chromosome which can be 'first to defect' can get itself into the egg nucleus rather than the polar body. We may hypothesize that such an action triggers similar attempts by the homologue in subsequent meioses, and when both members of a homologous pair try it at once, an extra chromosome in the offspring could be the occasional result. The fitness of the bearers of extra chromosomes is generally extremely low, but a chromosome which lets itself be sent to the polar body makes a fitness contribution of zero. Thus $P > S$ holds. For the model to work, an incident of 'defection' in one developing egg would have to be perceptible by others still waiting. That this would occur is pure speculation, as is the feasibility of self-promoting behaviour by chromosomes during a gametic cell division. But the effects do not seem inconceivable: a bacterium, after all, with its single chromosome, can do complex conditional things. Given such effects, our model would explain the much greater incidence of abnormal chromosome increase in eggs (and not sperm) with parental age.

CONCLUSION

Darwin's emphasis on individual advantage has been formalized in terms of game theory. This establishes conditions under which co-operation based on reciprocity can evolve.

Acknowledgements

For helpful suggestions we thank Robert Boyd, Michael Cohen, and David Sloan Wilson.

References and notes

1. G. C. Williams, *Adaptations and Natural Selection* (Princeton University Press, Princeton, NJ, 1966).

2. W. D. Hamilton, in R. Fox (ed.), *ASA Studies 4: Bisocial Anthropology*, pp. 133–53 (Malaby Press, London, 1975) [reprinted in *Narrow Roads of Gene Land*, Vol. 1, pp. 329–51].

3. For the best recent case for effective selection at group levels and for altruism based on genetic correction of non-kin interactants see D. S. Wilson, *The Natural Selection of Populations and Communities* (Benjamin/Cummings, Menlo Park, CA, 1980).

4. W. D. Hamilton, *Journal of Theoretical Biology* **7**, 1–52 (1964) [reprinted in *Narrow Roads of Gene Land*, Vol. 1, pp. 31–82].

5. R. Trivers, *Quarterly Review of Biology* **46**, 35–57 (1971).

6. I. D. Chase, *American Naturalist* **115**, 827–57 (1980).

7. R. M. Fagen, *American Naturalist* **115**, 858–69 (1980).

8. S. A. Boorman and P. R. Levitt, *The Genetics of Altruism* (Academic Press, New York, 1980).

9. R. Dawkins, *The Selfish Gene* (Oxford University Press, Oxford, 1976).

10. W. D. Hamilton, *Annual Review of Ecology and Systematics* **3**, 193–232 (1972) [reprinted in *Narrow Roads of Gene Land*, Vol. 1, pp. 270–313].

11. D. H. Janzen, *Evolution* **20**, 249–75 (1966).

12. J. T. Wiebes, *Gardens' Bulletin (Singapore)* **29**, 207–32 (1976).

13. D. H. Janzen, *Annual Review of Ecology and Systematics* **10**, 31–51 (1979).

14. M. Caullery, *Parasitism and Symbiosis* (Sidgwick and Jackson, London, 1952). This gives examples of antagonism in orchid–fungus and lichen symbioses. For the example of wasp–ant symbiosis, see ref. 10.

15. J. Maynard Smith and G. R. Price, *Nature* **246**, 15–18 (1973).

16. J. Maynard Smith and G. A. Parker, *Animal Behaviour* **24,** 159–75 (1976).

17. G. A. Parker, *Nature* **274**, 849–55 (1978).

18. J. Elster, *Ulysses and the Sirens* (Cambridge University Press, London, 1979).

19. S. T. Emlen, in J. R. Krebs and N. B. Davies (ed.), *Behavioural Ecology: An Evolutionary Approach*, pp. 245–81 (Blackwell, Oxford, 1978).

20. P. B. Stacey, *Behavioral Ecology and Sociobiology* **6**, 53–66 (1979).

21. A. H. Harcourt, *Zeitschrift für Tierpsychologie* **48**, 401 (1978).

22. C. Packer, *Animal Behaviour* **27**, 1–36 (1979).

23. R. W. Wrangham, *Social Science Information* **18**, 335–68 (1979).

24. J. D. Ligon and S. H. Ligon, *Nature* **276**, 496–8 (1978).

25. A. Rapoport and A. M. Chammah, *Prisoner's Dilemma* (University of Michigan Press, Ann Arbor, 1965). There are many other patterns of interaction which allow gains for co-operation. See, for example, the model of intraspecific combat in Maynard Smith and Price 1973 (ref. 15).

26. The condition that $R > (S + T)/2$ is also part of the definition to rule out the possibility that alternating exploitation could be better for both than mutual co-operation.

27. W. D. Hamilton, in J. F. Eisenberg and W. S. Dillon (ed.), *Man and Beast: Comparative Social Behavior*, pp. 57–91 (Smithsonian Press, Washington DC, 1971) [reprinted in *Narrow Roads of Gene Land*, Vol. 1, pp. 198–227]. Fagen 1980 (ref. 7) shows some conditions for single encounters where defection is not the solution.

28. R. Axelrod, *American Political Science Rev*iew **75**, 306–18 (1981).

29. R. D. Luce and H. Raiffa, *Games and Decisions*, p. 102 (Wiley, New York, 1957).

30. M. Taylor, *Anarchy and Cooperation* (Wiley, New York, 1976).

31. M. Kurz, in B. Balassa and R. Nelson (ed.), *Economic Progress, Private Values and Public Policy*, pp. 177–200 (North-Holland, Amsterdam, 1977).

32. M. Alexander, *Microbial Ecology* (Wiley, New York, 1971).

33. In either case, co-operation on a tit-for-tat basis is evolutionarily stable if and only if w is sufficiently high. In the case of sequential moves, suppose there is a fixed chance, p, that a given interactant of the pair will be the next one to need help. The critical value of w can be shown to be the minimum of the two side's value of

$A/p(A + B)$ where A is the cost of giving assistance, and B is the benefit of assistance when received. See also P. R. Thompson, *Social Science Information* **19**, 341–84 (1980).

34. R. Axelrod, *Journal of Conflict Resolution* **24**, 3–25 (1980).

35. In the second round, the length of the games was uncertain, with an expected probability of 200 moves. This was achieved by setting the probability that a given move would not be the last at $w = 0.99654$. As in the first round, each pair was matched in five games (ref. 36).

36. R. Axelrod, *Journal of Conflict Resolution* **24**, 379–403 (1980).

37. R. A. Fisher, *The Genetical Theory of Natural Selection* (Oxford University Press, Oxford, 1930).

38. J. B. S. Haldane, *Nature New Biology* **18**, 34–51 (1955).

39. W. D. Hamilton, *American Naturalist* **97**, 354–6 (1963) [reprinted in *Narrow Roads of Gene Land*, Vol. 1, pp. 6–8].

40. M. J. Wade and F. Breden, *Behavioral Ecology and Sociobiology* **7**, 167–72 (1980).

41. R. D. Alexander, *Annual Review of Ecology and Systematics* **5**, 325–83 (1974).

42. E. Fischer, *Animal Behaviour* **28**, 620–33 (1980).

43. E. G. Leigh, Jr, *Proceedings of the National Academy of Sciences, USA* **74**, 4542–6 (1977).

44. G. Akerlof, *Quarterly Journal of Economics* **84**, 488–500 (1970).

45. M. R. Darby and E. Karni, *Journal of Law Economics* **16**, 67–88 (1973).

46. O. E. Williamson, *Markets and Hierarchies* (Free Press, New York, 1975).

47. P. Buchner, *Endosymbiosis of Animals with Plant Microorganisms* (Interscience, New York, 1965).

48. W. D. Hamilton, in L. A. Mound and N. Waloff (ed.), *Diversity of Insect Faunas*, Symposia of the Royal Entomological Society of London, No. 9, pp. 154–75 (Blackwell Scientific, Oxford, 1978) [reprinted in *Narrow Roads of Gene Land*, Vol. 1, pp. 394–20].

49. E. O. Wilson, *The Insect Societies* (Belknap/Harvard University Press, Cambridge, MA, 1971).

50. M. Treisman, *Animal Behaviour* **28**, 311–12 (1980).

51. C. M. Yonge, in *Nature* **134**, 12–15 (1934), gives other examples of invertebrates with unicellular algae.

52. E. O. Wilson, *Sociobiology*, p. 273 (Harvard University Press, Cambridge, MA, 1975).

53. N. Geschwind, *Scientific American* **241** (3), 158–68 (1979).

54. D. C. Savage, in R. T. J. Clarke and T. Bauchop (ed.), *Microbial Ecology of the Gut*, pp. 277–310 (Academic Press, New York, 1977).

55. J. T. Manning, *Journal of Theoretical Biology* **55**, 397–413 (1975).

56. M. J. Orlove, *Journal of Theoretical Biology* **65**, 605–7 (1977).

57. W. D. Hamilton, *Journal of Theoretical Biology* **12**, 12–45 (1966) [reprinted in *Narrow Roads of Gene Land*, Vol. 1, pp. 94–128].

58. W. Henle, G. Henle, and E. T. Lenette, *Scientific American* **241** (1), 40–51 (1979).

59. See also I. Eshel, *Theoretical Population Biology* **11**, 410–24 (1977), for a related possible implication of multiclonal infection.

60. C. Stern, *Principles of Human Genetics* (Freeman, San Francisco, 1973).

'SEX ITSELF'

Pathogens as Causes of Genetic Diversity in their Host Populations

Seeking His secret deeds
With tears and toiling breath
I find thy cunning seeds
O million-murdering Death.
RONALD ROSS[1]

THE scientific Dahlem Workshops are a series with a long post-World War II tradition in West Berlin. They exist upon generous German funding via donors, foundations, and the Berlin City Senate. Conferees see them as a chance to spend a week with people they know, or know of, in distant places, to share ideas, hear the latest jokes and gossip, and to have a hardworking yet costless party. It was one of the meetings in 1982 that gave me my first visit to Berlin and it was to this conference I provided the paper of this chapter. It had as its subject 'The Population Biology of Infectious Diseases' and was set up by Robert May and Roy Anderson, doubtless in hope of a concentrated forum to discuss the new ideas in quantitative immunology they had recently originated; later, in 1986, I went to another Dahlem conference on sexual selection but that time did not provide a paper myself.

Nothing much in Berlin seemed to have changed between my two visits. During both the whole of the Western Sector seemed to me very full of corporate defiance and also of a rampant commercialism that went with it. Politically the conferences appeared designed in the same spirit, to be civic- and nation-level displays showing off the surplus cash of West Germany and the freedom of West Berlin to do as it pleased, including to defy all censorship. West Berlin was Hong Kong in Europe, and, like Hong Kong, sought to present a maximal contrast to communism across the way.

The Wall, the theory seemed to be, couldn't be high enough to hide the huge buildings of the 'West' or the huge ads you pasted on them. There were good, bad, and ludicrous aspects to all this. You could buy anything in West Berlin, talk about anything, criticize anything, and, if at the workshop, enjoy free lunches in rich tree-lined Dahlem amidst all its ponderous Berlin homes. And anywhere outside your hotel, especially downtown, you were bombarded by invitations to step into a booth and watch a sexy peep show. The peep shows, evidently, were another basic freedom of the West.

I think that our particular topic in 1982 was not one over which communism would particularly want to exercise censorship. Indeed in community medicine the Iron Curtain had always been a bit more translucent than it had been on other fronts although still, as on them all, one could not be there in Berlin and escape knowing that machine-gun-carrying Maxwell Demons (surely to us scientists they couldn't be just police) were still patrolling the 'Curtain' no more than the odd mile away and making it a valve through which 'cold' molecules from the East were never to be allowed to penetrate to watch sexy peep shows. More seriously, the barrier meant political thermodynamics was being defied for travel both ways. In the field of medicine, for example, it was easier—a little—for a Western virologist to go to watch the working of a (doubtless selected) vaccination clinic in the communist East than it was for a virologist to come from there to watch the British National Health Service fighting an outbreak of meningitis (in Hull): for on that other side they liked to maintain we were still—in matters of public health—in the age of Dickens and, to preserve this myth, the fewer really seeing how things were done (say, in Hull) the better. Even so, in this field, visits happened, advice and vaccine were exchanged, and the hospitals and clinics deemed watchable in East Germany—to judge from Solzhenitsyn's novel *Cancer Ward* concerning their equivalents in Russia—might not need to be very carefully selected. And yet no Eastern contributors came to our conference on health. One would have thought the 'selfish genery' of infectious bacilli and viruses to have posed only a most indirect challenge to anyone's dream of world order: but, as the potential thin end of a long wedge, the topic conceivably might be threatening for we need ever to be remembering that Lysenko and Stalin tried to ban the concept of genes in Russia entirely. Probably it would be more on account of the style and the spirit of the conference, and (if they had known about it) our photocopied pamphlet of irreverent Dahlem limericks produced at its end, that the East sent no representatives (though now that I think of it those limericks were from that other Dahlem Workshop I went to).

In the spirit of a soon-to-come evolution fashion show (real fashions of dress are always expensive), which I discuss more in Chapter 6, the whole Dahlem Workshop enterprise could be described as high-level Zahavian handicap—that is, an expense being assumed by the city of West Berlin to demonstrate its confidence, an effortless affordable potlatch party it could throw for the benefit of often very 'pure' (and thus needless) scientists and disciplines like me and mine. Actually in the case of the 1982 conference I describe, a lot of attendees were very serious applied medical men; however, this was not true of the second 'sexual selection' one that I attended, so the principle applies.

Anyway, how greatly in contrast to this exuberance of West Berlin was East Berlin as I saw it during my only brief excursion there, entering past those guardian 'demons' at Checkpoint Charlie on the last day of my visit. With its skyline so low and so different from the soaring tower blocks and its streets so lacking, anywhere, anything resembling the often tower-wide gaudy posters that adorned whole faces of the skyscrapers in West Berlin, much of East Berlin looked as though World War II had ended only a few months before. Exuberance—none of that. Heaps of rubble bulldozed to giant piles seemed jealously preserved like monuments—the bad memories of a person determined never to forget or forgive, of a person not wishing to recover, waiting, perhaps, for 'compensation'. I found that food was marvellously cheap but had to tramp long miles to find it. Then, with a restaurant located, I had to inch through what seemed some further miles of queue. Finally the food came dumped onto my single metal multi-dish plate. It was, in itself, excellent nutrition. The Pergamon Museum where I had hoped to see Schliemann's treasures from Mycenae and Troy proved to have secretive opening hours. Everything was firmly locked and no fool tourist except me was stirring anywhere near. Those haughty and blank doors offering no explanation left me enraged and I walked back to Checkpoint Charlie voting all strength to international handicap signalling and to commercialism if Dahlem Workshops on the one side and locked museums on the other could be taken as the typical outcomes of the different regimes.

If, when back from a Dahlem Workshop and again in the medical or academic sanctuaries of the West, you in your turn are invited by Silke Bernhardt, the efficient and handsome manager of the series, to help her in putting together another conference to follow or to branch off from the one you attended, the prospects will be quite rosy and roughly as follows: (1) A week's discussion of your favourite scientific topic will be organized for you. And organized is the word: not quite do the meetings have the bugle wake-up calls of a German Youth Hostel but there is something of that spirit—the restless German efficiency is palpable. (2) A book will be published in which you, as convener, may lead off with a free-wheeling review of everything you feel underappreciated. Does it go unrefereed? Of this I'm not quite sure, but anyway, once approved, your thoughts are guaranteed by the team in Berlin to meet with printer's ink in less than a year's time. Even if they are refereed you can, of course, with skill, still trash your enemies in the review, but, perhaps best of all, you have the fact that (3) only invitees may attend. Here lies the opportunity to be sure that your enemies don't attend, while as regards half-enemies who may have to (for there to be some show of balance), as the present convener you are still well positioned to advise Silke on the people suited to set up the next conference in your field: half-enemies, of course, can thus be subtly disparaged and the absent ones pushed still further into the intergalactic cold.

Apart from thus shaping the course of progressive science, how much does the effort invested in Dahlem Workshops affect Berliners and Germans generally? How much did we as a group meeting on infectious diseases, for example, help to topple communism—or, locally, help just to topple the Wall? Fortunately such questions needn't be our concern, but 'very little' I suspect is the real answer. To be more serious, however, science that is not connected with quick human benefits always has to be grateful for all pennies that fall from such sponsors as those of the Dahlem series. Whether as sparks from the clash of superpowers, as seemed the case with these meetings, or like wisps of horn-velvet floating from battles and displays of giants, as Medicis and Popes, Catherines and Fredericks and Drug Companies the Greats, or other great government institutions such as the Smithsonian in the USA strive to raise their prestige—and whether we are actual tines on such antlers (like, say, Sir Robert May) or just the discarded velvet that falls from them (like me), our part as scientists at these events is to provide our hurried but hopefully not stupid contributions. Honest writing for our discipline and yet at the same time a writing compatible with some part of the image our patrons wish us to project—a telling that science is useful, for example. Rare indeed are intellectual stars as independent as Michelangelo or Darwin, who can end telling the Pope to come to them, or keep just maintaining their stream of exemplary output while waiting for it to happen. Certainly the Dahlem Workshops, by organizing so much meeting, parting, thinking, writing, and drafting, all of it within such very compressed periods, have well proved their understanding of scientists' motivation and how to harness it. Library shelves bow under the printed proofs of this effort (you should look for a row of always greenish-blue and partly white book backs, if I recall): Dahlem sponsors have indeed attracted the best and have published at the very front, if not ahead, of each scientific field they have hosted. As conferences go, frequently I have heard the Dahlem events referred to by people as the most effective in moving a field onward of any that they have attended.

What did I contribute? The work on reciprocation with Bob Axelrod, as I told in the last chapter, had come as a fairly short interruption in my endeavour in the 1980s to understand sex. By the time of scribbling the paper of this chapter for the Dahlem Workshop in question, I had already split my 'sex' endeavour into two fronts, or rather out of my old problem, which I was beginning to call 'sex itself', I had coaxed a new branch, which, with much originality, I was calling 'sexual selection'. As I started to show in Chapter 2, however, I was approaching sexual selection from a fairly un-Darwinian angle where choosiness about mates was just another step in one's defence of one's descendant lines against their never-ending rain of parasites. Sexual selection for me had become more a matter of patching chinks in battered family armour for your son than of advising him what cuff-links and what lace he should wear at court, which was roughly how Darwin had had it—although I well knew how even useful armour tends to acquire its ornaments too.

It might have been justified for me to think at the time of this writing of some new name for the unusual mate-choice scene I had focused, and some 8 years later I did so. With the help of Bryan Hainsworth, a classics colleague at New College, Oxford, I cobbled out of old Greek an adjective 'sosigonic', which means very roughly 'conferring health on offspring' (see also Chapters 8 and 17). Sosigonic attraction it had been perhaps, as a first illustration, that drew Penelope to Odysseus, a tough and symmetrical red head if ever there was one. Penelope made him father to Telemachus, and perhaps it was a true love and the hope for more like the first that kept her faithful subsequently, in spite all those suitors besetting her while she was thought to be a widow. Meanwhile Odysseus, ever turning dire dangers into love affairs, redeployed the attractions that had captured Penelope to Nausicaa, to Circe (from whom three sons), and to Calypso (from whom two more—twins), and very likely to others still.

In Odysseus wit and resource backed his health and strength and all of them together served to relaunch his ship many times; my adjective 'sosigonic', intended to describe such as Odysseus, however, sunk at its first try. But perhaps I was justly jinxed. Only a little later I realized it was Alfred Russel Wallace who should have had the honour of naming this subbranch within the Darwinian theme;[2] but to detail how that comes about is jumping ahead. As will be seen, very little of the present paper, indeed almost only the final two pages, are about sexual selection, whether we name this the sosigonic variety, the Wallacean, or the something-elsan; by far the greater part of the paper has kept more to my trunk problem, 'sex itself', thus explaining my title. Because the background scenery of sex itself, and the theory I believe to be able best to fit with that scenery, is, as with most views, the more striking and beautiful the wider the panorama it is given to appear in, I mean now to redress a little the imbalance of the coming paper and to hint concerning some of the strange extremes that sosigonic mate choice may go to. If I am to use such images as of water slopping end to end in a bath, what, for mate choice, for example, do these choppy, rather slow waves imply?

As my search for potential evolutionary props of sex went on and as I came to focus more and more on parasites as my best hope, I thought at first about the most powerful agents in this line I knew, and thus especially about the acute epidemic diseases. I was looking for something rather the opposite of those agencies that had occurred to me during what I might now call the 'sosigonic' epiphany I described in Chapter 2, when I stared at the untorn 3-D phase diagram displaying its slow-wandering but still cyclic host–parasite interaction. That idea I could immediately see working best if my agents were numerous, small in their individual effects, and lagged in their onset in time; now for 'sex itself', it seemed, I needed almost exactly the opposite set of disease conditions: aspects of environment that changed not only powerfully but so rapidly as to give a negative heritability of fitness. Big fitness changes, near or actual reversals of advantage, must be occurring for the genotypes from generation to generation. With a bit of a stretch, as shown in the model of Chapter 2, this could be imagined to result from severe and very rapidly onsetting parasite selection pressures. In short, negative heritability of fitness means that the traits aiding your survival and reproduction in this generation are such as to harm your offspring if conveyed to them unchanged.

Sexual recombination, with its facilitation of all kinds of rapid rearrangement, seemed to offer a part palliation to such cruel regimes. While actual negative heritability, such as appeared in the main model of Chapter 2, appeared too much to expect in the real world (an unlikelihood to be confirmed for me in a later study by May and Anderson[3]), still a switching away from an advantageous combination together with its resultant linkage disequilibrium over a period of only a few generations didn't seem unreasonable. I find that in theorizing it is never good in modelling to disregard the extreme cases too soon, absurd as their states may seem. First, they are usually simpler to analyse and, by understanding them, you can fence your reality around a little more thoroughly and see more clearly what its limits are. Second, it is possible that the reality actually is more extreme than you imagined. It seemed to me good, therefore, at least to play with the idea that negative heritability might exist for some aspects of fitness: it was in this spirit that I began the model of Chapter 2. Actual constant negative heritability was so unsupported by evidence, however, and seemed bound to generate such crazy paradoxes in sexual behaviour that I am not surprised that others had chosen to dismiss it as a way sex could be stabilized:[4] certainly no example of a process provenly generating it in nature is known. But what if there was some way to 'stretch out' a negative heritability—substitute reversals of selection on interlocus combinations that were infrequent but never so infrequent that they would allow any clonal takeovers?

Really there was no way that asex, in the at best slightly imperfect state in which it normally first appears, could increase to the point of extinguishing functional sex in a reasonable population in less than a few tens of generations; so finding heritabilities that were always (or even just usually) negative didn't seem a necessity for the parasite theory to work. It was legitimate to imagine a kind of group selection working across species to reduce the chance that 'perfect' parthenogenesis would easily switch on, as was, shortly after this paper, to be argued by Nunney.[5] The more I thought about the realities of mate choice, including, of course, by use of plenty of introspection—what I was inclined to choose myself—the clearer it seemed to me that if I ever did discover nearly constant negative heritabilities they would have to be rare exceptions or/and would involve fitness contributions only from particular large-effect genes. The idea of preferences for manifestly sickly mates being a common rule of nature was just too fantastic and unsupported. I mention the idea, however, both in this paper and in the next because I have always rather enjoyed the idea that some species (most likely those with the tiny, expendable, one-shot males, which are sometimes routinely eaten by their mates during or after the unique act that they are allowed) may be pursuing an adaptive course that brings in this real ultimate of masochism—the giving to their loves that liberty with their bodies that alone can be utterly and incontrovertibly convincing as a demonstration of their lack of worth. Imagine how, if you have grown up naturally with the build of a Mike Tyson let us say, you might try to fool people you are a wimp; perhaps such a deception is even harder to imagine than its opposite that we sexual selectionists are so much more prone to discuss—being a wimp and trying to seem big and strong.[6] But imagine how the lovesick entreaty of a true undeniable wimp under conditions where negative heritability rules could be cast and how effective it would be: 'See how small, how sick I am—what a miracle I am here! But now that I am, I entreat you, give me my sole ambition, one copulation! If you doubt I am as sick as I pretend, just eat me. Gladly I sacrifice this gloriously miserable body just for you and my offspring.' Would it work? Try as I will, even focusing the most gory and kinky of the mantids, spiders, coccids, and midges that I know of—those groups where routinely the female does indeed eat her mate—I can't take it all quite seriously nor quite make it make sense. For a start this male should not be looking for the robust female most capable of eating him but rather for one who is, as nearly as possible, as substandard as himself.

More seriously, then, what I needed for 'sex itself' in the real world was for adaptations to be obsolescent on a timescale not longer than would be sufficient for a clone of asexual self-sufficients to complete its takeover in a case where fitnesses were held unchanging. This fairly modest extension to timespan would be enough to make chances to find something look much better. Cycles of varying intrinsic spans could well be contributing: fast ones could prop up 'sex itself' and select for 'scramble' mating—that is, they could propel a largely random selection of mates— while slow ones could be slanting Darwin's aesthetic sex choice towards making those Wallacean 'utilitarian in offspring' demonstrations that were to become my next interest. Whichever way the emphasis lay, this mate choice, to repeat, could be sosigonic: its choice would chase 'health for offspring' just as the word says—chase genes and combinations good not for ever but at least good for a few to many immediate generations. This is going to be mainly the topic for Chapter 6 and I will leave it at this point; but these ideas were pressing before my writing of the present one.

In effect, the Germanic efficiency in the Dahlem conference organization was causing a kind of pre-Copernican epicycle, yet another small 'retrogression' in these republications from my strict sequence of thinking them out. The present paper is in its true publication sequence relative to the next but was actually largely thought out and written later than the next one. Quickly composed along the lines requested by the workshop symposium where I was to present it and, along those lines, heavily weighting its enthusiasm against caution, I tried to show where I thought we had got to and where my parasite theme might be going. Normally I am an extremely slow writer and it almost never happens that I can write one paper so fast that it overtakes another. My main work at the time was the parasite/sexual selection topic and the writing up of that jointly with Marlene Zuk had begun and I did overtake it. After submission, both manuscripts went ahead fast with their respective journals, but *Science*, to which we sent the sexual selection one, if I remember, demanded two revisions before Mrs. Butz, the editor, thought our piece was acceptable, whereas the present paper, which had been requested by Robert May as a discussion paper for the conference, possibly suffered the opposite way in being criticized too little: it rushed headlong from being a hasty manuscript to being a published chapter in the conference book.

In the present republishings generally I try to alter as little as is consistent with the prose having clear sense. Thus I alter only my failures of previous proofreading and a few particularly unclear statements. Sometimes my copy editor also labours kindly to make me sound less weird and less 'un-PC' by altering a few words. Although there are thus

considerable numbers of minor textual changes in all the papers to make
for easy reading, I never alter the sense even if I now know the point I was
making to have been wrong. This applies to the ideas of the paper of this
chapter; however, the English has been more revised in this one than in
any other I republish either in this volume or the last, a consequence
necessitated no doubt by the original hasty writing. Bob May's instruction
had been a readable, speculative, and potentially controversial paper to
start discussion, with formalism to be avoided as far as possible. It sounded
easy but the time to do it in was extremely short. In the result I am proud of
the range of ideas that I still believe important that the paper covers, and,
on the whole, as I read it over now, I see little that I even would want to
change apart from the details of presentation.

On this positive side of what I include relative to previous chapters
(and leaving aside anything new the last two pages may have about sexual
selection) are the following: (1) the extensive evidence of *high heritability*
for both virulence and resistance; (2) a re-stressing of the *mobility–virulence*
connection; (3) being perhaps the first to point out the potential for
population partial subdivision—*metapopulation structure*—to protect
dynamic resistance polymorphisms; (4) a discussion of the kinds and the
unavoidable *costs* of being resistant; (5) hints from literature of the *wide
chromosomal scatter* of genes affecting resistance; and (6) perhaps my first
dim perception that *change per se* in so-called 'self-identity' molecules,
which previously I had tried to interpret as passwords, couldn't sufficiently
justify their existence—they must instead be functional tools. The last
point was later to grow into the amazement I still feel at the immense age
of many co-existing ancient variants of resistance genes. I began to see that
the variants are not merely variant of brand name, rather they resemble all
those occasionally used tools we tend to keep in our sheds and basements:
dusty for now we still resent being told by our spouse they're scrap and
rubbish; of course they are waiting to be used again! Right up to the present
(and this I could justly call a further point—(7)—in the paper) I strongly
continue the emphasis I gave to the potential of *molecular mimicry* by
pathogens to cause protein polymorphisms in hosts, although I remain
sadly uncertain still about how much of protein polymorphism is to be
attributed to this factor, as also on how much of metrical variation can be
attributed to the parasite-resistance variability more generally.

On the negative side of what I included I would now de-emphasize, as just suggested, my rather free use of the word 'password' and would lessen the emphasis I gave to what I was then calling 'concave fitness sets' (they should have been convex, anyway, I think, for mathematicians, who seem always to look piously upward at their diagrams where the rest of us tend to look down). Besides perpetrating much clumsy English, I also confess to several rather poorly thought out images and metaphors. A much better one for the population genetical dynamism I aimed to describe than the spikes in the fakir's bath was to occur to me later: this is the floating-ball image, which you will find in Chapter 16.

Although I use in the paper a metaphor of 'running in circles', I never referred to a 'Red Queen' in this paper even though the Lewis Carroll analogy had crossed my mind. At the time I regarded Lewis Carroll's RQ as a metaphor already committed—this lady was busy at Leigh Van Valen's tea party and therefore couldn't grace mine. Over there she was celebrating a rather different notion of fissioning and extinction of species that also lacks any real evolutionary 'progress'. Inspite of Van Valen's priority and his different use, the term, however, soon seemed to be kidnapped and brought to have its present and largely intrapopulation usage. It is possible, of course, that the chaotic circling of gene frequencies that I describe in this paper, as one example, may be connected to the 'getting nowhere' and the random mortality occurring in the twigs of Van Valen's tree.[7] A notion of *appelations contrôlées* for our concepts has seemingly not developed yet in biology, so I am very happy to go along the preference of others who would call my present theme a 'parasite Red Queen' paper (or, as I now often abbreviate it, a PRQ). I believe that for its time this paper was the strongest and most detailed account the idea had yet been given.

Quite possibly parasites will eventually prove to have more to do with speciation and whole-species extinction also than is currently credited. That species may split discretely more because of evolving incompatibilities in their anti-parasite techniques than for any other reason remains very speculative but my grounds for suspecting it, if anything, have strengthened since I wrote. For example, recently we have learned a very impressive lesson about one striking disease agent that chimpanzees tolerate and we don't. We see a very clear new reason not even to try to mate with a chimp even if a simian immunodeficiency virus (SIV_{cpz}) appears to be rare in them. But what about genuinely crossing—siring children from a chimp—if that could be done without a personal risk and without overwhelming public horror? If a fertile hybrid was possible from such a union, would this be an effective start, for example, towards transferring resistance to the HIV/SIV type of virus into the human population just as we once transferred to virus? If one could work with such an aim for as many generations as would be possible were it a case of developing a new resistance for mice or rabbits, probably it could be done, but there are already much easier routes. It is a mainstay of the PRQ theory that genetic resistance variation exists to almost all diseases, even those that, for all we know, have not been encountered before. Challenges like them, I suggest, probably have been encountered in the past. Thus one course to suggest in the present case is to marry a longstanding and still-healthy Nairobi prostitute;[8] another might be to tweak a hair or two from heads of attractive Caucasoids you sit behind on buses, apologize, of course, for the 'accident' but keep the hairs; multiply up DNA from the cells of the follicles obtained and in it search for that newly discovered gene that seems to give, surprisingly commonly in the Caucasoid race, resistance to HIV-1. The final step in the case of a positive result at the lab is then simply to take the same bus again and propose marriage.

Unrealistic as may be this social scenario, in essence something resembling this course may be the most practical AIDS sosigony available to Europeans; how much moreover appeals the further notion of encouraging a union between a handsome blond male Finnish resistor with the equally handsome and 'proven' resistor Nairobi prostitute: what a resounding defeat for HIV-1 would arise there! Back with the previous (and less wholesome) kind of experiment, my suspicion there, on the contrary, is that any line drawn out from a chimp–human hybrid, supposing it could propagate, would soon find itself generating such mismatched defence systems in other fields of disease, that, unless the F_2s and onwards could be cosseted through many generations of backcrossing with all the skill of a modern animal-breeding institute, it would end up a failure; in spite of any new AIDS resistance successfully gained, the other diseases exploiting the mismatched system components would finish the attempt.

These last ideas base themselves largely on an experiment already conducted for us, in the 'wild' of Europe, by house mice. In a striking study of parasitism in and near to the natural hybrid zone of the two European subspecies of house mouse in Germany (in an area near Munich) the geneality of 'hybrids' were found to be very unresistant to a nematoid worm that the 'pure' mice on either side of the hybrid line resisted reasonably well.[9] Publication of the paper showing this, however, was still 4 years in the future when I was writing for the Dahlem Workshop. Still more recent reviews of parasitism of hybrids have further shown that more often than not hybrids are less resistant than their unhybrid ancestors,[10] although, as we might expect from the phenomenon of 'hybrid vigour' as well as from the hybrid origins known for many parthenogenetic stocks, there are exceptions.[11] Perhaps high resistance is even largely causing the 'vigour' in the first generation while from this there follows on a rather hopeless dysgenesis in later intercrosses due to the mixed and incomplete systems that then come to be composed by recombination when the original cross was too wide.[12] Along with the Munich hybrid-zone results have also come recent contrary tests and findings[13] as well as a few more that are positive.[10,14] Perhaps we see vaguely a generalization here: first-generation hybrids (F_1s) are those that have the vigour and sturdy health of mules (for example); it is the later interbred (and therefore recombined) descendants from these that present the generally worse mixes,[12,15] and these perhaps deserve the more negative title of 'mongrels'. Think of wolfhound long hairs sprouting on dachshund legs or terrier nervousness and courage uniting, with disastrous results for postmen, with the power and bone-cracking bite of a German shepherd. All this, of course, from my point of view, needs to be translated into a talk of nonsensical, inconsistent splicing of component techniques intended for parasite resistance. But it must be remembered, still following the dog-cross analogy, that the good combinations that potentially make possible the really interesting new developments have to lurk there amongst the various new characters of the mongrels too, if only rarely; and that very often in order to become better you first have to become worse. I hope to come back to these matters of combinations as they affect the human setting in various ways in later chapters, and if a third volume ever follows this one, it, too, will come back to the importance of the theme raised here of valley crossing—deep and shallow—in evolution.

If F_1 status with its chromosomally balanced constitution can be fixed through parthenogenesis, then at least for a while all may go well with parthenogenetic hybrids. Probably this is exactly what is happening in the now legion of known cases of proven parthenogenetic hybrids between different species. While not denying such situations, my paper re-stresses, however, my general objection to the existence of common special advantages for heterozygosity as judged on a locus-by-locus basis. As will also be re-explained in later papers, in so far as such advantage exists it must clearly favour switches to parthenogenesis as I have just mentioned. These are most unfavourable to sex and are in fact its great danger; they provide excellent examples of 'evolutionary temptations' arising at a group level—in this case the issue is going for the very best fitness right now but in doing so losing flexibility for the future. But that heterozygote advantage at locus-by-locus level may not be as common as is often assumed is shown by the quite frequent failures to find evidence for it in careful studies of degrees of heterozygosity in wild populations plus some theory—not yet very adequate—about why not to expect it and plus many ambiguous cases.[16] Rarity of single-locus heterosis is also suggested by the fast improvement that the level of a disease resistance seems always to show when selected (see Chapter 6). There are plausible and perhaps even necessary alternatives to heterozygosity as mediating, *per se*, a prime advantage of outcrossing. Also, the advantage to heterozygosis that seems to arise in surveys of the average levels, and from correlations of these with fitness, in both animals and plants, may be an illusion.

Re-stressing my doubts about a wide importance of heterozygote advantage probably seems quite boring to most readers by this point, as also will seem the same stresses and others like it that I make in my paper. The doubts must be boring both to the genetically sophisticated reader who knows from independent sources something of the unresolving wrangle about heterosis (and this boredom in my case accentuated because I'm not attempting to bring forward anything new) and to the lay reader who will merely sense the matters he or she never heard of or wanted to hear, let alone caring that it's all controversial. The reason it is still worthwhile to mention the doubts may, however, be better understood if I say a little more about the particular meeting at Dahlem and the subenvironment I had been assigned to within it. The controversy about heterosis, in fact, is a very old one. As Darwin could well see but not solve, it is crucial to understanding outbreeding; thus it is crucial also to understanding 'sex itself'.

My particular group in the conference fell under the chairmanship of A. C. (Tony) Allison, famous for his work on the genetics and significance of the 'sickle-cell trait' in relation to malaria. Towards me at first, as our chairman, he seemed to convey a slight coolness, and if he hadn't much liked my discussion paper when he read it (perhaps on that first night of the meeting when it would have come to him in the general pre-circulation of the discussion papers that was occurring in the first few hours of the conference—super-efficient as always), I had reason not to be greatly surprised.

Sickle-cell anaemia, that hereditary disease common to West Africa, receives its name because of the partly collapsed and consequently sickle-shaped condition of the diseased red blood cells assumed. Allison's finding of how the sickling trait, which confers immunity to malaria (was it that the *Plasmodium* parasite simply can't grow comfortably within the misshapen blood cells or something more chemical?—I forget), derived from being heterozygous for a gene, which when present in its double (homozygous) dose makes the red blood so ineffective as a haemoglobin capsule that homozygous bearers all die in infancy, had become a classic of genetics. The story indeed had been a great new wonder of science progress that had been taught to me when I was an undergraduate at Cambridge. It was thought then that perhaps heterozygous advantages as now illustrated in this new sickling story would prove the biggest prop of polymorphism and variation generally throughout all diploid nature. Along with the intensive work on the structure of haemoglobin as a whole, which in Fred Sanger's lab in Cambridge would soon become the first protein fully sequenced for its amino acid building blocks (and shortly also the first even whose shape could be understood), the sickling story continued to unroll its details for a long time. The very newest at my time in Cambridge had been the discovery (though not by Allison) that only a single base-pair change in the DNA, resulting in just one consequent amino acid change, was enough to give abnormal shape to the molecule, this leading to the abnormal shape of the cell and finally to cell-breakage and the anaemia tendency.

It had all seemed early confirmation of the predictive potential of the newly discovered and wonderful 'code' of Watson and Crick. I could see how wonderful all this was, and was impressed; however, what was impressing me still more in this particular instance had been Allison's confirmation of some of the simple maths on the population side. He had shown that a relationship of gene frequency and fitnesses in West Africa had been exactly as predicted by J. B. S. Haldane's formula for cases of heterozygote advantage. Allison himself is a careful mason of science, and this together with his strong inclination to the medical and molecular side made him quite a far opposite type from me, in fact I think he was exactly the type most guaranteed to be scandalized by the style I had chosen in the rampage of speculative and jokey images I used in my discussion paper. But, in addition, I suspect Allison may have disliked the way my paper, via such images, generally poured cold water on heterozygote advantage as an important and common theme. It is almost essential in all my accounts of sex to pour such cold water. If heterozygote advantage were abundant my 'Red Queen' thinking on sex wouldn't work. If malaria was all that Africans suffered and the sickling trait the best that evolution could ever produce for them, Africans would do best to be parthenogenetic and all of them heterozygous sicklers: the production of the homozygous sicklers that sex dooms them to produce in every generation is obviously an enormous waste. The same theme can be found repeated very widely: diploid heterosis is bad news for sex. Returning to Allison, however, even if he was rather untalkative with me, he was certainly a good and fair chairman and deserves my thanks (as go also to Bruce Levin who mainly did the writing) for bringing together the very fair and useful report of our discussions in the subgroup, which was also published in the workshop volume.[17]

The advantage for me of the meeting was the chance to try to stimulate more practical scientists than me who would see how to make tests. The setting out of our ideas about these was supposed to be a principle objective in our report. We all agreed the ideas I had raised to be testable. But I am not aware that the suggestions our group made have led to much action. Perhaps the visions I had suggested were just too wild to catch notice; or, perhaps, for people deep in various practical fields, they were just too obviously improbable. What careful tests have been done have used so far mainly macroparasites—for example, the work of Curt Lively. I will show more of the general distaste for my reasoning that came from most quarters as we proceed into Chapter 16, wherein comes my main next statement on the 'sex itself' side; but I will also show there the many indirect indications now existing that many of the ideas of the present paper were right.

References and notes

1. One of the three stanzas written by Ross after, on 20 August 1897, the day he discovered ex-human *Plasmodium* stages establishing in stomach epithelial cells of *Anopheles* mosquitoes in India (J. O. Dobson, *Ronald Ross, Dragon Slayer* (Student Christian Movement Press, London, 1934)).

2. H. Cronin, *The Ant and the Peacock* (Cambridge University Press, Cambridge, 1992).

3. R. M. May and R. M. Anderson, Epidemiology and genetics in the correlation of parasites and hosts, *Proceedings of the Royal Society of London B* **219**, 281–313 (1983).

4. J. Maynard Smith, *The Evolution of Sex* (Cambridge University Press, Cambridge, 1978); May and Anderson (1983), note 3.

5. L. Nunney, The maintenance of sex by group selection, *Evolution* **43**, 245–7 (1989).

6. A good example of trying to seem big by sound are the South and East African so-called 'bladder' or 'bull' grasshoppers, in which the body seems inflated as in a child's balloon made into grasshopper shape (genus *Bullacris*: M. J. VanStaaden and H. Romer, Sexual signalling in bladder grasshoppers: tactical design for maximizing calling range, *Journal of Experimental Biology* **200**, 2597–608 (1997)). Almost for sure, however, full inflation and deep stridulations are not being achieved by the wimps in this species: the hint from the existence of a juvenile-like and wingless but sexual morph is that the destined wimps are opting early in development for a quite different strategy of mating—see the illustrations in S. H. Skaife, *African Insect Life*, rev. edn (Hamlyn/Country Life Books, London, 1979).

7. L. M. Van Valen, A new evolutionary law, *Evolutionary Theory* **1**, 1–30 (1973).

8. K. R. Fowke, N. J. D. Nagelkerke, J. Kimani, J. N. Simonsen, A. O. Anzala, J. J. Bwayo *et al.*, Resistance to HIV-1 infection among persistently seronegative prostitutes in Nairobi, Kenya, *Lancet* **348**, 1347–51 (1996).

9. R. D. Sage, J. B. Whitney III, and A. C. Wilson, Genetic analysis of a hybrid zone between *domesticus* and *musculus* mice (*Mus musculus* complex): hemoglobin polymorphism, *Current Topics in Microbiology and Immunology* **127**, 75–85 (1986).

10. S. Y. Strauss, Levels of herbivory and parasitism in host hybrid zones, *Trends in Ecology and Evolution* **9**, 209–14 (1994); L. Ericson, J. J. Burdon, and A. Wennstrom, Inter-specific host hybrids and phalacrid beetles implicated in the local survival of smut pathogens, *Oikos* **68**, 393–400 (1993).

11. S. Hanhimaki, J. Senn, and E. Haukioja, Performance of insect herbivores on hybridizing trees—the case of the sub-arctic beeches, *Journal of Animal Ecology* **63**, 163–75 (1994).

12. C. Moulia, N. Lebrun, C. Loubes, R. Marin, and F. Renaud, Hybrid vigor against parasites in interspecific crosses between 2 mice species, *Heredity* **74**, 48–52 (1995).

13. D. Shutler, C. D. Ankney, and D. G. Dennis, Could the blood parasite *Leucocytozoon* deter Mallard range expansion?, *Journal of Wildlife Management* **60**, 569–80 (1996).

14. J. R. Mason and L. Clark, Sarcosporidiosis observed more frequently in hybrids of Mallards and American Black Duck, *Wilson Bulletin* **102**, 160–2 (1990).

15. T. G. Whitham, P. A. Morrow, and B. M. Potts, Plant hybrid zones as centers of biodiversity—the herbivore community of 2 endemic Tasmanian eucalypts, *Oecologia* **97**, 481–90 (1994); K. D. Floate and T. G. Whitham, Insects as traits in plant systematics—their use in discriminating between hybrid cottonwoods, *Canadian Journal of Botany* **73**, 1–13 (1995).

16. R. C. Lewontin, L. R. Ginzburg, and S. D. Tuljapurkar, Heterosis as an explanation of large amounts of genic polymorphism, *Genetics* **88**, 149–69 (1978); B. J. McAndrew, R. D. Ward, and J. A. Beardmore, Lack of relationship between morphological variance and enzyme heterozygosity in the plaice, *Pleuronectes platessa*, *Heredity* **48**, 117–25 (1982); W. J. Libby and S. H. Strauss, Allozyme heterosis in Radiata Pine is poorly explained by overdominance, *American Naturalist* **130**, 879–90 (1987); J. M. Pemberton, S. D. Albon, F. E. Guinness, T. H. Clutton-Brock, and R. J. Berry, Genetic variation and juvenile survival in red deer, *Evolution* **42**, 921–34 (1988); J. B. Mitton and W. M. Lewis, Relationships between genetic-variability and life-history features of bony fishes, *Evolution* **43**, 1712–23 (1989); D. B. Goldstein, Heterozygote advantage and the evolution of a dominant diploid phase, *Genetics* **132**, 1195–8 (1992); E. Zouros, Associative overdominance— evaluating the effects of inbreeding and linkage disequilibrium, *Genetica* **89**, 35–46 (1993); M. H. Schierup, The effect of enzyme heterozygosity on growth in a strictly outcrossing species, the self-incompatible *Arabis petraea* (Brassicaceae), *Hereditas* **128**, 21–31 (1998).

17. B. R. Levin, A. C. Allison, H. J. Bremermann, L. L. Cavalli-Sforza, B. C. Clarke, R. Frentzel-Beyme, *et al.*, Evolution of parasites and hosts: group report, in R. M. Anderson and R. M. May (ed.), *Population Biology of Infectious Diseases. Dahlem Konferenzen 1982*, pp. 213–43 (Springer-Verlag, Berlin, 1982).

PATHOGENS AS CAUSES OF GENETIC DIVERSITY IN THEIR HOST POPULATIONS[†]

W. D. HAMILTON

Sex is likely to be the adaptation that enables large multicellular long-lived organisms to resist exploitation by specialized smaller shorter-lived organisms—that is, by parasites/pathogens. Antagonistic coadaptation between such species tends to entrain limit cycles or else repeating and largely non-progressive situations of countertransience of new defence and attack alleles. Models on these lines can account for (1) correlation of stable sexual reproduction with size and longevity and with biotic complexity of habitat, (2) abundance of protein polymorphism, (3) diversity of adaptive linkage values, (4) common linkage disequilibria in multilocus genotypes, and (5) 'good genes' mate choice and the excesses of sexual selection. Through parasites, frequency-dependent selection may account for much more variation than has been credited while immediate heterozygote advantage may account for much less. Through frequency-dependent selection, polymorphism based even on generally concave fitness profiles may be common.

A puzzle likely to occur to anyone hearing about evolution for the first time, and later very often forgotten, is that the rate of the whole process by natural selection must depend on the generation time. How, the listener then wonders, does anything manage to be as large and slow-breeding as an elephant? On the elephant's timescale of change, why do not bacteria of skin or gut, turning over generations a hundred thousand times faster, evolve almost instantly an ability to eat the vast body up? Worse still, among plants there are the aspen clones and redwood trees ...

The invention of multicellularity initiated a possibility of large body size and was a crucial step in evolution. The multicellular mode gives obvious advantages in competition with smaller forms (e.g. of plants for light in crowded habitats), and also in homeostasis (permitting colonization of new habitats), but at the same time, by rules of chemical logistics, this mode of

[†]In R. M. Anderson and R. M. May (ed.), *Population Biology of Infectious Diseases*, Dahlem Konferencen, 1982, pp. 269–96 (Springer-Verlag, Berlin, 1982).

growth entails an inevitable reduction of the intrinsic potential rate of expansion. Compounding the problem of slow growth, multicellularity must also have created many new food materials and many new avenues of exploitation for the organisms that remained micro. As cell adhesion becomes routine in body construction, it must become easier for organisms of similar size to attach themselves by similar means: in other words, the large host has a problem in defending many kinds of cells and in distinguishing at each cell surface a friendly co-clonal building block from an insidious foe. Unicells (and later other small organisms as well) that specialize in exploiting large hosts are, obviously, the precursors of pathogens and parasites.

A partial answer by multicells to the problem of evolution rate disadvantage can be to promote growth of appropriate specialist defender cell clones from among a varied array of such, using a model process of mutation and selection. This, of course, is well illustrated in the facultative response of the vertebrate immune system, especially on the antibody side: host clone is matched with pathogen clone and growth rates can be more comparable. Yet a gap remains and it seems that in general this answer is not enough. Mammals, for example, which have such a system most advanced, show no relaxation in their coreliance on something else, namely sex.

SEX

I believe that sex was the more immediate invention that enabled metazoans and large plants to forge ahead against their handicaps of inertia and invasibility (see Chapter 2 and refs 1–4). Sex also creates true species in an otherwise straggling mess of clones: if the idea about parasites is right, species may be seen in essence as guilds of genotypes committed to free fair exchange of biochemical technology for parasite exclusion. Also, if it is right, the effect of parasites on genetic diversity may be claimed to be vast. Parasitism has caused not only arrays of varying traits concerned directly with disease resistance, including perhaps much of the seemingly needless diversity of isoproteins,[5,6] but also the evolution of meiosis itself, setting up the basis of all Mendelian variation. Before giving further rationale to this derivation of sex, attention should be paid briefly to another, more obvious way in which large multicellular bodies might try to escape fast-adapting parasites.

DISPERSAL

Among potential benefits of multicellularity are greatly improved powers of movement. Large animals can swim, run, and fly much farther and faster than protozoans; even large plants, besides shedding pollen and seeds on the wind from greater heights, can join the muscle-power league through inducements attracting animal carriers. So one way to escape from locally adapting populations of unicells may be to outdistance them, move elsewhere. Such escape is likely to be an important motivation of animal and plant dispersal,[7] and, significantly from the present point of view, dispersal is often closely preceded or followed by sexual reproduction. Both activities tend to occur when local conditions worsen or when signs show they are about to do so.

SPATIAL VERSUS TEMPORAL HETEROGENEITY

Movement, however, is a very imperfect solution. Parasites become expert at clinging on. They can also often find ways of being even more mobile than their victims by riding on other animals, notably flying birds and insects. Such vectors the parasites harm relatively little.[8] Mobility of hosts carrying disease or mobility of vectors alone tend to homogenize disease problems over the whole species range, and, in the course of coevolution, it makes for strengthening disease virulence rather than evolution towards mutualisms (see Chapter 4 and refs 9 and 10). Even if there are no vectors and the host does not transport parasites with it, the host's movement, by continually providing new unrelated hosts to attack in any given place, may select for greater virulence in the pathogen (see Chapter 4). All this leads us to expect that most large multicellular species will be subject to at least a few pathogens that wander fast over the whole range and are virulent enough to apply significant, simultaneous, strain-specific selection to whole demes at a time. Local escape is not enough.

The view here of selection pressures for sex, spatially fairly uniform while changing over generations, is similar to that first stated by Jaenike[11] and is also similar to a view more recently emphasized by Hutson and Law[12] with respect to the problem of maintenance of recombination; other authors who have suggested parasites as a major factor for sex[2,3,6] instead emphasized the way new genes and combinations introduced hinder the transmission of parasites locally, especially from parent to offspring. Overall, however, it is the similarity of these various independently derived hypotheses on sex and recombination, all with disease in a primary role, that is most striking.

Population-wide selection would imply that the effective mean fitnesses determining demographic processes are more correctly estimated by geometric means of generational arithmetic means of the varying genotypic achievements than by overall arithmetic means (see Chapter 1). If fitnesses are very different between genotypes within each generation (selection is always strong), and the patterns of difference change fairly rapidly from one generation to another, then selection for the maintenance of sex, despite its halved efficiency compared to the alternative of parthenogenesis,[13,14] can easily be explained (see Chapter 2).

Sex is generally stable. It is especially so in large organisms, including those with low fecundity (e.g. elephants, humans, and palms). The low-fecundity examples are notable because other theories of sexuality predict stability in large high-fecundity organisms (such as trees) but not in low fecundity ones.[13,14] Such other theories, however, have tended to emphasize spatial rather than temporal heterogeneity of environment as the principal factor. Regarding fecundity, and also in their requirement of sibling competition, the other models are restricted as badly or worse, it seems to me, as my own model is regarding its requirement of strongly varying fitness profiles. Therefore I feel free at present to prefer my own. This, of course, does not exclude a role for patchiness of environment. Indeed, even without the sibling competition and 'lottery ticket' points made by Williams[14] and by Maynard Smith,[13] realistic as these are, on the scale of demes each patch could be cycling largely independently of others and the temporal model could still apply. However, migration between patches will then be a factor tending to dampen and eliminate the cycles. (For the latest theory on how migration tendency itself may be tuned under selection by risks of dispersal, deme size, and other factors, see Comins.[15])

Irrespective of the question of sex itself, the point may be raised that the 'multiple niche theory' has shown spatial heterogeneity to be able to preserve extra genetic variation even if some stage of the life cycle is totally panmictic.[16–18] The 'niches' of this line of reasoning could, of course, among the other possibilities, be endemic areas of disease—for example, marshes demanding adaptation to mosquitoes and malaria, while mountains are demanding adaptation to ticks and mountain fever. But this type of model applies only if adaptation or lack of it makes little difference to the numerical output from the niches. Social behaviour (e.g. territorial spacing) might bring this about in some kinds of species but, on the whole, I am doubtful of the breadth of applicability.[17,19] Another point against relevance of Levene-type models[18] is that species subject to such niche-based polymorphism should continually improve the ability to find and settle in habitat that matches genotype. This should lead to speciation and monomorphism, after which parthenogenesis could supervene.

SHAPE OF FITNESS AND ITS TEMPORAL CHANGE

My own preferred 'model' for the maintenance of sex through parasitism involves fitness profiles whose shape over the genotypes and over time may be said to resemble the shape and change of water that is kept slopping about in a bath (see Chapters 1 and 2). To put genotypes more explicitly into this picture, the bath can be imagined as designed by an Indian ascetic who has arranged long equi-spaced spikes to project from the bottom. These are three, four, or nine in number, depending on the model (and, incidentally, on the pain in analysis of the model) that is desired. The length of spike submerged by the water is current fitness. For the model to work best for the defence of sex and the reduction of parasite load, the following must apply:

1. The waves in the bath must be high; that is, fitness variance within each generation must be substantial—selection intense. Selection by parasites certainly has this potential (e.g. ref. 20); but the requirement becomes progressively moderated if the waves are slow on account of lags in feedback.

2. As the slop of the water goes up one side or another, concavity of the wave surface is helpful.

3. If waves sometimes hump high in the centre, troughs should also fall low in the centre; that is, high heterozygote variation helps but geometric heterozygote advantage over time favours parthenogenesis.

4. The tendency of waves to reverse is needed to keep the water mobile and within bounds. However, reversal in itself is bad for the mean fitness of a sexual species because when it occurs the most common homozygotes get low fitnesses. Parallel clones are likely to be yet more common and thus may fare still worse.

5. Finally, as regards the involvement of two loci or more, cornerwise slop in the bath (high fluctuations in linkage disequilibrium) should be moderated by the resulting splurge into low corners (maximal recombination), unless the timescale is such that the rise of a wave occupies several generations.

The reader will easily guess that the crudeness of imagery used in describing this model (including some ambiguity as to whether bathwater level is fitness or genotype frequency—though these, of course, tend to track one another under the required frequency dependence) reflects the crude and preliminary state of the ideas. Formal analysis and simulation of these problems have only recently begun.

Frequency dependence is obviously a key concept in the above 'model'. As it will be extensively referred to in what follows, the abbreviation 'FD' will be used in noun, adjectival, and adverbial senses, and 'FDS' will be used for frequency-dependent selection.

As stated in (2), common concavity of fitness profile (the water surface) is important in the model. Concavity implies profiles where the heterozygote ('het') is less fit than the arithmetic mean of fitness of the two homozygotes ('homs'); that is, $w(gG) < \{w(gg) + w(GG)\}/2$. (Abbreviations 'het', 'hom', and w for fitness will henceforth be used consistently.) Fitness concavity precludes het advantage within generations. Also, rather strongly, it tends to preclude het advantage in terms of arithmetic means (AMs) over generations. However, and perhaps surprisingly, lack of such advantage does not preclude polymorphism. With fitness profiles varying stochastically, it is sufficient to maintain polymorphism if geometric means (GMs) of genotype fitness over generations show het advantage.[21] (With Levene's spatial heterogeneity model, the requirement for polymorphism is less restrictive still: only harmonic means need show het advantage.[18] But in the context of sexuality there are other drawbacks for the Levene model, as already mentioned.) Even at the border of the requirement on GMs, polymorphism can still persist: in the two examples I have worked on gene frequency is central for a one-locus model but varies widely for a two-locus one (see Chapter 1).

Clearly in none of the models reviewed in this or my previous models of sexuality can highly fit heterozygotes be seen as an answer to the problem of sex (Chapter 1 and ref. 22): if such heterozygotes exist and can mutate to parthenogenesis, then parthenogenesis should supervene. Yet despite the rarity of clearly established cases of heterozygote advantage (refs 23–25, but see 26), population geneticists seem bemused by the possibility of abundant het advantage of the sickle-cell anaemia type and have given possible concavity of fitness and its consequences, especially its effects under FDS, very little attention. Likewise, temporal fluctuation of fitness values has been neglected with the exception of some attention paid to purely stochastic environments.[21,27] There are several reasons why concave fitness should be considered more widely.

One is that, in terms of viability at least, it is easier to imagine gene dosage multiplying rather than adding fitness effects. Multiplicative interaction here simply means that if, for example, one dose of a disadvantageous allele halves one's fitness compared to the normal, then being homozygous for the allele halves fitness again. The fitnesses of hom and het and hom then are in geometric progression, and so are concave.

Another point is that unless gene frequencies are usually outside of the mid-range (which for flat or central humped distributions could be taken to mean roughly as much time outside the range of 0.3 to 0.7 as inside), then in terms of numbers that have been present over a long series of generations, the heterozygote is the most common genotype. Thus all hostile biotic agents, including parasites, have had more opportunities to encounter heterozygotes and so should have adapted farthest towards exploiting these, so reducing their fitness relative to homozygotes. Whether variable or not over niches and sequential environments, fitness profiles would then be moulded more concave.

Other more direct reasons to expect concavity can also be suggested. One invokes antigen mimicry by pathogens. Only a phenotype that itself lacks an antigen altogether can respond to that antigen in a pathogen. Such a situation affecting a fitness profile is regarded as quite common.[28] Another kind of recessive benefit from a null character would come from sacrificing a molecule that somehow provides a cue or a 'door latch' for a pathogen to use. Recessive resistance to *vivax* and *knowlesi* malaria by Duffy-negative Africans (for details, see ref. 29) seems to be of this type.

There are, of course, also various arguments why hets might tend to be more fit than the mean of the homs, and some are as direct and immunological as the last one. Possession of a new counterattack molecule, for example, is likely to have beneficial effects that are more nearly dominant than recessive. Nevertheless, as mentioned, fully established cases of fitness overdominance, such as that of the human sickling trait in Africa, remain surprisingly rare, and this rarity is certainly a puzzle if direct heterozygote advantage is to account for a large part of genetic variability, as used to be assumed prior to the controversy about genetic loads and neutral alleles. Abundant genetic diversity in haploid bacterial populations further suggests that heterosis must be a factor that is far from universal[30] (but the involvement of sex or/and parasitism in bacterial diversity is less clear than it is for multicell). I feel that the expectation that the hets in known polymorphisms are likely to be more fit needs second thought, unless genotype frequencies are demonstrably very stable. Where fast-adapting parasites are involved, the argument about the frequency of hets in the evolution of overdominance[31,32] can easily be turned to a conclusion opposite to the usual one.

Current views on the multiplicity of alleles in histocompatibility complexes seem almost automatically to assume that hets are more fit. However, here there would seem to be yet another factor that would favour concave fitness. If a heterozygote requires two forbidden antibodies instead of one, it is open to attack by more host mimetic parasite strains, as is illustrated by its acceptance, as a rule,[33] of transplants from an increased range of donor genotypes. Immune response genes closely linked to (or actually being) genes for histocompatibility antigens, and these being codominantly concerned in antipathogen function, may offset this disadvantage. However, the existence of an overall advantage for major histocompatibility heterozygotes when all other genetic factors have been equalized (e.g. no prior outbreeding having applied over other loci), does not yet seem clear. Reason for such doubt is at least threefold: it can be *a priori*, empirical, and based upon observed patterns of genotype frequencies.

FREQUENCY-DEPENDENT SELECTION AND CYCLES

Constant concave fitness profiles always lead to gene fixation, but in cases where genotypes meet a higher incidence of their specialized parasites when more common (and hence decline in fitness) the variation of FD fitness profiles can prevent this. Sensitive and continued instability in the mid-range of frequencies makes dynamical behaviour—oscillations, long cycles, regular or not—especially likely.

A tendency towards permanent dynamical behaviour is inherent in host–parasite systems.[34–36] In a very general demonstration, Eshel and Akin[37] have shown how a wholly haploid host–parasite system, or a system where host and/or parasite is diploid with fixed allelic dominance in one way or the other, is likely to be permanently mobile: they show that all boundaries of the system can easily be unstable (i.e. the system pulls away from any state of near fixation) while simultaneously the one existing interior fixed point is also certainly unstable. This model has two strains of parasite a and b adapted to attack strains A and B of the host: the fitness of A increases monotonically with a decreasing frequency of a, and the fitness of B with a decrease of b (B reproduces better when less attacked); the fitness of a, however, increases with an increasing frequency of A, and the fitness of b with that of B (b reproduces better with more types it can infect and eat). I have simulated a model closely similar to this with a diploid host and haploid parasite, and with simple exponential functions for the FD fitnesses. Thus, fitnesses are functions such as $w(a) = e^{cp(A)}$, $w(b) = e^{cp(B)}$, $w(AA) = e^{2kp(a)}$, and $w(BB) = e^{2kp(b)}$, where p is the gene frequency ($p(A) + p(B) = 1$ and $p(a) + p(b) = 1$). The fitness of the het, $w(AB)$, can be set in a variety of ways; for example, identical to one or another hom (dominance), or as AM or GM between the homs. With any reasonable selection, parameters c and k and het fitness never set overdominant; this system always goes to a limit cycle with a period of not less than 10 generations. With slow selection, and especially if generation lags are added in the FD response as well, cycles lengthen indefinitely. If cycles go to such extreme gene frequencies that fixations occur, the implications of the model for variation and sexual selection may not change radically, provided that mutations at interacting attack and defence loci are not rare. A mutated defence or attack allele that again sets a species in motion after a pause need not be identical to the one that went extinct. Limit cycles here go over to a picture of a non-progressive coevolutionary pursuit, the process being ultimately stepwise and involving complete gene transiences in the population at every step.

In another simpler type of frequency-dependent model, the parasite population is not represented and fitnesses are simply made inversely dependent on the genotype frequencies in the host. Analysis of such models (see Chapter 2 and ref. 12) shows that at a certain threshold of intensity of selection, oscillations of period two begin and rapidly increase in amplitude with a further increase in the selection parameter. The long-term GM fitness rises with the amplitude of oscillation, and soon competing asexual strains can be excluded or at least kept to an extremely low frequency (Chapter 2). A very interesting feature of the haploid two-locus model (diploid two-locus versions have not yet been studied) is that when there is oscillation of the kind favouring sex, both the population level and the gene frequencies can be constant, and the intense FDS is then reflected only in the oscillation, positive to negative, of linkage disequilibrium (LD). The pattern of this model is much too simple to be expected in nature, but it suggests that gene frequency and demographic changes will not necessarily be prominent in the population processes engendered by parasitism. It also suggests that repeats of electrophoretic surveys of protein polymorphism at approximately generation intervals could be extremely interesting, particularly with respect to possible changes in the LDs that frequently turn up in single surveys.

Another interesting feature of the two-locus model is that oscillation onsets most readily, and sex has the greatest advantage over non-sex, when recombination between loci is maximal. Thus, if this model typified nature, chiasmata would be abundant and/or chromosome numbers very high. Other theories of sex and recombination, quite to the contrary, have difficulty explaining why recombination rates are not all very low.[13] With slower cycles (as obtained when a lag in the FD occurs) and above all when different parasites tend to engender cycles of different periods at different loci, it is probable that any degree of linkage can be shown to be selected for.[12]

The minimum-period cycles of the two-locus haploid model reflect the responsiveness of the genotype's fitness to its own immediate frequency. This is reasonable for highly infectious parasites that cause epidemics within one generation of the host. It further seems reasonable to ask what happens if the FD is accentuated still more by making the most frequent genotype, even when it is only marginally the most frequent, draw disproportionate attack from the parasite, so becoming very unfit while all other genotypes resist relatively effectively and to a similar degree. This was addressed by powering frequencies, renormalizing, and using the pseudo-frequencies so created in the fitness function, which is kept as a decreasing exponential. Indices in the range of 2 to 10 were tried. The effect of such modelling of what amount to strain-specific epidemics was to make various other patterns of cycle possible. Replacing the simple oscillations, some cycles of period four appeared, for example, but changes remained abrupt in terms of LD and of frequencies of particular genotypes. This applied also to chaotic[38] trajectories that could be obtained. In all cases correlations of fitness for particular genotypes from one generation to the next in these runs would be expected to be negative or near zero. When a lag was added along with or instead of the FD exaggeration through powering, however, cycles became slower and more rounded. This would make the long-term averages of intergenerational fitness correlations positive, as would also happen with the stately regular cycles of an Eshel-type model.

Slow-breeding parasites, therefore, are expected to create dynamical states in host populations such as will induce positive parent–offspring correlation in fitness; vagile epidemic-type parasites in contrast may create zero or negative parent–offspring correlations of fitness. These contrasted categories of parasite correspond roughly to what Anderson and May[39] call 'macroparasites' and 'microparasites' in their recent series of models of epidemiology. That the distinction is useful in epidemiology without genetics is suggested by the excellent fit to data obtained from the new models.[39–41] Nevertheless, there seems to be plenty of room for a genetical twist to be added. For the models outlined here, a change to treatment of genotype numbers and densities (best of all with the numbers escaping infection always accounted), instead of having genotype and gene frequencies as sole inputs from the last population state, would be a most desirable extension. If models made genetical like this prove necessary for predicting incidence patterns of other less-virulent diseases and to obtain improved fits to data generally, and if cycles are commonly involved, some important but neglected implications for the evolution of mate choice may follow from the genetical aspect. These implications would be based on the different parent–offspring fitness correlations mentioned above. Before considering this possibility in more detail, however, I will first briefly review likelihoods and evidence for the variation in genetic disease resistance, existence of which underpins the reasoning throughout this paper.

INNATE RESISTANCE TO PARASITES

Ways of resisting parasites are very diverse. However, likewise diverse are the levels of success: it must always be remembered that the other side is evolving and is often ahead. A comparison of large organisms to nations with fortified frontiers and with weapons and armies in readiness behind them[1,3] may help to give a useful hint of the complexity we can expect. Referring to vertebrates, even the depth and subtlety of the interplay of weapon systems of modern superpowers does not make an unreasonable comparison with the picture of the mammalian immune system that is emerging in journals of immunology.[42] With this in mind, the following suggested simple categories of antiparasitic variation must be considered a bare sketch of the possibilities. The first three categories below represent more passive and the remaining three more active aspects of defence.

1. Variants could build tougher physical barriers to the parasite (e.g. thicker cuticle).

2. Variants could change the structure of macromolecules in ways that do not lessen effectiveness in the host but which create problems for the handling by the parasites.[3,6] This implies that almost any rare new protein variant might have an advantage; it has already been suggested on these lines that parasitism may greatly help in explaining the diversity of isozymes.[5]

3. Variants could dispense with certain minor chemicals that serve as the cues or attachments by which parasites detect or enter their hosts.

4. New variants of active defence substances could be produced to overcome a pathogen's evolved resistance to previous defence substances. (Of course, such intrinsic antibiotics are not normally deployed all the time: they may be sequestered and released only when a parasite enters or feeds on cells.)

5. Variants could use new 'password' molecules, or combinations of such: every cell must produce the password to avoid being treated as an 'enemy' by cells that it contacts (Chapter 2).

6. Variants could recognise new, essential, and manifest molecules of a parasite and make lack of recognition trigger its counter attack. This is a response to a virulence strategy such as category 3 above on the part of the parasite.[43]

It will have been noted that the term 'recognize' in category 6 is vague and presumably implies that more than a single mutation must be substituted. The genetics and biochemistry of recognition by and of closely epigenetic molecules are matters of extreme interest, and their elucidation will have implications in fields that are seemingly very far from immunology and pathology. Thus studies of fusion[44] and mating behaviour,[45] social insect behaviour,[46] and speculation on 'green beard effects'[47] will all be affected. But, the main point here is that the probable evolutionary difficulty of choosing new key molecules for recognition gives strength to the recombinant password idea (Chapter 2); otherwise, it is pertinent to ask, why not keep inventing endless new complex password molecules rather than change around, by sex and by recombination, within a limited variety? Nevertheless, category 6 responses when pathogens have defeated those of category 3 seem to be what hosts are trying in the well-known gene-for-gene systems of crop plants and their parasites.[48,49] Possibilities of cycling in these systems have been shown.[36] The gene-for-gene system here is not at all the same as that outlined in connection with Eshel's instability proof, because in the plant pathogenesis, no matter how many loci and alleles are involved, there is always a universal virulent variety of pathogen or a universal resistant variety of host. Which of these is currently 'universal' against all varieties of the other is a matter of whether host or parasite temporarily has the upper hand (i.e. has the most recent mutation). The reason why a system may not fix for the most universal defence and attack genotypes is that defence and attack are costly: for recognition by the host to fail, it seems that the pathogen has to sacrifice, as in category 3, a useful tool, and for its part, the host has to answer this with a new, costly recognition of a different molecule on the pathogen to replace the ability now rendered obsolete. Therefore if, for any reason, the incidence of the parasite declines, varieties of host that are spending less on defence grow and reproduce fastest and again they rise in frequency. Along with the return of various grades of susceptibility, host density also rises and, finally, the combination invites a new epidemic. When this occurs, first it is pathogens with least comprehensive virulence that thrive but, later, as host resistance once more increases, more and more broadly virulent strains resume prominence. Such a verbal description of a cycle does not, of course, demonstrate that such a system must cycle—indeed, even the existence of the costs postulated above has been doubted—but it probably would. Cyclicity in host–parasite systems is hard to escape. A difference from the more certain 'complementary' type of cycle discussed in connection with Eshel's theorem is that the return of a former non-resistant host genotype may be said to be due to a physiological or 'static' cost. This is rather different from the selection by a lively 'biotic' cost, as in the Eshel-type model, when a differing pathogen supervenes and actively disfavours the genotype that had become common, causing the alternative to return to high frequency in its place.

If permanently dynamical behaviour is as characteristic of host–parasite relations as is suggested in this account, it may be that FD return based on static cost (i.e. on the inefficiency of being highly prepared for defence when no defence is called for, as suggested in the agrosystem gene-for-gene story above) will prove a more widespread agent of cycling than the complementary FD pattern suggested in connection with Eshel's result. Evidence of complementary resistance (i.e. those such that an individual cannot be resistant to both of two diseases at once) seems hard to find, and a review of literature in 1948 revealed at most one.[50] Complementarity implies negative correlations of strain resistance for two or more strains or species of parasites, whereas in fact, in the reported survey of mouse resistances, only correlations ranging from +0.17 to zero are recorded. More recent reviews also give little suggesting complementarity.[51–53] Nevertheless, a few examples can be found—for example, of mice to strains of a leukaemogenic virus,[54] and of plants to two kinds of insect.[55,56] Surveying the H-2 (major histocompatibility) range of genotypesin mice also reveals some evidence for complementarity of response to artificial antigens.[57]

On the other hand, innate resistances that are balanced by some static cost are likely to be very common. At the level of prokaryotes defending against antibiotics, existence of cost is well illustrated in the selective disadvantage of carrying plasmids when there is no antibiotic[58] (involvement of plasmids in bacterial 'sexuality' as well as in chemical protection is, of course, significant for the wider picture presented in this article but cannot be discussed here). As a more speculative example of static cost in a properly sexual animal, consider the opposition of lice (serious as vectors of diseases) and weather over the adaptiveness of human hairlessness. Conceivably the gene for male pattern baldness might have been, until extremely recently, involved in an evolutionary hassle of this kind. More realistic and currently topical, histocompatibility genotypes may supply examples. It is generally believed that many of the degenerative conditions connected with specific HLA genotypes are autoimmune in origin, and that such seemingly self-destructive genotypes persist in the human population because the phenotypes are hypersensitive to some significant alloantigen. The proper target antigen is probably on or in some important pathogen, but this antigen is mimetic to a molecule essential in the host. Examples of such mimicry are known,[59–61] but there are not yet enough, nor are the examples connected with sufficiently important diseases, for them to offer any convincing resolution of the problem of protein diversity. It is still very open, however, that it may be common that the price for a good immune starting point against a potentially lethal disease in early life is a slow, damaging attack, spontaneously or after sensitization by the disease, on self tissues in later life.[62,63] In this case, the genes concerned will increase when the disease is prevalent and decline when it is absent. Major histocompatibility antigens may fall in a category of 'password' defences and be at least a major syllable of the whole password or 'identity card' of the individual (see Chapter 2 and ref. 63). Frequency-dependent pressures from disease on the lines above could account for the extreme polymorphism of HLA, H-2, and homologous chromosome regions in other vertebrates. My own model on the password idea, however, suggests that linkage within the MHC region is much too tight for this region to supply ideal, whole passwords by itself. This doubtless largely reflects an inadequacy of the model, but one wonders about the possible role of minor HC loci and variable immunoglobulin coding regions as additional 'syllables'. And is histocompatibility mainly concerned with protection from slow diseases or acute ones? The seeming special connection with antiviral protection suggests acute, but among viruses there are those that are slow and that become latent, as well as those that are acute and ephemeral.

Other examples of genetic resistances with known or presumable static costs could be given. Two of special interest with a pattern similar to that of the gene-for-gene interaction of plants are the resistance to enteric disease caused by *Escherichia coli* in piglets[43,64] and the resistance to the dog heartworm filaria in the mosquito.[65]

In summary, certainly plenty of genetic variation in resistance exists,[66–70] and the resistance to almost any parasite or disease is readily increased by selection.[71] In many cases, genetic bases of such resistance variation are in process of being uncovered. Other papers in the Dahlem Conference volume have emphasized, for microparasites, the abundance of subclinical infections,[72] and, for macroparasites, the very clumped distribution of parasite numbers per individual.[73] Both facts hint at further large pools of relevant variation. Genetics, physiology, and epidemiology all lend strength to a view that the selectability of resistance, very different from what one expects for static, balanced polymorphisms, reflects a common occurrence of permanent dynamical behaviour in host–parasite systems.

SEXUAL SELECTION

Such permanent dynamical behaviour can be of many kinds. The water in the bath can slop fast or slow, in ragged chaotic waves or in a steady rhythm.

Sex thrives on movement: all kinds of it are better than no movement but oscillatory and choppy patterns help most consistently. In contrast, frigid parthenotes need and attend stasis. When cycles are slow, parthenogenetic mutants—at least, if they are perfectly fertile—could progress for periods on the order of half a cycle before recombinants from the sexual strain overtake them. In the long run, whether the cycles are long or short, an environment rich in parasites should make sexuality safe, but the average length of a cycle may profoundly affect the form of sexuality as it becomes manifested through sexual selection. If cycles are very short (e.g. a two- or three-point oscillation over the generations, see Chapter 2), then the parent–offspring correlation in fitness is zero to negative. Here a basis for the female choice of a mate who could minimize offspring losses through parasitism might be to pick the sickest male who looks as though he could still just effect copulation. Phthisic heroes in romantic literature are occasionally favoured but since authors of the time were often tubercular themselves there is room for doubt about realism. Generally females do not so choose; at best (in some insects, for example), they seem fairly indifferent. Perhaps oscillatory selections are always associated with enough slower cycles that while the parent–offspring correlation in disease resistance sometimes falls near zero, it does not descend below. It is also possible that in vertebrates mate selection is in a sense trying to adjust to this kind of problem by using MHC variability as a key to outbreeding, this as an alternative to looking at health more broadly. I refer here to the extraordinary findings of Yamazaki, Boyse, and others at the Sloan-Kettering Cancer Center[45] that mice of congenic strains prefer to mate with a mouse that differs from themselves at MHC, even when this is the only genetic difference in the whole of the genome. High rates of natural miscarriage that are being found in human married couples who are too alike at HLA[42] and similar implications from sizes of placentas in mice[74,75] add much extra interest to the Sloan-Kettering mate selection finding and also to tentative parallels that can be drawn to high multiple-allelic outbreeding mechanisms in plants. MHC-like phenomena in protochordates used both in adaptive graft rejection and for securing outbreeding are also relevant.[44] Even so, Yamazaki's work remains very surprising and perhaps needs to be repeated with even more careful attention to homogenizing family backgrounds of the animals, and perhaps cross-fostering. This would thus make the degree of 'innateness' of the effect really plain, for there are puzzling difficulties about what will happen if innate recognition for purposes of outbreeding undergoes evolutionary adjustment so as to provide innate recognition for the purpose of kin selection.[76–78].

On the matter of pregnancy wastage in mammals arising from too close a genetical similarity of mother and father, it may be remarked parenthetically that this seems regarded as further evidence of an advantage to heterozygotes at MHC loci,[42] whereas in my opinion the fact that more homozygous embryos are lost does not make an overall fitness advantage from this cause at all obvious. As adult females, the heterozygotes have a fertility disadvantage: they are less likely to produce fetuses they can react to and so retain. Population genetical analysis is needed here. Even permanent cycling seems a possible outcome of maternal–fetal incompatibility of this type, instead of the equilibrium with all alleles at equal frequencies that authors seem to assume—in other words, while simplest assumptions may imply an equal-frequency resting point of the system, stability of the point is another matter and needs proof. Regarding fitness factors other than those arising from the paradoxical 'histocompatibility' in gestation, it does, indeed, seem surprising if mice choose matings that tend to create het offspring which are less effective in resisting disease than homs would have been. But that 'less effective' is vague: following our ideas above about fluctuating selection there is often a good hom and a bad hom, and the het with its two alleles, bet-hedging the future as one might put it, may be the best most parents can hope for.

If long cycles predominate over short, the parent–offspring correlation rises asymptotically towards one half; in such a case it may be very definitely advantageous to choose a mate on the basis of his or her apparent ability to resist disease. It is suggested that various physical signs, similar to those used in medical checkups for life insurance, are indeed prominently displayed and attentively investigated in the course of animal courtship. Prominent but not completely honest, and often with great accretion of ornament, these signs of health have to be displayed in species where the slow disease cycles are important because an animal who hides them will not be mated except as a last resort. A margin of fakery is possible, but the choosing sex continually evolves new criteria, including criteria requiring expensive ornaments and stunts. These make faking harder. Even when male fighting is the dominant mediator of choice, display is still necessary, for a strong male who convincingly displays strength (reflecting his health and/or his freedom from parasites) will not have to waste energy and risk damage in fighting weaker males. But where fighting mediates sexual selection, showy epigamic characters involved in display will be less arbitrary and extreme than if direct preference mediated. This is because only preference can entrain runaway selection.[79,80]

Slow cycles tend to result from longevity of parasites and from various forms of lag in interspecies feedback. Along with other requirements underlying the potential effectiveness of mate choice, this suggests that effective disease agents will commonly be macroparasites and the infections they cause will be chronic and debilitating. Microparasites may sometimes give the right pattern (e.g. some viruses) but more often will cause acute diseases where the complete recovery of those that do not die will provide few cues for mate selection to work on. On the other hand, diseases that are acute and often fatal in juveniles and that then become chronic at various levels in those that survive the acute phase are ideal agitators of the suggested sexual selection. With this in mind, Marlene Zuk and I have examined literature concerning six chronic blood diseases in birds and have correlated their incidences with brightness of plumage and richness of song. The results have been favourable to the hypothesis so far: in the spectrum of passerines, the showy species are, on the whole, subject to more diseases in their blood (see Chapter 6).

Obviously this line of theory on sexual selection leads to many other testable predictions. If it is upheld, then some of the most spectacular variation of animals, as between males and females in polygynous species, and in some cases between various innate morphs within the sexes, will appear at least partly as an epiphenomenon of the cryptic variation in enzymes, antigens, and other proteins that was discussed earlier. This point of view does not exclude the connection between ecology and sexual phenomena that has very reasonably been urged in the past (e.g. refs 81 and 82). However, it is overdue for recognition that parasites and pathogens are environment, too, and that all of ecology is affected by their presence.

Acknowledgements

I thank I. Eshel, P. Ewald, L. Lee, G. Williams, and M. Zuk for much helpful discussion of topics in this paper.

References

1. K. Artz and D. Bennett, Analogies between embryonic T/t antigens and adult major histocompatibility (H-2) antigens, *Nature* **256**, 545–7 (1975).

2. H. J. Bremermann, Sex and polymorphism as strategies in host–pathogen interactions, *Journal of Theoretical Biology* **87**, 671–702 (1980).

3. H. J. Bremermann, Towards a theory of sex I: a new model, *PAM* **19**, 1–13 (1981) (Center for Pure and Applied Mathematics, University of California, Berkeley).

4. S. Kinne, *Diseases of Marine Animals: General Aspects, Protozoa to Gastropoda*, Vol. 1 (Wiley, New York, 1980).

5. B. Clarke, The ecological genetics of host–parasite relationships, in A. E. R. Taylor and R. Muller (ed.), *Genetic Aspects of Host–Parasite Relationships*, pp. 87–103 (Blackwell, London, 1976).

6. J. Tooby, Pathogens, polymorphism, and the evolution of sex, *Journal of Theoretical Biology* **97**, 557–76 (1982).

7. D. A. Levin, Pest pressure and recombination systems in plants, *American Naturalist* **109**, 437–51 (1975).

8. D. S. Wilson, *The Natural Selection of Populations and Communities* (Benjamin Cummings, Menlo Park, CA, 1980).

9. W. D. Hamilton, Altruism and related phenomena, mainly in social insects, *Annual Review of Ecology and Systematics* **3**, 193–232 (1972) [reprinted in *Narrow Roads of Gene Land*, Vol. 1, pp. 270–313].

10. R. W. Ewald, Host–parasite relations, vectors and the evolution of disease sever-ity, *Annual Reviews of Ecology and Systematics* **14**, 465–85 (1983).

11. J. Jaenike, An hypothesis to account for the maintenance of sex within popula-tions, *Evolutionary Theory* **3**, 191–4 (1978).

12. V. C. L. Hutson and R. Law, Evolution of recombination in populations experi-encing frequency-dependent selection with time delay, *Proceedings of the Royal Society of London B* **213**, 345–59 (1981).

13. J. Maynard Smith, *The Evolution of Sex* (Cambridge University Press, Cambridge, 1978).

14. G. C. Williams, *Sex and Evolution* (Princeton University Press, Princeton, NJ, 1975).

15. H. N. Comins, Evolutionarily stable strategies for localised dispersal in two dimensions, *Journal of Theoretical Biology* **94**, 579–606 (1982).

16. M. G. Bulmer, Multiple niche polymorphism, *American Naturalist* **106**, 254–7 (1972).

17. P. W. Hedrick, M. E. Ginevan, and E. P. Ewing, Genetic polymorphism in hetero-geneous environments, *Annual Review of Ecology and Systematics* **7**, 1–32 (1976).

18. H. Levene, Genetic equilibrium when more than one ecological niche is available, *American Naturalist* **87**, 331–3 (1953).

19. W. D. Hamilton, Ordering the phenomena of ecology (book review), *Science* **167**, 1478–80 (1970).

20. L. T. Webster, Inheritance of resistance of mice to enteric and neurotropic virus infections, *Journal of Experimental Medicine* **65**, 261–86 (1937).

21. J. Gillespie, Polymorphism in random environments, *Theoretical Population Biol-ogy* **4**, 193–5 (1973).

22. W. S. Moore and W. G. S. Hines, Sex in random environments, *Journal of Theo-retical Biology* **92**, 301–16 (1981).

23. E. Berger, Heterosis and the maintenance of enzyme polymorphism, *American Naturalist* **110**, 823–39 (1976).

24. L. L. Cavalli-Sforza and W. F. Bodmer, *The Genetics of Human Populations* (Freeman, San Francisco, 1971).

25. B. Clarke, The evolution of genetic diversity, *Proceedings of the Royal Society of London B* **205**, 453–74 (1979).

26. P. J. N. Hebert, D. C. Ferrari, and T. J. Crease, Heterosis in *Daphnia*: a reassess-ment, *American Naturalist* **119**, 427–34 (1982).

27. J. Felsenstein, The theoretical population genetics of variable selection and migra-tion, *Annual Review of Genetics* **10**, 253–80 (1976).

28. H. O. McDevitt and J. W. Benacerraf, Genetic control of specific immune re-sponse, *Advances in Immunology* **11**, 31–74 (1969).

29. A. C. Allison, Coevolution between hosts and infectious disease agents and its effects on virulence, in R. M. Anderson and R. M. May (ed.), *Population Biology of Infectious Diseases*, Dahlem Konferencen, 1982, pp. 245–67 (Springer, Berlin, 1982).

30. R. K. Selander and B. R. Levin, Genetic diversity and structure in *Escherichia coli* populations, *Science* **210**, 545–7 (1980).

31. P. M. Sheppard, Polymorphism and population studies, *Symposia of the Society for Experimental Biology* **7**, 274–89 (1953).

32. P. M. Sheppard, *Natural Selection and Heredity* (Hutchinson, London, 1958).

33. C. Stern, *Principles of Human Genetics*, 3rd edn (Freeman, San Francisco 1973).

34. R. M. Anderson, Vertebrate populations, pathogens and the immune system, in N. Keyfitz (ed.), *Seminar on Population and Biology*, pp. 249–68 (Ordina Editions, Liège, 1981).

35. I. Eshel, On the founder effect and the evolution of altruistic traits: an ecogenetical approach, *Theoretical Population Biology* **11**, 410–24 (1977).

36. J. W. Lewis, On the coevolution of pathogen and host (Parts I and II), *Journal of Theoretical Biology* **93**, 927–85 (1981).

37. I. Eshel and E. Akin, Evolutionary instability of mixed Nash solutions, *Journal of Mathematical Biology* **18**, 123–33 (1983).

38. R. M. May, Simple mathematical models with very complicated dynamics, *Nature* **261**, 459–67 (1976).

39. R. M. Anderson and R. M. May, Population biology of infectious diseases: part I, *Nature* **280**, 361–7 (1979).

40. R. M. Anderson and R. M. May, The population dynamics of microparasites and their invertebrate hosts, *Philosophical Transactions of the Royal Society* **291**, 451–524 (1981).

41. R. M. May and R. M. Anderson, Population biology of infectious diseases: part II, *Nature* **280**, 455–61 (1979).

42. A. E. Beer, M. Gagnon, and J. F. Quebbeman, Immunologically induced reproductive disorders, in P. G. Crosigniani and B. L. Rubin (ed.), *Endocrinology of Human Infertility: New Aspects*, pp. 419–39 (Academic Press, London, 1981).

43. B. I. Eisenstein, Phase variation of type 1 fimbriae in *Escherichia coli* is under transcriptional control, *Science* **214**, 334–9 (1981).

44. V. L. Scofield, J. M. Schlumpberger, L. A. West, and I. L. Weissman, Protochordate allorecognition is controlled by a MHC-like gene system, *Nature* **295**, 499–502 (1982).

45 K. Yamazaki, M. Yamaguchi, E. A. Boyse, and L. Thomas, The major histocompatibility complex as a source of odors imparting individuality among mice, in D. Muller-Schwarze and R. M. Silverstein (ed.), *Chemical Signals*, pp. 267–73 (Plenum, New York, 1980).

46. B. Hölldobler and C. D. Michener, Mechanisms of identification and discrimination in social hymenoptera, in H. Markl (ed.), *Evolution of Social Behavior: Hypothesis and Empirical Tests*, Dahlem Konferenzen, pp. 35–8 (Verlag Chemie, Weinheim, 1980).

47. R. Dawkins, *The Selfish Gene* (Oxford University Press, Oxford, 1976).

48. P. R. Day, *Genetics of Host–Parasite Interaction* (Freeman, San Francisco, 1974).

49. J. E. Vanderplank, *Genetic and Molecular Basis of Plant Pathogenesis* (Springer-Verlag, Berlin, 1978).

50. J. W. Gowen, Inheritance of immunity in animals, *Annual Review of Microbiology* **2**, 215–54 (1948).

51. G. J. Eaton, Intestinal helminths in the mouse, *Laboratory Animal Science* **22**, 850–9 (1972).

52. V. M. King and G. E. Cosgrove, Intestinal helminths in various strains of laboratory mice, *Laboratory Animal Care* **13**, 46–8 (1963).

53. D. L. Rosenstreich, Genetics of resistance to infection, *Nature* **285,** 436–7 (1980).

54. F. Lilly, Multiple gene control of viral leukemogenesis in the mouse, in H. O. McDevitt and M. Landy (ed.), *Immune Responsiveness: Relationship to Disease Susceptibility*, 279–88 (Academic Press, New York, 1972).

55. J. E. Hare and D. J. Futuyma, Different effects of variation in *Xanthium strumarium* (Compositae), *Oecologia* **39**, 109–20 (1978).

56. A. von Schonborn, The breeding of insect-resistant forest trees in central and northwestern Europe, in H. D. Gerhold *et al.* (ed.), *Breeding Pest Resistant Trees*, pp. 25–7 (Pergamon, London, 1966).

57. H. O. McDevitt and A. Chinitz, Genetic control of the antibody response: relationship between immune response and histocompatibility (H-2) type, *Science* **163**, 1207–10 (1969).

58. B. R. Levin, Conditions for the existence of R-plasmids in bacterial populations, in S. Mitsuhashi, L. Rosival, and V. Kremery (ed.), *Antibiotic Resistance* (Springer, Berlin, 1980).

59. D. P. Lane and H. Koprowski, Molecular recognition and the future of monoclonal antibodies, *Nature* **296**, 200–2 (1982).

60. J. N. Wood, L. Hudson, T. M. Jessell, and M. A. Yamamato, A monoclonal antibody defining antigenic determinants on subpopulations of mammalian neurones and *Trypanosoma cruzi* parasites, *Nature* **296**, 34–8 (1982).

61. J. B. Zabriskie, Mimetic relationships between group A streptococci and mammalian tissues, *Advances in Immunology* **7**, 147–88 (1967).

62. W. F. Bodmer and J. G. Bodmer, Evolution and function of the HLA system, *British Medical Bulletin* **34**, 309–16 (1978).

63. J. Dausset, The major histocompatibility complex in man, *Science* **213**, 1469–74 (1981).

64. J. M. Rutter, M. R. Burrows, R. Sellwood, and R. A. Gibbons, A genetic basis for resistance to enteric disease caused by *E. coli*, *Nature* **257**, 135–6 (1975).

65. P. B. McGreevey, G. A. M. McClelland, and M. M. J. Lavoipierre, Inheritance of susceptibility to *Difilaria immitis* infection in *Aedes aegypti*, *Annals of Tropical Medicine and Parasitology* **68**, 97–109 (1974).

66. A. R. Barr, Evidence for the genetical control of invertebrate immunity and its field significance, in K. Maramarosch and R. E. Shope (ed.), *Invertebrate Immunity: Mechanisms of Invertebrate Vector–Parasite Relations*, pp. 129–35 (Academic Press, New York, 1975).

67. M. F. W. Festing, *Inbred Strains in Biomedical Research* (Oxford University Press, New York, 1979).

68. F. Lilly and T. Pincus, Genetic control of murine viral leukemogenesis, *Advances in Cancer Research* **17**, 231–77 (1973).

69. H. W. Moon and R. H. Dunlop (ed.), *Resistance to Infectious Disease* (Saskatoon Modern Press, Saskatoon, 1970).

70. R. M. Williams and E. J. Yunis, Genetics of human immunity and its relation to disease, in H. Freedman, T. J. Linna, and J. E. Prier (ed.), *Infection, Immunity and Genetics*, pp. 121–39 (University Park Press, Baltimore, MD, 1978).

71. F. B. Hutt, *Genetic Resistance to Disease in Domestic Animals* (Comstock Publishing, Ithaca, NY, 1958).

72. F. Fenner, Transmission cycles and broad patterns of observed epidemiological behavior in human and other animal populations, in R. M. Anderson and R. M. May (ed.), *Population Biology of Infectious Diseases*, Dahlem Konferencen, 1982, pp. 103–19 (Springer, Berlin, 1982).

73. R. M. Anderson, Transmission dynamics and control of infectious disease agents, in R. M. Anderson and R. M. May (ed.), *Population Biology of Infectious Diseases*, Dahlem Konferencen, 1982, pp. 149–76 (Springer, Berlin, 1982).

74. W. D. Billington, Influence of immunological dissimilarity of mother and foetus on size of placenta in mice, *Nature* **202**, 317–18 (1964).

75. D. A. James, Effects of antigenic dissimilarity between mother and foetus on placental size in mice, *Nature* **205**, 613–14 (1965).

76. R. D. Alexander, *Darwinism and Human Affairs* (Pitman, London, 1980).

77. R. D. Alexander and G. Borgia, Group selection, altruism, and the levels of the organisation of life, *Annual Review of Ecology and Systematics* **9**, 449–74 (1978).

78. M. Ridley and A. Grafen, Are green beard genes outlaws?, *Animal Behaviour* **29**, 954–5 (1981).

79. R. A. Fisher, *The Genetical Theory of Natural Selection* (Oxford University Press, Oxford, 1930).

80. R. Lande, Models of speciation by sexual selection on polygenic traits, *Proceedings of the National Academy of Sciences, USA* **78,** 3721–5 (1981).

81. S. T. Emlen and L. W. Oring, Ecology, social selection, and the evolution of mating systems, *Science* **197**, 215–23 (1977).

82. V. A. Geist, A comparison of social adaptations in relation to ecology in gallinaceous birds and ungulate societies, *Annual Review of Ecology and Systematics* **8**, 193–207 (1977).

BRIGHT BIRDS

Heritable True Fitness and Bright Birds:
a Role for Parasites?

Victor came to the forest
  Cried: 'Father will she ever be true?'
And the oaks and the beeches shook their heads
  And they answered: Not to you.

AUDEN[1]

A LOT of the students attending my graduate-level course in the Museum of Zoology in Ann Arbor were interested in birds. The Bird Division of the museum was well known to universities in the USA and Canada and the Michigan Graduate School consequently drew bird-oriented people from all North America. Peter and Rosemary Grant, my colleagues in Biological Sciences who had joined the University of Michigan about a year after I did, studied birds exclusively and were busy every year in the Northern Summer with their now-famous study of evolutionary ecology—the actual year-by-year evolution—of the finches of the Galápagos Islands (those finches whose obviously recent evolutionary changes had so much excited Darwin).[2] At home in Ann Arbor where the Grants used their time for writing up, they kept a spectacular ginger cat. He, too, was a birder in his very distinct way but by use of claws rather than binoculars he kept the garden clear and gave them a break. The Grants always had several students working with them on the Galápagos and in Ann Arbor these usually attended my class sooner or later. Facing this class, my obsession with the topics of sex and sexual selection and my treating these topics as the most major challenges and sequels to all my earlier stuff—the kin-selection and sex-ratio themes—naturally brought me into many long discussions in which birds figured as our main examples.

Well, they were indeed very good examples, at least for the 'sexual-selection' end of my new interests. For the other end, birds actually don't have much in the way of alternatives to 'sex itself'. As to parthenogenesis, occasional fatherless turkeys of a particular line (white as it happens and, unfortunately for both biology and economics, all of them male) are about all that is known. Therefore for this side, birds were far from ideal. Among them could be found degrees of sexuality and of tolerance for inbreeding, but these were only topics of the foothills if you like, hardly the real thing. For asex's distinctive ecology, insects and other groups studied in the museum (molluscs, fish, even reptiles) were far better.

Yet birds really are as if made to exemplify effects of sexual selection. They provide many of the world's first discussed and most extreme examples. In North America, wild turkeys and sage grouse are examples while many other brilliant species—wood duck, cardinals, and scarlet tanagers—whistle, hover, and generally flaunt all manner of colourful plumes and effects nearby. A lot is known about bird anatomy, about the feather materials (did you know that black feathers are stronger than white?), about the patterns and energy of bird songs, and about what birds do with all of these matters mechanically, behaviourally, and all the rest. But, from my point of view, what is the point—the adaptiveness of it all, if there is any?

As it happened, sexual selection was rising as a fashionable subject on other fronts besides my own and, as always, bird studies were to the fore. The paper of this chapter, which Marlene Zuk and I succeeded to publish in a prominent journal, fuelled further the general enthusiasm. One of the other aspects, which I will return to later in this volume, concerns what we might call bird morality. Charming and humanoid as their social affairs may often seem to us, our planet's crew of these miniature and feathered survivors of the great reptiles are far from being either the marital or parental models of 'good' behaviour. Indeed, they are forcing us to think twice and twice again about many a dearly held human ideal. We are easily fooled, it seems. Lovers of art and nature in China, for example, picked those ducks we call 'mandarin' (*Aix galericulata*) as symbols of monogamous fidelity: yet in this notoriously varied duck family, where they could have done much better with, for example, geese and swans, they picked about the worst species. The male mandarin stays with his mate for only a fraction of one season.

Bird love, nevertheless, continues. Long before Marlene and I, even parasitologists, had liked them as a change from white coats and white worms, and this we saw reflected in the amount of work done on the parasites of birds of completely non-economic species. Possibly even those computer technicians who hated my cycles tearing their paper included some who, in May, would nip across the valley from North Campus to the Nichols Arboretum to 'tick' their glimpses of the 'spring warblers' flitting amid the exploding green leaves—those birds whose arrival tells of the final departure of the five or so months of the Ann Arbor winter. Possibly they'd even have been more forgiving and more interested if I'd explained how I was trying to chart new ideas about why one spring warbler (species or/and individual) is prettier than the next (though actually this did not in the warblers, as it turned out, correlate with the blood parasites). But in fact, of course, I hadn't thought any of this out at the time of my offending diagrams, rather the diagrams themselves were my entry points to new ideas on their way.

Lags that I brought in to give slower cycles, and thus the ink and paper of the U of M's plotter a marginal time to dry, were the stimulus setting me to wonder what weak and chronic agents, subject to what possible real lags, could fit the working of a similar scheme in nature. As I told in Chapter 2, it flashed on me at last that the parasites of the birds might have just such an effect. How often I'd seen those parasite life-cycle diagrams, those nature's black jokes about how one species comes to be connected to another—the parasite passing through the dragon fly to the fish, fish to heron, and then, in the heron's droppings and through the water, back to the dragonfly. Here surely could be the lags I suspected— one source of them anyway.[3] Mournful summer sounds floated to me from my far-off boyhood: they were in the garden at Oaklea in Kent where my mother's chickens languidly 'gaped' in their fenced yard in the heat. It was often these non-laying and unthrifty birds that mother killed for our meals and so it was she who showed me as a small boy the 'gape worms' that caused the condition—yellow threadlike worms embedded in inflamed tissues of the bird's inner throat. Much later I learned from books how earthworms, long-lived in the soil, could harbour the young of these worms, dormant and curled up like reminder knots tied in a handkerchief. There they could wait during their long quiescent phases for their chance to be eaten and to reinfect a bird. Then it was that I guessed that these perils latent in earthworms must be the key to why mother's chickens had often been distinctly aloof and unenthusiastic when I threw large worms I had dug up in the garden over the fence to them. Many studies of the behaviour of parasitized hosts suggest that earthworms coming onto the soil surface and lying there exposed deserve all the suspicion they seem to get from otherwise omnivorous chickens: in a natural situation the behaviour of invertebrates may be more being controlled by a parasite within, this pursuing its aim to be passed on, than by the self-preservation adaptations of the worm itself. A prey acting as though wanting to be eaten ought always to seem suspicious to a predator just as a 'friendly' wild fox trotting up ought to scare a human in all countries where rabies occurs. But all that's an aside: the main point is that having once thought of such enemies as these worms within worms, as it were, and of the obvious lags in natural selection that come by them, my mind 'went about' smartly like a wind-slanted yacht and with many bursts of spray and creaks of halyards set off on a new tack.

In parallel to such hits from my models I think it was particularly European hoopoes, as found embalmed in the quiet of the Ann Arbor bird library, that began my enthusiasm for specially bright bird species and what these might tell. Somewhere in that library I had seen a reference to a parasite bearing the evocative species epithet, 'upupae'; shortly somewhere else I saw another—it was the same parasite species name precisely. Therefore it appeared that at least two parasites had been noticed as found only in hoopoes! Possibly both these were cases out of Olsen's great grey parasite book,[4] the wonderful volume I bought for myself soon after first seeing it. It is full of creepy line drawings showing, for example, the elaborate multi-host cycles and the parasite stages where, among the simpler ones, you can see a gigantic and malaria-laden mosquito drilling the chest of a hapless man, injecting sporozoites and, moments later, drawing out again, along with the man's blood, the gametocytes of the *Plasmodium*, mankind's greatest killer of all. A few pages on (or was it back?) you could see the same man being bitten by a gigantic sandfly carrying *Leishmania*; elsewhere a dog-sized kissing-bug, *Triatoma*, delivering Chagas' disease to him; and so forth.

Why do those parasites in particular, among those many surreal and completely deadpan pictures shown by Olsen, seem so especially creepy to me? I think I know and will digress a little just to tell it. Something about the tube and the man's chest depicted brought back to me certain tubes and a pain in my own chest from 30 years in my past. It reminded me of how in Denmark Hill Hospital once two intern medics armed with a bike-pump-sized syringe, looking somewhat like the proboscis of Olsen's mosquito, had tried to draw off stale blood pooled in my right lower chest. It was blood flooded there from a wound after the explosion of a home-made bomb. The picture set me squinting in memory again down the pale dunes of my chest, still tented here and there at the time with plasters and bandages, towards a real surreal dreamscape that I shall never forget. Who let those two serious young whisperers loose on my anatomy it is now difficult to imagine; even to me then, a trusting 12 year old, they appeared to not quite know what they were doing and, in so far as I could make it out, it seemed the impossible. With unencouraging vagueness, measuring their distances with handspans and knucklengths out from certain landmarks such as the edge of my ribs, they found their spot. Then—well, imagine using a bike-pump-sized syringe to suck red-tinted junket through the wall of a rubber hotwater bottle in which, for some reason, the junket has first been allowed to set (that is, to clot in the case of my blood): you will then have an idea of their difficulties. As might be predicted, there would come into the barrel of the syringe a half inch of well-stirred red junket and a quarter inch more of a pink whey, and then nothing—even I could tell that the blood clot and perhaps also some of the lung substance of patient Hamilton had blocked their tube. Twice, I think, and again very like a mosquito looking around on a shirt sleeve for a gap between the threads, the two whisperers pulled up their rig and tried again in another spot, prefacing the new dig with another small but ineffective sting from a smaller syringe, which obviously applied a local anaesthetic to my skin and muscle. This local, however, hardly affected the major pain of the big needle piercing my pleuron. It seems to me now that I must have been both witnessing and feeling what it was like to be killed by rapier thrust several times repeated during that morning. And yet I watched and assessed it all in a rather detached, accepting way, merely longing to see them succeed—if rapier is the word, I had more the spirit that Hamlet had in Shakespeare's duel, I guess, than that of Laertes.

But another similar irrelevance that I can also at any time conjure from memory of the pages of Olsen is what I call the 'Blimp' drawing of *Giardia*, that strange protist whose shape and organelles makes a face so like that by which a certain cartoonist used to caricature a type of personality common in the days of the British Empire, a character, in fact, quite familiar to us by his presence at almost any grey-haired social occasion of the 1940s and especially at any my father might take his children to, he being with his roads very much an empire-builder himself though far in character from the style of Colonel Blimp. It was later that I learned that *Giardia* either had dispensed with or else never had had mitochondria, those little energy units that power almost every other nucleated cell on the planet. So after that I mentally assigned to Olsen's *Giardia*/Blimp a caption running something like: 'None of those damn newfangled mitochondria around when I was in Colon—we improvised, didn't need 'em.' Also, after the news about the lack of mitochondria I came to endure more stoically those dreadful creasing pains and the periods living for weeks afterwards with what the book aptly described as 'persistent foul-smelling, oily stools' after encounters with *Giardia* in Brazil. The disease came to me once after drinking from an innocent forest stream, another time after drinking from a glorious black-water lake of Amazonian Brazil. Surely, I thought, as the pains abated a little and I could think about the marvel of the absent mitochondria, I should feel privileged to be a habitat for such a wonderfully ancient organism! In truth, *Giardia* there in my colon might be lamenting another fading empire—that which held sway down the vast stretches of the pre-mitochondrial Archaean! But all this is irrelevant: they are instances only of such side images as fill weaker days while I struggle for further light on how parasites or whatever else might mediate sexual selection.

On stronger days, from such musings as those about my torn model plots and, later, the untorn and slower-cycling ones that followed, with Olsen's book open on my desk, it was only one floor up and 30 yards along to reach the bird library where I could dive and swim like a loon among shelves and shelves of past studies of birds, snatching from among them those few papers that dealt with parasites. At first I was looking for all kinds of macroparasites and for slow diseases (viruses were often good to read about) and was trying to gain an impression of whether strikingly sexually selected species were unusually prone. If they were, I wanted to see whether their loads were the host-specific ones that I reckoned ought be the main players. I was very keen to find close congeneric pairs of birds with one more showy than the other. The blue jay and the Canada jay became an early interest, in part replacing the hoopoe. The data they brought me fitted well, but unfortunately they were not congeneric. The curators of the Bird Division helped me to find some closer examples—the black duck and the mallard was one such pair and it may have been Bob Storer who suggested the two towhees of North America—but unfortunately these were geographically separated. But in any case data of a quantitative kind on the parasites of such close sets proved virtually impossible to find. It wasn't difficult to discover compilations of parasite species listed against their species of hosts; one tome, for example, gave me all this compiled together for ducks, geese, and swans. For any quantitative test such lists seemed to be of little use. Through more opportunities for study it is obvious that the abundant and widespread mallard will show a longer list recorded for it than will, say, the Kerguelen Island teal, confined as the latter is to a single tiny and seldom-visited oceanic island.[5]

University of Michigan librarians like to return books to the shelves themselves rather than let readers return them and sometimes only pretend to and get them lost. But in the Bird Division's library all shelves were well used and I could return the books myself. On the other hand I had the impression that older books and journals I was consulting in the Medical School's library were very little visited. So I just have to hope no asthmatic library assistant died of returning ancient books into the havens of dust whence my search for comparative parasite studies had extracted them; but then it is well known to readers that all the librarians who let you go to browse shelves for yourself are going to receive special blessings, along with the pure in heart and some others, from God, so I am confident their souls are both happy and asthma-free where they have gone, if they did die.

I sought in the dust mainly for literature on worms, protozoans, and viruses. The volumes that could be of any date (although according to size they had better be post-van Leeuwenhoek, or post-Pasteur, and the like). On the whole I passed over bacteria because as a group they seemed to provide relatively few diseases that were chronic rather than acute: for my sexual-selection story I was excluding the class of what I styled 'kill or recover' diseases because prospective mates would not be informed about them and many bacteria and also many viruses belonged here. Besides this preference for chronicity, I hunted also for specificity of parasites to single hosts and also those where lags could potentially affect cycling. Chickens, pheasants, turkeys, and grouse—technically the 'galliform' birds—were among my favourites of the time because, in addition to all the prior signs of high sexual selection (bright feathers, wattles, and suchlike), they were economically important and so had had much attention paid to their diseases. Ducks and geese had had much paid to them, too, but here the attention came about more through the incentive of hunting and, rather than from any chicken-wire perspective focusing the meat and eggs, the perspective of bracing and beautiful sites in the marshes. Gumbooted students begging offal from hunters at the edge of such shooting grounds seem to have met with—to judge from various publications that I scanned—willing responses, and perhaps many a duck hunter's wife came to sigh with relief later when her husband's 'bag' reached her already emptied of all its innards; meanwhile other wives, perhaps sighed oppositely when the parasitologists passed through with their 'bag' on their way to the lab (the sexism is inevitable here: it's just the way things were in the days when the data I scanned were collected). Anyway, the signs I could elicit from all that literature were sweet-smelling enough: it seemed to me that the neatly patterned and coloured birds did indeed have a tendency to host rather neat and special parasites too. But I needed something more than mere seeming and for a long time saw no way to any quantitative test of my idea.

I am not sure how I found my first list of blood parasites in the bird library; probably it was simply by following all index entries for 'parasites' in the journal volumes. Of course I excitedly scanned the first list I did find trying to assess, as usual, whether the brighter birds had more species and higher incidences. Again I had a positive impression. From early on in this search a fair protozoal and wormy crop for the cardinal, for example, was catching my attention. The papers usually stated that fixed numbers of microscope fields of blood-smear slides had been scanned and by this a standardized (and strictly always under-) estimate of the infection rate (or 'prevalence') was determined. So here at last I had quantification on the parasites of a kind, but the diseases studied seemed far from ideal in other ways. For example, there were no lags that I knew of for blood parasites apart from those possibly long stays of strains within individual hosts (the time a disease such as malaria spends in the insect vector is generally negligible compared with a human lifespan). A personal experience told me of one instance of the longer kind of stay; it was how slight 'tertian' fevers (fevers every other day) had affected me occasionally in the mid-seventies and really seemed likely to have been reawakenings of a *vivax* malaria infection I had caught in North Italy or West Turkey in 1960: yet the brief bouts dwindled and by the 1980s had ceased. So could this have been a 15-year lag (a half human generation), as would apply if the disease got back from me into a mosquito, which in fact it probably never did?

What was worse about the bird blood parasites was that they seemed, at least for the diseases of passerines, to be mostly high generalists. All the same, with all these shortcomings, here at last I had found data that might be put to a quantitative test. My first-discovered lists led to others. Generally, of course, search through citations led backwards, so by the end I had found plenty of old studies but had continually missed the best and most recent review that we might have used.

We? How so? When I told my class this new idea about slow cycling and how it would support permanent heritability of fitness and of its potential role for sexual selection I met with only a lukewarm response. I do not belittle the Michigan students for their scepticism, it is the stuff of science, and never elsewhere have I found a group that gave me a more honest and intelligent criticism on wild evolutionary ideas. Thanks perhaps particularly to the Grants, Peter and Rosemary, but also to Bob Storer and Robert Payne, ideas about bird microevolution were part of the air one breathed at U of M. Were I to list all of that class who listened to me raving on about the revelatory power of shabby hoopoe crests (I was highlighting the ectoparasites here: hoopoes wouldn't be able to preen them off, and nor could they preen the messed feathers they couldn't see; their best chance was to be genetically inedible to the parasites)—as I said, if I were to list them I would find myself citing a significant fraction of behavioural ecologists now teaching in America. There were post-docs as well in the year in question: I think that both Wallace Dominey and Paul Ewald were present, with the former rather favourably inclined to my idea, the latter less so, at least at the time (Paul's hummingbirds had almost no blood parasites if I remember, though I came to believe they might make up for this through seeming to host an exceptional variety of specialized mites).

The most pressing counterargument I had from the class was one that I was soon to hear again and again: that for any organisms with a highly developed immune system, as birds certainly have, parasite defence must be largely facultative; good resistance genes would be simply those building in the wonderful conditionalities of the system and, in so building, why should any clever new trick stay long enough in a polymorphism to fuel heritability? I partly agreed. But the evidence was that irreducible heritability did somehow remain and facultative responses had their limits, especially when other stress is present;[6] I think this is increasingly recognized so that I now less often meet that particular objection. Educable immune reactions are indeed at an evolutionary peak in vertebrates and perhaps they are as important for the achievements of this group as the seemingly equally open-ended and educable nervous system of the groups, the factor that is usually more emphasized.

To confirm how important the immune system is, just look at AIDS; and note also how for huge-brained humans it is still said that even the total of nervous tissue does not outweigh the sum of all tissues of the immune system. Yet from what I knew of my fellow humans, and even of my own family, it was clear that there remained behind the facultative stunts, an immense and seemingly innate variability in human healthiness. We are nowhere near all equally intelligent and we are nowhere near all equally healthy and there may be a subtle connection between these two by-themselves surprising facts. Texts on the medically available standard mouse strains told the same story as I read in others about chicken breeding; and yet these animals, mammals and birds, were said to be the twin peaks of the evolution of immune systems. Birds and mammals, I reflected at last, are also the peaks in another supposedly advanced character: homoiothermy (warmbloodedness). Could it be that their making (for reasons perhaps of developmental homeostasis) their bodies to be such ideal chemostats for microbes to grow in merely cancelled out with the general benefit of the facultative immunity and thus threw them back again, hardly better off than before, on the variation in innate immunity? All the books were telling me that variant genes were still there, did still give an edge against variants of the microbes pressing from all sides. What could be the edge? New passwords, old tricks suddenly revived . . . ?

So I argued and so tried to discuss with the class and just one voice spoke up in my support—Marlene Zuk's, a student I knew little of until that time. She brought up for discussion variation she had noticed in house finches in her native California. The bird was then solely an East-of-Rockies native and had not yet arrived in Michigan though it was on its way.[7] Marlene told that house finches sometimes had visible lesions on their legs but could not remember whether this was connected with their colouring; she also wondered why generally some bird species were so 'keen', so to speak, to become sexually exaggerated and others not. Her California birds were dull on the whole compared with the new array she saw in the woods of Michigan; California had nothing like a scarlet tanager. Might it be that the dry 'Med' climate of the West generated fewer mosquitoes and therefore less malaria? We discussed the exceptions: California quail tame as the rosy house finches and with their strange nodding top-knot. But this species, it turned out later, is unusually prone to diseases that were indeed generally the right kind, including a special *Haemoproteus* blood parasite that could cause serious epidemics, so it was a good exception.

So it happened that, as I found more lists of the haematozoa, Marlene and I formed an alliance. I asked Marlene if she would be bird-scorer in some test: we'd test statistically whether showy species were on average more parasitized with haematozoa than the rest. In our first try on one of the papers to our half-believing delight we found the slope we expected, though it was rather well short of statistical significance. As we added more haematozoa studies and combined the results, however, our first hint grew steadily stronger. I was unhappy with our technique because of the skewed incidence of the parasitism (statisticians later told me this didn't matter so much though I am still not sure why) but when I had spotted Goodman and Kruskal's alternative technique (see p. 225) I saw we could have a method that both dealt with the skew and enabled us to combine results from all our separate lists. In the interest of controlling what was shortly to be called taxonomic artefact, we cut down from considering all birds in our lists to considering only the passerines, even though this reduced the tendency of the tests to record significance. Eventually we were satisfied that the trend we had guessed at and had shown crudely in the first run was really there.

It would be boring for the general reader to include all back and forth that happened in evolutionary academia about the acceptance of our paper ('H&Z') after its publication. Yet equally it would be a pity to give no discussion. Therefore I have taken a middle course and put a lot of what I have noticed of the debate into a countercritical essay and include this as one of two appendices to this volume. It is slightly more technical in style than I aim for in the 'Introductions' themselves and you can skip it or read it as you please. Below I will just put a few of the points that seem of the most general evolutionary interest or else are important for chapters that follow.

By the time we were writing up Marlene and I had seen clearly not only how, if our assumptions were correct, we had found a source of permanently, positively heritable fitness but also how the downpassed benefits could be so substantial that the gains of acquiring them for descendants and the prices paid for them could include some fairly costly advertising. 'Genetic fitnessware' you might call what was being sold; knowing nothing else, would you buy a bundle from the shelf if you saw it in a shoddy packet?

I recall a discussion in the museum with Mark Kirkpatrick visiting us from the University of Texas about this time. Wallace Dominey was there and he and I stated our belief that females would evolve not only to appreciate bizarre characters, characters that would be costly in a direct sense for their sons, but they would even stump up yet more costs for effectiveness in the choosing. Mark Kirkpatrick, a strong believer at the time in the sufficiency of the Darwin–Fisher view of largely arbitrary sexual selection (indeed himself a pioneer in laying out the logical foundation of that view), brought against us the usual points. There was the predicted low heritability of strongly fitness-related traits and the seeming complete arbitrariness of the characters observed to be exaggerated. Amotz Zahavi's controversial views on sexual selection, which I had by now brought up in several runs of my class, were looming in the background of all our discussion. Mark had already published against these ideas: Zahavi, of the Hazeva Field Station in Israel, thus became an invisible third antagonist in the room (I should say that except on the subject of handicaps Zahavi and I have disagreed steadily about many other things, especially kin selection[8]). I note the absence of any Zahavi reference in my list of Chapter 5 but that Marlene and I do cite his 'handicap' idea in the present one and do so with what was for the time high approval.

Like Kirkpatrick and most others I had been having difficulty with the 'handicap' idea, in the first place not seeing from Zahavi's papers quite what he meant to include and exclude (a broken leg is clearly a handicap, for example: did he include that?) and, second, for all that I thought I did see, doubting that it could work. Not only had Zahavi not modelled his ideas quantitatively, not algebraically, nor numerically, but even the verbal explanation in his now-famous early papers had been extremely vague and brief. I could, however, easily see his point about costliness being a proof of honesty and my interest of the time was certainly showing me the worth that, to a mate chooser, honest advertisements could have; but, again, the limits in all this seemed vague and Zahavi's trenchant rejection of all Fisher, as of all bluffing (dishonest signalling), to say nothing of all kin selection, was, to say the least, offputting. Zahavi's initial published presentation of his idea was so confused that even after Alan Grafen had explained the Zahavi theory completely I give him (Alan) a good portion of the credit for what has become known as the Handicap Principle.[9] Credit in science is partly for effective communication, not just for right ideas; and if a vague sense of an idea is enough, credit for the Handicap Principle could slip back to the American economist Thorstein Veblen.[10] Likewise Darwin, not Patrick Matthew, gets the credit for evolution by natural selection because Darwin wrote his ideas clearly and persistently with extreme multiplicity of illustrations, not as a few paragraphs (clear though these paragraphs also were) of note F of an Appendix to a book on *Naval Timber and Arboriculture*.[11] What permitted Darwin's ability of lifelong concentration on one objective (launching of the study of evolution) as opposed to following all the varied normal occupations of a gentleman—his nicotine inhalation perhaps?[12]

Fisher's idea of a 'run away' process in sexual selection—the idea that the preference for a trait inevitably hitchhikes with, and thereby must exaggerate, the selection of the preferred trait itself—certainly works as Russell Lande, Mark Kirkpatrick, and others have shown, but it turns out to be a more frail process than Fisher and others have depicted it and than we, too, recently thought.[13] As for bluffing, even just thinking of certain early frights of my life, such as my mother's hens defending their chicks, or dogs radiating at me their ridged-back signals and then 'diminishing' and running if I approached, has taught me that bluffing certainly exists. The hen or the dog certainly suddenly looks larger: but it doesn't necessarily attack. Once my understanding of evolution had begun, bluffing in a wider sphere was confirmed to me by the impressive facts of Batesian mimicry. Incidentally, from Henry Bates on, and once anyone has seen and heard explanation of the good examples of Batesian mimicry, however can it be possible not to believe in the power of natural selection? Even the set of butterfly examples that Bates himself presented are enough and there

simply has never been any other remotely credible explanation for these
except for natural selection. Returning to the conjunction in Ann Arbor, it
is fair to say that that conversation became a kind of three-way wrestle
with Zahavi, me plus Wallace Dominey, and finally Mark Kirkpatrick in
the corners of a triangular boxing 'ring'. On the canvas that day Zahavi was
mostly the winner and remains so on the most major issue: numerous cases
are now known that bear out his insistence on choice for true worth. The
principle in biology of what I prefer to call the Zahavi–Grafen principle,
because Grafen quantified an otherwise vague notion, is no longer in any
doubt. Choices based on this principle can be either on behalf of the present
generation—that is, choice for the benefit of the chooser personally—or
for benefit of that individual's offspring, and evidence continually
accumulates for the high expense, so hard for weaklings to pay, involved in
most of the signals. Disfavouring Mark Kirkpatrick in the triangular
struggle, most studies that sought for pure arbitrariness in sexual selection
have failed to find it.[14] In the third corner Wallace and I stand together
looking dubious how much we have won—maybe we raise a half arm each.

Yearly the cases of positive heritability of fitness and of signal size
accumulate, showing us doing well in that respect, but the heritabilities just
mean that there must exist 'good genes' of some sort.[15] For the present this
favours just 'Wallacean' views generally, an adjective that conveniently
covers both the 'Dominey' and the 'Alfred Russel' versions I have
mentioned because neither evolutionist specified parasites in particular as
important agents of unfitness. Mere fitness heritability in the evolutionary
sense doesn't specifically favour my version and most still deny proof that
the varied genes that must underlie the fitness heritability are genes
concerned with infection and parasites. The concept of a more general
variability for 'good genes' has been around for a long time. Early and
again late in the 1970s both Robert Trivers and Amotz Zahavi
independently assumed this heritability in their arguments and so have
many others. Yet, if not my kind of good genes, what are these variable
entities in the gene pool coding and how are they staying so variable—in
short, why should I believe in variability other than that for which I know I
have evidence? Some say the variants are simply any unmutated genes that
still do what they were always evolved for, while their counterparts creating
alternatives are defective versions due to mutation; also, it is suggested, the
processes of sex and sexual selection are honed mainly for excluding/

avoiding the accumulation of such bad mutations. All this is a step better, to my mind, than claiming the processes to be connected with 'tangled banks'—that is, situations varying in space.[16] But still one has to ask, can there be enough of such degenerative mutation as to justify all the immense effort that sex implies? Also, isn't it a better answer for a species simply to dry up the flood of mutation at its source, if this really does so threaten; that is, why not evolve to lower mutation rates? These sides of the mutation-clearance version of sex theory seem little discussed and yet the evidence that the mutation rates could be very different, and sometimes are, is strong.

   As for honesty in the signals, I am not a complete believer in this because of the issue of bluff that I have already raised. On the other hand, I have long nailed colours to the mast that sexual selection will not be found to be wholly arbitrary, and have even pledged an unusual (for me) acceptance that whole-species selection (which means simply the process of species extinction) may have to be playing its part in the explanation of how taxonomic groups with rampant sex-signalling exaggerations survive and speciate as well as they do.[17] Yet again the highly ornamented and worldwide galliforms can provide examples. Sexual selection practically assisting such groups to resist the parasites that are long entrenched with their clades—and speciated many times in parallel within them—may play a part in this resilience. In the case of galliforms possibly we can be even more specific and point to coccidian gut parasites as prime movers. For Marlene and me it would be a triumph for the original idea if it comes to be shown that each feature of mate choice improves avoidance of a particular chronic parasite each species suffers from. Such an idea has been both discussed and hinted at by evidence.[18] A further tentative example here might be the exaggerated developments in the feathers surrounding the cloaca in both peacocks and birds of paradise. These seem rather obviously appropriate to multi-million-year hassles these birds have clearly undergone with diarrhoea-inducing coccidians. The *Eimeria* versions of bird coccidian parasites seem somewhat more virulent in general[19] than are the *Isospora* versions that passeriforms more typically suffer from.[20] The fact that in both groups species of *Eimeria* are the typical gut parasites even though birds of paradise are passeriforms and thus 'ought' to be hosting *Isospora* species is thus a further hint. Are these hind-end feather displays convergent and honest revelations of freedom from diarrhoea, or is the convergence in parasite genus and the manner of display merely a coincidence? In both pheasants and birds of paradise affording the female a view of the cloacal region and its feather tracts appears intrinsic to the display.

Similarly unfeathered patches and structures in birds, including bright legs, may be special for displaying varying conditions of the blood:[21] in this case it is chicken, turkey, and duck breeders who know already how they reveal diseases. But again there is a snag here and there are no statistics, and perhaps also too many ailments affect the blood, for the point to be very appealing. What is clear is that other theories offer nothing like such detailed predictions of 'meanings' for sexually selected exaggerations. So, as in the case of the predicted ecological correlations that I believe go with 'sex itself', favouring the Parasite Red Queen view, my best course is to try to give statistically supported and satisfying meanings for as many correlations as possible and especially for those about which the rival theories are silent. For sure, however, not all health signals can be expected to be disease-specific. I have mentioned, for example, that brilliant rainbow of Michigan birds that, contrasted to the rather dull set from California, was a factor drawing Marlene to the idea in its early stages. Intensity of infection by a feather mite in the house finch, and the extent of sores due to a virus, both of these measured just before a moult, correlate well and negatively with the degree of redness in the male's plumage after the moult: redness in house finches, it seems, 'reads out' parasite affliction generally,[22] while that the females like redness in their males had been proven previously.[23]

According to the mutation-elimination theory of sexuality sex should have no ecology. My expectations for the Red Queen view, on the other hand, often have an ecological twist and so far the data often accompany this. The Red Queen extension of that tries to predict the intensity and cues of mate choice ecologically and geographically, but for the 'sosigonic' view of sexual selection as I choose to call it,[24] the ecological, geographical, and pathological correlations tend to be a bit more obscure. One problem is, where do parasites flourish best, on islands or mainlands? Working on our paper Marlene and I gave some thought to birds on islands and decided that we found a preponderance of examples where showiness of island races was reduced. Probably for the most part, however, we were just accepting Ernst Mayr's authority on this matter,[25] rationalizing his claim to our theme on grounds that in colonizing an island a species might often leave parasites behind, especially those with the more complex life cycles that require presence of other species. This can happen.[26] But it is often also the case that when a host species reaches an island, and there is no close competitor, it becomes more abundant there than it was on the mainland: iguanid lizards on the Galápagos Islands are one example. Any parasite that colonizes along with an immigrant host, therefore, may find easier conditions for transmission. Intriguing examples are those reindeer that so astonishingly succeeded to colonize Svalbard, trotting there from Norway, it seems, in what must have been almost a hopeless venture—more astonishing even, perhaps, than Morioris attaining to the Chatham Islands in the Pacific, for the reindeer went across 400 miles (640 kilometres) of frozen Arctic Ocean. There the almost equally doughty first explorers from Russia and from Norway found them: reindeer smaller than mainland forms but proportionately with even larger horns— so not diminished but intensified sexual selection! As adults they have no predators but recent studies found that they took with them a plentiful supply of parasites.[27] On the islands these parasites, synergizing their detriment, of course, with the extreme weather conditions, the commonly marginal nutrition of the deer, and the absence of wolves, became major elements of mortality. Might the extra large horns reflect less need to emphasize to your wives your running ability, more need to show them how you could garner abundant calcium out of your forage—in spite of interference and haemorrhages due to your parasites?

Literature I have read since our joint paper has suggested to me that while sexual selection sometime does reduce in intensity on islands, there is perhaps an even longer list of cases where its signs become exaggerated. In this respect Robinson Crusoe Islands, already suggested in the press as a good haven where General Pinochet may escape not so much parasites but angry descendants of his victims, may well be a good place also for me to retire to, for, I hope, a quite different reason. On that island group also live mammals, birds, and plants that seem quite specially concentrated and specially evolved to test my theories, though it may be a moot point whether either I or the target species can survive long enough for the job to get done. The astonishing adventive (exotic) birds on Robinson Crusoe Islands, the old species and the young, have all come under pressure from the latest and weediest adventices—goats and brambles. These islands are more properly known as the Islas Juan Fernández although even in an Italian atlas I saw them marked as the 'Robinson Crusoes', following Londoner Daniel Defoe who wrote up the Crusoe story but was never nearer to them himself than the whole diameter of the planet. They stand out in the Pacific to the west of Chile rather as Svalbard does in the Arctic Ocean to the north of Norway. Hummingbirds have colonized them twice, and these events achieved by the world's most southerly 'hermit' species of the group must make a story as near incredible as that of those north-strolling reindeer or of the aboriginal Morioris. The first arrivals—obviously at least two individuals needed to arrive near simultaneously—had long ago evolved to a new species, which is now the most sexually dimorphic hummingbird found anywhere in the world; in other words it is one of the island exaggerations of sexual selection that I mentioned above. Even in the hummingbird jewel box of the Andean Neotropics there exists no more dimorphic pair. Much later came the second arrivals, offshoots again of the same rather dull and most extreme southerly mainland species. Again a minimal two must have arrived, for after their landfall they clearly did not hybridize with the established changeling cousins. The upshot is that now two species inhabit the islands. The second comers remain drab in both sexes and probably at the present point are no more than subspecifically distinct from the mainland form. The sexual dimorphism is modest, to the degree otherwise characteristic of their 'hermit' group. Thus the big puzzle is what happened with the first successful invasion, and the question I particularly ask myself, of course, is what parasites once suffered, and perhaps still suffer, each of those three populations—Mainlanders in Chile, Arrivals 2, and, finally and most interesting, the now fantastic and dimorphic Arrivals 1?

Ninety and more in age, and long after 'H&Z' has been consigned to some far limbo where the pages of symposia (see Chapter 17) on 'Wrong steps of science' lie forgotten and wait to be burned, eventually I'll return like Crusoe from my retirement on actual Crusoe's islands and reveal my findings. Whether I'll have then proved 'H&Z' right via the islands' hummingbird story or whether (these birds having gone too low for further study, or even being gone completely) I will have proved it by studying sex preferences of the goats plus their diseases (there are plenty of goat parasites, of course, and some do seem to follow inevitably to island populations), I am not sure: in any case it will be proved and I hope in time to intercede for the symposium volume. My line on 'sex itself' will also have been experimentally proved out on the island because using my principles I will have both re-sexualized and controlled the parthenogenetic bramble that also recently and ruthlessly invaded. The paper of Chapter 16 will sketch my background for the bramble story. With Chilean permission (available in principle I assume because they have already done what I intend on the mainland) I will be taking with me a rust fungus, plus also perhaps a few other special parasites of *Rubus* (there are good ones from my point of view, of great interest for their own sexuality). The goats and the brambles having been lowered to reasonable levels and the wonderful hummingbird brought back from the verge of extinction, finally I'll turn to my theories and rescue them likewise. In short, by the special and abundant parasites of all the three species, and perhaps also by watching the sexuality of the control agents I will have introduced, I will come back to show you how Marlene and I had been always always, in all matters of this chapter, completely right.

Notes and references

1. W. H. Auden, *Collected Poems* (Random House, New York, 1976).
2. J. Weiner, *The Beak of the Finch* (Knopf, New York, 1994).
3. There are many other kinds of lags for parasites besides, of course, the sometimes multiple 'vector' stages. Sometimes single vectors can store parasites for long periods. A tick has been shown able to starve for 11 years and at the end still to transmit the spirochaete for relapsing fever (E. N. Pavlovsky, *Natural Nidality of Transmissable Disease*, tran. N. D. Levine (University of Illinois Press, 1966), p. 23). *Ascaris* eggs can survive and infect humans and guinea pigs after 7 or even 10 years of dormancy in the soil (G. W. Storey and R. A. Phillips, Survival of parasite eggs throughout the soil profile, *Parasitology* **91**, 585–90 (1981); A. N. Brudastov *et al.*, Infectivity of eggs of ascarid (*Ascaris lumbricoides*) for guinea-pigs and human beings ten years after maintenance in soil [in Russian], *Medicinskaya Parazitologiya i Parazitarniye Bolezni* **39**, 447–51 (1970)).
4. O. W. Olsen, *Animal Parasites: Their Life Cycles and Ecology* (University Park Press, Baltimore, 1974).
5. Later I saw a way in which I could use these lists relatively unbiased by amount of study and with the help of Gene Mesher I succeeded to prove that showy ducks tend to more infestation with single-host parasites (relative to with generalists) than dull birds do, just as the theory of parasite-mediated sexual

selection would predict. This still-unpublished finding, which survived our re-examination of it at several taxonomic levels and partitionings, helped confirm my confidence in the theory's predictions. This study, however, came much later than the time I am describing.

6. L. Raberg, M. Grahn, D. Hasselquist, and E. Svensson, On the adaptive significance of stress-induced immunosuppression, *Proceedings of the Royal Society of London B* **265**, 1637–41 (1998); M. Zuk and T. S. Johnsen, Seasonal changes in the relationship between ornamentation and immune response in red jungle fowl, *Proceedings of the Royal Society of London B* **265**, 1631–5 (1998).

7. Later, the house finch became the subject of a series of neat studies initiated by Geoffrey Hill on the relation of parasites to redness and thereby to sexual selection. The latest I know is G. E. Hill, Plumage redness and pigment symmetry in the House Finch, *Journal of Avian Biology* **29**, 86–92 (1998), and this can serve as key to the others. For a negative result on a related finch, see G. Seutin, Plumage redness in redpoll finches does not reflect hemoparasitic infection, *Oikos* **70**, 280–6 (1994).

8. A. Zahavi, *The Handicap Principle* (Oxford University Press, Oxford, 1997).

9. A. Grafen, Biological signals as handicaps, *Journal of Theoretical Biology* **144**, 517–46 (1990).

10. T. Veblen, *The Theory of the Leisure Class* (MacMillan, New York, 1899).

11. W. J. Dempster, *Natural Selection and Patrick Matthew* (Pentland Press, Edinburgh, 1996).

12. Yet again at a tangent from this book's topic, Darwin was a lifelong cigarette smoker; Patrick Matthew in his book *Immigration Fields* (1839) inveighed against smoking.

13. A. Pomiankowski, The evolution of female mate preference for male genetic quality, *Oxford Surveys in Evolutionary Biology* **5**, 136–84 (1988); A. Grafen, Sexual selection unhandicapped by the Fisher process, *Journal of Theoretical Biology* **144**, 473–516 (1990).

14. R. V. Alatalo, J. Hoglund, and A. Lundberg, Patterns of variation in tail ornaments of birds, *Biological Journal of the Linnean Society* **34**, 363–74 (1988). For many other examples one has to admit so far, however, a semblance of complete arbitrariness—see, for example, the earwigs described in note 17 below. Mark Kirkpatrick possibly no longer holds the rather extreme position I attribute to him in the text, as is suggested by a recent article (M. Kirkpatrick and N. H. Barton, The strength of indirect selection on female mating preferences, *Proceedings of the National Academy of Sciences, USA* **94**, 1282–6 (1997)).

15. W. Dominey, Sexual selection, additive genetic variance and 'the phenotypic handicap', *Journal of Theoretical Biology* **101**, 495–502 (1983).

16. G. Bell, *The Masterpiece of Nature: The Evolution and Genetics of Sexuality* (University of California Press, Berkeley, 1982).

17. Extracting insects from beneath the bark of a dead, fallen jatoba tree (*Hymenaea*) in southern Brazil I once found what I thought was a species of earwig with six 'morphs' of its male, all the forms having weird and impractical-looking forceps while a quite different functional pair was on the supposed female. Presently slight but consistent differences showed me I had, however, collected six kinds of female also, each corresponding to a kind of male. Thus there were six species with what seemed to be exaggerated, different, and mostly functionally useless 'ornament-like' forceps in the males. All were obviously in the same genus, *Doru* (and all seemingly—this a separate puzzle—appeared disobeying the competitive exclusion principle by being in similar situations under bark in the same log). This log environment appeared to show a miniature of the cichlid-fish species-swarm situation in the African Great Lakes, which swarms likewise illustrate hyperspeciation associated with high sexual selection. Other parallels are Hawaiian *Drosophila* flies, Lake Baikal 'water shrimps', New Guinea bower birds, birds of paradise, and others. If parasites indeed make for high sexual selection, their presence seems beneficial rather than the reverse to species-level evolution—that is, they seem to encourage abundant speciation (pro and con this view of bright pattern in birds, see A. P. Møller and J. J. Cuervo, Speciation and feather ornamentation in birds, *Evolution* **52**, 859–69 (1998) and R. V. Alatolo, L. Gustafsson, and A. Lundberg, Male coloration and species recognition in sympatric flycatchers, *Proceedings of the Royal Society of London B* **256**, 113–18 (1994)).

18. C. Wedekind, Detailed information about parasites revealed by sexual ornamentation, *Proceedings of the Royal Society of London B* **247**, 169–74 (1992); C. Wedekind, Lek-like spawning behaviour and different female mate preferences in roach (*Rutilus rutilus*), *Behaviour* **133**, 681–95 (1996).

19. N. Hillgarth, Parasites and female choice in the ring-necked pheasant, *American Zoologist* **30**, 227–33 (1990).

20. D. C. Boughton and J. Volk, Avian hosts of *Eimeria coccidia*, *Bird Banding* **9**, 139–53 (1938).

21. B. Kouwenhoven and C. J. G. Van der Horst, Disturbed intestinal absorption of vitamin A and carotenes and the effect of a low pH during *Eimeria acervulina* infection in the domestic fowl (*Gallus domesticus*), *Zeitschrift für Parasitenkunde* **38**, 152–61 (1972); P. Yvoré and P. Maingut, Influence de la

coccídiose duodénale sur la tenseur en carotenoides du serum chez le poulet, *Annales de Recherche Vetérinaire* **3**, 381–7 (1972); M. D. Ruff, W. M. Reid, and J. K. Johnson, Lowered blood carotenoid levels in chickens infected with coccidia, *Poultry Science* **53**, 1801–9 (1974); U. D. E. C. Ogbuokeri, Pathogenicity, immunogenicity and control of several isolates of *Eimeria maxima*, and effects of several different *Eimeria* species on spermatogenesis of chickens, Ph.D. thesis, Auburn University, Alabama (1986).

22. C. W. Thompson, N. Hillgarth, M. Leu, and H. E. McClure, High parasite load in House Finches (*Carpadocus mexicanus*) is correlated with reduced expression of a sexually selected trait, *American Naturalist* **149**, 270–94 (1997); P. M. Nolan, G. E. Hill, and A. M. Stoehr, Sex, size and plumage redness predict house finch survival in an epidemic, *Proceedings of the Royal Society of London B* **265**, 961–5 (1998).

23. G. E. Hill, Redness as a measure of the production cost of ornamental coloration, *Ethology, Ecology, and Evolution* **8**, 157–75 (1996); A. V. S. Hill, Genetics of infectious-disease resistance, *Current Opinion in Genetics and Development* **6**, 348–53 (1996); G. E. Hill, Plumage redness and pigment symmetry in the House Finch, *Journal of Avian Biology* **29**, 86–92 (1998).

24. This could also be called appropriately the 'Wallacean' variant of Darwinian sexual selection (H. Cronin, *The Ant and the Peacock*, Cambridge University Press, 1992) though I think my word less cumbersome and likely to prove nearer to the true explanation for most secondary characters.

25. E. Mayr, *Systematics and the Origin of Species* (Columbia University Press, New York, 1942).

26. D. W. Steadman, E. C. Greiner, and C. S. Wood, Absence of blood parasites in indigenous and introduced birds from the Cook Islands, South Pacific, *Conservation Biology* **4**, 398–404 (1990); M. Petrie and B. Kempanaers, Extra-pair paternity in birds: explaining the variation between species and populations, *Trends in Ecology and Evolution* **13**, 52–8 (1998).

27. O. Halvorsen, Epidemiology of reindeer parasites, *Parasitology Today* **2**, 334–9 (1986).

HERITABLE TRUE FITNESS AND BRIGHT BIRDS: A ROLE FOR PARASITES?[†]

W. D. HAMILTON and MARLENE ZUK

―――

Combination of seven surveys of blood parasites in North American passerine bird species reveals weak, highly significant association over species between the incidence of chronic blood infections (five genera of protozoans and one nematode) and striking display (three characters: male 'brightness', female 'brightness', and male song). This result conforms to a model of sexual selection in which (1) coadaptational cycles of host and parasites generate consistently positive offspring-on-parent regression of fitness, and (2) animals choose mates for genetic disease resistance by scrutiny of characters whose full expression is dependent on health and vigour.

Whether mate choice could be based mainly on genetic quality of the potential mate has been a puzzle to evolutionary biologists.[1-3] Population genetic theory predicts that any balanced polymorphism for a selected trait ends with zero heritability of fitness, so that no one mate is better for 'good genes' than any other. However, females of many species act as if they are choosing males for their genes; thus 'good genes' versions of sexual selection have been frequently, albeit tentatively, suggested.[3-7] Here we propose a way out of the difficulty via a previously unconsidered mechanism of sexual selection, and give preliminary evidence for the operation of the principle in North American birds.

There may exist a large class of genes with effects on fitness that always remain heritable. The genes are those for resistance to various pathogens and parasites. The interaction between host and parasite (parasite here being interpreted in a broad evolutionary sense[8]) is unusual because it so very readily produces cycles of coadaptation. These cycles can ensure a continual source of fitness variation in genotypes.

To illustrate, imagine a host and parasite population, each with the two alternative genotypes H, h and P, p, respectively. For simplicity, assume the

[†]*Science* **218**, 384–7 (1982).

organisms are haploid, although diploid models can easily and realistically be made to work in a similar fashion. An H individual is resistant to pathogen type p, but susceptible to P, and vice versa for h individuals. The parasite, of course, flourishes in an individual host that is susceptible and dies (or is less productive) in a host that is resistant.

If a female chooses an H male when p will be the more common parasite genotype in the next generation, she is obviously getting a selective advantage, since her offspring will be more likely to be resistant to disease. As selection proceeds, both by the basic advantage of H when p is common and by the enhancement through any preference for H, the usual problem of variation damping out as all individuals become resistant might be envisioned; but meanwhile selection has been operating within the parasite population and has been favouring P. As the proportion of P individuals increases, the advantage of H falls; h then begins to increase in frequency and becomes the better genotype for females to choose.

Such a system usually has an equilibrium point where all four genotypes could occur together. But theory predicts that, given the pattern of host–parasite genotype interactions outlined above, this equilibrium point is unlikely to be stable.[9–11] If it is unstable, then a limit cycle, or at least a permanently dynamical behaviour of some kind, is instead the probable outcome. Cyclical selection affecting one locus in the host implies that there must be two generations per cycle where heritability is negative—those where advantage switches from H to h and vice versa. But it also implies that as the cycles lengthen the *mean* parent–offspring correlation in fitness must become positive, with 0.5 as the asymptotic upper limit. If the cycles are very short, then trying to choose mates for the 'right' genes for resistance is a perverse task; an animal might even do best to seek a 'worst-looking' mate. Despite a theoretical possibility (see Chapters 2 and 5), it is not clear yet if extremely short cycles (for example, period 2) are likely to occur in nature. Nevertheless, cycling could be relatively rapid under, in general, conditions involving intense selection pressure and pathogens that are short-lived and highly mobile and infectious (Chapters 2 and 5). On the other hand, weak selection, approximate equality of generation time of host and parasite, and also any lag in the feedback (such as long-dormant infective eggs or a long-lived vector) tend to create long cycles. Up to a point,[12] these should favour sexual selection.

Broadening an *a priori* case for cycles generally it may be noted (1) that epidemic rather than steady occurrence of disease can be involved without losing the tendency to cycle, and (2) that existence of two or more species of parasite differently virulent to host genotypes hardly differs conceptually from the case of two or more genotypes in an asexual parasite species. In any case, when several cycles of differing characteristic periods are in progress at once, care in choice of a mate should be eugenically rewarding. Then a male (for example) who is unmistakably outstanding in health and vigour offers females that mate with him an inherited healthiness in their offspring that is reliably well above average.

No direct evidence exists for coadaptational cycles of the kind suggested, but they have never really been looked for. Since the search requires study over several generations of a host, the cycles may be hard to show even if common. The possibility of cycles is suggested by differing varietal suscepti-bilities to particular diseases within host species,[13–24] by differing virulences of parasite strains,[13, 25–28] and by both kinds of variation within single species-pair systems.[29–33] It has been claimed that any domestic species can have its resistance to almost any disease rapidly improved by selection.[34] This again suggests that disease resistance may be permanently heritable in nature and speaks against the common occurrence of static equilibria due to fitness over-dominance.

Suitable parasites for tests of the theory are those that debilitate their host rather than either kill it or allow total recovery after brief sickness. Death of potential mates preempts the need for choice, while, if recovery is so com-plete that affected animals are indistinguishable from those which were never susceptible, a selecting mate has no usable basis for discrimination. Admit-tedly some kill-or-recover diseases may leave records of their severity in the form of stunted growth or shabby plumage, and thus may be able to provide cues of the required kind; but such diseases are likely to be too rapidly evolving, as mentioned above, to give the longish cycles favourable to sexual selection. The ideal disease is one that can be acute and cause heavy juvenile mortality, but persists in chronic form in survivors either as an infection actually latent or as prolonged aftereffects (such as autoimmune disease) throughout later life.

How could animals choose resistant mates? The methods used should have much in common with those of a physician checking eligibility for life insurance. Following this metaphor, the choosing animal should unclothe the subject, weigh, listen, observe vital capacity, and take blood, urine, and faecal samples. General good health and freedom from parasites are often strikingly indicated in plumage and fur, particularly when these are bright rather than dull or cryptic.[35,36] The incidence of bare patches of skin, which may expose the colour of the blood in otherwise furry or feathered animals, and the number of courtship displays involving examination of male urine[37] are of interest in this regard. Vigour is also conveyed by success in fights and by the frequently exhausting athletic performances of many displaying animals.[38,39] Display characters in polygynous species often seem to go far beyond the obvious ways of advertising health—huge tail feathers, for example, or wattles pigmented almost so as to conceal blood colour rather than reveal it. Already proposed processes of exaggeration through sexual preference may account for this.[40–42] (However, differing from Kirkpatrick,[42] if expression of display is conditional on sufficient reserves and these upon health,[7] there is less objection to females evolving preference for 'handicapped' males, the female's object being, of course, not handicap itself but a demonstration of health that cannot be bluffed. This roughly follows Zahavi;[43] but use of the word handicap and implication of an unconditional display character seem unfortunate. Even if females merely choose victors of combats, expensive unbluffable displays are still expected to evolve,[37] but, reinforcement through evolving preference being absent, the character should be less exaggerated and its expression remain wholly conditional. This may be illustrated in morphs of horned beetles (as discussed by Eberhard[44]).)

If susceptibility to parasites is as important in sexual selection as this idea suggests, animals that show more strongly developed epigamic characters should be subject to a wider variety of parasites (except for purely acute pathogens). In species where disease is relatively unimportant, or where only acute diseases occur, sexual selection should be less apparent and the animal less showy. Whichever sex does the choosing picks individuals with the fewest parasites and the highest resistance; the point is that such choice by one sex and advertisement of good health by the other is needed most in species where chronic parasites are common to begin with. Our hypothesis is contradicted if *within* a species preferred mates have most parasites. Direct evidence on this is very scanty and equivocal. Hausfater and Watson[45] found that egg counts of nematode eggs in yellow baboon faeces correlated positively with dominance rank. Freeland[46] gave male mice varying doses of nematode larvae and found that the level of infection correlated negatively with subsequent dominance. There are many recorded cases of emaciated animals proving to have heavy loads of parasites; such individuals could hardly be dominant. Hausfater and Watson hint that the nematode–baboon result might reflect greater food intake of dominant animals more than difference in worm burden or worm damage.

Our idea is supported if *among* species those with most evident sexual selection are most subject to attack by debilitating parasites. A test we have conducted supports the last point. We used data already in the literature on comparative parasitaemias of North American birds. In fairly comparable studies in different places, blood smears were taken from mist-netted and trapped birds. Fairly standard numbers of microscopic fields were searched for the presence of various haematozoa and usually also for microfilarial worms.[47] Data from two locations were substantial and particularly well suited to our analysis: Algonquin Park in Canada (Alg),[48] and Cape Cod (CC).[49] Four small surveys of birds in South Carolina and Georgia were included as one set (SCG);[50–54] these did not completely record microfilariae. Finally, a small set from the District of Columbia (DC) was included.[55] The DC set alone recorded *Toxoplasma*. Each of the six generic 'diseases'— *Leucocytozoon, Haemoproteus, Plasmodium, Trypanosoma, Toxoplasma,* and *Microfilaria*—was recorded when discovered as having P_{ij} cases of disease among n_j birds, where i is disease and j is bird species. Multiple infections were counted as one case for each disease involved. The diseases all seemed such as could contribute to the driving of our model. Most tend to be severe and fatal in young birds and then chronic or with long latent periods in the survivors; for some, hints of both variation in resistance[56, 57] and epidemic cycles[58] have been reported.

From the species lists, each sex of each passerine bird species was ranked by one of us who had not yet seen the disease surveys on a showiness scale of 1 to 6, with 1 being very dull and 6 very striking. Male scarlet tanagers were thus assigned a 6, while most male warblers rated a 3 or 4. No passerines were dull enough to rate 1 (chimney swifts, however, would fall here). The songs of male passerines were ranked similarly, on the basis of variety and complexity, by an expert in bird songs who was supplied with only the list of species and not their disease frequencies. Adult body weights for each species were also added to the data file. Since non-passerines were in very varying proportions both among themselves and with respect to the passerines, and were always a rather small minority, they were discarded from the analysis.[59] The passerines surveyed included 109 species and 7649 individual birds.

For each location (and for Algonquin juveniles and adults separately), we constructed ordered contingency tables for association of the display characters male brightness, female brightness, and male song, with the incidence of each disease. First we constructed disease expectations per bird: $X_{ij} = P_{ij}/n_j$. Then to spread the X_{ij} and give a discrete table when distributed over display ratings, we used, in most cases, 10 X_{ij} rounded to the nearest integer, so that the table set out 'expected cases per ten birds'.[60] From the contingency tables, Goodman–Kruskal gammas were calculated as the coefficients of association.[61, 62] Our choice of this in preference to more familiar multiple regression methods followed from the very non-normal distributions of the P_{ij}. For skewed distributions and for sparse entries in its tables, the Goodman–Kruskal method, on the null hypothesis of no association, leads robustly to a standard normal distribution of the following statistic: the gamma estimate divided by its standard error.[61] This is what we tested.

The tests lead us to reject the null hypothesis of no association. The associations are much more commonly positive than negative and often are positive individually at significant levels (Fig. 6.1). Coefficients for female brightness are less positive and significant than those for male brightness and song. These conclusions apply broadly across localities and diseases, although diseases are erratic in the trends shown in the different localities.

Toxoplasma was exceptional in giving all negative coefficients in the one set recording it; two of these were significantly negative at the 5 per cent level. Algonquin data (the most northerly set and by far the largest) gave the strongest evidence of association, while Cape Cod (the second most northerly

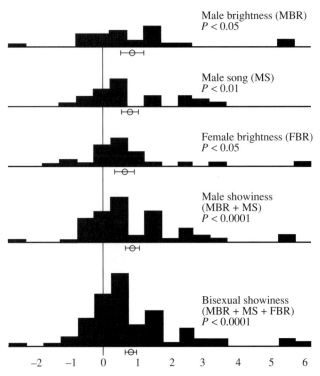

Figure 6.1 Distributions of measures of association of display characters with levels of incidence of six blood diseases in North American passerine bird species. Each of the first three histograms distributes the measures from 26 tables obtained from the following schema: 1 character × [5 diseases × (1 place × 2 ages + 2 places × 1 (all) ages) + 6 diseases × 1 place × 1 (all) ages]. The two lower histograms are accumulations of those above. Measure distributed is Goodman–Kruskal G divided by the standard error. It is grouped for intervals −0.25 to 0.25, 0.25 to 0.75, with 0, 1, and so on marked on the base line scale. On a null hypothesis of no association a standard normal distribution on this scale is expected. The bars show observed means ± standard errors.

and third largest set) gave no significant coefficients at all. The negative results for *Toxoplasma* are unexpected, but significantly positive results would not be expected either; this parasite is an extreme generalist regarding its secondary hosts, including birds, and the feedback from any one host species to the *Toxoplasma* gene pool must be correspondingly slight and coevolutionary cycles consequently are unlikely.[63] Values of gammas themselves are seldom high, 89 per cent being below 0.4. That they are on the whole mildly positive and that numerous values are needed before their trend is established conforms to theory, for it would be most surprising if our first-selected subset from among all disease organisms—namely, the subset of the blood parasites—were to provide all relevant activators of the cycles that underlie sexual selection. In our results it is the high gammas (two reach almost 0.8) that are puzzling but at present they are easily regarded as fortuitous.

The positive association found for female brightness might follow from imperfect sex limitation of gene expressions only selected for in males. However, we prefer the idea that in monogamous birds sexual selection aiming at genes for health can be working in both sexes at once. As with other suggestions for sexual selection under monogamy, this precludes the very large fitness differences that can occur among males under polygyny and also precludes extremes of a runway process.[40,41] Under any of these views sexually symmetrical ornament is expected to be relatively modest. This is so. However, there is also a widely reported phenomenon often involving bisexual ornament that seems explicable by our view and not by those alternatives where an advantage in natural selection merely primes sexual selection,[40,41] or where the whole process is imagined to work on arbitrary characters from the start.[42] Both bisexual brightness and degree of sexual dimorphism tend to decline on islands.[64,65] In our view this could happen because bird species escape from many of their parasites when they colonize islands.

Correlations of disease susceptibility (measured by X_{ij}) and the display characters (including body weight) indicated that in two of the data sets, log weight correlated more strongly with disease susceptibility than did any other character. These data sets (Alg and SCG) were also those that conformed best to our hypothesis for male display characters. Larger birds may provide a bigger target for the biting flies that carry protozoan diseases; or they may tolerate bites more because each bite is a relatively smaller subtraction of blood. Thus our result could be compatible with the predictions of classical sexual-selection theory, since greater sexual selection and showiness are often connected with larger body size[38,39] and this is easily rationalized through the superiority of large males in fighting. Still, if size increase occurring for any reason leads to more disease,[66] increased sexual selection through our suggested linkage is still expected and may entrain, among other things, further increase in size.

Other alternative explanations for the result could be offered. Evaluation of the many plausible and overlapping theories proposed about display and sexual selection, including our new one, is a daunting task and must be left to the future. For the present we conclude that eugenic sexual selection can work and may be common, and that our results hint at chronic disease as one agitator of the dynamic polymorphism that such selection requires.

Acknowledgements

We thank M. Perrone for special assistance with bird songs, and Wallace Dominey, Ilan Eshel, Paul Ewald, Peter Grant, Lester Lee, Trevor Price, and Richard Wrangham for helpful discussion of the work while in progress.

References and notes

1. J. Maynard Smith, *The Evolution of Sex* (Cambridge University Press, London, 1978).

2. G. Borgia, in M. A. Blum, and N. S. Blum (ed.), *Sexual Selection and Reproductive Competition in Insects*, pp. 19–80 (Academic Press, New York, 1970).

3. L. Partridge and T. R. Halliday, in J. R. Krebs and N. B. Davies (ed.), *Behavioural Ecology: An Evolutionary Approach*, pp. 222–50 (Blackwell, Oxford, 1984).

4. R. L. Trivers, in B. Campbell (ed.), *Sexual Selection and the Descent of Man 1871–1971*, pp. 136–79 (Aldine, Chicago, 1972).

5. P. J. Weatherhead and R. J. Robertson, *American Naturalist* **113**, 201–8 (1970).

6. A. C. Janetos, *Behavioral Ecology and Sociobiology* **7**, 107–12 (1980).

7. M. J. West Eberhard, *Proceedings of the American Philosophical Society* **123**, 222–34 (1979).

8. R. M. Anderson and R. M. May, *Nature* **280**, 361–7 (1979).

9. I. Eshel and E. Akin, *Journal of Mathematical Biology* **18**, 123–33 (1983).

10. J. W. Lewis, *Journal of Theoretical Biology* **93**, 927–85 (1981).

11. R. M. Anderson, in N. Keyfitz (ed.), *Seminar on Population and Biology*, pp. 249–68 (Ordina Editions, Liège, 1981).

12. Even if long cycles bring the mean offspring-on-parent regression near to $\frac{1}{2}$, there may be little mean variance in fitness for choice to work on because relevant genes are most of the time near to fixation. I. Eshel (personal communication) suggests that a product of the standard deviation and the parent–offspring correlation of genetic fitness variation would best reflect potential effectiveness of mate choice; thus the longest cycles do not necessarily give most power to sexual selection. But more theory is needed here.

13. J. W. Gowen, *Annual Review of Microbiology* **2**, 215–54 (1948).

14. J. E. Ackert, *Journal of Parasitology* **28**, 1–24 (1942).

15. M. M. Rosenberg, J. E. Alicata, and A. L. Palafox, *Poultry Science* **33**, 972–80 (1954).

16. V. M. King and G. E. Cosgrove, *Laboratory Animal Care* **13**, 46–8 (1963).

17. H. O. McDevitt and A. Chinitz, *Science* **163**, 1207–8 (1969).

18. C. E. Thorsen, in W. H. Taliaferro (ed.), *Immunity to Parasitic Animals*, Vol. 2, p. 913 (Appleton, New York, 1970).

19. H. W. Moon and R. J. Dunlop (ed.), *Resistance to Infectious Diseases* (Saskatoon Modern Press, Saskatoon, 1970).

20. G. J. Eaton, *Laboratory Animal Science* **22**, 850–9 (1972).

21. F. Lilly and T. Pincus, *Advances in Cancer Research* **17**, 231 (1973).

22. D. Wakelin, *Advances in Parasitology* **16**, 219–308 (1978).

23. D. L. Rosenstreich, *Nature* **285**, 436–7 (1980).

24. P. S. Brindley and C. Dobson, *Parasitology* **83**, 51–65 (1981).

25. R. D. Manwell and F. Goldstein, *American Journal of Hygiene, Series C*, **30**, 115–22 (1939).

26. J. E. Larsh, Jr, *Journal of Parasitology* **29**, 423–4 (1943).

27. C. G. Huff and F. Coulston, *Journal of Infectious Diseases* **78**, 99–117 (1946).

28. J. E. Larsh, Jr, *Journal of Parasitology* **37**, 343–52 (1951).

29. F. Lilly, in H. O. McDevitt and M. Landy (ed.), *Genetic Control of Immune Responsiveness: Relationship to Disease Susceptibility*, pp. 279–88 (Academic Press, New York, 1972).

30. P. B. McGreevey, G. A. M. McClelland, and M. M. J. Lavoipierre, *Annals of Tropical Medicine and Parasitology* **68**, 97–109 (1974).

31. J. M. Rutter, M. R. Burrows, R. Sellwood, and R. A. Gibbons, *Nature* **257**, 135–6 (1975).

32. P. F. Basch, *Experimental Parasitology* **39**, 150–69 (1976).

33. C. Dobson and M. E. Owen, *International Journal of Parasitology* **7**, 463–6 (1977).

34. F. B. Hutt, *Genetic Resistance to Disease in Domestic Animals* (Comstock, Ithaca, NY, 1958).

35. H. S. Peters, *Bird Banding* **1**, 51–60 (1930).

36. A. R. Jennings, E. S. L. Soulsby, and C. B. Wainwright, *Bird Study* **8**, 19–24 (1961).

37. B. E. Coblentz, *American Naturalist* **110**, 549–57 (1979).

38. M. S. Blum and N. A. Blum (ed.), *Sexual Selection and Reproductive Competition in Insects* (Academic Press, New York, 1979).

39. R. W. Wiley, *Quarterly Review of Biology* **49**, 201–27 (1947).

40. R. A. Fisher, *The Genetical Theory of Natural Selection*, 2nd edn, p. 151 (Dover, New York, 1958).

41. R. Lande, *Proceedings of the National Academy of Sciences, USA* **78**, 3721–5 (1981).

42. M. Kirkpatrick, *Evolution* **36**, 1–12 (1982).

43. A. Zahavi, *Journal of Theoretical Biology* **53**, 205–14 (1975); **67**, 603–5 (1977).

44. W. G. Eberhard, *Scientific American* **242** (3), 166–82 (1980); *American Naturalist* **119**, 420–6 (1982).

45. G. Hausfater and D. F. Watson, *Nature* **262**, 688–9 (1976).

46. W. J. Freeland, *Science* **213**, 461–2 (1981).

47. Light infections tend to be missed in these surveys. Since light infections might be prevalent in species in which sexual selection successfully combats parasites, such imperfect recording biases against our hypothesis.

48. G. F. Bennett and A. M. Fallis, *Canadian Journal of Zoology* **38**, 261–73 (1960).

49. C. M. Herman, *Transactions of the American Microscopical Society* **57**, 132–41 (1938).

50. P. E. Thompson, *Journal of Parasitology* **29**, 153–5 (1943).

51. J. W. Hart, *Journal of Parasitology* **35**, 79–82 (1949).

52. A. V. Hunninen and M. D. Young, *Journal of Parasitology* **36**, 258–60 (1950).

53. W. E. Collins, G. M. Jeffery, J. C. Skinner, A. J. Harrison, and F. Arnold, *Journal of Parasitology* **52**, 671–73 (1966).

54. A further good data set for South Carolina was overlooked until too late for inclusion: G. L. Love, S. A. Wilkin, and M. H. Goodwin, *Journal of Parasitology* **39**, 52–7 (1953).

55. P. W. Wetmore, *Journal of Parasitology* **27**, 379–93 (1941).

56. P. C. C. Garnham, *Malaria Parasites and Other Haemosporidia* (Blackwell, Oxford, 1966).

57. A. M. Fallis and D. O. Trainer, Jr, in J. P. Linduska and A. L. Nelson (ed.), *Waterfowl Tomorrow*, pp. 343–48 (US Department of the Interior, Fish and Wildlife Service, Washington DC, 1964).

58. V. T. Harris, *Wildlife Research: Problems, Programs, Progress* (Research Publication No. 104, US Department of Interior, Fish and Wildlife Service, Washington DC, 1972).

59. Inclusion of non-passerines probably would not have changed the overall trends. Some omitted groups would be against the hypothesis (owls) but others for it (grouse).

60. Such a table has 11 potential columns (0–10). But when a disease was rare sometimes only the first two columns, or even the first only, had non-zero entries; in such cases we used $100X_{ij}$ or, for one disease, $30X_{ij}$. Once several columns were present, the results were insensitive to the degree of multiplication.

61. L. A. Goodman and W. H. Kruskal, *Journal of the American Statistical Association* **58**, 310–64 (1963).

62. For tabulation and calculation we used the Osiris IV Data Management and Statistical Software System of the Institute of Social Research, University of Michigan, implemented by the Michigan Terminal System and the university computer.

63. A *post hoc* rationalization of negative toxoplasma results might be that a species successfully contending with several host-specialized parasites may have more chance of cross-resistance to newcomer generalist parasites from the same group. This point, however, emphasizes the need for proof that specialist parasites occur in sexually selected species, or, alternatively, proof that systems of multiple host and parasite retain the propensity to cycle.

64. E. Mayr, *Animal Species and Evolution* (Belknap/Harvard University Press, Cambridge, MA, 1963).

65. P. R. Grant, *Systematic Zoology* **14**, 47–52 (1965).

66. R. M. Geist, *Ohio Journal of Science* **35**, 93–100 (1935).

MAN OF THE
SAND-DUNE TEL

Parent–Offspring Correlation in Fitness under
Fluctuating Selection

Under the system [polygyny], the offspring, besides being equally
bright and brighter intellectually [than the offspring of monogamous
marriages], are much more healthy and strong.

LATTER-DAY SAINTS (c.1867)[1]

THE papers of this chapter and the next relate to Israel. The
young and ancient nation has been as important a contributor to modern
evolutionary theory as it has to modern fanaticism, with the two (plus
many other scientific achievements) making a curious mix in the present-
day country, like milk with creosote. The evolutionary side, which I regard
as the milk element because more useful to most people and allergenic only
for a few, makes it unsurprising that an important evolutionary symposium
was held in Israel in 1986. The paper of Chapter 8 was my contribution to
the symposium. That chapter will come back to the creosote streak as a
tangential issue of my experience there (fanaticism, of course, had no part
in the symposium itself). For the present chapter the point is that it is
unsurprising that Ilan Eshel, my co-author and a like-thinker of long
standing on many issues of evolution, as was mentioned several times in
Volume 1 of *Narrow Roads of Gene Land*, should live in Tel Aviv and
contribute to the mathematical side of an evolutionary paper. The Jewish
(which is not quite to say the Israeli) model of *Homo sapiens* seems to come
bundled with a cerebral coprocessor almost designed for handling genes and
probabilities in the mind, rather as the British model comes hardwired for
the invention and playing of ball games and for distributing them (together

with English language) around the planet. Neither national aptitude seems at first sight to have any plausible evolutionary background but probably they both do.

When I mentioned to Ilan I was having difficulty convincing people that substantial positive heritabilities of fitness could come out of coevolutionary cycles, he said the idea seemed likely enough to him. I then added, hoping to draw him in (as I had once drawn Robert May into a joint paper—Volume 1, Chapter 11), that there were two cases where such correlation, obviously, would not hold and, therefore, where doubters such as Mark Kirkpatrick could more possibly be right. One was a case like that of my *Oikos* paper in this volume (Chapter 2) where fitness decreases so immediately and fast when a genotype became common that one could obtain an alternation of optimal types, making fitness heritability negative. The other, I suggested, was when long lags in the frequency dependence and the conditions of strength of selection made cycles swing so slowly that there was time for gene frequencies to go so nearly to fixation for so much of the time that mean heritability must be almost zero. Ilan took longer to pronounce on this one and, when he did, he said he thought probably it might be so but it would be nice to have some analysis and show how it went. Counterpunching then with my crux and challenge, I said I would love to know about the intermediate cycle length that would be optimal for good genes' mate choice, if such should exist: it would give me a standard and would illuminate all my thinking about sexual selection! 'Standard?' Ilan frowned a bit and said he doubted anything could be so simple; surely it would depend on what sorts of cycle occurred and the fitness variance but probably—more cheerful again—we could get *something*, on the magnitude of heritability and its relation to the cycle length. The paper that follows shows what we did get. It consists of an analysis virtually all provided by Ilan, and, in the 'Introduction' and 'Discussion' sections mainly, the more diffuse stuff extending the line of 'Hamilton and Zuk' that was interspersed by me.

I forget when this conversation took place. Ilan is a scholar moving for the most part along an axis that includes California (Stanford) and Italy (Pavia) and his home in Israel but which consequently takes him easily through Britain and also through Detroit. He has been a likely invitee to the sorts of symposia that also invite me. We had been in correspondence from time to time since the early 1970s and he had also stayed at my home in Ann Arbor. Mathematician by training Ilan is an instinctive evolutionist: therefore evolution is the field in which his maths training has been applied. Seemingly he thinks very like me in most ways. Had I stood a few inches more to the right as I tightened the vice holding the bomb I was making in my father's workshop in 1949, so that shrapnel from its premature explosion entered my left ventricle instead of my right lung, Ilan with little doubt would have found and published most of the ideas for which I am well known, and would have done so to a schedule running no more than a few years later. Such, indeed, is often the true nature of 'priority' in science. At the same time he would certainly have provided the ideas with more sure logical foundations from the first than I did. On the other hand, groping, for my self-esteem, for a possibly negative side, he would have used more obscure titles to his papers and probably would have written no crazy essays (raised no 'oats', as I have called them) in support, like those I have used so often.

The fact is that Ilan independently discovered 'kin selection' only to find the ideas had very recently been published. Whether his ideas would have snowballed as fast as mine into a general appreciation might be a little more doubted, however, because I never thought him gifted in choosing titles for his papers. I imagine this lack of gift even now leads to much of his work being less appreciated than it deserves—maybe it's a penalty for publishing in a language not one's own to be a little more timid about titles, wondering just how they will sound. Yet who could guess, for example, that a paper of his entitled 'On the founder effect and the evolution of altruistic traits: an ecogenetical approach'[2] would contain the key early insights and predictions about disease virulence? Or guess that a paper he wrote with Ethan Akin on the 'Coevolutionary instability of mixed Nash solutions'[3] would be close in its theme to, and indeed almost making redundant, most of the findings about sex in this chapter and several others in this volume? For that key paper, if I had written it—as I would have been most proud to have been able—I'd have tried to pick out something both commensurate to its importance and at the same time, if possible, distinctive and enticing, though also, I fear, much more likely to border on frivolity and absurdity. For example, why not a hint of Rasputinesque scandal as implied by 'Lyapunov and the Red Queen: passion and instability'?—or, if one must be more sober, why not: '*Perpetuum mobile* in pursuit coadaptations'?

Ilan is also a philosopher and it is easy to see how he might easily have been a natural historian, too, had he not been first captured by mathematics. He feels strongly for the animals and plants around him, and, rather as I do, seems to take much of his evolutionary direction from this fascination. I suspect this was the reason for his early understanding of the idea of genetic altruism and, later, for his ready acceptance of a permanent heritability of disease resistance. That he could see a path of analysis ahead that would prove the ideas workable—that is, a path that would show, in the latter case, what sort of heritability it would be and how much—was secondary. In all humans the feeling of kinship to other animals simply has to run a deep course because it is there, built in, but some see this and jump lightly over it into a discussion of human attributes while others swerve—or, as they say about an unwilling horse in a steeplechase, 'refuse'.

If it is reasonable that human beauty is in very large part just hereditary health, as Marlene and I argued in Chapter 6, then such assessment is all the more true for animals.[4] For them no art of hairdresser or cosmetician, and no spoken lies enhancing the painted ones, can play much of a part. Beyond animals most of the ideas are reasonable for plants, too, in so far as these perceive other plants through their root contacts, through pollen received, through odours also blown in the wind, and in other ways.[5] Ilan sees and thinks on all this just as I do and, as to our animal interests, I noted other almost uncanny correspondences, such as that we both had personal stories about, of all things, the behaviour of persistently misanthropic crows. Crazy crows, crazy rememberers of them too, we were at one and all of us capable of the 'cultural distortions' or, as some might put it, 'just-so' stories, that can be built on such incidents. Might there not have been a remembering by the crows of harms done and then a generalizing of the events falsely—a crow's own 'just-so' story, in fact, this happening just sometimes to be wrong?

For my part I am not ashamed, instead proud, that a member of this greatly talented genus, *Corvus*, able in some cases to carefully fashion and use tools[6] in a very anthropoid manner, once falsely remembered me as an oppressor that I never was. Almost enjoying to be an injured innocent among crows is similar to my pride in name-dropping of my inward acquaintance with that ancient aristocrat of microbes, *Giardia*; likewise my pride to be so neanderthal—or is it *habilis*-like, as would be more African?—in my appearance that the male apes of two zoos picked me out, from all the crowd around me, as target for lumps of their hurled shit, and also rather as my school friends in more than one school picked me out to be named 'ape man' or 'cave man'. Such dis-affinity, indeed (to return to the crow story), as I experienced with a western American crow diving at me in Seattle, incidentally, led me to write in no mocking spirit in the introduction to Chapter 2 what I did about the 'crow's parliaments' reported to the museum curators Robert Payne and Robert Storer by old Michiganian ladies, and of their tolerance by the two men. When I had asked Payne personally why our marvellous and so sociable Ann Arbor crows (so different from English ones) weren't being studied by graduate students in his Bird Division he replied: 'Oh, we once tried but they're difficult.' There followed a typical Payne silence with no explanation. 'Difficult? How do you mean?', I said. 'Oh, you net them once and afterwards you never catch any—not just not the ones you caught, but none anywhere in the area, they seem to have learned what you tried.' Crows in Seattle, however, I thought later, certainly couldn't have known that twice in my boyhood, in Longsprings Wood, Well Hill, in Kent, I'd climbed to crows' nests in the great beech trees and had stolen one egg each time—or else can it conceivably be that a man, having done such a thing, gets that mean, crow-thief look in his eye that lasts all the rest of his life? But as for the primates and their turds, I swear that I have never once, even in childhood, wronged a gorilla.

As seems unsurprising on grounds that they both are in departments of Tel Aviv University, Ilan knows Amotz Zahavi well: but more surprising to me has been the mutual esteem between the two men, which I perceived was from both sides, with this seeming especially strange in the light of the open contempt in which Amotz holds all evolutionary modelling and especially that of kin selection, to which Ilan has contributed. Possibly Ilan's self-informed natural history plus such things as his sympathy for crows and their likes leads to acceptance by Amotz, these traits paralleling that intuition for behaviour in wild sources that Amotz himself so greatly developed.

The question I hoped our paper might answer—that of exactly what periods and what locus interactions under cycling would maximize the force of sosigonic sexual selection—isn't answered even yet for exactly the heuristic reasons Ilan originally gave me. My first idea had been groping for the impossible, as so often (but how I would still prefer, though, to be always groping for something—scorpions in a tree hole if need be—than never at all). Nevertheless in our paper we carried preparations for the answer to my original question to a point where anyone who has more definite ideas about their coadaptive situations can join us. Mark Kirkpatrick and Russell Lande once gave us Fisherian sexual selection set out with its proper maths and detail; Alan Grafen gave the Zahavian point of view similarly set out (while denying importance to the Fisherian one);[7] but the question of what regime is maximally powerful for sosigonic sexual selection still awaits its expositor. So are waiting also many questions about real life extremes of the process, whether these are the peacocks of the land, *Pavo cristatus*, and the peahens of the sea such as the copepod *Calocalanus pavo*—she with her train massive in proportion, if not truly as a peacock's, at least as a racket-tailed drongos's and all of it stretched out, vibrantly unstudied, in the Bay of Naples behind her.[8] Physical forces in all this I am treating as non-starters: but what parasites or other biotics drive such sexual paragons to their extremes? I still wait to know.

References and notes

1. Latter-Day Saints, *Journal of Discourses* **13**, 207 (c.1867–70).

2. I. Eshel, On the founder effect and the evolution of altruistic traits: an ecogenetical approach, *Theoretical Population Biology*, **11**, 410–24 (1977).

3. I. Eshel and E. Akin, Coevolutionary instability of mixed Nash solutions, *Journal of Mathematical Biology* **18**, 123–33 (1983).

4. Evidence for both these propositions begins to appear. A little caution is needed, however: just as intelligence is too complex to be fully summarized by any single dimension of measurement, so human beauty is more complex than just health, whatever that may mean at the start. Nevertheless broadly defined health clearly is a big component in human attraction: see, for example, T. K. Shackleford and R. J. Larsen, Facial attractiveness and physical health, *Evolution and Human Behaviour* **20**, 71–6 (1999) and S. W. Gangestad and D. M. Buss, Pathogen prevalence and human mate preferences, *Ethology and Sociobiology* **14**, 89–96 (1993). As an entry into the more extensive literature for animals, see K. Lagesan and I. Folstad, Antler asymmetry and immunity in reindeer, *Behavioral Ecology and Sociobiology* **44**, 135–42 (1998); A. P. Møller and H. Tegelstrom, Extra-pair paternity and tail ornamentation in the barn swallow *Hirundo rustica*, *Behavioral Ecology and Sociobiology* **41**, 353–60 (1997).

5. D. L. Marshall, Pollen donor performance can be consistent across maternal

plants in wild radish (*Raphanus sativus*, Brassicaceae): a necessary condition for the action of sexual selection, *American Journal of Botany* **85**, 1389–97 (1998).

6. G. R. Hunt, Manufacture and use of hook tools by New Caledonian crows, *Nature* **379**, 247–51 (1996).

7. A. Grafen, Sexual selection unhandicapped by the Fisher process, *Journal of Theoretical Biology* **144**, 473–516 (1990).

8. See the frontispiece of Chapter 10 in Vol. 1 of *Narrow Roads of Gene Land*, p. 352.

PARENT–OFFSPRING CORRELATION IN FITNESS UNDER FLUCTUATING SELECTION[†]

I. ESHEL and W. D. HAMILTON

─────

Cyclical selection that does not result in gene fixation may maintain a parent–offspring (P–O) correlation in fitness that is, most of the time, not far below one-half. This tends to happen when the following conditions are fulfilled: (1) slow cycling in allele frequencies occurs at many loci independently; (2) most alleles make only small differences to fitness; and (3) dominance in allelic fitness contribution is rare and overdominance rarer still. In contrast, drastic cyclical selection (dependent on large fitness differences without overdominance), because direction reverses so frequently, may give P–O correlations in fitness that are near zero or even negative. Also, if much fitness variance is maintained by heterozygote advantage, or is caused environmentally, the correlation is reduced in proportion to the ratio of relevant standard deviations.

P–O correlation in fitness that is almost permanently positive may help to explain mate choice in sexually promiscuous species. The most likely source of the multiple independent cyclical systems of selection required is host–parasite relations. To cause evolution of mate choice, and the subsequent manifestations of sexual selection, intermediate rather than extremely long cycles are most effective. A chooser's achievement for the offspring depends on the fitness variance in the sex chosen; in extremely long cycles this variance is on average close to zero.

1. INTRODUCTION

A crucial, almost paradoxical, problem haunting Darwinism from its very beginning is the source of maintenance of heritable variance of fitness. Natural selection, by definition, operates to destroy the unfit. In spite of this, plenty of non-environmental variation exists, and some of it certainly affects fitness. Knowledge of genes and the genetic structure of populations has provided a fairly satisfactory answer to a part of the seeming paradox. It has obviated Darwin's worry about blending inheritance, and it has helped to explain the

─────

[†]*Proceedings of the Royal Society of London B* **222**, 1–14 (1984).

─────

maintenance of variability of fitness in natural populations at equilibrium. Populations at equilibrium are, of course, not going anywhere under natural selection—but, does this matter? Perhaps they move only once in a while when a beneficial mutation occurs. Unfortunately, even if this is acceptable it leaves a major unresolved puzzle about *heritable* variation in fitness and about sexual selection. Traditional genetic models of population at equilibrium predict either no heritability of fitness or (with epistasis, recombination, and mutation brought in) a very small one (e.g. ref. 1; for alternative possibilities, see refs 2 and 3). Thus theory suggests that careful mate choice can achieve little, if anything at all, towards improving the fitness of offspring. A pessimism towards a strictly Darwinian view of sexual selection naturally arises from this point and seems reflected in many current writings: selecting females are assumed to be pursuing resources controlled by males rather than qualities of the males themselves (e.g. refs 4 and 5).

Assume, for example, a diploid one-locus *n*-allele model of viability selection with random mating. Let w_{ij} be the viability of an individual of type A_iA_j $(i,j = 1, \ldots, n)$. One can readily show that, in equilibrium, the frequencies $\hat{p}_1, \ldots, \hat{p}_n$ of the alleles $A_1, \ldots, A_n$, satisfy the set of equations:

$$\sum_{i=1}^{n} \hat{p}_i \, w_{ij} = \bar{W} \; (j = 1, \ldots, n), \tag{1.1}$$

where $\bar{W}$ is the average fitness in the entire population. Hence, a random female associating her genes with a specific gene A_j, has a chance p_i to pass the allele A_i to her offspring $(i = 1, \ldots, n)$ and thus to provide it with viability w_{ij}. The average viability of such an offspring will be

$$\sum_{i=1}^{n} p_i \, w_{ij},$$

which at equilibrium $(p_i = \hat{p}_i)$ is equal, as we see, to $\bar{W}$, independently of *j*. Thus 'choice' of any particular A_j-bearing male gains her nothing.

In other words, despite differences in fitness among individuals, there are no 'good genes' and 'bad genes' at equilibrium. The most fit will have offspring which are not more fit, on the average, than those of other individuals. Hence, a female cannot increase the fitness of her offspring by choosing an appropriate mate.

The theoretical finding is not surprising if we consider the meaning of equilibrium. Indeed, if offspring carrying a specific gene were better off on the average than others, then the frequency of this gene would increase instead of staying constant. Yet theory to the effect that females ought not to care about intrinsic quality in their mates is in disagreement with the apparent strong sexual preference observed in many natural populations and with the total of energy put at one stage or another into the adornment and performances of sexual selection.[1]

Actually any theory of sexual selection that at all resembles Darwin's original has, whether explicitly expressed or not, a necessity for the assumption of a positive heritability of fitness. Exceptions to this would be the theories of immediate advantage where the preference focuses qualities or attributes that are of significance for brood rearing. For example, selection for health in a mate purely so as to increase the chance that a mate survives through brood rearing and can participate fully in the parental care, would be excepted. So also would be selection of a mate purely for the good territory it holds, promising abundant food for the young. But these exceptions are neither Darwinian[6] nor are they general enough to account for all cases of sexual selection. Theories resembling Darwin's emphasize *genetic* endowment to offspring, and it is from this that a requirement of positive heritability follows. In some versions the endowment sought is imagined to be a general above-average fitness:[7] fitness, that is, in its everyday sense. In other versions a sexual selection episode starts with an advantage like this or else with some special utilitarian character that is under rapid promotion by natural selection. Later, through the polygyny which the preference has created, extra advantages to the preferred character are added.[8–10] All such models, which typically assume a constant environment (e.g. refs 11 and 12), have in common the difficulty that the variance on which they depend either dies away completely or at least exhausts its heritable component as the population approaches equilibrium. The basis of effectiveness of mate choice is thus eroded and whatever choice persists when it is gone (or reaches the low level that persistent mutation can justify) achieves, if anything at all, only a lowering of overall fitness.[13, 14] These results, already seemingly so unsatisfactory for the welfare of the species as a whole, are likely to be exacerbated by yet another cost that lies in the maintenance of the choosiness itself. A terminal state of generally lowered fitness along with a residual variation that has very low heritability seems to us inevitable in any genetic model that confines attention to sexual selection in a constant environment. Following such models species in which strong sexual selection had developed for any reason should show signs of having diminished their abundance. Perhaps many have; we do not discount the likelihood or the dysgenic power of the processes that are intrinsic in these models (e.g. ref. 13). However, the ubiquity of observations indicating strong sexual selection in stable abundant populations, and the fact that some animal groups have undergone considerable speciation with a mating system involving high sexual selection apparently intact throughout (e.g. bower birds), convinces us that a mistake must lie in the basic assumptions of the models. Changing this assumption, although it changes the model, by no means invalidates the previously important ideas (the 'runaway process' of Fisher; Zahavi's 'handicap principle'; how sexual selection may still operate under monogamy; and so on), but it does offer what seems to us a generally deeper and more plausible view.

While it is now widely accepted that environmental change of some sort is necessary to the evolution of sex (e.g. refs 11 and 15), we believe that it is also a crucial factor in the evolution of sexual preference. More specifically, redirecting old hints from papers by Haldane[16] and later writers towards still older problems, Hamilton has suggested that fluctuations in selection forces owing to host–parasite cycling may be the major basis of the evolution and maintenance of sex (see Chapter 2, and also Chapter 5 and references therein for parallel work). Under the assumptions that lead to this view, a situation may arise where many simultaneous host–parasite cycles affect many gene loci more or less independently. In this study we will show that when such simultaneous cycles occur and are mostly rather slow they can provide a permanent source of heritable variance of fitness in the population. Thus the cycles, via the heritability they establish, can serve as a basis for the evolution of sexual preference. This preference in turn, in this model, must be a relatively beneficial influence on health, although whether it pays its costs in this regard is a question we will not address.

As a first step we introduce some notation and analyse the advantage obtained by a choosing female (to take the predominant case) as a function of the heritability and the variance of fitness in the population.

If a correlate of fitness in a potential partner is being assessed during mate choice, it is far more likely to be, at the outset, a correlate of viability than of fertility. Direct correlates of fertility of a male would often be hard or even impossible to assess. Therefore, we assume for present purposes that viability is fitness. Assume that such fitness of a given individual at a given time is a function only of its genotype as expressed at the time in question. We are ignoring for the present the possibility of environmental variance in fitness. Assume, for simplicity, that this function is independent of sex. Let X_t be a variable, proportional to the viability of a random adult individual in generation t ($t = 1, 2, 3, \ldots$). Y_t is the viability of a random newborn offspring in the same generation. By $P(Y_{t+1}|X_t)$ we denote the conditional distribution of the viability of an offspring, born in generation $t + 1$ to a parent (say, father) whose viability is X_t.

The regression coefficient of Y_{t+1} over X_t is given by

$$B(Y_{t+1}, X_t) = \frac{\text{Cov}(X_t, Y_{t+1})}{\text{Var } X_t} = R(X_t, Y_{t+1}) \sqrt{\left(\frac{\text{Var } Y_{t+1}}{\text{Var } X_t}\right)}, \qquad (1.2)$$

where $R(X_t, Y_{t+1})$ is the correlation between the fitness of a father and that of its offspring. Such a correlation is often referred to as the heritability, in the broad sense,[17] of fitness in generation t. This value depends on the genetic set-up of the population in generation t, but also on the forces of selection in both generation t and $t + 1$.

From (1.2) it follows that a female who succeeds in obtaining a mate of fitness Δ standard deviations above the mean increases the expected fitness of her offspring, on the average, by

$$G(\Delta) = \Delta\sqrt{(\mathrm{Var}\ X_t)}\ B(Y_{t+1}, X_t) = \Delta\sqrt{(\mathrm{Var}\ Y_{t+1})}\ R(X_t, Y_{t+1}). \qquad (1.3)$$

Sexual preference of fit mates can therefore be selected for, only under conditions that guarantee the preservation of both substantial heritability and variance of fitness in the population. In the following sections we show that these two conditions are met when the population is simultaneously exposed to many, more or less independent, slow, environmental cyclings, each interacting with a different locus. Such cycling may be typical of a system composed of a host and many parasites (Chapter 6).

We start by studying the one-locus effect of fluctuating environment on the heritability of fitness.

2. HERITABILITY OF FITNESS OWING TO ONE-LOCUS EFFECT IN A FLUCTUATING ENVIRONMENT: A HOST–PARASITE CYCLE

Let us assume first a simple model of one locus with two alleles A and a in a host population that is afflicted by a parasite (or perhaps two). Assume that owing to some coevolutionary cycle arising from the parasitism the selection coefficients for genotypes at the A locus are fluctuating. As already mentioned host–parasite systems very readily go into cycles. The cycles could be of a demographic kind first postulated by Volterra, in which case perhaps one allele could be adaptive for low host densities and the other for high. Or the cycles could involve some genotype matching between host and parasite, in which case one allele tends to confer resistance to one parasite or parasite genotype and the other allele to another;[18] or there could be some combination of demographic and resistance cycling. The details need not concern us. We postulate merely that fluctuations in coefficients of selection occur and that in the consequent cycles changes in gene frequency are not so extreme that fixation will occur. Notice, however, that our focus here on the host–parasite relation as a source of fluctuating selection is only due to our belief in its being the most likely source; this is by no means a necessary assumption in the development of the model.

Let the viability of the three genotypes AA, Aa, and aa at generation t be proportional to $1 + \alpha_t$, $1 + \beta_t$, and $1 + \gamma_t$, respectively. Being interested in *relative values* we can assume, without loss of generality, that $\gamma_t = -\alpha_t$ (a normalizing factor can always be chosen so that the arithmetic mean of the two homozygote fitnesses will be 1). β_t is then the deviation of the fitness of the heterozygote from the average of those of the homozygote fitnesses. Let $p_1 = p_1^{(t)}, p_2 = p_2^{(t)}$, and $p_3 = p_3^{(t)}$ be the relative frequencies of the three genotypes after selection. Being interested in the initial advantage of female preference, we first assume random mating of males and females. The frequencies of the three genotypes AA, Aa, and as among newborn offspring in the next generation will be p^2, $2pq$, and q^2, respectively, when

$$p = p^{(t)} = p_1 + \tfrac{1}{2}p_2 = 1 - q. \tag{2.1}$$

Let us now calculate the distribution of the offspring fitness Y_{t+1} as a function of the father's fitness X_t. Note that for any father, there is a probability p that an allele passed by the (random) mother is A and a probability q that this allele is a. Hence, if $X_t = 1 + \alpha_t$, that is, if the father is of type AA, then there is a probability p that the offspring will be of type AA in which case $Y_{t+1} = 1 + \alpha_{t+1}$, and a probability q that the offspring will be of type Aa in which case $Y_{t+1} = 1 + \beta_{t+1}$.

In other words:

$$\left.\begin{aligned}
P(Y_{t+1} = 1 + \alpha_{t+1} \mid X_t = 1 + \alpha_t) &= p, \\
P(Y_{t+1} = 1 + \beta_{t+1} \mid X_t = 1 + \alpha_t) &= q, \\
P(Y_{t+1} = 1 - \alpha_{t+1} \mid X_t = 1 + \alpha_t) &= 0.
\end{aligned}\right\} \tag{2.2}$$

But, $P(X_t = 1 + \alpha_t) = p$, hence

$$\begin{aligned}
P(X_t = 1 &+ \alpha_t; = 1 + \alpha_t) \\
&= P(X_t = 1 + \alpha_t)\, P(Y_{t+1} = 1 + \alpha_{t+1} \mid X_t = 1 + \alpha_t) = p_1 p
\end{aligned}$$

and in the same way

$$\left.\begin{aligned}
P(X_t = 1 + \alpha_t;\ Y_{t+1} = 1 + \beta_t) &= p_1 q, \\
P(X_t = 1 + \alpha_t;\ Y_{t+1} = 1 - \alpha_t) &= 0.
\end{aligned}\right\} \tag{2.3}$$

Dealing with the other two father types similarly, we obtain the combined distribution of X_t, Y_{t+1} as given by Table 7.1.

By straightforward calculation we get the covariance:

$$\begin{aligned}
\mathrm{Cov}\,(X_t, Y_{t+1}) = (p_1 q + p_3 p)\,[\alpha_{t+1} + (q - p)\,\beta_{t+1}]\alpha_t \\
+ [(\tfrac{1}{2} - 2pq)\,\beta_{t+1} - \tfrac{1}{2}(p - q)\,\alpha_{t+1}]p_2\beta_t. \tag{2.4}
\end{aligned}$$

Table 7.1 Combined distribution of X_t and Y_{t+1}

| $X_t\backslash Y_{t+1}$ | AA | Aa | aa | |
	$1 + \alpha_{t+1}$	$1 + \beta_{t+1}$	$1 - \alpha_{t+1}$	Total
$AA\ 1 + \alpha_t$	$p_1 p$	$p_1 q$	0	p_1
$Aa\ 1 + \beta_t$	$(p_2 p)/2$	$p_2/2$	$(p_2 q)/2$	p_2
$aa\ 1 - \alpha_t$	0	$p_3 p$	$p_3 q$	p_3
Total	p^2	$2pq$	q^2	1

As a special case, if fluctuations of the environment affect mainly the homozygotes, leaving the viability of the heteroxygote close to the average, we get $\beta_t \approx 0$ and (2.4) becomes

$$\text{Cov}\,(X_t,\ Y_{t+1}) = (p_1q + p_3p)\,\alpha_t\alpha_{t+1} + O(\beta_t,\beta_{t+1}). \qquad (2.5)$$

In this case we also have:

$$\text{Var}\,X_t = [p_1 + p_3 - (p_1 - p_3)^2]\,\alpha_t^2 + O(\beta_t), \qquad (2.6)$$

$$\text{Var}\,Y_{t+1} = 2pq\alpha_{t+1}^2 + O(\beta_{t+1}). \qquad (2.7)$$

Denote by $x = p^{(t-1)}$ the frequency of the allele A among adults in generation $t - 1$. We know that the frequencies of genotypes among adults in generation t are:

$$\left.\begin{aligned} p_1 &= p_1^{(t)} = \frac{1 + \alpha_t}{W}\,x^2, \\[2mm] p_2 &= p_2^{(t)} = 2\frac{1 - \beta_t}{W}x(1 - x), \\[2mm] p_3 &= p_3^{(t)} = \frac{1 - \alpha_t}{W}\,(1 - x)^2, \end{aligned}\right\} \qquad (2.8)$$

where

$$W = 1 + (2x - 1)\alpha_t + 2x(1 - x)\,\beta_t. \qquad (2.9)$$

Hence

$$\left.\begin{aligned} p &= \frac{x}{W}\,[1 + \alpha_t x + \beta_t(1 - x)], \\[2mm] p &= \frac{1 - x}{W}\,[1 - \alpha_t(1 - x) + \beta_t x]. \end{aligned}\right\} \qquad (2.10)$$

Inserting (2.8) to (2.10) into (2.5)–(2.7), we obtain:

$$R(x_t,\ Y_{t+1}) = \frac{1}{2}\left[1 - \frac{x(2 - x)\alpha_t^2}{(1 - \alpha_t x)\,(1 - \alpha_t + \alpha_t x)}\right]^{-\frac{1}{2}} \frac{\alpha_t\alpha_{t+1}}{|\alpha_t\alpha_{t+1}|} + O(\beta_t,\beta_{t+1})$$

$$= \pm\frac{1}{2}\left[1 - \frac{x(2 - x)\alpha_t^2}{(1 + \alpha_t x)\,(1 - \alpha_t + \alpha_t x)}\right]^{-\frac{1}{2}} + O(\beta_t,\beta_{t+1}), \quad (2.11)$$

the sign being positive unless the selection forces alter their direction in the generation step in question. Hence, if the cycle is short, the heritability of fitness may be negative during a substantial part of the time. If, on the other hand, cycles are long and the selection forces are weak the term involving α_t^2 becomes negligible and (2.11) may then be written as:

$$R(X_t,\ Y_{t+1}) = \pm\frac{1}{2} + O(\alpha_t^2) + O(\beta_t,\beta_{t+1}), \qquad (2.12)$$

the sign being positive most of the time. Moreover, for any $-1 \leqslant \alpha_t \leqslant 1$ and $0 \leqslant x \leqslant 1$ one can readily verify the inequality

$$\frac{x(2-x)}{(1+\alpha_t x)(1-\alpha_t+\alpha_t x)} \leqslant \frac{1}{1-\alpha_t^2}, \tag{2.13}$$

while for selection forces not too drastically high (say $|\alpha_t| < \frac{1}{2}$)

$$\left[1 - \frac{\alpha_t^2}{1-\alpha_t^2}\right]^{\frac{1}{2}} \geqslant 1 - \alpha_t^2. \tag{2.14}$$

By using (2.13) and (2.14), (2.11) becomes:

$$O \leqslant \tfrac{1}{2} - |R(X_t, Y_{t+1})| \leqslant \alpha_t^2/2 + O(\beta_t, \beta_{t+1}). \tag{2.15}$$

Thus, even for such substantial selection forces as when AA is 50 per cent more viable than aa (implying $\alpha_t = 0.2$), R lies between 0.5 and 0.48.

We conclude that if cycling is slow and if the selection forces, determined by one locus without dominance, are not too drastically high, then the heritability of fitness, as determined solely by that locus, stays close to $\frac{1}{2}$ most of the time.

3. THE EFFECT OF DOMINANCE AND OVERDOMINANCE ON THE HERITABILITY OF FITNESS OWING TO THE ONE-LOCUS EFFECT

To study the general one-locus effect on fitness-heritability let us return to (2.4) and use the notation

$$\beta_t = \theta_t \alpha_t. \tag{3.1}$$

Here θ_t is the degree of dominance at generation t. $\theta_t = 0$ or $\theta_t = 1$ is the case of full dominance, $0 \leqslant \theta_t \leqslant 1$ is the case of subdominance. As a special case, if $\theta_t = \theta$ is fixed over time (2.4) becomes

$$\frac{1}{\alpha_t \alpha_{t+1}} \operatorname{Cov}(X_t, Y_{t+1})$$
$$= p_1 q + p_3 p + [p_1 q + p_3 p + (p_2/2)](q-p)\theta + (p_2/2)(q-p)^2 \theta^2 \tag{3.2}$$

and by a direct calculation from Table 7.1 we have

$$\frac{1}{\alpha_t^2} \operatorname{Var} X_t = [p_1 + p_3 - (p_1 - p_3)^2] - 2(p_1 - p_3)p_2\theta + p_2(1-p_2)\theta^2, \tag{3.3}$$

$$\frac{1}{\alpha_{t+1}^2} \operatorname{Var} Y_{t+1} = 2pq + 4(q-p)pq\theta + 2pq(1-2pq)\theta^2. \tag{3.4}$$

If the selection forces operating on the locus in question are not so drastic as to shift the genotype frequencies much from the Hardy–Weinberg distribution we have

$$p_1 = p^2 [1 + O(\alpha_t)], \qquad p_2 = 2pq[1 + O(\alpha_t)], \qquad p_3 = q^2[1 + O(\alpha_t)].$$

Therefore

$$\frac{1}{\alpha_t \alpha_{t+1}} \operatorname{Cov}(X_t, Y_{t+1}) = [pq + 2pq(q - p)\theta + pq(q - p)^2\theta^2] [1 + O(\alpha_t)] \quad (3.5)$$

and

$$\frac{1}{\alpha_t^2} \operatorname{Var} X_t = \frac{1}{\alpha_{t+1}^2} \operatorname{Var} Y_t [1 + O(\alpha_t)]. \qquad (3.6)$$

Hence, from (3.4), (3.5), and (3.6):

$$R(X_t, Y_{t+1}) = \frac{1}{2} \frac{1 + 2\theta (1 - 2p) + \theta^2 (1 - 4pq)}{1 + 2\theta (1 - 2p) + \theta^2 (1 - 2pq)} \frac{\alpha_t \alpha_{t+1}}{|\alpha_t \alpha_{t+1}|} + O(\alpha_t) + O(\alpha_{t+1})$$

$$= \pm\frac{1}{2}\left[1 - \frac{2pq\theta^2}{[1 + (q - p)\theta]^2 + 2pq\theta^2}\right] + O(\alpha_t) + O(\alpha_{t+1}), \quad (3.7)$$

the sign, as before, being positive as long as sign $\alpha_t =$ sign α_{t+1}.

In the case $0 < \theta < 1$ of subdominance one can use the fact that the expression

$$\frac{2pq\theta_t^2}{[1 + \theta(q - p)]^2 + 2pq\theta^2}$$

obtains its maximum when

$$p = \frac{\theta^2}{2 - \theta^2}. \qquad (3.8)$$

It then equals $(1 + \theta)/2$ and (3.8) can, therefore, be written as

$$\tfrac{1}{2} \geqslant |R(X_t, Y_{t+1})| \geqslant \tfrac{1}{2}\left(1 - \frac{\theta^2}{2 - \theta^2}\right) + O(\alpha_t) + O(\alpha_{t+1}). \qquad (3.9)$$

Moreover, the right side of (3.9) is indeed a lower bound of $|R(X_t, Y_{t+1})|$, obtained for the special frequency (3.8). However, in the case of subdominance, if the cycle is long, the value $2pq$ is likely to be very small most of the time and, as follows from (3.7), $R(X_t, Y_{t+1})$ is likely to stay even closer to $\tfrac{1}{2}$.

The situation is different in the case of overdominance, say $\theta > 1$ where $\alpha_t > 0$ (or $\theta < -1$, where $\alpha_t < 0$). In this case (3.7) still guarantees a non-negative value of $R(X_t, Y_{t+1})$ as long as the selection does not alter its direction. Yet for different values of p it may obtain any value between zero and a half. The value $R(X_t, Y_{t+1}) = 0$ is obtained for:

$$p = \frac{\theta - 1}{2\theta} = \hat{p}(\theta), \text{ say.} \qquad (3.10)$$

Values close to $\frac{1}{2}$ are obtained for p close to 0 or to 1. Note, however, that in the case of constant environment $\alpha_t = \alpha$, $\theta = \theta$, $(t = 1, 2, 3, \dots)$ the frequency $p = \hat{p}(\theta)$ of the allele A is globally stable. Hence when the cycle is long, as assumed before, and the selection forces are changed slowly, p is likely to stay near $\hat{p}(\theta)$.

We find, therefore, sharply divergent expectations according to whether A does or does not imply overdominance. If the cycle is long, then the heritability of fitness owing to a subdominant locus is likely to stay close to $\frac{1}{2}$ most of the time. On the other hand the heritability of fitness owing to an overdominant locus is likely to stay close to zero most of the time.

4. THE EFFECT OF MULTICYCLING: PRESERVATION OF BOTH HERITABILITY AND VARIANCE OF FITNESS

Until now we have restricted our study to the effect of one locus on the heritability of fitness. Denoting by X_t and Y_t the contribution of this locus to the general fitness of a random adult and a random newborn offspring, respectively, we have seen that the fitness heritability $R(X_t, Y_{t+1})$ attains values close to a maximum of half most of the time when the selection forces operating on this locus are moderate, when they are subject to the effect of a slow cycling, and when a polymorphism is maintained without overdominance. Unfortunately one must admit that such a locus can hardly be much of a source for a substantial variance of fitness in the population. Indeed, either the selection forces operating on this locus are very small or one allele must be close to fixation during most of each half of the cycle. In both cases the advantage $(G(\Delta) = \Delta R(X_t, Y_{t+1}) \sqrt{(\text{Var } Y_{t+1})}$ of a female attempting to choose a mate Δ standard deviations fitter then the average cannot be large (see (1.3)). The situation appears to be different when we regard, more realistically, the combined effect of many non-synchronized host–parasite cyclings, each having a moderate or minor effect on a different, non-epistatic, locus.

Denote by $X_t^{(i)}$ and $Y_t^{(i)}$ the contribution of the locus i to the general fitness of a random adult and random newborn offspring respectively,

$$X_t = \sum_{i=1}^{n} X_t^{(i)}, \qquad Y_t = \sum_{i=1}^{n} Y_t^{(i)}. \tag{4.1}$$

With independent effects of the various loci

$$\text{Var } Y_{t+1} = \sum_{i=1}^{n} \text{Var } Y_{t+1}^{(i)}. \tag{4.2}$$

With large enough n, this overall fitness variance in the population may not be small even if the component variances owing to individual loci are so. Moreover, with the law of large numbers, the overall variance may stay more or less constant even when each of the values $\text{Var } Y_{t+1}^{(i)}$ is changed from one generation to the next. To estimate the multilocus fitness-heritability

$R(X_t, Y_{t+1})$, we employ the assumption that the cycles are non-synchronized and loci are non-epistatic; therefore, for $i \neq j$

$$R(X_t^{(i)}, Y_{t+1}^{(j)}); \tag{4.3}$$

that is, the effect of locus i on the father's fitness is not correlated with the effect of locus j on the fitness of its offspring. Hence

$$\text{Cov}(x_t, Y_{t+1}) = E(\sum_{ij} X_t^{(i)}, Y_{t+1}^{(j)}) - \sum_{ij} EX_t^{(i)}, Y_{t+1}^{(i)}$$

$$= \sum_{ij} \text{Cov}(X_t^{(i)} Y_{t+1}^{(j)}) = \sum_i \text{Cov}(X_t^{(i)}, Y_{t+1}^{(i)})$$

$$= \sum_i \sqrt{(\text{Var } X_t^{(i)} \text{ Var } Y_{t+1}^{(i)})} \, R(X_t^{(i)}, Y_{t+1}^{(i)})$$

and

$$R(X_t, Y_{t+1}) = \frac{\sum_i \sqrt{(\text{Var } X_t^{(i)} \text{ Var } Y_{t+1}^{(i)})} \, R(X_t^{(i)}, Y_{t+1}^{(i)})}{\sqrt{(\sum_i \text{Var } X_t^{(i)} \sum_i \text{Var } Y_{t+1}^{(i)})}}$$

$$= C \, \frac{\sum_i \sqrt{(\text{Var } X_t^{(i)} \text{ Var } Y_{t+1}^{(i)})} \, R(X_t^{(i)}, Y_{t+1}^{(i)})}{\sum_i \sqrt{\text{Var } X_t^{(i)} \text{ Var } Y_{t+1}^{(i)}}}, \tag{4.4}$$

where

$$C = \frac{\sum_i \sqrt{(\text{Var } X_t^{(i)} \text{ Var } Y_{t+1}^{(i)})}}{\sqrt{(\sum_i \text{Var } X_t^{(i)} \, _i \text{ Var } Y_{t+1}^{(i)})}} \leq 1. \tag{4.5}$$

Thus, the fitness-heritability $R(X_t, Y_{t+1})$ is equal to the weighted average of the one-locus values $R(X_t^{(i)}, Y_{t+1}^{(i)})$, multiplied by the constant C, the weights of averaging being the geometric means $\sqrt{(\text{Var } X_t^{(i)} \text{ Var } Y_{t+1}^{(i)})}$ of fitness variances in the parent and in the offspring populations.

From section 3 and (4.4) it therefore follows that if overdominant loci (even under the cycling effect) are the main source of fitness-variance, then overall fitness heritability $R(X_t, Y_{t+1})$ must be close to zero.

If cycles are mostly short, then some of the values $R(X_t^{(i)}, Y_{t+1}^{(i)})$ are small, and some are negative and $R(X_t, Y_{t+1})$ must, again, be small (or even negative!).

However, if the selection forces operating on each locus separately are weak and if cycles are long so that they are changed slowly, one may safely use the approximation

$$\text{Var } X_t^{(i)} \approx \text{Var } Y_{t+1}^{(i)}. \tag{4.6}$$

In this case (4.5) becomes $C \approx 1$ and from (4.4) we get:

$$R(X_t, Y_{t+1}) \approx \frac{\sum_i \text{Var } Y_{t+1}^{(i)} \, R(X_t^{(i)}, Y_{t+1}^{(i)})}{\sum_i \text{Var } Y_{t+1}^{(i)}} \tag{4.7}$$

(the left side being always a little bit smaller than the right side owing to inequality (4.5)).

If, in addition, we assume no overdominance, we know that $|R(X_t^{(i)}, Y_{t+1}^{(i)})|$ $\approx \frac{1}{2}$, the sign being positive most of the time, and it follows from (4.7) that

$$R(X_t, Y_{t+1}) \approx \tfrac{1}{2}, \tag{4.8}$$

the approximation being also an upper bound.

5. DISCUSSION

This work concentrates on the limited question of the maintenance of a heritable variance sufficient for the evolution of sexual preference, treating specially the case where preference is based on a polygenic assessment. It has been shown above that, contrary to intuition, it is possible to have a selection process that does not exhaust the additive variance that it feeds on. With periodic reversals of the selection at many different loci, and with the timing of reversals uncorrelated, variance in fitness can be maintained at an almost steady value. In this situation the parent–offspring correlation in fitness is steady to the same degree at a value that may approach but does not exceed one-half. How closely the parent–offspring correlation approaches $\frac{1}{2}$ depends on, first, the average dominance of fitness effects at the loci, with over-dominance specially tending to reduce correlations towards zero, and, second, the average length of cycles, with long periods making for closer approach to the limit of $\frac{1}{2}$ except in the case where overdominance is common.

It was argued earlier (1.3) that $\Delta \sqrt{(\mathrm{Var}\ Y)}\ R(X, Y)$ shows the expected increment of offspring fitness for a female who obtains a male of fitness which is Δ standard deviations about the mean. Actually a female cannot be expected to gain as much as this for a given amount of effort (represented in Δ), because the phenotypic quality she selects always includes non-heritable effects from the environment. Assume that the female is choosing as a pheno-typic quality $Z = Y + \xi$ where ξ is the environmental effect, assumed inde-pendent of genotype, and Δ is now in units of standard deviations of Z. Then it can be shown that the gain to a choosing female is $G(\Delta) = \Delta \sqrt{(\mathrm{Var}\ Y)}$ $R(X, Y)\ (\sqrt{\mathrm{Var}\ X}\ /\ \sqrt{\mathrm{Var}\ Z})$. Since the correcting quotient factor is less than one and the rest of the regression is unchanged, a given amount of effort now achieves less, as would be expected. In other words, environmental vari-ation in the perceptible characters that correlate with fitness makes choosi-ness for those characters and consequent phenomena of sexual selection less likely to evolve.

With this point made, the way is clear to appreciate a more interesting diminution of the advantage of choosing. Since, excluding the case of com-mon overdominance, $R(X, Y)$ rises toward $\frac{1}{2}$ as cycles lengthen, it might seem that the power of sexual selection would always be greater for longer mean periods of cycling. However, this neglects the fact that if the cycles are long, gene frequencies almost inevitably go very extreme with consequent reduction of the variances. In fact, in reasonable models $G(\Delta)$ must tend to zero as the mean cycle becomes very long, and this, combined with the fact that if all cycles are of period 2, $G(\Delta)$ must be negative, implies that this function has a maximum for some intermediate median period. With the mixture of cycles occurring in real life, there is therefore some mean periodicity that maximizes the advantage that females can get for their offspring by a given effort of mate selection.

The problem of characterizing cycle mixtures that would maximally promote sexual selection in the above sense will not be pursued here. Such a study would require more detailed specification of a model with attention paid to the non-linear dynamics that fixes the nature of the cycles, whereas here we have only been concerned with very general assumptions about cycle length.

Although parasite coevolution was suggested earlier to clarify the underlying idea, our result does not require that parasites be the cause of cycling. Other kinds of biotic interaction might cause cycling. So also might fluctuations of the physical environment. Physical examples here would be dry to wet cycles occurring on a timescale of a few to many generations, hot and cool periods similarly, or changes owing to the sunspot cycle. However, such physical fluctuations do not seem to be likely to be as important as the biotic factors for two reasons. First, they do not slacken their pressure when alleles approach fixation. Consequently, it is relatively hard to maintain polymorphisms under regimes of this kind (see Chapter 1). Second, even if non-biotic 'cycles' occur and are buffered against fixation in some way that does not involve overdominance to the degree that destroys the parent–offspring correlation (see ref. 19 for possibilities), it remains hard to believe that there can be many such cycles operating at one time. Our result involving a parent–offspring correlation that is both steady and substantial requires many cycles co-occurring—in effect, polygenic cycling. If species were subject to just a few drastic cycles it is likely, first, that such cycles would have been noticed, and, second, that sexual selection based on them would show dramatic switches in preferred traits at particular points of the cycle: again, a phenomenon never observed. In contrast to the potentially fluctuating physical factors, parasites, on the other hand, are very numerous. Every species of substantial body size (e.g. any bird or mammal) has many parasites, and a high proportion of them are quite host specific. (The characteristics of parasites relevant to coevolutionary genetical cycling are described in Chapter 5.)

While any numerous large free-living animal such as a bird or mammal is likely to have a numerous set of parasites, the same becomes less likely to apply as we review primary parasites themselves as potential hosts of hyperparasites and so on down. The very fact of being a parasite tends to ensure a life that is somewhat secluded biotically, although some exception must be made for ectoparasites and species with complex life cycles and free-living stages. Altogether parasites are likely to experience fewer more drastic cycles or changes of selection that are dependent on genotypes of their hosts. If the parasite's life cycle is fairly long, as usual in macroparasites, this may well select for maintenance of sex, but it will not be conducive to sexual selection occurring on the lines discussed here. These predictions seem to fit roughly with the facts of parasite sexuality (for a discussion see ref. 20).

Returning to hosts in general, another prediction is that we expect sexual selection to be most highly developed in the part of the host range where host-specific parasites are most numerous. This would normally be the centre of the range. Examples of this are easily found. Unfortunately they are also open to explanations not involving parasitism.[21] Far more testing of the ideas in this paper would be a direct search for genetical evidence of polygenic cycling.

Acknowledgement

We are grateful to J. Seger for helpful discussions.

References

1. P. D. Taylor and G. C. Williams, The Lek paradox is not resolved, *Theoretical Population Biology* **22**, 392–409 (1982).

2. E. Mayr, *Population, Species and Evolution* (Harvard University Press, Cambridge, MA, 1976).

3. R. Lande, The maintenance of genetic variability by mutation in a polygenic character with linked loci, *Genetical Research* **26**, 221–35 (1975).

4. S. T. Emlen and L. W. Oring, Ecology, social selection and the evolution of mating systems, *Science* **197**, 215–23 (1973).

5. T. R. Halliday, Sexual selection and mate choice, in J. R. Krebs and N. B. Davies (ed.), *Behavioural Ecology: An Evolutionary Approach*, pp. 180–213 (Blackwell, Oxford, 1984).

6. S. J. Arnold, Sexual selection: the interface of theory and empiricism, in D. Bateson (ed.), *Mate Choice*, pp. 67–107 (Cambridge University Press, Cambridge, 1983).

7. A. Zahavi, Mate selection—a selection for a handicap, *Journal of Theoretical Biology* **53**, 205–14 (1975).

8. R. A. Fisher, *The Genetical Theory of Natural Selection* (Oxford University Press, Oxford, 1930).

9. P. O'Donald, *Genetic Models of Sexual Selection* (Cambridge University Press, Cambridge, 1980).

10. R. Lande, Models of speciation by sexual selection on polygenic traits, *Proceedings of the National Academy of Sciences, USA* **78**, 3721–5 (1981).

11. J. Maynard Smith, *The Evolution of Sex* (Cambridge University Press, Cambridge, 1978).

12. J. W. F. Davis and P. O'Donald, Sexual selection for a handicap: a critical analysis of Zahavi's model, *Journal of Theoretical Biology* **57**, 345–56 (1976).

13. O'Donald 1980 (ref. 9), p. 32.

14. M. Kirkpatrick, Sexual selection and the evolution of female choice, *Evolution* **36**, 1–12 (1982).

15. G. C. Williams, *Sex and Evolution* (Princeton University Press, Princeton, NJ, 1975).

16. J. B. S. Haldane, Disease and evolution, in *Symposium sui Fattori Ecologici e Genetici della Speciazione negli Animali, Supplemento a La Ricerca Scientifica*, Anno 19°, pp. 68–76 (1949).

17. D. S. Falconer, *Introduction to Quantitative Genetics* (Ronald Press, New York, 1960).

18. I. Eshel and E. Akin, Coevolutionary instability of mixed Nash solutions, *Journal of Mathematical Biology* **18**, 123–33 (1983).

19. J. Felsenstein, The theoretical population genetics of variable selection and migration, *Annual Review of Genetics* **10**, 253–80 (1976).

20. G. Bell, *The Masterpiece of Nature: The Evolution and Genetics of Sexuality* (University of California Press, Berkeley, 1982).

21. I. Eshel, Sexual selection, population diversity, and availability of mates, *Theoretical Population Biology* **16**, 301–14 (1979).

CHAPTER 8

AT THE WORLD'S CROSSROADS

Instability and Cycling of Two Competing Hosts with Two Parasites

*Tell me (said he) since thou art here again in the peace and
assurance of Ullah, and whilst we walk, as in the former years,
towards the new blossoming orchards, full of the sweet spring as the
garden of God, what moved thee, or how couldst thou take such
journeys into the fanatic Arabia?*

C. M. DOUGHTY[1]

'AND did those feet in ancient time/Walk upon England's
mountains green?/And was the holy lamb of God/In England's pleasant
pastures seen?' So wrote William Blake in the opening lines of a famous
poem. A short and unmetaphorical answer to his question is: 'No; but as a
boy he did walk over very similar rocks.' Jesus grew up in a chalk landscape,
one born out of the soft white limestone built in the same milky
Cenomanian Cretaceous Sea as reared eventually, at what was once its
western extremity (just as the site of Nazareth was at its eastern), the small
hills here in southern England. These make the South and particularly the
North Downs, as we call them; the latter almost certainly were the
'mountains' Blake was thinking of when he wrote the poem, for it was
among them, near Shoreham in Kent (or else among others similar that
were of the same formation near the Sussex coast), that the artist was
accustomed to take refuge at times from London and the 'dark satanic mills'
that also enter the poem.

When I saw Nazareth in Israel in the spring of 1985 it was indeed
green and quite like the North Downs. Its hillside outside my bus window

and the buildings above the slope (we did not enter the town) reminded me especially of the chalky slant of Great Lines going up from Chatham to Gillingham, beside the Medway River in Kent. This slope I had seen many times during my National Service when I was in the British Army's Corps of Royal Engineers and undergoing training in Chatham and Gillingham. Facing Nazareth I wondered if those rare, un-British and steppe-evoking plants that grow on Great Lines, such as the purple star-thistle, might be on the slope here too. It is very possible; *Centaurea*, the genus of the star thistle, has many species in Israel.

The meeting that had put me into the bus from which I saw this hillside of Jesus' childhood was perhaps the most extraordinary and instructive I ever attended. A principle feature was the nomadic programme that brought 30 or so scientists not only to pass by Nazareth, but to carry us daily to or past one such historic venue after another. Our talks were interspersed with the moves. All three of our organizers, Eviatar Nevo, Daniel Cohen, and Daniel Zohary, acted as expert tour guides as the buses rolled. Between them they supplied research-standard information on anything and everything we visited and passed—information at times too deep and too abundant to be absorbed. Whatever might be your preferred '-logy' of the landscape, at least one of our guides could expound upon it— geo-, eco-, zoo-, palaeonto-, some of archaeo-, and I am sure others I have forgotten. The gap they frankly admitted to was that of the New Testament–Roman–Mohammed–Crusader span, indeed all that which begins with Nazareth and shortly plunges into the Diaspora—exactly the times that, for most people, provide the most powerful lures of the Holy Land. But our guides were not 'most people' and because none of we conferencees cared greatly for the post-prehistory part either, it was not greatly missed. The meeting began in and toured Jerusalem; it brushed the shores of the Sea (what an exaggeration!) of Galilee, it saw Nazareth from afar as described; and so on. As to fortresses, we visited two, Masada and the huge castle at Acre that remained an island of Christendom long after all other gains of the Crusades had been washed away by the resurgent tides of Islam. For the most part we were content to pass these more historical sites just as we passed Nazareth, being all the more enchanted by them perhaps because of the decent haze of distance that enwrapped their divisive myths.

Nearer at hand the natural beauty of the land fell on me like a wolf on a jerboa: I think I was killed and rose again and in my new state could a little understand for the first time the passions that all successive invaders seem to have had to call this key strip of land eternally their own. We must have been there at the best time of year: surely it could not be more beautiful at any other. Except in the stoniest deserts everything was flowering or else green under a brilliant but not brassy sky. In the riot of young wild plants Danny Zohary pointed out to us the progenitors of our flax, wheat, barley, oats, and lettuce as well as many a herb condiment that must have begun right here to make the neolithic meats less dangerous,[2] to expel human worms, and forestall malaria. Likewise the fruit trees: at the road sides were the carobs, olives, figs, true sycamores (really another fig in case this name should confuse the English), pistachios, and date palms. With an almost impossible freshness and beauty considering their root space, the wiry bushes of caper dangled and flowered in cracks of desert rocks. Zohary's best evidence[3] tells that many of these plants may have been first cultivated here or elsewhere around the Fertile Crescent or possibly in Egypt. The patchily productive land of the Levant has, of course, a record of occupation stretching vastly earlier than the advent of the Israelites or even of the peoples they displaced; indeed it is likely that all of us, even the farthest Australians and Fuegians among us, share the progenitors who moved through this strip of non-desert that makes an immediate threshold to three of the world's continents.

What is the plural of 'Exodus'? Exodi? Whatever, it happened many times. Here had passed *Homo erectus* as well as, later, that 'R. & D.' prototype, *H. neanderthalensis*, which became so nearly, but seemingly not actually, our European ancestor.[4] Here, finally, came the youngest sibling: *sapiens*. He joined for a time (even it seems together or alternating in the same caves) the earlier and stouter (though equicranial) cousin. Later still, and in the higher *sapiens* levels as shown in the digs, it is evident that plant cultivation was practised in the land thousands of years before Abraham drove in his flocks. Clearly by the time of the writing of the Bible the conquering Hebrew pastoralists had been themselves already deeply infused with the culture of the townsmen and cultivators around them, an infusion that was to continue for millennia before they themselves came to contribute in a similar way to new waves of thought and writing, mostly far to the north and west. In this manner the great influence the Diaspora had on equestrian pastoralists and their predecessors in the far reaches of

Europe came to lie in the skills of born townsmen—Jewish clerks, merchants, or, at the very least, cultivators, or else as literate bailiffs managing cultivators on behalf of illiterate equestrian lords. Emphatically it was not the skills of pastoralists, in spite of the fierce traditions derived from their past tribal life, that the Jews kept for themselves.

Corresponding to the change of Jewish lifestyle over the millennia, our seminar, though at least as nomadic as Abraham had been, didn't sleep us in black tents but instead in kibbutzes and student dorms, many of them extremely modern. To our engorged minds the timespan from Abraham to Augustus seemed only a passing flash—a flash within the long tale, for example, that was told to us by the caves of Skhul under Mount Carmel and beside that briefer 'tell' (the occupation mound) of pre-biblical Jericho. But even this 'briefer' is to be measured in thousands of years. Most impressive to me on this far-prehistoric side were the waving mimic fields of wild barley and, more rarely, of wheat or oats growing between the oak trees on rich slopes. So alike were they to small planted fields that it was difficult to realize that they were wholly natural and that the same tall annual grasses had waved first to finches, antelopes, and to empty wind for hundreds of millennia before the glances of newly ex-African *Homo* ever fell upon them. We—I mean our ancestors—saw them there first, I suppose, like the graceful dancers of the wind they were and still are; but eventually saw them also with a more ordinary hunger for the sweet milk of their swelling seeds.

Eibi Nevo told us of the extreme micro-site differentiation of these 'fields', where almost one side or another of an oak tree could make a provable difference to the isozyme gene frequencies he detected. Inbreeding and random drift, though both obviously occur, are not enough to explain the differences he finds; instead these are correlated with various physical and soil factors—shade by the oak tree or the full sun, damp soil or dry soil.[5,6] It can be imagined, however, that I pricked my ears at a somewhat different slant asserted by Daniel Zohary during these discussions. He noted that incidences of pathogens in the sites also correlated with the isozyme frequencies. Inexhaustable Eibi has not neglected this resistance adaptation[7] but it has been less his central theme than abiotic factors.[5,8] Daniel also told me of a remarkable proportion of the genome of the crop progenitors he was studying that gave signs of being devoted to codes for disease resistance: he even gave a figure for the proportion, but I have unfortunately forgotten it and I noted no reference to a paper. It was high, and no doubt because of it and other kinds of variation they possessed, wild populations in the Levant region are now of great interest to plant breeders eager for sources of new genes or for retrieval of those carelessly lost. Very important for their programmes are genes that can be bred into modern crop plants to help them resist the epidemics of fungal, viral, and other pathogens that transfer and evolve so swiftly within our vast genetic monocultures.

It may be, therefore, that the natural dynamics of disease resistance in these wild 'mimic' fields of Israel, as they become known for their own wild pathogens, will be among the first to test some of my speculations about the uses of sex. Even the seemingly adjustable inbreeding of the hermaphrodite plant is favourable to the testing because, in principle, it can be used to separate the roles of heterozygote advantage (but beware here of a common illusion—high heterozygosity doesn't have to mean a high background of heterosis!) and of new multilocus homozygous combinations. Really it isn't my extremely simplistic models that need testing for no real situation can resemble them closely. The tiny 'fields', the real patches among the oak trees, that I was seeing in Israel obviously make an ensemble quite opposite in kind to the unitary population I was trying to model. Rather it is the background ideas of those models that matter and the fact that they can protect alleles in spite of their unpromising, undivided design.

The 'metapopulations' of Ladle, Johnstone, and Judson[9] and then of Judson alone[10] are far more realistic. Those models came much later than the paper of this chapter; my manuscript didn't entirely neglect the metapopulation (that is, the notion of a patchwork of partially isolated subpopulations) although I modelled this only in a most artificial way. I saw the idea of a population of populations in the background as an insurance against locally lost genes (p. 298) or, under the actual conditions I modelled, of species being lost from an ecosystem. The extra condition for this to work is, of course, that over generations the populations in a patchwork desynchronize their host–parasite variations, and in the paper I record my concern for the condition for desynchrony to be determined by theorists. The final work on the paper was done after the trip so possibly I was influenced by the deme complexes I had been seeing scattered on the Levantine hillsides, but I think the idea was actually of longer standing and those parts were already in the manuscript before I travelled. Virtually all lasting populations in the world are metapopulations—in other words, they are patchworks. In Israel, besides the scattered and clumpy (above-ground) grasses that we saw, Eibi Nevo pointed out to us many swarms of the mounds of his scattered but un-social (and underground) mole rats that he has shown to have a somewhat similar propensity for intense local variation though, of course, the animal is non-hermaphrodite and is less rapidly changing than are the annual grasses. Mole rats can live several years. Fascinating as the mole rats are, however, it is fairly clear that the understanding of their 'wild' genetics is due to progress more slowly than is the case with the wild wheat and barley. These grey, velvety, and toothy sausages are occasionally pests of fields, I gathered; nevertheless their affairs and depredations obviously can never compete with the interest that half the world has in securing its main food supply.

It was not only the wild life of Israel that seemed to me so beautiful. The inanimate landscapes themselves seemed to have a special demonstrative clarity, even the sky had its special blue. Perhaps I'm saying 'demonstrative' here again because Eibi several incarnations back in his varied life had been a geologist and he was nothing if not an explainer: in the setting of a desert cliff, for example, his words could stand us on the shore of that Cenomanian Sea I mentioned at the beginning and could add the pterodactyls circling there and crawling on the sand. But all that was now sealed in the rocks: as to the present, I have never walked in desert mountains quite as barren as those we passed through at the edge of the Negev nor elsewhere ever scuffed my shoes on desert stones shattered into such sharp pieces by the sun or had seen stones, which, when so cracked, expose such rich though lifeless patterns within. In terms of scenery set for the second day of creation, the Sonoran of Arizona now seemed to me a verdant Eden compared with some mountain walls of the Dead Sea or the valleys of the Negev behind them.

Close under the heights of the Masada Fortress, however, looking across the so-called sea, the mud, water, and light on the mountain wall on the Jordan side, like the chalk rock of Nazareth, gave me another quite unexpected revelation, more to do with life and today.

I went once in my teens to an exhibition of the Royal Academy of Arts in Burlington House in Piccadilly, London. I think it was a bicentennial or something like that of the Academy and it included pictures by artists from way back in its history—Constable and Romney I remember and inevitably, of course, Turner. But the picture that most caught my attention was not by any of these. Low and central in its frame you saw painted in side view what I thought at first to be a sheep but soon realized, from the title on the frame, was a goat. With a wreath of flowers circling its neck and, as I remember, with colourful ribbons attached to many other places in its shaggy wool, the animal stood bogged above its knees in a level plain of mud. Its head was bowed as if in some terrible sadness and it seemed near to collapse: this far it had come into the mud and it was stuck fast there and was hopeless. Beyond it, a level grey-streaked white merged finally into a brilliant band of haze, above which, near the top of the picture, rose a precipitous wall of mountains, all pink and blue and chevronned with ravines that seemed still to ripple before the viewer's eyes with the heat of a fading desert day. The mountains seemed as much the subject of the picture as the goat: the artist had painted a distant inaccessible wonderland, almost another planet—had started a sci-fi book cover, it might be, but then had messed it because, as anyone could tell him, he needed a central something that wasn't a bedecked goat. Instead, why not have a hero zapping alien robots with a ray gun perhaps, or, if he wished to be more historical, why not just David killing Goliath?

I returned several times through the crowded gallery to try to understand the picture—to pull out, as it were, a festering tooth it had created by its tragic and cruel beauty. The picture by itself wasn't quite all I had to understand it by, there was a clue: 'And the scapegoat shall go over into a land uninhabited' said a caption on the frame and these words, like the picture itself, stuck to my mind like a burr. The artist's name was beside the frame too—William Holman Hunt—an artist I had never much valued and knew best for his *The Light of the World* in St Paul's Cathedral in London, a picture showing Jesus crowned with thorns, lantern in hand, knocking in a bluish and ivied darkness at a rustic door. The picture was soppy for me, a damp allegorical, but it was enough to suggest that Hunt's *Scapegoat*, which was far better, also must be allegory. Later in Brazil and Arizona I came to see evening mountains resembling those of *The Scapegoat* picture but none was ever quite right: now on the shores of the Dead Sea I knew that I saw exactly the background I had remembered from Burlington House. The wall of Jordan across the lake and the salty flats and crusts on this side were if anything more exceptional, more other-worldly, than the

painting had made them. I knew also that I stood on the lowest dry point of
our planet's surface and I faced a far-off mountainside half way up which,
perhaps, lay the level of the world's ocean. Salt and death were on all sides;
everything around me was extreme. A bulky black wingless grasshopper put
out red and doubtless poisonous blood from his cowled neck and at the
same time pretended death as I plucked him off a desert bush. Were there
bushes, too, on the far mountains? I could not see them. There must be
bushes there; there were probably trees under the haze as well. But the goat
of Hunt's painting had never, never been going to reach them, no more
than could I if I tried to cross the crusted mud and the dried brine and the
lake beyond—perhaps this was the 'land uninhabitable' as Hunt had
interpreted it, simply the Dead Sea. Whatever, the goat had been driven to
death; well might it be sad!

Presently one of my companions, when I told him of the
remembered picture, enlightened me further about the story—or was it
about the ancient Jewish custom that he told, a custom that I gathered still
persists to some extent in stricter sects? Later, following his lead, I looked
up the tradition of the 'scapegoat' in the Bible (Leviticus 16:20). I still
don't know what Holman Hunt did mean exactly, if indeed he meant
anything, because an artist is certainly not obliged to 'mean'. Was his
intention, for example, anti-semitic? Or was he being more purely
allegorical, in a variant on the more typical Christian allegory that is drawn
out of Isaiah—about how our sins make others suffer, and we know it and
are sorry and place wreaths and throw out marvellous compliments to those
we have oppressed even while we still send them to die? Anyway, partly
through Hunt's painting that vision of the Dead Sea became a part of the
rebirth I mentioned and such images combined to enfold all the other
living wonders that I outlined above. The world's deepest dry-land place
brought me somehow inside Earth itself and into the presence of its
elemental forces, mostly so hostile and so pre-life like the brine and blazing
sun, but then also rising through all the quirks and strangeness towards
these extreme oddities of human behaviour, including the bedecked goat
(or at least the thought of one) and, then right there where I stood,
appeared three tourists bobbing high on the lake's surface and trying to
look fulfilled while their modern lakeshore hotel seemed almost to bend
over them, quivering in the heat, eager to attend shortly to their sunburn
and salt.

The fortress of Masada, seen on that same day I think, inspired also, but in the evening the Dead Sea revived memories of the painting and made me very sad, as often happens after an extreme emotion. I guess I felt like the goat itself and, on another side, I thought quite unmetaphorically about all those animals and plants of the region that were caught between the forces Hunt had depicted and of how they, knowing nothing of whether Jews or Arabs were rightful owners, or what ownership meant, instead just moved away and away from the centres of the quarrels as the burgeoning populations spread, with sweet spring by sweet spring being taken from them and even the deep, dead water under the land that created the springs being sucked away. Especially I was remembering what I had seen as we drove towards the valley over the heights just outside Jerusalem where there was so much scraped white limestone and so many white buildings rising and where there had seemed such a hopeless race by the races to populate, expressing the inexorable biology of our species. Was it just for this, the desperate animal imperative of reproduction, that *H. erectus* and then *sapiens* had wandered out of Africa? One might have hoped the Jews to have remembered better the other deep biological imperative that is more special to their own traditions, a part of which is that up-and-wander spirit itself—that of the pastoralists they once were proud to have been, and which is so much closer to the world of outreaching imagination they had learned to teach. Well, one could understand the desperation to have a place of their own after their experiences in Europe; but the alternative in Europe, too, had been simply to ease off their own increase and thus to become less feared.[11] Over my lifetime it has saddened me that they should have lowered themselves into the company of former oppressors, re-enacting, under reversed roles, the old and predictable story.

What about my paper in all this? I should come to that. Writing it had been an unusual task for me in two ways. One was that I had chosen to study an ecological and population-dynamical interaction instead of a population-genetical one. The other was that, compared with my usual sloppy standards in works that are purely my own, I carried the mathematical analysis unusually far. In explanation of the second point, I guess I was a bit fearful of that Jewish facility for logic and maths that I mentioned in the introduction to the last chapter and hinted at again in this one; I was heading for a kind of homeland of all of this, and I had better stretch myself and do an adequate job if I was not to seem a fool. What I achieve in the paper analytically is, by the standards of, say, Samuel Karlin (promoter of our symposium and the editor, with Eviatar Nevo, of its volume in which my paper appeared), very elementary; still, it was an effort for me and a relief to find that I could really prove most of what I had suspected. But why was I trying to prove anything about just a numerical interaction of species? It wasn't, as it stood, evolution—why invade a rather unimportant corner of a territory of Robert May?

I attempt to explain this in the paper and will not repeat the arguments here. The ultimate objective is still sex and the 'how' of its support, but I was trying mainly to satisfy myself a little further about the likelihood of an intrinsic abundance of cycling in the living world—or rather to satisfy myself about what I might more broadly call the vibratory tendency to be expected whenever an ecological interaction includes one or more 'pursuits'. By 'pursuit' I mean one species 'seeking' another and that other species 'fleeing'. Of course it's not just literal 'pursuit' and 'fleeing', though those could well happen, too, on the actual ground. Rather it was the more general situation where one species needs another for its own life, most commonly as its food, and the other is trying to evade—trying not to be food, habitat, or whatever else the pursuer is after. I knew the answers for the simplest situations since the earliest pioneers of interspecies population biology, Alfred Lotka and Vito Volterra, had provided them.[12] I knew also about Robert May's general claim that adding more random links to a web of interacting species gave less chance, not more, that the total situation can be in perfect stasis. Thus if several Lotka–Volterra cycling systems are put somehow into arbitrary contact with one another, May's results predict them to be less likely to reach an equilibrium. But how will they depart from any initial unstable balance and what will the end result be? Will the combined systems cycle less or more wildly than their binary fundamental units, and, above all, does it become less or more likely that something has to go extinct? And what about synchrony of cycling? In this paper I added just one interaction to link the two halves of my mini-ecosystem, one of them potentially symmetrical—a competition of the two hosts. I knew I would be unable to answer my question about the final dynamics fully analytically. But it would not be difficult in a tentative way to parallel the stability analysis—that routine drill conducted out in 'flat plains' of math relations where so-called 'linear' approximations can be applied. By a simulation I could show experimentally something of what was happening in the mountainous terrain that encircled the plain. There I could play with ideas that might possibly steer my embryo ecosystem away from or towards paths of extinction. I wanted, in short, more of a feeling for communities and especially for their behaviour when they were far from a potential equilibrium and contained pursuit interactions.[13]

May's approach seemed only to look for stability: all else was treated as failure and was thrown away; in short, the unstable systems were expected to simplify by at least one species although this outcome was not proved and might not happen. If a species is not lost, and yet nor is any unchanging situation stable, then obviously an interesting permanent dynamic must result. It seemed quite likely this would involve changing selection pressures that would be favourable to sex if the genetics was there as well. Indeed it seemed to me this intermediate outcome of non-stability but non-simplification might be common: why could there not be bounded cycling with no elimination?[14] I have to admit that in the outcome of this particular paper May was the winner: most cases of cycling that I studied would, if unaided by extra assumptions, have lost at least one species. The extra factor I brought in to prevent this was potentially realistic but at this stage very untried: a metapopulation with low levels of migration. As you will see in the paper I simply postulate multiple demes to be there in the background—like Cadmus's soldiers I conjured them from the ground, but unlike soldiers I wanted them an unruly bunch, to march all out of step so that, as a whole, they could provide constant input to ailing populations. That such behaviour was possible for metapopulation was shown later in the models first of Frank[15] and then by Ladle *et al.*[9,10] The paired subsystems of the present model do indeed go 'out of step' but this is due to host competition and is an effect only weakly connected with the desynchronization in a metapopulation (it must be remembered that in the present paper I am treating only one deme). Given, however, that the further work confirms that desynchronization is usual in all reasonable models, we obtain easily, as this paper begins to show, a shimmering vibration of frequencies throughout most of the living world. Not only does the metapopulation never lose genes or species but often, as shown in many of the cyclic states produced by the present model, it is clear that the individual demes never lose them either, or at most do so temporarily.

The ultimate object was not species but polymorphic genes—that tinker's bag of tricks that species carry along and freely interchange. As I explain in the paper, and here touch on again, I had had all along the idea that the results I got for species interactions would translate into similar results in a gene-frequency version of the model. With a few caveats one would simply substitute the word 'allele' for the word 'species'. At the time I was working on this I had not foreseen that I needed to be very careful in any such translation and that parthenogenetic clones, to a lesser extent than genes but with this offset by their great efficiency, would be strongly protected too (I will return to mention this problem in Chapter 16). Anyway, as regards the gene changes in sexuals, something like the 'translation' mentioned has been done though so far with no attention to sex.[16] But the reader will be justified to be a little puzzled why, after a long 'Red Queen' introductory discussion about coevolutionary genetics and sex, I launch into what may seem a very limited 'Red Pawn' of a model about population demographic interactions. For the moment I just have to ask you to believe me that the antics of the 'Pawn' that I describe are a prelude for wider and more complex dances that my subsequent papers are going to show performed by the Red Queen. What is more interesting to begin to discuss right here is the question of how the set of dance-like motions arising in an ensemble of connected pursuit situations[17] comes to divide itself between genetic 'neo' Red Queen motions (that is, the dynamic of alleles and genotypes within species if they are sexual) and what one may properly call classical Red Queen manoeuvres at the species-in-community level. I call these 'classical' because they are much closer in spirit to, though still far from identical with, the ideas for which Leigh Van Valen originally invoked Lewis Carroll's character.[18] Both processes might be said to be playing their parts in helping to keep the world 'green' in spite of all the combined activities of its herbivores and parasites—see Chapters 4 and 5.[19]

When populations of parthenogenetic species (or rather the clones or clades of 'museum' species, because strictly the species concept doesn't hold for parthenogens) are studied genetically it usually turns out that they consist of mixtures of from several to many genotypes and that these clones (1) are not each others' closest-related clones among the sets known and (2) are found to be undergoing dramatic changes in relative frequency if they are studied over time.[20] Aphids, water fleas, and rotifers have provided well-studied examples and these situations may come near to illustrating what I am actually simulating in a simplistic way in the paper. They illustrate the parthenote clades coming as near as possible to having their cake and eating it—having the variety but also efficiency—and through this keeping (as I conjecture) their frequency-dependent antagonists at bay. That their play is not fully successful compared with using recombinatorial sex in the very long term seems to me clear from the fact that, with very rare (and perhaps still questionable) exceptions,[21] obligately inbreeding, and purely parthenogenetic species and lines, whether plants or animals, seem to die out before achieving any substantial evolutionary advances, very few of them achieving even genus level of diversification let alone anything higher. In fact, most only achieve species level in the eyes of rather generous-minded taxonomists (others referring to their rather ill-defined assemblages as microspecies, agamospecies, Jordan species, and the like).[22,23] In the short term, however, their success, at least in terms of present abundance and the occupation of difficult habitats, is striking.[24] What is happening with the varied clones in these populations is essentially that each is 'hiding' among others that are different enough to give parasites a hard time swapping from one to another. Their specific enemies (mainly parasites, I believe) are wasting their effort searching in the multitudes of the inedible for the few individuals they can attack. All clones thus get a special advantage when they are rare and this is why they do not die out and why many are present. This idea links, of course, with the very widespread similar advantage that is believed to be a major factor in diversifying all living systems, whether the types involved are parthenogenetic or sexual.[25]

If you have ever searched for a particular species of plant or animal in a tropical forest you will understand hiding by being rare. The best naturalist in the local village may tell you, 'It occurs, I can find it'; and yet if he hasn't seen the species recently even for him it may take days to find just one specimen. The same is probably true when it is a parasite that is searching: like the local naturalist, the parasite knows what it needs and that it is there, but the target is hidden in an immense diversity of other things. Let us fix on a gravid female butterfly seeking a particular species of tree or bush. The butterfly perhaps has various chemosensory advantages over the village *mateiro*, but it's only an insect with a distance vision that is quite mediocre. The butterfly has to be very close to a plant before it even has a chance to use sight to discriminate the often extremely subtle shape differences of tropical forest leaves. Most butterflies I see working in this way in the dim understorey of a forest seem to be almost randomly alighting on any likely looking leaf, tasting it momentarily with their feet (your foot is a tongue if you are a butterfly). A female ithomiine butterfly, heavy of abdomen and longing to find her plant, may go like this for hours from leaf to leaf: for as long as you watch you may not see her lay even one egg. When she does find the plant it may be small and suitable only for one or two, so that she soon resumes her search. Long-winged, slow-flying, highly poisonous and often semitransparent butterflies are generally looking for members of the potato family, Solanaceae, though for a few it can be the periwinkle relatives in the forest instead. In the case I described of one egg laid on a tiny plant they might be acting like predators of those plants but, more usually, by the definition that they eat their host but usually don't kill it, caterpillars are parasites. We don't know yet why evolution forces specialism on parasites but many possibilities are being discussed by biologists—too many for comfort. My favourite recent suggestion is a recent one of Whitlock.[26] Given that specialism occurs, however, it is easy to see how biodiversity can feed on and breed its own increase—and that this process also ends up helping to keep the forest green. Despite its truly huge variety of kinds of small herbivore (orders of magnitude more than a temperate forest will have), the leaves in a tropical forest are no more tattered by insects and fungi than are those of the temperate forest.

Obviously all the above raises a question of whether, if my general idea is right, being in a tropical forest might in the end relieve rather than increase the need to have sexuality to help control one's parasites. Could the simple hiding-by-rarity and enemy time-wasting themes—the million wrong leaves interspersed between the two that are right—be enough to ensure a comfortable survival for everyone? As a general way of protection the idea asks for a further question about how a species or individual could begin using such a course. Getting to be rare goes against the grain for life and perhaps has disadvantages at least as great as those attached to being common and maintaining the 'wasteful' operation of sex for parasite control. One way to get to be rare, obviously, is to bud off, become a new species, from some corner of the common one you presently are but want to abandon. Budding geographically, on an offshore island say, is a good way of doing this as Ernst Mayr has emphasized but, of course, there can be no planning. Given you have come far enough that the mate preferences necessary to begin a new species have already appeared, how do you persuade all your old parasites that you are so different that they should decide to stop eating you? A difficult problem, but not worse than others that beset the study of speciation. I think the factor of a limited escape from the ubiquitous force of parasitism can only help with both problems at once.

In the paper of Chapter 5, I suggested that species may separate not primarily because of the need for modular adaptation to the physical environment nor even for such adaptation to the biotic environment as it is normally imagined (defining limits, say, to the range of prey species that the size of a predator makes efficient to be taken, with speciation rendering these ranges discrete), but rather because of the chance evolution of, in different parts of the parent species' range, radically different techniques for resisting parasites. It seems likely that incipient resistance may start by such radically different paths that eventually it becomes disastrous to mix and re-associate their elements—the F_2 hybrid may be, as it were, receiving a mixture of machine-gun parts manufactured for models from different armament companies. As I mentioned in the introduction to Chapter 5, suggestive though still sparse evidence for such an importance of incompatible defences exists, its present best example being the extremely poor resistance to a nematode shown by hybrids of the two subspecies of the domestic mouse in Europe.[27] In any case, this is digression: the main point, again, is simply that parasitism hints at neglected advantages for the superabundance of speciation events that have occurred on our planet and also at routes that, at least from a distance, look easy to travel in that general direction. Let me pass on to discuss whether we should expect less sexiness when species richness attains its pinnacle, as in tropical forest: can the hiding within such huge mixtures make sex unnecessary? Seemingly, no.

The facts suggest that sexiness does not reduce in tropical forest; it even increases, though much less than the species diversity does. Tropical trees are somewhat more sexual than temperate counterparts, at least in so far as this can be judged from the variability of their isozymes[28] and from what is known of their breeding habits.[29] For the big trees this is perhaps not so surprising: if it had been one of these species you had asked the naturalist of the village to find, the trip in the forest is likely to have been briefer although not necessarily in distance (big trees are his landmarks, he'll know where to go—and, by the way, if he's at all an Amerindian, will expect to lead you to them by his dark, slippery paths at a run). Massive, conspicuous plant bodies like big trees may be easy targets for their insect enemies, which often have the extra search advantage of flying over and looking down on the canopy and probably also of being more sensitive to the gusts of a tree's special scent.[30] I am not yet aware of comparable information on sexuality and polymorphism levels of understorey shrubs and herbs but I predict that the excess of genetic diversity over the temperate equivalents may prove to be less the smaller and more scattered the plant. From my own observations on breeding systems of the river-bank annuals of the Amazon, I see little difference from that of temperate weeds: ability to self-pollinate is common while, as to extremes, I have found neither examples of parthenogenetic seeds nor of separation of the sexes. A ploughed field in Oxfordshire would yield about the same; in fact, one could almost summarize by saying that the sand and mud banks of the Amazon are indeed ploughed fields and the Amazon is the plough (a resemblance from which local people take good advantage). But among the large perennial arborescent 'weeds' of Amazon river and sand banks, a situation dramatically different from that of the smaller weeds exists and it is one I am not sure can be well paralleled in the temperate zone: among these very conspicuous tree-like weeds separation of sexes is very common. Likewise, in the case of two hermaphrodite species of the extremely flooded areas that I tested, I found no ability of the showy flowers to self-pollinate.[31]

Plants of all these big and highly 'apparent' species,[32] easy to find because they are largely confined to river banks, often have extremely tattered leaves. It seems as if the marginal and flood-prone places they inhabit in company with only a few other equally tolerant pioneers denies the hiding strategy. Their habitat is linear and within it they exist unmixed and at higher density, a considerable contrast to conditions in the fully two-dimensional and far more species-diverse and uniform *terra firme* forest nearby. Thus the pioneer species probably have to endure a larger than usual pressure of potential parasites attempting 'jumps' from another host species and finally succeeding. Completely contrary to my starting idea, which was that the far-out-from-forest species on the flooded margins would be low in sexuality because they support few parasites and mainly 'struggle for existence' against physical forces, I now suspect that sexuality will always be high in such species; in fact that none can abandon sex and survive in these unmixed and easy-to-find situations. Often I see these species (cecropias and the Amazonian willow as examples) skeletonized nearly to leaflessness. Correspondingly I suspect that genic diversities in these pioneer plants may be amongst the highest known.[33] These plants' richness in secondary defensive compounds is already established. There is a parallel, here, I believe, to the reindeer, which likewise for a while puzzled me by its obvious high sexuality until I learned how abundant the parasites of reindeer actually are. The environment may indeed be harsh when, if ever, reindeer parasites are outside their hosts, just as the Arctic is harsh on everything living, but for a reindeer parasite one problem is already solved in a way it is not for parasites of deer in forest thickets: its host is large, common, homoiothermic, and, most important, has almost nowhere to hide (see also Volume 1 of *Narrow Roads of Gene Land*, Chapter 7).

All this is again a long digression from the talk I gave in the Israel symposium but I hope it shows some of the background of my thoughts and also what followed. There is another way in which all these digressions are also relevant to my trip, a bit sad for me to mention but important to be understood. Amongst patterns of planetary biodiversity *Homo* is in an atypical line: we are anything but biodiverse—a monospecific genus and even tribe (to some) and this is not common. On the other hand, the physical diversity of human races compared with that seen in another group—butterflies say—might have persuaded a Martian taxonomist who hadn't yet started to focus on human interfertility that the forms of *Homo* had already reached species status. It may be that when we emerged to the point where we could begin to observe ourselves scientifically, we were actually not far from having undergone a new radiation that would have returned our group towards a biologically more typical condition of several or many species to the genus. Our Martian, had he arrived around the start of our now passing millennium, might indeed have observed many towns and other crowded habitats where he could see racial forms existing well mixed and yet still remaining quite distinct and little hybridized: and, at least for an Earthling taxonomist, such mixing without hybridity is the usual criterion of the different forms being species. Yet, clearly, if a radiation was about to happen, it hadn't quite done so in prehistory and it hadn't either, quite, during the Renaissance when came the first stirrings in the global human pudding: the Polos marched east from Venice, Columbus west from Genoa, Vasco da Gama sailed south, and so forth; and all of these by their immense journeyings began to push the great spoon of change. Journeys end in lovers' meetings: also, it has to be said (especially when travellers go armed or in ships), they quite often end in rapes too.

Thus soon by their own sperm as well as by the ideas that their achievements stimulated, the explorers and colonists started to mix things up. Incipient crystals of what may have been almost different species were stirred back into solution and this was happening, for the first time since the early days of our species in its confined world in Africa, on a truly whole-species and planetary scale. Europeans did not like all that they saw in their travels. The caste system of India, for example, almost a species radiation in itself and like to that which our all-seeing Martian might have noted in African great lakes for cichlid fish, was frowned on by British invaders and partly broken up. The invaders, however, showed themselves at the same time quite happy to insert a two-tier caste system of their own in its place, which, as might be expected, and like many similar social inconsistencies around the world, didn't work. In spite of countercurrents in which many remote peoples suffered or were outright extinguished, right up to the present a spirit of panhumanism has increased in pace and eventually has swept all before it (Chapter 9). India, China, and Japan eventually all lowered their barriers. Almost all others did too. As a species, or rather as we are directed to be a species by the intellectuals who write for us, we are opting not for the tropical (Brazilian) *terra firme* route in the forest as above (which route's best instance had been again caste-chequered India as it existed prior to the British Raj). Instead we are choosing the 'river-bank pioneer' route—that of remaining a single and insistently outbreeding and multiresource-using gene commune covering the entire world. In a few places and situations, however, the new panhumanism didn't sweep away the older patterns of incipient separation: one was South Africa; another Germany, especially in the 1930s and early 1940s. Yet another was and still is Israel, starting from the first stirrings of the Zionist idea in mid-nineteenth-century Europe (in which the universal racist theories of the time still had a seminal part[34]), and continuing with the underlying concepts in the state's constitution as it came to be formed. There are scattered other instances persisting, of course—other crystals that have resisted being stirred back into soup; every reader will know of some.

It is sure to seem ridiculous to many for me to connect the human phenomena being mentioned to the kinds of biological motivations I have given above for speciation. It will have to be admitted, however, that some traits a zoologist or a well-informed Martian taxonomist would recognize are present. For example, such differing emphases on food and who handles it as are observed in all the more orthodox sects of Judaism in Israel today (all matters, be it noted, relevant to dangerous microbes food may carry, and concerns of hygiene and diet related to this), and these are exactly what we would expect to arise during a speciation process, the main (but moderate) differences from biological situations being that Judaism had come to use a cultural element of inheritance to replace what genes once had been doing more slowly. Indeed, combining this observation with the way elsewhere in the world other caste-like groups likewise strongly emphasize differing diets and notions of hygiene, such a situation bears out even the predictions of the particular 'parasite-oriented' view of speciation I favour. Other instances of traits of incipient biocultural speciation are in differing standards and laws existing for 'us' and 'them' (these resembling how most of us see as perfectly natural the different laws applied to 'us' and to, say, macaques and cows). Perhaps most worrying to panhumanists are yet other hints of incipient speciation that lie in a stated total and unchangeable belief in the supposed revelations of an ancient holy book, a book that includes not only the hygienic and dietary instructions just mentioned—which were sensible enough, indeed advanced, at their time of enunciation—but also specific discouragement of outmarriage, and, perhaps most worrying of all, injunctions for genocide where necessary against competing groups.

In the world of animals, ants perhaps provide *Homo*'s nearest equivalent for typical broadness of niche. If an unspecialized ant species had a Bible I'd expect to find in it extremely similar injunctions about food, ant genocide, and so forth, as I find in the actual Bible, and I would have no difficulty to suppose these as serving each ant colony well in its struggle for existence. Very probably something like such sets of instructions exist effectively encoded in ant genes but, for sure, we wouldn't think it a good step for our species to read out this information and then re-encode it as a kind of written senior cultural wisdom (in evolution history, of course, ants are greatly our seniors).

In short, from a humanist point of view, were those 'species' the

Martian thought to see in the towns and villages a millennium or so ago a good thing? Should we have let their crystals grow; do we retrospectively approve them? As by growth in numbers by land annexation, by the heroizing of a recent mass murderer of Arabs,[35] and by honorific burial accorded to a publishing magnate (Robert Maxwell), who had enriched Israel partly by his swindling of his employees, most of them certainly not Jews, some Israelis seem to favour a 'racewise' and unrestrained competition, just as did the ancient Israelites and Nazi Germans. In proportion to the size of the country and the degree to which the eyes of the world are watching, the acts themselves that betray this trend of reversion from panhumanism may seem small as yet, but the spirit behind them, to this observer, seems virtually identical to trends that have long predated them both in humans and in animals.

After these thoughts it will come as no surprise to the reader to hear that the part of our nomadic meeting in Israel that I enjoyed least was the day we and I spent touring in Jerusalem. Unquestionably this is one of the world's most beautiful cities: its setting, its walls, great buildings, the aura of its history, these are all stunning. Legacies of the past, of so much of our traditions, jut at the visitor from every wall and corner. Yet neither in the tourists and nor the inhabitants did I seem to see much enjoyment of it all. A tension was palpable and there were few smiles. Machine guns hung on shoulders of almost civilian-dressed young men and women in the streets rather like book-filled backpacks might dangle on students walking the Diag of the Campus in Ann Arbor. As a group of mostly grey-haired and obviously harmless tourists, we merited hardly a glance from these Platonic guardians of the city and so much of being ignored was quite welcome. But it was not quite so comfortable to find oneself taking on, it seemed, a practical invisibility to other Jerusalemites—for example, to black-garbed, long-haired men who strode the streets seeming to see only walls and buildings around them, or sometimes to see not even those; and who would pause for and notice only others who were attired similarly to themselves. Caused to feel so invisible, it came almost as a relief when, at the Wailing Wall and having it explained to us by one of our guides, I noticed the guide being stabbed himself by a clearly angry glance from one of the black-clad men. He then lowered the book from which he had been reading aloud in order to sweep all of us—all of our teeshirted, sandalled, and gawping group of scientists, such as might have taken the scheduled day trip out from any similar meeting in Europe or America—with a long cold stare, a look that seemed to tell us clearly enough that in his opinion we should not have been there; in his eyes we were frivolous and irrelevant in such a place.

I know very well that we could have experienced the same degree of unwelcome in holy places in the holy cities of Islam or in parts of Tehran, and actually my most vivid similar experience to that I have just described was a reaction from an elderly male Scot in Edinburgh, who overheard me joking to a friend and laughing while walking the morning pavement of a Presbyterian Sabbath near to Princes Street. It was the same turn, the same stare: you are acting wrongly, you are polluting my thoughts. In the case of the Israeli at the wall there was also: 'you understand nothing of this', not even if that fool of an Israeli guide tries to explain it to you. Fair enough: I certainly didn't understand much. But for sheer fanaticism the Middle East on the whole seems to have it over Edinburgh and even over Northern Ireland. I am sure I am not the first to have wondered what it is about that part of the world that feeds such diverse and intense senses of rectitude as has created three of the worlds' most persuasive and yet most divisive and mutually incompatible religions. It is hard to discern the root in the place where I usually look for roots of our strong emotions, the part deepest in us, our biology and evolution; even a recent treatise on this subject, much as I agree with its general theme, seems to me hardly to reach to this point of the discussion.[34,36]

What can one find, in fact? Could it have something to do, once again, with the peculiar self-sufficiency of pastoralists who survived well for thousands of years alongside their softer and more opulent Middle Eastern agrarian neighbours? By their very ability to persist on unenvied desert diets and to use only camel-borne and shoulder-borne possessions, they were not only surviving but threatening and looking down on the occupants of town and villages. Or might it be that the region tended to include the very first to experiment with the agrarian lives, together with the fast rise to high density that these new styles of living permitted, so that they carried with them into their new lives a stronger imprint from the transition they had undergone (what had gone wrong in it, or right, and why?) and thus carried along also specially unfaded memories of a tribal past? There had been no time for the group selection that was later to teach people more tolerant and integrative ways, no time to learn how not to enshrine as unchangeable all the convictions and doctrines that had been so useful for survival during the phase of being hinterland tribes. Adding to this possibility perhaps a macromutation for imagination or for those 'logic' neurones for the brain or the like that I proposed semiseriously in the last chapter to have occurred in Abraham or in some Sumerian ancestor, giving him good reason to feel superior as well as self-sufficient as he drove his flocks west.

In such a picture possibly the scene might have been set for all that was to follow. Jesus and Mohammed were, of course, both claimed (plausibly enough) to have descent from Abraham and through them the Old Testament memes for intolerance may have passed down through all the major splits . . .

I am always happier if I have at least a conjecture and these rather vague ones, which I break off at this point, can serve me for the present. Of course I would dearly like to see improved offerings about a historical phenomenon[36] that seems to me very much in need of some sort explanation—why is the Middle East so fanatical? I should add that the above-described sensations on our tour applied almost only to Jerusalem. More democratic, scientific, or open-minded men than our guides themselves, for example, could hardly be imagined. But it seems to me one of the most paradoxical of all the many impacts Judaism has had on the present world that it is the cousins (at least in a broad ethnic sense) of the very people who claim, on the authority of ancient writings, a right to deprive other Middle Easterners of their land and freedom, who generate from among their intellectuals the arguments that convince the rest of us that panhumanism is a worthy ideal and that the course that follows from it is *Homo*'s best hope for the future. A long list of examples could easily be given: I could start, for example, with Karl Popper and his book *The Open Society and Its Enemies* and I could end with books of Richard Lewontin and Stephen Jay Gould. It seems to be particularly these men of Jewish origin (not, however, orthodox followers of the faith as a rule) who seek to persuade us that race differences are non-existent and everyone can mate with everyone in confidence of equally capable children. If Zionism is to escape the thought 'They mean panhumanism only for some', then these writers must include severe criticism of Israel on the same grounds that they applied criticism to South African apartheid and to racial and class separations in the USA. Demographic competition and overt racism of which religionism is just one kind have to end in Israel and the country has to become a true democracy, shaking off the present theocratic restrictions in its constitution.[37] If this doesn't happen, genuine panhumanists and genuinely democratic states are justified to treat Zionism and Israel with the same reserve, sanctions, and suspicion as we treat all the similar inward-turned '-isms' and personality cults of other antidemocratic states. Meanwhile, in the works of those who pretend to a belief in panhumanism while financially or otherwise supporting Israel on its present course, we are justified to say that we detect the taint of hypocrisy.

What I have written here is certain to draw accusations of racism on me when this book is read. I will end, therefore, by pointing out that, contrasting so far as I can see with any philosophy whatsoever of the past, I am presenting in this volume of my papers concerning sex the first biological rationale for panhumanism in its most basic, single-species sense that has ever been offered. I am proud to call myself the first anti-racist who has a good evolutionary reason—and even mathematical models—for being so. My view advocates what, it appears, *sapiens* of the distant past never advocated—neither with regard to his cousin *neanderthalensis* (else the latter's genes at least would be with us still whereas it is claimed they are not[38]), nor with regard to *erectus* in South East Asia (which species was long contemporary, it seems, with our own, at least in Java). Most experts are telling us again that with *erectus* neither genes nor mitochondria came down into the younger species once the stocks had split apart in Africa. A common current moral stance, which I call crypto-racist, says in effect, 'I see, of course, the beauty and the equal status of all those outsiders; still, given the differences—the clashes and all that—I wouldn't want my daughter to marry one.' My theory says, to exaggerate it a little: 'Clashes? That's all just culture, and you'll soon adapt. Health is the most vital thing and haven't I sufficiently proved why we are all good for each other? Look for whatever you like, marry, and don't think twice.'

I myself have plenty of ideas on how to proceed, some exaggerated, some serious. If you bet on AIDS as the curse of the twenty-first century, then let my International Marriage Bureau (IMB) advise you. A first-class Nairobi prostitute will be found for you, a proven uninfectable by the HIV-1 virus.[39] She'll probably provide you with the broadest-based protection for descendants. But in case you're finicky as to the profession, we may be able to find you a very respectable Finnish lady of excellent resistance instead.[40] If you tremble rather for the coming of the multi-drug-resistant tuberculosis, IMB will seek out the right East European, survivor of that other kind of holocaust that region experienced: he or she may quite possibly be a not-too-orthodox Jew and also just possibly a mild case of Tay-Sachs.[41] Again, if instead you fancy a change from the cold north (or south) and like to imagine descendants as rich plantation owners in the swampy, falciparous tropics, then we'll find you a West African whom we will certify on grounds of several genes at several loci as being your very best mate against malaria;[42] resistance to other, mainly tropical ailments such as hookworm can probably be thrown in as well. IMB already has many other suggestions and at the rate of present progress of medical and genome knowledge soon will be able to propose a somebody for everything, excepting possibly the new diseases at present being inducted into our species by live vaccines, xenotransplants, and the like. Because of the sometimes great phyletic distance of the hosts involved in such completely new developments, those diseases will need time for true resistor genotypes to manifest: they will come but more slowly, possibly only after several generations (meaning a scale of 100 or so years), especially if complex combinations have to be put together. But for those who haven't yet decided which epidemic it is best to fear, the first golden rule is very simple: follow your senses of beauty and love wherever they lead. Unfortunately for IMB's chance to charge you for telling you all this, the crucial message is being given to you all the time straight out of the back of your skull. But because scientists are always supposed to spell out their sources even when they are just common sense, let me give the credit for this not to IMB but to John Lennon and especially to his song 'Love is all you need': in short, love, marry, and re-create human beauty through whatever human form you adore—though, please, for the rest of our sakes, not too much of it, not too many.

Of course I am being a bit frivolous again here and this is why I bring in my Bureau (yet to be incorporated, I am afraid; if it was perhaps I'd be rich). Clashes of culture do matter in marriages and so do other more personal clashes of psychology. But 'health for children'—or 'sosigony' as I have tried unsuccessfully so far to christen it—matters a great deal too. It matters more, it seems to me, than most realize. Some of the more endogamous of religious sects are already badly burdened. This may come about through a lack of understanding of what they are doing genetically, especially about how they shut in mutation and exclude new resistance;

and I am not just writing here of the well-known major bad gene mutations of endogamous sects. To repeat, so far as I know I am the first to have proposed sosigonic gene exchange as positively and rationally supporting the panhumanist programme. What has been emphasized in this field in the past has been rather the opposite—health *dangers* of race hybridity.[43] Such dangers certainly can exist but probably the Rhesus blood-group problem of some Europeans gives us the measure of them. In other words, they seem for the most part minor. When we changed course in our approach to racial and divisions-of-labour events of speciation—that is, before the Polos went travelling in Asia and roughly at the time when the Martian I supposed above formed his opinion about our being already a species complex, it seems we were still, over the world, well short of any major genetic barriers. All this discussion, of course, goes far beyond the paper that follows but I write it out here to show where my ideas were tending both before, and as they have become changed by, the dramatic days I spent in Israel.

I almost wrote 'dramatic *first* days in Israel', which would be in one sense true but in another not. Never in Israel as Israel before, I had been in Palestine—was a baby in Haifa. While my mother kept me and my year-older toddling sister practically shut within the British garrison's married quarters during the shortly pre-World War II days, as my mother later told me, and this because of danger from both Zionist and Arab terrorists opposed to the British Mandate, my father was working for the British Army on the plans for an intended Haifa–Baghdad road. The road was never built: the war came instead and then Israel. I found it difficult to persuade both Eibi Nevo and Ilan Eshel that any notion to build a Haifa–Baghdad road existed but a map in my father's file drawers scrawled over in his own terrible but unmistakable handwriting and showing the proposed lines, proves this. How strange to be there again, not quite in Haifa this second time but high above it on Mount Carmel where the new university was, and that I should be visiting an Institute of Evolution set up by a man, Eibi Nevo, whose own father, an early twentieth-century Jewish immigrant from Lithuania, I could almost imagine to have felt a duty as a Zionist to shoot a uniformed British officer, such as my father, if the opportunity arose! (Eibi, however, has told me his father was not such an immigrant as ever would, and also that it was always a small minority of Zionists even at that time who advocated violence.)

The day in the University of Haifa was almost the last of our meeting. Perched on a mountain summit, the glassy main university building seemed a bit unreal, like an alighted spaceship or a project for a time machine. I am glad it isn't a time machine or else some fanatic, after reading this essay, might go back to bomb the Officer's Mess of Haifa 10 years earlier than it was indeed bombed, with much loss of life, by Zionists just after the war, and thus he might have ended the sapper captain in sand-coloured shorts and Sam Browne sword belt, the man who was to determine in large measure the course of my life. Retrospectively he might have undone a course that has led to this writing as well as, farther back, to all that sociobiology, those 'selfish' genes, and many other claimed ills of our changing culture that I have had a part in. But I suppose, had a Zionist time-traveller had such power, he would almost for sure have gone deeper in time and headed for bigger game—why not bomb, let us suggest, the mess tent of Pompey the Great when he was camped at Jerusalem, or of Emperor Hadrian (who, incidentally, invented the name Palestine for the country he had just caused to cease to exist as Judaea), or even, travelling another 700 years, plant it in the tent of Nebuchadnezzar? What different spins such bombs might have brought to the history of our planet!

Anyway my idle thoughts here mean little; there is more point to say about Eibi, knowing both the man and his work, which has been so massively and honestly favourable to the biology I admire (and all of this of the type that is sometimes reviled under the name of the dread 'adaptationist programme'), that he is one whom I would not fear personally as this time-traveller if he became one. Naturalist and historian at heart, he would be far more likely to pull out a notebook than a bomb, and Hadrian, like-minded about efficiency and in having an interest in everything, probably would have taken notes on his visitor also (I read somewhere that Hadrian even wrote reminders to himself about topics he wanted to raise with his wife at breakfast). Instead I would fear much more others, such as he whom I saw turning to us science tourists at the Wailing Wall. Indeed, such is Eibi's own drive to be objective about everything that I think of him as the spirit least likely of all I know to be upset by what I have written here—less in fact than will be many who are completely unconnected with Israel as a country. For many, I fear, my approach to Israel is too simplistic, too irreligious, and just not of the right kind.

References and notes

1. C. M. Doughty, *Wanderings in Arabia Deserta* (Duckworth, London, 1908).

2. J. Billing and P. W. Sherman, Antimicrobial functions of spices: why some like it hot, *Quarterly Review of Biology* **73**, 1–46 (1998).

3. D. Zohary and M. Hopf, *Domestication of Plants in the Old World* (Clarendon Press, Oxford, 1993).

4. G. P. Rightmire, Deep roots for the Neanderthals, *Nature* **389**, 917–18 (1997); R. Ward and C. Stringer, A molecular handle on the Neanderthals, *Nature*, **388**, 225–6 (1997).

5. B. P. Forster, J. R. Russell, R. P. Ellis, L. L. Handley, D. Robinson, C. A. Hackett, *et al.*, Locating genotypes and genes for abiotic stress tolerance in barley: a strategy using maps, markers and the wild species, *New Phytologist* **137**, 141–7 (1997); L. L. Handley, D. Robinson, C. M. Scrimgeour, D. Gordon, B. P. Forster, R. P. Ellis, *et al.*, Correlating molecular markers with physiological expression in *Hordeum*, a developing approach using stable isotopes, *New Phytologist* **137**, 159–63 (1997).

6. K. Kato, Y. Mori, A. Beiles, and E. Nevo, Geographical variation in heading traits in wild emmer wheat, *Triticum dicoccoides*. 1. Variation in vernalization response and ecological differentiation, *Theoretical and Applied Genetics* **95**, 546–52 (1997); E. Nevo, I. Apelbaum-Elkaher, J. Garty, and A. Beiles, Natural selection causes microscale allozyme diversity in wild barley and a lichen at 'Evolution Canyon', Mount Carmel, Israel, *Heredity* **78**, 373–82 (1997); E. Nevo, Genetic resources of *Hordeum spontaneum* for barley improvement (paper presented at The International Conference on the Status of Plant and Animal Genome Research, San Diego, 12–16 January 1997); E. D. Owuor, T. Fahima, A. Beiles, and A. Korol, Population genetic response to microsite ecological stress in wild barley, *Hordeum spontaneum*, *Molecular Ecology* **6**, 1177–87 (1997); S. Somersalo, P. Makela, A. Rajala, E. Nevo, and P. Peltonen-Sainio, Morpho-physiological traits characterizing environmental adaptation of *Avena barbata*, *Euphytica* **99**, 213–20 (1998).

7. G. L. Sun, T. Fahima, A. B. Korol, T. Turpeinen, A. Grama, Y. I. Romin, *et al.*, Identification of molecular markers linked to the Yr15 stripe rust resistance gene of wheat originated in wild emmer wheat, *Triticum dicoccoides*, *Theoretical and Applied Genetics* **95**, 622–8 (1997); F. Tzion, G. Sun, V. Chague, A. Korol, A. Grama, Y. Romin, *et al.*, Identification of molecular markers linked to the YR15 stripe rust resistance gene of wheat introgressed from wild emmer wheat, *Triticum dicoccoides* (paper presented at The International Conference on the Status of Plant & Animal Genome Research, San Diego, 12–16

January 1997); T. Fahima, M. S. Roder, A. Grama, and E. Nevo, Microsatellite DNA polymorphism divergence in *Triticum dicoccoides* accessions highly resistant to yellow rust, *Theoretical and Applied Genetics* **96**, 187–95 (1998).

8. E. Nevo, Genetic diversity in nature: patterns and theory, *Evolutionary Biology* **23**, 217–46 (1988); H. Pakniyat, W. Powell, E. Baird, L. L. Handley, D. Robinson, C. M. Scrimgeour, *et al.*, AFLP variation in wild barley (*Hordeum spontaneum* C. Koch) with reference to salt tolerance and associated ecogeography, *Genome* **40**, 332–41 (1997).

9. R. J. Ladle, R. A. Johnstone, and O. P. Judson, Coevolutionary dynamics of sex in a metapopulation: escaping the Red Queen, *Proceedings of the Royal Society of London B* **253**, 155–60 (1993).

10. O. Judson, Preserving genes: a model of the maintenance of genetic variation in a metapopulation under frequency dependent selection, *Genetical Research, Cambridge* **65**, 175–91 (1995); O. P. Judson, A model of asexuality and clonal diversity: cloning the Red Queen, *Journal of Theoretical Biology* **186**, 33–40 (1997).

11. Increase of Ashkenazi Jews in eastern Europe in the span of the nineteenth century is said to have been almost fourfold (S. Jones, *In the Blood: God, Genes and Destiny* (HarperCollins, London, 1996)). This implies a doubling about every generation. Very surprisingly this fact seems almost never to be discussed as part causative background to the holocaust, an omission that continues even when claims of group competition are the focus (K. MacDonald, *Separation and its Discontents* (Praeger, Westport, CT, 1998)).

12. F. Scudo and J. Ziegler, ed., *The Golden Age of Theoretical Ecology* (Springer, Berlin, 1978).

13. Symposium volumes are not a good way to get your work noticed! Two similar theoretical mini-ecosystems involving two prey/hosts and two predator/parasites have been studied since my paper was published, and neither cites it.

    One study (J. Vandermeer, Loose coupling of predator–prey cycles: entrainment, chaos, and intermittency in the classic MacArthur consumer-resource equations, *American Naturalist* **141**, 687–716 (1993)) is a close analogue to mine but gives most of its attention to describing the cycling, bifurcation, chaos, and intermittency possibilities displayed in the non-linear dynamics of the author's four-species system. Its results basically fit with my verbal description. because of the competition of 'prey', it also finds the cycling of the two pairs moving maximally out of phase.

    In the other study (J. D. Van de Laan and P. Hogeweg, Predator–prey coevolution: interactions across different timescales, *Proceedings of the Royal*

Society of London B **259**, 35–42 (1995)), a pursuit situation (termed 'predator–prey') is set up between 'quasi-species' (equivalent to clustered asexual-strain mixtures) with the pursuit–avoidance evolution being conducted in a circular phenotype space traversable by mutation. The initial quasi-species pair spontaneously splits into two pairs, which then (as the most robust among many possible patterns) conduct a long-term periodic dance of fusing and parting while at the time all morph frequencies undergo rapid short-term changes.

Neither study includes sex or Mendelian genetics but both confirm the almost unavoidable dynamic behaviour within mini-ecosystems involving exploitations. Thus both provide background support for the 'Biotic Red Queen' view of sex.

14. I. Eshel and E. Akin, Coevolutionary instability of mixed Nash solutions, *Journal of Mathematical Biology* **18**, 123–33 (1983).

15. S. A. Frank, Spatial variation in coevolutionary dynamics, *Evolutionary Ecology* **5**, 193–217 (1991).

16. J. D. Van de Laan and P. Hogeweg, Predator–prey coevolution: interactions across different timescales, *Proceedings of the Royal Society of London B* **259**, 35–42 (1995).

17. R. H. Whittaker, Evolution and diversity in plant communities, in *Diversity and Stability in Ecological Systems*, pp. 178–96 (National Bureau of Standards, US Department of Commerce, Springfield, PA, 1969); D. A. Levin, Pest pressure and recombination systems in plants, *American Naturalist* **109**, 437–51 (1975).

18. L. M. Van Valen, A new evolutionary law, *Evolutionary Theory* **1**, 1–30 (1973).

19. See also S. Pimm, *The Balance of Nature* (University of Chicago Press, Chicago, 1991).

20. M. A. Mort and H. G. Wolf, The genetic structure of large-lake *Daphnia* populations, *Evolution* **40**, 756–66 (1986); P. D. N. Hebert, Genotypic characteristics of cyclic parthenogens and their obligately asexual derivatives, in S. C. Stearns (ed.), *The Evolution of Sex and its Consequences*, pp. 175–95 (Birkhauser, Basel, 1987); A. Gómez, M. Temprano, and M. Serra, Ecological genetics of a cyclical parthenogen in temporary habitats, *Journal of Evolutionary Biology* **8**, 601–22 (1995).

21. O. P. Judson and B. R. Normark, Ancient asexuals—scandal or artifact—reply, *Trends in Ecology and Evolution* **11**, 297 only (1996); O. P. Judson and B. B. Normark, Ancient asexual scandals, *Trends in Ecology and Evolution* **11**,

A41 only (1996); S. J. Schrag and A. F. Read, Loss of male outcrossing ability in simultaneous hermaphrodites—phylogenetic analyses of pulmonate snails, *Journal of Zoology* **238**, 287–99 (1996).

22. M. Mogie, *The Evolution of Asexual Reproduction in Plants* (Chapman and Hall, London, 1992).

23. How purely parthenogenetic some of these forms are is often a moot point. In some evidence of Hardy–Weinberg ratios (proportional association of alleles within loci) in the diploid genotypes indicates that they certainly do sometimes cross. Even when such ratios are not approached there may still be evidence of occasional crossing, or at least of the new forms having originally been created from crosses of pre-existing clones (S. Asker, Progress in apomixis research, *Hereditas* **91**, 231–40 (1979); A. J. Richards, *Plant Breeding Systems* (Allen & Unwin, London, 1986); J. T. Crease, D. J. Stanton, and D. N. Hebert, Polyphyletic origins of asexuality in *Daphnia pulex*. II. Mitochondrial DNA variation, *Evolution* **43**, 1016–26 (1989)). Obviously in so far as such systems persist, they can have advantages similar to those of the more regular cyclical parthenogenesis as occurring in most aphids, water fleas, etc.

24. M. F. Dybdahl and C. M. Lively, Diverse, endemic and polyphyletic clones in mixed populations of a fresh-water snail (*Potamopyrgus antipodarum*), *Journal of Evolutionary Biology* **8**, 385–98 (1995).

25. J. B. Gillet, Pest pressure, an underestimated factor in evolution, in D. Nichols (ed.), *Taxonomy and Geography: A Symposium*, pp. 37–46 (London Systematics Association, London, 1962); J. Connell, On the role of natural enemies in preventing competitive exclusion in some marine animals and in rain forest trees, in P. J. van der Boer and G. R. Gradwell (ed.), *Dynamics of Populations*, pp. 298–310 (Centre for Agricultural Publication and Documentation, Waageningen, 1971); D. H. Janzen, Herbivores and the number of tree species in tropical forests, *American Naturalist* **104**, 501–28 (1971); C. K. Augspurger, Seed dispersal of the tropical tree, *Platypodium elegans*, and the escape of its seedlings from fungal pathogens, *Journal of Ecology* **71**, 759–71 (1983); C. K. Augspurger, Offspring recruitment around tropical trees: changes in cohort distance with time, *Oikos* **40**, 189–96 (1983); C. K. Augspurger, Seedling survival of tropical tree species: interactions of dispersal distance, light gaps, and pathogens, *Ecology* **65**, 1705–12 (1984); P. Becker, L. Lee, E. Rothman, and W. Hamilton, Seed predation and the coexistence of tree species: Hubbell's models revisited, *Oikos* **44**, 382–90 (1985); C. Wills, R. Condit, R. B. Foster, and S. P. Hubbell, Strong density- and diversity-related effects help to maintain tree species diversity in a neotropical forest, *Proceedings of the National Academy of Sciences, USA* **94**, 1252–7 (1997).

26. M. C. Whitlock, The Red Queen beats the Jack-of-All-Trades: the limitations on the evolution of phenotypic plasticity and niche breadth, *American Naturalist* **148**, S65–S77 (1996).

27. S. Y. Strauss, Levels of herbivory and parasitism in host hybrid zones, *Trends in Ecology and Evolution* **9**, 209–14 (1994); C. Moulia, N. Lebrun, C. Loubes, R. Marin, and F. Renaud, Hybrid vigor against parasites in interspecific crosses between 2 mice species, *Heredity* **74**, 48–52 (1995).

28. M. D. Loveless and J. L. Hamrich, Ecological determinants of genetic-structure in plant-populations, *Annual Review of Ecology and Systematics* **15**, 65–95 (1984); J. L. Hamrich and M. D. Loveless, Associations between the breeding system and the genetic-structure of tropical tree populations, *American Journal of Botany* **74**, 642–2 (1987); M. D. Loveless and J. L. Hamrick, Distribution of genetic variation in tropical woody taxa, *Revista de Biologia Tropical* **35**, 165–75 (1987); Nevo (1988) in note 8; J. L. Hamrick and D. A. Murawski, Levels of allozyme diversity in populations of uncommon neotropical tree species, *Journal of Tropical Ecology* **7**, 395–9 (1991); D. A. Murawski and J. L. Hamrick, Mating system and phenology of *Ceiba pentandra* (Bombacaceae) in Central Panama, *Journal of Heredity* **83**, 401–4 (1992).

29. K. S. Bawa and P. A. Opler, Dioecism in tropical forest trees, *Evolution* **29**, 167–79 (1975).

30. J. D. Nason, E. A. Herre, and J. L. Hamrick, The breeding structure of a tropical keystone plant resource, *Nature* **391**, 685–7 (1998).

31. Murawski and Hamrick (1992) in note 28, however, reported that *Ceiba pentandra*, formerly a common riverbank giant tree in Amazonia, self-pollinated frequently although at rates very variable from tree to tree.

32. P. P. Feeny, Plant apparency and chemical defence, in J. W. Wallace and R. L. Mansell (ed.), *Biochemical Interactions between Plants and Insects*, pp. 1–40 (Plenum, New York, 1976).

33. They were not outstandingly high for a *Cecropia* in Mexico (E. R. Alvarez-Buyla and A. A. Garay, Population genetic structure of *Cecropia obtusifolia*, a tropical pioneer tree species, *Evolution* **48**, 437–53 (1994); B. K. Epperson and E. R. Alvarez-Buylla, Limited seed dispersal and genetic structure in life stages of *Cecropia obtusifolia*, *Evolution* **51**, 275–82 (1997); A. A. Garay and E. R. Alvarez-Buylla, Isozyme variation in a tropical pioneer tree species (*Cecropia obtusifolia*, Moraceae) with high contents of secondary compounds, *Biotropica* **29**, 280–90 (1997)), the only species of the genus studied genetically so far. This was a forest light-gap species, however, to which type the diversity/concealment factor may apply at least in part (one light-gap group perhaps

being of similar apparency to the big tree that it replaces). Comparing their findings with those for other pioneer species, the authors expressed surprise at the diversity they found within individual light-gap patches.

34. K. MacDonald, *Separation and its Discontents* (Praeger, Westport, CT, 1998).

35. Baruch Goldstein. Modern Islamic political ethics, it has to be said, include similar and even more widely agreed upon 'fundamentalisms' as is shown in the shelter and pension Saudi Arabia gives to Idi Amin, apparently on the ground that he adheres to Islam. For the most part Amin, when president and dictator of Uganda, was obeying the Koran (4: 92–95) in avoiding mass murder of co-religionists whilst promoting mass murder of others. Goldstein doubtless believed himself obeying the explicit instructions of the Tanakh in the killing, within the 'Jehovah-given' towns, of opponents of Israel, of all ages and sexes, as in Biblical times (Deut. 7:3 and 20:16).

It has been urged to me that Goldstein was a member of a small and very extreme group whose actions the majority of Israelis abhor, and, in parallel to this, definitely I would not wish to think British morality judged by the acts of Tom Hamilton, who a few years ago unaccountably burst armed with a rifle into a British primary school and massacred children before finally killing himself. So far as I know, however, no single Briton has praised Tom Hamilton (an adoptee by the way—sharing his name but, I am glad to say, no Y clone with me!), whereas at least to an outsider, Goldstein's act, and the public praise it receives from his sect, appear as just the 'ultra-violet' of a spectrum of anti-Arab racism (including numerous other killings) that occur far more commonly than news media represent. All this concerns acts for which, seemingly, Israeli perpetrators receive either no censure or else astonishingly light sentences; see, for example, C. C. Smith, 'Shooting and crying', *The Observer* (31 July 1988); for further examples and general critique see I. Shahak, *Jewish History, Jewish Religion: The Weight of Three Thousand Years* (Pluto Press, Boulder, CO, 1994) and N. Chomsky, *Fateful Triangle: The United States, Israel and the Palestinians* (Pluto Press, London, 1999).

36. K. MacDonald, *A People That Shall Dwell Alone* (Praeger, Westport, CT, 1994).

37. Shahak (1994) and Chomsky (1999) in note 35.

38. Ward and Stringer (1997) in note 4; T. F. Bergström, A. Josefsson, H. A. Erlich, and U. Gyllensten, Recent origin of HLA-DRB1 alleles and implications for human evolution, *Nature Genetics* **18**, 237–42 (1998).

39. K. R. Fowke, N. J. D. Nagelkerke, J. Kimani, J. N. Simonsen, A. O. Anzala, J. J. Bwayo, *et al.*, Resistance to HIV-1 infection among persistently seronegative prostitutes in Nairobi, Kenya, *Lancet* **348**, 1347–51 (1996).

40. F. Libert, P. Cochaux, G. Beckman, M. Samson, M. Aksenova, A. Cao, *et al.*, The Delta CCR5 mutation conferring protection against HIV-1 in Caucasian populations has a single and recent origin in Northeastern Europe, *Human Molecular Genetics* **7**, 399–406 (1998).

41. B. Spyropoulos, P. B. Moens, J. Davidson, and J. A. Lowden, Heterozygote advantage in Tay-Sachs carriers, *American Journal of Human Genetics* **33**, 375–80 (1981); S. J. O'Brien, Ghetto legacy, *Current Biology* **1**, 209–11 (1991); W. W. Stead, The origin and erratic global spread of tuberculosis—how the past explains the present and is the key to the future, *Clinics in Chest Medicine* **18**, 65 (14 pages) (1997).

42. A. V. S. Hill, Genetic susceptibility to malaria and other infectious diseases: from the MHC to the whole genome, *Parasitology* **112**, S75–S84 (1996); A. V. S. Hill, Genetics of infectious-disease resistance, *Current Opinion in Genetics and Development* **6**, 348–53 (1996).

43. N. Stepan, *The Idea of Race in Science: Great Britain 1800–1960* (Archon books, Handeb, CT, 1982).

INSTABILITY AND CYCLING OF TWO COMPETING HOSTS WITH TWO PARASITES†

W. D. HAMILTON

—

Competition between hosts fortifies (and complicates) the tendency to oscillation in host–parasite systems. A model on the lines of standard community ecology using host–parasite dynamics plus host competition shows that, as competition increases, (1) the fixed point of the system becomes less likely to be stable, (2) the departure from the fixed point remains as likely to oscillate, and (3) permanent cycling (or irregular quasi-cycling) readily follows.

The presence of parasites complicates but does not eliminate the 'competitive exclusion principle' for a pair of hosts. When competition between hosts is strong, the double system easily loses either one parasite or a host–parasite pair through extinction. Low levels of immigration strongly reduce cycle amplitude and prevent extinction.

These findings are discussed in relation to the theory that sex is an adaptation to combat the rapid coevolution of parasites. Extinctions of genes (or local extinctions of population) due to parasites do not necessarily undermine the episodic quasi-repetitive selection processes that support sex.

1. COEVOLUTIONARY PURSUIT AND FLEEING BY SEX

Among possible interactions of species in an ecosystem there is a general kind that may be called coevolutionary pursuit (CP). A parasite with host (*p–h*) and a predator with prey (*p–p*) are examples of it. In a square matrix which shows for a local community the effects that current abundance of each species has on the growth rate of every other, pairs of species involved in coevolutionary pursuit produce pairs of entries of opposite sign: a CP is in fact characterized in a 'community matrix'[1] by this '+\−' interaction, which simply reflects that abundance of the pursued species (host or prey) is positive—increases growth rate—for the pursuer, while abundance of the pursuer species (parasite or predator) is negative—decreases growth rate—for the pursued.

---

†In S. Karlin and E. Nevo (ed.), *Evolutionary Processes and Theory*, pp. 645–68 (Academic Press, New York, 1986).

At least since the work of Lotka and Volterra (see Scudo and Ziegler[2]) it has been realized that the $+\backslash-$ interaction causes oscillatory population phenomena. The oscillations following perturbation may be damped and disappear or may grow. If they grow, the species system may end in some compact finite cycle where both species persist in repetitively changing numbers. Or else the oscillations may grow without limit until one species, most likely the host or prey, hits zero. Even when this happens, however, the system may still allow all species to exist if the local deme where extinction occurs is part of a larger population that from time to time reseeds patches with migrants.

On the basis of its abundance and inherent tendency to cycle, it has been suggested that the CP situation, and especially its *p–h* variant, can explain the existence of a very widespread and seemingly inefficient system of reproduction. This system is sexuality (refs 3–5, and see also Chapters 2 and 5). Introducing this idea, of course, at once steps outside the confines of the community matrix approach to ecology because the model underlying that approach makes no provision for variation within species. However, it is clear that the dynamical findings must still apply in a broad way even when there is variation. Also, reinterpreting 'species' in the community model as 'varieties' can make interesting suggestions about the advantages sex might have in a broader model—a model where sex was being really tested during both interspecies and varietal changes. This paper, therefore, will look in the community dynamics model for robust sources of cycling that might support sex over parthenogenesis.

The effects of having sex will not be addressed in the models here: the cycles I discuss will involve changes in population numbers only. Sexuality, of course, is not primarily concerned with numbers of individuals; it is concerned rather with numbers of genes and gene combinations. Hence there will be important questions left unanswered in this paper about what changes of combination need to arise during population cycles for sex to be useful; whether such changes as do arise increase or reduce the mean density of population; and whether the changes reduce the cycling tendency or exaggerate it. None of this will be addressed directly. Following May[6] it will be assumed that if cycles can emerge from population-density dependence they can also, in parallel ways, emerge from genotype-frequency dependence. Following a paper of my own (see Chapter 2 in this volume; but for cautions on the specific model of this chapter see ref. 7), it will also be assumed that when a fair amount of dominance and epistatic interaction contributes to the fitness of genotypes, not only does cycling remain possible, but, when it occurs, it gives a sexual population a good chance to outcompete any mixture of asexual clones that might be derived from it, despite the advantage that clones have in reproductive efficiency. The chance is especially high when the selection is strong, as can easily be the case with disease.

Of course, if clones are not efficient in reproduction, as might be expected with first mutations to parthenogenesis out of long-established sexuality (and as seems often to be the case, e.g. ref. 8), the chances for sex to resist inroads of parthenogenesis is better at the start. However, there would seem to be many marginal situations where, if low initial fertility were their only handicap, clones could establish themselves (e.g. by colonizing islands where competition is lacking). In such places they could then improve efficiency by further selection (as has sometimes happened with asexual lines under artificial maintenance). Finally, they could reinvade the sexual areas in which their matriarchs originated. The rarity of observation of such courses argues, in my opinion, that a fairly short-term strong selection is supporting sexuality in most species.

In my model of 1980 (see Chapter 2) there were no fluctuations of population numbers at all. Cycling of genotype frequencies arose from pure frequency-dependent selection of a somewhat artificial kind. In contrast, one could alternatively imagine density-dependent dynamics following roughly the lines of the model given below, but with feedback based on numbers of genotypes taken separately, rather than on numbers of all individuals taken together, as in the pure ecological version; further one could imagine that the cycles of p–h-selected genotypes 'drag' the frequencies at other polymorphic loci along with them, so producing continual changes in parts of the genome that have nothing to do directly with the CP and the parasites. If epistatic fitness effects are involved, such repetitive changes can give sex an advantage. Unless prevented by the structure of the model, overall population numbers would in such a case cycle in some rather complex fashion but would probably not vary so widely as they did before genotype variation was brought in. This would be due to a tendency of a more complex dynamic to dephase automatically the genotype cycles within each species. Smoothing of overall changes through such dephasing is revealed in the simulation runs of the model below (section 4) if we accept to think of the two hosts as genetic varieties of a single species.

In wider perspective, and looking in particular at Jaenike's and my own version of the parasite resistance idea about sex, as long as the cycling tendency is not too strongly damped, and stable cycles are not too closely confined, the numerous p–h interactions that must apply to almost all eukaryotic species can be seen as a recipe for eternal turbulent genotype change. In effect, the pursued species, if sexual, is using segregation and recombination to 'run away from' the coevolutionary advances of its various pursuers. Such advances come particularly fast in the case of pursuing parasites, it is suggested, because of the usual smallness and their consequent shorter generation time (see Chapter 5 and ref. 4). To combat their transmutations sexual species are continually putting back together the yet unextinguished components of defence systems that served them in similar phases of similar cycles in the past. Doing so, they 'flee' more effectively than can asexual species: hence arrives an advantage to sex that may exceed the advantage that parthenogenesis has in efficiency.

'Fleeing' and 'pursuit' as described so far are largely a matter of running in circles. However, even if old alleles sometimes go extinct and completely new advantageous ones arise by mutation, it is still probably true that pursuit situations are more apt to make gene transiences common, as is needed to protect sex, than are other agencies of natural selection. This may amount to saying that some older ideas about sex, such as 'Muller's ratchet', may look stronger if attention is focused particularly on pursuit situations, especially the host–parasite one. Even if not quite running in circles here, the system is still most of the time dodging around in a small evolutionary area. Sex on this view has a basically defensive conservative function and the macroevolutionary developments that arise from it are initially only side-effects.

The reasons why the parasite–host relation, of the two major classes of CP, is more likely to be important for sex can be summarized as follows:

1. Parasites tend to be short-lived and consequently can rapidly evolve new modes of attack.

2. Parasites tend to be specialists and predators to be generalists; it is harder to imagine how a generalist can get into cycles.

3. For any one host or prey, p–h relations are much more numerous than p–p. (On the severity side, predation events may seem more decimating and parasites in contrast often very trivial; but, in fact, debility due to parasitism probably precipitates capture by a predator much more often than is recorded—see, for example, refs 9–11.)

4. The coevolution of host and parasite is more likely to take place in discrete orthogonal steps (e.g. create a new 'password' surface molecule), whereas p–p relations are more likely to be a matter of adjustment of position in a continuum (e.g. choose a particular compromise between fleetness and strength). Cycles are more likely to be robust and permanent when mutation to an intermediate position between two initial morphs is not possible.

None of these four factors leans unequivocally in favour of the p–h rather than the p–p variant. For factor 2, for example, it may be objected that the possibility of cycling when there is a generalist predator (or parasite) has hardly been studied theoretically. A first glance at what has been done, however, is not at all encouraging. Lotka[12] has pointed out that a predator with two prey can maintain its numbers on one of them while it drives the other extinct: to the benefit of both pursuer and pursued extinction is less likely when dependence is on a single species. Here we do not seem on our way towards a recipe for eternal cycles. However, given again metapopulations made up of many demes with reseeding by migrants following extinctions, or when given special refuges from parasitism or predation when at low density (perhaps through genotypic or learned switching of the pursuer's attention to the more abundant prey or host), it is possible to imagine systems with generalists that at least preserve all their species. It is, of course, another step to demand that on top of that they also cycle, or at least give conditions of selection causing frequent gene transiences in the way that would be favourable to sex, but this possibility might repay study.

Such a line towards dynamics of generalist pursuers will not be followed here. The simple connected community that I propose to look at does potentially cover the one-parasite–two-hosts situation as a special case. However, my primary objective is to understand more about the cycling induced by specialist parasites. I hope to show that they still cause cycles when their hosts are not ecologically isolated but interact among themselves—in other words, that Kolmogorov's limit cycle[2] arising from CP does not disappear as soon as it is involved in a more speciose and interconnected situation. In addition, I hope to show reason to believe that a general result on cycling that is implied in a recent work by Eshel and Akin[13] is not substantially weakened when absolute densities of morphs replace relative densities (i.e. replace gene frequencies) as the fitness-determining variables. The particular relation holding between hosts in the model studied will be competition.

2. THE MODEL

Let there be two host species, Ho_1 and Ho_2. Let each have a parasite which has at least a minor degree of specialization. Call these Pa_1 and Pa_2. If the population numbers of these four species in a particular generation are H_1, H_2, P_1, P_2, let those of the next generation be given by the relations

$$H'_1 = H_1F_1, \qquad H'_2 = H_2F_2, \qquad P'_1 = P_1G_1, \qquad P'_2 = P_2G_2, \qquad (1)$$

where all the factors F_1, F_2, G_1, and G_2 are, in the most general case, functions of all four population sizes: for example,

$$H'_1 = H_1F_1 (H_1,H_2,G_1,G_2). \qquad (1a)$$

Suppose that the system has an interior equilibrium point $(H_1^*,H_2^*,P_1^*,P_2^*)$; that is, a point such that $H_1^* = H_1^*F_1$ with $H_1^* > 0$ and likewise for the three others.

The first step towards describing the stability of this equilibrium is to evaluate the Jacobian matrix of the linear approximation of the system that applies near to the equilibrium point. This matrix, formed from evaluations of all the partial derivatives at the equilibrium point, will be written:

$$\mathbf{J} = \begin{bmatrix} \underset{a_1}{-} & \underset{b_1}{-} & \underset{c_1}{-} & \underset{d_1}{(-)} \\ \underset{b_2}{-} & \underset{a_2}{-} & \underset{d_2}{(-)} & \underset{c_2}{-} \\ \underset{e_1}{+} & \underset{f_2}{(+)} & \underset{g_1}{(-)} & \underset{h_1}{(-)} \\ \underset{f_2}{(+)} & \underset{e_2}{+} & \underset{h_2}{(+)} & \underset{g_2}{(-)} \end{bmatrix}. \tag{2}$$

In this, we have by definition, for example, that

$$e_2 = \left. \frac{\partial \{P_2(G_2 - 1)\}}{\partial H_2} \right|_*, \tag{2a}$$

where $|_*$ means evaluation using $H_1 = H_1^*$, etc. Note that the particular slope instanced here refers to the effect of the second host on the second parasite and, this being appropriate host for that parasite, the slope will be positive. This is indicated by the $+$ sign just above e_2 in the matrix. The expected sign is given likewise for all the other symbols in the matrix, with those that might reasonably be zero-bracketed.

The next step is to form $\mathbf{J} - \lambda \mathbf{I}$ (i.e. to subtract λ from every term in the main diagonal), and then to find all four zeros for λ for the determinant:

$$|\mathbf{J} - \lambda \mathbf{I}| = 0. \tag{3}$$

Assume the four roots found numerically or otherwise. It then follows that if one or more of the roots of this equation is such as to imply mod $(\lambda + 1) > 1$, then the fixed point $(H_1^*, H_2^*, P_1^*, P_2^*)$ is unstable; if, on the contrary, the opposite inequality holds for all of the roots, then the fixed point is stable (see ref. 1). Any case where the largest modulus is equal to 1 would need further analysis to determine stability; however, such a case is exceptional and if it should hold we can see that stability and instability must be both near at hand. Changes of the parameters of the model in such a state are likely to have large effects for or against stability. Thus any stochastic event can be expected to have longer aftermath than under other conditions. Even if fluctuations reduce, the aftermath changes will probably be similar to effects of cycling from the present point of view. This case will not be discussed further.

In general, the expansion of (3), the characteristic equation of the system, is a quartic, which generally will not solve neatly. One case where it does factorize into a product of quadratic expressions, however, is broad enough to give guidance on some interesting questions. This is the case where the two subsystems, Ho_1 with Pa_1 and Ho_2 with Pa_2, are 'identical twins' in all their parameters, including those that imply cross-effects onto twins. Because the subset having this symmetry is more restricted than one could wish, cases where roots are found numerically from the general Jacobian will be discussed and also some simulations of dynamics of the complete system will be reported. In this way it is hoped to give some idea of how robust the conclusions drawn from the symmetrical case may be. Simulation is especially needed, of course, in those situations where instability holds because no useful analytical tools for showing the existence of simple cycles seem to be available once more than two species are involved. These unstable cases are crucial in the present study.

Formally the intended symmetrical subset of matrices is defined as having $a_1 = a_2 = a$, $b_1 = b_2 = b$, ..., $h_1 = h_2 = h$; in other words, we simply drop all subscripts in **J**.

When this is done, the characteristic equation derived from **J**, after factorization, is

$$\{\lambda^2 - (a_+ + g_+)\lambda + a_+g_+ - c_+e_+\} \{\lambda^2 - (a_- + g_-)\lambda + a_-g_- - c_-e_-\} = 0, \quad (4)$$

where

$$
\begin{aligned}
a_+ &= a + b, & a_- &= a - b, \\
g_+ &= g + h, & g_- &= g - h, \\
c_+ &= c + d, & c_- &= c - d, \\
e_+ &= e + f, & e_- &= e - f.
\end{aligned}
\quad (4a)
$$

Signs and relative magnitudes in these expressions and their implications for the eigenvalues (the roots of (3)) will be discussed below. It is worthwhile to look first, however, at another subset where simple quadratic factors can be found and where symmetry of the subsystems is not required. Suppose the subsystems are not linked with one another at all (i.e. all those elements in **J** with letters b, d, f, and h, of whichever subscript, are set at zero). Then obviously a quadratic characteristic equation appears for each of the now separated subsystems:

$$\lambda^2 - (a_1 + g_1)\lambda + a_1g_1 - c_1e_1 = 0, \quad (5a)$$

$$\lambda^2 - (a_2 + g_2)\lambda + a_2g_2 - c_2e_2 = 0. \quad (5b)$$

Comparing these to the factors in the previous equation it can be seen that there is a generic similarity. Instability depends on the relative magnitude of certain sums and differences of effects on growth rates in the linked systems, and on simple within-subsystem effects on growth rates in the individual subsystems when unlinked. In (4) a_+, for example, can be described as the effect on joint growth rate of hosts of perturbing one of the host populations alone, while in (5a), a_1 is implying the effect of change of the first host population on its own growth rate—in this case there is no effect from the other. Similar observations hold for all other symbols in the first factor.

Similarity of form allows the conditions for stability to be discussed at least in preliminary stages by selecting any one of the four equations given or implied above and dropping all of the subscripts. Subscripts can be reinserted when there is need to go into more detail.

In this light, the roots are all of the form:

$$\lambda = \{(a + g) + \sqrt{(a + g)^2 + 4(ag + ce)}\}/2 \qquad (6a)$$

or equivalently:

$$\lambda = \{(a + g) + \sqrt{(a - g)^2 + 4ce}\}/2. \qquad (6b)$$

Thus there is a pair of complex roots, with implied potential for oscillatory behaviour at least near to the fixed point, if

$$(a - g)^2 < -4ce. \qquad (7)$$

Clearly ce must be negative if there is to be any chance of oscillatory departure. A glance at the matrix $\mathbf{J}$ shows that this is just what the CP relationship tends to ensure, except for some situations that might arise from the composite nature of c and e in the linked case (i.e. the case arising from (4); note that, unlike with (5), dropping subscripts in (4) is not equivalent to dropping them in matrix $\mathbf{J}$).

We now make some assumptions about magnitudes of the various gradients set out in the Jacobian:

1. With the exception of a and b (where any relation of magnitudes may hold), the within-subsystem effects (measured by c_i, e_i, g_i; where $i = 1, 2$) are much larger than the across-subsystem effects (measured by d_i, f_i, h_i; where $i = 1, 2$). This means that c_+ and c_- both have the same sign as c_1, and likewise for the other letters with the sign subscripts except for a.

2. The effect on growth from one's own species numbers is much less in parasites than it is in hosts, so that $a \gg g$. This is usually reasonable because for parasites host finding is more crucial than the parasite's own density.

3. Now a repeat of the almost definitional assumption about parasitism: hosts benefit their parasites but parasites harm their hosts. Given assumption 1, this means $c < 0$ and $e > 0$, and therefore $ce < 0$. This is a necessary condition for complex roots as noted above.

Under these assumptions, looking at (6a), (6b), and (7), we see that:

(a) If p–h interactions are generally powerful compared to competitive effects, then oscillations are likely. This is the expected generalization of Lotka–Volterra theory.

However, if parasite species interact strongly within and between themselves in much the same way that host species do (i.e. if the sub-matrix of **J** that has g and h is broadly similar to the submatrix that has a and b), contrary to assumption 2 above, then p–h interactions have less need to be strong for oscillation to occur. Then with $a = g$ we can see that quite small host–parasite effects could cause oscillatory events. This seems unlikely to apply to most parasites. It could more reasonably apply to predators apart from the fact that predators are unlikely to be con-fined to such a simple two-prey scenario.

(b) In any reasonable model the fitness of hosts on their own will show negative density dependence in the following sense. At low host numbers we may expect that a will be positive (numbers grow faster the more are present to reproduce), while if the equilibrium is at high density, then a will be negative. The latter case, in making a_+ likely to be negative, favours sta-bility. An equilibrium anywhere near to a condition where a is zero, is almost certain, through the expected smallness of g, to imply complex roots. Moreover, the equilibrium is likely to be unstable if $-ce$, measur-ing the power of the p–h interaction, is relatively large. As host density at equilibrium falls below that which gives $a = 0$, a rises, making strong p–h effects less necessary to attain instability.

Summarizing (a) and (b), if parasites have large effects on mortality of hosts and especially if they depress equilibrium levels far below carrying capacity—or if they interfere competitively among themselves, as discussed in the next section—then the internal equilibria of the system(s) are unlikely to be stable, and any departure will be oscillatory. In such cases, with reason-able assumptions, limit cycles or more complex dynamic attracting sets are expected to follow, as discussed by May[1] for the case of the unlinked sub-systems.

3. EFFECTS OF HOST COMPETITION

In this section to help fix ideas we depart from previous generality and make specific assumptions about how each host population affects the other. Later, a little thought will show that similar results should hold under other reason-able assumptions.

Assume that in the absence of parasites the recurrence relations for the two hosts are

$$H_1' = H_1[1 + s\{1 - (uH_1 + vH_1)/K\}]$$

$$H_2' = H_2[1 + s\{1 - (vH_1 + uH_2)/K\}], \tag{8}$$

where $0 < v < 1$ and $u + v = 1$. Here v is the parameter linking the sub-systems, and it can be called the coefficient of allocompetition. Its value shows the proportion by which each host's growth rate responds to numbers of the other species relative to its responding to numbers of its own. If $v = 0$, the equations are simply those determining discrete logistic growth: this case, of course, has the numbers of each host growing towards K as the asymptotic upper limit. As v is raised, each population becomes more responsive to inhibitory pressure from the other host. At $v = 0.5 = u$ each species is responding to all individuals equally without discrimination of species. If $v > u$, each is more inhibited by numbers of other species than by numbers of its own: this could happen with interference competition.

Increasing allocompetition is intuitively destabilizing. One way of seeing this more formally is to note that as v passes upwards through 0.5 the two straight isoclines, $uH_1 + vH_2 = K$ with zero growth for H_1, and $vH_1 + uH_2 = K$ with zero growth for H_2, switch across each other, changing a stable (improper) node at $H_1 = H_2 = K$ into a saddle. The species with greater numbers then drives the other one out. At $v = 0.5$ the species coexist but do so at best in neutral equilibrium.

Increasing allocompetition is likewise sure to be destabilizing in the model where each host has a parasite. This is easily shown formally using the specific competition model implied in equation (8). Parasites being present, these equations each need an additional term that will be a function of parasite numbers and, probably, of host numbers. It will represent the reduction of host numbers or host reproduction that the parasites cause. We need not specify the form of such terms and will simply call them $-P\text{term}_1$ and $-P\text{term}_2$.

To complete the system we would also need to describe the functions G_1 and G_2 in (1). The following argument does not require that these be specified. However, one thing will be said about them. It is simplifying and reasonable to assume that v does not appear in either of the Pterm expressions or of the G expressions. These are not concerned with interhost competition and therefore presence of v would not be expected. The point of this assumption will appear shortly.

We may abbreviate (8) by replacing the square-bracket expressions by symbols $\tilde{F}_1$ and $\tilde{F}_2$; then after including the Pterms we may form the partial derivatives like (2a) that give a_1, b_1, a_2, and b_2 for the Jacobian:

$$a_1 = \frac{\partial\{H_1(\tilde{F}_1 - 1)\}}{\partial H_1}\bigg|_* - \frac{\partial P\text{term}_1}{\partial H_1}\bigg|_*$$

$$= \frac{\partial\{H_1\tilde{F}_1 - 1\}}{\partial H_1}\bigg|_* - \frac{\partial P\text{term}_1}{\partial H_1}\bigg|_*.$$

For compactness the last term of these expressions will be given the symbol $P\text{slope}_1^*$. Then, performing the differentiation in the first term:

$$a_1 = s\{1 - (uH_1^* + vH_2^*)/K\} - usH_1^*/K - P\text{slope}_1^*.$$

With the assumption of symmetry we can drop subscripts and obtain:

$$a = s(1 - H^*/K - usH^*/K - P\text{slope}^*).$$

A like procedure for b_1, assuming that $P\text{term}_1$ contains H_1 but not H_2, and after assuming symmetry, gives:

$$b = -vsH^*/K.$$

Hence, $a_+ = a+b = s - P\text{slope}^*$, which is independent of v.

Recalling the assumption made above which implies that v is not involved in any other elements of the Jacobian, we can turn back to (4) and see that the '+' quadratic factor as a whole is also independent of v, so there is one pair of eigenvalues which does not change as the degree of interhost competition is changed. If either of these roots gives $\text{mod}(\lambda + 1) > 1$, the system is unstable irrespective of interhost competition or lack of it. A similar argument can be applied regarding the independence of interparasite competition, if any exists.

Note that the constancy of a pair of roots depends on the assumed symmetry and is lost if this symmetry is dropped. The reason for constancy is easy to see if we think of the model being started symmetrically (i.e. with $H_1 = H_2$ and $P_1 = P_2$). Obviously these equalities will continue to hold as the model runs, and so it behaves as if each subsystem was running on its own; and with the assumptions above the way it runs does not depend on v. Thus one of the necessary stability conditions, that given by the '+' factor, is the same as the stability condition for an isolated subsystem.

If both '+' roots are inside a unit circle centered at $(-1,0)$ in the complex plane, then the '−' roots must be examined to determine stability.

As v increases, starting at zero, b decreases also starting from zero, while a increases by an exactly corresponding amount. The trace of the matrix **J** increases, making it generally more likely that there will be an eigenvalue with modulus greater than one. In fact, under reasonable assumptions, it is virtually certain that by the time $v = 0.5$ (full competition) the system is unstable. Under symmetry this follows because

$$a_- = a - b = s(1 - H^*/K) + (v - u)sH^*/K - P\text{slope}^*.$$

The first term is always positive, and the second is positive as soon as $v > 0.5$. The last term, concerned with the detrimental effects of parasitism, is also positive under fairly reasonable assumptions—these being to the effect that the ability of parasites to take advantage of increasing host numbers shows diminishing returns as host numbers grow, so that the damage rates from a fixed number of parasites diminishes. If all three are positive, then a_- is positive. Instability then follows from (6), and likelihood of oscillatory departure is implied by (7).

In summary, competition always moves to destabilize the dual host–parasite system when symmetrical and we can reasonably suppose that it does so also in asymmetrical cases. Upon destabilization, as noted in the last section, there is an oscillatory departure.

4. FATE OF OSCILLATIONS: PROTECTION FROM EXTINCTION

The main method I have used to investigate the non-linear dynamics has been simulation. The usual device, once away from the linearized region, of plotting the patterns of flow in relation to isoclines is not easy to apply in this case because of the dimensionality. For the stimulation I used CP equations that are discrete versions of the differential equations 4.4 of May.[1] In these equations the host's response to density was generalized to the form of (8); that is, the term H/K in May's system became in mine $(uH_1 + vH_2)/K$ for the H_1' equation and $(vH_1 + uH_2)/K$ for the H_2' equation.

If the subsystems had stable cycles or equilibria when considered alone, then at low levels of the coefficient of allocompetition in the situation was not much changed except that cycles might occur where previously they had not at all, and that a stable point could be pushed into small cycles by the other subsystem if that one cycled intrinsically. Other interesting but also expectable features were that the cycling in the two systems tends to be out of phase, perhaps permanently and at least most of the time, and that when two periods synchronize out of phase they do so at some period intermediate between the periods of the separate parts.

As allocompetition is increased, at least one of the cycles moves to a state where most of the time both its host and its parasite occur at very low numbers, and where brief upsurges of numbers are followed by crashes. Eventually a crash may, with the formulas of the discrete system, bring a host number below zero—in effect cause an extinction. Or it may stay positive but go so low that the population would in practice have to die out. Extinction of the host is, of course, followed by extinction of the specific parasite. In the simulations the host numbers also often do not go to inviable lows but the parasite numbers do so. Then the parasite must be considered extinct. Once relieved of it, of course, the host recovers strongly.

Following an extinction various things may happen, but obviously none is entirely satisfactory if the hope is that cycling may be common in complex ecosystems as they actually exist. The difficulty of one or another subsystem losing its parasite or going wholly extinct under the added destabilizing influence of competition becomes rapidly worse as v approaches 0.5. At and above that value extinction appears inevitable. All this amounts to saying that parasitism does not enable a system to escape from the competitive exclusion principle even when parasites are most of the time depressing numbers far below the joint carrying capacity. When niches overlap only partially

($v < 0.5$) we know from the previous findings that the model can create cycles where none would exist if the host species did not compete. Simulations show that these cycles can be very complex. Often they are apparently permanently irregular especially when of large amplitude. Complexity in itself, however, need cause no dismay for a theory that cycling backs sexuality; it is probably good in the sense that it must make it harder for the parasite to adopt any regular system of morph changes to track its host. On the other hand, the fact that competition can easily destabilize a situation too much and cause extinctions is adverse to the theory.

What might dampen the cycles and, without extinguishing them, hold numbers from going too low? An obvious possibility is any constant input of migrants into the cycling population. Simulation shows that this is indeed very effective in damping and simplifying cycles. In the specific model derived from May's 4.4,[1] a constant small input of the hosts alone, or of hosts and parasites, pushes the cycles away from both zero lines very effectively. Counterintuitively, the hosts-alone input is more effective in raising minimum and mean parasite levels than it is in raising minima and means for the hosts. As the constant input is increased, the cycles reduce their amplitude and also their complexity further. Finally, and of course simultaneously, they become a single point—the stable equilibrium.

The moderating effect of immigration is not just a theoretical possibility. A laboratory demonstration that a steady input of 'clean' hosts could stabilize a cycling *p–h* system and prevent extinctions was given by Scott and Anderson,[14] using guppies involved with the monogenean parasite *Gyrodactylus*. Compared to my simulation reality turns out to be complex, as one would expect, and it was found that some levels of immigration in the experiment could increase the amplitude of cycling. However, conditions are far from being those of the simulation and it is not clear that a comparison is very valid. For example, generation times of host and parasite are very different, and the period of cycling is actually shorter than the normal generation of hosts.

Either in theory or reality, where could a steady influx of migrants come from? And, in order to stabilize, how steady must it be? On the second question, the only direct answer that will be given here is that making the migrant input a Poisson variate instead of a constant does not much affect the issue. This is seen in the simulations. But this doesn't address the problem of a likely periodic inconstancy. An attractive answer to the first question where the system can be thought of as a set of demes is that the influx to individual demes comes from the population as a whole—the immigrants are settling from some sort of migrant cloud. For this to work in the sense of fairly constant input it would have to be that the cycles in the various demes are out of phase. An important further question therefore emerges as to whether migration in itself will have a synchronizing or desynchronizing effect. Mostly, this question will have to be left for future work, but some very speculative comments are given below.

1. Physical conditions may set the parameters of population behaviour differently in different demes. From this it will result that:

   (a) Demes have different natural periods of oscillation and will not readily synchronize.

   (b) Some demes will be favourable to one species and some to another. High fluctuation correlates with high reduction by parasites and not with high productivity. Thus the main demes supplying the migrant cloud are likely to be relatively constant. (The factor of protection of species of genes from extinction implied here has obvious kinship to the multiple-niche polymorphism of Levene and others; see, for example Hedrick *et al.*[15])

2. Migration rates of the various species are all different from one another. This implies that an influx of hosts, say, will not tend to be followed by an influx of parasites at a time appropriate to reinforce the way the deme has reacted to the extra hosts. Hence the recipient deme may go into a mini-cycle or an extended one and so be brought more out of phase with the deme that is the donor of the influx. However, this deme may be doing the same thing back, and it is not quite clear that the effect necessarily dampens cycles: experimentation or, better, analysis is needed.

3. Factors such as time taken to travel, iteroparous reproduction, and deposition of dormant propagules all deserve to be discussed, but generally it is again not clear how their influence will go.

Altogether there seems reason to hope that a deme structure spread over a diversity of habitats and with migration between will be able to (1) prevent overall extinctions, (2) prevent even local extinction, and (3) still allow intrinsic cycling in some demes, which in turn, through their migrants, will impose cyclical pressures on those demes that are more intrinsically stable.

As is relevant to the parallel to the problem of sex, cycling in these statements may refer to numbers or to gene frequencies or to both.

5. 'CYCLES' THAT INVOLVE EXTINCTION

If in the course of cycles the numbers of a species go extremely low, then there is much more variability about what will happen next. The situation becomes truly stochastic. One possibility, as already discussed, is that the element goes extinct. If it does, uncertainty as to the future—including the future of the vanished element—continues. Two agencies might bring the lost element back: (1) immigration from elsewhere and (2) mutation.

Mutation cannot apply in a case of a lost species but it can revive a lost allele or at least produce one that is effectively the same.

Stochastic sequences of extinctions and revivals of lost elements are qualitatively different from deterministic cycling. However, they are bound to be repetitive and non-progressive in rather the same way. The period of change will probably be longer and less regular. This will not give such consistent short-term support to sexuality as cycles do (see Chapter 5). However, changes of selection, emphasizing gene combinations, probably still come very fast under this regime of extinctions and re-establishments concerned with resistance compared with what is expected in the non-CP processes of adaptation. Judging from the literature of parasite and parasitoid ecology, local extinctions and establishments of species occur commonly.

In the case where we are concerned with the extinctions of alleles and their reappearance later by mutation, the suggestion that the CP, and especially its p–h relation, is the main supporter of sex may be considered as a version of the long-prevalent Weismann–Muller–Fisher view in which the principal achievement of sexual reproduction is bringing 'good mutations' together into the evolving stock as fast as possible. In my present version of this theory, the abundance of pursuits coming from parasitism would supply the sufficiently rapid succession of 'good mutations' and consequent gene transiences shown needed for that theory to work. Likewise, once originally 'good mutations' had become outmoded due to spreading innovations in the coevolutionary partner, sex could be considered to be working for efficient elimination of bad elements, both outmoded old and mismutated new, much in the way it is supposed to work in the 'Muller's Ratchet' idea (for a review of this and other sex theories, see ref. 16).

In an extension of this view of the role of mutation, it is not even required that the new mutation in a host is like the last one to go extinct. In the current environment it may be that a completely new allele may serve as well as or even better than a revival of the old. All that is needed is that it confers resistance against upcoming types of parasite. With this suggestion we may seem to leave anything that could be described as cycling behind. However, it remains true that there is unlikely to be any other pressure for novelty that is as constant and prolific as the coevolution of parasites. It is arbitrary novelty that arrives for the most part, but that does not matter; in fact, the present theory may be said to flourish under a general aimlessness of evolution. If, for example, it is a case of something like new passwords being needed (see Chapter 2), we are still in an evolutionary realm that is essentially unprogressive and not too unlike the case of true cycles. At the same time this is a realm where it is much easier to think of *frequent* suitable variants for the theory occurring than it is with most trends of evolution. So here, as in the cases of true cycles, we see possibilities for the kinds of gene combinations that sex could be busy making up and taking apart and at least we can imagine it working on the scale of generations with combinatorial efficiency that would make sex able to compete with the intrinsic, purely reproductive efficiency of parthenogenesis.

References

1. R. M. May, *Stability and Complexity in Model Ecosystems* (Princeton University Press, Princeton, NJ, 1973).

2. F. M. Scudo and J. R. Ziegler, *The Golden Age of Theoretical Ecology: 1923–1940* (Springer, Berlin, 1970).

3. J. Jaenike, An hypothesis to account for the maintenance of sex within populations, *Evolutionary Theory* **3**, 191–4 (1978).

4. H. J. Bremermann, Sex and polymorphism as strategies in host—pathogen interactions, *Journal of Theoretical Biology* **87**, 671–702 (1980).

5. H. J. Bremermann, The adaptive significance of sexuality, in S. Stearns (ed.), *The Evolutionary Significance of Sex*, pp. 135–94 (Birkhauser, Basle, 1987).

6. R. M. May, Non-linear problems in ecology and resource management, in G. Looss, R. H. G. Helleman, and R. Stora (ed.), *Chaotic Behaviour of Deterministic Systems*, pp. 389–439 ((North-Holland, Amsterdam, 1983).

7. R. M. May and R. M. Anderson, Epidemiology and genetics in the coevolution of parasites and hosts, *Proceedings of the Royal Society of London B* **219**, 283–313 (1983).

8. R. V. Lamb and R. B. Willey, Are parthenogenetic and related bisexual insects equal in fertility?, *Evolution* **313**, 774–5 (1979).

9. D. Jenkins, A. Watson, and G. R. Miller, Predation and red grouse populations, *Journal of Applied Ecology* **1**, 183–95 (1964).

10. R. K. Murton, *The Wood Pigeon* (Collins, London, 1965).

11. J. C. Holmes, Impact of infectious disease agents on the population growth and geographical distribution of animals, in R. M. Anderson and R. M. May (ed.), *Population Biology of Infectious Diseases*, Dahlem Konferenzen, 1982, pp. 37–51 (Springer, Berlin, 1982).

12. A. J. Lotka, *Elements of Mathematical Biology* (Dover, New York, 1956).

13. I. Eshel and E. Akin, Coevolutionary instability of inner Nash solutions, *Journal of Mathematical Biology* **18**, 123–33 (1983).

14. M. E. Scott and R. M. Anderson, The population dynamics of *Gyrodactylus bullatarudis* (Monogenea) within lab populations of the fish host *Poecilia reticulata*, *Parasitology* **89**, 159–94 (1984).

15. P. W. Hedrick, M. E. Ginevan, and E. P. Ewing, Genetic polymorphism in heterogeneous environments, *Annual Review of Ecology and Systematics* **7**, 1–32 (1976).

16. G. Bell, *The Masterpiece of Nature: The Evolution and Genetics of Sexuality* (University of California Press, Berkeley, 1982).

BISHOP WYKEHAM ON EVOLUTION

Discriminating Nepotism: Expectable, Common, Overlooked

He said Bruno was a terrible heretic. I said he was terribly burned.
He agreed to this with some sorrow.

JAMES JOYCE[1]

THE trip to Israel of the last chapter was made while I was still living in the USA. About this time Sir Richard Southwood alerted me to a vacancy for a Royal Society Research Professor in Britain. I applied, was offered the post, and shortly I moved with my family to Oxford. There had been a similar chance to return to Britain a few years previously but, although I went so far as to visit England to discuss the idea, I did not pursue it. Too many of my family were in phases of schooling that made the change inadvisable and perhaps also I was too contented in my work in Ann Arbor.

By 1984 two aspects of life in the USA had made me feel differently. One was the level of violence—this was felt even in a seemingly rural university town like Ann Arbor. Part of the problem perhaps was not being quite far enough away from Detroit, but some was local. The level itself hadn't appreciably changed for us but my family was being more affected by it or at least by its threat. Remembering my own childhood in Kent, and contrasting this with how in Ann Arbor we couldn't even let our young daughters cross the park at the side of our garden to play with their friends a half mile away, was frustrating. Of course I had no illusion about England being free from the problem: even in my own boyhood it angered me when I realized that my sisters were less free than I was to wander as they pleased in the woods and fields around our home. But like other aspects of the progress of civilization, rising levels of violence in Britain lagged a few years behind those of the US.

The other main incentive for the move was also a 'rising level' and this had changed in the 2 years, and was about as depressing for me as the violence. I see it now as the beginning of what came to be called 'political correctness', but then it just seemed an increasing intolerance about how others lived and spoke, a rising pressure to conformity. Mountain men in remote valleys might escape the new social climate but one couldn't in Ann Arbor. An early tentacle of the trouble for me was the growing witch-hunt directed at tobacco smokers.

Never a smoker myself, I have always been fairly indifferent to whether and where others smoke around me. I think this is part of a general stoicism about minor nuisances of life that arise out of activities that seem pleasurable to others. Human music from portable radios drowning bird song in public parks is another example. Even in crowded Britain there are still country places where I can go to hear bird song. Even there, I admit, the drone of the nearest motorway may dull the notes for me a little; they may also reach me through air detectably scented with smoke from where gardeners are burning weeds; but neither the drone nor the smoke are things I notice very much. Back with possible causes of lung cancer, the main issue today, however, if cancer-causing molecules are in the smoke of those weed bonfires, which were a pretty constant accompaniment of my childhood, I still treat them as too scarce to matter—scarce and beneath notice: my ancestors lived with far more of such molecules than I do. Still those ancestors somehow continued my line and made no protest except perhaps in building each new chimney on their houses a little better than the last so as to have more smoke-free comfort. Likewise I believe in letting people die young, or middle-aged, or old as they choose, especially if their dying is unselfish—abandons no complete dependants, for example—and if it is inexpensive to the state. Theoretically, of course, by that last condition, it could be that my own tolerance is selfish because, with less spent nationally on health care for the elderly due to more dying young, it is possible for my taxes to be less; but I don't think that even subconsciously I am thinking of that. People should be informed of risks they are taking and I don't doubt the accumulated evidence that tobacco smoking impairs health and (at least for some) endangers life. Therefore I thoroughly approve the health warnings on tobacco advertisements. But it is surely very obvious that a lot of other advertisements should carry such warnings too—advertisements for alcohol, for example. It is also obviously wrong to target ads promoting either of these types of danger at children—that is, at persons at lower ages than a culture has decided should be allowed to use the drugs.

But as for so-called 'passive smoking' in enclosed spaces (as opposed to those whiffs of bonfire smoke out in the country and weed fires in my mother's garden), if evidence for the supposed ills of this passive smoking were examined with half the casuistic logic and statistical skill so often devoted to trying to show that human intelligence is not heritable,[2] I believe it would be realized that there is virtually no evidence for these ills at all, or else that they are quite trivial.[3] Where, for example, are the reports on identical twins reared apart—the one twin having gone to a non-smoking household and not smoked him/herself and the other to a smoking household and likewise not smoked him/herself? Leave alone where are the studies controlling for all the other 'heritable IQ' confounding factors that 'IQ' critics would have us believe in? Almost all the same factors can be adduced as misleading in the case of effects of passive smoking.[4]

Perhaps I am just insensitive; even, I suppose, I might be attracted to smoke. Certainly I am not distressed by a little of it any more than I am by the smoke of bonfires. In the presence of barbecue smoke I have to admit to being sometimes quite fumotaxic—that is, tending to turn and move up windward like a platypezid fly does towards a bonfire. These insects, by the way, with their nominative elegant flat feet and special silvery-velvety reflectances on their black backs painted as if by puffs of their favourite pollutant itself, should be bred as mascots by the tobacco companies; for these flies actually hold their lek mating swarms in bonfire smoke whenever they can. If ever moved to imperil your neighbours and the planet by burning weeds in your back garden again, in spite of all the current fuss, take the chance to watch the platypezids dancing in the smoke—flying right into the big puffs and sometimes tumbling out again with a female embraced. A weed fire is obviously a big event for platypezids. But then— who knows?—the opposition will come back at me on this one, suggesting perhaps they, too, don't really enjoy it and are just vying to impress the females by their contempt for platypezid tracheal cancer.

Were barbecue smoke as well known a signal for a human sexual orgy as weed smoke seems to be for platypezids, I might be even more fumotaxic than I am, and doubtless some actual orgies at camp-fire feasts indeed lurk there in my ancestry. The fact is that smoke qua smoke often is for me a friendly and human sign: I like it in moderate doses but even I have to admit that it can be too much. How many campfires burning wet sticks in windy hollows have I myself huddled over just in Britain alone—a lung full of smoke breached for every calorie my chilled limbs draw from the fire, while, to judge by fellow campers, my watery red eyes stare harpy-like back at them from my blackening face like theirs at me! How many home fires also have my ancestors likewise huddled by, striving for warmth again and for the hot and harmless food that the fire brings them, watching, perhaps, from under some bleak overhang of rocks, grey clouds scudding on the heights of glaciers high above; or else out, on the plains, sitting choking on the bitter acrolein of mammoth meat burning on the hot stones—or, yet again, just squatting over dry weed fires in teepees built from butchered mammoths' very ribs! Those are the backgrounds that, without any shadow of doubt, I have come from and there I see the origin of my smoke indifference and my smoke laughter; others seem to have come to their humanhood by more delicate paths than these though I don't know quite how or where. The point is that if any species besides the smoke-loving genus of the platypezids should have evolved resistance to the effects of passive and even active smoking, that species is without question *Homo sapiens*. Comparisons to rates of cancer in obligately smoke-puffing rabbits (poor things) are certainly going to be quite unfair on this problem: rabbits never invented fire, and had burrows to hide in while the grass-fires skimmed overhead.

As for the smoke of those modern and miniature weed fires, the cigarettes we make from *Nicotiana tabacum*, things that you put half into your mouth and draw smoke through (strange business, I have to admit), many people, especially check-in clerks at airports, seem confused and even annoyed about my indifference so sometimes I end up explaining and apologizing unnecessarily, saying that I am, perhaps, an 'active passive smoker' and assuring them over and over that I *really, but really* don't care which section of the aircraft they put me in. In punishment for levity on a subject they are trained to see as a serious social issue, the clerks seat me on the edge of the two zones, which means only too frequently in a row over the wing—exactly where I had told them was my highest priority to avoid, indeed a position I would almost willingly exchange for a non-wing window seat that put me beside a broken-down Chernoby1 reactor in full puff, so strong is my desire to look down and see the world as a whole for a few moments, and thus all these human passions in their proper perspective.

Actually I am not really an 'active passive smoker' and I do avoid blue wisps and clouds in a room if I can; they are beautiful and have taught me many a superb lesson on turbulence, but I am more at ease with them on the whole when they are, like the world through the aeroplane window, seen from a distance. How trivial, though, is all that seen in the perspective of the fires of our ancestry and compared with the fuss that so many make. It is actually no joke to state that I am an 'anti-anti-smoker'. I still feel, as I felt in those last years in Ann Arbor, that America was seeing a rather unpleasant and interfering pressure in modern life—almost a witch hunt—that bodes ill for our crowded future. Opposite to how it had been in the previous and rather similar hue and cry about witches and evil eyes, Britain is lagging the US in this mania—but, as usual nowadays, it seems she is trying to catch up.

While the 'anti-smoke' pollution of human freedom was one factor pushing me to Britain in the early 1980s it still wasn't the most significant event telling me I was getting seriously out of touch with America and was going to find it difficult just to be myself if I remained there. Rather it was the experience of being telephoned by a lady academic to talk to me about a letter I had sent to her department recommending one of my graduate students for an advertised post. She told me that phrases I had used were very harmful both to the student's chances and to my own reputation. Because I had taken great care in writing the letter and considered it to contain, in fact, the strongest support I had ever given to any student, I was immediately dismayed and puzzled. I was soon enlightened by her where the trouble lay. I had written to the effect that the student was 'exceptionally strong on the theoretical and statistical side and with an ability especially remarkable in view of her sex and non-mathematical background'. The offending words will be obvious to anyone today: my informant made it clear to me that I ought to have treated the student as if I had never noticed its gender.

Hundreds of experiences have taught me that women are not as good on average at maths and spatial problems as men. This matter seemed really to need no statistical tests although plenty have been recorded. In a related field, day-to-day orientation, many women seem almost to boast about their tendency to get lost in towns and buildings: very few men do. The entire sweep of history as I knew it suggested to me the same things, and the difference continues even right up to the present in spite of all the recent equalizing forces that have been applied. Even in rats females have been shown to be less good at spatial and route-finding tasks than males although why this should be is still largely a mystery.[5] As always, of course, the overlap of abilities is wide and there are innumerable women who are much better mathematicians and visualizers than innumerable men—I could name many who are better than me. For sure also, plenty of good women engineers, physicists, and the like are qualifying where there exists the will to enter these activities. Still, the average difference remains for me striking and is confirmed through a multitude of channels. Aptitude tests performed on small children are just the start of them. Nowadays some parents make a special point of trying to override any cultural bias that may have existed in erstwhile courses of infancy and childhood and I, too, have made my own small intrafamily attempts, out of interest, to see if the tendency could be overridden (diagrams of theorems of Euclid pinned inside the hood of one daughter's perambulator formed a part of one experiment). But neither I nor others who have tried it, I think, can record much success (my daughter, however, did pass advanced-level school maths).[6]

Anyway, consequently I didn't believe the theory that the difference comes wholly from cultural factors, and I assumed that, in spite of all the outcry of a few based so far as I could see on very flimsy evidence and excuses, most intelligent people of both sexes would interpret my letter the same way as I meant it. Holding this expectation, I had wanted to be sure my student would not be automatically devalued by a reader of my letter who might interpret it as a more vague complement: '"Exceptional", he says, but he means no doubt "as women go"—who wouldn't leave that unsaid of his own student?' I had added the phrase to forestall this, to make clear that I was aware of the average difference and nevertheless gave unqualified praise. I assumed any 'search' committee would understand this and that it would raise, not lower, my student's chance. I still can hardly credit that what I wrote would be interpreted in any other way. I suppose that if everyone believed what to me was an obvious untruth—that there weren't and never had been any average psychometric differences between the sexes—then my words could be seen as odd and inappropriate. And yet still, surely, they would be harmless. But in the gathering new climate—in which, by analogy, I suppose you are not supposed to mention in a letter that you happen to have noticed that a student has the handicap of being blind—they were not harmless: some acid remarks from the person on the phone made this plain when, amazed, I tried to explain what I had meant.

My dismay from this event was soon compounded by the fact that the student, when I explained to her how this letter seemed to have backfired, showed dismay herself—and, in fact, appeared to agree with the caller. She likewise seemed to feel disparaged by the words in this best letter I had ever written (I had showed it to her when the topic arose; later she told me that she had not felt herself disparaged but that I was harming myself—that was not how it seemed at the time). The end result was that I decided, rather as I had done at Cambridge when I had been told there was no possible connection between genetics and social anthropology, that I should try to pursue my career as a misogynist somewhere else with the corollary to this decision that I needed to be careful about taking on more graduate students generally if my best efforts were going to harm them in ways I had become too socially blind to see.

My student didn't get the job; but then neither did she get the others that followed for which I put out a more 'PC' version of the letter. In short, what I had written seemed to make little difference. Perhaps people had been reading it the way I intended. They were hard days for jobs in all circumstances and she did find a good one eventually.[7]

If I was leaving, in Ann Arbor, one world I found hard to understand, the one I was entering turned out not so transparent to me either. Meeting Oxford academics I felt a bit like Gulliver just departed from the friendly giants of Brobdignag amongst whom he feared to be stepped upon and arriving at the land of Laputo where he found himself politely but harmlessly ignored. Especially during my first weeks in Britain in 1984, when I had moved ahead of the rest of my family (they needing to remain for the completion of the school year in Ann Arbor), arriving in Oxford was like landing on another planet. During those weeks I had full board and lodging in New College in Oxford. I remember four things from my earliest impressions of my new environment especially vividly.

The first was having breakfast alone in the Senior Common Room at New College and seeing the glass of orange juice I had just set down slide away from me across the table top. It was no message to me from the spirit world, not even from the deceased or the still-living dons whose pencilled portraits looked down on me from the walls; I soon realized the sliding glass to be just the combination of an unlevel floor and a superb polish on the table top. In fact, I thought, cheering a little as I solved the mystery, I had done problems like this at school: the coefficent of friction, a glass stands, forces normal to and along the plane . . . find the tilt and so forth. Now, perhaps, I was privileged to see where some of the problems of my school physics had come from (ours were often taken from the Oxford and Cambridge Board's examination papers). Spirit-propelled the glass, of course wasn't; but the combination of a medieval floor (or actually an all-periods one—it looked to have been remade many times) and of enough servants to polish a table top so perfectly has continued to seem to me quite socially magical.

Back with physics and in the same room, it seemed to me at first some other of the more dynamical problems of my school days might have originated right in this room, too, but in this second guess I was quite definitely wrong. At the end of the room by the window stood a neat contraption of brass pulleys and an inclined model rail track, quite reminiscent of an antique 'Fletcher's Trolley' that had been used at my school to show the quantitative rule of acceleration under gravity. I thought this apparatus might be a historic science specimen, perhaps some eighteenth-century Fellow had made it, a colleague of Fletcher's, a dreamer, perhaps, who thought Newton's mechanics ought to become important in education right away in Oxford instead of in 200 years time . . . So I myself dreamed, but the first time I attended a Senior Common Room after-dinner 'dessert' occasion, a quite different and still-current function for the apparatus appeared. The small tables at which we sat were arranged in a horseshoe with the tables for the most part close enough for cut-glass decanters of port and Madeira wine to be passed easily from table to table. But at the gap of the horseshoe there was a problem and it was to bridge this that the rail track with pulleys was designed: the contraption meant that the decanters could be still passed by means of the inclined railway without anyone rising from his seat. But why not instead of this, you may wonder, close the horseshoe to a full circle? I wondered, too, and at first thought that it must be due to the risk of a waiter jogging an elbow as he stepped into the ring, or that the thing was to be a sort of 'dumb waiter' so that the college staff could be sent home as a social evening progressed. Later, however, it was explained to me that the horseshoe would often be set facing a fire that warmed all participants in the dessert. Perhaps this fire would have been so hot burning up all that British coal of our past two centuries and which we now lack, as to be uncomfortable to pass near.

The third thing that impressed me as a social matter also and this time completely on a side of life that I always find quite difficult, as earlier paragraphs and these introductions as a whole have probably shown. I found myself having considerable trouble following the conversations of my colleagues at New College, still more in contributing anything that could fit in and be understood. If anything, an even greater difficulty occurred in understanding and laughing appropriately at jokes. But because there had been a similar though lesser trouble at first with the jokes across the Atlantic, it seemed that perhaps my present problem was more with my recent training than with specifically Oxford humour—or perhaps, alternatively, it was just, once again, with my slow and still mammoth-rib-tented mind.

Language and what was said in my new environment appeared to me lacking in what I think of as American directness. Directness has to be good for truth, you may imagine, and therefore I must be criticizing the New College style. But this is not quite the case; I came to realize as I stayed at Oxford that this whole matter is itself subtle—one might even say the whole matter is rather indirect. For example, the American directness can be applied very heartily to political correctness once that idea has got to you. I had realized this in Ann Arbor when I was phoned about my letter. Here in Oxford all such PC issues—the supposed unisex brains, the smoking issue, etc.—seemed a lot more complicated than they had in Ann Arbor and they also had to be talked about in more oblique ways: you might believe in PCness or you might not, but you didn't discuss your view directly and, above all, not at lunch time. Actually in New College when I joined there was a major extra reason why you didn't discuss such issues with women at lunch time and this was that there weren't any—or not women who were present as fellows of the college (you could bring them as guests provided they weren't your wife). Given the extended familial environment we aimed to provide for our undergraduates, this seemed even to me quite a strange situation because we were already at that time admitting about a third of women among the undergraduates. But I should add that there are plenty of women fellows at New College now: we catch up and variegate. The sexes now mingle all along the tables, even at the high table in the hall; in a similar expansion our one-shot-heart-failure Wessex cream, which used to be the only sauce for desserts, is now variegated from time to time by yoghurt.

Over the years I have grown used to the more roundabout ways of thinking and writing that are typical in Oxford. I expect this is already evident in my style in this book, which probably seems as complicated to Americans now as my fellow dons' conversations and their humour did to me when I was newly returned to Britain in the spring of 1984. You might suppose I would have been prepared for Oxford through having been an undergraduate at Cambridge University, and that the systems cannot be very different seemed detectable in the way Cambridge was by far more frequently mentioned than any other university after Oxford itself (which came top of the mentioned universities by maybe about 99 per cent). While I was at Cambridge, however, I had never got near even to a Middle Common Room, leave alone to a Senior Common Room, and was something of a recluse as well. My two subsequent experiences of university life at staff level at Imperial College London and at the University of Michigan had done almost nothing to prepare me: the three places seemed like vertices of an equilateral triangle, with, at Oxford, an ocean as deep and wide as the Atlantic still separating me from my start in London. As for coming to Oxford as opposed to trying to get to Cambridge, Oxford has certain advantages. Oxford is about as far from the sea as you can get in Britain but that means it is close to a lot of places too—a journey from Cambridge to the Welsh border town of Hay-on-Wye, for example, to buy old text books would be so tedious you would need a hotel in Hay. Also thanks to the damper climate further west in England, more species of moss can colonize my car than I could hope for in Cambridge.

The fourth novelty at New College was the unexpected discovery of the source of a saying that I had often heard. I would not have guessed William of Wykeham, who founded the college in 1379, to have been a likely source for an evolutionary insight that was to predate Darwin by about 500 years. I found he had uttered, or at least had implied, what amounted to two very pertinent thoughts on human evolution, one of them very modern and one rather old.

Wykeham's motto was 'Manners Makeyth Man' and he made it that of the college. During my lifetime I have seen many aspects of human life suggested as being unique to our species and have watched these fall away one by one as more is discovered about animals generally and especially discovered about our nearest relatives, the apes. Tools, blank-slate learning, trade, recreational (and face-to-face) sex, even language—all these had to some extent suffered by such comparisons. But as I pondered Wykeham's neat phrase so lightly tossed to us by this rich, self-made, and notably non-aristocratic man of the fourteenth century, I was surprised to realize that it stood up well (along with, perhaps, what is often regarded as almost the opposite of manners: the lofty art of stone-throwing[8]) against almost all that I knew from the evolutionary point of view. For humans, manners certainly existed and they were important and had probably been so as far back as we could imagine; and yet I knew of no non-human species that could be said to observe anything like manners. To convince me that chimpanzees, for example, have manners I need to be shown some behaviour that is at once social and arbitrary, is taught and learned positively in one group, and viewed with distaste, or even anger, when it is performed in another. Chimpanzees react seemingly automatically with suspicion and hostility to animals perceived as not of their group: was this the counterexample I needed? Could their reaction be to just 'the bad manners' of the outsider? Perhaps in a sense yes, but what is seen to happen seems out of scale for what we consider manners. With chimpanzees the strange individual is often violently attacked. I would be much more persuaded it was a case of manners if this was a more minor hostility or disgust arising towards minor social transgressions in animals that are already accepted into a strange group. As a background I should state that eventual acceptance of strange females is a normal part of the dispersal and outbreeding pattern of chimpanzees; males so far as I know are never accepted no matter how 'good mannered' or ingratiating they try to be.

If William of Wykeham showed he had picked out something uniquely human for his motto, thereby expressing a perception that chimpanzees could hardly understand even if we had words with which to explain it to them, then it is fair to add that another aspect of his legacy, another idea that he enshrined at New College, probably wouldn't puzzle them at all. He built a school for boys in the city of Winchester 40 miles to the south and then he built New College in Oxford to be its sequel for the further education of the boys that Winchester would have started. He provided in effect a complete production line to turn out literate and religious men into the indefinite future.[9] Along with the buildings and the rules he also endowed scholarships to help needy but talented boys to be educated at both places. His will stated, however, that certain of the scholarships and the places for fellows at New College were to be restricted to 'Founder's Kin'. This was the part of his bequest that I think chimpanzees would understand well. Nepotism is the warp and weft of their lives and the bishop's bias would seem to them utterly natural; it would only surprise them that the whole of his enterprise hadn't been restricted in the same way. Wykeham was a bishop so he was not supposed to have descendant kin but, whether or not he died truly without any, he certainly did have a surround of family, including a married and fertile sister;[10] therefore 'nepotism', so often used figuratively, is a more than usually exact word in his case.

But as the generations rolled after the foundation, the concept of Founder's Kin became more and more complicated and dubious. Finally it became corrupt. Through the simple fact of every one being born of two parents and through the slowness of the medieval change of population, numbers of boys with a just claim to some degree of kinship mostly about doubled also every generation, and, of course, the growth in such a proportion would go on with less and less in the way of chance variation as generations replaced each other. Assuming three generations to a century let us suggest, then, that Wykeham's indirect kin increased every century by a factor of 10 (not far from 2^3), though it could easily have been more. By the early nineteenth century this gives him about 10 000 kin via descent from his close family in the fourteenth, a number that quite possibly represented a large part of the population of Winchester city.

A story emanating from Winchester seems to reflect not only a popular and humorous understanding of sociobiology but also understanding of the idea that by the end of the nineteenth century was to become codified into Francis Galton's famous 'regression to the mean'. For Galton, at first regression was just a statement of a rule of quantitative human heredity he had discovered but both he and others soon saw it to embody a very important and general statistical concept that could be useful throughout science. Anyway, back to Winchester, the story says that when there was doubt about whether a pupil at Winchester could be eligible for a scholarship as Founder's Kin, the headmaster hit him on the head with the flat of a trencher—that is, with one of the wooden plates used in the school refectory. If the trencher broke the boy was to be judged Founder's Kin. In other words, the popular estimation was that the talents of Founder Wykeham had already regressed out of sight: his descendants were even below the mean—'numbskulls', 'boneheads' in short, and were implied less intelligent that the average ordinary scholarship boy.

That ordinary scholarship boy with no kinship to claim might have been, but never was, the novelist Thomas Hardy. As cities, Hardy was familiar with both Winchester and Oxford. Of all Oxford colleges New College could perhaps be called the most 'Wessex'-oriented in its traditional intake. Thus it is likely to have inspired Hardy's most personal interest and perhaps also his longing. He never went to a university but his novels suggest by many signs that he would have liked to. Perhaps a sardonic and poetic appreciation of all three of Wykeham's 'Manners', Founder's Kin, and even that folk version of Galton's 'regression' enshrined in the trencher joke, lurk in the background of various scenes of his last novel *Jude the Obscure*. One tragic and key scene seems to be set in an existing tavern, The Turf, which nestles still under a wall that is bounding both to medieval Oxford as a whole and, within that wall, to New College. In the tavern, Jude, the self-taught country youth, competes in learning with well-bred ill-mannered Oxford students who are likely fair depictions of Founder's Kin as they had become in the nineteenth century.

In the spirit of my paper that follows, gentler and more effective methods than the trencher for present-day use at Winchester could be suggested. Even in Wykeham's time bloodhounds must have been available. A kennel of elite 'Wykeham Hounds' could have been kept at the school with the animals trained on the intimate clothing of the closest and most genealogically bona fide representatives of the bishop's lines (which means generally, of course, the most inbred). More surprisingly, recent results suggest also that an appointment of a human Dame Kin Sniffer at Winchester School could have worked also and she could have replaced the kennel with perhaps a slight increase of dignity. Women have long been known to have a better sense of smell than men, and the recent evidence is that this extends to being better at distinguishing degrees of kinship. But before anyone jumps up to create this sniffer post for any institution—in Winchester or elsewhere—I had better mention that the matter is complicated and the research into its fundamentals still in a preliminary stage. An unsettled question is whether the sniffer should be Founder's Kin herself or not. As to the possibility of such testing having been discovered and used in earlier times, the signs are that before the invention of the contraceptive pill (because the best results seem to come if a sniffer has been on it for a few weeks before performing her tests), the decision she gave would have been too variable in its accuracy—there is evidence that her hormonal variation during the menstrual cycle could create subconscious biases, pro-kin or against.[11] I will come back to this strange situation in a minute.

More seriously, it is an interesting scientific question whether the fellowships reserved for Founder's Kin had an inhibitory effect on New College scholarship. I suspect it did; it hasn't been, in fact, a college of distinction proportionate to its size. Just before the time when in 1871 the system was reformed and all fellowships thrown open to full competition, 21 of the 70 fellows at New College had been Founder's Kin. So far as I know none of these men became well known, or at least weren't so internationally. At a dilution of the bishop's genome of about 1 in 65 536 (arguing as before but assuming the equivalent of one extra generation to take account of descent from the bishop's sister and not from himself, implying 16 nominal generations in all) and because a human is believed to need code for about 60 000 proteins to unfurl his anatomy and drive his body, each Founder's Kin could proudly still carry about one blueprint for one 'Wykeham' protein out of the 60 000 Wykeham himself had carried.

Would a Founder's Kin have a higher proportion than this of the bishop's memes—that is, his ideas—even his 'Manners' if you like? Possibly, and especially in so far as fathers and sons may indeed have gone to New College to receive the imprint of its sandstone and tutorial smoke into a continuing family 'memotype'? (I omitted to mention, by the way, in my anti-antismoking diatribe that pipe smoke combined with ivy, and this specifically to be encountered at Oxford, has been conjectured by at least one professor[12] to be an education in itself). But I still think that, after the revolutions and counterrevolutions that affected the religion and the government of the entire country (all them as it came to pass with a rather ferocious focus around the city of Oxford), and after many other social changes, including waxing and waning of various Founder's kin family fortunes, this cannot have amounted to much. With about a thirty-thousandth of just one episcopal protein left and, we may allow, just a little more of Wykeham's Manners, we arrive at something like the boor student who spouted Latin against Jude in The Turf and also at the warden who so politely and yet so unkindly rejected him. After the four further generations that bring us from that time to the present, a Founder's Kin aspirant now has only a millionth of the bishop's genome, less than enough to specify code for even the smallest enzyme. Thus substantial as they may have been for the first two or three generations, the virtues tracked by the Founder's Kin principle of support are clearly at the present time convergent with the virtues of homeopathy, which, as regards faith in infinitesimal essences, can be viewed as a kind of sister belief system to genealogy. Of course, special provisions for Founder's Kin by long-deceased benefactors exist for many venerable institutions around Britain as well as for New College and, differing from our decision made at the college a full century back, in many cases the provisions seem to be still applied,[10] a fact which is surprising to me on the same basis as I am surprised at irrational public faith in homeopathy.

Perhaps the idea of the bloodhound is after all the best one: perhaps when a bloodhound can no longer detect any candidate with any resemblance to the best genealogically proven Founder's Kin of his period, it is time to bring the whole Kin tradition to a close and to devote the money to helping the Judes instead. I think that roughly this is what New College has done.[13]

Even if it weren't for all the halving dilution over many centuries that has made Bishop Wykeham's genome, by the twentieth century, to disperse as utterly as I describe, and even if we were talking instead about detecting fragments of essences of, say, John Ruskin, influential writer and critic but not ancestor (again childless), who worked in Oxford merely in the late nineteenth century, I am sure my suggested technique isn't going to be tried today. My 10 years in England before the discovery of the human ability to detect products of the hypervariable histocompatibility system (which provides the background to the Kin Sniffer idea) had been enough time for the Political Correctness that I had fled from in the USA, along with that country's standards of violent crime, to catch me up across the Atlantic. Today the act of advertising for a Dame Kin Sniffer—essentially female and preferably, as I said, a woman already on the pill—if not actually illegal will be sure to annul such positive PC points as Winchester School had gained by admitting occasional girls (though they are Teachers' Kin, as I gather, so the wheel perhaps turns again) to its Sixth Form and New College by letting women into its Senior Common Room. At New College, belatedly, they may be said to replace the Founder's Kin and with great improvement, because they have undoubtedly been selected on merit.

In such various ways, then, my ideas for updating the mystiques of Winchester School and New College fall flat. Some may unkindly tell me that, in practice, all my ideas were doomed through the Wykeham scholarships (for instance) having fallen to (or stayed at) a value of only a few pounds per term each. But here they would be missing a major point— how mystiques thrive on being almost potty, and some of my suggestions are surely that. A concept as vague as Founder's Kin becomes almost like a religion and can always have new roads ahead of it. Meanwhile for my taste, feminism and the PC vogue that just now are visiting Britain from America can be considered at least moderate improvements on the transportation and transmadeiration engines of New College's past and all that those implied; and maybe PCness and the like look not so bad when compared to the overstretched matter of the Founder's Kin. As Bishop Wykeham in life—versatile, worldly, conformist—so his college: 'New' ever since so named relative to Oriel College in 1379, New College is new once again.

The job at Oxford was wonderful—it was no teaching and all research. The university landed me like a returned space shuttle and New College sheltered me for the time being in the time-machine hangar I have just described. The Zoology Department gave me a few square metres more of floor space than I had had in the museum in Ann Arbor. Looking out from the building I had some gains and some losses. A hedge of a spiny berberis below my window put on colours my first autumn like an oiled pool: it was a delicacy not really British (the bush is Chilean, I think) but the colours shone well in the British rain and proved more nostalgic for me than had been the strident reds and yellows of maples and sumachs that I had come from. There in the museum it had been honey locusts below my window and these turned only to a drab leguminous yellow. But now I remember even those trees with nostalgia. Mastodon-repelling spines, necessity to their ancestors, were in the late twentieth century in Ann

Arbor as politically incorrect as admitting you noticed a person's gender and, as I told in Chapter 3, the spines had gone. But the trees' fine humour, especially jokes (these being of a rather mathematical kind), were OK and retained, and, above all, the trees had been always shady and pleasant to step out under on hot summer days. The jokes were best seen in the late spring when their leaves as they expanded played with quaint indecision on whether to be once or twice divided into leaflets. Often a leaf that I spotted as I walked up the slope to that studentfoot-bridge that crossed East Huron Avenue would bring me to uncontrollable guffaws through quaint achievements of both states together in the one leaf: even artist Paul Klee had never made better two-dimensional surprises than these, so it seemed to me.

Other trees that I knew from my windows had humour, too, but generally this lay more in their names. There was that pride of India tree in the Nichols Arboretum, for example, while across my new Oxford lawns I found a tall tree resembling an ash but larger-leafed and of a family, Simaroubaceae, with members traditionally even bitterer than the ash's bitter cousin and confamilial the olive. In its name outclassing even the pride of India, this tree was called the tree of heaven (TOH). Then again, back in Ann Arbor, diagonally from my window (and right at the side of East Huron Avenue for anyone who might care), stood a fine tall specimen of the Kentucky coffee tree (KCT). The TOH was once pinnate (that is, its leaves were once divided into leaflets), like an ash and like the simpler leaf versions that the honey locusts (HL) had: but the KCT was always twice pinnate, so the two nicely bracketed the indecision of the HL—and, now that I think of it, the KCT may also have the largest bipinnate leaves in all North America.

The world, for both these much-loved middle-distance mistresses of my middle years and idle moments, has moved swiftly and sadly on. The KCT of Ann Arbor became in one instant the first Kentucky Fried Coffee Tree that I have known (a name I herewith suggest for a café chain in Britain or in America—though it has to be a good one if to survive for the reasons I gave in Chapter 4) when one summer afternoon it was split and killed almost before my eyes by lightning under a green tornado sky. Over here just months ago in Oxford, my berberis hedge was pulled out and the tree of heaven itself escaped by less than a branch length the uprooting storm of the founding of a new science building that is now rising. This building sadly will soon obscure the whole tree from me, even supposing it survives, and already I can't see if it has. As the Oxford (or really Italian) ragwort became a patron flower in my mind to the Italian restaurants of King's Cross (Volume 1, Chapter 2), so is the TOH now patron to the Chinese take-aways of Oxford—a traveller like them from a remote land; emphatic, daring, primarily feminine, food for far-flung shantung silk caterpillars and their like . . . Whence comes your strange name, TOH? Is even the 'willow pattern plate' really a 'tree of heaven plate', the tree bowing long leaves (not leafy shoots) over the house, wilting the leaves out of the tree's sympathy for the parted lovers? To me, the tree in that famous design looks more like my tree, this TOH I saw from my Oxford window, than like a weeping willow. And how often I see others of its kind; sad like this with soot thick on their leaves in hot summer days but then, after rain, revived and happy and a little washed—strong angels in the urban decay of London, Boston, Chicago, Bratislava. Wherever invading, wherever rejoicing in the usual absence of silk caterpillars and other pests of its homeland, I see it reaching out the wings of its long leaves towards me from whatever damp wall or edge of a railway siding gives it a footing, calling to me from its blackened greenery that where man's works are failing, nature can still bless. Heavenly in this way, the tree has certainly become for me; but in the town where I write this, whimsical images of the kind I am typing aren't enough, so I hope someone will explain to me properly sometime the name's origin—if possible using Oxford English Dictionary's full etymologese. But until I receive this explanation I can continue my own fancies.

One 'angelic' hint the tree did pass to me is of a more scientific kind but can't be connected with the name because it is part of a long (and perhaps tall) story that no one except me believes. I give only the barest summary here, not expecting it to be believed. Flowers of the TOH, appearing in July, are insignificant—they are small and greenish but in spite of this are quite attractive to bees. The tree for its leaves is also dull in autumn and neither the quiet 'berberis' nor the flaming 'sumach' colours come to it, nor even the flat yellows of honey locusts. The young fruits, however, in late summer to early autumn become for a time the tree's real glory and possibly even the reason for its name. Yellow and orange, long elliptical flakes of fruits are held massed among still dark-green and frond-like leaves with each twig holding them like a flowering palm in miniature.

Perhaps you will think such colouring on the fruits is easy to understand, these being like the rowan berries and colour simply to attract birds to disperse them. But you are then wrong for no bird touches them; they are thin and later dry out and are destined simply to be blown away.

While some may notice I am hinting at a scientific puzzle of why dry windborne fruit should colour as they mature, others may spot the betrayal of an important theological secret known to few. There are times when British suburban streets are going to be strewn with free Keys of Heaven . . . but I think St Peter must long ago have grown wise to this and, after a first lucky few, changed the lock . . . Better then for me to try using them to unlock a quite different and scientific door, that which may open to the origin of all non-photosynthetic plant coloration. My idea, which was partly inspired exactly by TOH fruit colours in the first place, is that all such colours are older than, and quite opposite in their intent, to that of the colours of flowers and fleshy fruits. Originally I believe all bright plant colours hadn't the least 'adaptive purpose' to suggest food to a bee or a bird but rather signalled the opposite. They signalled the taste you'll get should you try to smuggle a key past St Peter in your cheek—simarouba gall with a faint mockery of a smell of roast beef to deter extinct mammals like the elephant-like dinotheres as well as insects; and the colours were to warn of these. First plant colours (other than chlorophyll, and this includes the autumn colours) were, in fact, 'Don't eat me' or 'Eat at your peril' signals. But I must come back to that later.

Returning yet again to the novelties of my new life in Oxford in 1984, it is worthwhile to tell how, to right and left beyond the tree of heaven and among the various towers and spires that in those days were visible from my window, it was the scattered parts of the Bodleian Library that became the most important institution that I came to use in Oxford after the Zoology Department itself and New College. Once I had understood where all my relevant fragments of this library, which pervades Oxford much the way a sumach at least attempts to pervade a lawn by its suckers, lay scattered in the city, and also how thin the fragments I had to tread gently on the toes of the librarians whom I found to be more like sensitive priests than like that mixture of students and dedicated ladies—or I should say dedicated persons—I had become used to in Ann Arbor, this great institution proved excellent for all my needs, except perhaps for a few aspects of my growing interest in parasitology.

Shortly before Oxford I had received an invitation from David Fletcher in Athens, Georgia, to contribute a chapter to a volume he and Charles Michener were editing. It was supposed to bring together the knowledge now accumulated on animal abilities to recognize kin and the uses they made of them. Because the editors seemed to want an overview from me that would either introduce or else summarize the book as a whole, I was hesitant. I felt I was out of touch with kin selection as it had developed recently and I knew of many people who understood the later refinements better than I did. Like having a bad 'stitch' in a cross-country race, my concentration on sex had dropped me back in the field and it was a pain to try to run let alone to catch up. Alan Grafen, for example, was about to publish an exceptionally clear account of the theoretical side of kin selection:[14] David Queller likewise was more authoritative than I in updating the theory; finally, as for the data aspect, there, too, I wasn't fully abreast. In short, there seemed various people who could do an overview better. But my admiration of Charles Michener was very great.

Most would agree that the world's three greatest bee scientists in the period since I began my research had been Charles Michener in the USA, Warwick Kerr in Brazil, and Shoichi Sakagami in Japan (of course, I don't mean here just 'honeybee' bee scientists, the names for that subfield might be different). Michener stood a little above the others, at least for me. He had worked in Brazil and he and Kerr had been co-inspirers of my visit there in 1963. Ever since I had first encountered his work, the stream that continued out of his own hands or out of those of his students has been among my most exciting scientific reading; I stayed amazed at his ability to 'keep up' and discover ever more novelty in this great group. (Much the same, incidentally, has applied to Kerr and his group in Brazil.) Michener's most recent shock piece for me had been a stingless bee out of buried amber—amber that was even from the twilight of the dinosaurs— and yet was a perfect social worker bee in all its details. It is indeed so little changed from a bee that flies today in Amazonia that it sits in the same genus. Hence many details of its lifestyle can be fairly confidently predicted—if predicted is quite the right word for a honey-maker that died about 74 million years ago. Stephen Jay Gould was right that in some cases some animals do most amazingly not change.

Longer ago, and beginning about 10 years before the invitation to contribute to Fletcher and Michener's volume, what had become particularly appealing to me out of Michener's Kansas group had been the research on the ability of tiny 'sweat' bees to recognize their kin. Sweat bees get their name because they often fly to sip sweat from a person's skin, although this was not a very distinctive name for them in Brazil because numerous social stingless bees, a quite different group (and that of the fossil bee just mentioned), do so also and much less timidly. Anyway, one would have thought sweat bees to be a most difficult group in which to investigate kin recognition and yet the Michener team had managed to do it in great detail.

I mention the Kansas work in my paper that follows but much more about it, of course, is to be found in other chapters of the Fletcher and Michener volume. The theme of kin recognition in bees has continued to pour out its novelties, including ever more from the inexhaustibly surprising honeybee. But this is all running ahead; at the time I told Mich that I would try to write something but it might not be quite the overview that they wanted. What I did write was, as will be seen, quite idiosyncratic even if it has a remnant of the 'overview' idea that leads me to give at least a mention to some chapters out of the rest. I was doubtful when I sent off the first draft to Fletcher whether they would want it. Was my paper too full of naive social errors like my letter for my student? Would its reception be the story of my previous 'dedication' of that paper to Washburn, a work contributed out of admiration but seen only as a tug into the mud? As it turned out the editors seemed quite pleased. Maybe the inclusion of a limited amount of controversial, free-wheeling ideas aids the sales of a book without too much lowering its tone; perhaps, too, curious, unshakeable, silver-haired Michener was of an age not to be much bothered about 'correctness'.

The paper, then, is lightly written and it aims to carry you along, scientist or not, with little effort and little more needs to be said here to explain it. One of the areas where I think its topics continue exciting today is in the extensions and the chemical details of kin recognition in bees, as has already been mentioned, but also in other animals—coelenterates, sponges[15] even slime moulds (Myxomycota).[16] Another is mate choice based on histocompatibility in mammals.[17] This is an un-kin preference at the present but it surely implies a foundation for choosing kin innately and this, if used, as sometimes seemingly it is,[18] has potential to lead to some paradoxically antisocial and antigenomic forms of co-operation and to peculiar countermeasures that these engender.[19] This really is one of the currently exciting open ends of kin-selection theory. Because sex and the meshed gear trains that lead from it to exogamous mate choice all come together here I continue to watch this field with great curiosity and there is a lot happening.[20] Already it is clear that recognition systems have been invented several times with signals and receptors made up from different themes of organic chemistry, all of them offering, however, a range of variation that is either automatically self-recognizing in simple cases[15] or that has grown parallel with another system, so although the signal and receptor molecules are different they still are always able to match shape against shape like a lock and key.[21]

Suppose you put your name and address in the phone book so that your friends can find you, but suppose that old 'friends' out of your wicked past, and whom you hoped you had left, search in the same phone book. Suppose they discover your address and draw you back into the world you thought you had escaped and back even into deeds that become injurious to your new and real friends . . . Such a seeming outline for a TV thriller is oddly not far from describing the risk arising when deepest identities of your gene loci are posted publicly in skin scents and urine, thus laid open to being read by what might be described as the Mafia of the genome. Just as certain internal codes of the Mafia may force from us a grudging admiration for their peculiar 'honour among thieves', as also a fascination for how blackmail and protectioneering are made into a way of life, so the consequences of genomic alliances and of squabbles within the body are also instructive and astonishing.[22] In a mother–fetus pair in which classic relatedness enters, they become still more astonishing and even fraught with potential danger for the whole diploid participants.[23] In the struggle over resources here, it appears that even a kind of behind-lines commando of cells may be dispatched out of the fetus into the love-hated land of the maternal uterus with the mission to forcibly regulate affairs in the latifundial young master's interest—and 'master' here by the way isn't quite just another of the thoughtless chauvinisms of my writing, or at least it shouldn't be seen so by an evolutionist. Her theory should tell her that the male fetus is bound to struggle more selfishly with his mother about provisioning than a daughter fetus should struggle over hers.[24] Evidence confirms, in fact, that the male fetus does struggle more. Note here that the oppressive 'latifundium', the initiator in this struggle, whether male or female, is the fetus, not the sometimes supposed all-powerful mother. But the X chromosome of a heterogametic male may be abetting the fetus in its greed, although an X in a mother opposes this.

Historians of the acts of such genetic Mafiosi will find their work still further enlivened if analogous intrigues can be shown operating beyond the placentas and blood vessels and achieving similar sinister objectives later, out in the full light of day. Such intrigues have been speculated, but only recently have the first clear cases been documented.[19,25] They concern those superkinship traits first speculated by me in Volume 1 (Chapter 2)[26] and later styled 'green-beard' traits by Richard Dawkins. I discussed their possibility, and for the most part dismissed this as low. As I mention in the paper of this chapter (p. 351), we know that non-identical twins are a little aware about their genetic relatedness by seemingly direct means of estimation (that is, other than by knowing that they are twins);[27] therefore with a slightly higher level of expectation than I had in 1964, I anticipate in the present paper how it may turn out that exceptionally and even 'foolishly' mutually helpful friends will be found more close in their *HLA* haplotypes (that is, those genes they carry in that code their tissue-transplantation type) than can be explained either by genealogy or random chance. Real members of the Mafia, of course, are not so foolish as to use anything like an actual green beard to inform about membership (instead it's supposed to be how you shake hands, or kiss, isn't it?). In the case of a genetic Mafia gene, however, a member of the organization just has to let out some public sign, like one crooked finger that can be felt in a handshake that just has to be there in some form or other.

This digit in itself doesn't pin a crime but its presence is like a motive proven and thus can raise suspicion: the digit is that the receptor gene, that coding for the perception of the green beards, has to be closely linked on the same chromosome as the gene locus for the green beard itself. Thus in the case of the finding by Pakstis and colleagues[27] about non-identical but same-sex twins that I cite in my paper, we need not (at least on the evidence given) worry too much about a Mafia being at work in this effect because the gene set used to show that the twins were successfully recognizing closer and lesser degrees of sibship was scattered widely in the genome. If it had been otherwise—if the only genes whose extra similarity in the pair correlated with the twin's opinion that they were specially close or distant were in the region of HLA (the human tissue-transplant-controlling locus)—our suspicion of a green-beard effect would be greatly raised. It is known that mice, rats, and (as I mentioned earlier) people[11,28] do sniff airborne clues to tissue-transplant (histocompatibility) type. It is therefore of great interest to learn that the human genes controlling olfactory function have recently been found to be both numerous and to occur widely scattered on the human chromosomes, with none so far detected on chromosome 6, the chromosome of MHC—the gene complex that encodes the tissue transplantation antigens. Thus the human smell system looks as though it was specifically designed to avoid the possibility of 'green-beard' effects arising involving the sense of smell. It is not, of course, actually 'specifically designed'[29] but rather it is probable that when olfactory genes did arise on, or were transferred to, the chromosome where existed one of the best polymorphic systems for producing 'scent' versions of green-beard signals (actually always more probable than visual ones), the results were in the long run disastrous and the lines where mutation or translocation occurred died out due to the detriments to whole organisms or/and populations of the extremes of selfishness.

In the remarkable case of a seeming bona-fide green-beard gene recently discovered in the red fire ant by Laurent Keller and Kenneth Ross, the 'sniffer' gene seems to be inseparably linked to the gene creating the 'smell' that is detected.[19,25] This is rather as one half of the theory predicts; but then why isn't the other half of the theory working, the half that says that the green-beard gene complex should be killing its host species off and for this reason isn't found? Unfortunately for *Homo sapiens* and his outdoor affairs in the southern USA, red fire ants are the very opposite of a dying species; in their new areas of colonization they have been enormously increasing. This may be the case of a kind that I mentioned briefly in Volume 1 (pp. 344–5), where I suggest that an adaptation that is definitely selfish and destructive at one level of grouping of genes may provide a benefit at a higher group level.[30] The modest numbers of 'green-beard' executions of queens that occur due to the gene may facilitate a type of organization that is proving very beneficial at the colony level and above— including beneficial to the overall density of population and thus creating the conflict with human interests. If this interpretation is right it shows two ideas I should have brought together. Had I done so I might have been less dismissive of the likelihood of green-beard effects generally; perhaps cases like this will prove to be not so uncommon.

As shown in the paper of this chapter the mouse ability for such detection of genetic products of MHC has been familiar for some time but evidence, especially that showing the prevalence and importance of detection in wild situations, has strengthened greatly since I wrote.[31] The demonstration of a human ability to perceive MHC (or HLA as the human version is called) is very recent and still needs more confirmation. But as the claim stands a striking twist already is that which I have already hinted in my suggestion about kin-sniffers for Winchester School: not only are females better at detection, just as in mice (and also best of all, it seems, at the fertile stage of their menstrual cycle), but their ability declines and their preference even slightly reverses if they are taking the contraceptive pill. Because the mode of action of the pill is a mimicry of pregnancy in women, the Swiss authors conjecture that a sosigonic attraction to non-kin exists when a conception is in prospect and that, after an onset of pregnancy, this switches to an attraction to kin—which, too, can be adaptive because kin provide the most guaranteed support base for the coming birth event.[11] As for the olfactory interest being sosigonic—that is, with the aim to chase after genes beneficial to the health of descendants— the team has recently added definite evidence of this for mice.[32]

Work of all these kinds tends to receive a sceptical and even antagonistic hearing from most geneticists and also from physiological scientists generally. Medical biologists and geneticists working on the immune system particularly seem to dislike the idea. Although the research is being pushed ahead by outsiders to the medical field, insiders cannot complain that it started with outsiders because it was one of theirs, E. A. Boyse at the Sloan Kettering Cancer Research Institute in New York, who first suspected and proved the phenomenon in mice. Among the others, however, there seems almost a fear of a contamination, as if the 'clean' and 'objective' field of genetics and physiology was in danger of being smeared with something unpleasant—all this talk about 'altruism', 'selfish genes', 'green-beards', and the like that they had long hoped would go away and they would never seriously have to attend to. The above fields are not alone, of course, in disliking the idea of a contamination by 'sociobiology'. Within this volume we have already reviewed reactions of this kind that were both naive and extreme. One example was from the parasitologist who confessed himself much opposed from the first to the idea of sosigonic selection for showiness in birds; another that I recall was in a letter that

came to unintended hands and then was passed to me. It was a botanist
protesting at a manuscript he had reviewed and rejected. His comments
were explicit that the work he had dismissed appeared to him an attempt to
make a bridgehead in plant science for sociobiological ideas. He looked on
the attempt like a Spanish doctor of the time of Columbus might have
looked on a first case of syphilis in Spain (if that was really when the
disease came to Europe, which actually I doubt)—he would say it was a
case of measles and hope never to see it again. Fortunately reactions are not
always so irrational. In the case of findings of the unexpected and
seemingly chemical aspects of kin recognition they usually take the form of
minute examinations of every particle of the evidence, and pedantic
criticism of a standard, which if applied, as I said earlier, to the evidence for
ills of passive smoking would have lead to almost none of such evidence
being published.[33]

Because I have been consulted several times by editors puzzled about
conflicting reports from referees of manuscripts, I have seen at first hand
various instances of this kind of antipathy, starting with the objections to
the early pathbreaking discoveries of kin recognition in ground squirrels,
macaques, and tadpoles by Warren Holmes, Bruce Waldman, and others.[34]
The phenomena were indeed very surprising. Blobs in the jam jars of my
childhood, identified to me by such names as 'tadpole' and 'pollywog', were
turning out actually to 'care' who their siblings are and about whether they
want to be with them. It seemed as unlikely (I'd once have said) as that
those sappy aphids on the rose bushes would be able to do harm or good to
either near kin or far.

How vastly more complex is the non-human world than we think in
our first impression of it, and how much more similar, in fact, to the twisted
wood of our own species. Just as eventually I learned to think better about
aphids, thanks to Shigeyuki Aoki, so eventually I learned to think better
about tadpoles too. A faint memory of what I had seen when the occasional
tadpole in my crowded jar grew stunted and died may have stuck within me
and prepared me just a little for reading much later about the tiger
salamanders when news of these broke. Perhaps by then there had been
added an early but still post-Waldman hint in some journal that I do not
remember that it might, actually, not be such a good idea for a tadpole to
fraternize in its pond with the big boys of its own kind, especially if food
was short. Even with those hints, reading of salamandery mires and ponds

of the Grand Canyon Rim[35] brought memories of my first read of Arthur Conan Doyle's *Hound of the Baskervilles* and of my first astonishment and horror also at that story of Kuru, the spongiform brain disease that the Fore natives of New Guinea caught from each other by ritual cannibalism, causing them, as the original reports put it (and, I think, using terms of the Fore people themselves) 'to die laughing'. The tiger salamander, it seemed, was a strongly kin-cognisant amphibian of the plateau ponds. It had gone farther with its ability to know its kin than the average Waldman frog or toad and had entered a system of rather sharply separated growth schedules and decisions about what to eat, using as cue for both, in rather sinister fashion as it turned out, the proportion of the non-kin that it sensed in its neighbourhood. Bringing to mind the word 'tiger' in the animal's name, maybe you can guess the rest of this story; in any case you may if you wish look up and see the pictures of these 'Tadpoles of the Baskervilles' in David Pfennig's papers.

Such cases aside, and returning to biases of our field, science is probably no more distorted by the wistful long-shot investigations and by over-ready acceptance of papers trying new and challenging, but possibly wrong, facts and theories, than it is by opposite kinds of wistfulness—those accepting only the papers that splash more weak concrete to hold together popular, established, and moralistic beliefs. The end is as with earthquakes: final re-adjustments are the more drastic the more the second type of those biases has made stresses in a system of thought to accumulate. It is true for morals as well for science. It might have been true (might still be true) that any of the sets of experimental data I have tried to help to be published, whether those showing kin recognition, or those showing bird coloration, or plant kin selection, or credibility of the oral-polio-vaccine theory of AIDS,[36] were then or still are flawed in their design. The beauty of science as a whole, however, is, as I state repeatedly in these memoirs, that it is self-correcting: all errors sooner or later are pulled out of the growing structure of truth and discarded. But it is harder, and asks for more forests to be pulped, to sort out the mess that comes with a Kuhnian revolution than to sort out a single, too-hasty paper. And the former revolution can even cost lives,[37] maybe a huge number of them.[36]

While writing the above paragraphs I was trying to think of arguments I would use against myself if I was the 'other side' in this matter —what I could bring, for example, to support anti-smoking campaigns, to

favour the absurdities of 'political correctness', and so on. I eventually
thought of one rather sweeping explanation that might apply to all
controversies of this kind. It is a little short of a general justification for
irrational gospels—at least it is as we first view it—though perhaps at some
stage it can come near to being so.

There may exist truths that the very nature of human
communication almost forbids us to entertain. Imagine a book manuscript
in your hands and you read it and find that it has made an incontrovertible
case that knowledge is dangerous and that all books ought to be burned—
the alphabet no longer taught. But the real point is that someone has asked
you to try to think of a free-enterprise publisher for this book. Will you
succeed in finding one? An older and similar question is whether a
democratic country should allow to come under its mantle of protection for
free speech the demagogue promoters of politics who would immediately
abolish both free speech and democracy if gaining power. As yet another
instance, can newspapers in a country with a constitution claiming it to be
'self-evident' that 'all men are created equal' publish uncontroversially a set
of arguments and data showing with no shadow of doubt that the
proposition is not only not obvious but not true? Or, what is probably still
more acute for the same country, can it publish this in the face of all the
obvious benefits that the 'untrue' proposition has so far brought to it? These
questions are clearly coming close to some of the matters I have been
discussing in this essay. We know how the above ambiguous 'equality' is
usually interpreted by thoughtful analysis, how it is believed to have meant
'created equal under a just legal system'—that is, the (please note) persons
that the proposition intended to refer to are to have equality of
opportunity, of legal redress, and so on. But, already, that is all a lot more
complicated and doesn't sound so much like anything that is 'self-evident';
and most significantly, perhaps, it has already lost some of it demagogic
appeal.

Maybe seeking absolute truths in some fields as opposed to being
content with mystical and euphemistic half-truths lying vaguely in the
same line is like promoting full literal free speech that includes free speech
for fascists and communists, or like envisioning that actually impossible
trend of genome evolution that ensures that fitness advantages achieved
through intragenomic conflict—gametic drive, green-beard altruism, and
selfishness, and all that—will be as unopposed by genes of the rest of the

genome as are the genes that confer fitness advantages to the entire genome—or, as a still more acute contrast, those that confer benefit to the whole group or whole species. In reality an absence of opposition within the genome simply won't happen, or at least if it does happen it will hardly be observed by us because the species in which such 'free selfish action' happens will soon be extinct. Moreover the genome that is failing to oppose its 'outlaws' of this kind is equivalent to a genome without regulator genes; otherwise how could those regulators, whose effect is to interfere with green-bearders, be excluded without having other regulator excluders coming in to wreck all the rest of the genome's self-beneficial organization? In short, a genome that is perfectly 'fair' to all kinds of fitness advantage, whether disruptive or not, couldn't be anything like what we would today call an organism—it could be a leaderless gang of DNA bits and pieces, a minute, weak-membraned proto-cell of the primordial ooze perhaps, nothing more.

The fact that a realistic genome can't prevent green-bearders being resisted[38] and thus couldn't permit, using our analogy, perfect 'truth' or 'free speech' in the genome is just as well for us because allowing such 'truth' would be like licensing a Mafia within us to do as it pleases. Imagine a green-beard superlocus (that is, a chromosome region with DNA space for several elements) that has linked to its green-beard detection function a particular element coding for murderous spite directed at anyone seen not displaying a green beard (such spite supposed capable to affect, amongst the rest, even the green-bearder's own siblings). It is obvious that a population in which a linked complex of this kind is present and spreading would soon have no individuals lacking green beards, all having been killed by the green-bearders. In itself that might not matter—the health of a resource-limited population could be unchanged. We need think only of an additional intrinsically harmful gene also linked in the green-beard complex—say, a recessive that happens to be in the same chromosomal inversion—and then we see that by the green-beard gene's increase to 100 per cent (that 'drive' of green-beards to 'fixation' as evolutionists put it), our focal species is greatly weakened and very likely to become extinct. This is due, of course, to the linked harmful gene becoming universal along with the beards. It is important to realize that selection of the type discussed here would normally continue to gene fixation whether their effect for the whole organism, population, or species happens to be good, bad, or indifferent. In other words, as I explain for one case in my discussion of 'Samuelson's error' in the paper, unless there are additional assumptions, such systems do not reach polymorphic equilibria—they reach neither neutral states nor intermediate stabilizations.

Might it be, then, that in a meme-dominated and societal analogue of the genomic evolution I have just described (which analogue we can take to be the pan-humanist culture of today) the seekers for a total truth about, say, racial intelligence, about male–female differences, about the harmlessness or even therapeutic outcomes of smoking, and so on, are like those genes inseparably linked into drive complexes and functioning (unknown to themselves) to 'defend' their complex in its basically divisive work—defend it against inhibition from outside influences? This would be similar to the worst accusations being made against those scientists who investigate the above socially sensitive matters, which are to the effect that the scientists are secretly on the side of some form or other of parasitic 'drive' in society and that their proclaimed concern about truth is a pretence: what they really want, the view says, is the triumph of fascism, of capitalism, the Mafia, or whatever. Following the same line might it be fair also to say that the champions of 'no difference' in race or sex, or intelligence (and, to go again to that more trivial level, of the idea that we will all get cancer if we allow smokers the smallest licence), are the guardians of a greater 'untruth' that allows people to live together in mutual harmony, implying that these critics really deserve to be praised as our protectors even when they are factually wrong? I think there is a valid analogy here and I think also it is roughly how the self-appointed guardians choose to present themselves—leaving aside, usually, the step of frankly admitting that they are promoting factual untruths when they know that they are.

Yet answering both the above questions with a 'yes', and thus disapproving the activities of the first set (the scientists) against those of the second (the social dogmatists), would be extremely foolish.

In effect, 400 hundred years ago similar arguments to those I have just given for the first set were at work telling us of the dangers of letting Giordano Bruno and Galileo publish their ideas. Likewise, until quite recently, similar arguments were also praising the social benefits that would come from witch-hunts. In the one case moral chaos was predicted if Copernicus's and Galileo's challenge to Genesis and to full authority of the

Bible was allowed to pass; in the other a boost to health, economy, and divine approval was promised to flow from the elimination of evil eyes and supposed empowerments of the devil. But the boot in these cases is not just on the other foot, it's on quite someone else's. Amazingly in my own lifetime there have been two book-burning philosophies, if they can be called philosophies, plus attempts at effecting them. Both came from extremes of the political spectrum. These extremes might at first seem completely opposed to each other but really were not.[39] Both moves came from philosophies that could be described as righteous and totalitarian.[40] The more recent book-burning, which occurred in Mao's China, emerged from the extreme Left, the side of the political spectrum where most of our 'guardians' presently sit and the side where, so far as I noticed, there was a conspicuous lack of protest throughout all that incredibly damaging series of acts that went on in China. Persuading people to vote for programmes that have at first unstated totalitarian goals, just as Hitler's Fascists persuaded Germans to vote at a time when they were shortly to become promoters of the other earlier recent book-burning episode—this, I admit, to be slightly more of a danger; however, such danger actually increases from the other side, too, if our present self-appointed 'guardians' are given free rein and they also happen to be wrong. These 'guardians' write sometimes undisguisedly proposing their own brand of complete social control. Moreover, should they prove wrong about race and class differences and be merely successful in opposing all measures to conserve intelligence in democratic states,[41] then they simply cause those states to become more prone to foolish votes, which eventually will allow demagogues and totalitarians to stroll undisguised through all the usual barriers set up to them by democracy.

An alternative, and I think more probable, interpretation is that our 'guardians' are themselves like the drive genes, or at least they can be likened to genes that defend a driving segment of chromosome against distant inhibitors. In this analogy we must accept that they are in favour of a 'truth' they believe in. But the reality of this truth is simply power. To state this more plainly, the 'guardians' are demagogues seeking to persuade—falsely—that the interpretation I just abandoned was the right one. They seek to persuade us that their aim is not power or popularity but instead to be guardians and to promote improved states of society. For me, however, their inclination to direct a high proportion of their writing directly at the

book-reading but non-science-trained public is a very damaging sign and
so is the unscrupulously selective reading and sloppy scholarship we see
embodied in their output.[42] Their books typically have far shorter reference
lists than those of their opponents, and usually they do not respond to
challenges from scholars they criticize.[43] These facts weigh heavily in
favour of those insisting on more factual if less-popular versions of 'truth'
as being more genuine guardians.

The parallel of avowed communist academics to drive agents is, in
fact, strong. The motivation may be desire for power, for money, or for
popularity. If communism had indeed triumphed in the world as it seemed
it might be going to when all the critical activity I have been addressing
was at its maximum in the 1960s and 70s—that is, when American
Academia, revolting against the Vietnam War, seemed almost in the
preliminary stages of inciting a new American revolution—they might
have seen themselves ending both as popular heroes and as top professors,
one following the other. On the other hand, it is harder for me to caste a
man like Philipe Rushton,[44] taking an example from the other side, in a
similar light. If 'driving' at all, Rushton has to be admitted to be promoting
a segment of the pan-human chromosome that is very distantly situated
from his own locus, Ontario, supporting a locus situated at the far end of
Asia. I say this based on his claim that East Asians are the race most
advanced in the mental prerequisites for permanent resource-limited
civilization.

Any human science not aiming for factual truth in human social
matters is as inevitably doomed to bring costly accidents in the long run as
would be an unfactual science of technology. Take eye surgery: would you
rather have the best science-trained man from Guy's Hospital in London
do your cataract operation, would you call in a witch doctor to do it, or
would you rely simply on swallowing a text from the Koran? If all cultural
beliefs are equally true, the three treatments are to be equally trusted.
Better still, think of aeronautics. For this technology the thread of
understanding that runs through Archimedes, Galileo, Newton, and
Prandtl is essential education for doing what we now do: Richard Dawkins
neatly summarized the alternative by writing 'Show me a cultural relativist
at 30 000 feet and I'll show you a hypocrite'.[45]

Back with sociology, by letting a supposedly rational and
academically endorsed database to drift far apart from the common-sense

observations of ordinary intelligent people, the 'science' promoted by demagogues makes our institutions increasingly unstable, like ships that are being slowly overloaded. The existence of a body of obvious untested nonsense enshrined in academic departments not only creates disrespect for the 'discipline' and departments themselves where the so-called knowledge resides but, by reflection, it damages respect for academic knowledge as a whole, encouraging a pseudoscience that ordinary people see as no better justified than other kinds of nonsense that appeal to them more (racism could be one example).[46] If we want to avoid painful revolutions of society, like those in the twentieth century that afflicted many parts of the world due to Marxism (painful both in getting into belief in the untruths and then getting out of them), we have to stop using just incantations and manifestos to settle truths about society just as we generally know better than to let them start to dominate the technology of aeronautics.

My analogy between real human democracy and the 'democracy of the genome' seems to me apt and useful but I admit that, whichever way one looks at it, the analogy is loose and it is hard to assign all the parts of the comparison unequivocally. To take an extreme view of this, what if it is true, for example, that our science isn't far enough advanced to warn us that what we are doing with particle supercolliders (say) sets us teetering on the edge of initiating some unimagined chain reaction that will make our planet into a supernova? If that is really the situation, then burning the books at least for now might really be the best thing for us to do—not only for our species but for all life. But if we are going to be cautious about that sort of possibility we should begin the book-burning right now and retreat as fast as possible not just from civilization but from being human; let us start using incantations and political correctness to keep the aeroplanes aloft after all. And let us merely sigh as one by one they crash until none is left. On the grounds that the technology of chimpanzees is more unlikely to be competent to set off a supernova, it is probably sufficient for us to retreat to our former status of being an uncommon anthropoid once again. But if, against all this, the more probable extinction threat for us is a comet on a collision course with Earth then, of course, we should do exactly the opposite: burn no books at all, not even those that seem to us the most anti-rational (in science wrong ideas have to slowly die, as I said, and truth is healthiest in the absence of anything inhibiting the free competition of ideas). We need, in short, to strive as fast and hard as we can to be ready with means to deflect the comet when it comes. That is going to need the best of all the science we have built up.

It is still very possible, of course, that all such effort will not be a success and will carry us into rather than away from some great dangers. The present proliferation of nuclear weapons to new nations occurring as I write poses a much more realistic problem than the supernova possibility. Giordano Bruno was more a poet, mystic, and alchemist than he was a scientist but by his self-sacrifice he stays representative of that Renaissance fever that was busy resuming, in his time, the modes of thought of ancient Greece[47]—mixing it a bit still to the occult, admittedly, but then so did Newton and so did the Greeks. Even as a philosopher, Bruno was nowhere near to Galileo's standard and failed to die because any wings he had been able to engineer, either of thought or artefact, had failed him. He wrote, however, a poem invoking the Daedalus–Icarus myth and in it he predicted his own probable fall to a fiery fate. Exactly so he did fall—into a fire lit for him by the Roman Catholic Church in the Campo dei Fiori in Rome. Through his death as an (albeit imperfect) hero of the Renaissance, his spirit and his reputation as a true prophet live to this day and with what pleasure I saw, during a recent visit to Rome and to the Campo de Fiori where he was burnt and where they still sell flowers, a white chrysanthemum fixed where someone had climbed to put it, lighting the bronze bas-relief depicting his trial on the plinth of his statue. A similar thought, then, to that of Bruno, I claim, applies to fears about the overdevelopment of simple and honest science, whether it is that of particle physics, nuclear technology, aeronautics, or sociology. If sociology or social science wants that title—science—that they so often seem to despise, let theirs be true.

Here is the second half of Bruno's sonnet in which, imagining himself soaring on 'Daedal's wings' in 'Philosophic Flight' (the poem's title), he foretells his fate:

> Where wilt thou bear me? O rash man,
> Recall thy daring will: this boldness waits on fear.
> Dread not, I answer that tremendous fall,
> What life the while with this death could compete,
> If dead at last to earth I must descend?
> Strike through the clouds and smile though death be near
> If death so glorious be our doom at all![48]

References and notes

1. J. Joyce, *A Portrait of the Artist as a Young Man*, p. 271 (Penguin Books, Harmondsworth, 1992; 1st published 1916).

2. L. J. Kamin, *The Science and Politics of IQ* (Penguin Books, London, 1974); L. J. Kamin, *Intelligence: The Battle for the Mind* (Pan Books, London, 1981); S. Rose, R. C. Lewontin, and L. Kamin, *Not in Our Genes* (Pantheon, New York, 1989).

3. Anon, *The Economist*, 71–3 (20 December 1997).

4. The same criticisms, that correlations don't show cause, also apply to a lot of the evidence supposedly confirming how smokers harm themselves. While I accept that it is proven that smoking harms various aspects of health, including causing (in some) lung cancer, I see also much backsliding, following this demonstration, so that an incredibly sloppy acceptance of evidence has become normal (see R. Matthews, Smoke gets in your eyes, *New Scientist* **162**, 18–19 (29 May 1999), although I would put the case for biased samples of studies getting to be published more strongly).

   In particular, two issues seem to be neglected. First, the possibility of a huge variability in humans' ability to withstand health damage due to smoke seems researched much less than it deserves even though there is already striking evidence (e.g. T. F. McNeil, T. Thelin, and T. Sveger, Psychosocial effects of screening for somatic risk: the Swedish alpha-1-antitrypsin experience, *Thorax* **43**, 505–7 (1992)). Second, there is gross neglect of the idea (first raised by R. A. Fisher in the 1950s) that due to unlucky inheritance or social background there may exist unhealthy people, who, due to their condition, experience above-average craving for nicotine stimulation and who also, and quite independently of the tobacco, are destined by their middle to late ages to show up in ill-health and death statistics, whether or not they are allowed to smoke. Obviously such sloppiness I refer to would never be allowed if instead of the proposition 'smoking causes lung cancer', the claim was being made that 'low IQ (or bad genes) cause low socioeconomic status'.

   It is worth adding to this point mention of the two issues the possibility of a connection. Genes essentially for poor health, may, in my view, underlie a lot of social/mental inadequacy and it may be this feeling of inadequacy that entrains search for alleviation through drugs; finally, in the case of some personalities and drug reactions, perhaps it is this restless search that settles on the stimulus of nicotine. The main point here is not that any of this has much evidence but that there are many plausible possibilities besides direct smoke-caused illness that could explain the correlations seen, and choice between them has important bearing on the justice of the present anti-smoking witch-hunt; all such alternatives should be considered carefully.

5. The greater facility with spatial problems might be connected with greater cranial capacity. I do not know about rats and other mammals but, for humans, a difference in cranial capacity favouring males is real even after body-size corrections have been applied, as was re-confirmed recently by R. Lynn (Further evidence for the existence of race and sex differences in cranial capacity, *Social Behaviour and Personality* **21**, 89–92 (1993)). Because there are many psychometric characters in which the sexes do not differ and some also (e.g. language) in which women excel, what is done by men with the extra capacity is unclear. But handling and storing spatial images seem strong possibilities. Most who work with microcomputers have noticed how our need for RAM and storage space seems to jump up by an order of magnitude when we start manipulating graphics and pictures: word processing, in contrast, is much less demanding.

6. A new popular summary of all this that seems to me fair and very largely bearing out my point of view, including with recent evidence, is D. Blum, *Sex on the Brain: The Biological Differences Between Men and Women* (Viking Penguin, London, 1997). Another very relevant briefer contribution on sex differences as they affect the 'gender gap' at work is K. Browne, *Divided Labours: An Evolutionary View of Women at Work* (Weidenfeld & Nicolson, London, 1998).

7. There is no doubt that the women's liberation movement has opened many new doors to women and I must admit their new successes compared with male achievements in most fields since the movement gathered speed have continually surprised me. In so far as this reflects greater fairness in opportunities provided for the two sexes I applaud the success of the movement. But in so far as the movement has persuaded women generally that they ought to strive to imitate achievements of men, it is not at all clear to me that the movement has resulted in greater summed happiness for women. This is because progress has been largely achieved by inducing women to become hormonally closer to men and thus adjusting their ambitions and paths of development in male ways. For some this brings women to the places they always have wanted to be. But not all have wanted: great numbers plunging or urged into the course of masculine striving suffer reduced chances of motherhood and, by this and in related ways, I believe are brought into an unhappy conflict with their biological nature.

Geneticists have long realized that the Y chromosome that makes a human male serves almost solely as a switch for developmental pathways and that the development to an adult male is later channelled by hormones produced by the gonads or by other tissues that follow from the primary switch. Search for genes for specific male characters induced in the brain or elsewhere have largely been unsuccessful. At the same time it has become clear that a lot of genes for striking male secondary characters reside on the X chromosomes where their activation is subject again to the mentioned switches. It follows that having the female XX genetic constitution is probably no bar to developing almost any male character desired if treatment, either psychological or by chemicals, begins early enough; which is to say that the sexes are basically (that is, genetically), equal apart from the switch.

But I think it is a fraught decision both for parents and individuals whether to press development always towards a male pathway. Of course, very similar things could be written about males trying to be more femaloid and parents trying to make them so, but this is less common and less pressurized. The latter, however, contrary to the present cultural trend, is most likely to be the course of the future for our species: first, the trend has already been slowly in progress through the past 5 million years to judge by morphology, and, second, civilized living conditions seem more and more to demand it.

It is foolish to imagine, however, that such a trend is going to happen much beyond where it stands now simply as a learned or culturally induced phenotypic plasticity—induced, for example, by such a course as giving boy toddlers more dolls to play with. Women who believe the convergence course for the future is desirable must make the effort to create babies to promote the trend; and they must do so by men who are currently not the most masculine types.

Anyway, none of this bears on what I regarded as the obviously untrue proposition enshrined in current gender 'political correctness': that was that there is no difference in average ability of the sexes. The love of complex abstractions, facilitating mathematical ability, is one of the obvious fields of difference favouring males. Typical average sex contrasts of this kind characterizing promiscuous species, and involving larger, more combative males, are bound to be with us for some time. Some of the differences (for example, linguistic skills) favour females. Again, Blum's book (see under note 6) is an easily read recent summary.

8. P. M. Bingham, Human uniqueness: a general theory, *Quarterly Review of Biology* **74**, 133–69 (1999).

9. J. Buxton and P. Williams (ed.), *New College Oxford 1379–1979* (The Warden and Fellows of New College Oxford, Oxford, 1979).

10. G. D. Squibb, *Founder's Kin: Privilege and Pedigree* (Clarendon, Oxford, 1972).

11. C. Wedekind, T. Seebeck, F. Bettens, and A. J. Paepke, MHC-dependent mate preferences in humans, *Proceedings of the Royal Society of London B* **260**, 245–9 (1995).

12. S. Leacock, *Moonbeams from the Larger Lunacy* (Lane, London, 1916).

13. Of course there are various ways in which regression to the mean may be impeded and one is that a social class in the population may inbreed. Random maters within that class then regress to their class mean, which may be different from the whole population mean, and members of such a class, because of the multiple potential pathways of inheritance, retain high chances of carrying genes of a particular ancestor. This undoubtedly has occurred in the British upper classes but the degree to which it could have made Founder's Kin persistently more 'worthy' of university training is questionable. First they are major themes of this volume that inbreeding leads to ill health and that the inclination of the natural human psyche is to outbreed *for health*. These factors will undoubtedly have worked against pure intraclass inbreeding. Second, the British population has been continually infused with able immigrants from abroad. These will have raised the average scholastic ability of the general population (e.g. Thomas Hardy) and, in so far as not admitted to be potential spouses for FK, will have relatively lowered the deserts of those as university entrants.

14. A. Grafen, A geometric view of relatedness, *Oxford Surveys in Evolutionary Biology* **2**, 28–89 (1985).

15. O. Popescu and G. N. Misevic, Self-recognition by proteoglycans, *Nature* **386**, 231–2 (1997).

16. M. J. Carlile, The genetic basis of the incompatibility reaction following plasmodial fusion between different strains of the myxomycete *Physarum polycephalum*, *Journal of General Microbiology* **93**, 371–6 (1976); E. B. Lane and M. J. Carlile, Post-fusion incompatibility in *Physarum polycephalum* plasmodia, *Journal of Cell Science* **39**, 339–54 (1979).

17. D. Penn and W. Potts, The evolution of mating preferences and MHC genes, *American Naturalist* **153**, 145–64 (1999); R. Ferstl, F. Eggert, E. Westphal, N. Zavazava, and W. Müller-Ruchholtz, MHC-related odors in humans, in R. L. Doty and D. Müller-Schwarze (ed.), *Chemical Signals in Vertebrates*, pp. 205–11 (Plenum Press, New York, 1991).

18. E. Svensson and F. Skarstein, The meeting of two cultures: bridging the gap between ecology and immunology, *Trends in Ecology and Evolution* **12**, 92–3 (1997).

19. L. Keller and K. Ross, Selfish genes: a green beard in the red fire ant, *Nature* **394**, 573–5 (1998).

20. A. D. Richman, M. K. Uyenoyama, and J. R. Kohn, Alleleic diversity and gene genealogy at the self-incompatibility locus in the Solanaceae, *Science*

273, 1212–16 (1996); N. Zavazava and F. Eggert, MHC and behaviour, *Trends in Ecology and Evolution* **12**, 8–10 (1997).

21. J. Klein, *Biology of the Mouse Histocompatibility-2 Complex* (Springer-Verlag, New York, 1985).

22. S. Ohno, *Sex Chromosomes and Sex-Linked Genes* (Springer-Verlag, Berlin, 1967); W. D. Hamilton, Extraordinary sex ratios, *Science* **156**, 477–88 (1967) [reprinted in *Narrow Roads of Gene Land*, Vol. 1, pp. 143–69]; D. Haig, Genetic conflicts in human pregnancy, *Quarterly Review of Biology* **68**, 495–532 (1993); N. Ellis, The war of the sex chromosomes, *Nature Genetics* **20**, 9–10 (1998).

23. Haig (1993) in note 22.

24. We are close here to the true root of male chauvinism (the male baby has more to gain by being big and strong than has the female baby and is thus more selfish against its mother). But we also are close to the root of female chauvinism—the fact that the next baby may have a different father plus that, perhaps, neither has been sired by the seeming social father. In these facts begin all the human social conflicts of sex.

25. A. Grafen, Green beard as a death warrant, *Nature* **394**, 521–3 (1998).

26. W. D. Hamilton, The genetical evolution of social behaviour, I and II, *Journal of Theoretical Biology* **7**, 1–16, 17–52 (1964) [reprinted in *Narrow Roads of Gene Land*, Vol. 1, pp. 31–82].

27. A. Pakstis, S. Scarr-Salapatek, R. C. Elston, and R. Siervogel, Genetic contributions to morphological and behavioral similarities among sibs and dizygotic twins: linkages and allelic differences, *Social Biology* **19**, 185–92 (1972).

28. C. Wedekind and S. Füri, Body odour preferences in men and women: do they aim for specific MHC combination or simply heterozygosity?, *Proceedings of the Royal Society of London B* **264**, 1471–9 (1997); D. Penn and D. Potts, Chemical signals and parasite mediated sexual selection, *Trends in Ecology and Evolution* **13**, 391–6 (1998).

29. R. Dawkins, *The Blind Watchmaker* (Longman, Harlow, 1986).

30. W. D. Hamilton, Innate social aptitudes of man: an approach from evolutionary genetics, in R. Fox (ed.), *ASA Studies 4: Biosocial Anthropology*, pp. 133–53 (Malaby Press, London, 1975) [reprinted in *Narrow Roads of Gene Land*, Vol. 1, pp. 329–51].

31. W. K. Potts, C. J. Manning, and E. K. Wakeland, Mating patterns in semi-natural populations of mice influenced by MHC genotype, *Nature* **352**, 619–21 (1991); W. K. Potts, C. J. Manning, and E. K. Wakeland, The role of

infectious disease, inbreeding and mating preferences in maintaining MHC diversity: an experimental test, *Philosophical Transactions of the Royal Society of London B* **346**, 369–78 (1994); Penn and Potts (1998) in note 28.

32. C. Wedekind, M. Chapuisat, E. Macas, and T. Rülicke, Non-random fertilisation in mice correlates with the MHC and something else, *Heredity* **77**, 400–9 (1996); T. Rülicke, M. Chapuisat, F. R. Homberger, E. Macas, and C. Wedekind, MHC-genotype of progeny influenced by parental infection, *Proceedings of the Royal Society of London B* **265**, 711–16 (1998).

33. W. Trevaskis, Up in smoke, *New Scientist* **150**, 52 (1 June 1996).

34. B. Waldman, Toad tadpoles associate preferentially with siblings, *Nature* **282**, 611–13 (1979); W. G. Holmes and P. W. Sherman, The ontogeny of kin recognition in two species of ground squirrels, *American Zoologist* **22**, 491–517 (1982); H. M. H. Wu, W. G. Holmes, S. R. Medina, and G. P. Sackett, Kin preference in infant *Macaca nemestrina*, *Nature* **285**, 225–7 (1982).

35. D. W. Pfennig, M. L. G. Lob, and . P. Collins, Pathogens as a factor limiting the spread of cannibalism in tiger salamanders, *Oecologia* **88**, 161–6 (1991); D. W. Pfennig, H. K. Reeve, and P. W. Sherman, Kin recognition and cannibalism in spadefoot toads tadpoles, *Animal Behaviour* **46**, 87–94 (1993); D. W. Pfennig and W. A. Frankino, Kin-mediated morphogenesis in facultatively cannibalistic tadpoles, *Evolution* **51**, 1993–9 (1997).

36. J. Cribb, *The White Death* (HarperCollins, Sydney, 1996).

37. B. Glass, W. Gajewski, A. Petrament, S. M. Gershenson, and R. L. Berg, The grim heritage of Lysenkoism: four personal accounts, *Quarterly Review of Biology* **65**, 413–79 (1990).

38. See M. Ridley and A. Grafen, Are green beard genes outlaws? *Animal Behaviour* **29**, 954–5 (1981) for some cautions and clarifications on this topic.

39. N. Weyl, *Karl Marx: Racist* (Arlington House, New Rochelle, NY, 1977); G. Watson, *The Idea of Liberalism: Studies for a New Map of Politics* (Macmillan, London, 1985).

40. The idea that certain persons or religions, or political movements have pontificial right to say what nature is like and that all then 'ought to' or can be made to 'believe' what they tell, is, of course, extremely old. But, recently, it has come especially persistently out of the extreme left of politics. I will write more about the probable demogogic cause of this phenomenon in the introduction to Chapter 12: briefly the left is 'for the common people' and to appeal to any people you do best to say what they want to believe. It is a bit unfair for the present context but for the record here let me show an extreme example out of the history of the Lysenko movement in Soviet science—in

the course of which, of course, at least pretending to believe certain nonsense you were told about genetics suddenly, at the very meeting I quote from, became a matter of life and death, or at least of a long visit or not to Siberia. This is verbatim from the report of a meeting of the Lenin Academy of Agricultural Sciences of the USSR, the one at which complete dominance of 'Soviet genetics' by Trofimov Lysenko came to be achieved:

> The Michurinian biological science will continue creatively to develop Darwinism, unswervingly and determinedly to expose the reactionary, idealist Weismann–Morganian scholasticism, which is divorced from practice, to combat the servile worship of bourgeois science which is unworthy of Soviet Scientists, and to emancipate researchers from survivals of idealist, metaphysical ideas. Progressive biological science repudiates and exposes the false idea that it is impossible to govern the nature of organisms by creating man-controlled conditions of life for plants, animals and microorganisms.

> > Hail the progressive Michurinian biological science!
> > Glory to the great Stalin, the leader of the people and
> >   coryphaeus of progressive science!
> > (*Stormy, prolonged and mounting applause and cheers. All rise.*)

(I. D. Kolesnik, From the Session of the Lenin Academy of Agricultural Sciences of the USSR to J. V. Stalin, L. A. of A. S. of the USSR (ed.), *The Situation in Biological Science*, Proceedings of the Lenin Academy of Agricultural Sciences of the USSR (Foreign Languages Publishing House, Moscow, 1948).

The sentence beginning 'Progressive' is the one most relevant here and the point of it for now is that it cannot be shrugged off as just an effect of the dire personality in Stalin; if anything the effect runs the other way. It was such wild wishful thinking on the part of many intellectual communists that gave the demogogues Stalin and Lysenko their start. The idea of a malleability of both humans and nature which it is wrong to doubt seems to have been stronger in the USSR than elsewhere even before the revolution, perhaps being a reaction to the extreme rightist regime that was so long in power. That the idea is certainly older than Communism in Russia could be shown by various parallels to the gist of the above in works of pre-communist authors, the best known being Prince Kroptkin (see D. P. Todes, *Darwin without Malthus: The Struggle for Existence in Russian Evolutionary Thought* (Oxford University Press, Oxford, 1989)).

41. R. Lynn, *Dysgenics: Genetic Deterioration in Modern Populations* (Praeger, Westport, CT, 1996).

42. R. B. Joynson, *The Burt Affair* (Routledge, London, 1984); R. Fletcher, *Science, Ideology and the Media: The Cyril Burt Scandal* (Transaction Publishers, New Brunswick, NJ, 1991); J. P. Rushton, *Race, Evolution and Behavior, a Life History Perspective.* (Transaction Publishers, New Brunswick, NJ, 1995).

43. J. P. Rushton, Race, intelligence, and the brain: the errors and omissions of the 'revised' edition of S. J. Gould's *The Mismeasure of Man* (1996), *Personality and Individual Differences* **23**, 169–80 (1997).

44. J. P. Rushton, The equalitarian dogma revisited, *Intelligence* **19**, 263–80 (1994).

45. R. Dawkins, *River Out of Eden* (Weidenfeld & Nicolson, London, 1995).

46. P. R. Gross and N. Levitt, *Higher Superstition: The Academic Left and its Quarrels with Science* (John Hopkins University Press, Baltimore, MD, 1998).

47. A. Cromer, *Uncommon Sense: The Heretical Nature of Science* (Oxford University Press, Oxford, 1993).

48. G. Bruno, The Philosophic Flight, in M. Van Doren (ed.), *An Anthology of World Poetry*, p. 547 (Cassell, London, 1929).

DISCRIMINATING NEPOTISM: EXPECTABLE, COMMON, OVERLOOKED[†]

W. D. HAMILTON

The ideas and discoveries presented in this book help us to see biology as a whole. They link across disciplines—for example, psychology to immunology —and they also link across the gap that we imagine to separate the human from the rest. Animals are even more like us than we thought. It turns out that they care about kinship as much or more than we do. In this chapter, I first review the probable reasons for this caring and the limits to which its evolution can go. After that, I present probable reasons for the long neglect of nepotism as a worthwhile subject of enquiry. Why did we not see this unity before? Thoughts about the evolution of human attitudes have to be speculative; but beyond even speculation, the chapter ends with some personal opinions about how the new matters of identity and relatedness and recognition, once joined to ecology, enable us to see ourselves reflected in several new ways and so change our world view. These closing opinions are expressed, of course, entirely independently of other authors.

New genes contributing to altruism advance in frequency provided that the altruistic acts they cause create, on average, benefits sufficiently greater than losses, and provided the benefits go to close-enough relatives. 'Sufficiently greater' and 'close enough' are quantified in a now well-known approximate criterion:

$$br - a > 0.$$

Here a is loss to altruist, b gain to beneficiary, and r relatedness. In what follows the relation above will be referred to as the kin-selection criterion or merely as the criterion. It is equivalent to a statement that the inclusive fitness effect of the action must be positive.[1] What is happening is that through the enhanced reproduction of others due to altruism, more copies of the causative gene are created than are expected to be lost to the altruist.

[†]In D. J. C. Fletcher and C. D. Michener (ed.), *Kin Recognition in Animals*, Ch. 13, pp. 417–37 (Wiley, New York, 1987).

Because of common ancestry, a relative has an increased chance of carrying the gene or genes. Their presence is never so certain as in the altruist itself but the criterion says that the gain ratio (b/a) of the act may more than compensate.

Various limitations and potential inaccuracies in the criterion have been shown. However, accepting slow Darwinian evolution in the adaptations concerned—evolution mediated by genes of small effect not obligately expressed in their bearers—the criterion works well at least to a point where crudities of current ability to measure fitness and not the failings of theory, must be the overriding limitation in any test.[2–4]

Following a Neodarwinian view, then, complex traits of altruism can be built up by accumulation of genes that satisfy the criterion. Each adds to a set already there and this set—to which, in effect, the new gene is applying for admission—is a part of the 'environment' over which gene effects must be averaged in determining whether selective forces will allow it to rise in frequency. Interaction with some of the genes already there may be epistatic but this is taken into account in evaluating average selective effects.[5] Thus in the context of inclusive fitness (as with many other criticisms directed at Neodarwinism) there seems little cause to worry about limitations of one-locus theory. The problems raised by complex epistasis in polymorphic situations are, of course, interesting, but their importance is probably only in proportion to the occurrence of non-neutral polymorphisms generally.

The set of traits involved in the adaptive suicide of the worker honeybee when it stings a vertebrate attacker of its nest can be taken as example of the claims just made. Some traits are structural (e.g. barbs on the sting and features promoting abscission of a sting apparatus[6]) and some are behavioural (e.g. ferocity[7]). The queen whose safety the acts of suicide serve is carrying the same gene unexpressed, and it is through her breeding success that the genes are maintained. The worst pathologies of the kin-selection criterion arise when genes for social behaviour are unconditionally expressed (i.e. expressed by every individual of a given genotype). For example, a dominant gene for suicidal altruism has to exterminate itself in one generation if unconditional, irrespective of recipient relatedness; even if its benefits are always directed to colonal relatives, these themselves have to die as they express the gene. Results with generally the same slant, often concerning multiplicative fitness interactions and genes of potentially large effect, have been discussed by Cavalli-Sforza and Feldman[8] and others. In situations approaching eusociality, long before caste differences are apparent, there are numerous asymmetries that make conditional expression the natural course,[9] and, in my opinion, genes with small effects are likely to be by far the most important. Multiplicative fitness interaction is indeed plausible, but if the effects are small its selection criteria converge with those of additive interaction. Conditionality, although mentioned, was insufficiently emphasized in my previous work.[1,10]

In the case of the genes modifying behaviour there is good evidence of a differance at a racial level. However, this is easily understandable from the differences in predation to which bees are subject in different continents;[11] there is no reason to expect polymorphism in behaviour within any one area—or, at least, no more reason to expect it than there is for any non-social character. Regarding variability in general, the theory differs only slightly from the classical Neodarwinism which it extends. Within its sphere of application, compared to the classical theory, it implies that less variability will be due to gene differences and relatively more to environmental conditioning. It has to assume that genetic differences are present from time to time, during the spread of new mutations through the population. However, these transiences will usually go to completion. Most of the time the various morphs, acts, etc., will be observed seeming carefully to aid only relatives that are more likely to carry the same causative genes, whereas in fact all members of the population carry them. Thus, contrary to a common misconception, the theory does not imply that differences in social behaviour have to be genetic differences.

All this, of course, applies not only to altruism. A very similar argument works for the opposite of altruism and implies restraint on the evolution of extremes of selfishness and spite. Genes will not go ahead if they take fitness away from relatives for too little self profit.[1,10,12]

It should by now be clear that, if kin-selection models are a fair image of reality, closeness of relatedness is a crucial factor for determining adaptive courses of action in social situations. Thus an ideally adapted social organism would know the relatedness of every individual around it and would make its behaviour finely conditional on the relatedness perceived. However, it does not follow from this that ability to discriminate degrees of relatedness automatically implies that kin selection is the model relevant to its origin. In fact, since even earlier than Darwin, it had been realized that most organisms tend to avoid closely inbred matings. The reasons must have to do with the function of sexuality and this is not yet quite resolved (see, for example, refs 13 and 14, and Chapter 5); but whatever the function is, here must be another set of reasons for discriminating. Some animals certainly do use discrimination for purposes of mate selection. Japanese quail, for example, evidently use an early imprinting of their chick companions towards obtaining, much later, preferred degrees of consanguinity in their mates.[15] In general, however, ability to discriminate, even when its usage is obscure, is prima-facie suggestion of the importance of kin selection in nature. Even if started by other pressures, it is likely that discrimination, once present, diversifies the sociality of a species in both group-beneficial and group-harmful ways. Inevitably, as ability to discriminate refines, new avenues for social behaviour open, and the manifestations can be both positive (e.g. eusociality) and negative (e.g. infanticide).

Writing on the kin-selection criterion 20 years ago I pointed out that where substantial increments to inclusive fitness could occur through discrimination, the ability would be expected to evolve. Four pages were given to the topic together with a summarized principle, corollary to the kin-selection principle, as follows: '*The situations which a species discriminates in its social behaviour tend to evolve and multiply in such a way that the coefficients of relationship involved in each situation become more nearly determinate*[10]. The focus on the 'behaviour of a species' in this statement now seems to me odd. The way of expressing the matter is also indirect and, probably, was cowardly (i.e. aiming to divert from the main point and to avoid sounding racist). Examples that I had been able to find at that time, however, were no more than faintly suggestive that a widespread phenomenon might exist in animals other than humans. I knew, of course, of the keen interest which *Homo sapiens* always has in relatives. It did not seem appropriate to discuss this in the paper because culture was so obviously a confusing factor; human and introspective information, however, had been very important for germinating the idea that a quantitative criterion for social transactions generally might be found. Apart from humans, my best encouragement at that time for the evolution of discrimination lay in examples of birds learning to recognize their own young at exactly the rearing stage where they were beginning to mix with the young of other parents.[10,16] Beyond this I only had various scraps of evidence at more remote levels, suggesting a racial or locally nepotistic kind of discrimination in various animals. In retrospect the interest of these scraps is that, even at that time, they had been long known, and, by furthering an idea that animal and human cases might run parallel, they could have suggested discrimination within the range of close relatives. One might wonder why, even while a theory was lacking, such evidence was not sought. I shall return to this point later.

Since 1964 evidence for discrimination has expanded greatly, particularly during the past few years. Cases have been found where not only do we not know how animals discriminate, but we can hardly even guess why they do so or what use can possibly be made of the ability either in mating or in any other actions that plausibly affect inclusive fitness. Many of the new cases are discussed in the book in which this paper first appeared. Waldman's discovery of discrimination in tadpoles may be picked out here as an early surprise.[18] Not only the occurrence but the remarkable fineness of the discrimination is well established in this case: tadpoles distinguish even their half-siblings from their full-siblings. The list of anurans known to discriminate is growing, yet the function of such abilities in the wild remains quite unknown.

Altruistic or selfish acts are only possible when a suitable social object is available. In this sense behaviours are conditional from the start. Conditionality upon assessment of relatedness of a potential recipient is, in fact, just one of several ways that the behaviour may be refined. For example, an animal might well change its behaviour according to the age or state of health of the interactant concerned. In what circumstances, in general, would discrimination be expected to be most developed?

First, it is obviously expected to be highly developed where potential gains to inclusive fitness are great and where along with this goes a high cost when acts are misapplied (see also ref. 19). This implicates the species that are already fairly social. Adaptive co-operation brings about, ultimately, acts of successful reproduction that would not otherwise have occurred; as we have seen, it is important to an individual participating in causing such extra reproduction that it should involve relatives, not just any animal in the population.

The second and third factors bearing on high discrimination interact with the first one; however, to see them we have to move round the benevolent group statue of co-operation and look at the scowls and concealed daggers (or at best blankness) expressed on the other side.

Wherever extensive preparatory behaviour towards reproduction occurs, selfish acts, in the form of usurping the preparations made by other individuals, can be very rewarding. Examples here are egg dumping that parasitizes the brooding efforts of another bird of the same species[20] and the usurping of nests in the nest-building Hymenoptera.[21] Other examples are legion: think of any object on which an animal expends much preparatory effort; then study the natural history, either in the literature if this exists or in the field. Very likely you will find that this object is sometimes usurped. On the principles outlined, if an animal is going to parasitize, it should, other things being equal, parasitize the most distant relatives, so that discrimination of degree is expected.

The third and final consideration I will mention is that positive powers are most expected where inactive context-driven (and therefore cheap) discriminations are least likely to be effective. For example, an aphid that has just made a dispersal flight might be expected to change to a more selfish and damaging exploitation of the host plant it reaches. After its dispersal any companions found feeding on the same plant are unlikely to be its siblings. Therefore it need not 'consider' any more the effects that its actions may have on the welfare of neighbours. It should become more purely selfish. If there is a change in behaviour, this aphid is performing a *de facto* discrimination based on context. But the same crude method could not be used where, say, even before any migration, there might be many mixed clones and where it might be advantageous to distinguish sister from non-sister aphids and react to them accordingly. To effect discrimination here (supposing it is needed: one sees no compelling reasons why it should be in aphids, but then one sees none for tadpoles either) requires something like a chemical comparison of the target aphids with the self or with a certain set of highly probable siblings, or with the mother before the offspring lost contact with her: all this, with the necessary receptors and neural connections, requires a much more detailed specification in the genome than the reaction triggered by the context of a new plant. In aphids, such examples are hypothetical as yet, but in light of what is now known for toads and bees, it would not be very surprising if real examples were to be found.

A common situation in which all the conditions may be fulfilled is that of broods fathered by more than one sire. In cases of mixed insemination it will be impossible to know which are full siblings in the brood and which are half siblings unless there are fairly refined powers of discrimination. The more obvious courses of social learning—knowing who the mother is, the family context, etc., as well discussed by Alexander[22]—will not serve here. Yet the difference between the relatedness coefficients (0.5 and 0.25) is considerable, and brood mates are often in situations where both high altruism and easy self-profitable selfishness might be advantageous. Thoughts on these lines lessen the surprise that a ground squirrel that seems to have a high level of promiscuity[23] also proves to have a ready ability to discriminate relatives in various ways, at least up to the level of telling full from half siblings.[24] With these squirrels it is actually known that they give alarm calls that benefit others and are dangerous to the giver, and that the calls occur in social contexts where it is especially likely that kin will be benefited. Thus the calls are altruistic, and, in a general way, the expectations of kin selection are confirmed. Honeybees, on the other hand, have long seemed a contrary case to the ground squirrels with the regular mixture of full and half siblings rather ignored,[25,26] despite the existence of well-developed kin discrimination in other bees,[27] however, at least some evidence of kin discrimination has now been found in the honeybee also.[28–30] In general, a good test of the thesis of this chapter will be to compare discrimination abilities between close species that are monogamous and those that are promiscuous, given that the broods can do something social in the first place. It should be pointed out that although the discrimination has to be a positive adaptation if it exists at all, it need be nothing mysterious and its arrival in a series of small steps is easy to imagine. As a minimum one might suggest an array of odoriferous chemicals whose production and quantitative mixture depends on a set of varying genes. These genes may be polymorphic for quite other reasons[19] but in the end, via the odoriferous products (possibly wastes) that they contribute, they effect a chemical signature. Next—and this could happen quite separately—there needs to be an ability to habituate to familiar odours: with respect to the chemicals we have suggested, much of the time it would be a matter of habituating to the particular blend of odours that characterizes self. Finally, behaviour needs to be made conditional in a graded manner on the degree of familiarity of the odoriferous environment in which an animal finds itself at the moment when it is in a position to do something social. The last step is the only one where kin selection has to be involved.

It is easier with most animals to imagine all this happening for odours than for visual stimuli. For example, a vertebrate cannot, without a mirror, habituate to its own face and yet this is the place where humans most expect to find the landmarks for individual and kin recognition. The fact that even humans, who are normally thought of as relying on vision much more than on smell, can fairly easily recognize both their mates and their kin by smells imparted to clothing[31] suggests that an old mammalian method is still available to us, and that the method has remained in spite of our about 90 per cent monogamy and in spite also of some probable reversal of evolutionary 'policy' on nepotistic kin discrimination that occurred in recent human evolution, as will be discussed later.

In the above suggestion of how to get discrimination out of habituation, the reader may have noticed that if it works at all, the proposed mechanism has no reason to stop at discriminating full siblings from half siblings. Among full siblings, for example, there are those who happen to have a more similar shake out of chromosomes from the parents and those who have less. (For any specific relationship, as specified by pedigree connections, the mean relatedness is fixed—barring problems brought in by the sex-linked part of the genome—but variability about this mean depends, roughly speaking, on the number of freely recombining units in the genome; that is, on the 'recombination index'.[32]) Rather surprisingly there is already evidence that human discrimination, using in this case undoubtedly many media besides smell, can discriminate among full siblings.[33] So long as the genes involved in all parts of the discrimination are well spread among the chromosomes, most genes apart from the 'non-discriminator' allele that is being replaced, benefit from any action that fulfils the kin-selection criterion on the basis of the estimated r. Thus we need not expect that other elements in the genome will evolve to suppress the effect. Such a system of measuring similarity differs sharply from one that reacts to a particular trait—that is, a 'green-beard trait',[34,35] as will now be discussed.

The system outlined above was based on a minimal kind of 'social learning'.[22] Prior to application in behaviour, learning, such as self-habituation, can be done by the individual in total isolation. Could there be any mechanism of kin recognition that does not need learning of any kind? In fact, there do seem to be effects describable as kin recognition which are like this, but at the same time there are also reasons to believe that they cannot evolve into nepotistic patterns of any complexity.

Absolute innate discrimination of genotypes, differentiating them as to whether they do or do not overlap with self at one particular locus, is already well known in all those plants whose outcrossing is regulated by the one-locus, multiple-allele, self-incompatibility system.[36,37] The reaction happens between the diploid stigmatic tissue of the whole plant and the haploid gametophyte arising from the pollen grain that has alighted on it. Here it is a case of seeming failure to co-operate when the two tissues are 'related', resulting in actual death of the gametophyte. The failure happens whenever the gene in the pollen grain is the same as either one of the genes at the incompatibility locus in the receiving plant. In essence, it is as if the pollen grain has a chemical 'green beard' by which it is recognized, and if the host finds it to be at all similar to its own, the germinating pollen grain is rejected. But is it necessarily the reversal of usual kin co-operation that it seems? Maybe it is the pollen grain that auto-destructs, or at least grows non-competitively when it perceives by the 'green beard' of its host that it has landed on a close relative. In this case the death is altruistic—an evolutionarily 'taught' reaction based on the fact that if a closely related pollen grain does race ahead and claim an ovule, the resulting seed is practically doomed through its homozygosity (or its lack of differing genome). Thus the genes of the pollen grains propagate better by giving to their probable twins in the ovum that they might have entered, the chance of joining the more suitable gene partners that are brought in by a different pollen grain. At the back of this speculative but not unreasonable interpretation are the still-mysterious advantages of sexuality and outcrossing. These advantages, of course, also underly the very existence of kin selection as we know it, because without them there would be no mendelizing genetics. But that line of thought leads off towards longer digressions than can be pursued here.

The above example will probably be considered a cheat of a 'green beard', not really what Dawkins meant. But given that this pollination phenomenon exists (e.g. in cherry trees) it is possible to imagine the following. When cherry tree roots collide underground they could measure their genetic overlap by the same type of mechanism. Then they could either co-operate by physically uniting their root systems (natural root grafting between neighbouring conspecific trees is common in some species[38, 39]) or else they could keep separate and compete. Note that the reaction of pollen and stigma is unlikely to carry over to roots in quite the same form because (1) tissues of colliding roots are both diploid whereas pollen is haploid, and (2) even within the floral parts the pollen reaction is often very localized[37] (e.g. injecting pollen into the ovary sometimes bypasses the inhibition). However, the idea of reactions developed primarily for optimal outbreeding being switched for use in nepotism is generally reasonable and perhaps is the most likely course where genetic markers are being used. There is actually a difficulty about how genetic polymorphism can evolve if nepotistic identification is its main use; this is explained by Crozier.[19] Strengthening the idea of switching, Crozier refers to the remarkable finding of Smith[40] on the kin-biased mating reactions of male sweat bees.[41] These bees are the same species as studied by Buckle and Greenberg[27] in their work on discriminative guarding.

There are, of course, also other possibilities—for example, one cherry tree could fasten on the other parasitically—but for the present discussion we assume that if they unite, each is able to defend itself against harm.

Suppose that a root connection is made and that subsequently one tree is dying for want of water, to which the other has limited access. Suppose the other supplies the dying tree with water at the benefit-for-cost ratio far more generous than the limit set by the kin-selection criterion based on pedigree relatedness. The helping tree as a result sets less seed and the helped sets more seed—but suppose the 'more' is not enough to fulfil the criterion except at the recognition locus and possibly at loci very nearby. Suppose this happens again and again in instances of such paired colliding trees. If now, elsewhere in the genome, a mutant gene arises that blocks the outward sap flow at this benefit-for-cost ratio, then that gene is advanced; this is easily seen by the kin-selection principle (a claim by Samuelson to the effect that such an altruism-blocking gene would not supervene completely, but rather would be carried to an intermediate ratio dictated by the starting conditions, is an error[42]). Thus as regards its actions as a potential donor, the genome evolves to suppress a 'green-beard' transfer. Effects of this kind would therefore be transitory and be unable to evolve into proper adaptations. If such aid is given regularly through root grafts, it is likely to be established on some basis either of typical kin selection or else of reciprocation.

Potential plant examples like those suggested above may seem remote, even if valid theoretically, to animal-oriented readers of this volume. The following examples probably won't seem much better, even though they are animal and even though the animal in my first case is—well—woman. There is good evidence that a minor part of human sterility may result from too great a similarity of the partners for genes in the major histocompatibility complex.[43] It appears to result from failure of the fertilized ovum to implant properly. The blocks evidently are at a different stage, but have some sort of parallel to the genetic incompatibility systems of plants. Whether we admit anything 'green-beardy' about this or not, the parallel itself is the more impressive when we learn that a system almost certainly homologous to the major histocompatibility complex reaches back to our own sessile ancestors (or slightly more accurately, our remote 'aunts'), the tunicates,[44] and is probably used by them to prevent coalescence of adjacent individuals. Possibly it reaches even to the coelenterates.[45] On lines of the parallel, referring again to the human zygote and to the possibility that 'altruism' is involved in a decision not to attach and grow, I note here, first, the complete reversal of the normal effect of histocompatibility, and, second, the seemingly completely maladaptive outcome of the failure to implant that even arguments about altruism will not easily explain. Surely both the potential mother and the fertilized ovum should agree that it is better to produce some offspring, even if more than ideally homozygous for HLA, than to produce no offspring at all. Perhaps it is only the unusually maladaptive extremes of such early fetal loss that come to notice. (Were the effect not rare, heterozygotes might be expected to be much in excess at HLA loci just as they are at incompatibility loci in plants; so far this does not seem to be the case.) In any case there is at least an interesting hint of a rather innate recognition ability, and this case leads on to notice of yet another related phenomenon on the borders of immunology: the house mice which show mating preferences based on unlikeness of certain genes in the histocompatibility complex.[46] This effect is very surprising, becoming perhaps the more so when it is known that it is mediated through odour coming from urine. We cannot do more than note here what appears to provide the basis for a minimally learned discrimination system of the kind already outlined, and that it seems to have a potential to become 'green-beardy' in so far as the effect is concentrated in the small *H–2* region of one chromosome. But, once again, for the reasons already given in the hypothetical case of the trees, we do not expect anything describable as an innate kin-recognition adaptation, to be much used for regulating social behaviour other than mating.

One of the novel outcomes of trying to rationalize social behaviour by arguments of inclusive fitness is the realization that not all elements of the genome are expected to 'want' the same thing, and that they may be making contributions to strategies that are actually contradictory within the one genome.[47] The unlinked gene that suppresses the effects of a 'green-beardy' one in the previous paragraph or which acts in the case of supposed interactions of cherry-tree roots is an instance of this. Both these genes could be on autosomes. More regular instances of intragenomic conflict can be expected between the sex chromosomes and autosomes. Since sex chromosomes are usually a single pair, and only very occasionally a major part of the genome, and since the rest of the genome goes symmetrically with sex and submits to the very fair and uniform process of meiosis, the effects of conflict of strategy between autosomes and the X chromosome are not expected to be very apparent. The Y, being usually inert, does not come into this argument much except reflexively in explaining the inertness. Such manifestations, if ever found, would be discriminatory and so deserve to be mentioned here. Only one example will be given. The behaviour of female birds during brood care might be expected to be slightly biased towards assisting their male offspring.[25] This is because they pass on to sons a very substantial chromosome, the X (if we can extrapolate from those few cases where its size and genetic richness are established), whereas they pass on to daughters only a small inert Y. Their male mates might be expected to attempt to compensate for any bias by preferentially feeding the female young. There do seem to be some hints of such biases;[48] but as we would expect from the outweighting of sex chromosomes by autosomes, they are slight effects found only in some species. It is too early to say whether the inclusive fitness rationale is important.

Perhaps the most interesting thing to come out of the realization of possible conflict within the genome is a philosophical one. We see that we are not even in principle the consistent wholes that some schools of philosophy

would have us be. Perhaps this is some comfort when we face agonizing decisions, when we cannot 'make sense' of the decisions we do make, when the
bitterness of a civil war seems to be breaking out in our inmost heart.

With evidence now at hand, most of it rapidly accumulated over the past
10 years and summarized in this volume, it begins to seem that some ability
to recognize kin and to react accordingly will be found in any social animal if
looked for carefully enough. This potentially implicates almost every animal
species, because almost every species is social at least to the extent of mating.
If we admit the lower-level phenomena concerned with tissue graft rejection,
recognition may encompass much of the Metazoa over again. Sponges, sea
anemones, and corals know their kin[49] and graft, avoid, and fight accordingly;
slime moulds (Mycetozoa) fight or do not fight lethal chemical battles according to how they differ in genes, and these genes reflect kinship.[50] Such universality raises a puzzle. At the higher end we have known that our species is
inclined to nepotism for as long as we have known anything; we are steeped
in its existence and have known of it since, as infants, we began to have any
ideas at all. Then we have had it all imposed again (if with many variations)
from the coded lore of every human culture. Even this is not all that should
sensitize us to the issue of relatedness. The idea that animals throw back pale
or distorted images of ourselves and we of them is also ancient—far, far older
than Darwin. Why, then, have we not long expected to find that animals have
ways of discriminating kin? Why were the first recent papers about actual
cases reviewed with almost open hostility and suspicion? The reason must
have to do with the fact that at least in civilized cultures, nepotism has
become an embarrassment.

We admit certain aspects of nepotism, such as parental devotion to offspring, because it is obvious that even high civilizations cannot do without
them. I believe that even the most extreme communists would react with
horror, like everyone else, to an idea that people's babies should be routinely
switched around in the natality wards of hospitals, as a matter of state policy,
even though, if put into effect, this policy would immediately bring to reality
some of their most ardent dreams. For example, apart from justice in the ran-
domization of opportunity, the policy would, of course, in the space of one or
two generations, settle the question of nature and nurture in human ability
once and for all. Experiments rather on these lines although less extreme
have actually been tried. One of the most daring and recent has been that of
the Israeli kibbutzim; but I know of no such experiments that have not had to
retreat from the original ambitious aim. Adoption does succeed. It succeeds
extremely, surprisingly, nobly—as an outstanding credit to the above-animal
nature of human—when it is considered in the light of some of the issues and
facts raised in this book. Yet even this doesn't succeed wholly. Virtually no
one chooses adoption except as a last resort.[51]

The same spirit that approves of adoption in principle opposes nepotism rather definitely once outside the circle of closest relationships. When we hear that a son has inherited his father's business we think that this, at the least, is only natural. What else to expect: perhaps actually it is a 'good thing', an expression of strong positive spirit in the nuclear family which we approve. If it is a nephew that inherited the business we are little less certain. If it is a son of a cousin we may mutter something about 'blatant nepotism, unjustified class privilege'. It is extremely hard, perhaps impossible, to know how much of this attitude is natural in us and how much is a product of our teaching. It is difficult to know how much of it is sincere, even in oneself. For sure, large parts of such attitudes are taught. One has only to think of how whole cultures have been swayed towards or away from racism within living memory to see that. For sure also, what we express as ideal is not always what we do, and what we do usually errs on the side of being more nepotistic. Such hypocrisy over nepotism is understandable on evolutionary principles. A world where everyone else has been persuaded to be altruistic is a good one to live in from the point of view of pursuing our own selfish ends.[21] This hypocrisy is the more convincing if we don't admit to it even in our thoughts—if only on our deathbeds, so to speak, we change our wills back to favour the carriers of our own genes. However, when all this has been said, a case remains, I believe, that a major reversal of natural selection acting upon the more extreme forms of nepotism took place in recent human evolution. Since the apes we may have become intrinsically more ambivalent.

I shall try briefly to explain why I think this has happened. Possibly gear-wheels will have been heard grating in this chapter—actually a little back from here—and if the chapter seemed sketchy already it is now going to seem worse. In excuse I must plead that I am trying to explain something that is sure to be controversial in a short space . . . that is, to explain the puzzle of why people don't want to know about animal nepotism and why nepotism in general has become an embarrassment.

The new pressures which I believe came to affect the issue arose out of the diversity of human activity as it developed, contrasted to the activity of apes. One, the major new pressure, comes from our division of labour. Another comes out of our increasing intimacy with plants and animals, arising ulti-mately out of the need to control them in securing our food supplies.

If it makes sense at all to talk about the efficiency of a community—that is, if the community is integrated enough to have 'functions' that can be measured as more or less efficient—then the ideal road to division of labour is probably that taken by the cells of the embryo or by the worker honeybees of a hive. Here, every individual carries the code both for all the varied pre-liminary paths of ontogeny, and for the detailed acts of living. All these are to combine together for the wellbeing of the superorganism. Sometimes, especially in the social insect examples, the animal does actually perform several of the activities in its lifetime. In these cases division of labour is all a matter of going to the right part of a general programme carried by every individual. It works well so long as storage of very long stretches of unused script is not too much of a problem and the individuals are closely enough related that antagonized interests do not cause the system to break down. In the case of cells the latter condition is clearly met; in the case of honeybees it is met, to judge from their behaviour, by a rather narrow margin—met with the help, perhaps, of severe threat to the non-co-operative colony from the outside. In humans, the relatedness in a self-supporting group of any size isn't close enough and division of labour has come about from the start in a different way.

Here there may come a 'just a minute' from the reader. The human system is indeed different, you may say, but this has nothing to do with relatedness. Rather it is because humans are not, and never were, anywhere near to having the kinds of innate programme of development that cells and honeybees have. Instead the information for diverse human activities is stored in culture from which appropriate parts are taken up by learning—taken up into the most broadly impressionable nervous system that is known. And, you may add, your argument was astray earlier because, using your own standard of emphasis on the individual, what says that the net outcome of human division of labour needs to be efficient—what unit is supposed to benefit if it is?

Well, I agree with these criticisms but not entirely. My answer to the last one is that human communities and cultures do replicate sufficiently in their own right[52] for there to be appreciable evolution towards efficiency, although I am not sure how this could be measured. The earlier criticism is much harder. The impressionability of humans is exceptional but it is not unlimited, and downloading a culture-specific set of beliefs and norms into a 'blank' memory cannot be the whole story. I must be brief here because this is digression: I will claim that it should be obvious to anyone who has sat in a classroom or played games on a school playing field that all people are not equally teachable in all skills. If the obvious experiments based on infant adoption have been done by those who believe otherwise, then their results have not been reported. Therefore, if done, the results were probably negative. Adoption studies that have been reported, on identical twins reared apart, indicate much heritable idiosyncrasy in personality and abilities. Furthermore, *a priori* it is rather difficult to imagine any 'blank slate' arising by evolution. Admittedly there has been a steady progress in the more flexible and learned approach to adaptation and it is slightly easier to imagine a general mentality that could become a disc jockey or an inventor than it is to imagine a physique that could become a high jumper or a weight lifter—but it is not much easier. The scenario for the origin of division of labour that I give below has at least a plausible slant and story. I postulate a mechanism basically analogous to the way interspecies mutualism is expected to arise out of the goalless meandering of interacting species in an ecosystem.[52]

The pressure begins when our ancestors, probably still confined to Africa, expand their exploitation outwards from the narrower niches of apes. It is assumed that brains permit them to understand the past and to imagine the future, that they understand tit for tat (see Chapter 4 in this volume), and that they easily recognize particular alien groups. Let us imagine two tribes, one of the savannah, one of the forest, which sometimes met along the forest edge. For a long time, if they take notice of each other at all, they snarl and fight. Their traditions—we suppose they already have language—admit to no intertribal common ancestry; indeed they would feel insulted at the thought of it for each tribe regards the other as barely simian. Then gradually a change comes. When 'we—the people' go to the forest now we take grass seed and meat with us; these we find easy to gather and scavenge out on the plains. We leave them around. Gradually we have found that instead of throwing their spears at us, 'they—the ape pigs' give us spears; and spear shafts, of course, are what we went to the forest for anyway. And when they howl and mouth at us in that crude way of theirs across the glades, well, now it doesn't sound such a bad sort of uproar when you're in the mood for it. At times they stand up almost like people and one may even like to watch the way some of them—the sows—wave themselves about. One could almost . . . In short, we have the impression that they are greeting us in their crude way and they want us to come back. Of course they wouldn't want to come to us bringing their spears and bow staves and fruit. The sun would roast that pitiful pale skin of theirs right off their backs if they came into the savannah—unless we camped where there were a few more trees, then they might come. But if they did, we would want to make sure they can see that those trees are not a forest they can move into; they're our trees. So is it our forest for that matter—it's just that we are not choosing to live there for the present. Those creatures are nothing to us except a clever kind of monkey or pig that can be useful when it decides not to be vicious. So long as they realize that, I suppose, there would be no harm in having them actually around our camp sites; they climb trees so much better and could bring us fruit before the parrots take it . . .

By some long, long 'rationale' such as this (in reality spread over hundreds of thousands, if not millions, of years, and with the trends I have set out realized, in part, in actual changes of genetic disposition, and at the same time confused much by intermarriage), *Homo economicus* can be imagined to come into existence. More and more complexity would be added as the story goes along. Domestication of animals and the invention of agriculture occasion huge strides. The invention of metal-working has been another even more recent incentive to diversity. Here, in particular, it is in itinerant smiths of recent recorded times working in quite alien cultures, half admired, half despised, and in any case so much needed that their position is relatively secure, that one can see a hint of what has been going on.

Definitely on all fronts it has become imperative not to bristle with hostility every time you encounter a stranger. Instead, observe him, find out what he might be. Behave to him with politeness, pretending that you like him more than you do—at least while you find out how he might be of use to you. Wash before you go to talk to him so as to conceal your tribal odour and take great care not to let on that you notice his own, foul as it may be. Talk about human brotherhood. In the end don't even just pretend that you like him (he begins to see through that); instead, really like him. It pays.

As civilization gathers pace and more and more diverse activities open up, it becomes, of course, more and more impossible that a single human type, either in physique or mental disposition, could do all the tasks equally well. And not just tasks but roles. It is not just the physiques and dispositions for farmers, bankers, and watchmakers; civilization needs a more subtle mix, with ingredients drawn from a diversity that was present and used even before the *economicus* stage. (A possibility that integration of diverse gene-tical aptitudes may have been part of the group-living of primates in much earlier arboreal times is raised by the polymorphism found for retinal pigments in *Saimiri sciureus*, a New World monkey.[54]) Civilization needs leaders deter-mined to stand on top and willing to try novelty; it needs sheep content to be lowly and to be told what to do; it needs introverts to be seers and inventors, and extroverts to keep everyone talking to each other. It is a symbiosis of aptitudes.[55] It is like the mutualism of a lichen made vastly more complex. It is emphatically not a matter of bringing all the needed diversity out of a single genome, any more than this is being done to achieve the complexity of an ecosystem (a reverse conjecture, compatible with this but trying to under-stand ecosystems and many of their functions by comparison to human economies, was developed by Ghiselin[56]).

With this in mind and also looking into the great black pit of non-co-operation that opens at our feet out of the so-called Prisoner's Dilemma once civilized interdependence has risen,[57] then surely it is not difficult to see why selection for nepotism has gone partly into reverse and why we are confused about it.

This outline of the present situation and the implied possibility that, once we have caught up with the extra complexity and have it properly figured out, it might be adaptive to pursue nepotism again wholeheartedly, may not seem very likeable. Certainly if it is anything like true, it raises worries about the future of civilization that must be addressed sometime if we are sincere about preserving some human attributes that we now believe to be good. However, I do not intend to face the worries here. Such a discussion would have to con-sider hypocrisy—that which may be latent in my own view about and also which is fairly obvious in the alternatives that are usually preferred. There is indeed here an obvious demagogic value in a talk of generosity rather than of sordid trading; also in a general shouting about brotherhood; also in denial of 'natural' nepotism . . . However, this is a large subject, and it overlaps many minefields in the human subconscious. I will leave it and turn to the other lighter and more likeable factor that I see as countering nepotism in recent human evolution.

We have grown oddly fond of animals as we have of plants also. We like them for their own sake; we like to have them around as pets and we like flower gardens. As we isolate them and bring the species close to our gaze, as we look after them almost as if they were our children, we inevitably see their lives with a degree of sympathy that we never gave them before. And then we begin to realize their similarity to ourselves, and beyond this to glimpse the unity of all living nature.

Why we should develop this liking for plants and animals is hard to say. At first sight to connect pets and flowers with the domestication of other species which are being ruthlessly exploited once they are controlled, does not seem promising. However, it is likely that the truth lies this way. Arbitrary and useless as may seem a Mexican's love of plants that will cause him to hang a *Sedum morganianum* in an empty tin can from the porch of a building made out of the sides of packing cases and old corrugated iron sheets, it is difficult to imagine that this love is not connected with the immense contribution that Mesoamericans have made to the crop-plant flora of the world. Out of the same land also came the first ideas of a botanic garden where every kind of plant would be grown; now, of course, botanic gardens all over the world store plants and genes to protect our food supply from disaster, to say nothing of storing plants whose potential use is as yet unknown. The contribution to domestication of the indigenous peoples of South America is not so great but it exists, especially in the more civilized areas, so that when one finds a tribe in the Amazonian forest with no domestic animals save the dog, and yet with parrots, seriemas, pacas, monkeys, deer scrambling and strolling about the house as pets, one can imagine that the sympathies that make people inclined to this seeming waste of effort have, in other places and times, paid off for them. Meanwhile, if famine strikes the pets are probably sometimes eaten— as are the guinea-pigs that are kept half as pets by Andean Indians, even without famine.

Here, by the way, we are not looking at the first roads to domestication. Nor is the first road in the Middle East, nor even human. Not far through the forest from huts where Brazilians keep their pets I have watched *Azteca* ants attacking and cutting up a colony of wasps of a species that nests, normally peacefully, inside the nest of ants—and at least in that area, never nests elsewhere. In the case of the attacked colony a recent flood had probably interrupted some of the ants' foraging trails. The *Azteca* have undoubtedly been keeping their 'pets' for millions of years before people came to the Americas. Older still, perhaps back in the sunset of the dinosaurs, must be the origin of similar events I have watched in a primitive ant *Hypoponera punctatissima* with its myremecophilic Collembola. But this is digression.[58] The social insect developments would be on a much more innate and less-flexible basis than those of humans. My main claim for the present about the keeping of pets and plants on the road to domestication is that in the human (if not in the formicine) kind of approach it helps greatly if you are interested in and even love the living things around you.

If you have been saying all your life of the people in the next valley that they are little better than pigs—they eat things that pigs do, gorging with the manners of pigs while also they copulate pig-like in the streets of their villages (one personally hasn't seen this but everyone knows it is so), and so only deserve to be speared on sight like pigs—then it may at last trouble you acutely to realize that you love your own pig and are determined that one in particular will not be put to death and eaten. Once this troubling begins, and once there is enough social surplus to make pet-keeping and flower-growing possible, so that the idea can be constantly present and its effects kept not too costly, then the process can go rushing ahead in leaps and bounds—in fact, perhaps, has to rush on until it reaches the point where it says: 'These animals are actually much better than us; they are sort of divine.' Once this stage is reached, then we know how to rationalize and react if it happens that 'they' go on spearing us even when we tried to stop doing it because of their being too pig-like. It can then transpire that actually they are worse than animals, wicked and perverted, fallen from natural grace (as indeed are some of our own tribe, come to think of it). Now their unnatural wickedness can be the excuse for our harmful acts. Yet one hopes, when this stage is reached, having once seen them as part of a great living scheme which we might enter gently and mould to our advantage, as we did with our animals, we do not throw spears with quite such abandon as before, but do it more defensively. This is all a bit fanciful, but I believe it has at least a grain of truth, and if this is the case it may help to explain one of the facts that I mentioned early in this chapter: the shock and indignation aroused by the discovery of nepotistic discrimination in animals. We had become proud of what we can do, but also sorry about it and nostalgic for a supposed golden past. In certain moods we idealize the state of nature and consider ourselves 'fallen'.

There is little reason to expect that *Australopithecus* kept pets; *Homo neanderthalensis* at least may have loved flowers.[59] Just possibly this is connected with the fact that in australopithecine times the hominids in Africa had less morphological diversity than exists in hominids today (i.e. within *sapiens*), whereas after *neanderthalensis*, rather as if the disappearance of that near species came to be regretted, the diversity of human races fans out and, with one or two known blots of extinction, the fan persists down to the present. The possibility of genocidal interaction among groups of early hominids has been suggested[60,61] and is plausible both theoretically[62] and by comparison with other broad-niche social species, including ants[63] and primates.[64] Some aspects of human behaviour seem to require an antecedence of strong kin-group selection, and for this to work it is almost necessary that groups be mutually hostile as well as genetically closed. Intra-amicable and relatively non-nepotistic groups of Oceania[65] are possible examples. Unfortunately, as to distant patterns of group selection in the main stock it is unlikely that the fossil record can reveal much direct evidence.

As for one of the recent blots of extinction, I would like to think that the sudden reversal of the policy of the white settlers towards the native Tasmanians, when it was realized—unfortunately too late—that they were nearing extinction, was an expression of a new ethic of conservation and of respect for all forms of life. Started from a small origin along with pet-keeping and horticulture a very long way back, suffering vicissitudes but never extinguished during the explosions of domestication and agriculture, this ethic has been very rapidly expanding in the past hundred years.

It may seem insulting to the Tasmanians to suggest that the emotions that tried to save them were the same as those now trying to save the giant panda. (Maybe as they on Flinders Island, so the panda in our zoos expresses its proud contempt for what I write here by refusing to reproduce.) But it is surely better to be saved by patronizing emotions than not to be saved at all. It seems to me that this is a sounder start towards a more peaceful and varied phase of humanity than the impossibly idealistic visions that are commonly advocated. In essence, loving everyone as oneself is as unthinkable, and as contrary to the grain of all life as a programme of deliberate baby swapping would be. Yet if we are content with a more-realistic objective of a stable human 'ecosystem' that has integrity and beauty equivalent to that of the natural ecosystem out of which it grows, then the first steps are not far to seek. We must cease to pretend with cries of 'brotherhood' and other nonsense, that there is one ideal way of life, one dominant culture, one right or inevitable programme of evolution for our species, whether claiming a way of a noble savage, a Marxist, a Christian, an Englishman, or any other. Then, still countering less-liberal views, there follows one rule that most different cultures should be willing to enforce, however varied and individualistic their practices may be in other ways: that no race or genotype has the right to unlimited numbers, and yet every race has a right to be protected against extinction.

References and notes

1. W. D. Hamilton, The genetical evolution of social behaviour, I, *Journal of Theoretical Biology* **7**, 1–16 (1964) [reprinted in *Narrow Roads of Gene Land*, Vol. 1, pp. 31–46].

2. R. Michod, The theory of kin selection, *Annual Review of Ecology and Systematics* **13**, 23–55 (1982).

3. R. Boyd and P. J. Richerson, Effects of phenotypic variation on kin selection, *Proceedings of the National Academy of Sciences, USA* **77**, 7806–10 (1980).

4. J. M. Cheverud, A quantitative genetic model of altruistic selection, *Behavioral Ecology and Sociobiology* **16**, 239–43 (1985).

5. R. A. Fisher, *The Genetical Theory of Natural Selection* (Oxford University Press, Oxford, 1930).

6. H. R. Herman, Sting autotomy, a defensive mechanism in certain social Hymenoptera, *Insectes Sociaux* **18**, 111–20 (1971).

7. C. D. Michener, The Brazilian bee problem, *Annual Review of Entomology* **20**, 399–416 (1975).

8. L. L. Cavelli-Sforza and M. W. Feldman, Darwinian selection and altruism, *Theoretical Population Biology* **14**, 268–80 (1978).

9. M. J. West-Eberhard, The evolution of social behaviour by kin selection, *Quarterly Review of Biology* **80**, 513–30 (1975).

10. W. D. Hamilton, The genetical evolution of social behaviour, II, *Journal of Theoretical Biology* **7**, 17–32 (1964) [reprinted in *Narrow Roads of Gene Land*, Vol. 1, pp. 47–82].

11. M. L. Winston, O. R. Taylor, and G. W. Otis, Some differences between temperate European and tropical African and South American Honeybees, *Bee World* **64**, 12–21 (1983).

12. W. D. Hamilton, Selfish and spiteful behaviour in an evolutionary model, *Nature* **228**, 1218–20 (1970) [reprinted in *Narrow Roads of Gene Land*, Vol. 1, pp. 177–97].

13. G. Bell, *The Masterpiece of Nature: The Evolution and Genetics of Sexuality* (University of California Press, Berkeley, 1982).

14. W. M. Shields, *Philopatry, Inbreeding and the Evolution of Sex* (State University of New York Press, Albany, 1982).

15. P. Bateson, Optimal outbreeding, in Bateson, P. (ed.), *Mate Choice*, pp. 257–77 (Cambridge University Press, Cambridge, 1983).

16. M. D. Beecher, Signature systems and kin recognition, *American Zoologist* **22**, 477–90 (1982).

17. P. A. Buckley and F. G. Buckley, Color variation in the soft parts and down of Royal Tern chicks, *Auk* **87**, 1–13 (1970).

18. A. R. Blaustein, M. Bekoff, and T. J. Daniels, Kin recognition in vertebrates (excluding primates), in D. J. C. Fletcher and C. D. Michener (ed.), *Kin Recognition in Animals*, pp. 333–57 (Wiley, New York, 1987).

19. R. H. Crozier, Genetic aspects of kin recognition: concepts, models and synthesis, in D. J. C. Fletcher and C. D. Michener (ed.), *Kin Recognition in Animals*, pp. 55–73 (Wiley, New York, 1987).

20. Y. Yom-Tov, Intraspecific nest parasitism in birds, *Biological Reviews* **55**, 93–108 (1980).

21. G. E. Bohart, *The Evolution of Parasitism among Bees*, Forty-first Honor Lecture (The Faculty Association, Utah State University, Logan, 1970).

22. R. D. Alexander, *Darwinism and Human Affairs* (University of Washington Press, Seattle, 1979).

23. J. Hanken and P. W. Sherman, Multiple paternity in Belding's ground squirrel litters, *Science* **213**, 351–3 (1981).

24. W. G. Holmes and P. W. Sherman, The ontogeny of kin recognition in two species of ground squirrels, *American Zoologist* **22**, 491–517 (1982).

25. W. D. Hamilton, Altruism and related phenomena, mainly in social insects, *Annual Review of Ecology and Systematics* **3**, 193–232 (1972) [reprinted in *Narrow Roads of Gene Land*, Vol. 1, pp. 270–313].

26. R. E. Page and R. A. Metcalf, Multiple mating, sperm utilization and social evolution, *American Naturalist* **119**, 263–81 (1982).

27. G. R. Buckle and L. Greenberg, Nestmate recognition in sweat bees (*Lasioglossum zephyrum*): does an individual recognize its own odour or only odours of its nestmates?, *Animal Behaviour* **29**, 802–9 (1981).

28. M. D. Breed, Nest mate recognition in honeybees, *Animal Behaviour* **31**, 86–91 (1983).

29. W. M. Getz and K. B. Smith, Genetic kin recognition: honeybees discriminate between full-sisters and half-sisters, *Nature* **302**, 147–8 (1983).

30. M. D. Breed and B. Bennett, Kin recognition in highly social animals, in D. J. C. Fletcher and C. D. Michener (ed.), *Kin Recognition in Animals*, pp. 243–85 (Wiley, New York, 1987).

31. P. A. Wells, Kin recognition in humans, in D. J. C. Fletcher and C. D. Michener (ed.), *Kin Recognition in Animals*, pp. 395–415 (Wiley, New York, 1987).

32. M. J. D. White, *Animal Cytology and Evolution*, 3rd edn (Cambridge University Press, Cambridge, 1973).

33. A. S. Pakstis, S. Scarr-Salapatek, R. C. Elston, and R. Siervogel, Genetic contributions to morphological and behavioral similarities among sibs and dizygotic twins: linkages and allelic differences, *Social Biology* **19**, 185–92 (1972).

34. R. Dawkins, *The Selfish Gene* (Oxford University Press, Oxford, 1976).

35. M. Ridley and A. Grafen, Are green beard genes outlaws?, *Animal Behaviour* **29**, 954–5 (1981).

36. D. Lewis, Genetic versatility of incompatibility in plants, *New Zealand Journal of Botany* **17**, 637–44 (1979).

37. D. de Nettancourt, *Incompatibility in Angiosperms* (Springer, Berlin, 1977).

38. B. F. Graham and J. H. Borman, Natural root grafts, *Botanical Review* **32**, 288–92 (1966).

39. J. E. Stone and E. L. Stone, The communal root system of red pine: water conduction through root graft, *Forest Service* **21**, 255–62 (1975).

40. B. H. Smith, Recognition of female kin by male bees through olfactory signals, *Proceedings of the National Academy of Sciences, USA* **80**, 4551–3 (1983).

41. C. D. Michener and B. H. Smith, Kin recognition in primitively eusocial insects, in D. J. C. Fletcher and C. D. Michener (ed.), *Kin Recognition in Animals*, pp. 209–42 (Wiley, New York, 1987).

42. P. A. Samuelson, Complete genetic models for altruism, kin selection and like-gene selection, *Journal of Social and Biological Structures* **6**, 3–15 (1983). The statement on p. 8 'The ultimate proportion of A_2's and a_2's are not predictable, since this depends on the happenstance of initial $[N_{ij}(0)/N(0)]$ ratios', is not correct. For any gene frequency at the locus of A_1, a_1, the fate of the locus A_2, a_2

must follow by the same reasoning that Samuelson correctly uses for the first locus. The fate is determined by Average $(l_{12}, l_{22}, > \text{Average } (l_{11}, l_{21})$. Therefore a_2, the suppressor allele, goes to fixation, and, as it does so, expressed altruism in the population vanishes. Linkage between the two loci does not stop this.

43. A. E. Beer, M. Gagnon, and J. F. Quebbeman, Immunologically induced reproductive disorders, in P. G. Crosignani and B. L. Rubin (ed.), *Endocrinology of Human Infertility: New Aspects*, pp. 419–39 (Academic Press, London, 1981).

44. V. L. Scofield, J. M. Schlumpberger, L. A. West, and I. L. Weisman, Protochordate allorecognition is controlled by a MHC-like gene system, *Nature* **295**, 499–502 (1982).

45. R. Lubbock, Clone-specific cellular recognition in a sea anemone, *Proceedings of the National Academy of Sciences, USA* **77**, 6667–9 (1980).

46. E. A. Boyse, G. K. Beauchamp, and K. Yamazaki, The sensory perception of genotypic polymorphism of the major histocompatibility complex and other genes: some physiological and phylogenetic implications, *Human Immunology* **6**, 177–83 (1983).

47. W. D. Hamilton, Extraordinary sex ratios, *Science* **156**, 477–88 (1967) [reprinted in *Narrow Roads of Gene Land*, Vol. 1, pp. 143–69].

48. S. T. Emlen, Cooperative breeding in birds and mammals, in J. R. Krebs and N. B. Davies (ed.), *Behavioural Ecology: An Evolutionary Approach*, pp. 305–39 (Blackwell, Oxford, 1984).

49. J. E. Neigel and J. C. Avise, Histocompatibility bioassays of population structure in marine sponges: clonal structure in *Verongia longissima* and *Iotrochota birotulata*, *Journal of Heredity* **74**, 134–40 (1983).

50. M. J. Carlile, Cell fusion and somatic incompatibility in myxomycetes, *Berichte der Deutschen Botanischen Gesellschaft, Berlin* **86**, 123–39 (1973).

51. M. Daly and M. Wilson, *Sex, Evolution and Behavior* (Willard Grant, Boston, 1983).

52. W. D. Hamilton, Innate social aptitudes of man: an approach from evolutionary genetics, in R. Fox (ed.), *ASA Studies 4: Biosocial Anthropology*, pp. 133–55 (Malaby Press, London, 1975) [reprinted in *Narrow Roads of Gene Land*, Vol. 1, pp.329–51].

53. D. S. Wilson, *The Natural Selection of Populations and Communities* (Benjamin/Cummings, Menlo Park, CA, 1980).

54. J. D. Mollon, J. K. Bowmaker, and G. H. Jacobs, Variations in colour vision in a New World Primate can be explained by polymorphism of retinal photopigments, *Proceedings of the Royal Society of London, B* **222**, 373–99 (1984).

55. W. D. Hamilton, Selection of selfish and altruistic behavior in some extreme models, in J. F. Eisenberg and W. S. Dillon (ed.), *Man and Beast: Comparative Social Behavior*, pp. 57–91 (Smithsonian Press, Washington DC, 1971) [reprinted in *Narrow Roads of Gene Land*, Vol. 1, pp. 185–227].

56. M. T. Ghiselin, *The Economy of Nature and the Evolution of Sex* (University of California Press, Berkeley, 1974).

57. R. Axelrod, *The Evolution of Cooperation* (Basic Books, New York, 1984).

58. This note partly explains the digression. I want to mention an animal named for

me and even more to express my loss from the death in 1986 of O. W. Richards, who gave me a job at a time when no other biologist that I knew thought that animal nepotism was a 'worthwhile subject of study'. The species of the *Azteca* is not known; the wasp is *Stelopolybia hamiltoni* O. W. Richards. The collembolan is some species of *Cyphoderis*, as Richards told me. It was O. W. Richards also who knew, and also spelled for me, *Platyarthrus hoffmannseggii*, which had made a white dust as of mealy bugs among the black *Lasius* ants in forked garden soil; who knew *Leptinus testaceus*, when I only described it, beetles blind as the white woodlice but no symphiles of the bumble bees in whose nest they were scrambling, instead parasites of the mice that had been there before. So with many others I learned from the man whose pockets, even whose dinner jacket, never lacked a specimen tube, and whom now I can consult no more.

59. R. S. Solecki, *Shanidar: The Humanity of Neanderthal Man* (Allen Lane, London, 1971).

60. R. Pitt, Warfare and hominid brain evolution, *Journal of Theoretical Biology* **72**, 551–75 (1978).

61. H. Szarski, Why did the human brain cease to increase 100,000 years ago, *Bulletin de l'Académie Polonaise des Sciences, Classe II, Séries des Sciences Biologiques* **29**, 381–3 (1983).

62. D. R. Vining, Group selection via genocide, *Mankind Quarterly* **21**, 27–41 (1982).

63. E. O. Wilson, *The Insect Societies* (Harvard University Press, Cambridge, MA, 1971).

64. I. Eibl-Eibesfeldt, *The Biology of Peace and War*, transl. by E. Mosbacher (Thames and Hudson, London, 1979).

65. M. D. Sahlins, *The Use and Abuse of Biology: An Anthropological Critique of Sociobiology* (University of Michigan Press, Ann Arbor, 1976).

LAND OF
THE RISING SUN

Kinship, Recognition, Disease, and Intelligence:
Constraints of Social Evolution

Fortune is a blind god,
Flying through the clouds,
And forgetting me
On this side of Jordan.
ANON

O N A H I G H hill top that overlooks Kyoto city, ancient capital of Japan, I watched free-living Japanese macaques for the first time. One of a crowd of scientific tourists, I had poured from a bus onto the flat, gravelled open space surrounding a kiosk and café that crown the hill and found the monkeys all around me. They gave us brief glances and moved quietly out of the way if we came near but much more they seemed to be watching the kiosk. Obviously they were free-living on this hilltop but right in this place could not be called exactly wild; however, the steep slopes over the edge of the small plateau were densely wooded and possibly the troupe obtained most of its food from there. In the kiosk you could buy drinks and small food items; I remember being surprised to see canned *guarana*, that caffeine-rich fruit drink from fruits of a liane whose black seeds in the Amazon forest look down on you like pupils in blood-shot eyes. I knew the drink well but thought it had remained endemic to its native Brazil. I don't remember whether some items in the shop were intended for feeding to the macaques. But even if feeding them is forbidden it obviously happens, and to protect both tourists and the café's wares the building is enclosed from eaves to ground with stout chain-link fencing wire.

When our party arrived the macaques were sitting, wandering, grooming, nursing infants, and, as I said, just watching. For these first few minutes I saw indeed the peaceful species I had read about. Then someone threw something out from inside the wire mesh of the kiosk, a scattering of small objects—peanuts, I think—and within a second all had changed. My memory gives what I can only describe as a swirl of grey fur and red faces rushing to the spot where the first lucky animals were already grabbing the prizes from the ground and scampering away with them pursued by others. Most of the later arrivals rushed into the grey generalized fight at the centre, but some sprang onto the vertical cage wire of the café and fought for places near to the crowd on the ground. Yet others flung themselves up and onto the roof where, spacing themselves along the eaves, they eagerly watched from above, perhaps trying to spot where the nuts or whatever were rolling to concealed places. Perhaps they would memorize these and search for them later. Because those on the vertical wire were well in view, I watched them the most and it was here that I saw one action that has stuck in my mind ever since. A female came along the wire in a swift horizontal hand-over-foot progress and as she passed another who was holding a baby she seized the clinging young one out of her arms. Hardly pausing in her progress along the wire, she swept it to her mouth, bit it, and threw it to the ground. The baby screamed, the mother after a momentary unsuccessful swipe and grimace at the passer, jumped down into the crowd to rescue it. The perpetrator ran on and up onto the roof.

These, then, were the peaceful, constructive, social primates that had been pictured to me in the ethological papers of Professor Kinji Imanishi and his followers. One of his group had come with us in the tourist bus to be our guide and he now stood back watching us watch the monkeys. As if he had known what our first impression would be his eyes seemed to me sad. Imanishi's group had recorded how the monkeys are socially patterned by their matriarchs, how they largely peacefully inherit rank from them, and how they follow and are protected by leader-like young males in dangerous situations. They had told how troupes produce, very occasionally, geniuses who discover novel and advantageous habits that the rest of the troupe then learns—the washing of food items before eating them in the fashion of an American racoon was one such discovery.

I am confident that all these observations are correctly recorded and that the social organization of the Japanese macaque is indeed very complex, perhaps even humanoid to a degree unusual for monkeys. In the troupe I was observing, after peace had been restored, I saw a trait that is even stranger perhaps then the food washing, or the hot-spring bathing I had already read about, its very strangest feature being that it had no utilitarian function at all, thus making it almost a frontal challenge to the 'sociobiology' of old Bishop Wykeham that I mentioned in the last chapter. In short, it seemed a possible case of 'Manners Makeyth Monkey'. The habit appeared confined to a few youthful males, who, in the periods of relative peace in the group, would gather and play with small piles of stones—a bean-sized gravel that they would heap together, toss lightly, re-gather, stir, then, perhaps, gather into a handful and carry to another site where the play would continue. The activity seemed not at all social and this probably disqualifies it to contend with the Bishop's motto. It seemed to me more like a compulsive, lonely play activity of a shy child, reminding me likewise of pointless routine actions I have developed myself from time to time, especially in my childhood: 'Don't keep doing that, Bill—are you batty or something?', my father would say as I made odd clicking sounds or for hours kept bouncing a ball into a corner to watch the spin. I didn't see any old macaque ticking off these young males for what they were doing, but then why did they almost furtively move their play site from time to time? Apparently the habit had been 'invented' by one young male and had spread to the others.

At last, after the peanut shower, peace again prevailed and I wandered slowly among the animals, staring at many a red face and being very coolly stared at in return. I noted how easy it was to tell individuals apart and this principally just by the faces, just as with us—indeed one couldn't escape the thought that they looked more like Europeans than Japanese. Pausing by a particular player with his stones, I reflected that what was remarkable in all the published portrayals of this most northerly of all monkeys was not exaggeration of these constructive and idiosyncratic features that the Imanishi group had recorded but rather that they had left completely off the record all other traits that might be called 'dark' from a humanistic point of view; indeed they explicitly denied such other sides existed. I had seen no mention in any account I had read, for example, of a single instance of the explosive and spiteful strife such as that I had just watched over the peanuts.

As if reading my thoughts, Dr Itani, who had been the first recorder of the strange stone play that had arisen in this group alone, came towards me. We started a conversation and he told me that he thought it a great pity about the feeding by tourists and the presence of the café; both made the monkeys behave unnaturally. He said that because the place was so near to Kyoto, such feeding was inevitable—the visitors liked the monkeys to be visible and not too shy. I asked if there were wilder groups to study. Shrugging, he said, of course; he had studied those and he was no longer particularly concerned with this one; when he had been working elsewhere, he had tried to minimize the unnatural influences.

But troupes with no cafés on which food lusts can become focused, no overcrowded spaces around them,[1] no tourists to provide 'fallen' models for imitation—would such troupes really be much more peaceful and uncompetitive than what I was seeing? Was it just our own sin—original as called, compounded, etc.—being reflected back to us by these macaques at Arashiyama? Or was the violence and the selfishness actually more deeply original than the books admitted? Perhaps where food gathering is more arduous, less surplus energy exists to be diverted into such strife. Perhaps wilder troupes living in places where stress from the sheer capriciousness of tourists never touches them and any stress they experience comes more from the harshness of nature are truly more peaceable and co-operative. I doubted, however, that they were much more peaceable or that selfish acts among such wilder group were rare. All my readings in primatology, wild and captive, and even my own observations elsewhere, have suggested otherwise.

My intention in describing this visit to the Arashiyama hilltop and troupe is not just to pass on random notes of a biological tourist. The descriptions the Japanese behaviourists had given of these monkeys are closely connected to the reason why the Kyoto symposium of 1986 was being held and why I was invited. Hence they are connected to the origin of my paper of this chapter, and also connected to those philosophic and comparative issues about 'truth' on which I closed my introduction to Chapter 9.

The symposium marked the approximate mid-swing of a pendulum in Japanese evolutionary and ecological studies—or maybe not so much a pendulum as the click of a ratchet. Through critical decisions that seem to have started at several centres within Japan, Japanese ethologists were busy transferring from what may be described as a 'panglossian' account of

behaviour and ecology to what I would call a realistic one. Or, as many of
the Japanese attendees expressed it to me, they were transferring from a
tacit, though in some cases reluctant, allegiance to the views of prophet-
biologist, Kinji Imanishi, to the acceptance of objective evolutionary
science. Respect for authority is usually high in Japan and custom seems to
affect the paths of thought more strongly than it does in the West.
National levels of education are high, however, and Japanese scholastic
endeavour has been becoming ever more linked with that of the rest of the
world. Educated, progressive people cannot be persuaded that black is
white indefinitely, and especially not if the traditional acceptance seems to
be leading nowhere. Then the overthrow of custom seems to occur as
swiftly and as unanimously as it was being resisted before. A social example
long ago in Japan was the switch from being a pro-natalist society before
and during World War II, when they experienced consequently a great
growth of population, to becoming an anti-natalist one just after the war
with abortion suddenly made legal and free for all. As a result of this
change, birth rates plummeted and overpopulation was checked at once.
During the change, it is worth mentioning, parenthood continued to be
very responsible and the general attention to child care admirable;
moreover, so far as I know, no hospital clinic performing abortions was ever
bombed, nor a doctor performing the operations shot.

Imanishi first printed his disbelief in the existence of competition
between organisms before the war but his series of books and his rise to
power in the university system, in which, by being appointed professor in
Kyoto, he soon attained a dominating position, began after the war. It was
the same period I mentioned above, one of radical national reconstruction.
By 'dominating position' I am far from implying any malign drive to
disempower opponents (such a drive as had, say, Trofimov Lysenko when
he became top plant breeder in Stalin's Russia).[2] Imanishi seems indeed to
have been as a man almost as benign as his beliefs and, in fact, might be
considered a strange case of thriving by not competing himself, just like his
theory believed. Nevertheless, his persuasive writings plus the ready 'faith'
that others gave to him, seemingly stifled competing lines of thought. His
ideas reached out to all educated people from his popular books whether or
not they also encountered those books at a university. The culture as a
whole, from housewives to professors and on to demobbed kamikaze pilots
(classes which, by the way, must sometimes have overlapped—the last-

mentioned were certainly not the brainwashed fanatics that are sometimes presented to us[3]), all wanted to believe in the biological 'facts' Imanishi was presenting. As if in response, as his career progressed, he ranged onward to paint his 'ecology' with still freer brushstrokes and, as one might also put it, a more Picassoesque licence to depart from reality. From pre-war work on niche separation among mayfly immatures in streams and on the ecology of trees on mountain sides, he moved, after the war, to study feral horses and thence to the Japanese macaques; finally, he confined himself purely to didactic writing based on his reading and synthesis of this with his (supposedly) carefully observed biological examples.[4]

In bomb-shattered post-war Japan, Communism and American-style Capitalism, the religions of the victors in the war, circled each other warily. But the rapidly changing culture came to no point of Macarthian or Lysenkoist explosion; rather it is as if somewhere, hidden discretely perhaps in the mists over the Sea of Japan, the alien spirits coupled and gave birth to the hybrid that, in the sphere of sociology and understanding of evolution, was to take over.

There was a great need for all people to pull together for the process of rebuilding a bombed-to-bits and disheartened country. In a rather similar way in the early years of Bolshevik Russia, a revolutionary demagogue and writer, Prince Kropotkin, died honoured by millions for dreaming up a biology where animals of the same species strove only against their inanimate environment and never against each other. Darwin, Kropotkin insisted, had been wrong. The facts of nature showed that only survival of the fittest species could ever make any sense. Even that competition was dubious for the prince: for example, he became vague even about the antagonisms that might be involved in food chains. One book tells how 'the falcon' (an unhappy exception in being, like many top predators, rather solitary animals) lives by 'brigandage' and because of this was 'decaying throughout the world'.[5] Well, as if obedient to his prophesy, it is true that in 50 years falcons really were almost dropping from the skies of almost all the first world nations; but I think now we would all agree that their difficulties were due to organochlorine pesticides and certainly not to 'brigandage'. Should we admit from this example that Kropotkin had the knack of the truest of prophets in being right for the wrong reasons? I think not because, while still quite unrepentant in their carnivorous habits, kites and falcons are coming back and this is undoubtedly because DDT and like pesticides have been banned. To illustrate, in November 1999 I watched five red kites soaring over the Chilterns in the mere 5 minutes it took me to drive under them on the motorway to Oxford; 20 years ago I would have had to go to Wales to see perhaps two, and those in a special spot. Nevertheless, it does seem that in spite of such obvious nonsense as got into his books, Kropotkin's *Mutual Aid* provided to the revolution in Russia—and for the whole reconstruction period that followed it—an effective biological propaganda and a suitable model for social consumption. He provided not facts but a religion that persuaded people, and that worked. Likewise exactly, it seems to me, in post-war Japan (and quite independently of Kropotkin so far as I know), Imanishi prescribed another very Kropotkin-like biology for similar circumstances in Japan.

Does truth by faith always replace truth by evidence in such reconstruction periods? Does telling, and even believing in, social untruths—lies, as one might put it, of the people, by the people, and for the people—sometimes become almost an intellectual duty, with reason to be ignored? If Edward Wilson's *Sociobiology* had been fully spelled out by those great Neodarwinists, R. A. Fisher, J. B. S. Haldane, and Sewall Wright, during the immediate aftermath of World War II, would Europe have been ruined again, Europeans floundering still in strife like Yugoslavia is now and riven all over again by the old national discords?

To judge by the writings of some of the critics of *Sociobiology* one might think so. Would the 'economic miracles' of Japan and Germany never have taken place? Possibly the answer might be 'yes' to all these; but, if so, the devil has a price he demands of us sooner or later for all unscientific prophesy, a bill that comes to us eventually in two forms. First, there is the cost of the failure of the predictions themselves; for example, crime is predicted to wither if we behave as directed by a false hypothesis and crime in the actual event doesn't wither, instead thrives along with the genes and memes that create it, with these last making the change hard to reverse—a situation that I believe is the sad case in Russia today. Second, there is a cost in increased load to society of the prophet's own memes and genes—as (looking to wider issues of the more distant past), for example, through the enlargement of priesthood—doctorhood—bureaucracy that the revered prophets create, as also, sometimes, through personal contributions of the prophet's own direct or collateral family. The last-mentioned personal type of addition must be minor for a long time regarding the effect of any one prophet but cumulatively may slowly become important, as seems visible in the inclinations to religiosity and fanaticism that plague India, Iran, and Israel today, all these places being dominated by religions that explicitly link fecundity very directly with the practice of religion (all successful religions, however, undoubtedly do the same in less explicit ways).

My criticisms of Western popularist biologists and my decision, in the last chapter, that they must be judged to be more like 'drive memes' themselves, rather than be seen as 'guardians' of drive prevention, apply closely to Imanishi. In a thoughtful, personal review of Imanishi's work and influence, a former admirer, Yosiaki Itô, admits at the end to more potential danger from a popular misuse of Imanishi's philosophy in the form of extreme political movements than of any similar misuse becoming based on sociobiology.[4] I also have pointed out in Volume 1 of these memoirs how uncritical belief in group harmony has often provided a sure and followed signpost towards totalitarian and warlike policies.

How objectively Imanishi ever looked at the freshwater stream organisms from which he gained his reputation, or looked years later at his macaques, I find it hard to know. For both he managed to see a complete absence of competition and instead saw some sort of a pact sealed in heaven to attain optimal productivity. Stream detritus and filter feeders are superficially promising for this view: their food can be imagined flowing past them; they have no control over what comes and that which they fail to catch goes on where they cannot follow it; all they need, it might seem, is their crevice in a stone and beyond that maybe would do well something like a Tanuki statuette at that crevice's door (see below) to invoke Good Luck—good things to happen to come by. But what if Imanishi had made even so small as change as to base his research on caddis in streams, instead of on mayfly larvae? For those, with silk-built and silk cemented houses so often, so obviously, crowded on the stones (where they are sometimes so abundant that they can, in parts of Japan, be collected as human food), I would have thought the matter of competition for space at least was plain . . . But once he had focused a highly evolved quasi-rational species like that I saw thronging the hill top of Arashiyama, it is still more hard to see how he could have failed to perceive pairs at least, if not whole groups, to be in very direct competition.

This is not to say that Imanishi's influence on the study of monkey behaviour *per se* was all negative. There are aspects of his ideas I heartily endorse, and without doubt he founded a very energetic school. For example, differently from most animal behaviourists in the West at the time, he believed that it did no harm to anthropomorphize one's animals and to give names to mark them as personalities, thus increasing one's insight. He also believed, in contrast to the 'snapshot' accounts popular in the West, it was necessary to follow the personalities for long periods so that snapshots could extend to become sagas often over several generations—like Russian families in *War and Peace*. Kinji Imanishi and Jane Goodall were twin revolutionaries when it came to this naming and personification, but also to that unabashed feeding that brought the animals into sight where they could be watched and so their thoughts imagined. I have even wondered whether Goodall can have been inspired in her methods through reading the earlier Japanese accounts (the earliest of all, unfortunately, up to about 1965, had been in Japanese).

By 1986 Japan had made itself again a world power and despite the destruction of so much of its industry, even a leader in technology in many fields. By a high average intelligence, by adopting among themselves Kropotkin–Imanishi style of cohesive action, by repressing (perhaps) all elements of imagined 'falconic brigandage', excepting such forces as they could aim outwards from their shores (as towards whales, for instance, these living very far off and in pan-competed seas), and by a degree of voluntary effort that may be unprecedented in civilized history, Japan achieved its economic miracle. Doubtless a criminal underground meanwhile was quietly taking advantage of all the mutual trust, patriotism, and hard work in the way such criminal subcultures always do; and doubtless also more legal but still selfish entrepreneurs flourished relative to the less-selfish ones. Doubtless in everyday life other kinds of individual selfishness successfully propagated a little extra at all levels in the trusting atmosphere. But these changes have not been particularly apparent on the surface and the Japanese remain far above the average of nations in their respect for law and orderly behaviour. I cannot but wonder, however, whether by the 1980s when our conference occurred there had arrived a national perception that through their teaching that selfishness would die out without special intervention, they might instead be helping it to flourish. A rash of exposures of corrupt businessmen and politicians in the 1970s and 80s might, for example, have suggested this. In any case, long before the symposium of 1986 that I attended, attitudes towards the truths of animal behaviour had indeed started to change.

Beginning in 1983 a special research project involving more than a hundred ecologists and ethologists was run under the sponsorship of the Japanese Ministry of Education, Science and Culture on 'Biological Aspects of Optimal Strategy and Social Structure'. The Kyoto symposium of the same title that I attended was the terminal flourish in this project. A dozen or so evolutionary biologists from the West, myself among them, all of us having been active in developing and testing the newer kinds of Darwinian social thinking, were invited. My suspicion is that the title was a little euphemistic and that simpler, more truthful ones could have been 'The Positive-Sides-of-Sociobiology-and-ESS Symposium' or even 'The Anti-Imanishi Symposium'. Anyway, that was how it seemed to work in practice: the displacement of Imanishi as prime spokesman for the 'Three Es' (Ecology, Evolution, and Ethology) in fact occurred in Japan: it was conducted in an atmosphere of exemplary unselfishness and mutual aid by a whole generation of young biologists, these aided by just a few of my own generation or older men such as Yosiaki Itô whom I came particularly to admire.

This first visit to Japan was for me at least a stunning cultural experience and one that became greatly enhanced by several trips after the symposium to give talks in other universities and to experience Japanese countryside and natural history. The longest such side trip I made was with Naomi Pierce and Andrew Berry (both of whom were working at Oxford at the time) to Taiwan. Shigeyuki Aoki, discoverer of soldier aphids (see Volume 1, p. 266), was our guide and our purpose was to see some of the world's most dramatic soldier-defended galls. Another trip was by me and Mary Jane West Eberhard, with Yosiaki Itô this time as our expert guide, to see social insects and some other lines of insect research on the island of Okinawa. Itô was one of the organizers of the conference, and ever since I had met him in Oxford the previous autumn had done a lot towards the elimination of some of my more stereotypical misconceptions about the Japanese. These, perhaps, had been conditioned in me in part by an uncle, a New Zealand land surveyor, who had been a prisoner, throughout World War II, of the Japanese in Changi Prison Camp on Singapore Island. A character or a physique more fitted than Itô's to break down stereotypes of any kind is hard to imagine; but hard to imagine also is how such a lighthearted and yet simultaneously stern rebel could come to be serving at the centre of an organized revolution. Perhaps because of Itô and the Okinawa trip, and perhaps especially because of a certain morning when Mary Jane and I visited (I for the first time) a coral beach, my whole experience of Japan that time is likened in my mind, dreamlike, to a long swim in a warm and strange sea—far out from land I was but in a buoyant and unthreatening water, thinking deeply on the great unity of mankind.

Of sharks during that metaphorical swim I had no fear, but the swim was not quite free of contacts with what I might call sharp corals—if I may so liken some of my moments of embarrassment arising from misunderstanding of Japanese customs and sensitivities. Everyone was helpful and had an obvious (an even slightly disconcerting) expectation that the visitor would not understand their world and would be helpless to live in it on his own, so that their assistance to us was vital. Unfortunately it was perfectly true; the visitor wasn't able. As we arrived at the university hostel in Kyoto each of us was issued with a card to use in the event of our getting lost during any excursion. We were told to show it to any passing taxi driver and that it said in Japanese something like 'Please drive this poor man to the door of the . . . Hostel'. Sometimes, however, the man couldn't avoid asking something of an ordinary passerby or shopkeeper for himself: then the efforts of the stranger to extricate and to help, even in the face of an insoluble language problem, were extraordinary and endearing.

An employee of the Shinkansen 'Bullet' Train might look angrily at me if I caused delay for one second through not having my ticket ready or by not standing between the right pair of white lines on the platform; but when, after leaving the ride, a similarly uniformed and grim brother of that same official[6] had finally managed to understand my sign language and a desperately emphasized bus ticket, and saw the name of my destination, he seized the heaviest of my suitcases and ran with it for a hundred yards to the correct long-distance bus that was already puffing its great diesel in readiness for departure. There, having frowned severely at my tip offered for his efforts, he departed with bow and smile, clearly well pleased by his contribution to the efficient working of his country. Perhaps this man's job in the bus station was indeed simply to wander about and look for helpless and lost *gaijin* tourists like me; but I still have to say that he helped with a physical feat of instantaneous and disinterested exertion that I have never encountered elsewhere.

Atmospherically, and especially when one is off the main road and into the abundance of its steep hills, Japan is just like the colour prints out of the nineteenth century. Its sky is strong and deep, and pours down on you alternatively very hot sunshine, warm mists, and very heavy rain. Both before and after the rain the high black wooded hills stand streaked with white scarves of mist, just like the mountains in the famous woodcuts. Lines of the land, too, like the people, seem especially purposeful: a Hokusai descendant travelling today would be more likely to divide his picture by the bold diagonal of the jib of a crane, or by a crowded motorway, or by an electric pylon viewed from an unusual angle, than he would be to use the old traditional objects and artefacts—the band of mist, slanting cherry tree, or a tilted log that workmen, above and below, are hand-sawing into planks, but otherwise the idea would be the same. The flats and the mountains of the land are still there, but among them again our reborn artist is likely to choose as foreground for his distant Mt Fuji not that old trio of peasants in conical hats, instead some weird tractor trimming tea bushes on a steep slope or a quite different one floundering sprayfully but efficiently across the mud of a tiny rice field. But then again, if you spot this artist at all, this Hokusai the Second is more likely to be noticed for digitally electronic cameras that swing from his shoulders, and more likely to be riding a motor bike than to be on foot, satchelled and with his sketchbook or carrying hand-carved wood blocks.

The rice fields and the tea plantations on the slopes are still there. Tea bushes, for example, made the dark stripes on far low hills or else flashed past the train or car close by like groups of resting furry green caterpillars, while, all along that grim yet very colourful megapolis that stretches down almost the whole length of the eastern plains of Honshu where the train speeds, you see lambent and paler green polygons flooding all spaces between buildings. Rice, I noticed, had still to be planted in every wet plot, the plots sometimes filling such small spaces between buildings that little can be gained by not cultivating the plants by hand. I was told that growing rice has been almost a national religion ever since the privations of the war, and that, incurring an uncharacteristic diseconomy, the government still tries to avoid dependence on imports of this most vital food.

Deep as yellow manioc is for Amazon caboclos, or wheat and bread for Europeans, an image of rice must be crystallized deep in the Japanese psyche. Any lack of these familiar basics in any culture implies hunger. But perhaps even here a revolution is coming. In the hills of Mishima late one afternoon, I watched a red sun light up the topmost cone of Fuji soaring over the haze of the great rice- and town-filled plain I had come from, and I bought for my family what seemed to me like the discarded 'crystals' of a culture that was almost past—they were extracted from the string-tied packages of porcelain rice bowls in a shop that sold nothing else. Going at about four bowls to an English pound, they were the only cheap souvenirs I have ever found in this land of the mighty Yen, either on that trip or any other. I had just that day visited Motoo Kimura at the Japanese Institute of Genetics and in the street with the shop I wondered whether his very institute might be gestating some change for the land that this shopkeeper knew about. Did he know that genetically engineered horse chestnuts soon to be put out by the institute, perhaps, were about to take over as the national staple, or nettle-seed porridge—foods that he could see were sure to need different vessels? Unless foreseeing a coming inutility, why sell hand-painted eggshell-china rice bowls at such a low price? Its a fantasy of mine for sure but, with so many surprises seen already in Japan, no new one seemed impossible.

So much for my wilder and wider impressions of the country: what about our conference? As commonly for me it is through a small item by a person I didn't know that I remember the meeting most vividly. Itsuko Yamamoto, a young scientist from the University of Kyoto, spoke and showed a film about her observations on the birthing of female and male racoon dogs, *Nyctereutes*.[7] Birthing by a *male* racoon dog—will that stir a sleepy eyelid of my reader just as little? Well, it's almost birthing by a male. And it has to be said that Mr Nyctereutes, the male 'tanuki' as the Japanese call him, is a 'new dog' in the most modern sense of 'new man' without any question. The light conditions that would allow Ms Yamamoto to shoot her video and yet not be disturbing to the animals had to be very low so that the male I watched in the film was just a thick-set furry shadow, almost a black ball rolling about, but, ball-like and mobile as he was, he contrasted sharply with the stationary curved shadow of his recumbent 'wife'. He fussed to and fro, into and out of the box the zoologists had given them; he brought more straw, licked his mate front and rear ends, and, as she increased the moans of her parturition, seemed to fuss ever more frantically—lay beside her, got up, fetched yet more straw . . . The female, too, was astonishing. Her cries during this process were so humanoid that I found myself having an irrational urge to rush up to either the screen or to the projector to ask if I could help with something. Finally, when the first pup was fully born I could see the male licking a new black spot on the straw as if he could never stop. 'Vocalisation by the newborn seemed to induce more active male direct care', as Yamamoto put it in her report, 'and the first pup was intensively licked dry by both male and female. After a while the female began to give birth to a second pup. Meanwhile the male continuously licked the first pup while huddling . . . with it.' In this wise Mr Nyctereutes played mate, midwife, and—almost—mother.

As Yamamoto's paper tells, in one pair the litter was of five with one pup stillborn, and two others soon to die. To the other pair she observed four being born and all were still alive at the time of our meeting. Fecundity is clearly high compared with us humans but so is early mortality: with us it is as if the mother's physiology detects the least viable in her 'litter' while they are still in her womb and rejects them (see also Chapter 3 of Volume 1 and Chapter 12 to come) whereas in racoon dogs the same elimination of sex's worst combinations and worst mutants seems to happen after birth—it would happen as soon as possible according to the evolutionist's expectations (Volume 1, Chapter 3). It is very clear from the behaviour, however, that once born a pup is treated as a very precious asset by both of its parents. As Yamamoto's summaries show, after the birthing is complete the mother spends only a little more time in the box than the male does and, later, it is always the male who spends more time there. Possibly this is unique among mammals. Part of the reason for it probably lies in racoons' food being of small items—fruits, insects, and other small animals, which the male cannot carry in bulk back to his mate in the way other canids carry meat. It is best, Yamamoto suggests, for the milk provider to go out for her own food while the male guards and warms the litter, as she recorded him doing for 60–80 per cent of his time. Yamamoto's film went bang for me like popped balloons and with it went some other Wykehamish mottoes that I had vaguely dreamed up for our species. 'Travail Makeyth Woman', for example, was obviously a non-starter for uniqueness. Worries over travails of Woman weren't uniquely making Man either; if it made 'Man' then it had to make 'Racoon Dogge' too.

So should the species of that rounded and furry shadow I watched behaving so well on the video be chosen as the ideal masculine mascot for the feminist movement? I doubt this or at least I'd advise caution; biologists can perceive elsewhere some hints that other humanoid traits may be concealed up this animal's furry sleeves. The Japanese seem to have long loved the tanuki: dog, racoon, humanoid—all merge in those statuettes for good-luck that stand outside thousands of shops in Japan, some of them in towns but still more in the rural areas. Often the statuette lurks by doors to small wayside shops and eating places, a thick-set, jolly figure, a black animalized Silenus, his cheerful and short snout beckoning you, and while he squats so erect and humanoid he is also obviously very 'erect' also in a quite other well-known sense as well. What catches the eye most of all, however, is just below the primal male organ—a truly gigantic and well-filled scrotum. Exaggerated these parts may be in the statuettes but they may well be founded on relative realities of the animal. Because it is now agreed among evolutionists, on good evidence,[8] that large testes go with voluminous ejaculate and that this in turn accompanies promiscuous sexual habits, and because, as I say, folk representations of animals often have at least a comparative truth to them, we had better be careful about imagining that the racoon dog male is a model of monogamous fidelity. Does tanuki society suffer, perhaps, a 'single-mother' problem that Dr Yamamoto hasn't yet detected?

Because of its introduction into western Russia for the value of that furry back and sleeves that I had detected rather vaguely in the video, the tanuki has made a huge jump out of its original East Asian range and is now spreading in eastern Europe. He (and, of course, she) arrived recently, for example, in the great woods of eastern Poland and has reached also to Finland where, sad to say, he (and perhaps still more in this case she—in proportion as he still, in Europe, is spending more time being the 'house-husband' than she being housewife), is getting a reputation for killing domestic and game birds. But the arrival now makes it possible for European biologists to participate in unravelling whose mascot, if anyone's, these ambiguous humanoids deserve to be. For the present the Japanese choice is to make the male be a mascot and a pet simply to Lady Luck, and this seems a safe one and, as always, an ascription difficult to disprove.

The symposium pot-boiler that follows is designed, like that of the last chapter, to be easily read. Once again I had been asked for an overview of kin selection and once again I chose to evade duty and to sketch where I felt the centre of interest in social evolution was moving. Why, for example, are the great potential advantages of close relatedness in social behaviour, which I had explained in my first papers, so often neglected in evolutionary social developments? The answer I give is connected with my new ideas about sexuality and also with my ever stronger conviction about the importance of parasites. To present this key issue yet again, why does a mother, for example, give birth to offspring only half related to her instead of creating ones fully related? For sure an individual mother can't normally choose but over evolutionary time the options are undoubtedly there for her. My answer is that otherwise those diseases, which, after the ups and downs of local natural selection of their own, have succeeded to crack the 'code' of your generation, will find they have cracked also that of all our offspring as also all your further descendants. All down a susceptible mother's descendant line these diseases will have their way unchallenged. As another example, why, as a worker honeybee, do you put up with sisters related to you by only about three-eighths of their genome when you could (during previous incarnations of your genes in queens) have made them related by three-quarters? Here it is your mother indeed who set up the situation but her decision is likely to be mostly in your interest as well as hers, as we see by thinking through the inclusive fitnesses of the case. If she hadn't gone for the orgy of polyandry over the trees, philandering with (fatally for them) 15 or so out of a great comet-tail of competing drones that attended her, it is more likely that you wouldn't be working to raise even siblings because you would be dead—you'd have ended, for example, as a stinking liquid sack of 'foul-brood' bacteria dangling out of a wax cell instead of being the well-formed adult worker that you now are. In your life as a larva with half sibs pressing around you sired by the dozen or so fathers and partitioned away from you only by the delicate wax walls, it is much less likely that any single strain of a deadly bacterium could make a quick culture medium of the whole brood comb. In short, the theory suggests that polyandry gives the average strain of bacterium less chance to spread and this makes everyone safer.

These explanations are probably regarded by most biologists as still speculative but increasingly they seem accepted at least as a hypothesis worth testing. The idea of the mixed genotypes in a population stopping an epidemic has, of course, been understood for some time by agronomists and plant pathologists (probably in animal husbandry scientific circles, too, though on that I have done less reading). Even as I prepared to ignite and send up my first balloons of advertisement concerning this anti-parasite heterogeneity and its importance in the problem of sex, I was finding plenty of current interest in very disparate applied journals about the so-called 'multiline' plantings of crops, the aim of which was similarly to check disease spread (and ultimately thus reduce reliance on chemical control agents). A parallel notion that I mention in this chapter's paper has its roots considerably farther back—to the 1960s at least;[9] but it would be surprising if it had not been kicked around at some time earlier still—inevitably it would have been by Haldane in the Red Lion Pub, I would imagine. I guess I first heard the parody from physics that I use in the paper, that 'Nature abhors a pure stand' (it's 'vacuum', of course, in physics), when I was an undergraduate doing botany at Cambridge but I don't remember hearing anything then that connected the idea with parasites. Ten years later a kind of 'multiline' approach to the great multispecies diversity problem had appeared in independent publications by Joseph Connell and Daniel Janzen.[10] In applying the same idea at the intraspecies and variety level to the high polyandry of large-colony social insects, the first mention may be mine in the coming paper. Very shortly afterwards, however, and certainly without having seen my piece (which is unsurprising given where this was published), Paul Sherman, Tom Seeley, and Kern Reeve[11] conjectured exactly the same resolution for the puzzle of polyandry of the social insects and gave a better survey of the evidence. In 1998, what we may perhaps, in view of the parallelism, now call the multilineal notion of polyandry against parasitism, has entered a new phase of lively controversy about its evidence and predictions.[12]

The main basis of criticism has been that it is hard to see enough destruction of wild social insect colonies by disease, leave alone to see this destruction proven to be correlated with low polyandry. Here I only comment that it is easy in any census of a wild population to overlook failing colonies; really, for honeybees, for example, the fate of every swarm needs to be traced and something like a doctor's death certificate produced for every failing colony. It needs to be a careful one, that doesn't overlook the fact that although invaded and finally destroyed by a honey-badger, the badger dared to attack only because the colony was weak due to its parasites. As a parallel, in early days of kin selection it seemed similarly paradoxical that so many multiple-foundress nests of the wasp *Polistes* showed colony growth rates that certainly didn't, via arguments of inclusive fitness, justify acts of joining; but when more careful surveys took into account the rates of complete failure of single-foundress nests, usually due to predation in this case (though, again, we must be aware of the subtle disease influences that easily underlie this), the paradox disappeared.

As instanced on a small scale by the near simultaneity of the publications of my version and that of the Sherman group for the role of polyandry, separate origins of ideas are common throughout the history of human thought—Newton and Leibniz separately inventing the calculus is one well-known example. Such trends put (or ought to put) a damper on too much credit to first discoverers, especially when, as with my writing just a few lines on honeybee polyandry in my paper here, a first 'discoverer' often scores just one hit by one shotgun pellet out of the great blast of his misses; hence more credit is due to those who put together the painstaking evidence for how the same subject really deserves widespread attention.

So it may turn out eventually with my humorous yet semi-serious remarks that I place at the end of this paper about what may be going to happen when unlimited cloning becomes available to humans. That such a technique must be well on its way with no obvious barrier was clear enough to me when I wrote. For some time I had been giving an opening salvo and some humorous remarks on this in seminars, with the theme of how my work, as yet theoretical only, certainly had an applied aspect: I told that I was racing to show what the human male sex was good for in evolution before new cellular techniques brought together with feminist man-hatred sealed my sex's doom. 'That is, the sex of half of you in this audience', I would say ominously, drawing usually as many nervous sighs as laughs, probably because the idea that the male was a cheating and relatively useless sex (at least for the immediate future) had not yet sunk in. After it has, and the average person reacts with horror at the idea of humans being able to clone themselves and being allowed to do so to any great extent, it is probably because of a vision of youthful Howard Hugheses in multiplets strolling out from some palatial home in the USA. Cloning labs are perhaps imagined as built on at such a home's back—in other words, people are thinking primarily in terms of what *male* millionaires might dream of and pay to have done. And I can well imagine unusual men with enough wealth and power who might like to try this: Howard Hughes himself might be one and his phobia for germs might make him good at handling the sterile glassware that was necessary.

If this sort of wish for descendance is a danger at all, however, it is more likely, given equal access to the technology, to emerge from the likes of Germaine Greer—women so much involved in the battle against men that they have never thought of babies by the normal means until it is too late. The essential point, though, is that men need women more than women need men: women survive better, are more stable, more happy lacking men than the reverse. A 'Hughes' self-cloner will not simply want for sexual gratification if he causes a male world for himself but he will unavoidably need, for a very long time to come (much longer than it will take to perfect the technique already being demonstrated in the Edinburgh sheep), to have the service of willing surrogate mothers to produce the offspring. A female self-cloner, on the contrary, will not need men at all, judging by the hint from Dolly, and if of lesbian inclination she might live sociably and happily in the complete absence of men. Thus as I hint in much fewer words than this at the end of the paper, I would suspect communes of women pooling resources and all cloning together, or even reproducing 'sexually' through fusions of their eggs (to which I'll return below), are a considerably more likely route on the female side than are go-it-alone millionairesses.

Two factors make this course more likely. First, it's more egalitarian: within the commune at least everyone can have a share of whatever reproduction is going on. Second, let us first note that hormonal control of human physiology and development is advancing very fast in parallel. Once absence of the Y chromosome has done its stuff in early sexual development, making by its default a female fetus (embryonic femaleness is certainly this way determined in mammals), and after first traces of ovaries and a womb have been formed, it will be open to us, or at least to endocrinologists, to masculinize female offspring in later development to any extent desired. I would expect a commune able to competently perform nucleus substitutions to be competent, and willing also, to experiment to varying degrees on this side. Whether the result is achieved by applied drugs, or by extreme genetic combinations, or by accidents, the feasibility of a hormonal course is already evident. Careful biometry is hardly needed, just plain eyesight and a TV screen, to show that the body styles of women Olympic athletes are approximations to men's. Add to this their doctors' notes on cessations of their menstruation during training and sometimes for long periods after it ends, and it becomes very obvious that one is seeing an interplay of self-induced hormonal effects. It is thus very likely that if extra hormone treatments are applied early enough, almost all the special attributes of male brains and male physique (bar the primary ones) can be simulated in women likewise—on the mental side would come better maths, better spatial abilities, more daring, worse verbal and literate skills, less social sensitivity, and so on. Really this isn't just prediction of a 'simulation' either: we, the two sexes, are basically the same in our genes, just apart from that relatively small number of functional switches that are set in the small and largely inert Y chromosome. Indeed it is already becoming clear that genes that actually code for maths and spatial abilities are not even among those on the Y (I am less sure about those concerned with 'daring' but there, too, it is much more likely to be the dosage of the Xs, than absence of a Y-produced substance, that is entraining women to more caution).

Actually, genes proven to affect intelligence are turning out to be distributed to a puzzlingly high degree on the X chromosome itself.[13] Easily accounting for what sex differences exist in mentality, women have more of these intellect-related genes due to their double dosage of X but the fact seems to be that, as with the genes for male physical secondary traits, they transcribe these less—that is, they less often mobilize them in early development to build neurones and their connections, to create the potential for masculine moods and the like. Maybe this is because women traditionally came to need such attributes less as they became dependent on men for their food supply (the male's speciality in meat provision coming at the first) and also for their protection, these dependences being based in turn on the growing dangers of giving birth to large-headed and helpless babies needing long years of care. Instead, perhaps, it was better for women to channel their resources towards good health and therefore the immune system. Perhaps also the little tried and highest flights of intelligence were inclining bearers to many risky and unproven choices as well as to good ones, and these could yield only moderate gains of reproductive success even when the moves turned out to be good ones, whereas they could readily cause disasters when they didn't. For men, advantages on the success side when it is a success can be much greater[14] while the disadvantages coming to males on the disaster side may be only a little worse than those facing females all the time—in both cases a limit is simply set at zero reproduction.

This difference in ability to translate success into offspring undoubtedly is the most important cause of all the psychic differences of the sexes. But another crucial difference that is becoming equally important for understanding the differences is that men inherit the genetic capacities of their minds, high or low as they may be, mainly from their mothers in proportion because the genes, as stated, tend to be concentrated on the X chromosome. Stupidity, it seems, like colour blindness, comes to men from their mothers—mothers who are generally themselves much less afflicted because of the averaging influence of their other X chromosome. But the same applies if we substitute 'genius' for either 'stupidity' or 'colour blindness' (the word 'afflicted', however, perhaps need not be changed for either case). As to what men do with the genes of intellect (or lack of it) that they inherit, and these in 'switched-on' mode, it is in the nature of the sexual-selection system of the human line that males are obliged to show whatever they have received in the best possible light. This has been an unavoidable part of the advance of outstanding brain, an advance achieved against great suffering imposed both on their own and the opposite sex.[15] Our great but cruel Mother Nature feels nothing for this being herself not just 'colour' but totally blind[16]—she neither foresaw nor sees now the suffering she is causing. Even if she weren't blind, however, I am afraid I easily imagine her watching all that goes on very coolly—preening herself, glancing round for applause, when she has a new success in her long-running absurd experiment in maleness but hardly uttering one sniff for all those sons who have to fall in their droves through her unfairness. Bursting their hearts in ill-health and rejection and solitude, dying by battle—that must be the counterpart of the successes. And hardly a tear she sheds either for those deaths of daughters in childbirth in whom simian hips jam hopelessly with the new bulbous heads that she caused all of us of both sexes to crave.[15] But why accord to Mother Nature this illusion of sight and knowledge—aren't we all really just a step in this doddering blind woman's unplanned progress, a step that has turned out (inevitably?, by good luck?) the greatest constructive experiment yet in the Universe.

Returning to my previous scenario, that of the imagined reproductively self-sufficient communes of lesbians, the next point to make is that with sufficient hormonal control available, all of the above and the production of all male attributes, with the exception of gonads and penises, are available within a community of technologically competent women, as is also the completion of the switch now in progress for *Homo* to Caesarean birth. Female attributes would doubtless likewise be available as developments from male fetuses but, first, they might be harder to put in place due to the awkward continuing influence of the Y (I'm not sure about this; how much does the Y actually do by itself after it has worked the fundamental switch that will provide the male gonads?) and, second, male fetuses need wombs in which to develop so that numerous surrogate mothers are still needed.

The frightening implication for me as a male who has tended to appreciate his penis and testicles is not principally that of female-to-female cloning giving rise to a wholly feminine communal scenario, such as I have sketched. All the arguments and facts in this volume should eventually convince you, I hope, that it is the general fate of clones, after their creation, to become steadily less healthy. If they do so, the health of clonal humans will eventually reach the point where anything, even having old-fashioned babies shared genuinely with a partner, will come to seem a best choice even to the greatest of egotists. If lack of conviction conveyed in my arguments for the immediate effects of the Parasite/Red Queen connection has been your problem, then, at least when you look to the very distant future, you will probably still hope that after such experiments have been watched failing there can be still enough men somewhere on the planet—perhaps on some far South Sea island—to bring us back to what we are now, to being bisexual. On these lines it has seemed to me that a much more fatal threat for real men—men like me with balls, penis, and Y—lies in a certain ambiguity of that phrase 'by a partner' that I just used above: onset of female-to-female sexual breeding—fertilization by female egg nuclei inserted into other eggs—is the greatest danger for men. So far this one is much less discussed and yet the same kind of technique that has just opened the way to the creation of Dolly the sheep is quietly laying this very feminine baby on our doorstep too.[17]

The techniques are certainly going to be used at least for our dependent animals and plants. The drive towards nameable and standard crops that began in the Neolithic Revolution and brought us our smooth fields of wheat and barley (all so inbred and so standard and thereby so perfectly machine-harvestable), that long before my birth brought our millions-strong clone of a tree now spread on several continents, a tree which a Mr Cox, for example, once reared from a 'pippin' of an apple he had enjoyed, a drive that has made already almost square and box-ready tomatoes, our chickens so tender and so over-ready for the oven that they collapse under their own weight if reared to adulthood—all of this vast trend that has been and is still going on is surely hardly likely to pause in its stride now just because people are feeling shocked for a week or two at a prospect of clonal Howard Hugheses. Cows with no bull in their field have been familiar for a very long time and this is just one aspect of the unnatural inbreeding and genetic restriction that are both already well accepted. Few seem even to notice how, under modern practices, the cows themselves are seldom seen in the fields either and in some cases might hardly know how to stand up in fields and what to do if they came. Thus it is clear that not many of us are going to turn even a hair if bulls should quietly cease to exist, which will be the likely outcome of both of the breeding systems mentioned above. If our milk and beef stay cheap, why should we care? This will be the general reaction.

When I planned my introduction to this paper I was not thinking of these issues: Dolly exploded on me halfway through and has left me a little dazed and amazed, not because of the novelty but because of all the fuss. But I had planned from the first to say something in this essay about my fantasy on humans that concludes the paper to show that I still didn't think the ideas I had floated so wholly fantastic. The event at the Roslin Institute in Scotland, where Dolly was created, and the fuss proved they weren't. There was even some local Taiwanese colour bearing on the thoughts I gave at the end of the paper that I wanted to weave in, all of it a bit trivial but stemming from the same trip. There was to be mention of Puli, 'Centre of China' as it called itself on a roadside placard—in English—on the way in, and there was Mrs. Chung, our hostess, with her competence and wonderful kindness, and Sun-Moon Lake with all its honeymoon hotels so near its shores as well as those pure white cauliflower-like 'hotels' of the aphids we had gone to see, hotels (or galls actually) in which half of the residents might be soldiers but every single one of them (making half the population of each gall) a sterile amazon. Finally, I meant to tell—and to be quite truthful am doing so in this very sentence—how one of the human honeymoon hotels was wholly occupied by a *convention of women*. They were teachers, I think, or they said they were, no word of 'reproductress' or 'amazon' did we hear but then none of us knew Chinese—and all were streaming in and out of the main door but under their colourful and varied flower bed of parasols, they all looked to my eyes terrifyingly the same—much more so, I am ashamed to say, than had even the macaques on Arashiyama . . . Therefore, you see, I could imagine it all happening there already, in Puli, 'Centre of China'. So I was going to end this piece. But with the arrival of Dolly the sheep to scare people, I feel I can rest my case; no more fantasies are needed.

Looking at the whole matter of cloning once again in a more immediate, experimental light rather than contemplating such vast changes and forebodings as male extinction as I have outlined above, what actually is the harm if we let some likes of Hughes or Imelda Marcos have their clones if they want them? I have to confess that I personally would take great interest in the situation and I wouldn't feel that our world was morally falling apart as I watched. The health issue (and perhaps an issue concerning premature senescence too) is bound to force wisdom concerning clones in the end. No animal anywhere near as large physically as we are has managed to be parthenogenetic, as far as we know, and on the plant side there are only a few trees that can be so through seed. I assume, therefore, that the health issue is going to control any excesses of cloning eventually, just as it will also control the equally unnatural practices we have already installed in our species (for example, the use of routine Caesareans for birth) or which we are attempting to install (for example, xenotransplants). So, too, in the long run it is almost certain to force us to lower various anti-natural banners we have tried to raise in the name of humanism or religion (anti-abortion as well as anti- all of the following: -infanticide, -euthanasia, -eugenics, -population control).

What we already have of cloning in our species obviously causes no harm; indeed, twinning seems to me like a gift God has given us, the 'fraternals' and the 'identicals' so neatly paired, to help our self-understanding and our differentiation of the roles of nature and nurture. Curiously, considering the present furore about cloning, I have never heard it suggested that either identical twin or quadruplet of a nine-banded armadillo loses rights or dignity in its life through being one of a set two or of a four with equal endowment. Of the un-uniqueness of self in these sets admittedly something more might be said, but that more with interest than with alarm because, out of all I have read, such genetical un-uniques themselves, almost without exception, tell us that they enjoy their situation. Instead of mumbling on about human dignity, we should certainly think much more about things we already do to reduce it in our distant cousins. Surely it is obvious that twins and healthy clones lose only the tiniest jot of dignity, if at all, compared with a man who is forcibly kept alive after he has expressed his wish to die, or compared with a turkey too heavy and weak to stand up without breaking its legs.

References and notes

1. C. Russel and C. M. S. Russell, *Violence, Monkeys and Man* (Macmillan, London, 1968).

2. B. Glass, W. Gajewski, A. Petrament, S. M. Gershenson, and R. L. Berg, The grim heritage of Lysenkoism: four personal accounts, *Quarterly Review of Biology* **65**, 413–79 (1990).

3. I. Morris, *The Nobility of Failure* (Holt, Rinehart and Winston, New York, 1975).

4. Y. Itô, Development of ecology in Japan, with special reference to the role of Kinji Imanishi, *Ecological Research* **6**, 139–55 (1991).

5. D. P. Todes, *Darwin without Malthus: The Struggle for Existence in Russian Evolutionary Thought*, p. 130 (Oxford University Press, Oxford, 1989).

6. Some parts of a European's 'Japanese' stereotype, I am afraid, still persist and in particular the uniformity (to us) of Japanese faces seems quite difficult to deal with.

7. I. Yamamoto, Male parental care in the Raccoon Dog *Nyctereutes procyonoides* during the early rearing period, in Y. Itô, J. L. Brown, and J. Kikkawa (ed.), *Animal Societies: Theories and Facts*, pp. 189–96 (1987).

8. R. V. Short, Why sex?, in R. V. Short and E. Balaban (ed.), *The Differences between the Sexes*, p. 3–22 (Cambridge University Press, Cambridge, 1994).

9. J. B. Gillet, Pest pressure, an underestimated factor in evolution, in D. Nichols (ed.), *Taxonomy and Geography, A Symposium*, pp. 37–46 (London Systematics Association, London, 1962).

10. J. H. Connell, On the role of natural enemies in preventing competitive exclusion in some marine animals and in rain forest trees, in P. J. van der Boer and G. R. Gradwell (ed.), *Dynamics of Populations*, pp. 298–310 (Centre for Agricultural Publication and Documentation, Waageningen, 1971); D. H. Janzen, Herbivores and the number of tree species in tropical forests, *American Naturalist* **104**, 501–28 (1971).

11. P. W. Sherman, T. D. Seeley, and H. K. Reeve, Parasites, pathogens and polyandry in social Hymenoptera, *American Naturalist* **131**, 602–10 (1988).

12. B. Kraus and R. E. Page, Jr, Parasites, pathogens, and polyandry in social insects, *American Naturalist* **151**, 383–91 (1998); P. W. Sherman, T. D. Seeley, and H. K. Reeve, Parasites, pathogens, and polyandry in honeybees, *American Naturalist* **151**, 392–6 (1998); B. Baer and P. Schmid-Hempel, Experimental variation in polyandry affects parasite loads and fitness in a bumble bee, *Nature* **397**, 151–4 (1999).

13. G. Turner, Intelligence and the X chromosome, *Lancet* **347**, 1814–15 (1996); but see also E. B. Hook, Intelligence and the X chromosome, *Lancet* **348**, 826 (1996) and G. Turner, Intelligence and the X chromosome—Reply, *Lancet* **348**, 826 (1996).

14. See, for example, R. D. Alexander, *The Biology of Moral Systems* (Aldine de Gruyter, New York, 1987) and M. Ridley, *The Red Queen: Sex and the Evolution of Human Nature* (Viking, London, 1993).

15. W. R. Trevathan, *Human Birth: An Evolutionary Perspective* (Aldine de Gruyter, New York, 1987).

16. R. Dawkins, *The Blind Watchmaker* (Longman, Harlow, 1986).

17. Such ideas about female-ruled sexless and neo-sexless futures as I raise in this introduction seem to be far from threatening right now, even as remote wishes, but this may be more because the public is still in a state of shock over the advent of mammalian cloning *per se*. I do not think the present dread of and distaste for cloning will necessarily continue in all groups. A survey of just 79 adults recently found no case of acceptance that human cloning should be allowed and no woman 'wish(ing) to see a world without men' or 'men becoming biologically redundant'. Those possibilities (although I have discussed them) are obviously very extreme ideas to present to interviewees so soon after the advent of the new technique.

KINSHIP, RECOGNITION, DISEASE, AND INTELLIGENCE: CONSTRAINTS OF SOCIAL EVOLUTION[†]

W. D. HAMILTON

It is both expected by theory and found in practice that co-operation evolves most easily between genetically related individuals.[1] Colonies of social insects, for example, are mostly single families of siblings, and in those cases where they contain distantly related reproductives they rarely achieve typical levels of co-operation and colony lifespan. However, there are exceptions to this. For example, the honeybee's seemingly extreme preference for half sibships in its colonies, implied by the multiple insemination of its queens, is still a strange exception but not so exceptional as it once seemed. At first it appeared that the trait might be shared only with a few highly social ants,[2] but this has not been upheld.[3-7] There is, in fact, a growing opinion that the high relatedness of $\frac{3}{4}$ that arises in full sisterships of Hymenoptera due to their male haploidy has not been the most critical factor for the common evolution of eusociality in the Hymenoptera, even though theorists have on the whole upheld the idea that the exceptional relatedness, if present, can strongly favour eusociality (refs 8–11, but see also 12 and 13).

Including polygyny (i.e. multiple queens) along with the problem of multiple insemination, a puzzle remains as to how some weakly intrarelated colonies can be as co-operative as they are. Part of this puzzle is in long-known cases (e.g. the tropical multiqueened polybiine and ropalidiine wasps, and the slave-maker ants) but part comes from more recent evidence on relatedness in surveys of electrophoretic alleles.[3,5,14] Apart from the slave-maker case, which perhaps more resembles the problem of the fostering of the cuckoo, the examples prompt a thought that reciprocity—at least as a subsequent development—may be a factor in many social insect species. Cases of inter-species tolerance or co-operation reinforce this possibility (e.g. ref. 15) as do

[†]In Y. Itô, J. L. Brown, and J. Kikkawa (ed.), *Animal Societies: Theories and Facts*, Ch. 6, pp. 81–102 (Japan Scientific Societies Press, Tokyo, 1987).

various features of the pleometrosis of wasps,[16–19] and also those cases of polygyny in ants where the genetic diversity is known to be both high and stable.[20] However, if such reciprocity occurs within species, its origin is likely to have occurred after eusociality had been initiated by kin selection, because this is an easier route to travel than arriving at reciprocity without kin interaction, especially when the animals are highly mobile (see Chapter 4 and ref. 21).

HIGH KINSHIP WITH LOW CO-OPERATION

Besides weakly intrarelated colonies co-operating too much, social evolution theory has another puzzle that is similar but at the opposite extreme. This is the subject of this paper. Some highly related colonies seem to co-operate too little. Specifically, the question is why assemblages of genetically identical but separate individuals (i.e. clones) should show, on the whole, less co-operation than is found in the sibling and half-sibling assemblages of social insects.

With many clones there is circumstantial evidence that the asexual reproduction has not existed for long enough for a social development to evolve. Parthenogenesis has a scattered phylogenetic distribution. This is mostly at the level of species within genera and of strains within species, suggesting that in terms of macroevolution, events of parthenogenesis are short-lived and contain somehow the seeds of their own death. Another factor is that most parthenogenesis is in 'weedy' opportunist species, while in general the adaptations of such species, with their emphasis on rapid reproduction, do not involve the environment manipulations that could be diverted into co-operative acts.

When writing on this subject 20 years ago, thoughts on the above lines made up my main excuse for the non-occurrence of conspicuous co-operation within clones. However, since then the picture has changed rather dramatically, at least for a few cases: there are now known examples of highly developed altruism, even of sterile castes, within separated clones. Such examples, together with others that have been known but little recognized for a long time, are listed in Table 10.1.

Avoiding the problem of the expected short life of purely parthenogenetic strains, all the examples turn out to have regular sexual reproduction occurring along with, or else interspersed with, the amictic generations. Equally interesting, in three of the five different cases it turns out that the co-operation mainly serves the colony's need for defence. This finding links with other accumulating evidence for the importance of defence in the evolution of social insects generally.[16,19,22–25]

In the case of aphids, Aoki[30] in Japan was the discoverer of the 'soldier' caste or phase and he has remained the main student of this phenomenon with few other followers so far. From his rate of discovery of new species

Table 10.1 Sterile morphs within separated clones

Group	Morph/function	Reference
Myxobacteria	Stalk cells/disperse spores	26
Acrasiales	Stalk cells/disperse spores	27
Coelenterata		
Acanthopleura elegatissima	Soldier polyps/acquire and defend space for colony	28
Trematoda		
Dicrocoelium dendriticum	Brain worm/affects host behaviour for survival of clonal sibship	29
Insecta		
Aphididae (> four origins)	Soldier juveniles/defend colony	30, 31
Chalcidoidea	Soldier larva/defends clonal parasitoid sibship	32–34

showing some aspect of soldier behaviour, it is obvious that there must be a great many more examples yet unknown. It is also now clear—from Aoki and Kurosu's[35] observations—that the phenomenon is not confined to the Far East. The case of these aphids illustrates how easily arguments *a priori* go astray in biology. In my case, it was possibly correct to look at *adult* aphids and rationalize, as I had been doing, that they are such specialist insects so devoted to the use of plant phloem sap and to rapid reproduction that they can have no preadaptations to do anything positive for one another.[2] But even this slighting view of the potential altruism of adults may have been excessive, for there are hints that a mother aphid may, for example, chemically influence the plant she sucks in a way that is favourable to her descendants living on it.[36] In any case, it turned out that very striking altruistic adaptations found by Aoki concern very *juvenile* aphids, which are much more like generalized Homoptera in their morphology and therefore more likely to be able to discover adaptations to a new role. The role to which they did adapt turns out to be, in essence, wielding swords (e.g. *Colophina*) or else daggers (*Pseudoregma*) in the colony's defence. In yet another genus, *Astegopteryx styracicola*, the force of soldiers[37] is so numerous that, considering also the huge gall produced by the aphids' host tree and inhabited by the colony, it might be conjectured that some defence is being extended to the tree as well. The stated population of 100 000 aphids in a gall, with about one half of these being soldiers, makes this *Astegopteryx* a rather highly evolved social insect when compared with colony sizes found in species we accept as highly social in ants and termites.

In spite of the above discoveries, dramatic as they are, the actual number of castes and the degree of differentiation of morphs are not impressive compared with ants and termites, and the number of cases known to involve any such altruism remains very low.[38] Because the members of the aphid colony are potentially of identical genotype, for which theory says that any kind of altruism that pays more than it costs should be selected, we may again ask why the trend has not gone farther. More generally, the same question may be asked about a more inclusive set of species—the traditional social insects with the species of Table 10.1 added. Amongst all these, why is the number of morphs so small, and only the degree of the co-operation effected so slight, compared with conditions found in the assemblage of cells that makes up the metazoan body?

In the human body the number of distinct cell types that are easily recognizable by morphology also runs to dozens. Considering separated clone or family members, counting the number of morphs that can be brought out of a single genome or closely related set, aphids again may be the record holders, but the morphs here are obviously more concerned with sex and stages of the complex seasonal cycle than with simultaneous division of labour or with any adaptation that could be considered altruistic. Being alate admittedly is often 'altruistic' and equally could be considered part of a division of labour within the family;[39] but apart from this, the aphid soldier is the only explicit instance of altruism. In any case the maximum number of morphs seems to be about one dozen. Altogether, the achievements both of separated clonal colonies and of the traditional social insects in the direction of division of labour are very meagre. I believe that the contrast can be explained largely by two factors:

1. The issue of cheating in the co-operation: cheating becomes much easier when the co-operators are separate individuals.

2. Disease. Against this no purely behavioural strategy is likely to work except to become less social, which is contrary to the assumed trend. Soldiers *per se* cannot fight disease. Sex and genetical diversity possibly can, and this implies another possible limiting factor, which will be dealt with in detail later.

The above two factors are actually linked: indeed in a sense there is really only one—*cheating*. This is to say that the force preventing division of labour comes from evolution in an unproductive, unco-operative side to life. This always gains its living at the expense of more-productive and co-operative behaviour. The first factor above was being cheated by individuals that are not yet distinct even at the species level and whose parasitic behaviour is almost certainly facultative. The second aspect was being cheated by individuals whose evolutionary path diverged in the extremely distant past— cheated by lines that for the most part remained microbes when metazoans began to build large multicellular bodies. Disease microbes are usually obligate and specialized, but not all are so, and there is evidently a continual possibility for the free-living to switch into parasitic niches. Thus there are many disease microbes that do not have evolutionary continuity as pathogens with each other. But remarkably, on the host's side, the basic methods for defence against pathogens may have been continuous over vast periods. Emphasizing the connection of the two apparently distinct problems facing co-operators, it is beginning to seem that *analogous tools* are being used for protection at all levels. In fact, at least for the chordate line, there is a strong hint that *homologous tools* may be being used both in the containment of disease and in recognition of self and kin at the level of whole animals. Recognition, of course, is the key to success with the first factor, the conspecific cheaters.

RECOGNITION

There are really two problems: recognition of kin for kin-selected co-operation, and recognition of previous co-operators (and defectors) for reciprocity. I will discuss these in turn.

How does an animal know who its kin are? In more general terms, how are kin, whether cells or individuals, to be distinguished from alien invaders that ought to be attacked or excluded—or at least monitored to watch what they are doing? Speaking in everyday images rather than in chemical practicalities, one way, of course, is to observe who is born out of whom and never let any individual out of your sight—pictorially, keep hold of mother's apron strings, to which, it can be hoped, all other true siblings are holding on too. This is fairly literally the way of zooid colonies of Bryozoa, Siphonophora, and various other groups. Judged at least from the ability of such zooids to differentiate their morphology, it works well.[40] It is largely also the way of the cells of the metazoan body, and in fact the method of zooids keeping in contact or growing out of a common stem really makes the chemical recognition mechanism potentially identical for the two cases. But what actually is the danger if the individuals separate?

Suppose an aphid morph were to specialize in a role where it uses food less for its own growth than to manufacture chemicals, which, when put back into the plant, influence it in ways beneficial to other members of the colony. What is to stop another aphid strain or species from specializing in settling always near to colonies possessing this morph, never contributing to the altruistic processing of the plant itself but simply imbibing the fruits of the altruism it finds? Clearly this strain (assumed equal in other things but with no slowing of its growth due to production of the altruists), would have a reproductive advantage. Such strains would become a load on the productive colonies, so reducing the success of the species that had invented the co-operative morph. Aoki's discoveries strongly suggest that such specialist forms may be found to exist if they are looked for in the right way. One form of plant exploitation that should have a good chance of revealing them is gall formation, because, at least when gall morphology is reasonably advanced, its very nature tends to exclude parasites. However, in the early stages, galls are as exploitable as anything else. It is thus particularly fascinating to find that one of Aoki's suggested evolutionary routes to the soldier morph involves a migrant form whose original adaptive function was to wander off, like a self-propelled cuckoo egg, to exploit the galls being set up by other mothers.[41] On this interpretation the evolution of soldier aphids is not so much a case of swords being beaten out of ploughshares as of swords being beaten out of burglars' jemmies. Aoki's alternative explanation—presented in the volume in which this essay first appeared[31]—is only a little different and preserves the same slant, the issue being changed to parasitism of an already dug well hole rather than of a ready furnished home. Both his suggested routes are excellent illustrations of the essentially opportunistic nature of evolutionary innovation. Elsewhere I have drawn attention to how in the ancestral line of a honeybee queen a soldier's sword has been refined back into a murderess's stiletto.[42]

So far there is no indication that aphids can discriminate clone members. The ability to do this would greatly facilitate co-operative phenomena: free-riders, and, still better, malignant cuckoos, could be ejected.[2] Obviously, it is very desirable to know more about polyclonality in aphid colonies, and a particularly valuable check on kin-selection theory will come when it is possible to contrast colonies that have soldiers with others that don't. Ideally the comparisons will involve species that are either close in ecology in respects other than degree of mixing, or close in phylogeny, or both.

If it is ever possible to look at measured degrees of sociality along with measured abilities of kin recognition more generally, it will almost certainly turn out that the two are strongly connected. Spiders and other arachnids, for example, with very few exceptions, have shown no abilities for kin recognition as yet, and, correspondingly, there is very little morph differentiation in the semisocial and primitively eusocial species.[43,44] Vollrath,[45] for example, found only a slight and overlapping size difference between the reproductive and non-reproductive individuals in *Anelosimus eximius*, the most eusocial species yet described. Admittedly the situation in *Polistes* is not much different. The traditional social insects, on the other hand, all have kin recognition to some degree. Always, so far as is known, this recognition is based on olfaction. The way in which it is done varies from species to species, and so far there does not seem to be any single homologous system uniting the phenomena in the way the major histocompatibility complex bids to unite the chemical-recognition phenomena in chordates. Interesting themes in kin recognition of social insects are the degree to which acquired odours are used[46] and the degree to which genotype-specific chemicals are blended together into a colony odour before it is used.[47]

Colony recognition using blending and learning prevents the schisms within the colony that are likely if phenotype matching based on unmixed individual odours is used. At the same time, too much learning opens a way for parasites, for if a blended colony odour exists the parasites may find a way to smear the same mixture over themselves. Probably there are various compromises reached between the possible extremes. At one extreme there would be a high regularity of ejection of strangers and, combined with occasional intracolony schisms, ejection of unusually segregated true colony members. At the other extreme, intracolony life would be very harmonious provided it is free of the parasites that in this case could more easily get in. There are probably parallels here with the defences of the metazoan body. For humans there are hints that allergy-prone humans may be also good resisters of disease.[48]

The pros and cons of various methods of kin recognition cannot be considered in detail here, fascinating as the subject is. In any case, the importance of kin recognition in facilitating division of labour among separated individuals should not be overstressed. There is another more direct physiological reason why separate individuals have a difficulty. Even if they should want to co-operate, it is harder for them to communicate about who should do what. After separation, the well-tried chemical methods have all suddenly become unavailable. A possible exception to this is the situation of colonies whose members still keep close together in wet environments. With the cellular slime moulds (Acrasiales), for example, water-borne molecules are still used,[49] but achievements are on a very small scale. Air-borne chemical messages have, of course, been invented in numerous groups, and in the social insects they have reached a fair degree of variety and informative power,[50] but the combination of requirements for volatility and consequences of turbulence, which, when present, erratically varies and mixes up chemical signals, must inevitably restrict the complexity of what can be said. The same effects will restrict the ability of a receiver to detect fraudulent messages. Cellular slime moulds seem surprisingly indifferent about mixing with alien colonies or even species. Possibly their habitat makes it difficult a priori for strains specializing in parasitism to succeed because the secluded sites create barriers for dispersal: in that case mechanisms to prevent heterogeneous strains fusing in the coenobial phase are unnecessary. But, as expected, it is beginning to appear that parasitism between cellular slime moulds is not completely absent.[51] In the case of mycetozoal slime moulds that do not have nuclei forming separate modules, recognition, rejection, and sometimes lethal reactions of one colony on the other, are very striking.[52]

Returning to the problem of keeping individuals informed about colony needs, without which even willing and honest co-operators will end up taking inappropriate roles, it is obvious that sight and sound are far better media than olfaction. Later still, building on vision and hearing, symbolic language gives a further leap in the ability to decide needs and to convey instructions. Chemistry and odour provided the primeval, natural media for regulating kin interaction. Sight and sound, although doubtless also important for the old kin system in some species, grew up as the speciality of the more 'modern' approach to co-operation—namely, reciprocation. This touches a huge subject. To discuss the complexities that language, for example, introduces into such communication is quite beyond the present scope. However, principles of reciprocation and how they differ from those of kin co-operation are relevant topics, and to these I now turn.

Co-operation by reciprocation[53] potentially supervenes out of kin-based co-operation but then, once established, it can continue even when relatedness has fallen to zero (see Chapter 4). It has rather severe conditions for its evolution, but once these are attained there is greater freedom. When a particular run of reciprocation begins, it is crucial that all future interactions be between the same two partners. This can be ensured by two methods. The first may be characterized pictorially as 'keeping a finger in a partner's buttonhole'. Obviously this corresponds closely to 'keeping hold of mother's (or sibling's) apron strings' in the case of kin recognition. The principle is not to let the other interactant out of your sight. The other method is carefully to memorize the interactant so that it can be recognized as an individual upon re-encounter. Clearly this second method opens the way to much more complex systems of reciprocation. Achieving individual recognition of just one interactant probably requires less information storage than learning the set of traits sufficient to categorize a relative of given degree, and, as a bonus, none of the calculus implicit in an estimation of relatedness is needed. However, if an individual is going to have reciprocative relations with many interactants at the same time, the information to be stored goes up proportional to the number, which is not at all the case with categorical kin recognition. Thus it is not surprising to find that it is the relatively largest-brained of the vertebrates, humans, who appear to be life's current champions at individual recognition (e.g. ref. 54) nor is it surprising that humans have quite large regions in the brain where damage seems to impair recognition in particular.[55] In general, a fairly large brain would seem to be a necessity.

DISEASE

It is now time to turn away from these highest levels of co-operation and from the enemies that are of similar size and sophistication to their hosts and go back to that major set of enemies that operates at the level of cellular co-operation—that is, back to pathogenic microbes. This may seem far removed from the intraspecies struggle we have just been discussing, but it will be argued that, while the struggle at the lower level has seen some strategic victories for the large host, the war has never been won, and that effects of its continuance are manifested at all levels of sociality, often in surprising and subtle ways[56] (see also Chapter 6). Tiny and efficient, pathogenic microbes thrive by doing the limited things they can do very fast. For this it suits them to carry little surplus luggage in the way of a DNA programme that it might only occasionally use. Hence they do best if the growth conditions are uniform. This in turn implies a trait that is actually found: that pathogens are inclined to specialize on one tissue or group of tissues within a large animal's body. Extrapolating still onwards from such specialism, we expect pathogens to have preferences also between individuals—to prefer, say, in the case of one strain, living in the liver that is produced by genotype X to living in the liver produced by genotype Y. There is constantly accumulating evidence that such preferences also exist. As I and other authors have argued (refs in Chapter 5), a likely and most important outcome of specialization by parasites at the genotypic level is sex—a perpetual need of the large host to create newly different body environments, and perhaps specifically to create for its offspring new self-recognition chemical passwords, substances whose absence on a cell can be used to trigger the body's counterattack on invaders. Being a specialist is certainly no panacea for all the problems of being a parasite. The more specialized a parasite is the farther it can expect to have to go to find its next suitable habitat. Thus another factor making for the success of parasites generally is the bunching or dispersion of suitable host genotypes.[57-59]

These thoughts now bring us face to face with a crucial conflict that is affecting all levels of the organization of life and which consequently constrains possibilities at all levels of co-operation. It turns out that the very conditions that can make social co-operation most harmonious and best able to resist parasitism by conspecifics—that is, the condition of being a colony of clonally identical individuals—is the very one that will make the same colonies most exploitable by pathogenic microbes. I believe that this at last shows the reason why soldier-forming aphids are not more abundant and why species having them have not carried their division of labour farther: species that try this path eventually become diseased and parasitized to extinction. Obviously this line of reasoning leads to a testable prediction as follows.

If average relatedness within colonies is assessed for a range of aphid species by electrophoresis or other method, it will be found that the relatedness correlates not only with the presence of colony-benefiting adaptations such as a soldier caste or long post-reproductive lifespan,[36] but also with incidence of diseases and of low-mobility parasitoids. In testing the truth of this, it will obviously be desirable to control for density and for colony size, because these almost certainly also affect pathogen success.

An old empirical saw of plant ecology borrows from a saying in physics and goes: 'Nature abhors a pure stand.' From the researches of plant pathologists, and also from entomology (many insects, especially the small ones, act just like diseases to their plant hosts[60]), an important reason for the truth of the proverb is becoming clear. Dynamics of wind dispersal, as also of some other kinds of dispersal (e.g. through random contact), often imply that there will be critical densities of susceptible types that must be reached if an epidemic is to occur. Thus, assuming the problem of distant dispersal (between colonies) is solved, the rate at which a disease can build up within a colony, and even whether it can build up at all, is likely to depend on the degree of uniformity among the genotypes present. For the honeybee, a glance at any beekeeping manual reveals the serious, decimating nature of the diseases affecting this highly social species. The more compact and numerous a colony becomes the more it is almost like a metazoan body in the opportunities it offers to pathogens. Even the honeybees themselves behave as if they know the extreme importance of disease and hygiene. Their hives are always tidy; the combs themselves are so clean that we have no hesitation to eat the honey extracted or even consume them whole. We also know how ruthlessly and rapidly the bees throw diseased larvae out of the hive, and even that the rate at which they do so can actually be crucial in determining whether a hive will completely succumb to a disease like foul brood or not.[61] Workers killing their siblings in this manner are performing a role equivalent to that of the T-lymphocytes that kill diseased cells in the mammalian body before the cells can liberate more virus.

Rather as facultative responses from a complex, ingenious immune system seem to be never quite enough to solve completely the problem of disease for the mammalian body, so the hygienic instincts of nurse bees are evidently not enough to prevent disease in the honeybee colony. One may suspect that insofar as defence systems do succeed, colonies quickly evolve to be larger and still more homeostatic until disease problems catch up. All this implies that *sexuality has to remain the essential back-up system of defence against disease, both for the body and the colony.*

Without special defence, the conditions for microbial growth in the large homeothermic body would be almost ideal although this might be offset a little by the tissues being so varied. In the colony there are no such differences between 'tissues' at the individual level. Fixing attention on the honeybee again, the hive as a whole is indeed homeothermic (as are nests of most highly social insects), while nothing like an immune system for detecting and combating microbial enemies is known to exist. It is likely, therefore, that an extra emphasis on genetical diversity will be needed in the honeybee hive to overcome a susceptibility of what is otherwise an unusually 'pure stand' that is being kept under unusually constant conditions. The special potential uniformity comes, of course, from the haploidy and the unvarying genetic contributions of each male. I believe that the high polyandry of *Apis* queens has evolved so as to offset these various factors of uniformity that make the colony an ideal habitat for microbial growth. Other very highly social insects have rather similar tendencies to promiscuity to the honeybee,[3,7] and there seems to exist a general tendency for highly social species to carry more genetical variability.[62] The same evolutionary favouring of colonies with diverse genotypes might also be a factor in the polygyny of tropical wasps: however, I am unaware of evidence that these groups suffer more from disease, and in fact have personally seen more colonies die from disease in temperate single-queen vespine wasps than I have in neotropical polybiines.

Disease is, of course, also expected to affect behaviour in more direct ways than those considered above. Behaviour to avoid contagion was well discussed by Freeland.[56] Obviously, this factor is going to lead to a motive for exclusion of strangers that will confuse with their exclusion on grounds that they are potential cheats at the intraspecific level. This builds up a double ground for expecting colony members to exclude individuals of their own sex, but, on the other hand, there is going to be an opposition of the factors with regard to unlike-sexed strangers. The colony needs new genes to combat potential disease in its offspring. It does not want, however, in acquiring the new genes, to catch new diseases. This conflict seems the more paradoxical because sex, according to the thesis adopted here, is ultimately aimed to reduce disease: it also contributes to causing the problem that it exists to cure. There is not space to discuss this paradox except to state in summary that one would expect sex and dispersal to be carrying genes around on at least the same scale of distance as the media of contagion, whatever they are, carry the pathogens. Thus outbred sex is expected to correlate with the existence of far-moving independent agencies of disease transmission, and inbred sex with a preponderance of diseases transmitted by direct contact of conspecifics.

The seemingly probationary period that canid and primate groups in particular apply to intending joiners, during which the alien animal follows the group but does not join, could serve primarily as a quarantine.[56] The preference which social insects seem to have for open-air mating might be interpreted on similar lines: here it would not be a matter of quarantine but of having as few colony members touch the stranger(s) as possible. Spacing out behaviour in other colonies (e.g. aphids[63]) might also occur partly to avoid contagion.

The effects of disease on social behaviour can be summarized on an arbitrary timescale:

1. Long term: disease causes outbred sexuality. Where asexual reproduction becomes established, disease causes eventual extinction. Clonal colonies persist longest where (a) they remain small and (b) sexual generations are interspersed.

2. Middle term: particular problems of disease, through influence on degree of inbreeding and polyandry, sexual selection, etc., regulate the pattern of genetic diversity found within colonies. This genetic diversity determines the course of kin selection and kin recognition.

3. Short term: disease causes evolution of grooming, behaviour to avoid contagion, rejection and diseased offspring, etc.

INTELLIGENCE: A WAY OF ESCAPE?

The necessity of good memory for reciprocation in the advanced form that allows interactants to separate and re-join has already been mentioned. Memory is, of course, not the same as intelligence but some of the same neural equipment can serve for both. Or at least, there is overlap—if one function grows, there is likely to be a correlated growth in the other. It is hard to imagine reciprocative relationships of the parting–rejoining kind in an animal where there is not some intelligent appreciation of distant future possibilities. For example, it may be of the type: 'If I help him do this now, then later—though it may be weeks—he may help me do that.' Or else: 'If I explain it to them, then they'll see, too, that we could work together.' What is in common here is a vision of the distant goods that co-operation might be expected to produce, and the ability to experience such visions obviously creates numerous new possibilities of action. With humans it is commonplace to experiment mentally even with kinds of co-operative ventures that have never been tried before. This is also done, of course—and done every minute—for the effects of one's own individual actions in the purely physical world, whether it is hitting a golf ball or planning the roof beams for one's house. In general, it is the accuracy of forecasts *in increasingly untried situations, increasingly distant in time*, that is the measure of the possession of intelligence. Applying such methods to social activity is much the same except that the reasoning must cover simultaneously the points of view of the other intended interactants—it must estimate their enthusiasm for outcomes, their probabilities of defection, and so on. The attempt to formalize all this, of course, is the Theory of Games. This, again, is a huge subject outside the scope of this essay, but, because we are here concerned with the emergence of the rational goal-seeking beings that are presupposed by the theory, it is at least pertinent to remark that the game-theoretic framework assumes the problem of recognition to be already solved—for example, I am not aware of discussion of what happens if the payoffs are sometimes paid by mistake to wrong parties. Yet this might well be expected to happen as evolving inter-actants struggle to cope with ever-increasingly complex game situations.

Beyond simply showing the goods that co-operation might create, intelligence opens other very important doors for social progress. It suggests, for example, that it might be worthwhile to contribute to the maintenance of a system of laws that aids co-operation in the most general way, so that its fruits come more within reach. Unfortunately, beneficial as it is to have co-operation, and for it to be law-bound and policed so as to prevent cheating, it is usually also advantageous to the individual to cheat on his contributions to the policing if he can (i.e. not be a policeman, and to avoid any necessary taxation). This is a big problem. It grows in importance as relatedness within groups diminishes, and (what is much the same thing) as selection working between groups also diminishes. However, the advantages of social co-operation are so great and the preconditions for it so well built in humans that progress continues, if intermittently; and, in general, the facilitatory laws and police forces and religions appear and thrive and help it along. The laws seek ever new ways to stop cheating. They work especially well, of course, to stop cheating against whatever class it is that happens to have the power to be making the laws; but also in the long run, in the average expression that comes after many civil revolutions of one kind or another, they seem to succeed in working fairly democratically, while simultaneously they adapt to achieve their ends at diminishing costs in the policing itself.

Religions for their part seek ever new and more seductive ways to exhort people to act co-operatively and not cheat. Thanks largely to the ingenuity and impressionability of the human mind (i.e. to its intelligence) both sides of the process keep pace with the forces against them, and, as all this works, civilization progresses and even accelerates into ever more mazily complex patterns of division of labour and ever greater productivities of co-operation. The laws and the religions are themselves being generated, of course, by memetic[64] rather than by genetic processes of evolution. These processes are extremely rapid by standards of ordinary evolution. But it is well to remember that they are so on the basis of a pattern of human impressionability that was established by ordinary genetical natural selection as it acted in a social primate living in the savannahs of Africa; now their task has become the quite paradoxical one of making the descendants of this primate happy and orderly while living, for the most part, in utterly changed circumstances— such as in the heart of cities. The contrast with the environment where we came from is strange indeed. If we saw swarms of men and women crouched on the outside ledges of St Paul's Cathedral and of the Bank of England as evening falls in London, and if they were clutching one another, gibbering, and grooming each other's hair and skin, while their bowler hats and umbrellas and handbags lay scattered on the pavement below them, where there would be lions and leopards prowling, waiting to prey on anyone who might fall down, we might have less reason to be surprised than by what we actually do see—crowds going to the trains all hatted and dressed and seemingly so calm. A poem by T. S. Eliot said of this scene of the rush-hour crowds: 'A crowd flowed over London Bridge, so many, I had not thought death had undone so many.' In the light of the vanished savannahs, spoiling Eliot's poetry perhaps, but with a similar intent that merely looks a few million years further back, the evolutionary biologist might well reinterpret the last phrase as 'I had not thought death had undone *enough*'. In other words, it is certainly inevitable that such new physical and social circumstances, including the religions and laws that support them, create new genetic selection pressures; but these pressures have been operating for such a short time compared to the explosive growth of the new memes and the changes of environment that it is not at all clear what the trends of selection are, and whether they are stabilizing or destabilizing for the civilization that generates them. There are probably examples of both kinds.

One example of a new trend whose existence is easy to see, but whose direction is hard to interpret, is that created by the victory of modern medicine over disease. Inevitably this is already leading in all civilized countries to a relaxation of the selection that, in all other species, has always eliminated badly mutated and badly copied DNA. The same competence to nurture every baby born up to adulthood, and to protect everyone against infectious disease is also countering the ancient process of coadaptation that resists the antagonistic changes of pathogenic microorganisms. The current human enthusiasm for cures, hospitals, and doctors, on the part of both medical scientists and the public, seems to reflect faith that such relaxations can raise no problem because the memetic evolution—that is, the production of better and better cures—can go on for ever and can more than keep pace with any degeneration of the human genome due to mutation. However, it is not clear to me that this faith is justified. It is possible to imagine the current trend leading to an increasingly delicate balance, a metastability. If a stage is reached where almost every individual in the population carries a lethal although treatable genotype at at least one locus, such that effects of the genes have to be offset by medication derived from a complex technological industry (severe diabetes could be one example), then a war or civil disruption could initiate a chain reaction in which industries became successively paralysed by sicknesses among their workers. Unless drugs could be sufficiently stockpiled (and actually were so) this would lead to a situation where almost every one would die except perhaps for members of a few religious groups that strongly restrict both medical intervention and exogamy of their members.

The ageing of society that is happening so rapidly within our lifetimes, while not itself, of course, a genetical change, can only bring such a metastability closer. With the degree of interdependence assumed above, civilizations would have become superorganisms to the point where, in a sense, they lost title to their prefix of 'super': they would begin to suffer from sudden death like metazoans. Correspondingly, they would probably have to find ways to compete with each other and reproduce like metazoans and *Astegopteryx* aphid colonies.

Whether the trend towards medical dependence at all ages, plus the rapid and interacting trend towards a more aged society, really threatens such disasters, is a question of 'futurology,' and, once again, is beyond the present scope. But it seems to me that questions like this need to be considered more than they are. They are not dismissed by those calculations made during an

earlier controversy over eugenics that showed that policies concerning particular diseases could have little immediate effect. First, to allow any pool of deleterious genes to increase, as by genetic counselling, is not a satisfactory
long-term policy—the pool has to overflow some time at a rate determined
by mutation.[65] Second, the problem must be considered as one of the genome
as a whole. Possibly I am just odd in not having a heart that leaps with joy at
every new achievement of medicine and surgery. I admit to being far from
ecstatic, for example, when I read in a newspaper that a 'hole in heart'
mother has had a 'hole in the heart' baby, and that this new defective heart
has been, like the mother's, patched up. Instead of delight I see an uninviting
vista of more holes in hearts to come, and of an increasingly hospital-bound
population stretching down the ages. Of course, I admit to walking on two feet
myself, and to riding a bicycle, both of which are metastable activities requiring the proper functioning of thousands of my neurones, as well as of my eyes
and organs of balance. But the question for us as individuals is: do we want
our descendants to become like the cells of organs, organs which in turn comprise some larger entity—that is, totally dependent on the functioning of our
civilized system as whole, mere cogs in its wheels? Is it inevitably to this end
that the human pattern, like the metazoan and the social insect patterns
before it, had to evolve? Lastly, if we should not want to evolve this way, are
we still able to halt the trend that is already in progress, and if so how?

If the trend to make civilized societies into completely interdependent
superorganisms goes ahead, there is one fairly obvious streamlining of the
process that can probably be effected in the fairly near future. In the light of
topics already raised, it deserves to be discussed. This is the elimination of
sex.

If, as there is reason to believe, the existence of sex is linked to disease,
then the elimination of disease makes sex unnecessary. There will no longer
be evolutionary support for the process of mixing and matching and
separating genomes, whether the main function of that process turns out to
be correcting errors of the genome,[66] or, as others believe[67,68] (see also
Chapter 5), the function is the creation of new or repeat combinations that
defeat the changes of infectious pathogens. Such lack of support by selection
will not, of course, in itself, cause mutations that eliminate sex; with the sexual system left to its own devices there would be a long time to wait for it to
disappear. But in an increasingly intelligent and technological world it is no
longer necessary to wait for mutations. Science will probably soon offer us
the means to clone ourselves directly into offspring, *in utero* at first, and *in
vitro* later, if we so wish. There is probably no reason why this should not be
done with cells of either sex, but if, for efficiency, one sex was to be eliminated, it would doubtless be the one that has always been the less serious and
utilitarian of the two that would go—the one not provided with a womb, and
the one that, at a later stage, would probably be less competent at keeping

watch over the germinal 'test tubes' and less intrinsically interested in the product that comes out. In other words, males would have become both ultimately and proximately useless, and they would be dropped. Elimination of males, it may be noted, would also be more humane because evidence indicates that males are less good at living without females than *vice versa*, and atavistic sexual urges would probably cause less heartache in a world of pure females than in a world of pure males—although admittedly, for either sex, these urges could no doubt be dulled by adding suitable antiaphrodisiacs to the cocktails of drugs everyone would be receiving already.

If the above scenario seems threatening and unlikeable as well as fantastic to some readers, as it does to me,[42] it may be a suitably inconclusive end to this discussion to reflect briefly on some difficulties that stand in the way of the extreme case just sketched—that is, the cloning and the loss of a sex. These difficulties at least seem guaranteed to slow the coming of the medical utopia to beyond our own lifetimes. And they also reflect back to the point where I started.

For reasons already given, everything ought to work very peacefully in a clonal civilization once it is established. There might be some problem about aptitudes (see Chapter 9), but this can probably be overcome by making sure to pick not just one but at least a few high-quality generalist genomes in the first place, and by diversifying computers and robots to take the brunt of the new superhumanity's less-pleasant tasks in any case. Everything should quickly become very harmonious: it would be much as in the Communist dream, but with a more sound scientific basis: the functions of policemen and lawyers (mostly males anyway, so this would fit) could start to wither away.

The problem is, what genomes or genome to pick? Even within a nation, even within a village, how could we possibly agree? I don't know the answer to this and am, in a sense, glad that I don't know because I don't want it to happen. But to take a worst view, we might notice a hint that emerges not out of theory but out of what actually happens in a certain human and (as I think) quite laudable intervention that is in the same general field. This intervention is the human artificial insemination by donor to relieve infertility. In this activity, medical students seem to be the favoured donors, because— well, so far as I can tell, simply because they happen to occur just down the nearest corridor in medical institutions. This hints that, if it can happen at all, the cells of the final world organism may eventually be cast in the semblance of some very self-assured, competent, and doubtless, for this case, female, medical student.

SUMMARY

- The understanding of social evolution still has puzzles about how weakly interrelated societies manage to be as co-operative as they are, and about why more highly related societies aren't more co-operative. Possible ways of resolving these puzzles have been discussed.

- It is suggested that colonies of separated clonal individuals usually have difficulties of communication, and that these affect both the efficient integration of activities, and the detection of non-co-operative cheaters. The attenuation of chemical communication when individuals became separate made a barrier to progress that was hard to overcome until the advent of communication by language. However, using air-borne phero-mones, the social insects made moderate advances before this.

- Cheating remains a problem throughout the living world and is very far from resolved in humans. Its reduction is aided, however, by high powers of recognizing individuals. Synergistically further benefit accrues if there can be added the same high powers of communication as assist co-operation.

- Because co-operative societies have to have contact between their members, contagious disease in particular and parasite attack in general, is always a serious threat. This factor works in favour of communities with weak intrarelatedness and correspondingly high levels of intracolony genetic variability.

- Probably it is disease combined with the difficulty of co-ordination already mentioned that prevents clonal colonies of separated individuals differentiating to the same degree as do cells of the metazoan body.

- Humans will soon probably be in a position to form co-operative colonies of separate clonal individuals if they so choose. However, kinship theory predicts a difficult problem about the choosing.

References

1. W. D. Hamilton, *American Naturalist* **97**, 354–6 (1963) [reprinted in *Narrow Roads of Gene Land*, Vol. 1, pp. 6–8].

2. W. D. Hamilton, *Journal of Theoretical Biology* **7**, 1–52 (1964) [reprinted in *Narrow Roads of Gene Land*, Vol. 1, pp. 310–82].

3. C. K. Starr, in Smith, R. L. (ed.), *Sperm Competition and the Evolution of Animal Mating Systems*, pp. 427–64 (Academic Press, New York, 1984).

4. P. Pamilo and R. Rosengren, *Biological Journal of the Linnean Society* **21**, 331–48 (1984).

5. R. H. Crozier and R. E. Page, *Behavioral Ecology and Sociobiology* **18**, 105–15 (1985).

6. R. F. A. Moritz, *Behavioral Ecology and Sociobiology* **16**, 375–7 (1985).

7. R. E. Page, *Annual Review of Entomology* **31**, 297–320 (1986).

8. W. D. Hamilton, *Annual Review of Ecology and Systematics* **3**, 193–232 (1972) [reprinted in *Narrow Roads of Gene Land*, Vol. 1, pp. 270–313].

9. Y. J. Iwasa, *Journal of Theoretical Biology* **93**, 125–42 (1981).

10. J. Seger, *Nature* **301**, 59–62 (1983).

11. A. Grafen, *Journal of Theoretical Biology* **122**, 95–121 (1986).

12. E. Kasuya, *Research in Population Ecology* **24**, 174–92 (1982).

13. P. Pamilo, *Behavioral Ecology and Sociobiology* **15**, 241–8 (1984).

14. R. Gadagkar, *Proceedings of the Indian Academy of Sciences* **94**, 587–621 (1985).

15. A. P. Bhatkar, *Experientia* **35**, 1172–3 (1979).

16. Y. Itô, *Journal of Ethology* **1**, 1–14 (1983).

17. Y. Itô, *Zeitschrift für Tierpsychologie* **68**, 152–67 (1985).

18. Y. Itô, *Research in Population Ecology* **27**, 333–49 (1985).

19. Y. Itô, *Monitore Zoologico Italiano* N.S. **20**, 241–62 (1986).

20. K. G. Ross and D. J. C. Fletcher, *Behavioral Ecology and Sociobiology* **17**, 349–56 (1985).

21. J. J. Bartholdi, C. A. Butler, and M. J. Trick, *Conflict Resolution* **30**, 129–40 (1986).

22. N. Lin and C. D. Michener, *Quarterly Review of Biology* **47**, 131–59 (1972).

23. D. L. Gibo, *Canadian Entomologist* **106**, 101–6 (1974).

24. G. J. Gamboa, *Science* **189**, 1463–5 (1978).

25. H. R. Hermann (ed.), *Defensive Mechanisms in Social Insects* (Praeger, New York, 1984).

26. J. W. Wireman and M. Dworkin, *Science* **189**, 516–23 (1975).

27. J. T. Bonner, *On Development* (Harvard University Press, Cambridge, MA, 1974).

28. L. Francis, *Biological Bulletin* **150**, 361–76 (1976).

29. W. Wickler, *Zeitschrift für Tierpsychologie* **42**, 205–14 (1976).

30. S. Aoki, *Kontyu* **45**, 276–82 (1977).

31. S. Aoki, in Y. Itô, J. L. Brown, and J. Kikkawa (ed.), *Animal Societies: Theories and Facts*, pp. 53–65 (Japan Scientific Societies Press, Tokyo, 1987).

32. Y. P. Cruz, *Nature* **294**, 446–7 (1981).

33. Y. P. Cruz, *Journal of Experimental Zoology* **237**, 309–18 (1986).

34. W. M. Wheeler, *Journal of Experimental Zoology* **8**, 377–438 (1910).

35. S. Aoki and U. Kurosu, *Journal of Ethology* **4**, 99–104 (1986).

36. W. D. Hamilton, *Journal of Theoretical Biology* **12**, 12–45 (1966) [reprinted in *Narrow Roads of Gene Land*, Vol. 1, pp. 94–128].

37. S. Aoki, *Kontyu* **47**, 99–104 (1979).

38. G. F. Oster and E. O. Wilson, *Caste and Ecology in the Social Insects* (Princeton University Press, Princeton, NJ, 1978).

39. W. D. Hamilton and R. M. May, *Nature* **269**, 578–81 (1977) [reprinted in *Narrow Roads of Gene Land*, Vol. 1, pp. 377–85].

40. E. O. Wilson, *Sociobiology: The New Synthesis* (Belknap/Harvard University Press, Cambridge, MA, 1975), especially ch. 19 and table 19.1.

41. S. Aoki, *Kontyu* **47**, 390–8 (1979).

42. W. D. Hamilton, *Quarterly Review of Biology* **50**, 175–80 (1975) [reprinted in *Narrow Roads of Gene Land* , Vol. 1, pp. 357–67].

43. R. E. Buskirk, in H. R. Hermann (ed.), *Social Insects*, Vol. 2, pp. 282–393 (Academic Press, New York, 1981).

44. L. Aviles, *American Naturalist* **128**, 1–12 (1986).

45. F. Vollrath, *Behavioral Ecology and Sociobiology* **18**, 283–7 (1986).

46. M. Breed, *Animal Behaviour* **31**, 86–91 (1983).

47. A. Mintzer and S. B. Vinson, *Behavioral Ecology and Sociobiology* **17**, 75–8 (1985).

48. J. Newell, *New Scientist* **110** (28 May), 27 (1986).

49. J. T. Bonner, in E. Sondheimer and J. B. Simeone (ed.), *Chemical Ecology*, pp. 1–19 (Academic Press, New York, 1970).

50. E. O. Wilson, in E. Sondheimer and J. B. Simeone (ed.), *Chemical Ecology*, pp. 133–55 (Academic Press, New York, 1970).

51. D. R. Waddell, *Nature* **298**, 464–6 (1982).

52. M. J. Carlile, *Bericht der Deutschen Botanischen Gesellschaft* **86**, 123–39 (1973).

53. R. L. Trivers, *Quarterly Review of Biology* **46**, 35–57 (1971).

54. P. Garigue, *Canadian Journal of Economic and Political Science* **22**, 301–18 (1956).

55. N. Geschwind, *Scientific American* **241** (3), 180–99 (1979).

56. W. J. Freeland, *Biotropica* **8**, 12–24 (1976).

57. J. E. Van der Plank, in J. G. Horsfall and A. E. Dimond (ed.), *Plant Pathology*, Vol. 3, pp. 229–89 (Academic Press, New York, 1960).

58. J. Harper, *Population Biology of Plants* (Academic Press, New York, 1977).

59. B. R. Trenbath, *Annals of the NY Academy of Sciences* **287**, 124–50 (1977).

60. P. W. Price, *Evolutionary Biology of Parasites* (Princeton University Press, Princeton, NJ, 1980).

61. N. Rothenbuhler, *American Zoologist* **4**, 111–23 (1964).

62. R. C. Berkelhamer, *Evolution* **37**, 540–5 (1983).

63. J. S. Kennedy and L. J. Crawley, *Animal Ecology* **36**, 147–70 (1967).

64. R. Dawkins, *The Selfish Gene* (Oxford University Press, Oxford, 1976).

65. W. D. Hamilton, review of *Human Diversity* by K. Mather, *Population Studies* **19**, 203–5 (1965).

66. H. Bernstein, H. C. Byerly, F. A. Hopf, and R. E. Michod, *Science* **229**, 1277–81 (1985).

67. H. J. Bremermann, *Journal of Theoretical Biology* **87**, 671–702 (1980).

68. H. J. Bremermann, in S. Stearns (ed.), *The Evolutionary Significance of Sex*, pp. 135–94 (Birkhauser, Basel, 1987).

BEING RARE AND SUCCESSFUL

Parasites and Sex

Time flies like an arrow;
Fruit flies like a banana.
J. ULEHLA[1]

FROM the last, this chapter goes forward by one year to 1988. Around that time evolutionary interest in sex was at a peak. Begun in the 1960s and put into perspective, along with a summary of a lot of the relevant evidence, in 1982 by Graham Bell's massive review,[2] the debate seemed to reach a high point. In 1987 and 1988 two multiauthor books came out on the subject,[3] one of them containing the paper of this chapter. After that the activity declined a little. The fever to organize 'sex' meetings likewise began to abate. I ceased to hear Bruce Levin argue with Bryan Clarke about three times a year concerning what J. B. S. Haldane might have or might not have foreseen if only we knew and I ceased to meet all the other popular players at conferences held scattered all over the evolution 'hemisphere'. Had John Burdon Sanderson in fact come back from the dead in the mid-1980s to tell us all what he had thought, he wouldn't have needed even a home; air-borne and hotel-borne all of the time, he could have circled the globe like a restless Erdös or albatross, dropping in at Albuquerque, Honolulu, Kyoto, Bangalore, and Berlin (only mentioning the more exotic sites I alone went to), the oceanic islands not of actual evolution (though in the case of Hawaii the island was that as well) but of the study of it; in so doing he could have told us of all calculations he had discarded in pubs as well as what he had meant at Pallanza. Instead, I think that to follow the sex-meeting trail after 1990 he would have needed a university post to anchor in and a grant to pay travel like the rest of us.

By the end of the period, even without Haldane, we had chewed and pulled the matter about in all directions like a litter of fox cubs gnawing a tough dead goose, but even then we could hardly decide what to leave and what to swallow. Most of us still can't. As I write in 1998, 10 years later, I personally feel the matter already had been fairly well resolved, and this volume as a whole will show what I think was the outcome; but then I had thought pretty much the same in 1988. In contrast, I think most see the puzzle as having been little changed—a goose much bitten about and a mess made but something important and definitive still to come. But most will also admit, I think, that somehow the furore of interest has drained away. In 1988, even a third multiauthor symposium book came out on sex—that including my paper of the next chapter. But because the papers for that were requested at a level suited to senior schoolchildren—and thus required to be broad and readable—my paper in that volume was not intended as a serious summary, as you will see.

The abatement of interest must have come, then, not due to any consensus but because we evolutionists were committing ourselves increasingly to particular theories and becoming less inclined to listen to others—or we were declining to listen, at least, until there was much more evidence. But perhaps in this again I speak mainly for myself. I always admired James Thurber as a humorist and recall how somewhere vividly he presents his personal policy in debating with people—how, during an argument, he found it best not to listen to his opponent at all because this enabled him to keep his mind calm and keenly focused on what he was going to say when his opponent stopped talking. It was as if we had all read Thurber.

There are, of course, several other reasonable theories besides the one I favoured: I still do think about them briefly but on the whole have followed Thurber's advice. This may sound very un-Popperian and even a dishonest approach to science,[4] and in a way it's true; but, as I have said in earlier introductions, one of the wonderful properties of science is its self-correction: at the same time there is a lot to be said for everyone doing their own advocacy. Very possibly the truth comes faster that way—the courts, at least, seem to think so. Consider Gregor Mendel and how well he started with those two fine papers, even forgiving in them that small fudging he added or selected—the step of presenting the best he had got, which probably was hardly considered reprehensible in those prestatistical days. Then look further at how, after those papers, he completely failed to follow through after his discovery. Quite likely this was because he had not the benefit of Thurber's advice—was listening far too much to other people, listening to spurning, incomprehending, and outright erroneous remarks. Had he not listened to the Swiss botanist Karl Nägeli, for example, who had recommended *Hieracium*—hawkweed daisies—for his further studies, he himself would probably not have started on that cryptically sexless group of plants and would not have become confused. Well, let Mendel divert from his triumph as he must, I hope the above sketch roughly sets scene for you for the present paper—the scene as I saw it around and just after 1988. The arguments were by then all in place and I do not know any radically new theory to have arisen since that time. We drew back and experimented, or, the lazier of us like me, generally awaited more evidence.

The book that included the present paper that I wrote jointly with Jon Seger was edited by Richard Michod and Bruce Levin. The parallel multiauthor book of a year earlier than ours was of about the same size and ambition—to summarize current views—and that was edited by Stephen Stearns from Basel. I think I remember being asked by both sets of editors some time in 1985 to contribute to their books and that initially I declined both of them, being too busy with the big model whose results in progress obviously couldn't be ready for publication soon enough. I felt too busy even to summarize my previous models and results into a paper. Rick Michod and Bruce Levin therefore suggested I might team up with Jon Seger (whom I think they had already asked) and that, between us, we might write something on the parasite angle. Again I wasn't keen; I think I already had entered the commitment to write the paper that is reprinted in Chapter 12. Therefore I suggested to Jon, who at that time seemed to be thinking about sex on much the same lines as I was, that he should take on the whole assignment. He agreed, but when he sent in his paper (it was done as about the first of his new work, I think, from within his new job in the University of Utah) the editors didn't like it because it was more on sexual selection than 'sex itself' and also, rather strangely, because it was too novel—it involved a very neat new model on sexual selection. You might think the editors would have loved that—a new model from Seger tends to have exactly the attributes of my title. But only at this point did they make it quite clear that they were determined to have a review of frequency-dependent ideas relating to sex, not new additions nor evidence.

Others might have been too irritated to continue but Jon is inexhaustibly courteous. His already written offering having been sent off to another journal (*American Naturalist* I think it was) he immediately started writing the book chapter again. This time he insisted that I be re-involved to the extent of making additions and alterations to go into the second and, hopefully, final draft. So it came about and a joint manuscript (really 90 per cent Seger), and not really about frequency-dependent selection as a whole but rather about the parasite possibility within the idea, was dispatched in May 1987. Although, once again, contrary to editorial guidelines, Jon had included three more new models, our brief review and Jon's lucid style were liked and the paper was accepted by the editors.

I contributed so little, mainly just examples and parts of the discussion, that I hardly feel justified to include it as a paper among mine in this volume of reprints. But Jon had made a neat mini-review so that I was also loathe not to be involved. Should you have been unhappy with the earlier introductions to the problem of sex because of versions too vague, too biased, too contorted of English, or too frivolous, you would do well to try again for a proper understanding by reading this one: many students have said to me unexpectedly and rather pointedly, long after we have first discussed 'sex', that recently they read 'Seger's paper in that book'; and how they now understand better what I have been on about for so long. A little huffily after this, I will leave the chapter to speak for itself. It is worth adding here, however, that if the reader would like to learn more about the other theories and about the aspects of 'frequency dependence' we left out, a very good extra source is the paper by James Crow, which is also in the Michod/Levins volume. To my mind Jim Crow stands at present as perhaps the least-sung hero of evolutionary population genetics of the twentieth century. Everything he has done, including his famous text of population genetics with Motoo Kimura, has been important for general progress in evolution theory. His long-term ally, Kimura, was slightly more of a sung hero, it may be said, but this is partly for the positive reasons of his excellent quantitative and mathematical reasoning and partly also negatively, arising from his ingenuous interpretations, his brilliant tools, combined with naive application to real life. Jim Crow held back from endorsing all these (no one, however, has held back from endorsing the usefulness of Kimura's tools themselves). Anyway, back in the 1960s, these two had been the key to the whole re-opening of the sex topic and, at least for me, Crow retained, very differently from my own advocacy, perhaps the most balanced overview on the topic all along. To repeat, his review in the Michod/Levin book will be found characteristically brief and clear.

There are a few points to make about our paper. The phrase 'Evolution of sex', which is so usually given to this problem as in the titles of all three of the books I have now mentioned, is misleading; the phrase should really be 'Evolutionary *maintenance* of sex'. This is because what we are almost all primarily concerned with under the heading is why sex doesn't disappear. Margulis and Sagan, in the book of Chapter 12, were saying something serious about the original evolution of the sexual system; others also doubtless have evolutionary ideas about that remote origin, but these people tend to be a different class of evolutionist—microbiologists, palaeontologists, and straight origin-of-lifers . . . As Margulis and Sagan explain, the origin must have occurred way back in the Proterozoic eon for which they give the wonderfully broad dates of 2500 to 570 million years ago (it's the late Precambrian I suppose, roughly, in my antique terms; but nowadays it has become recognized that evolution and geology went on in other places besides Celtic Britain plus that outlier in the Jura Mountains). During this immense period must have occurred for the first time those meetings and partings of cells that began to assume vague similarities to the processes we now know as fertilization and meiosis. In other words, pairs of cells having fused, and having entangled their chromosomes within a joint nucleus, separated and re-emerged eventually—miracle of miracles—with balanced and yet different combinations of their hereditary material. Margulis, in that next-to-come book, skates gracefully to this ancient crux with her title 'Sex: the cannibalistic legacy of primordial androgynes'; meanwhile I plod the periphery, dragging my heavy phrase 'maintenance of sex' behind me like a ladder, peering at this and that feature of ecology. Really this title for what I actually study will never do. It echoes with workman's feet and the slap of paint on an old door. Even tweaking the 'maintenance' side a bit—so that (think of a cover design) it conjures visits to the shops with blank faces or where rubbery gadgets hang in the window—will not do either nor catch the right neurones. In short, editors and publishers cannot love it: hence that less correct but more general 'Evolution of sex' it evidently had to become in all three volumes.

The events in the Precambrian were, however, tremendously important and it is a pity that no one in either of the two more serious books was asked to address this even if any account would have to be very speculative. For the descriptive side, priestess Margulis provides her druidic sketches, as I watched at the meeting of that coming volume I have mentioned. She and Dorion Sagan point out that freeliving unicells do sometimes attempt, under conditions of starvation, the cannibalism of conspecifics;[5] most importantly it is known for the attempts to become static and to fail. Such the ragged failures of the Precambrian—what marvels may almost have happened, and, had they happened, what different chemical feet we might have stepped off with, what different scenes we might be seeing today! And what different novels by different intelligent beings might be being written. Or else, perhaps, no marvellous chance occurred; some piffling non-recombining chromosome system might have come and no novels at all.

But when that happened which did happen (you are right, it is meant to sound like the Bible) we can regard those events as pregnant with the future of all multicellular life on Earth. Boring and disregarded, the 'drawn' games tended to play to empty stadiums. They were the lesser teams probably; all the excitement of their times, amongst the unicells, would have focused the successful cannibalisms—the big boys, gladiators, spiked, thumbs turned down. It was at those rare drawn games, however, that recombination and reduction division started. At others not so drawn (nor so conspecific) started other subtle and pregnant victories, those interweavings of fusions and divisions that were the first successful pathogeneses and parasitisms—the sneaks in and the bursts out; in short, viruses and other intracellular attacks. These also, most importantly, were originating in the same era although, because bacteria and even viruses themselves (as also even computer assembly-language life forms) can have parasites, these probably began much earlier.

Even the Romans, I think, never thought of staging actual cannibalism for their gladiatorial shows. But what a scene we may imagine there in the Precambrian! An acute unicell of the time (she'd be female, of course, the very notion of masculinity is supposed but not yet invented), who happened to have a ticket to what was to be one of the dull 'drawn matches', might have glanced up, bored, to notice shadowy forms bending behind her. As inconceivable to her as life forms of Jupiter would be to us, she'd see for the first time huge creatures and dry creatures—monsters of open air, giants, compilations of unicells all tumbled together like Vigeland Oslo statues, domed brows at the top furrowing anxiously at the play. In the dark back rows pentadactyl hands would rest on objects stranger still to our unicell's bewildered vision—objects that, by all the thermodynamics of the inanimate, ought never even to exist: bows, for example, ready for hands to grasp and pull, and books (and I would like to think even this one) ready for interpretation through the eyes she can see in these monsters but cannot understand. Through her own eyes (I have to endow these for this mythic occasion) our unicell might see people, elephants, and redwood trees. How anxiously these unborn beings would watch every move of the game—and how frantically they would cheer for each equalizer but groan for each next goal! At the end, if a draw, wild excitement in the shadowy figures would reign at what to our lonely unicell would seem merely a wasted ticket—the game she watched where no one ever ate anyone. All those ghosts of the unborn would have been willing for those key happenings that, a billion sun-cycles of the planet in the future, would permit their birth: it was not a win but a draw—a DRAW!—that they wanted. And how they would love that calm parting in which the teams troop from the field not having exchanged just their players' shirts but their arms and heads and legs.

All this amounts to saying that if there is one event in the whole evolutionary sequence at which my own mind lets my awe still overcome my instinct to analyse, and where I might concede that there may be difficulty in seeing a Darwinian gradualism hold sway throughout almost all, it is this event—the initiation of meiosis. The idea has helped me to see one plausible root out of which one may grow DNA repair, gene exchange against parasites, mutation, or whatever. But, failing my own energy to create an account in which disciplined thought can replace all the above whimsy, I much need an account by some other evolutionary mechanist or a set of them—Tim Cavalier-Smith, perhaps, on the one hand or Michael Doebeli and Nancy Knowlton[6] perhaps on another more tactical. It is on this snag of meiosis that, at the moment, I still base my uncertainty whether I expect the Universe to be full of life or empty of it apart from us. Possibly the likelihood of a drawn game being resolved in the neat and symmetrical way that eventually came to be a meiosis—that is, the draw that led to the creation of the eukaryotes—is so unlikely to happen again that a finite Universe would give no chance for it. If it hadn't happened then a world lacking siblings as closer relatives than sexual offspring would never have begun the building of our big and indefinitely capable bodies. We would have forever remained as multispecific and intrinsically selfish tissues within, at best, stromatolites—that is, unicellular parts and patches within those mixed communities that often stand, like oozy hybrids from coral reefs mated to mushrooms, around the shallow margins of warm seas, sites where they've been standing, it seems, since far into the Precambrian era. Superorganisms these are, if you like, but I would say they have stayed without one haziest multispecific philosophical idea ever developing between them in all the immense time. And these stromatolites were seemingly the best fate of our game watcher when the games were always wins.

Just as the previous two chapters could be considered a tidying away of old issues of mine—relatedness, sociability, and insect themes—so this one and the next two can be considered to clear the ground and set up a base camp for the next assault; in fact they prepare for that big simulation model on sex that was actually already in progress and, as mentioned, about 90 per cent deterring me from the paper of the present chapter. Various things that Jon Seger and I say in the chapter indicate the coming endeavour. I can't remember if it was Jon saying these things or my additions as I revised his draft. For example, we ask and partly answer several questions, addressing in a tentative way how the very simple models involving coevolution at one or two genetic loci, as in this paper and some of my preceding one, may strengthen in their support of sex if the number of loci involved in the tussle is increased. We (or rather Jon) in the paper forthwith increase the number as far as eight and uncover hopeful signs. But they remained only signs and together we did not follow them up. In general, the selection regimes in this paper were far too weak and the cycles generated far too long to have any hope of defeating a full twofold extra efficiency of parthenogenesis. The strength of selection assumed on a single locus by a disease was a reasonable one (about a 5 per cent higher fitness for the resistant genotype) but it is reasonable to expect much higher differentials when all factors (or all resistances) are being considered together. The bird that unhappily compounds being 5 per cent more sick due to each of several of the less good alleles it happens to carry at each of several loci, probably doesn't get a nesthole—or, if it doesn't need one and could potentially nest in a bush, still quite likely it doesn't breed.

My work with Robert Axelrod and Reiko Tanese was modelling all this with resistance involving up to 14 loci and the result was proving dramatic, so that, almost by the time of this paper, I could answer that there were indeed 'great advantages over a system that involved only one or two loci'; I could also state (see the question a few sentences below) that it was easy to find reasonably realistic states of parasite coevolution in which a recombining sexual population would permanently outbid an equivalent asexual population of the same size, all this in terms of the genotypic diversity each could maintain. In Jon's eight-locus model in the paper he incidentally confirmed what we knew, that having reversible mutation was a strong preventative to alleles going extinct even under the unhelpful haploidy he had assumed; but much more interestingly he found that too much of mutational aid doesn't close but widens the reproductive gap in favour of asexuals. In this paper we leave this difference uncommented but the effect comes up again dramatically in the paper of Chapter 16—see Fig. 16.1. There the effect is explained and reason is also given why the lower mutation rates are the more realistic ones when resistances are concerned.

The later papers certainly bear out the expectation we voice in this one that modelling more realistic situations was going to give situations that would be 'frighteningly complex' (p. 442). Jon's seemingly chaotic cycling out of three-allele weak selection was probably the basis for the statement. But our hopeful suggestion in the following sentence to the effect that 'Artful simplifications will undoubtedly be the key to making their behaviour understandable' has so far not been realized. I suspect that the hope as written was mainly Jon's and may have been a nudge intended as much for me as for the reader of the paper, reflecting that by the time of this paper I was almost in open rebellion against taking out more time to look for 'artful simplifications' and against trying to make any of the detailed behaviour, as opposed to manifest outcomes, 'understandable'— and I was willing to settle for very partially understandable too: Jon, I think, wouldn't. Probably, in short, I was already trying to hide lacunas of intellect under such excuses as the proven unpredictability and difficulty of analysis of chaos. This condition, it seemed to me, was looming ever more commonly in all the host–parasite dynamics I had been experimenting with, especially when with multiple loci—or at least it was something so like chaos that it seemed to make little difference whether one dubbed it by that name or not because in either case it appeared utterly analytically intractable. Jon, however, is mathematically of sterner stuff, and I think still quite strongly objected to all my 'complicated' simulation models, such as those that would follow. He would like to take steady and fully understood steps into the unknown. Even at the 7000-mile separation we endured as we wrote—he in Salt Lake City and I in Oxford—I had to imagine, as I contributed my vague touches, Jon's eyebrows shooting to improbable heights on his tall face and a succession of those phrases he usually brings out for the furtherance of decent precision in a project, phrases that may have begun out of the 1960s and 70s slang but that acquired with him a special scientific slant. 'No way', for example, is one I imagine he would make to some suggestion of mine lacking even any hint from a model, and probably wrong. But then also I remember such as 'It is the case that . . . , is it not?' and especially 'Well, if at all, *let's do it right*'.

But saying that I was despairing of fully understandable formal analysis is not saying that I expected to get results with just no understanding at all. In my opinion we were in an equivalent realm of difficulty to that of the three- and more 'few'-body problems of Newtonian physics, which are harder to understand than the statistical answers emerging when the 'bodies' are very numerous and bounce properly like molecules in a gas. But before simply assuming we could jump to many loci in the sex problem and to an assumption that multiple polymorphism will be automatically self-preserving under the selection we were proposing, we had to give at least simulation accounts of some of the 'few-locus' versions and hopefully find good hints in them (and reasons if possible) to the effect that the resistance variability involved would be permanent. To my mind this is a weakness still present in the quantitative genetics theories I was drifting from—they could not easily show why the variability they assumed was permanent.

In 1988 we were, as I hinted at the beginning of this introduction, at about a high point of the favour that would be given to the host–parasite idea of 'sex itself' by theoretical geneticists, or indeed by any evolutionists, for years to come. Graham Bell of McGill and John Maynard Smith of Sussex had swung across in this direction in 1987 from espousal of other possibilities.[7] Curt Lively[8] and Steve Kelley and Janis Antonovics[9] plus their colleagues were bringing in very favourable field and experimental findings. Even on the front of the extensions of the idea to sexual selection most things looked bright: Andrew Read's studies published in 1987 and 1988, for example, seemed to confirm my and Marlene Zuk's finding on the parasitism of the North American birds with considerably better phylogenetic control than we had had, and I knew Marlene had uncovered more of our correlation for South American birds, though these were not published until several years later.[10] But, as I say, it was a high point: a deep hiatus was to follow of which something has been written on the sexual-selection side in the introduction and appendix of Chapter 6 and about which I will write more on the 'sex itself' side in the introduction to Chapter 16.

If both sides of the idea are now becoming more respectable again it is probably mainly through causes quite extraneous to my near world of wrangling theorists. First, one must speak of the recent popularizers working both through books and the mass media—this can change popular perceptions and thus indirectly attack the prejudices of academia. Second, there is the steady stream of favourable data for both sides of the general idea of the involvement of parasites that has been emerging (far more than from anywhere else) from labs, especially in Scandinavia and Switzerland. And, lastly, there is the rapid accumulation of medical information that not only gives detail to the immense variability of resistance to infection that I had presupposed as probable in all my sex theorizing, but also begins to reveal the immense sweep of one source of support: molecular mimicry. All biological mimicry, whether of the classic stuff of Bates's butterflies or this new kind based on the face-to-fold recognition of molecules that, through sequence matching, is becoming itself easy to recognize[11] (but see note 12) is strongly favourable both to polymorphism and to frequency dependence of selection.

There are a few statements of uncertain validity in this paper that I now see. Did Jon know something about parthenogenesis of plants in Utah, unknown to me, that encouraged him (us) to write the 'apomictic' part in the following: 'Annual plants are more often self-pollinating or apomictic than are perennials'? Did Donald Levin's paper, which we cite, also back such an 'apomixis' statement?[13] The 'self-pollinating' part is OK but now it seems to me that apomixis (parthenogenesis via seeds) is actually quite rare in annuals. From the top of my head I can think of only three cases: *Erigeron annua*, *Aphanes microcarpa*, and *Chondrilla junceoides*. These are true seed apomicts, although I could, with a brief look for the names, add several viviparous grasses. I may indeed have seen some of these near Salt Lake City in the midst of what Nancy Moran once pointed out to me as an immense flora of annuals that inhabits the semi-desert around the salt lake—strange species that live, if 'live' is the word, in suspended animation in dusty 'seed banks' for most of their long 'lifespans'. But the well-known predominant apomicts like the *Opuntia* prickly pears are obviously better parallels to our perennial, apomictic, large-flowered, and European rain-soaked brambles (say) than they are to our tiny and green-flowered parsley piert (*Aphanes*). Does it matter? If annual apomicts are scarce, this is scarcely the most major mystery of sex. Clearly they can have more or less fixity of type as they choose among various gradations in persistent sin—that is, use the far more adjustable methods of fixing variation through self-pollination.

Jon and I, known to each other since my early visits to Harvard in the 1970s, never met during the writing of this paper. Perhaps this was as well, or as well that I did not go to Utah, because every aspect of the history and physical situation of Salt Lake City makes it for me a home of a strong brand of preposterous and yet romantic nonsense, and I would only too easily add to it. One probably needs a sterner mind than mine, such as Jon's, to withstand these, and, even when guarded by the university, not fall into parallel preposterity oneself.

References and notes

1. Yura Ulehla (see Vol. 1 of *Narrow Roads of Gene Land*, Chapter 3) claims not to be the author of this blank verse nor to recall who was. Nevertheless, it seemed to me somehow characteristic that it reached me from a Czech—scion of that nation that provided Jerome K. Jerome for English literature. In Manaus, Brazil, likewise, it was a Czech shopkeeper, who, noticing I was travelling with an Italian, told me in mixed English and Portuguese how one Italian could always make 10 Englishmen laugh whereas on the contrary even 100 Englishmen could not make one Italian sad. Combining Jerome with this, my estimate became that actually neither 1000 Englishmen nor a stifling, noisy equatorial town could sadden one Czech. One Englishwoman might— but just the one must try it, not 10.

2. G. Bell, *The Masterpiece of Nature: The Evolution and Genetics of Sexuality* (University of California Press, Berkeley, 1982) and review by W. D. Hamilton (Unravelling the riddle of Nature's masterpiece, *BioScience* **32**, 745–6 (1982)).

3. S. C. Stearns, ed., *The Evolution of Sex and Its Consequences* (Birkhauser, Basel, 1987); R. E. Michod and B. R. Levin (ed.), *The Evolution of Sex: An Examination of Current Ideas* (Sinauer, Sunderland, MA, 1988).

4. I should add here, however, that I have not knowingly misled on anything in any paper. A wise scientist never ignores evidence, of course, his theories soon forsake him if he does; but in science, as in law, advocacy is not a matter of lying. You put the best case that you can for the theory you believe a likely winner; but likewise you must do nothing to impede others to work similarly for the theories they favour. (In case of interest, both my grandfathers were lawyers.)

5. L. Margulis and D. Sagan, Sex: the cannibalistic legacy of primordial androgynes, in G. Stevens and R. Bellig (ed.), *Nobel Conference XXIII. The Evolution of Sex* (Harper and Row, San Francisco, 1988).

6. M. Doebeli and N. Knowlton, The evolution of interspecific mutualisms, *Proceedings of the National Academy of Sciences, USA* **95**, 8676–80 (1998).

7. G. Bell and J. Maynard Smith, Short-term selection for recombination among mutually antagonistic species, *Nature* **328**, 66–8 (1987).

8. C. M. Lively, Evidence from a New Zealand snail for the maintenance of sex by parasitism, *Nature* **328**, 519–21 (1987).

9. S. E. Kelley, J. Antonovics, and J. Schmitt, A test of the short-term advantage of sexual reproduction, *Nature* **331**, 714–16 (1988).

10. M. Zuk, Parasites and bright birds: new data and a new prediction; in J. E. Loye and M. Zuk (ed.), *Bird–Parasite Interactions*, pp. 179–204 (Oxford University Press, Oxford, 1991).

11. G. Strobel and S. Dickman, Prime suspects lined up in MS mystery, *New Scientist* **146**, 16 only (1 April 1995); H. Baum, H. Davies, and M. Peakman, Molecular mimicry in the MHC: hidden clues to autoimmunity?, *Immunology Today* **17**, 64–70 (1996); V. Apanius, D. Penn, P. R. Slev, L. R. Ruff, and W. K. Potts, The nature of selection on the major histocompatibility complex, *Critical Reviews in Immunology* **17**, 179–224 (1997); S. Roy, W. McGuire, C. G. N. MascieTaylor, B. Saha, S. K. Hazra, A. V. S. Hill, *et al.* Tumor necrosis factor promoter polymorphism and susceptibility to lepromatous leprosy, *Journal of Infectious Diseases* **176**, 530–2 (1997); R. G. Feldman, G. T. Rijkers, M. E. Hamel, S. David, and B. J. M. Zegers, The group B streptococcal

capsular carbohydrate: immune response and molecular mimicry, *Advances in Experimental Medicine and Biology* **435**, 261–9 (1998); A. Sahu, J. O. Sunyer, W. T. Moore, M. R. Sarrias, A. M. Soulika, and J. D. Lambris, Structure, functions, and evolution of the third complement component and viral molecular mimicry, *Immunologic Research* **17**, 109–21 (1998); G. R. Vreugdenhil, A. Geluk, T. H. M. Ottenhoff, W. J. C. Melchers, B. O. Roep, and J. M. D. Galama, Molecular mimicry in diabetes mellitus: the homologous domain in coxsackie B virus protein 2C and islet autoantigen GAD(65) is highly conserved in the coxsackie B-like enteroviruses and binds to the diabetes associated HLA-DR3 molecule, *Diabetologia* **41**, 40–6 (1998); Z. S. Zhao, F. Granucci, L. Yeh, P. A. Schaffer, and H. Cantor, Molecular mimicry by Herpes Simplex Virus—Type 1: autoimmune disease after viral infection, *Science* **279**, 1344–7 (1998).

12. C. Roudier, I. Auger, and J. Roudier, Molecular mimicry reflected through database screening: serendipity or survival strategy?, *Immunology Today* **17**, 357–8 (1996); M. A. Atkinson, Molecular mimicry and the pathogenesis of insulin-dependent diabetes mellitus: still just an attractive hypothesis, *Annals of Medicine* **29**, 393–9 (1997).

13. It has come to my notice that the 'dance floor' image as part of the imagery of Red Queen thinking was seemingly first used by R. H. Whittaker (Evolution and diversity in plant communities, in Brookhaven Symposia in Biology, *Diversity and Stability in Ecological Systems*, pp. 178–96 (National Bureau of Standards, US Department of Commerce, Springfield, PA, 1969)), not by Donald Levin, as I had it in Volume 1 of *Narrow Roads of Gene Land*. Whittaker seems to have implied only ecological interactions of *species*, however, whereas Levin added a dance of the *genotypes* within them. To cap these I am proud to promote my own version as having included heterosex *couples* pairing off like minichromosomes (Chapter 10 of Volume 1). But, of course, you can't carry this kind of imagery too far. If you try to go further with this, the Red Queen one, we will need pickpockets among the dancers in my version, these having their own dance on the same floor with their victims, or, if you like, more literally, vector mosquitoes having their mating swarms over the dancers' heads, and within them, in their stomachs, the malaria parasites having theirs, and all this in the air above; and so on—altogether a very wild carnival but one very suitable, it must be said, for all tropical streets and jungles where carnivals and parasites and sex all attain their summits and all do sometimes mix together in many-level orgies.

PARASITES AND SEX[†]

JON SEGER and W. D. HAMILTON

━━━

Parasites of many kinds have long been recognized as important regulators of population size (e.g. ref. 1), but only during the last decade or two have they been widely viewed as the protagonists in fast-paced (and long-running) evolutionary thrillers involving subtle features of the biochemistry, anatomy, and behaviour of their hosts. On this view, their power as agents of evolution derives from their ubiquity and from the great amounts of mortality they can cause (which are also the properties that make them effective agents of population regulation) and, just as importantly, from their imperfect (but improvable) abilities to defeat the imperfect (but improvable) defences of their hosts. Thus each party is expected to experience the other as a change-able and generally worsening part of its environment. In principle, prey and predator species have the same kind of relationship. But predators usually have generation times as long or even longer than those of their prey, while parasites may have generation times many orders of magnitude shorter than those of their hosts. If this asymmetry allows parasites to evolve improved methods of attack much faster than their hosts can evolve improved methods of defence, then the hosts' best defence may be one based on genotypic diversity, which, if recombined each generation, can present to the parasites what amounts to a continually moving target (e.g. refs 2–14 and Chapters 1, 2, 5, and 8).

In this chapter we discuss the idea that parasites may often play an impor-tant part in the maintenance of sexual reproduction. First, we distinguish the problem of maintaining full-fledged sex from that of maintaining genetic recombination in a species that always reproduces sexually. Then we describe the kinds of arguments and evidence that have been advanced to support the view that parasites may be uniquely able to generate the large selective differences that are required to pay the 'cost of sex' in most species. Finally, we discuss a few of the many specific predictions that can be derived from different versions of the host–parasite hypothesis. Whatever weaknesses this hypothesis may have, untestability does not seem to be one of them.[15]

[†] In R. E. Michod and B. R. Levin (ed.), *The Evolution of Sex: An Examination of Current Ideas*, Ch. 11, pp. 176–93 (Sinauer Associates, Sunderland, MA, 1988).

We use the term 'recombination' to mean the creation of new gametic associations of existing alleles at different loci, through the mechanisms of crossing-over and reassortment of chromosomes. Because the nucleotide positions within a single functional gene can be viewed formally as separate loci, some forms of intragenic recombination are included in our definition. We use the term 'sex' to mean nominally biparental reproduction involving differentiated male and female individuals or reproductive functions. This definition includes hermaphroditism, and even selfing, although selfing is in some respects a partial retreat from sex (as discussed more fully below).

Felsenstein[16,17] argues that all ecologically motivated theories for the evolution of sex and recombination fall into one of two categories, depending on the cause of the maladaptive linkage disequilibrium to be lessened by recombination. Theories belonging to the Fisher–Muller category invoke random genetic drift, while those of the Sturtevant–Mather category invoke selection that periodically changes direction. Models of host–parasite co-evolution show a generic tendency to cycle, or to move incessantly in some other, more complicated way, because any changes that increases the average fitness of one species tends to lower the average fitness of the other (e.g. refs 18–20). The idea that sex is a major weapon in the war against parasites would therefore seem to be an instance of the Sturtevant–Mather theory. But genetic drift caused by finite population size can also give rise to varying frequency-dependent selection mediated by parasites and thereby to an advantage for sex. Thus, in its most general form, the host–parasite coevolution hypothesis seems to be simultaneously an instance of both of Felsenstein's categories.

COSTS OF SEX AND RECOMBINATION

The problem of sex and the problem of recombination are closely related, but they are not simply two names for the same thing. Given that a population reproduces sexually, there still remains the vexing question as to why its recombination rates do not evolve downward towards zero (e.g. refs 17, 21–23), and, more generally, there remains the question as to why its recombination rates should have equilibrium values other than zero and one-half (e.g. refs 24–27). But the conditions that favour recombination, given sex, may be much less stringent than those that favour sex itself, given fully viable asexual mutants (e.g. ref. 28).

Under outcrossing, half of a population's parental investment will go into males (or, more generally, male reproductive functions). If males do not themselves rear offspring, then an asexual form that produced entirely female progenies would have a rate of increase that was twice that of its sexual counterparts, and it would drive them to extinction in very few generations.[29] This is the twofold cost of sex. It is sometimes referred to as the cost of meiosis, but it is less a consequence of meiosis than it is of the sex ratio, in species where only the females rear offspring. If males contribute as much as females do to the rearing of offspring, then there is no cost of sex in this ecological sense, because an all-female asexual clone would have no reproductive advantage over an equivalent sexual species. Nonetheless, a new partheno-genic mutation arising within the sexual species could increase, if the sexual males unwittingly paired with the parthenogenic females and helped to rear their offspring. But as this mutation increased in frequency, the number of available males would decrease, and in the end there would be a new all-female clone with no reproductive advantage over an equivalent sexual species whose males mated only with their sexual conspecifics (e.g. ref. 30). Selfing reduces the cost of sex by permitting the evolution of strongly female-biased patterns of reproductive allocation, but it also reduces the potential benefit of recombination by creating extensive homozygosity. In many respects, partial selfing can be viewed as a continuously adjustable approach to asexuality.

It is sometimes argued that obligately outcrossed species lack the genetic variation that would easily allow them to give up sex, and that except for this constraint, many would do so. This presumed inability to experiment with asexuality is often viewed as a product of group or species selection, on the assumption that sex may permit the long-term survival of a population, despite its short-term disadvantages. But many species of cyclically or facul-tatively parthenogenic animals (and self-compatible plants) could easily go literally (or effectively) asexual, if selection were pushing them in that direc-tion.[28,31] It follows that outcrossed sexuality, with its attendant sex-ratio penalty, must somehow be paying its own way in these species, and that a satisfying general explanation for the prevalence of outcrossed sex should not appeal to long-term group or species selection, but should instead identify short-term benefits of sexual reproduction that give it something like a compensating twofold advantage over asex.

Nonetheless, strong development barriers against an easy switch to asexual-ity could exist in some taxa, and these barriers could have evolved through group selection. A sexual lineage that easily gave rise to viable asexual forms might often find itself driven to local extinction through competition with its own asexual forms. If these in turn were doomed to early extinction through inflexibility, then the lineage might not leave many descendants, compared to an equivalent lineage in which there happened to be no developmentally feasible route to asexuality. But the fact that asex and selfing are viable altern-atives in some taxa shows that short-term advantages to sex must exist, at least in those taxa. And there is no reason to suppose that such advantages exist only in taxa showing lapses from exclusively outcrossed sexuality. Thus, even if we knew that group-selected barriers to asex had evolved in some taxa, the problem of the twofold cost would still be with us.

These arguments are well known, and they are discussed in several chapters in *The Evolution of Sex*. We have rehearsed them here to emphasize that a full explanation for sex requires that it often has large and persistent selective advantages over asex. The attraction of parasites is that they seem likely to be able to generate such advantages.

PARASITES AND POLYMORPHISM

There can be no benefit in reducing linkage disequilibrium unless ecologically significant genetic polymorphism actually exists. Thus we need to ask, first, whether host–parasite interactions are expected to cause the accumulation of such polymorphism, and, second, whether there is evidence that they do so.

Theory

Host–parasite coevolution can be viewed metaphorically as an 'arms race', in which each side is continually searching for new and improved methods of defence or attack. If the new methods are unconditionally better or worse than the existing methods, then the mutations giving rise to them will either sweep to fixation or be lost, and there will be no tendency to accumulate polymorphism. But if better methods of defence or attack tend to cost more than their alternatives (in that they drain resources away from reproduction), then a 'better' defence will be of net benefit to a typical host individual only if the host is likely to be attacked by a parasite against which no weaker (and cheaper) defence will work. The parasite faces a similar dilemma, and thus the two species find themselves playing an evolutionary game that is closely related to the well-known 'war of attrition'.[32,33] Similarly, if the differences among phenotypes are qualitative, such that particular defences simply work best against particular attacks, and vice versa, with no differences of intrinsic cost, then host and parasite can be viewed as a pair of coupled multiple-niche models, in which each species provides a variable environment for the other. In either case, the relative fitness of a given defence (or attack) will depend on the frequencies of the different attacks (or defences) currently being employed by the other species.

Under these circumstances a polymorphic equilibrium may exist (e.g. refs 34 and 35), but if the full genetic and population dynamics of both species are included in the model, the equilibrium is likely to be unstable (e.g. refs 9, 18–20 and 36–46; see also ref. 47). The boundaries are usually unstable as well, which implies that the two species will engage in some kind of permanently dynamical 'chase' through the gene-frequency and population-size planes. As one host genotype increases in frequency it favours the increase of the parasite genotype best able to exploit it, whose subsequent increase lowers the fitness of that host genotype, allowing a different host genotype to increase, which favours a different parasite genotype, and so on.

Levin[3] applies Whittaker's famous coevolutionary metaphor[48] to the special case of host and parasite, who sweep back and forth across the evolutionary 'dance floor'. This captures the sense of lively, coupled movement, and almost makes it sound like fun. Or do they tend to bump into the walls? In particular, do the dynamics tend to keep the host species polymorphic? If not, then the host will usually have little to gain from recombination.

In the simplest one-locus models without mutation or migration, the two species either circle endlessly in a neutrally stable orbit determined by the initial conditions (if the model is cast in continuous time), or else they spiral outwards towards the boundaries (if the model is cast in discrete time). Figure 11.1 illustrates the dynamics of one such model. In most models of this kind, as in this one, the fitnesses of the host genotypes depend only on the current frequencies of the parasite genotypes, and vice versa. But the current state of the parasite population reflects the recent history of the host, so the fitnesses of the host genotypes depend, indirectly, on their own frequencies over many previous generations. The same is true of the fitnesses of the parasite genotypes. Thus any genotype that was common in the recent past is likely to suffer relatively low fitness at present, because of the evolutionary change that its commonness induced in the other species. In effect, the current position of each species is a 'memory' of the recent history of the other, and so the fitnesses of the genotypes within each species appear to exhibit negative frequency dependence with a time delay, even though there is no explicit intraspecific frequency dependence in the model.

In general, there are no stable internal equilibria in models of this kind unless they incorporate some form of explicit intraspecific frequency dependence, density dependence, or heterozygote advantage. Thus in a finite world, their cyclical dynamics would be likely to degenerate into irregularly spaced episodes of monomorphism, punctuated first in one species, then in the other, by the reintroduction of the lost allele and its rapid passage to fixation. This would not seem to be very favourable for sex, because any given locus would tend to be monomorphic for long periods of time. But there are several ways to rescue the situation.

First, there is the appeal to mutation and migration. The tendency to spiral outwards can be arrested even in the simplest discrete-time models by remarkably small inputs of genetic variation uncorrelated with the current state of the population (Fig. 11.2). A low rate of mutation or migration gives rise to a stable limit cycle near the boundaries, and as the rate is increased the cycle shrinks inwards towards the central equilibrium point, which is finally stabilized at rates above a certain critical value (see the legend to Fig. 11.2).

Second, there is the appeal to multiple alleles. The two-allele model of

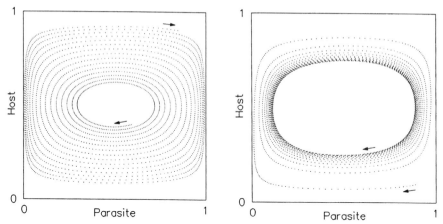

Figure 11.1 (Left above) Gene-frequency trajectory of a simple one-locus host–parasite model. Each species has a single haploid locus with two alleles, at frequencies h_1 and $h_2 = 1 - h_1$ in the host and p_1 and $p_2 = 1 - p_1$ in the parasite. Hosts and parasites encounter each other at random, so the probability that any given host individual is attacked by parasite type 1 is proportional to p_1. Parasite type 1 is most successful on host type 1, and parasite type 2 is most successful on host type 2, while hosts are most successful against parasites of opposite type. Thus the expected fitness of a type 1 host is negatively proportional to the frequency of type 1 parasites: $W(\text{host } 1) = (1 - s)p_1 + p_2$, where s is the penalty, to the host, caused by successful parasitism. Conversely, the expected fitness of a type 1 parasite is positively proportional to the frequency of type 1 hosts: $W(\text{parasite } 1) = h_1 + (1 - t)h_2$, where t is the penalty, to the parasite, caused by the host's successful defence. The fitness of type 2 hosts and parasites are constructed in exactly the same way. Given these four fitnesses, it is easy to write down the recurrence equations for h_1 and p_1. For $0 < s < 1$ and $0 < t < 1$, the central equilibrium at $(0.5, 0.5)$ is unstable, as are the boundaries. The case illustrated here is $s = 0.05$, $t = 0.15$. Each point shows the gene frequencies of parasite and host (p_1, h_1) in one generation, and the entire trajectory is 2500 generations long.

Figure 11.2 (Right above) Gene-frequency trajectory of a simple one-locus host–parasite model with mutation in the parasite. This model is exactly the same as the one described in the legend to Fig. 11.1, except that the parasite species experiences a mutation rate m (which can also be thought of as a migration rate). If the mutation rate exceeds $0.25st/(2 - s - t + st)$, then the central equilibrium at $(0.5, 0.5)$ becomes stable. Lower mutation rates give rise to stable limit cycles, such as the one illustrated here for $s = 0.05$, $t = 0.15$, $m = 0.0005$. As in Fig. 11.1, the points show the joint gene frequencies of parasite and host in successive generations, and the total length of the trajectory is 2500 generations.

Figs 11.1 and 11.2 always cycles in a highly stereotyped and regular way, but the equivalent three-allele model has very complex dynamics that depend more strongly on parameter values and initial conditions than do those of the two-allele models (Fig. 11.3). Without mutation or migration, the three-allele gene-frequency trajectories eventually become stuck near the boundaries, as in the two-allele model, but the time to quasifixation may be longer than in the two-allele case. This suggests that the relatively chaotic dynamics of a highly multiallelic system might be less inclined to drive particular alleles to

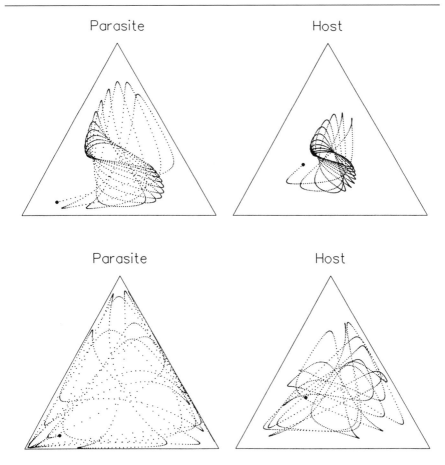

Figure 11.3 Gene-frequency trajectories of a simple three-allele host–parasite model. This model is the three-allele generalization of the model described in the legend to Fig. 11.1. Each of the three parasite genotypes successfully attacks one of the three host genotypes, and suffers a fitness penalty *t* when associated with either of the other two host types. Like-wise, each host suffers a fitness penalty *s* when successfully attacked by the one parasite type that can penetrate its defences. There is no mutation in the model shown here, although mutation in the parasite has essentially the same kind of stabilizing effect in the three-allele case as it does in the two-allele case shown in Fig. 11.2. Separate gene-frequency trajectories for parasite and host are shown here, because a phase diagram can show only the frequency of one allele in each species. The fitness parameters *s* and *t* have the same values as in Figs 11.1 and 11.2, and the trajectories are again 2500 generations long (but here every other generation is plotted, rather than every one). The upper pair of trajectories begin at a point chosen to be similar to the starting point of Fig. 11.1; the lower pair are the continuation of the trajectory shown in the upper pair. Thus 5000 generations are shown in all. All meaningful values of *s* and *t* and all starting points apparently lead eventually to chaotic gene-frequency dynamics, in the absence of mutation or migration. Both species may eventually become stuck near the boundaries most of the time, but each continues to make at least occasional passages out through the middle of its gene-frequency space.

very low frequencies than are the more regular dynamics of a system in which there are only two possible allelic states in host and parasite.

Third, there is the appeal to multiple loci. The argument here is that although most loci may be monomorphic most of the time, enough of them will be polymorphic enough of the time to give an advantage to recombination. In effect, this is the Fisher–Muller view of multilocus evolution: a favourable new mutation is almost guaranteed to be out of linkage equilibrium with alleles at other loci, and this may retard its progress to fixation. The argument applies to every kind of transient polymorphism, no matter what its cause, so parasites retain a special importance only to the extent that they are a frequent cause of adaptive gene substitutions (see Chapter 8).

Finally, there is the appeal to biological complexities not represented in the simplest models. These range from molecular and physiological details of the way in which host and parasite interact (e.g. mechanisms of acquired immunity), through their life-history patterns (e.g. parasite life cycles involving more than one host species), to their population structures (e.g. subdivided host populations with limited migration among subdivisions). In many of the cases studied to date, the addition of realistic detail has reduced either the occurrence or the severity of cycling (e.g. refs 42, 43, and 49), but in some cases it has increased the tendency to cycle (e.g. ref. 45).

One very important detail is the generation time of the parasite.[9] If this is much shorter than that of the host (as is often the case), then the parasite population may more or less fully adapt, within each host generation, to the current distribution of host genotypes. In the limit, each host generation faces an array of parasite attack strategies that depends only on the distribution of genotypes in the previous generation of hosts. This greatly shortens the effective time delay in the frequency dependence experienced by host genotypes, and thereby gives rise to shorter cycles more likely to produce sustained polymorphism at loci controlling the host's defences. Indeed, under various kinds of simplifying assumptions it all but eliminates the need even to model the actual dynamics of the parasite population, which can be represented as a simple phenomenological frequency dependence of the genotypic fitnesses of the host population, perhaps with a short time delay. This is the route taken by many 'host–parasite' models, especially those that focus on the possible advantages of sex and recombination in the host (e.g. refs 25, 42, 43, and 45, and Chapters 1, 2, and 5). In some models this process of abstraction is carried even farther, and the parasite population is represented as a regime of externally imposed alternating fitness differences among the various host genotypes (e.g. ref. 50 and Chapter 7).

In Chapters 1, 2, and 5, I and my coauthors explore one-locus diploid and two-locus haploid models with phenomenological negative frequency dependences of the form

$$W_i = \exp[d(1 - xp_i)],$$

where d is a parameter that sets the strength of the frequency dependence, x is the number of different genotypes (two or four), and p_i is the frequency of genotype i, either in the present generation or in the previous one. May and Anderson[42] examine similar models in which the fitnesses are derived from standard epidemiological models for infectious microparasites such as viruses and bacteria. Parameter values corresponding to mild frequency dependence tend to give stable polymorphic equilibria, but more extreme values give rise to two-point cycles and, in some cases, although at generally unrealistic severity, of selection, to higher-order (but still jagged) cycles and finally to chaos (see also ref. 45).

In the two-locus haploid models, these cycles may involve alternating co-efficients of linkage disequilibrium, where an excess of coupling gametes (AB and ab) in one generation is followed by an excess of repulsion gametes (Ab and aB) in the next, without any gene-frequency changes at all! Here the advantage of recombination is easy to see, because it allows the common (favoured) genotypes in one generation to produce, among their progeny, reasonably large numbers of the rare (disfavoured) genotypes that will be favoured in the next. As a result, the geometric mean fitness of a sexual population can be much larger than that of the corresponding asexual population. In effect, the sexual population 'hedges its bets' by retarding the rate at which its genotype frequencies respond to selection. These models are extremely artificial, but they show that intense frequency dependence can generate temporally varying linkage disequilibria of the kind that may strongly and persistently favour recombination.[24,25,46]

But if the frequency dependence is not so extreme, so that the host's genotype frequencies cycle only weakly or not at all, then recombination may actually be selected against. To see why, consider again the two-locus haploid model with four genotypes (AB, Ab, aB, and ab) and suppose that the fitnesses of these genotypes are identically frequency dependent according to a scheme such as the one mentioned above, so that there is a stable poly-morphic equilibrium with all four genotypes present in equal frequencies. Then there will be no linkage disequilibrium and no advantage (or disadvantage) to recombination. But if the pattern of frequency dependence is made less symmetrical, so that the equilibrium genotype frequencies are not in linkage equilibrium, then recombination will tend to generate too many of the genotypes with the lower equilibrium frequencies, and modifiers of recombination that tighten linkage will tend to be favoured by selection (e.g. refs 17 and 23). Indeed, a mixture of four asexual clones would easily keep itself at the equilibrium frequencies, and, given the sex-ratio advantage discussed above, such a mixture would overwhelm even a non-recombining sexual population.

In all of the models discussed so far, the negative frequency dependence of fitnesses is caused by the presumed complementary specificity of the interactions between hosts and parasites. From the point of view of an individual host, it is bad to be common because the parasites best able to evade your defences are themselves likely to be common. With two alleles at each of two loci, no genotype can remain rare for long. We need to ask whether any qualitatively new behaviour is likely to arise in more complex models that permit real rarity, which is to say, in models with many alleles at each of many loci.

Given an overall pattern of negatively frequency-dependent genotypic fitnesses and a very large population size, it is clear that (1) many alleles could be maintained at each locus, and (2) a system that involved several loci in the determination of the host's defensive phenotype might have great advantages over a system that involved only one or two loci. Under these assumptions the number of functionally distinct host genotypes might be very large. In an infinite world this would make no difference; a cloud of clones, each at low frequency, would still defeat an equivalent sexual species, as long as the dynamics of interaction between host and parasite did not give rise to vigorous cycling. But in a finite world there are limits to the number of genotypes that can be maintained, even in the absence of cycling, in either a sexual or an asexual population. In particular, rare asexual clones are always at risk of going extinct.[51] Rare sexual genotypes may also disappear, of course, but if their constituent alleles remain in the population, then they can be recreated in subsequent generations. As is often remarked, only in a sexual population can every individual be unique. With respect to defence against parasites there may be no particular advantage in being unique, but there may be great benefit in being very rare.

Will a recombining sexual population actually maintain the multilocus allelic diversity required to give it an average genotypic diversity greater than that maintained by an equivalent asexual population of the same finite size? Intuition suggests that it should, but many intuitively reasonable arguments about recombination have turned out to be wrong. Here one might also imagine that in a system with many loci, each individual locus is so unimportant that it could easily slide into monomorphism if unconstrained by linkage to other, functionally related loci (e.g. refs 52–54). To explore this question we constructed the simplest possible Monte Carlo simulation of a finite population subject to frequency-dependent selection with respect to an eight-locus, two-allele haploid genotype. Each of the 256 possible genotypes is assumed to determine a unique phenotype, with a fitness inversely proportional to its frequency. For a wide range of population sizes, rates of mutation or migration, and strengths of selection, more genotypic diversity is maintained under sexual than under asexual reproduction, and this is reflected in higher average fitnesses of the sexual populations. The disparity in genotypic diversity (and hence average fitness) between sexual and asexual reproduction becomes very large for population sizes on the order of 100. Some typical results are shown in Table 11.1, with the model described more fully in the legend.

In summary, the dynamics of host–parasite interaction tend to give an advantage to rare host genotypes, under the usual assumption that coevolution between host and parasite tends to produce complementary attack and defence phenotypes. But this advantage of rarity may or may not cause large amounts of variation to accumulate in the host, depending on many details of the life histories, population structures, and genetic systems of both species. The relative generation time of the parasite, its mode of transmission, its average virulence, and its effect on the population density of the host have all been identified as important variables, as has the host's ability to mount immune reactions. But polymorphism can apparently be maintained if there is some degree of complementarity between the genotypes of host and parasite, and if there exist one or more complicating factors sufficient to prevent runaway cycling of the kind that leads to effective monomorphism.

Before looking more closely at the ways in which parasite-induced polymorphisms might give an advantage to sexual reproduction, we will briefly

Table 11.1 Equilibrium genotypic diversities and average fitnesses in an eight-locus simulation of frequency-dependent selection*

N	m	Reproductive system	Average number of genotypes	Average genotypic diversity	Average fitness
1024	10^{-3}	Sex	222	131	0.99
		Asex	68	34	0.97
	10^{-5}	Sex	201	108	0.99
		Asex	18	16	0.94
256	10^{-3}	Sex	96	56	0.98
		Asex	21	13	0.92
	10^{-5}	Sex	60	37	0.97
		Asex	8	7	0.86
64	10^{-3}	Sex	25	15	0.93
		Asex	8	6	0.82
	10^{-5}	Sex	8	7	0.85
		Asex	3	3	0.66

*The model species is either hermaphroditic or asexual. There are 256 possible genotypes (eight haploid loci, each with two alleles). The fitness of each genotype is $W_i = 1 - p_i$ and around this the distribution of progeny sizes is made Poisson. N is the total population size, and m (or migration) is the mutation rate per locus per generation. Each set of conditions was run to approximate equilibrium (1000 generations), and then the number of genotypes present, the genotypic diversity, and the mean fitness was calculated every 20 generations, for the next 200. Each number given in the table is the average of these 11 figures, averaged again over four independent runs. Genotypic diversity is calculated as $1/\Sigma p^2$. Results were highly consistent within and between runs.

mention (without attempting to review) the various lines of evidence indicating that such polymorphisms exist.

Evidence

Complementary 'gene-for-gene' systems appear to be fairly common in certain crop plants and their fungal pathogens (e.g. ref. 55; see refs 7, 11, and 56–58). These systems motivate most of the fully coevolutionary models that have been published to date. Barrett[58] argues that these systems are usually more complicated and less symmetrical than is commonly believed and that the equivalent systems in undomesticated species are even messier. Thus a one-to-one relationship between genes in the host and genes in the parasite is an extreme instance of relationships that are probably more often one-to-many, many-to-one, or even many-to-many (i.e. fully polygenic on both sides). But even though the genetics of these systems are usually more complicated than the simple gene-for-gene hypothesis would suggest, phenotypes still tend to exhibit complementary specificity.

Variation in innate resistance to protozoans and helminths has been documented for several animal species, especially mice (see Chapter 5, and refs 42, 43, and 59–64), but in only a few animal systems is there yet any evidence for the complementarity that motivates the models discussed above (e.g. ref. 65). Several genetic complexes assumed on functional grounds to affect disease resistance are notoriously polymorphic (e.g. HLA in human beings), and there is epidemiological and other genetic evidence that different genotypes may vary in their susceptibility to different diseases (for entries to the literature on HLA, see refs 66–70). But showing that a system is polymorphic, or even that there is variation for resistance, is not the same as showing that the polymorphism is maintained by frequency-dependent interactions with particular species of parasites.[15] It may be difficult to imagine what else could be maintaining all that polymorphism, but as yet there seems to be little direct evidence, even of the limited sort that exists for crop plants and their fungal pathogens.

At the phenotypic level there is abundant evidence of negative frequency dependence, mainly from experiments on grasses (e.g. refs 71 and 72; see ref. 11). Although these experiments show clearly that individuals may be fitter when surrounded by unrelated nearest neighbours than when surrounded by close relatives, for the most part they say nothing about the mechanisms generating the frequency dependence. There are, however, a few experiments showing that mixtures of inbred lines may suffer less damage from pathogens than do monocultures,[73] and suggesting that this may be one reason for their superior yields.[74,75]

Many patterns in the geographical distribution of plant breeding systems and animal parthenogenesis conform to the general expectation that sex and recombination should be most valuable in stable, biotically rich environments, and least valuable in physically harsh, disturbed, or otherwise biotic-ally impoverished environments (e.g. refs 3, 4, 10, 11, 28, and 76; but see ref. 77 for a critique). This association is consistent with the view that parasites are most troublesome in biotically rich environments (e.g. in the tropics), but it is also consistent with the view that sex is an adaptation to straightforward competitive and prey–predator interactions.

Annual plants are more often self-pollinating or apomictic than are per-ennials.[3] This is consistent with the idea that perennials should be more troubled by parasites than should annuals, because they are easier to find and have longer generation times. But perennials should also be more troubled by competitors, so again the comparative evidence tends to be ambiguous.

The strongest evidence of complementary coadaptation between hosts and parasites in nature comes from the work of Edmunds and Alstad[78,79] on the black pine leaf scale, a homopteran that infests Douglas fir and several species of pines in western North America. Scales show limited dispersal, and they are completely sedentary once settled on a host. Adjacent trees often differ enormously in their total load of scales, but most trees become more seriously infested as they grow older. Through a series of reciprocal trans-plantation experiments, Edmunds and Alstad have shown that the increased infestation of older trees is explained mainly by the local adaptation of their indigenous populations of scales, and not by a general weakening of defences with age.

Scales are haplodiploid (males are haploid, and females diploid). Alstad and Edmunds[80] show that where two adjacent trees touch each other, the sex ratio tends to be lower than it is on the opposite sides of the same trees. Alstad and Edmunds interpret this as evidence of 'outbreeding depression' caused by gene flow between the two populations of scales, each of which is better adapted to its own host than to the other; males, being haploid, are expected to suffer worse from the effect than are females. Pines and firs defend themselves with extremely complex and individually variable mix-tures of terpenes and other toxic compounds (e.g. ref. 81), so it is possible that scales benefit by adjusting their own defences to the particular mixtures produced by their host trees. In principle, this hypothesis could be tested experimentally.

On balance, the existing evidence is favourable to the idea that parasites are often a cause of polymorphism at loci controlling certain aspects of the defences of their hosts, but it is not yet decisive as to the generality or the importance of the phenomenon. The main problem is that different kinds of evidence tend to come from different systems—genetics here, population biology there, physiology somewhere else. When the chain of causation has been tied together at all these levels for even one system, the fragmentary evidence from other systems will probably seem more coherent, and thus more compelling, than it does at present.

POLYMORPHISM AND SEX

Given polymorphism, there remains the question as to how it favours sex and recombination. As Felsenstein[16,17] has emphasized, recombination accomplishes only one thing: the reduction of linkage disequilibrium. If this is to be advantageous, then there must be epistatic fitness interactions between loci whose linkage disequilibria periodically change sign, either because of drift or because of changing patterns of selection.

Complementary attack and defence interactions of parasites and hosts could generate epistasis on the fitness scale, but we are not aware that this has ever been demonstrated, even in the well-studied gene-for-gene systems. To the extent that rarity *per se* is favoured, epistasis is almost guaranteed, because particular combinations of alleles may be very rare even if each of the constituent alleles is itself fairly common. This would seem to be, at least in principle, a special strength of the host–parasite hypothesis.

Fluctuating disequilibria can easily be generated by random drift, giving rise to the 'Fisher–Muller' version of host–parasite coevolution, as exemplified by the simple model discussed earlier and illustrated in Table 11.1. Because host–parasite models have an innate tendency to cycle, the perturbations caused by sampling in finite populations may also set off spiralling gene-frequency changes that generate additional, selectively induced linkage disequilibria, even where the interactions are not of a form that would sustain such cycles in the absence of stochastic perturbations (see Fig. 11.2). These selectively induced linkage disequilibria would also change over time, depending on the phase of the cycle at which the population found itself, propelled by a combination of random and deterministic forces. No such effects are seen in the stochastic model described above and illustrated in Table 11.1, because the parasite population is represented implicitly by a simple fixed pattern of negatively frequency-dependent genotypic fitnesses in the host. A finite-population model with evolving parasites would be very difficult to analyse, but might prove interesting. In any event, such a model would apparently be one in which randomly induced and selectively induced linkage disequilibria were inseparably entwined; it would therefore be simultaneously a Fisher–Muller model and a Sturtevant–Mather model, in Felsenstein's taxonomy.

In an infinite population governed entirely by deterministic dynamics, cycles giving rise to changing linkage disequilibria can also be sustained, as emphasized by Hamilton (see Chapters 2, 5, and 8), but the interactions between host and parasite need to be stronger than they do in the case of a finite population. Extreme parameter values are needed to generate twofold fitness advantages for sexual reproduction in the simple two-locus models studied to date, but more complicated multilocus or multiallele models are likely to produce large advantages for sex under reasonable assumptions about the fitness differences associated with different host genotypes.[42,82] In principle, several independent mechanisms, each of which produced a small advantage for sex and recombination, could be combined to produce a cumulative advantage of almost any desired size. This argument applies to all kinds of mechanisms, not just those defending hosts against parasites. But the members of a typical species probably face parasitic threats from many quarters and the defences involved seem likely to be at least partly distinct from each other.

Where sex involves active mate choice it can do more than reduce linkage disequilibria. In theory it can actually generate linkage disequilibria, but more plausibly, it can change gene frequencies. If hosts and parasites are engaged in coevolutionary gene-frequency cycles of intermediate length and severity, then much of the time there is likely to be heritable variation for fitness within the host population (Chapter 7). Hamilton (see Chapters 1, 2 and 5) has argued that under these circumstances, females in polygynous species might benefit from attempting to choose mates that were relatively free of parasites, and thus relatively likely to have genotypes conferring above-average resistance to the currently dominant strains of parasites. Kirkpatrick[50] describes a three-locus model in which female choice for a 'showy' male trait that reveals parasite burdens can be driven to fixation, under an externally imposed regime of alternating selection at the locus controlling resistance. Such a pattern of female choice might pay part of the cost of sex in polygynous species (which tend to suffer the full twofold cost because they typically have no male parental investment), but it is not clear how large the benefit might actually be, because the best resistance genotypes are being favoured by natural selection anyway.

Regardless of the extent to which female choice could help to pay the cost of sex, it provides an opportunity to test specific hypotheses that arise as implications of the more general hypothesis that host–parasite interactions generate heritable fitness differences. For example, Hamilton and Zuk (see Chapter 6) and Read[83] show that brightly coloured bird species tend to carry more genera of blood parasites than do duller species, as might be expected if sexual selection tends to be relatively strong in species that are relatively prone to infection. Like most comparative studies, this one cannot rule out alternative schemes of causation, but these can be examined experimentally, and several such experiments are now under way or soon to be reported (e.g. ref. 84 and see refs of Chapter 17).

SUMMARY

Selectively important linkage disequilibria involving loci that affect the interactions between hosts and parasites could be caused either by selection or by drift (or both), and they could vary either in time or in space (or both). Thus parasites, as agents of selection, are not tied even in principle to any particular category of models for the evolution of sex and recombination. As Bell[11] points out, they could play as important a role in Tangled Bank models (which emphasize spatial variation) as they do in Red Queen models (which emphasize temporal variation). The asymmetry would appear to lie in the greater dependence of Red Queen models on a role for parasites, because it is difficult to imagine what other selective agency could provide sufficiently large and rapidly changing fitness differences, involving epistatically interacting loci.

There remain many interesting theoretical issues to be explored, particularly those involving realistic details of host and parasite life histories (both of which can be very complicated), in the context of fully coevolutionary treatments of the dynamics of both species or, even better, several species of hosts and parasites (see Chapter 8). These models will be frighteningly complex. Artful simplications will undoubtedly be the key to making their behaviour understandable.

But the main need, as we see it, is for more evidence concerning the actual interactions between hosts and parasites, at both the individual and the population levels. In particular, it seems important to know much more than we do about the costs of various attack and defence mechanisms, ideally for both members of a pair of interacting species (but see refs 85 and 86, for bacteria and viruses), and it also seems important to know how far we can generalize from the complementary genetic systems of plants and their fungal parasites.[58] Without such evidence there are too few constraints on models intended to answer questions about the amounts of ecologically significant polymorphism that might be maintained by host–parasite interactions. Many of the relevant experiments are ecological and evolutionary in scope, so they will require large population sizes and large numbers of generations. Levin *et al.*[15] consider the kinds of experimental systems that are most likely to prove both tractable and useful for these purposes.

By contrast, there seem to be few limits (other than imagination and knowledge of natural history) on the number of potentially testable comparative predictions that could be generated from the basic premise that host and parasite may be engaged in a fast-moving coevolutionary struggle. For example, if infections tend to spread in epidemic fashion within large social insect colonies, then we might expect social species to engage more frequently in multiple mating, and to have higher rates of recombination, than do their solitary relatives[13,87] (see also Chapter 10). For similar reasons, butterfly species that live at low population densities might be expected, other things being equal, to distribute their eggs singly or in small groups among a large number of host plants, rather than piling them together on a few plants, where there is greater risk that an epidemic could take hold among a large group of relatively homozygous and genetically similar siblings. We expect that many interesting new predicitons will soon be made concerning the ecological and demographic correlates of mating systems in various groups of plants and animals. Comparative studies will not address the quantitative questions that arise from the abstract theory, but they may nonetheless derive a great deal of power from the way they exploit distinctive features of the biologies of particular groups of organisms.

After this was written, Burt and Bell[88] reported that excess chiasma frequency in the males of 24 species of undomesticated mammals is positively correlated with age of sexual maturity. Excess chiasma frequency is defined as the total number of chiasmata in excess of one per bivalent. The raw correlation is very strong ($r = 0.88$), as is the partial correlation taking out the effect of body size ($r = 0.69$). Neither excess chiasma frequency nor age of maturity is correlated with chromosome number, and excess chiasma frequency is negatively correlated with litter size. Burt and Bell interpret this pattern as evidence 'that crossing-over may function to combat antagonists with short generation times but does not function to reduce sib competition'.

The hypothesis is similar in spirit to the one mentioned above, concerning the recombination rates of social insects and their solitary relatives, with lifespan in mammals playing the role of colony size in insects. In each case, the factor expected to promote recombination is one that is expected to make the species a relatively easy target for fast-evolving pathogens. Social insects appear to have higher chromosome numbers than do their solitary relatives,[89,90] but Burt and Bell[88] find no evidence that chromosome number is related to age at maturity in mammals. It will be interesting to see whether this apparent inconsistency between the two groups of organisms can eventually be resolved, and whether similar patterns can be found in other groups.

References

1. R. M. May, Parasitic infections as regulators of animal populations, *American Scientist* **71**, 36–44 (1983).

2. J. B. S. Haldane, Disease and evolution, in *Symposium sui Fattori Ecologici e Genetici della Speziazone negli Animali, Supplemento a La Ricerca Scientifica,* Anno 19°, pp. 68–76 (1949).

3. D. A. Levin, Pest pressure and recombination systems in plants, *American Naturalist* **109**, 437–51 (1975).

4. R. R. Glesener and D. Tilman, Sexuality and the components of environmental uncertainty: clues from geographic parthenogenesis in terrestrial animals, *American Naturalist* **112**, 659–73 (1978).

5. J. Jaenike, An hypothesis to account for the maintenance of sex within populations, *Evolutionary Theory* **3**, 191–4 (1978).

6. H. J. Bremermann, Sex and polymorphism as strategies in host–pathogen interactions, *Journal of Theoretical Biology* **87**, 671–702 (1980).

7. H. J. Bremermann, The adaptive significance of sexuality, *Experientia* **41**, 1245–54 (1985).

8. H. J. Bremermann and J. Pickering, A game-theoretical model of parasite virulence, *Journal of Theoretical Biology* **100**, 411–26 (1983).

9. R. M. Anderson and R. M. May, Coevolution of hosts and parasites, *Parasitology* **85**, 411–26 (1982).

10. G. Bell, *The Masterpiece of Nature: The Evolution and Genetics of Sexuality* (University of California Press, Berkeley, 1982).

11. G. Bell, Two theories of sex and variation, *Experientia* **41**, 1235–45 (1985).

12. M. V. Price and N. M. Waser, Population structure, frequency-dependent selection and the maintenance of sexual reproduction, *Evolution* **36**, 35–43 (1982).

13. J. Tooby, Pathogens, polymorphism, and the evolution of sex, *Journal of Theoretical Biology* **97**, 557–76 (1982).

14. W. R. Rice, Parent–offspring pathogen transmission: a selective agent promoting sexual reproduction, *American Naturalist* **121**, 187–203 (1983).

15. B. R. Levin, A. C. Allison, H. J. Bremermann, L. L. Cavalli-Sforza, B. C. Clarke, R. Frentzel-Beyme, W. D. Hamilton, S. A. Levin, R. M. May, and H. R. Thieme, Evolution of parasites and hosts group report, in R. M. Anderson and R. M. May (ed.), *Population Biology of Infectious Diseases*, Dahlem Konferenzen, 1982, pp. 213–43 (Springer, Berlin, 1982).

16. J. Felsenstein, Recombination and sex: is Maynard Smith necessary?, in P. J. Greenwood, P. H. Harvey, and M. Slatkin (ed.), *Evolution: Essays in Honour of John Maynard Smith*, pp. 209–20 (Cambridge University Press, Cambridge, 1985).

17. J. Felsenstein, Sex and the evolution of recombination, in R. E. Michod and B. R. Levin (ed.), *The Evolution of Sex*, pp. 74–86 (Sinauer, Sunderland, MA, 1988).

18. C. Person, Genetic polymorphism in parasitic systems, *Nature* **212**, 266–7 (1966).

19. B. Clarke, The ecological genetics of host–parasite relationships, in A. E. R. Taylor and R. Muller (ed.), *Genetic Aspects of Host–Parasite Relationships*, pp. 87–103 (Blackwell Scientific, Oxford, 1976).

20. I. Eshel and E. Akin, Coevolutionary instability of mixed Nash solutions, *Journal of Mathematical Biology* **18**, 123–33 (1983).

21. M. Nei, Modification of linkage intensity by natural selection, *Genetics* **57**, 625–41 (1967).

22. M. Feldman, Selection for linkage modification. I. Random mating populations, *Theoretical Population Biology* **3**, 324–46 (1972).

23. M. W. Feldman and U. Libermann, An evolutionary reduction principle for genetic modifiers, *Proceedings of the National Academy of Sciences, USA* **83**, 4824–7 (1986).

24. B. Charlesworth, Recombination modification in a fluctuating environment, *Genetics* **83**, 181–95 (1976).

25. V. Hutson and R. Law, Evolution of recombination in populations experiencing frequency-dependent selection with time delay, *Proceedings of the Royal Society of London B* **213**, 345–59 (1981).

26. L. D. Brooks and R. W. Marks, The organization of genetic variation for recombination in *Drosophila melanogaster*, *Genetics* **114**, 525–47 (1986).

27. L. D. Brooks, The evolution of recombination rates, in R. E. Michod and B. R. Levin (ed.), *The Evolution of Sex*, pp. 87–105 (Sinauer, Sunderland, MA, 1988).

28. J. Maynard Smith, *The Evolution of Sex* (Cambridge University Press, Cambridge, 1978).

29. J. Maynard Smith, What use is sex?, *Journal of Theoretical Biology* **30**, 319–35 (1971).

30. M. K. Uyenoyama, On the evolution of parthenogenesis: a genetic representation of 'the cost of meiosis', *Evolution* **38**, 87–102 (1984).

31. G. C. Williams, *Sex and Evolution* (Princeton University Press, Princeton, NJ, 1975).

32. J. Maynard Smith, The theory of games and the evolution of animal conflicts, *Journal of Theoretical Biology* **47**, 209–21 (1974).

33. J. Maynard Smith, *Evolution and the Theory of Games* (Cambridge University Press, Cambridge, 1982).

34. C. J. Mode, A mathematical model for the co-evolution of obligate parasites and their hosts, *Evolution* **12**, 158–65 (1958).

35. J. H. Gillespie, Natural selection for resistance to epidemics, *Ecology* **56**, 493–5 (1975).

36. S. D. Jayakar, A mathematical model for interaction of gene frequencies in a parasite and its host, *Theoretical Population Biology* **1**, 140–64 (1970).

37. P. Yu, Some host parasite genetic interaction models, *Theoretical Population Biology* **3**, 347–57 (1972).

38. S. Rocklin and G. Oster, Competition between phenotypes, *Journal of Mathematical Biology* **3**, 225–61 (1976).

39. D. Auslander, J. Guckenheimer, and G. Oster, Random evolutionarily stable strategies, *Theoretical Population Biology* **13**, 276–93 (1978).

40. J. W. Lewis, On the coevolution of pathogen and host. I. General theory of discrete time coevolution, *Journal of Theoretical Biology* **93**, 927–51 (1981).

41. J. W. Lewis, On the coevolution of pathogen and host. II. Selfing hosts and haploid pathogens, *Journal of Theoretical Biology* **93**, 953–95 (1981).

42. R. M. May and R. M. Anderson, Epidemiology and genetics in the coevolution of parasites and hosts, *Proceedings of the Royal Society of London B* **219**, 281–31 (1983).

43. R. M. May and R. M. Anderson, Coevolution of parasites and hosts, in D. J. Futuyma and M. Slatkin (ed.), *Coevolution*, pp. 186–206 (Sinauer, Sunderland, MA, 1983).

44. S. A. Levin, Some approaches to the modelling of coevolutionary interactions, in M. Nitecki (ed.), *Coevolution*, pp. 21–65 (University of Chicago Press, Chicago, 1983).

45. R. M. May, Host–parasite associations: their population biology and population genetics, in D. Rollinson and R. M. Anderson (ed.), *Ecology and Genetics of Host–Parasite Interactions*, pp. 243–62 (Academic Press, London, 1985).

46. G. Bell and J. Maynard Smith, Short-term selection for recombination among mutually antagonistic species, *Nature* **328**, 66–8 (1987).

47. J. Maynard Smith and R. W. L. Brown, Competition and body size, *Theoretical Population Biology* **30**, 166–79 (1986).

48. R. H. Whittaker, Evolution of diversity in plant communities, *Brookhaven Symposia in Biology* **22**, 178–96 (1969).

49. H. J. Bremermann and B. Fiedler, On the stability of polymorphic host–pathogen populations, *Journal of Theoretical Biology* **117**, 621–31 (1985).

50. M. Kirkpatrick, Sex and cycling parasites: a simulation study of Hamilton's hypothesis, *Journal of Theoretical Biology* **119**, 263–71 (1985).

51. M. Treisman, The evolution of sexual reproduction: a model which assumes individual selection, *Journal of Theoretical Biology* **60**, 421–31 (1976).

52. R. C. Lewontin, The interaction of selection and linkage. II. Optimum models, *Genetics* **50**, 757–82 (1964).

53. R. C. Lewontin, Models of natural selection, in C. Barigozzi (ed.), *Vito Volterra Symposium on Mathematical Models in Biology*, Lecture Notes in Biomathematics No. 39 (Springer, Berlin, 1980).

54. I. Franklin and R. C. Lewontin, Is the gene the unit of selection?, *Genetics* **65**, 707–34 (1970).

55. H. H. Flor, The complementary genic systems in flax and flax rust, *Advances in Genetics* **8**, 29–54 (1956).

56. P. R. Day, *The Genetics of Host–Parasite Interactions* (Freeman, San Francisco, 1974).

57. J. A. Barrett, Plant–fungus symbioses, in D. J. Futuyma and M. Slatkin (ed.), *Coevolution*, pp. 137–60 (Sinauer, Sunderland, MA, 1983).

58. J. A. Barrett, The gene-for-gene hypothesis: parable or paradigm?, in D. Rollinson and R. M. Anderson (ed.), *Ecology and Genetics of Host–Parasite Interactions*, pp. 215–25 (Academic Press, London, 1985).

59. J. C. Holmes, Evolutionary relationships between parasitic helminths and their hosts, in D. J. Futuyma and M. Slatkin (ed.), *Coevolution*, pp. 161–85 (Sinauer, Sunderland, MA, 1983).

60. J. M. Blackwell, Genetic control of discrete phases of complex infections: *Leishmania donovani* as a model, in E. Skamene (ed.), *Genetic Control of Host Resistance to Infection and Malignancy*, pp. 31–49 (Alan R. Liss, New York, 1985).

61. D. Wakelin, Genetic control of immunity to helminth infections, *Parasitology Today* **1**, 17–23 (1985).

62. D. Wakelin, Genetics, immunity and parasite survival, in D. Rollinson and R. M. Anderson (ed.), *Ecology and Genetics of Host–Parasite Interactions*, pp. 39–54 (Academic Press, London, 1985).

63. D. L. Wassom, Genetic control of the host response to parasitic helminth infections, in E. Skamene (ed.), *Genetic Control of Host Resistance to Infection and Malignancy*, pp. 449–58 (Alan R. Liss, New York, 1985).

64. A. Sher, R. Correa-Oliveira, P. Brindley, and S. L. James, Selection of the host for resistance: genetic control of protective immunity to schistosomes, in M. J. Howell (ed.), *Parasitology—Quo Vadit?*, Proceedings of the Sixth International Congress of Parasitology, pp. 53–7 (Australian Academy of Science, Canberra, 1986).

65. W. H. Benjamin, Jr and D. E. Briles, Evidence that the pathogenesis of *Salmonella typhimurium* is dependent on interactions between *Salmonella* and mouse genotypes, in E. Skamene (ed.), *Genetic Control of Host Resistance to Infection and Malignancy*, pp. 239–44 (Alan R. Liss, New York, 1985).

66. L. P. Ryder, A. Svejgaard, and J. Dausset, Genetics of HLA disease association, *Annual Review of Genetics* **15**, 169–87 (1981).

67. G. Thomson, A review of theoretical aspects of HLA and disease associations, *Theoretical Population Biology* **20**, 168–208 (1981).

68. W. F. Bodmer, HLA today, *Human Immunology* **17**, 490–503 (1986).

69. W. F. Bodmer, Human genetics: the molecular challenge, *Cold Spring Harbor Symposia on Quantitative Biology* **51**, 1–13 (1986).

70. P. W. Hedrick, G. Thomson, and W. Klitz, Evolutionary genetics: HLA as an exemplary system, in S. Karlin and E. Nevo (ed.), *Evolutionary Processes and Theory*, pp. 583–606 (Academic Press, Orlando, FL, 1986).

71. R. W. Allard and J. Adams, Population studies in predominantly self-pollinating species. XIII. Intergenotypic competition and population structure in barley and wheat, *American Naturalist* **103,** 621–45 (1969).

72. J. Antonovics and N. C. Ellstrand, Experimental studies of the evolutioanry significance of sexual reproduction. I. A test of the frequency-dependent hypothesis, *Evolution* **38**, 103–15 (1984).

73. J. A. Barrett, The evolutionary consequences of monoculture, J. A. Bishop and L. M. Cook (ed.), in *Genetic Consequences of Man-Made Change*, pp. 209–48 (Academic Press, London, 1981).

74. M. S. Wolfe and J. A. Barrett, Can we lead the pathogen astray?, *Plant Diseases* **64**, 148–55 (1980).

75. M. S. Wolfe, J. A. Barrett, and J. E. E. Jenkins, The use of cultivar mixtures for disease control, in J. F. Jenkyn and R. T. Plumb (ed.), *Strategies for the Control of Cereal Disease*, pp. 73–80 (Blackwell Scientific, Oxford, 1981).

76. M. T. Ghiselin, A radical solution to the species problem, *Systematic Zoology* **53**, 536–54 (1974).

77. M. Lynch, Destabilizing hybridization, general-purpose genotypes and geographic parthenogenesis, *Quarterly Review of Biology* **59**, 257–90 (1984).

78. G. F. Edmunds and D. N. Alstad, Coevolution in insect herbivores and conifers, *Science* **199**, 941–5 (1978).

79. G. F. Edmunds and D. N. Alstad, Responses of black pine leaf scales to host plant variability, in R. F. Denno and H. Dingle (ed.), *Insect Life History Patterns: Habitat and Geographic Variation*, pp. 29–38 (Springer, New York, 1981).

80. D. N. Alstad and G. F. Edmunds, Selection, outbreeding depression and the sex ratio of scale insects, *Science* **220**, 93–5 (1983).

81. K. B. Sturgeon, Monoterpene variation in ponderosa pine xylem resin related to western pine beetle predation, *Evolution* **33**, 803–14 (1979).

82. D. Weinshall, Why is a two-environment system not rich enough to explain the evolution of sex?, *American Naturalist* **128**, 736–50 (1986).

83. A. F. Read, Comparative evidence supports the Hamilton and Zuk hypothesis on parasites and sexual selection, *Nature* **328**, 68–70 (1987).

84. M. Zuk, The effects of gregarine parasites, body size and time of day on spermatophore production and sexual selection in field crickets, *Behavioral Ecology and Sociobiology* **21**, 65–72 (1987).

85. B. R. Levin and R. E. Lenski, Coevolution in bacteria and their viruses and plasmids, in D. J. Futuyma and M. Slatkin (ed.), *Coevolution*, pp. 99–127 (Sinauer, Sunderland, MA, 1983).

86. B. R. Levin and R. E. Lenski, Bacteria and phage: a model system for the study of the ecology and coevolution of hosts and parasites, in D. Rollinson and R. M. Anderson (ed.), *Ecology and Genetics of Host–Parasite Interactions* (Academic Press, London, 1985).

87. P. W. Sherman, T. D. Seeley, and H. K. Reeve, Parasites, pathogens and polyandry in social Hymenoptera, *American Naturalist* **131**, 602–10 (1989).

88. A. Burt and G. Bell, Mammalian chiasma frequencies as a test of two theories of recombination, *Nature* **326,** 803–5 (1987).

89. P. W. Sherman, Insect chromosome numbers and eusociality, *American Naturalist* **113**, 925–35 (1979).

90. J. Seger, Conditional relatedness, recombination and the chromosome numbers of insects, in A. J. G. Rhodin and K. Miyata (ed.), *Advances in Herpetology and Evolutionary Biology: Essays in Honor of Ernest E. Williams*, pp. 596–612 (Museum of Comparative Zoology, Harvard University, Cambridge, MA, 1983).

CHAPTER 12

THE HOSPITALS ARE COMING

Sex and Disease

Are God and Nature then at strife,
That Nature lends such evil dreams?
So careful of the type she seems,
So careless of the single life;

TENNYSON[1]

A COMPARISON of this chapter's paper with that of the last will show what I meant about Jon Seger and clarity but still I hope that it achieves some of what I intended—that even if its notions are hazy as they flash past, it can at least be a downhill and suggestive read. It is a good example of a 'wild oat', as I defined this very technical concept in Chapter 6 of Volume 1, and this and five other 'conference' papers already included in the present volume show how my prediction in those early pages about my increasing 'promiscuity' with age has come true. One may well ask who wants to read so many mere essays concerning a scientist's work, or will look forward to still more to come; but from my own point of view, who wouldn't be tempted to write them if the writing, not to be too critically reviewed, bought you a costless magic carpet for travel to Uppsala, Berlin, Jerusalem, or Kyoto—all places where you had never been?

On the other hand, focusing the present chapter, did I really need a magic carpet to take me to the dairy lands of northern Wisconsin? I knew already its flatness, its faint morainic hills, its corn and cows: all these were sure to be very like the hills and corn and cows I had been seeing near Ann Arbor for years and, anyway, I had already visited the University of Wisconsin campus at Madison. In the same introduction in Volume 1 as the 'wild oats' were defined, however, I mentioned 'honour' as another side of the temptation to sow my oats into the literature, and here in northern Wisconsin there was a kind of April Fool honour lurking. I could guarantee a surprised look from a colleague when I said, no, I wouldn't be here that week (when he wanted me to do something—this was another benefit). No indeed, I could say, I had been invited to The Nobel Conference. It was best to pass on quickly after this, otherwise I had to explain that actually it wasn't a stage of the selection process for a great prize, instead a boarding school in northern Wisconsin held a meeting every year and called it . . . by which time, of course, it was my colleague who would be busy and moving on. So, why did I go?

The Gustavus Adolphus school and the conference sounded unusual and the school certainly deserves marks for initiative, if not caution, in deciding to expose their scholars to extremists and peddlers of unmarketable ideas like me. But what drew me across the Atlantic, more than anything, was actually the chance to be at work head to head again with Bob Axelrod in Ann Arbor for a week or so before the conference. This would be on the big simulation project on support of sex we had started about a year previously, a project already mentioned and which you will read most about in Chapter 16. At the time of the Nobel Conference, Bob, aided by Reiko Tanese, was attending to the programming and computation side in Ann Arbor while I was providing some of the biological steering from Britain. Over the Atlantic the Hermes of e-mail was busy with our back and forth but we both knew that we could make much faster progress if we could experiment together, even if only for a few days. Without the chance to visit Ann Arbor on the way I might not have accepted the invitation.

As to how the conference series came to be called the 'Nobel Conferences', the main clues seem to lie in the name of the boarding school plus American immigration history in the northern borders of prairies and just west of the Great Lakes. The school was set up by and mainly serves an immigrant Scandinavian community and it is named after a great if somewhat internationally expansive Swedish king. An ancestor of my father-in-law had been a gunner in his army, so perhaps my children should pray for the king's soul just for the mercy that he did not send his gunner to set up too many dangerous emplacements. In the spirit of this Adolphus, then, one of whose substantial achievements in Sweden had

been to inaugurate the public education system, and having conceived the
idea of expanding the minds of more senior pupils by contemporary trends
in world ideas, the school decided to ask the Nobel Foundation if they
might name the intended series of conferences after the dynamite inventor
and creator of the well-known prizes, just as they had named their school
after the previous and gunpowder-toting king. They received gracious
permission. Whether also some funds towards running costs and the little
volume that is published from each symposium flowed in consequence I am
not sure. The Nobel Conferences of Gustavus Adolphus College had
seemed able to attract outstanding speakers in a series of very diverse
topics: economics, cognition, evolution, genetic engineering, and so on.
Not only the pupils of Gustavus Adolphus school itself but some from
neighbouring schools attended the meeting and parents were present in
force. There was a kind of a school open-day atmosphere spiced with
gowns, hoods, and pomp; finally, all of it was sprinkled with the latest semi-
baked scientific and moralistic ideas from us the speakers.

Given the general hoodedness of the occasion, it was not a
conference where we speakers would normally expect startling revelation
from our peers. But I greatly appreciated meeting for the first time Peter
Raven, a famous botanist and director of the Missouri Botanic Garden, and
I found him unexpectedly supportive to my ideas about parasites as the
main key to sex. Later I recalled a reason not to have been quite so
surprised. This was an early paper he had written with Paul Ehrlich on the
evolutionary purpose of all those plant 'secondary' chemicals whose
hexagons and other more runic patterns scatter the pages of specialist
textbooks of bio- and organic chemistry—an endless stew of shapes and
combinations. Ehrlich and Raven had suggested that all of the substances
with which we scent, flavour, and de-worm—and sometimes even
terminate—our lives, are evolved in the plants not for the purposes that we
put them to but rather to be their own weapons in their neverending battle
against exploiters, with these ranging from fungi to giraffes and then back
to bacteria. In short, it turns out, surprisingly, that Socrates' experience
with the alkaloids in hemlock (*Conium maculatum*) in ancient Athens may
be the nearest to telling us about their real, original function. Basically they
are all evolved as protective poisons. After my recalling all this, Peter
Raven's favourable view of the Parasite Red Queen idea seemed to me
less surprising; indeed in a sense I had just been giving him back his own

idea of long ago while adding a few notes of my own about genetic recombination.

I also enjoyed at the meeting Lynn Margulis's superbly illustrated tour through the earlier aeons and more minute crannies of life on our planet as she expanded that strange title I mentioned in my introduction to Chapter 11. Likewise, in the sense that I also mentioned there, her talk gave more speculative justification to the title of the book of this conference than anyone else's did. Likewise, however, it exposed yet again to me her seeming increasing preference for mysticism over mechanism and for rhetoric over reality, these permeating the way all her data were discussed.

What is worth updating out of the miscellany of thoughts that I included in my own essay? What temptations did I try to throw to the new-minted minds before me and, to mix my metaphors still more, is it worth re-baiting here any of the hooks that I did throw? I am painfully aware that all my baits and hooks are of kinds that human fish are apt to touch once and then flee from for ever more; such is the intrinsic nature of the theme of natural selection. As for the fisherman who casts the hooks, I guess I can only seem a gloomier figure crouched beneath a still more green and rainsoaked umbrella today than I seemed at that Nobel Conference. The matter is serious and I shouldn't joke: really I have a duty to put back some of the baits, to give them another whirl out there with you—my presumably somewhat less-young present public. Like prophet Amos in the Bible, we scientists can't turn away from our roles either. It doesn't mean scientists have to be windbags or moralists; nor do they have to believe themselves born to pass on truths. Yet in a way their plight is even more inescapable than that of Amos the unwilling prophet: scientists study the mechanism of the Universe and, to do so, all use under some generosity or law of their culture parts of a social surplus that others have accumulated. They ought, therefore, to let their providers know of any dangers that they find, in the course of their studies, affecting society. Even though he was supported by company shares passed to him by his parents, and even though out of embarrassment over what he has found he delays for 20 years, Darwin knew that he was morally bound eventually to publish his results and that he must do so properly and without any hiding. If the public supports scientists explicitly in their study like it does me, the obligation to be truthful, and not to be unfairly self-advantaging by means of the support, is all the stronger. If the telescope that the Royal Society, with behind it the taxpayer, buys for me, for example, enables me to find an asteroid headed for Earth, I must tell this—I cannot with any justice just turn to introspective religion even if I think death is inevitable for us all and actually the best thing. So, too, with the more social issues whose fundamentals I have actually worked on: if this-and-this policy continues to be done then the mechanisms of the Universe I know of show that-and-that will follow. After the best truth one can see has been told, people can choose their course for themselves.

The editors of the conference volume preceded each of our contributions with a few paragraphs of explanation about who we were and in some cases they also inserted warnings for their readers; it was hinted that they should watch whether the supposed experts might be stepping beyond what they knew. I found myself represented as 'a persuasive writer … difficult not to agree with … , but history should not bias your judgement', and readers were asked to consider whether Hamilton's critics might be right when they 'suggested his arguments were not applicable to all sexual species', continuing: 'Are all the sexual members of the microcosm plagued by diseases? Do plants that are free from diseases violate Hamilton's hypothesis?'

My answers to these, as you can guess, are a loud 'Yes' for the first followed by an almost deafening 'No, because there aren't any' for the second. The editors had asked me which other speakers I esteemed most highly and seemed to note carefully the rather cautious remarks I made, so I imagine they had also obliquely sought opinions about me. This made me wonder when I saw the book if it might have been Lynn Margulis who primed the questions; two ideas suggested it. One is her earlier expressed contempt for the 'puerile numerologists' of evolution theory (meaning, of course, such as me), who, she evidently supposes, never look at organisms either by microscope (see Volume 1 of *Narrow Roads of Gene Land*, p. 267) or in the field and therefore simply fail to see how inevitably sex must 'emerge' out of all the world's orgasmic togetherness, although 'emerge' for what purpose is never clear. The other is the way microbiologists are so apt to deny that their subjects suffer diseases, even though when looked for they seem always to be found. Admittedly 'disease disease agents' tend to be more mobile, variable, and intermittent than members of microbial mutualisms whose occasional long-term importance Margulis has justly emphasized for us. Anyway, these ideas suggested to me the input of a microbiologist. Frankly I believe there is a tendency for microbiologists themselves not to look enough at wild microbes down their microscopes or else they would know the wild-disease scene better. When not primed by outsiders to become defensive, microbiologists often readily admit to having noticed huge die-offs in their populations that they don't know a cause for. In those cases where they have managed to focus on the dying organisms before they disappeared, they have often found viruses to be massively involved.[2] Sea water, for example, can be virus soup; it can have from 1 to 250 million virions to a millilitre.[3] If you think of it, even 1 million is a lot to be in a teaspoonful, and beyond this it is salutary to remember that a virion isn't a virion unless some host cell's machinery was subverted—always to the death—in order to produce it.

I am even conjecturing that this recently discovered abundance of marine viruses, doubtless connected with the generally favourable environment provided by sea water for disease spores generally, may be connected with the fact that parthenogenesis is much more rare in the sea than it is in freshwater (especially contrasting to life in small and temporary lakes and ponds) or on land. Mutualist viruses are, admittedly, not an impossibility at the level of the whole organism. But apart from retroviruses, which, through their strong tendency to gain vertical transmission with immediate entailed interest in the survival of the host (and there may be several possibly important mutualisms fanning from this[4]), I can hardly think of a single case of mutualism being proven except for those seeming to occur among the viruses themselves.[5] At the level of the host cell, however, every virus that emerges has to have been that cell's parasite by definition.

That small organisms offer fewer niches for parasites and hence tend to be hosts to fewer kinds is, of course, a fact I don't contest, but its main outcome is simply to highlight the correlation of high sex with large size, a correlation which I consider one of the strongest circumstantial points in favour of the Parasite Red Queen argument and mention as such in various of my papers. As to the big organisms, the trees you pass as you walk by a wood may indeed look healthy enough; but peer carefully among them and you will always notice some that are being suppressed; how healthy do these ones look? How easily the eye of optimism passes over these lost tree souls of the wood and their diseases! Look even at a leafy twig of the most successful of the victors and then count, in full summer, the proportion of leaves findable that show absolutely no sign of attack by insect, fungus, bacterium, or virus. Large organisms in the wild are never completely healthy.

Back with the microbes, it is those groups that, by the evidence of their taxa, could once perform meiosis in their evolutionary past but never do so now—nor show any sexual fusions—that worry me much more than the very small and seemingly primitively asexual ones (prokaryotes). But even for those I find ever more ways by which I can see them fit under the Parasite Red Queen hypothesis. I learned recently that the spores (black usually, as expected for needed ultraviolet protection) that customarily float highest in our atmosphere belong to the so-called 'fungi imperfecti'; that is, to the group for which the sexual stages have never been seen even though, as implied by their very inclusion in the fungi, all the species certainly are descended from sexual ancestors. Asexuality accompanying disparate dispersal rates between hosts and parasites has come as a theoretical prediction from Red Queen models[6] and examples illustrating the idea extend all the way from the cloud-borne spores just mentioned to earthly dandelions, while, in the other case of being far more sedentary than their parasites, there are the snails and the wingless cave crickets and stick insects. As to existence of some unparasitized plants, as implied by the Gustavus Adolphus editors above, presumably here the critics meant plants larger than green independent unicells. Well, I challenge anyone to show such an unparasitized multicell plant species to me; that is, show me any unparasitized population that is not either controlled by humans (under glass, existing strictly *in vitro* or the like) or else, *ex vitro*, not being treated heavily with manufactured pesticides. In agriculture such externally applied chemicals substitute, of course, exactly for those heterogeneous internal or structural natural pesticides and deterrents that, as Raven and Ehrlich were first to point out, are normally manufactured within every natural plant's own body. Just as expected we find these chemicals most in those tissues that are most crucial for the plant to defend (they are abundant, for example, in inner bark, the key artery of the tree, and are also made more in long-lived expensive evergreen leaves than in 'expendable' deciduous ones[7]).

But why am I suggesting that anything is gloomy in these particular rejoinders to critics; why is my image of the angler catching no interest?

The best entry to the serious human side that follows in the rest of this introduction, and affects all potential resolutions of the problem of sex currently considered in evolutionary biology, is to reprint verbatim the last exchange in question time just after my talk at Gustavus Adolphus College. Observing exactly the spirit of the editors' opening instruction to the schoolchildren, let us look carefully at the last of those questions that came to me:

AUDIENCE QUESTION: Just how much damage has medicine inflicted on the genetic reserves of humans by its attempts to keep genetically unfit individuals alive and producing children?

MAYNARD SMITH: Speaking as someone who would be blind and useless in a hunter-gatherer society without my spectacles, I'm extremely glad that people are keeping people like me alive and even allowing us to reproduce.

HAMILTON: I'm all in favour of keeping John Maynard Smith alive, or others like him, but I think there is a potential problem in that we get better and better at treating more and more people with genetic defects. I think there is a possibility we might build up a situation where every one of our descendants has several lethal genes which require medication all of the time. If there were to be some breakdown of civilization, an earthquake or other kind of crisis, such that the medication could not be provided, then we might have an unstable, escalating crisis in medical care. If everyone is a diabetic, for example, the operatives in factories making insulin had better not join in any general strike or they may end killing themselves along with the rest of us. In the midst of such a crisis, the only survivors would be some lucky people on some South Pacific island that never had medical attention and so were still competent. I do think there is going to be a growing problem along these lines.

MAYNARD SMITH: Don't you think that in the timescale in which medical treatment is going to lead to an increase in the frequency of deleterious genes, we are going to see techniques for actually changing the genes themselves (that is, eugenics)? If we can actually transform deleterious genes into beneficial genes in the germ line, then I see no reason why in a hundred years' time we wouldn't do just that. Maybe I am too optimistic, but I see that as a sensible response to your idea.

HAMILTON: Such technical eugenics would indeed seem very

desirable. However, techniques are not going to arrive by magic, nor are corrected genes going to insert themselves into embryos by magic. Perhaps we should think right away about being more open-minded about research on early human embryos than we are now if we want to go that route. Some may decide that such research is even less desirable than going along with old-fashioned natural selection.

As to how the audience question came up without being already answered, as it is now at least in sketchy fashion, in the paper you are about to see, this is because I had written for the talk more than I could get through in my own-hour address at the meeting. Because the four pages relevant to the question came at the end of my paper, I hadn't reached this last section during the talk. But the audience member appeared to have sensed where my lecture was going. If the reader glances at the four pages now she will see that technological advances that I expected to happen 'soon' have virtually happened already (1997), this only 10 years after the publication of this conference lecture. At least, they have happened if we can take it that it's not expected to be any more technically difficult to clone new humans out of 'adult' non-germ-line cells than it has been to clone new sheep. In this sense I have proved a successful prophet and the theme of human parthenogenesis that I often raised as a joke with my graduate class in Ann Arbor has become possible reality.

But what has cloning to do with the question asked? Notice that the discussion has transferred from optical aids such as spectacles and an implication of other possibly more 'somatic–genetic' modes by which the effects of bad genes might be corrected in phenotypes, to corrections that might in the future be made in germ-line cells—that is, in those cells that go on to form the gametes and then through them to contribute (half by half) to whole individuals of the next generation. The questioner at the meeting evidently wanted to focus 'eugenics' in the old Galtonian sense, or at least to do so with regard to what is often referred to as 'negative eugenics'—the elimination of genes agreed to be defective. He did not intend to question me in Maynard Smith's sense—that of somatic genetic engineering. I had meant to cover the import of his question in my talk but had run out of time.

Now recall all the public fuss that greeted the publication of the new technique for cloning a sheep, recall the media pronouncements, speech by the Pope, comments by experts of all sorts, and remember how that fuss appeared to be simply reflecting a public horror at the idea that the tinkering with 'natural' processes that went on to create Dolly the sheep might shortly be applied to humans. If Professor Maynard Smith, a present doyen of modern evolution theory, believes that in a hundred years we are going to be able to correct an abnormally mutated gene in a human germ line he needs to say how we are to deal, along the way, with this kind of public reaction. People want to be *natural* and want so far as possible to be naturally healthy: the most common ideal, perhaps, is to have a physique like a gymnast or a football player whether you actually are or ever intend to be in that line of work. The best I can imagine is that Maynard Smith believes that experimentation of the kind that created the clonal sheep— those physical extractions of cells and eggs and transposition of chromosomes—should be pushed ahead and extended to monkeys and then extended thence also to various non-human germ-line cells, with this done with great care and detail in preparation for the switch in which we will suddenly achieve all such transfers perfectly in humans, without losing a single embryo.

Can the corrections he imagines really be done without a preceding tinkering *in vitro*, with lots, probably hundreds, of lost embryos generally of the kind that led up to the cloned sheep? Perhaps I am dropping back in my understanding of what is already possible; perhaps Maynard Smith and others who seem so happy with the present course of genetic engineering and with its promise to dispel all the worries of the older eugenicists, already see way past those usually inevitable 'research and development' phases to uncomplicated procedures for correcting germ-line genes, news of whose possibility hasn't yet reached me and that we would all happily accept—just as we would accept the taking of the wondrous gene-correcting pill. But certainly I don't see that far. Even the prospect of 'human artificial chromosomes' (HACs)[8] that are now being proposed as possible germ-line additions, and which are the nearest seemingly yet imagined to be able to provide something like the effect of a gene-correcting pill, will do nothing to stop the slow deterioration of the tens of thousands of other genes that the HACs provided in the germ line have not been engineered to carry. Indeed, HACs seem at first a possible path— the path of *ad hoc* HACs as one might poetically put it. When available HACs might be added progressively until all the old genome had become redundant, and perhaps even at some stage the old chromosomes might be engineered to be turned off by some particular HAC so that the beneficial wonders of the inserted team would not be interfered with. But what is happening to the new engineered human chromosomes as they in turn make their way down the generations? The rain of mutation doesn't falter: the *ad hoc* HACs also are mutating. As they become unwieldy (as they surely will have to) or become very numerous (see below) we shall soon need a second generation of HACs to put right the now-defective genes of the first HACs. And so on. In short, what seems inevitable in this course, once it is begun, is a radical replacement of human genomes, in all their varied and (it may be) always slightly defective glory, by a standard version that is being kept up to scratch through continual efforts in labs and libraries. Then, of course, it's goodbye, too, to any notion that your genes carry those positive and special traits of humanity that are latent in you alone and worth your efforts, through love and marriage, to propagate. People as we know them, in their present infinite variety, will have ceased to exist.

Leaving aside technical details, in the line of the HACs or otherwise, let us imagine for a moment that the optimal procedure has indeed been achieved: you are given a pill to take for 6 weeks and in consequence throughout your ovaries or your testes that longstanding curse of your ancestors, the bad gene of which you carry a copy (the gene of cystic fibrosis or amaurotic idiocy, let us say), will have been put back to (or be at least in sufficiently paralleled to) its normal form. Your descendants will never have to worry about the disease again; if it does come again it will be due to a new mutation, which now, in your line, is no more likely to happen than in any other. If your descendants marry carefully they need not worry until after this re-mutation. Even if they marry carelessly, their offspring can, of course, again be fixed, providing there is enough cash (which is to say social surplus) flowing from somewhere, and this in itself is not too unreasonable to suppose. All this applies equally to all of us, or at least it will apply if a sufficiently liberal public-health service continues. Such a cure would indeed be wonderful; yet I see nothing in the literature that even begins to discuss how such a point is going to be reached: germ-line interference has become virtually a taboo subject today. All the recent books about the genetic and reproductive marvels predicted for the human future are completely silent about it; in some of the many books the word 'mutation' is not even indexed.

Until such technical germ-line solutions are attained (if they ever can be for all defects—and I have already hinted at what I see as the snag of indefinitely continued correction via HACs or the like), what we have to contemplate is a drawn-out messy affair of tinkering with oocytes, egg cells, primary spermatocytes, and such like. Hopefully for progress, God at some point may decide no longer to forbid us what everyone seems certain today that he does forbid; namely, tinkering with fertilized ova and early embryos and finally cleaning up the lab bench by letting some of them die. In other words, to attain our ultimate goal we will need a very different public attitude from that which recently greeted the cloning results at the Roslin Institute, and also from that which has accompanied all the other forays by technology into the human germ line and into early human embryology. My guess is that if anything as un-messy as the pill technique imagined above (or merely no more messy than the more stopgap HAC technique) can be developed, it will need early on an attitude extremely different from the present one. Even with early terminations allowed, it is quite likely that sufficiently effective techniques will never be developed. In any case, it is on these issues that John Maynard Smith and various other prominent figures of modern genetics and evolution should be giving us realistic sketches of their imagined procedures before gloss-painting for us their genetic utopias that suppose real genetic cures for genetic diseases that are supposed to be arriving. To my mind even the matter of providing phenotypic patches is being gloss-painted as easier and more benign than it actually is.

I would very much like to be reassured. Even one disease that could be put right in the germ line by a 'pill' pinpointing the wayward gene would be a vast achievement. I personally would not fuss too much about the 'messiness' of the techniques providing the path to this triumph. Intrinsically I seem (though, possibly more than I realize, it may be in the rational evolutionary point of view I have developed) not to care nearly as much as most people about early human embryos: for me, tiny embryos that seem to be in fish-like stages of human development, or earlier, are genuinely fish-like or even more primitive. Indeed, I believe additionally that they experience less of pain, fear, and danger than fish experience. Because of this I would genuinely be happier 'terminating' such early human embryos in a Petri dish than I would be 'terminating' an adult fish in an aquarium. The whole evolutionary rationale I have developed by my life's work (that is, perceiving nature through the notion of 'inclusive fitness and the like—see Volume 1) adds up to an idea that dying for the good of others similar before or to follow, is probably much more natural for a human embryo than it is for an adult fish.

We can see plant embryos, for example, doing this all the time. Have you ever wondered what has happened to the other two embryos that once quite obviously existed alive behind the other two 'eyes' of a coconut? Related palms usually have either three or six ovules per flower and commonly most or all of them get fertilized. A religious evolutionist might believe that God, for some good reason, originally guided three embryos into the gynoecium of the palm flower; but she would then have to admit that, at least since the genesis in evolution of the genus we now call *Cocos*, He must have ceased to intend that all three should survive, because now from the three ovules there comes always only one seed. Which of the three original embryos survives? I strongly suspect it is the one that is the most vigorous from an early stage of its growth in the green fruit. For many embryos finding themselves with bad genes, or experiencing unlucky accidents in their development, eventual death is well nigh unpreventable. Then why not make it come sooner for the benefit of siblings, so less material is wasted? I am convinced, having identified many parallels to support the idea, that such defective embryos hardly care about their fate, or, in so far as they do care, find it actually sweet—once the battle of who is the best of the set has been decided—to be in a position of lending aid to a sibling. Thus they willingly die, possibly even, unasked, they auto-destruct.[9]

Back with the adult fish in the aquarium we see a major contrast; there is almost no reason why any fish will want to auto-destruct. All my study of natural selection and of natural history tells me that adult, or even quite small fish that are already able to move and to avoid threatening stimuli, will feel some analogue of fear as I approach with my scoop or my ether jar: they really try to avoid my interference with them even when my aim is just to scoop them from the water. In short, my point with both fish and coconut embryos is that the limits of fear, worry, and pain are proportional to what fear, worry, and pain have evolved with purpose to attain in normal situations. Illustrating normality, I don't fear every day, for example, that a meteorite is going to fall on my head: the event is unlikely and to be all day glancing upwards would be very distracting and probably quite ineffective, while living in a bomb shelter is very inconvenient. Illustrating attainment, many will have noted in wildlife films how, when finally downed by the lion, the wild antelope seems to cease to feel pain at exactly that moment when further struggle to escape has become pointless. Evolutionarily all this is understandable: when death is certain the avoidance reactions associated with pain and fear no longer have any point.

These thoughts, of course, don't tell us that the killing of a human embryo should be a light matter. There are the feelings of the parents to be considered: how easily can they create a replacement, how far are they themselves persuaded by my arguments above, as opposed to imagining an infant wandering alone in limbo or tormented in hell? And so on. But basically my work can provide a rationale and some systematic guidance for the humane treatment of animals as well as humans. More even than pain, fear is the point of it all. Brave spirits have seen this for some time. For example, Miriam Rothschild has written perceptively that it seems to be almost impossible to bring a farm animal to a slaughterhouse without the animal showing signs of fear. Perhaps that is just the smell of blood lingering there; blood-smells are obviously something that prey herbivores in the wild ought to be sensitive to in the interest of their safety. In that case perhaps a painless drug making the animal lose its sense of smell before the journey could be used. In any case, with the problem of this fear in mind, Miriam Rothschild arranged a striking experiment that may have revealed the very kindest way to cull sheep for butchery. It was to provide a marksman with a silenced rifle, and let him, from a hide at the side of a pasture, fire through the heart of each of a few sheep that needed to be culled from the flock. The sheep so killed simply lay down; around them their neighbours continued feeding as if nothing had happened. Excepting those hit reeling from shock and alarm for just a few seconds after the heart had been pierced, no sheep during the whole operation even behaved nervously, let alone with apparent fear; it was as if the unfeared meteorite I referred to above had struck—as if I was dead, while in the eyes of the crowd around me I had simply chosen to sit down. After dusk the dead (seemingly still sleeping) sheep were carried away. Unfortunately Miriam's procedure at present seems impractical for culling on a large scale.

With treatments that might be tried on human embryos, such as rectification of defects in the human germ plasm, troubles for me begin much later than the stage of the early embryo, and come in quite different ways. If that wonderful tinkering of a germ line in an experimental animal, to put a particular defective gene back to normality, ever comes to pass roughly two general avenues for treating our own species will open up.

The achievement with cystic fibrosis by itself, to take that example again,[10] would affect only a small minority of the population. Thus the first of my major difficulties in imagining any generalization is that we need to lay out all the slightly different methods necessary for the several hundred known genetic ailments—haemophilia, phenylketonuria, Parkinsonism, and so on. Second, and attended with vastly more difficulty as it seems to me, we must also set about characterizing and dealing similarly with the much longer list of all the *unknown* genes. Is this course really even possible? High priority would obviously not at first be given to genes with small but non-crippling effects—to genes for defects of eyesight, say, which only need correcting lenses, as in John Maynard Smith's comment earlier. If the work on all the known severe defects is successful, we might come to the slight ones eventually. Instead the really great trouble for me is that the list of the 'rest' of the defects is nearly endless.

We have above 3000 million base pairs of DNA taking care of coding for our perhaps 100 000 genes. The intentions to correct all of those seem to me akin to the communists' dream of a state able to take care of every detail within a vast nation's economy: all the 100 000 genes are capable of mutation into non- or wrongly functional forms. Not only are they capable of this mutation but they have a characteristic rate of random change and miscopying. The rate applies to you at this very moment: cosmic protons are zipping through your room as you read; when they hit you they do trivial damage along a streak of somatic cells that will be easily replaced; but sometimes instead they will lay ionized tracks through your testes or your ovaries and the 'free radicals' these create, knifing into your DNA, potentially create effects due to last in all your descendants. I write this partly to scare a little, and to force to take notice those of my readers who (unlike me) are not postreproductive, but also because the warning can lead on to my next point.

A key issue in all current biological theorizing about support for sexuality is whether organisms have an overall rate in their germ line that amounts to more or less than one bad mutation per genome per individual per generation. The point for evolutionary sexists like me is that a claim exists[11] that the 'mutation-clearance' theory of sex becomes a sufficient explanation if the rate is more than one per genome per individual per generation; there will then be no need for additional recourse to parasite-escaping coadaptations or to any other factor. The point here, not just for us evolutionary sexists but for more ordinary human worries, is that, whether the real rate is an order of magnitude above or below the 'Kondrashov threshold' of one (see Chapter 16), it still implies for us a truly depressing future if we are going to continue with the current enthusiasm for phenotypic curing (as by spectacles, phenylalanine-free diets, and the like) of every defect in the germ line as we identify it and as it occurs—in other words, if we just allow to continue that degradation of the human genome that the questioner at the Gustavus Adolphus meeting was alluding to.

An amount of mutation setting all of our 'lines' back all of the time, even at a rate of 0.1 per genome per individual per generation, means that although you can have a 9 out of 10 chance that nothing bad has happened to you—in your germ line—in your lifetime, very little such hope can apply to your extended family at a distance of cousinship; or, if you prefer, very little can apply to all your descendants at the distance of the ensemble of your great grandchildren. Thus even an optimistically low estimate (such as would please us of the Parasite Red Queen persuasion[12] that our theory of sex is winning, and on the same grounds disappoint most of those of the Deleterious Mutation camp) has a profound significance for how we should view feasibility of both the 'germ-line engineering' or the 'medical-fix' versions of eugenics that are proposed by John Maynard Smith. So far as I can judge from the above interchange at the meeting, JMS seems to regard the 'germ-line engineering' option as optimal eventually but meanwhile has no qualms about the 'medical-fix' line of treatment being carried as far as possible. As I have indicated I am much more pessimistic about both, the one because it's so damaging to population health and the other because it's so far off and even of dubious theoretical feasibility. So far as I can judge out of somewhat equivocal passages, Steve Jones, a prolific and popular writer on the theme of our genetic future, countenances only the 'medical-fix' type of solution for reaching his utopia.[13] This course is often referred to as 'somatic genetic engineering' because inserted DNA is to be targeted to the somatic cells only; its procedures are obviously not genetic engineering in the sense of changing Mendelian (that is, heritable) genes and therefore it is not eugenics—indeed as regards the original meaning of the word is the reverse.

Returning to the central theme, if humans turn out to be near the Kondrashov limit—that is, if on average every gamete has one bad mutation created during the lifetime of its producer—it is obviously not going to be nearly enough to test a baby for the subset of the few hundreds or so of well-characterized genetic defects for which its family history indicates it may need treatment with the wonderful pills that have been developed. There are certainly tens of thousands more possible mutations that the baby could have that have not been characterized well enough yet even to be named as genetic diseases. Thus, if we are taking the germ-line gene-engineering path, which I regard as the only one that will avoid the steady build up of instability to civilization and that will avoid the slow withdrawal of all our attributes as free-living organisms (as I mention in my reply to my questioner), then really for every baby proposed to be born we need some course as drastic as the following:

> Take a gamete from the father and an ovum from the mother;
> extract, stretch out, and scrutinize the entire length of the DNA,
> chromosome by chromosome, all 23, taking great care not to cause
> any further damage during the process. Finally compare entire codes
> found with those specified in the coming 'Handbook of the
> Standard Genes of *Homo sapiens*'.[14] Where the comparison of the
> actual with the normal shows sequences in the gametes to be
> definitely non-standard they must all be corrected by our supposedly
> already perfected genetic technology.[15] When all is done, the DNA
> must be coiled again into its chromosomes and the whole of each
> corrected single set slipped back into the readied and empty nucleus
> of an ovum of the woman who is to bear the child.

Surely in this sketch of a procedure it can be seen why I suggested that our horror at the techniques used to make Dolly the sheep is relevant to all our idealistic schemes of eugenic reproduction as well, whether they are asexual or not. Obviously such a method of producing babies would be not only an immense, and (for the present almost unbelievable) technological feat, but its use on a wide scale will indeed represent a complete finale to evolution by natural selection within our species. The method arranges that all possible advantageous changes are to be eliminated along with the definitely bad ones: we are effectively accepting that the definition of *Homo sapiens* has become fixed from the time of the publication of the imagined human genome 'Handbook'.

In the same sci-fi-like spirit that gives us the method for correcting mutation-bearing gametes we might go further, and could, I suppose, proceed right to the point where certain conformations that are non-standard by the Handbook might be 'passed' for use as having possibly a good effect, as foreseen by some ethics committee of molecular, medical, and developmental biologists of the time. Because we know that almost all mutations are harmful, the ethics committee pronouncing on this in the said future period will have to be very cautious, especially if they still have the same scruples as at present, under which, once back inside the key zygote cell, the DNA and the zygote it creates, are considered to be fully human and inviolable for their generation (once again, my second note in the recipe for a genome check refers). This, I presume, is also the attitude of the Pope as well as of many 'right to lifers' and assorted other intellectuals, Christian and otherwise. On the bright side of this gloomy picture will fall cases where the modification that the ethics committee elects to let pass indeed turn out to be good ones—better than expected. Even on the slightly dimmer side, where a particular modification doesn't turn out well, then providing the change is not so bad that the offspring fails to reach sexual adulthood, our technical wizards will always be able to make amends by putting everything back to 'standard' for next by treating the gametes that this individual produces.

To me all of this scenario is, from a practical point of view, absurd, and because there are simpler ways to proceed that involve no pain and vastly less fear, it is rather pointless to consider it. Yet in all the talk of a coming genetic utopia, other than this account, I have not yet seen any discussion of how corrective genetics is supposed to work on a scale that is really commensurate to the problem of the continual genome-wide mutation that is certainly occurring all the time. My outline may seem to you crazy, but, facing an uncertain rain of somewhere between 0.1 and 5 fitness-lowering mutations per genome per generation, and perhaps averaging 10 per cent in effect, and needing to maintain human genetical competence as at present, what is *your* idea about how to proceed? How else is it to be done; what, really, is John Maynard Smith's (or Steve Jones's) idea? In short, I am very sceptical that the problem is being taken seriously by those who promise us a coming genetic utopia.

The public loves and obeys demagogues; indeed, that's what being a demagogue means—a persuasive person. A few admittedly seem to succeed in the narrow trade of being demagogue doomsters but most demagogues are successful by giving either slightly or hugely false promises about rosy futures. There ought to be a scale on which we could rank soothsayers by their successfulness in their art—and I mean not in the first place in how correct their predictions are but rather how well they achieve notice and popularity in spite of making untrue predictions. We should think of marks by which we could scale the failed outcomes achieved by persuasive prophets in the past as well those of present practitioners. If a is the value to the public of what actually happens and s the value of what a demagogue predicted was going to happen from a course he persuaded to be adopted, then an index of demagoguery (that is, of falsity of prediction) could be calculated as $10(s - a)/k$, where k is the greater of s and $-a$ (s and a are being allowed negative values as well as positive—for example, should the discussion concern how to prevent bombings by terrorists, it might be that some explosions are virtually certain to occur). The units of this index might be named simply *demogs*, and the prophet who gets his predictions perfectly right will be given a score of zero demogs. This is a complement to him from my point of view, even if his message was one of terrible gloom, because every one knows what is coming and can optimally prepare. A prophet who predicts something good is going to happen from a course of action when actually absolutely nothing good does, or the reverse, will get a score of 10 demogs. This isn't quite the limit achievable for my index because trial substitutions in the makeshift formula will quickly show that if you predict something good and then something terrible happens you could attain a score of more than 10; nevertheless my scheme is on the whole a kind one (and perhaps too kind) for bad prophets. For example, if you predict that an approaching asteroid will not only miss our planet but, by its near passage, will improve the wheat crop in Russia (value $+1$) but then actually the asteroid hits Earth (value $-100\,000$), you score only 10.00001 demogs. This suggests that very little more contempt descends on you than if you had predicted a miss that would have no beneficial effect. Your worst case would be if you had said the asteroid would hit but it would turn out to be a blessing because you knew somehow that God, counting the event as being his mistake, would send all the killed persons to heaven as a compensation ($+1\,000\,000$); but then, when it did hit, it turned out

He had actually been angry at Earth all the time and all the unrepentants you had encouraged by your cheerful and do-nothing prediction, went to hell ($-1\ 000\ 000$). Your extra punishment then would be to bear a label of -20 demogs, which—in hell, I presume—would get you some well-deserved extra snarls.

My scheme for attaching these quantitative reputations to demagogues might work quite well for politicians, too, because they often do imply predictions and sometimes even have to commit themselves to a time limit—for example, predicting that a certain good will eventuate even during their term in office. Unfortunately for the case of the genetic demagogues I wanted to discuss, it is going to be almost impossible to pin the labels because the prophets we are thinking of will be dead long before we can assess whether their rosy predictions actually materialized or not. Therefore these partly frivolous paragraphs (in which, by the way, I am forbearing to bother you with the definition of a *cassandra*—the unit of pessimistic predictions) needs to have a serious tail pinned. Appearing as optimist, euphemist, humanitarian, evolutionary cosmetician, a Dr Pangloss the Younger, et cetera, et cetera, does actually make one extremely popular. And popularity brings immediate success; appearing, on the contrary as a pessimist, as a stern person, a doomsayer, generally has a quite opposite effect on the public. The ancient Jews so loathed listening to stern and gloomy Jeremiah that they lowered him into a hole in the ground, hoping, I suppose, that he would there either shut up and plead to be let out or at least would have his voice muffled. But the Bible holds him a prophet: he turned out right. We should ask ourselves, therefore, whether we are being misled by people who just want to be popular and to sell books. Are we letting kindly and cheerful men stop tears in our children by handing them lozenges of honeyed heroin? I say 'kindly' and 'men' here with an intent, first, because I think they really do mean to be kind and desperately want to convey messages of hope; second, because those I have in mind seem all to be males; and, third, because I think of women, instead, as being on the whole far more realistic as well as usually more genuinely and insightfully concerned about the fate of future generations. In any case I strongly recommend at least some lighthearted rating of prophets in terms of a measure resembling my demogs whenever you hear of utopias and genetic engineering being mentioned—also that you ask the mentioners to be quite specific about the steps they envisage leading to their goal. What are the techniques and how are they to be developed—in terms of practise on embryos, for example? And how to fix not just one or two but 100 000 genes, each mutating in many ways?

Short of the germ-line-added and permanent HACs already discussed, other techniques that are already practised with domestic animals and plants—for example, the importation of new genes into cell lines by using innocuous retroviruses or like means—may soon provide once-a-lifetime treatments for well-characterized genetic defects of individual human sufferers. But at the moment I don't see how these methods come near to being applicable to germ-line cells. The service of the genes to cells and to chromosomes is just too haphazard: if genes go to the wrong place, as some almost certainly will, then over the generations we will encounter all the problems of multiple and of insufficient copies. Once again I warn optimists that American parents of the hereditarily damaged will sue perpetrators for many lifetimes' worth of damages—and, for once, I agree that the litigation may be deserved. Such problems don't, of course, deter the plant and animal breeders from targeting the germ line because they can have a high proportions of failures without being either sued or called mass murderers on account of them. In agriculture, the practical advantages of the genes' insertion when it does work can greatly outweigh the losses when it doesn't; but, as stated, for the duration of the current climate, the problem with *H. sapiens* looks extremely different: some ethicists won't allow us any elimination of our failures without calling us murderers.

I will be happy if the publication of this volume goads someone to explain what other route geneticists see for reaching their promised goals, routes that, in my reply to John Maynard Smith, I doubted to exist. What have I overlooked? The best course I can think of has a degree of messiness in the research and in the intrusion of the act of making up the zygote that makes the technique leading to Dolly the sheep look like a chemical dip for sheep ked, done in a tarpaulin trough in a field, in its comparison to what will be necessary. With perhaps only a moderate messiness added, I can imagine intra-uterine procedures, which, had they been available 70 years ago, might have forestalled, for example, Maynard Smith's need to use spectacles (or my own for that matter); yet I cannot imagine those procedures except in such almost sci-fi scenarios of inspecting chromosomes as I have outlined above. Only such can eliminate the need

for a re-treatment for every defect for every generation of offspring.
Presumably Maynard Smith can imagine the techniques or else perhaps he
foresees unlimited advances in the more external aids, like his spectacles,
which he would not feel to be bothersome—better and more lenses (as
needed for the accumulation of more mutated defects), xenotransplants
into our descendants of lenses extracted from farmed eagles perhaps . . . Or
else what?

In the relative gloom of my own lack of such vision and in that still
deeper darkness of a thought that all the present optimism on these issues
(seen by me as so false, rating about 7 demogs on my scale or more) I see
something extremely like to the dreams of communism that were so
omnipresent in my youth and, like those dreams, so equally hopeless—
silvery dreams of the centralized, all-singing, all-protecting, all-predictive
economy. Seventy years of murderous hate- and gloom-filled history in
Eurasia, massacres and starvations on a scale dwarfing even those of Hitler,
have shown plainly enough how such dreams simply do not work. In the
recurrent gloom of watching (perhaps this time from the plinth of a 'sure
foundation of unyielding despair', as Bertrand Russell once expressed the
stoic's position), very similar mistakes being now proposed, I take up the
stand that natural selection—or semi-natural selection if you will allow
this phrase—is still necessary and will be necessary for us for a very
long time.

Let us back off a little now from my futurism in the last few pages to
gaze for a few minutes on the old and tried method that has brought us to,
and has kept us at, our human standard, holding us there even after the
main rush of our progress—that great up-pouring of nervous material into
our foreheads that happened through the Tertiary and Quaternary
geological periods—had ceased. The sentry on the bridge of life is as
merciless as she is great; there is no denying that finding oneself on the
wrong side of her who guards the integrity of the human genome is the
cause of a large part of the sorrow of the world.[16] But reflect that at least
she takes no pleasure in the suffering she causes and she doesn't have to be
cruel: you can ask her to shoot early and shoot straight if you want to—or,
like Miriam Rothschild, hiring her own sharpshooter, your parents at least
could so entreat her. In my opinion it is those who insist on the
prolongation of the courses and the effects of bad genes, those that promise
an eternal 'treatment' of effects, who are cruel to humans. I also cannot

avoid suspecting even that Mammon may be moving behind the cheery prophets of today, quietly pulling strings and inciting many an absurd promise. Indeed, how could the immensely powerful medical and pharmaceutical interests, busy with ever more and more profits derived from procedures in our ever-enlarging and ramifying hospitals, clinics, and pharmacies, fail to urge us onward in the courses that benefit them—fail to promise easy treatments for everything if we will only just pay? Instead of a blind acceptance of this, I ask you to consider whether there are not ways of overseeing and guiding selection that will be eventually far kinder, and whether there is not a different philosophy of the inevitable that can at least reduce the sorrow so often coming from ill chance.

One basic principle underlies both the Parasite Red Queen theory of sex and the Mutation-Clearance theory. This is that, when sexuality is present, bad, or just unfashionable genes can be eliminated several at a time. It is this that inspires my image in some of my papers of some genotypes falling into 'mineshafts' that occur on an otherwise rather level landscape of possible living states. We can take it that we all have a few bad genes; without too many, however, we don't become mineshaft victims. It is to say that our bad stuff, whatever it is, isn't generally enough to make us unmarried, infertile, or dead and thus lost from contributing to the future population (unless, perhaps, we ourselves choose to be lost by our free will, simply deciding to have no children). Put differently, having one or possibly several mutated genes (or on the Red Queen view simply again some 'out of fashion' genes—see Chapter 18), may not matter for a healthy life; but having mutations in numbers above a certain threshold brings on rather quickly the kinds of failures just mentioned. What it is like today to live (and perhaps also to die) at around this threshold of healthiness has been movingly and bravely told, including the mentioned decision to have no children, in an article in a Sunday paper by an 'only daughter of an only daughter' (these were daughters sick, more or less, in both of the generations) a few years ago. Ironically (or could it have been by an editor's intent?) the immediate previous page of the same issue of the newspaper carried an article entitled 'When your husband is a hunk'.[16] This was by another woman, healthy and good-looking enough in herself to judge from the photo, who describes being married to a male dancer and model. This second writer provides an equally vivid picture of what life is like at the top end of a 'good-gene' spectrum. Such life has its snags, too, as she describes (such as having phone numbers slipped by competitor women into your husband's pocket from the other side as you walk leaning on his arm), but clearly these snags are less. Alas, I don't believe that medicine will ever make us all or even most of us like that 'hunk' couple; but, due to the disaster scenario of the re-onset of natural selection that I have already predicted, medicine is also going to have great difficulty making us all to be like the first unfortunate woman too. Illustrating my point, even at her best I think this brave woman would admit she was in no condition to run the machines she needs for her life to be continued, leave alone to build such machines. But scientific medicine can send a lot of us in that direction and, over the generations, it certainly is trying to do so now.

Repeating the above, and as I explain in more detail in the papers of Chapters 16 and 18, the part played by sexual recombination in the living scene is the bringing of bad genes together into single zygotes so that they are dropped to zero fitness *several at a time*. I would like to be able not just to italicize but underline that twice. If recombination is disallowed, as it is in the clones of a non-sexual species, then badly mutated genes are isolated from one another and more individuals have to die as they are cleared from the population one by one; J. B. S. Haldane was the first to point out how, under parthenogenesis, one 'genetic death' (as he termed it) is bound to occur for every bad copy ever occurring if the population is to maintain its status quo. That is to say, one death must occur for every even mildly deleterious mutation that has entered the germ line of a living body. In other words, one elimination per mutation becomes a logical necessity under parthenogenesis if average fitness (in the wild unmedicated environment) is not to sink indefinitely. Obviously the parthenogenetic course is a much less-efficient way of eliminating bad genes than that which is empowered by sex when present; therefore if we are willing to ignore in our arguments the cost of sex in reducing purely reproductive efficiency compared with the efficiency in clones (which arises due to the inevitable evolution of maleness), it becomes extremely easy to find models in which, in a world with mutational and/or environmental change, sexuals succeed best. Many people, I think, delude themselves that there is no evolutionary problem of sex by speculating along these lines; for sure, many have pointed to such models in publications and a surprising stream still continues, all of it seemingly ignoring the snag of the immediate twofold inefficiency that comes from the production of males (see various previous chapters including 10 of Volume 1 and especially Chapter 16 of this volume).

Now notice that our image of stripping out the DNA from gametes and subjecting it to codon by codon examination has, of course, at least the same sort of inefficiency as that of asexual reproduction, because that, too, would not be permitting the sexual process to play its recombinatorial part in the elimination. Assuming, however, that we are allowed to reject—in effect to condemn—poorly endowed gametes (I think that even the Pope might not criticize us males for allowing rejection of at least some of our billions of excess sperm) we could still bring in sex's trump card and improve our efficiency a little by conducting a preliminary scan of gametes, searching for those where recombination has done some of the work for us. In other words, we would be looking for gametes with as few as possible of the non-by-the-book genes that we want to see eliminated. But the work of reading all the DNA in multitudinous gametes to find these best haploid genotypes is still so daunting that this course would almost certainly not effect an economy of the procedure in practice; what we really need here is almost the genetical equivalent of the famed Maxwell Demon of physics— and this means a demon who is, in effect, just as elusive and impossible.[17]

In our previous state of nature, before the great religions took hold with other theories, the naturally aborted fetus and the malformed neonate, which the mother's instinct was undoubtedly to let die, would have far higher than average probability to be the 'mineshaft' victims. These defectives, I insist, *it is the evolved function of the sexual process to eliminate*. This idea, as I say, is the common standard of both major theories of sex that stand today, and the plain fact is that we are going to have to struggle almost impossibly hard against Nature if we seek long-continued preservation of such individuals. While various sophistries are possible about how it is not in the best interests of mutation-burdened embryos or fetuses to be let die, surely for parents and siblings it is quite obviously the best thing that can happen. If calculations are needed to back up what the human heart, down all the ages, has always told, the new calculus of inclusive fitness can readily supply them. Obviously it is best for mineshaft victims themselves, as also for the family, if bearers of unusual loads can be detected and disposed of through maternal (physiological) rejection reactions already in the uterus, or else through induced selective abortions, all of it at the earliest intra-uterine stage possible; best of all, death would occur at a stage of development so tiny that not even the most officious priest or 'right-to-lifer' would even notice it, and, of course, a lot of miscarriages are exactly of this nature—commonly these underlie those mere pauses in the otherwise regular sequence of the menses. The extreme worst outcomes for ill-starred zygotes (as also for their parents and siblings) are those cases where defects that are bound to cause death or else to bring on severe disease before adulthood do not reveal themselves until long after birth.

A decision threshold of great significance to my mind (and in the light of all the above) comes around the time where a new infant starts to be well bonded with everybody close to them. This threshold is definitely not set at the moment of birth. Especially if a baby is at all noticeably deformed, it is normal for there to be period of ambivalence on the part of the mother. If the baby can survive in spite of a visible problem, and can be expected to proceed to a meaningful life, then once again the outcome seems not so bad and the mother normally warms to it and bonds. A not-uncommon course for such genetic victims is to become non-reproductive 'helpers' within extended families. There their presence may even increase summed inclusive happiness of the various members while at the same time their role is best for their own inclusive fitness also (I am thinking, for example, of those with Down's syndrome). So placed, the unlucky starters can also even become the Akhenatens, Vincent Van Goghs, Isaac Newtons, Stephen Hawkings, and, sometimes by their amazing cultural contributions, they can add happiness to the whole world in the ways creative people have even if sometimes at the expense of great sadness of their own.[18]

In spite of these often cited examples, however, I strongly reject the idea that parents should welcome ill-health in their offspring because the handicapped child's achievement, great or small against its obstacles, should somehow elevate the spirits, the humanity, and what not, in all concerned: in fact this view illustrates quite well that I meant about the sentry of the bridge of life being less cruel than many moralists are. While, of course, I immensely admire achievements of Newton and am the more astonished at them by the fact of his having been an underweight weakling at birth and expected to die (as may also have been the case with the Marfan's syndrome mutant Akhenaten[18]), I am still left imagining what Newton might have been had he the same brains without that (possible)

birth handicap. Perhaps with a 5-kilogram birth weight he would not only
have discovered all he did but would have led a happy life and raised a
large family like Johann Sebastian Bach—and in the course of this other
life been nicer to his friends and supporters and more forgiving to coiners
than he was. While I admit that it is surprisingly easy to produce examples
of creative gifts in handicapped people, I am still waiting for statistical
evidence that any real correlation exists, and am doubting rather strongly,
meanwhile, that handicap actually causes gift. It might be so, but even in
that case a question of happiness remains. Maybe I will begin to believe it
when I hear that parents commonly state a preference for their baby to be
born physically defective in order that they may have a chance to be
parents to a Newton or the like—or just, more humbly, to become good
parents to a new member in the Deaf Nation. If they don't prefer physical
handicap, this is presumably because long ages of natural selection printed
a message in them that reads 'more defect, less inclusive happiness'.[19] A
wheelchair-bound offspring trundling on the pavement to become Stephen
Hawking must surely be an immense consolation to its parents; but such a
thing happens only to exceedingly few, as everyone knows. Because of my
belief that most parents prefer not to bear a defective child, and because of
the generally low chance of creating a Hawking (or a Toulouse-Lautrec to
bring in one other example), I admit that I doubt even the common
assurance of parents that looking after their defective child has been a
'rewarding experience'. Before accepting such a claim I would like to see
more objective and statistical tests, rating such possible matters as their
depression, frustration, early divorce, and all similar likely consequences.
On similar lines never have I yet heard that sperm of genetically
handicapped men being requested (or suggested) at clinics performing
artificial insemination—nor have I heard of parents who are encouraging
their handicap and offspring to reproduce more than their other, normal
children. Indeed in the case of the more severe hereditary defects I would
consider parents who did so very anti-social because it is obvious that each
wheelchair or bed-bound victim must add to the burden of those fit and
taxpaying individuals who ultimately have to provide for them to survive.
Hence the remarks that handicapped offspring are 'blessings in disguise'
and so on to their parents seem to me to be messages merely intended to
put a tough face on an unavoidable burden.[20]

For all these reasons I am unapologetic about discussing the courses that I see as likely to reduce human congenital mutational handicap and proposing to measure success in these courses by appeal to abatement of fear and to some sort of population average of 'inclusive happiness'—full details of this statistic have yet to be worked out (I haven't tried much) but to my mind they are obviously less difficult than working out the details of attaining Steve Jones's utopia. Through the common (perhaps Jonesian?) view that natural selection is obsolete, it seems to me that we are making a grave mistake and soon will find ourselves pushing agonizingly and pointlessly against forces that will eventually prove tectonic. Following this image, what we are trying to do is like trying to recover lost territories of India by pulling them out from under the Himalayas! By kidding ourselves about some weird kindness to embryos, to neonates, and the like now, we are actually being very unkind to numerous far more sentient persons of the future—principally and firstly I mean to families and secondly (to bundle it all together) to every one. In addition we are showing ourselves to be very muddled.[21]

This is the situation that comes inevitably if we place a too high and indiscriminate a value on human zygotic life, no matter whether sentient or conscious, or neither. When as a young man I felt drawn to but decided against the manifest 'humane' attractions of communism, which many of my contemporaries seemed so wholeheartedly to embrace, it was for very similar reasons to those I explain now. I was stirred by the ideal of communism, and even by its battle-cries, but soon felt that a lie was hidden somewhere, that the demagogues were busy, and that communism's notions of what humans were or could be persuaded to become were, again, pitting themselves against almost tectonic forces of human nature as I knew it. Besides using my own intuition of human nature I was also, of course, reasoning about human nature on the lines first argued by Darwin. Since his times we have all gone much further, especially during that great spate of new understanding that came after the 1960s. I believe that history during my life has proved me justified in my scepticism about communist visions of human nature (or my pessimism if you prefer it). I believe I will be proved right in the present case, too, and, it seems worth saying so, let the hate mail come as it must. But it is long for me to wait to be right on this one. Perhaps I should get myself more attention to my theme at once, at least with most people if not with the intellectuals, if I predict how Europeans and Americans will increasingly lose out in the Olympics and other international sports, losing to all the native Africans and South Sea Islanders whom our too-skilful doctors have so far left alone, . . . how we see it happening in the European football teams more every year . . . But let me for now leave this one out; let me try out one more personal tack—in fact, switch to a specific description of the deaths of three particular 'personalities' I have known.

As I have probably made plain enough already (and as is sure to bring me the hate mail when this is published), the human neonate has, in my opinion, virtually no 'soul' until reaching an age when it starts its own ability to bond with particular adults. This onset, whose delay after birth (so different from ducklings, for example[22]) seems to me to have been provided to us like a gift from God, appears to be set about the time of the first smile of true recognition (the very earliest smiles of a baby are, as is well known, automatic and can be evoked by crude smiling models of a face, as was shown by John Bowlby in the 1950s). If early signs of a coming severe handicap can be detected during that early postnatal period, I hold that the kindest thing for the family in which the defective child is to be dependent, and for the child itself, may well be to kill it. It is well known, of course, that such an act can't normally be performed without very persistent psychological consequences for the parents, especially for the mother. But these reactions can still be turned to advantage for what I am calling inclusive happiness, as examples I am about to give will suggest: in effect, the happiness of the family, with or alternatively without the defective, is the key to the decisions that I envisage and from there the benefits branch outwards to the whole of society. And it seems to me likely that without the legal and social disapproval, the psychological consequences will probably be less severe anyway.[21,23]

Seven babies were born to my mother. The next sibling following me (the second born) was a boy, whom my mother called 'Jimmy'. We other children seldom heard about him because he died a day or two after his birth due to a congenital bowel obstruction. When Mother did speak of him it was with unusual brevity and always by that name 'Jimmy', never 'my lost one' or anything like that. Also she spoke of him with an obvious but fatalistic sorrow. Really there was not much to say about him except his name; and she often added that his dying had allowed her to give 'more to the rest of you'. Had 'Jimmy' been born today I imagine he might easily have been saved by some intestinal re-section or the like; but I know well from my mother's attitude to all these matters that she would never have allowed it if she could have prevented it. Although all of us were born in maternity homes or hospitals, in the present technical climate that says what can be done should be, I am sure she would have preferred giving birth to her babies alone and at home if she thought such an operation had been likely to have been forced on her in a hospital: she was a doctor herself and well knew the likely situations.

In my memory, as on these pages, those inverted commas always surround 'Jimmy' however he appears; and I think of the commas as being my mother's too. For all of us he was so faint an image that my mind hardly even accords him much of a shred of a soul; he is the name only—a name more nebulous to me than an abstract noun, a reaction of a woman, my mother, to life's accidents. Indeed most of all he exists for me exactly as that—an illustration of my mother's thoughts and morals.

Utterly different from this 'Jimmy' has been the case of my other brother that died. He was our youngest sibling: Alex. He died in a climbing accident during his first week at university and everything I can remember about his personality I have preserved fiercely ever since the accident. I think other members of my family have done the same. It seems to me that I remember almost every detail of that bold, open, friendly personality that I had already known for 18 years—a boy eager always for information, the one who told me which pop singers were decent poets and I should listen. Because of his death I believe I and Alex have remained in a brotherly symbiosis even more than we would have done had he lived. Through his unabated teenage daring he is even a danger to me to this day—and even slightly, also, he is a danger for the population of Oxford. This is because I took over and still ride his bicycle and I ride it so far as I am able with all the speed and daring that I remember in him. He lost one bike and became forced to save and buy the one I ride by having been too trusting (he had left the other unlocked one day as he wandered among the beech woods on Polhill in Kent). Out of my respect for that trust he had in humanity, and that refusal to think ill, I leave his second bike also unlocked almost everywhere I go in Oxford too—and so far no thief has taken it nor student even 'borrowed' it. Alex comes into me as I step astride its battered but still un-repainted frame of 30 years; I bend to the same low handle bars as he did, though I admit that those and the steel of the frame are about all that remains of the original bike: wheels, chains, sprocket sets, and much more have been replaced five times over and tyres, it must be, over 20 times. Even the steel stem of the saddle of the original once snapped clean off under me with its rust as I rode down Park Street— just opposite, incidentally, the museum where Huxley once battled Bishop Wilberforce . . .

How different, then, these two brothers, 'Jimmy' and Alex, are to me. And why can't others see such differences as I do—and certainly such as my mother saw? What sickness is come into our Christian religion, so unrelated to any life's secret that Jesus Christ ever taught, to try to make such matters seem otherwise? In this spirit, returning to the original theme, how fiercely I think one must defend and nurture each baby that a family wants to keep. But in contrast to this how ambivalent it is *right and natural* for all of us, inside and outside a family, to remain towards neonates that parents themselves are uncertain about. Only a 'monkish darkness in human thought' (as C. M. Doughty once put it about beliefs in the unicorn's horn), as well other grotesque fruits of celibacy,[24] and, above all, certainly nothing ever recorded by Jesus, have to underlie the modern and so-called 'civilized' tendency to imagine that embryos have souls.

As you can see from this, it is certainly not any supposed sacred or 'emergent' quality of diploid human cells that makes me look askance at Roslin Institute techniques as they might become applied to human reproduction. Rather it is just the hopelessness of getting anywhere—for a start, anywhere near to the degrees of skill needed to make lasting improvements in defects we all have and are still continually acquiring.

All evidence that I know suggests to me that blastulas are even more thoughtless and less subject to pain, than, say, are *Eudorina* algae, simplest of multicell creatures—green dots that may well swim in the rain tub outside your door. Leave alone should early human embryos be compared to the smallest free-swimming fry of a fish! Those eudorinae, elegant raft-like beings that I choose out of many because they have about the same number of cells as a human blastula, are above all motile; therefore, like other species of their general kind and like tiny fishes, any *Eudorina* individual certainly can try to avoid a noxious stimulus by turning and swimming away from it. Ability to perform such an act by itself puts *Eudorina* to a level where basic elements of pain and fear must be present. A human blastula, in contrast, is created immobile; it can't do anything except wait for what comes to it, whether this coming is good or bad.

Exactly the same holds for the later human fetus almost right to the time of birth (although I expect here to receive many protests from women who will tell me that at least a later fetus has its ways to make some of its wishes and states of happiness known—a kick being just one of them; well, I'll admit to that one). But, obviously, as the stages of development pass and more and more organs of the fetus acquire ability to withdraw from or reach towards the stimuli that affect them, one needs to worry more and more about its possibilities of pain. But for much longer than this I think one still need not worry at all about the other great factor in the arguments about human and animal rights: fear of death. The onset of this idea will not be until several years after birth. To be feared, death needs either to be witnessed by an intelligent and self-aware organism, or else it needs to have been carefully explained to it, which latter demands even more special conditions.

Freya, the dog my family had in Ann Arbor, was a cross of Alsatian, Malamut, and wolf. The wolf part, as we were told in the Farmers' Market where we found the puppy displayed, came at one-eighth in the mixture; the Alsatian part (or German Shepherd as Americans call this breed) had obviously the most of the phenotype, but the Malamut gave the build, the thick fur, and a particular pattern on the face. One day, desperate to follow me, Freya jumped the fence by our gate as I walked out across Fuller Road on my way to work and was run over by a car. Her bones were broken in many places and her skin so ripped from her muscle all over her trunk that I could feel it slowly filling with fluid like a loose bag as I picked her up. She seemed not in great pain but instead was very worried that I was angry and, of course, she could not stand up. Half an hour later she lay calmly in my arms licking my face and hands while the vet, who had told me that her case was hopeless and that even at the best of survival (if that were possible) she would never walk again, gave her a lethal injection. During that day and afterwards I cried for Freya more than I had done even about Alex my brother when I had heard that he had fallen from the cliff. Even now Freya stays enshrined in my mind in much the same part as he does; indeed she seems to me to have been a character quite similar to Alex in many ways, including that quiet friendliness she had for almost everyone who came to our house and an eternal eagerness for any new adventure. Finally she is similar to him in the stage of her life—a little beyond puppy hood—when she died; and even the way she died was similar. In both cases, an excess of trust seemed to have played a part: with Freya the trust was in me—namely, that I would surely know when it was safe to cross a road; while Alex, for his part, fell roped to a courageous but less-experienced climber whom he hardly knew and who was first to drop.

If one is going to kill a baby, clearly it is only tolerable for it to be killed painlessly, as Freya was. That is not very difficult. If possible a neonate should be killed also, as Freya was, in a full spirit of love. I am glad I have never been in a situation where I have been even tempted to kill a neonate, although, stepping out just a little in this direction, I have told an attendant doctor about the wishes of both my wife and myself that if our baby should come out severely handicapped, and unable to live without strong medical intervention, then we wanted no heroics. I also have little doubt that if trying to make out on Robinson Crusoe's Island alone with my wife, I would indeed with my own hands kill a defective baby and am fairly sure that I should feel sorrow that this had been necessary but no sense of wrong-doing. I believe that my wife would have agreed to the act. Likewise my mother, if realizing for herself that 'Jimmy' was formed so as to be unable to feed, would have killed him, too, rather than watch him slowly starve. To my mind the case of a baby whose best course for itself and the family is to die, and the case of a family dog in the condition I described, are not actually so different. I realize it will estrange many from me to read this but I simply ask such people to be patient and to think about it, and to try to do so without preconceptions.

Up to the point where we lost her, our dog Freya was like a fourth child and indeed lived a life parallel to (though in her development, of course, much faster than) our 'American' baby, Rowena, who was also born in Ann Arbor. The wolf in Freya gave her strange traits that perhaps made her seem still more child-like. For example, she never barked: perhaps it was the wolf also that endowed her not only with this but also with a great steadiness and gentleness with children, including with our Rowena; no more 'domestic' breed of dog I have owned has ever seemed to me so utterly trustworthy. It's been hard for me to forget about another near death Freya suffered, when, again, she followed me without my knowledge and took a deadly straight line to catch up with me as I skied across the frozen Huron River on my way to work. Before I even saw she was following she had gone into an unfrozen section that I had carefully skirted. She wasn't silent that time but crying desperately to be helped, scrabbling at a smooth ice edge at the side of the hole where a swift black porridge of flowing ice made up the current of the river, threatening, at the least relaxation of her efforts, to sweep her under the sheet and thence all the way—drowned immediately—to the distant dam below Gallup Park.

To come anywhere near her would certainly have been to fall in myself, so I skied frantically for home to fetch a rope that I could use to tie myself to a tree and then go lying on the ice to the edge to try to help her. I feared I had no chance to be in time; but, just as I reached the steps up from the ice into our garden, a huge wet lump hit me from behind and knocked me into the snow. In some scrabbling kicking frenzy in the hole she had solved that problem for herself—and how it still, still saddens me that, later, she couldn't likewise solve her problem with the car! What to do in ice holes doubtless had been written over and over in her grey-wolf genes, all along her chromosomes; how to deal with speeding metal, two tons and four wheels of it at a time, had not. Did she realize during that brief time in the freezing river that I had been running not to leave her, but to get help? I like to think so, and in the joyous rumpus after she hit me and we rolled skis, poles, satchel, and wet fur all together in the snow suggested certainly her celebration of a problem we had solved together: 'Sorry I worried you', her joyful whines seemed to say.

Once up on the ice again, and just as before, it is worth noting, she had run towards me again completely silently like a wolf. So that now, due to that steady silence that held during all our walks and that I still so well remember, it is not really accurate for me to write of her as wrote one ancient Roman on a tombstone he erected for his dog: 'This stone records the death of my most faithful companion/Now the sound of him is heard only in the silent pathways of the night.' But his thought had been a true one and Freya indeed sometimes does run in exactly those pathways that his beautiful words evoke. Probably, however, not only the barking but the actual pathways and all the other details with his dog were different: that Roman, for sure, had no occasion, as we had, for example, to remember his dog pulling his baby on a sled along snowy woodland tracks of Peach Mountain . . . or her tying his own legs and his skis together until he must fall by circling him in her sled harness . . . All this has become Freya's immortality and I hope I add to it here a little, as only humans can, by writing about her.

Almost exactly like this it can be, then, with babies who happen to die and with those that need to and do die. My mother's 'Jimmy' simply was born and then died: she certainly didn't kill him; he died very quickly for himself. But I believe that in the desert-island scenario I sketched, seeing him as likely to live only with a great handicap, she almost certainly would have killed him, or at least would have 'exposed' him as is the more common action and as is illustrated (without reproof) even in the Bible—this is the case in the legend of Hagar and Ishmael. Anyway, those who must die or who need to die like 'Jimmy' must know only love up to the last moment and no fear; and it is surely very obvious that the parents' love after the last moment should be then passed on to a replacement baby or to some other target that equally deserves it. Freya deserves to be more remembered and more loved than 'Jimmy' to my mind, a statement I think even my mother would well understand even though she did not know this particular dog. 'Jimmy' was human, Freya not: maybe you think that should make a great difference, but at the times of their respective deaths Freya had accomplished far more heroic and 'humane' feats in her life for all of us to appreciate than 'Jimmy' ever did, though this, of course, leaves no fault implied to 'Jimmy'.

Now to turn from all this to a more cheerful topic, even to one that could be hilarious if people would only treat it so: why not let humans clone themselves in moderation if they want to? I'm afraid I just can't get quite serious about all the fuss about human cloning. Just as I find our equal-aged clones of two—our twins—fascinating to learn from and feel them to be provided for us, again, as if as a gift from God to teach about ourselves, so I would find an unequal-age clone of two (a father, say, with his clonal son) fascinating to watch also. How much would such a father feel the need to dominate the son, for example—and would there be extinction of the usual Oedipal reserve? An extinction on both sides perhaps; but if not, then on which? And so on. I admit I don't care for the idea of either such pairs or such multiples being too abundant—confusing and monotonous are the words that most come to mind. I discuss elsewhere[25] (and see Chapters 16 and 18) my belief that the problems of health and happiness will soon prevent any overenthusiasm in this direction if human cloning should ever come to be tried on a more than a trivial and experimental scale. But it will not surprise you after the thoughts I expressed in Chapter 1 that I would find a father–son clone less worrying to observe than a mother and clonal daughter.

The problem for me is that the techniques that make cloning possible are going to increase various other dangers for our species that come to us in a disguise of humane progress. The goal towards which we are drifting is quite the opposite from what we think: we believe we are defending the rights of individuals, where we actually are very busy preparing their destruction and making individuals to become as individually helpless as cells in superorganisms. Technological superorganismic masters are soon going to be far beyond any individual's ability to control and, in fact, they'll be paying less and less heed to whatever we, as *individuals*, may be saying and wanting. Indeed I believe that only these superorganisms have any chance of being able to perform that magic that John Maynard Smith and Steve Jones need to fulfil their visions. Simple pills that fix mutated germ plasm and leave individuals as intact and as free as they are now, as in my sci-fi caricatures of situations above, are too far off at present, even for 'let's pretend' stories.

But what chance have these suggestions of mine to be heard when pitted against the vast clamour of medical-oriented industries and the outpourings of sentimental media, the exhortations of sundry megareligions? How can they be heard when a heroic 'hospital' film runs on the TV almost every night? I am very pessimistic. I think of a stone tossed into the pool of a sweet syrup spread all between us by our ethical demagogues and then I think of the few brief waves from this stone I have thrown spreading and drying. I think also of the phrase that is bound to come after a first reading? 'This man has to be like Hitler to write like he does; why, even Hitler may have loved his dog if he had one!' Writing his history in Ancient Greece, Thucydides added that see-if-I-care disclaimer in his opening paragraphs, telling how he intended to stick to the truth about the recent war with Persia and not to be swayed by Athenian wishfulness and pride—and how he well knew that his truth was going to be disliked. As I state in the paper of the chapter, the currents of our First World for the present seem far too firmly set. Like those 'volunteers' who step forward for a medical experiment in a prison, we have already railroaded ourselves towards our various trial functions within the superorganisms of the future. I think that not even a stream of disasters of the kinds I have mentioned—the asteroid, the epidemic, the worldwide general strike—is going to make any majority in our democracies see that I am right so that our course could be changed in time. Much as an epileptic unfailingly forgets (or else never knew) his actions during a seizure, so we easily forget the unseemly actions that happened during the political crises of our own past, even if they are fairly recent. Like, for example, that different ethos that Britain had during World War II, when killing almost any whole and adult male German was considered a high virtue. We in Britain forget and deny this even though it was exactly that ethos that enabled us to win and thus to return to our present peaceful and antimilitary stance. I ask myself, then, what hope is there for any practical evolutionary philosophy of individual value in a culture that is so at ease with and seemingly hardly notices a 15 per cent rate of births by Caesarean section, one which, by this sign and by others, is approving that its women should evolve (or hormone-induce themselves to be) as narrow-hipped and as athletic as men, all of this done, seemingly, in a spirit of a need to 'catch up' with men in an Olympic record book?

Perhaps I am wrong, or at least am wrong about how far and fast such an ethos may be carrying us towards the coming state that I have called the Planetary Hospital. Again an asteroid impact may serve as a model to fix our thoughts, and to suggest what we are risking, for our descendants, by our acceptance of the Mephistophelean gifts of modern medicine. The disaster, however, could be of many other kinds: nearest and most trivial of which, and also perhaps a kindly warning, is the 'millennium bug', while the one actually most likely and by far more serious are the risks of world war and of pandemic. Perhaps we will survive the former even when, after the war, every baby has to be cut from its mother; and, perhaps after the asteroid, scalpels dusted with iridium will prove sterile no matter what other gunk they have picked up out of the rubble and with that we shall get through—a few of us—as this new species, *Femina sectipartum*. Case-hardened in our bodies by natural selection, like the birthing scalpels that, in that age of rubble, we will have beaten from fossil car springs (I am borrowing here images from Russell Hoban's moving and, it seems to me, realistic *Riddley Walker*[26]), we will start anew in a bright world with many of the old constraints of our evolutionary topology forgotten. I don't doubt that very human brains would still be left to all that survive the new holocaust and what was left for us to start from would now have almost no limits to its expansion—no more limits at least than a proto-peacock galliform had for its tail. With the problem of the size of the pelvic aperture gone due to the complete ubiquity of Caesareans, hips would narrow and there would be no longer any constraints on women winning in the Olympics or in anything else, just as, among deer, the does normally can outrun the heavier stags. As I noted more than once in Volume 1, this new *Femina sectipartum* would not be the first example of a radical topological innovation occurring to reproductive anatomy during evolution. A topological 'cut' that occurred in some ancestor of those flat, red, smelly companions of less-clean beds and lodgings around the world (such as the one where I slept in Galway town—as related in Chapter 4) seems to have been quite successful as a converse to the Caesarean concerning a way of injecting the sperm.[27]

At this point in writing my first draft, I seemed to hear a bitter feminine laugh at my 'bed bug' allusion (for that is the animal I refer to). I have tried to follow the laughter's thought: 'How like a male, even how psychopathic, even in an insect,' she was saying, 'to carry it that far!' Shortly with relief I reflected how, in the matter of Caesareans, we human males just have to be blameless; Caesareans respond to a female evolution problem and can't be our fault . . . But is it true? Males have been and have stayed in the lead in the most striking long trend of our species: head size.[28] And because the fetal head was to become that pivotal and eventually too-tight key causing most of the trouble in human childbirth,[29] we with the larger heads, in a certain sense, have caused this trouble in our species too. But, again, was that really the start of it? Even more fundamentally the problem began because women found themselves for one reason or another liking the men who had the more fetal appearance and therefore the larger heads, and these were the kinds of heads that indeed had the capacity to hold most brains: their liking had been fundamentally for intelligence and is part of why we are such a breakthrough and a success. A liking for large and fetal-style headedness was a choice that quite clearly women failed to abandon even after many were finding themselves doomed in many resulting struggles of childbirth. Selection by this death in childbirth, without a shadow of doubt, has produced the present wide feminine hips: if you had not thought of this before I am sorry to shock you, but every time a man looks at that shape of the female torso that so attracts him, it has to be faced that he is looking at consequences of hundreds of thousands of adult womens' deaths—in a sense a vast graveyard shimmers behind that subtle and, to us, so beautiful hourglass curve. With one excepted factor, which is probably much less important and which I will touch on in a moment, there is absolutely no other way in which this widening of the feminine pelvis, so different from that of the male, could have happened.

This selection through death during obstructed childbirth has continued right up to the rational present, as you may read movingly told in many a Victorian novel. For the pre-Victorian era you may read further in Jane Austen how all her heroines seek suitors who are kind and decent, but—and alas for many a daughters' fate, though it isn't discussed—they want them to be intelligent as well. Male preference for wide hips, on the other hand, since it began (obviously in the far Palaeolithic to judge from the figurines), may have been working a little to save women. In other words, it may have worked to leave aside as unmarried spinsters all the exceptionally narrow-hipped and more masculine types, those who could never have a baby safely anyway. Evolving secondarily to appreciate wide female hips, after such deaths began to become serious in the *Homo* line, men could almost be said to have tried to save women a little from the consequences of their reckless preference while against such efforts women have continued, willy-nilly and for their undoing, admiring large male heads. Doubtless more than the heads what they have been really admiring are the wild, sporting, joking, mathematical, musical, and, be it said, often the more complexly and engagingly deceitful minds that also tend to inhabit the larger heads . . . Out of all of which back and forth, in summary, surely our one and only dependable message is that the trends of the two sexes are quite as inextricably entwined as the functions of heart and lung have become entwined in a single body. In the evolutionary sense it was the human sexes *together* that 'willed' for us to become this most rational ever of all our world's species. But, most sadly for the women, it has been they that, especially in the modern world, have ended paying the highest price for the wonderful developments that were brought about.

Please believe me that this is the truest and deepest (and, it must be, of course, as written by a man, the most unbiased) account you have yet read about how our so-called 'man's world' of today came to be. It is probably of little use to ask for this to be believed, coming from a man. We males today meet only accusations of patronage if we try to tell of our admiration and our thanks for sacrifice, or try to bring forward how we endeavour to return to women, out of deep gratitude, the very best that the minds they give us with such trouble can produce.[30] The main thing to be decided now is something far more pragmatic: simply, from where we are, where do we go next?[31] Is it right that we casually give away a birthright of our ancestry without discussion? The birthright is nothing less than the right to be free, a pair in paradise, forming their own island, giving birth naturally.

Returning to the more general problem of maintaining the health of future generations in spite of the efforts of modern medicine, it should now be clear why over the years of my working life I have been slowly developing a paranoia for hospitals. The growth of hospital buildings and of all the infrastructure of clinics, pharmacies, the increasing domination of our universities by medical schools associated with huge hospitals all over the developed world, the unending love affair the public have (or are browbeaten to have) with medicine through the TV screen plays on doctors and on hospital life, accelerant birthing by Caesarean section, the indefinite increase of pharmacies even in the Third World, the accompanying increase of our everyday pills that tumble from every bathroom cabinet—all these matters combine to tell us how we are heading. Happy in our progress for the present, like children on a fairground roller-coaster, prattling of the genetic castles to come, we skim onward, uninformed, into the bowels of a great Planetary Hospital of the new millennium[32] (and see Chapter 18). My dread is even worse. It concerns how, by the time our joyride ends, we will realize that we have come into the concrete heaven of our TV dreams but only at basement level: all the higher levels, those with the views, the nice chairs, and the broad desks, will not be those where we play our ultimately drab games as hospital heroes—the patients, the nurses, the doctors—but rather will be assigned to 'Administration'. We ordinary humans will go 'up' to 'Administration' only when, occasionally, we find ourselves healthy and skilled enough to be wielding pliers and circuit testers: we'll be invited only to install the new diagnostic and administrative computers or equipment or to fix the old ones. We will have, in effect, roles akin to those of the glial cells that look after the needs of the neurones in our own bodies. Even those of us that still have minds to think with will no longer have any idea about what the 'Hospital' is doing.

Maybe this nightmare won't happen, but I am not clear what better outcome can possibly emerge from our present course. On the lines of those genetic cousins to Maxwell's Demon whom I attempted above to imagine scanning our DNA, end to end, telling us all our errors, fetching out the

codes from the standard gene book to correct them, and, then, following
this thought to the realization that such demons are practically and perhaps
theoretically impossible to employ (this, again, rather in the manner that
James Maxwell's prototypes are impossible in physics), it seems to me that
perhaps the only alternative open to us, other than that of a return to a
stern regime similar to that of our past, is indeed to abandon our 'selves'
into the arms of just such a much larger organization as I have tried to
visualize. Maybe a global human organization can indeed evolve to manage
us (or 'emerge to manage us' if you like—that seeming the more
fashionable phrase for such one-off processes), and at the same time this
superhuman progress can actually make us become those swarms of
labouring demons whose combined and emergent consciousness indeed
understands the 'Handbook of the Standard Genes of *Homo*'. A genuinely
superhuman organization like this imagined and exemplary Planetary
Hospital is perhaps able continually to recreate and maintain us. So it will
be for such of us, at least—mites within its bowels—as the organization
thinks it needs; for the rest it will soon find excuses for their dying out, just
as we do for our domestic animals that already we have forced into the
plight that I predict.[33] The 'thinking' part of such a superhuman system is
what I am referring to as the 'Administration' levels of the 'Hospital', while
a current example of an 'organ' for this thinking part would be its library:
that will become eventually, no doubt, a massive set of electronic databases
and treatment algorithms.

In the background of my thoughts as above have been the organelles
employed by the cells of all multicellular creatures. Increasingly electronic
upper and 'admin' levels in my 'Hospital' would look after us in much the
same way as the genome of the eukaryote maintains a remnant genome,
sufficient for the purposes that the overarching nuclear-ruled whole
requires, in the form of its one-time bacterial allies now going by name
such as mitochondria and chloroplasts. Were the bacteria perhaps lured
into the eukaryote cells by chemical temptations analogous, in some
remote sense, to our 'hospital' TV dramas? Were promises held out to them
how grand, dramatic, and safe life would be once they are 'inside', promises
of how wonderful would be the biochemical machines they could play
with, and what dizzying and expert communication they would see flying by
them on all sides? Were they told how they would be borne aloft in moving
palaces—humble denizens of soil and slime no longer nor even of merely

stationary stromatolites. There in the new palaces they would be fed, advised, and cosseted by the wisest of guides and companions—chromosomes out of the nuclei. Wow, all this for free! A millionaire's life on a cruise ship and all for nothing. So the offer to the symbiont may have seemed at first.

Perhaps many of you admit to a faint realism in these metaphors but do not find them at all frightening. If so there is really nothing more to say: let it be so for you. But then you ought to confess that the metaphors do not fit at all with that fashionable idea of high value for human individuality—that which seems the core of the fuss, for example, about cloning and the sheep Dolly; indeed you ought to be clear that they imply quite the opposite of such value. Even asexuals (such as the prokaryotic ancestors of mitochondria and chloroplasts in their former free life in the Precambrian certainly were) always seem to care somewhat about individuality as is shown by their typical evolution of genetic diversity as coexistent clones in local populations.[34] Sexuals still more always care deeply about diversity—it's the very essence of that expensive commitment they are insisting on. But the ancient prokaryotes becoming mitochondria and chloroplasts had eventually to forget all that; they ended up losing their competence, and, almost, their diversity, completely.

Is there not some middle course? Yes, but it is simply that which I have been advocating: you have to grasp the whole of the nettle where sits the glorious rose. Guardians of the vaunted individualism in our species, champions of an adduced right to be genetically constituted 'anyhow', with a supposed equality continuing despite the 'anyhow'—these guardians criticize any hints of an attempt to maintain genetic standards in *Homo*.[35] That such criticism tends to come from people committed to extremes of leftward politics seems not surprising when we reflect that, from the first, Marxist communism always seemed to be happy with the idea that individual wills and ideas could and ought to become merged into more collective versions; however, the idea that this state would be compatible with individually free-form and free-living bodies seems not to have been thought through clearly. Indeed the position adopted on the far communist wing with respect to everyone being able to learn everything if they are just properly taught, including learning the principles and views of communism, seems to have driven the movement's intellectuals into an orgy of sloppy and tendentious scholarship that has for some decades been applied in claiming to show up errors in hereditarian ideas.[35-37]

Results showing that virtually all human traits carry substantial heritable portions in their variability—contra Stephen Jay Gould, Leo Kamin, Richard Lewontin, Steven Rose, and others similar—can fairly be said to have accumulated over the twentieth century in a steady trickle. The refutations of this trend have been negligible and very contrived whereas the counter-refutations become stronger and more detailed each time around,[28,38] and also become less and less commonly answered.[39] There can be little doubt that identification of specific and localized genes will soon follow—some are already claimed. By the same standard, results with the same import for variation in intelligence can be said to be accumulating in a steady stream also.[28]

Finally, using that stream metaphor yet again, the evidence that aspects of health—the main topic of my Gustavus Adolphus presentation—are heritable can be likened to a mounting torrent. Pick up any issue of either of the new journals *Nature Genetics* or *Nature Medicine* and there is a high chance that a new description of a heritable human resistance to a disease will meet your eye or you will read of how an autoimmune disease is strongly connected with a variant human genotype and how a wild (and often in itself mild) pathogen interacts to produce the situation.[40] Glancing at random in the same way through the equally new journal *Structure* in the Radcliffe Science Library in Oxford, seeking something else, recently my own eye was caught by an article giving a biochemical interpretation very similar to the image that I used in the present paper only as a quaint illustrative parable—that is, my shapely sops-to-Cerberus idea or the logs-to-alligators. I was astonished. This macromolecular field, it seems, is so rich, so fast progressing, and so full of so many surprises in detail, that virtually you can tell your 'just-so' story first and then, out of the infinite variety of ploys and counterplays occurring in host–parasite coevolution, search for the facts that fit. In some rather trivial ways the case is different because caterpillars and even rats[41] eating, of all things, runner-bean plants or their seeds is not what you are likely to first think of when I give you a metaphor of a frail canoe of the vertebrate body threading a swamp of disease and of alligators attacking it. But such quasi-external feeders of ours as scabies and house-dust mites are disease in humans, or cause it, while, as to the sizes involved, alligators relative to the frail canoe of my image may have it about right for rats trying to eat bean plants if not merely the beans on them. It would be better for me, I admit, if the *Structure* article just cited had discussed the counteradaptations to amylase inhibitors in rabbits, for which I am sure some similar story will eventually be told. A more important difference is that I was thinking of proteases as being the alligators' teeth while my intended victim was to be a vertebrate. Instead, in the *Structure* article, amylases were the primary aggressive tools and it was a plant (or a seed) that was the victim. In any case, a particular point here is that the massive revelation of connections of genetic variability with autoimmune diseases on the one hand and with infectious diseases on the other will probably soon turn out to have more connections with all the above-mentioned variability in intelligence than we have yet suspected.

The abundance of the genetic variability in intelligence has always been a surprise; it is a little hard to see why we are not all selected to the upper limit—in fact, selected in just the way the intellectual left, ignoring or arguing away the evidence, pretends to think we have been. If we imagine, however, that detailed correspondences of neuronal epitopes and bacterial or viral epitopes may exist commonly (of the kind suggested by recent investigations of Sydenham's chorea, for example), we see an almost unlimited possibility for genetics, infectious diseases, and psychological idiosynchracies to be bound inseparably together. This complex will very likely include parts of that emergent cerebral attribute we call intelligence. In cases of Sydenham's chorea, infection with a *Streptococcus* of low virulence has been found to be connected, via subsequent autoimmune attack on a very localized region of the hypothalamus, with a compulsive behaviour disorder.[42] Further evidence along similar lines comes from the way high intelligence combines with high rates of certain types of disease and neurological and psychological disturbance in certain endogamous groups, such as the Jews.[43] Again, if we really are to help people's aspiration concerning their children in the next millennium, these matters must definitely be recognized, not swept under the carpet. There are soon going to be attractive but impermanent (and ultimately disappointing) ways to help people both in respect of health and/or intellect, as by 'gene therapy' and possibly later by 'HACs'; and there are also going to be ways to help them that are only a little harder in their confrontation to our immediate longings but which ultimately can offer much greater and more permanent rewards—and these as a bonus can have hereditary high health thrown in as well.

No matter what new name we may choose to give to this genuinely most benign route, many will still see it for what it actually is: eugenics. Associated with this, the many will also think of all that went wrong partly under the name in the 1930s and 40s and the word certainly must come with a dread ring in various ways. But those at first naive and ultimately ghastly policies that brought the bad name should not be made too much of. They were never espoused by Francis Galton, who coined the word eugenics: to him the idea held out an *escape* from cruelties of natural selection. So do I see it. Thanks to Herculean efforts of geneticists and evolutionists in the second half of the twentieth century, we have now gained enough understanding of the general intricacy of the human genetic situation to start something far more effective than anything that could have been started in Galton's time and much more humane even than he imagined. We are at last able to apply many alleviations in ways that will not end mortgaging the whole human genome to doubtfully stable technology. It is simply required that we can accept a few steps running counter to a current and much advertised ethos—an ethos, however, whose creation was all of it based on false premises from the first and never was natural to the human mind. By infringing our rightly esteemed individuality just a little, we can save individual mental and physical competence of individuals for the future *and* increase human genius and freedom of all good kinds; and all this can be done in totally uncruel ways. Alternatively, we can let individuality continue to have unfettered reign as we tend to do today and our ghosts-to-come can stand by and watch the individuality of our descendants slowly dying, all of it happening in the course of our moving into what I have imaged in this essay as the Planetary Hospital. We doom ourselves, in fact, to watch the beginning of a great oppression in which each of our lines becomes gradually enslaved through its physical and mental incompetence. The best hope if our present course is continued, however, is an ironic one—that within about 500 years we won't be caring any more about what we are losing; psychically and physically because we will have become ciphers and be without choice. We will have no choice but to take the drugs and accept the operations, with all this happening within, and as parts of, those greater entities whose evolution we have set in motion. A new life form will be present; but it won't be humane and it won't be loving. Is it the destiny we want?

To many readers much of the above may seem quite 'spacy' and speculative. I will end this introduction, therefore, by offering another more immediate and observable prediction—and the start of this one will probably be observable at least within the lifetimes of some persons alive today. If here, in the so-called West of the world, we can't get our minds around the advantages for our descendants' happiness of being just a little tougher in such matters as those of supposed rights (and souls) of embryos, and of supposed rights of all adults to reproduce without restraint, I believe similar scruples will not stop East Asians and perhaps some other cultures from adopting the more rational and fruitful policies I have outlined. To some extent we already see this happening—note, for example, the recent striking statistics on attitudes to negative eugenics[44] and also note the government policies adopted in Singapore and in China, all of it denounced by so much Western criticism.[35] If the West doesn't but the East does adopt such policies, I predict that Eastern cultures will pull increasingly ahead of the rest of the world in the directions that we also should be going. In my opinion those Easterners will deserve all the increased health, intellect, and happiness that they get.

In what else has the passage of 10 years proved me right or wrong in my article? Quite often the expansion of a successful idea obliterates some of the steps leading to it; in fact, it often takes something like a request that you explain yourself to a schoolchild, as at the Gustavus Adolphus symposium, to bring all of the important stages back. Although in the paper I coauthored with Jon Seger (Chapter 11) we mentioned 'multiline plantings' as an agricultural technique, this aiming to illustrate the importance of dispersion and findability by parasites as part of the pressure to genic diversity in the world in general, we didn't mention this factor as a part of the pressure to local species diversity that it obviously also has to be. We instanced biodiverse habitats as ones likely to be benign to parasites generally as well as offering maximum new opportunities for them to effect host transitions. But it was Jon, I think, who insisted to mention that not just the parasite interactions with hosts but also the very numerous competitive and predatory relationships in biodiverse habitats could help to explain why use of sex was so specially consistent there. He wanted us to implicate a dynamic of opposing species of all kinds, not just a dynamic between varieties of important parasites. This was pretty much how I thought of it myself but perhaps because of criticism and claims of 'goodness of parasitism'[45] and the like, I had tended to become more and more monomaniacal about the parasites. The neglect they obviously received from ecologists played its part, too, and Jon by no means cured me of this monomania, I'm afraid. As regards biodiversity *per se* and whole species impacting on a given host in the ecological struggle, this has been mentioned repeatedly all through my line of models since the beginning of this volume. Species or varieties, parasites, predators or competitors: they are just the choices in what one bets on in a composite line-up of ever changing Red Queen 'pursuits' of one kind or another. I have come to emphasize the parasites because of their usual strong specificity, the numbers of their species compared with numbers of host species (the 'many on one' factor tending to give differing periodicities and thus demanding recombination), their ubiquity in the environment, and finally the importance of rectifying generally their (then) state of neglect. Sometimes to avoid seeming too monomaniacal (though don't be fooled, I remain so!) I use the word 'exploiters' as this can include predators as well.

Following in part from my own pottering observations of ecology in the wild and in part following a theme of Joseph Connell[46] and Daniel Janzen[47] (these authors in turn stemming from a 1960s' forerunner in Gillet[48]), the frequency-dependent nature of all 'exploiter' pressures on species seems to underlie not only the variation within species and the temporal and spatial handling of this variation within the gene pool—in effect what I have been studying in my own theme of sex—but to a large extent the world's multiplicity of species may depend on this frequency-dependent interaction itself. A feedback between the numbers of host-dependent exploiters (plus, one must certainly add, a very positive extra effect coming from the ever-improving searching skill of the exploiters over evolutionary time), and the numbers of species of hosts on which they depended, seems to me the main explanation of the vast biodiversity on our planet. In the present paper I was trying to explain my ideas to schoolchildren; therefore I mentioned frequency-dependent host-finding and (over longer spans) host-species-transference, as steps in my argument for sex, or at least as the preliminaries and corollaries of that argument. Usually I took this to be an obvious ecological background that hardly needed stating; but writings by Wills[49] have reminded me that the ideas are important and seemingly are still often overlooked, in which light, then, even in this paper they probably receive too little attention (see p. 517).

The 'multiline' planting technique that tries to duplicate artificially a small part of the usual natural variability of resistance in a stand of a species, so as to become an agricultural answer to disease-spread problems, is touched upon on pp. 522–3).[50] This mention comes within a paragraph on costs of resistance and here I only need add that in spite of some back and forth over whether costs of resistance really exist,[51] the modern data seem on the whole to be falling (as will be discussed further in the Appendix to Chapter 16—see p. 827) the way we 'RQueeners' would wish. My own view can be broadly characterized as insisting that costs for hosts must exist or else all hosts would be resistant and parasites unlikely even to exist—surely an absurd conclusion. An almost identical argument tells me that there must be costs to being a virulent parasite, too, otherwise all hosts would be extinct also, or at least would be universally parasitized—which, of course, makes an even stronger biological *reductio ad absurdam* than does the previous implication.

Coming to the genetical resistances I used as illustrations for my theme, an example I brought in early in the essay was prematurely chosen. The paper I cited was a poorly founded example of a supposed HLA genotypic resistance to the AIDS virus, which even by the time of my publication had been largely withdrawn by the authors. Similar claims quickly followed, however, and there is now quite a list of genetic variations disposing for or against developing AIDS,[52] including a surprising latest one, which, at the time of writing, seems to show a seemingly preformed genetic resistance at moderate frequency in Europeans while absent in Africans.[53] Because of HIV-1's status as a virus that seemingly recently jumped to us from African monkeys, one might have expected exactly the opposite. With such monkeys as source and doubtless from earliest times being butchered by African hunters, one would expect Africans to have experienced numerous assaults and perhaps brief successful inroads by the virus from time to time and thus have been at least a little selected for resistance.[54] In any case, in order to update my article, it is only necessary to substitute the reference to the new paper for my reference 5, and at the same time to insert 1 per cent in place of 8 per cent as the frequency of Europeans who cannot catch AIDS: wording as in the paper could then stand. For the rest, several HLA alleles conferring improved and worsened prospects of progress to AIDS following HIV-1 infection are now, as I said, additionally known.[52]

As to showing that fuel for my dynamical Red Queen process was abundant via review of other 'resistance' items in genomes, I don't know why I didn't give more human examples—or at least show the likelihoods of them—following my statement of a 'need to look for other systems with which this one [MHC] could be interacting' (p. 525). Quite a number were already well known besides the protease inhibitors that I chose to mention. the classic blood-group system, ABO–Lewis-secretor, confers various slight resistances and just by itself has its elements well spread around the genome. This system could well have been mentioned; so now could other polymorphisms, such as one to which Peter Lachmann drew my attention from his work within the 'complement' component of immune defence;[55] as also could have been mentioned the minor histocompatibility systems.[56] My enthusiasm for the elaborate polymorphisms in protease inhibitors about which I had recently read, evidently took my attention from the others; but the undeniable summary of all such examples had to be that variability with high potential to affect resistance, and often clearly dispensing it, was plentiful and widely scattered throughout genomes in people, mice, and birds.

Notes and references

1. *In Memoriam*, Canto 55. A. Tennyson, *The Works of Alfred Lord Tennyson* (Macmillan, London, 1893). The poem *In Memoriam* was published in 1850, thus 9 years before *The Origin of Species*.

2. J. A. Mayer and F. J. Taylor, A virus which lyses the marine nanoflagellate *Micromonas pusilla*, *Nature* **281**, 299–301 (1979); T. F. Thingstad, M. Heldal, G. Bratbak, and I. Dundas, Are viruses important partners in pelagic food webs?, *Trends in Ecology and Evolution* **8**, 209–13 (1993); W. Magliani, S. Conti, M. Gerloni, D. Bertolotti, and L. Polonelli, Yeast killer systems, *Clinical Microbiology Reviews* **10**, 369–401 (1997).

3. S. Hara, K. Terauchi, and I. Koeke, Abundance of viruses in marine waters: assessments by epifluorescence and transmission electron microscopy, *Applied and Environmental Microbiology* **57**, 2731–4 (1991); O. Bergh, K. Y. Borsheim, G. Bratbak, and M. Heldal, High abundance of viruses found in aquatic environments, *Nature* **340**, 467–8 (1989).

4. O. Bagasra, *HIV and Molecular Immunity: Prospects for the AIDS Vaccine* (BioTechniques Books/Eaton Publishing, Natik, MA, 1999).

5. J. I. Cooper and F. O. MacCallum, *Viruses and the Environment* (Chapman and Hall, London, 1984).

6. R. J. Ladle, R. A. Johnstone, and O. P. Judson, Coevolutionary dynamics of sex in a metapopulation: escaping the Red Queen, *Proceedings of the Royal Society of London B* **253**, 155–60 (1993).

7. P. D. Coley, J. P. Bryant, and F. S. Chapin, III, Resource availability and plant anti-herbivore defense, *Science* **230**, 895–9 (1985).

8. R. Taylor, Superhumans, *New Scientist* **160**, 25–9 (3 October 1998).

9. One path to test the idea of semi-voluntary deaths within sibling groups of palm zygotes would be to study early competition in a related palm, the babassú, *Attalea speciosa*, in Brazil. The gynoecium of this has six ovules and it is not uncommon for all of them to form seeds, although ripe fruits in which one or two have already aborted are more common. Nevertheless strong competition between the surviving seeds after germination, such that only one can survive, is certain and is ensured by the fact that all the six (or a lesser number) are bound together in a massive 'shell' that prevents any separation without destruction. Sometimes several seeds are destroyed by bruchid larvae (this is the same family of beetles as provides the larvae that bore into our garden peas) but commonly, also, several are intact during and after fruit fall and dispersal. The seeds germinate simultaneously and seedlings extend and root out from their 'pockets' within the massive and extremely strong babassú 'nut'.

   Because eventually only one babassú palm can survive from these single-fruit seedling sets, experiments in which the demography of the natural groups of sibling competitors is compared with artificial groups of non-siblings, formed by seed substitution, could check my predictions. Some tests are as follows: (1) in the sibling groups less rapidly growing siblings will quickly die; (2) in non-sibling groups the struggle for survival will be more prolonged; (3) in consequence of (2), the struggle will also be more wasteful in that the leader and eventual winner will grow and attain maturity less rapidly than will winners from sibling groups. For information on similar kinds of conflictual interactions in plants and among their offspring, see B. Furlow, Flower power!, *New Scientist* **161**, 22–6 (9 January 1999), and also U. Shaanker and K. N. Ganeshaiah, Conflict between parent and offspring in plants: predictions, processes and evolutionary consequences, *Current Science* **72**, 932–9 (1997).

   Still earlier than such 'abortions' and 'miscarriages' at fruit and whole-plant levels, it is interesting to note that analogous conflicts, and sometimes doubtless self-sacrifices, occur in seed plants amongst multiple embryos within single seeds (as, for example, in pines and other conifers). Moreover, from what we can reconstruct from comparative evidence in nearest cousins they have been institutionalized in flowering plants since well before their intitiation. It seems that these and their sister group, the few-but-weird Gnetales (W. E. Friedman, The evolution of double fertilisation and endosperm: an 'historical' perspective, *Sexual Plant reproduction* **11**, 6–16 (1998)), have been not only the experimenters with new plant leaf styles that we have long known (the 'broom'-like habits as in the 'Mormon tea', *Ephedra*, of the western USA, the leafy *Gnetum* lianes of the tropics—top marks for a leaf design pregnant with the future but few for their branching and strength—and the amazing *Welwitschia* 'two-leaved turnip' of the Namib Desert whose strap-like leaves extrude for the whole life of the plant like toothpaste) but experimenters in nutrition administration and early death too. In short, in the ancestor of the two groups a double fertilization generated genetically identical competing embryos—twin potential babies. A struggle between them was pointless genetically and instead, obedient to kin selection, one twin fertilized ovum began to devote itself regularly and faithfully to protection and nourishment of the other twin. The 'altruistic' twin embryo has evolved to become that important and almost defining characteristic of the modern seed—its endosperm, a store of food flesh, as we might put it, of a self-sacrificial twin embryo (E. L. Charnov, Simultaneous hermaphroditism and sexual selection, *Proceedings of the National Academy of Sciences, USA* **76**, 2480–4 (1979)). Now we find it being eaten not only, as intended, by the young plant during its germination, but also in other circumstances by more remote beneficiaries as well—by us, for example, every time we consume a slice of bread.

A technical objection to the above might be that the endosperm in flowering plants isn't—now anyway—truly a genetical identical of the developing embryo. An extra gametophyte nucleus is added in course of the parallel, non-embryo fertilization. But I agree with William Friedman and others that the antecedent state is likely to have been the formation of a real duplicate embryo and that only later (rather on the lines of the pure genetic imprinting effects discussed by David Haig) the mother sporophyte (the plant we see) evolved in cahoots with her daughter gametophyte (that is, the post-meiosis tissue, or 'embryo sac', within the flower) to add another nucleus to the auxilliary and sacrificial embryo. This would be selected, doubtless, because giving more 'maternal' control over an often too strong resource-acquisitive power in the new tissue (which, by the way, sometimes even involves haustoria—fungoid cell extensions—reaching into and among the maternal cells, presumably to grab nutrients when they are not being offered). Especially in cases where neighbouring ovules can be fertilized by other fathers (or neighbouring flowers pollinated by them—that is, generally, under what may be called 'Haigian' conditions, which, in the case of flowering plants, are especially likely in big, close inflorescences and/or under wind rather than insect pollination), the dedication to altruism in an identical endosperm-like tissue might often over-reach the interests of both gametophyte and 'whole plant' mothers and produce irregularity in the seed.

The extra maternal genetic input would be needed because, in effect, the suicide twin had proved too successful by its fanaticism. When bringing both paternal and maternal genes to its task with equal weighting, this twin would be prevailing too strongly against the 'mother' consortium, making the early seeds bigger at the expense of those from later pollinations, which would become deprived or aborted. Thus became provoked, I suggest, the maternal evolutionary countermeasure of the extra nucleus whose prepared gene-translational biases could re-emphasize maternal optima. And there finally you have it—Freud re-rendered for flowers!

10. This gene may, by the way, have been lifesaving in certain circumstances and this fact rather than mutation may explain how it has reached the frequency it has. It may protect against typhoid or other enteric pathogen (G. B. Pier, M. Grout, T. Zaidi, G. Meluleni, S. S. Mueschenborn, G. Banting, *et al.*, *Salmonella typhi* uses CFTR to enter intestinal epithelial calls, *Nature* **393**, 79–82 (1998)). But even allowing for a possible future collapse of civilization, hygiene and good clean water provide obviously better ways to combat such infections.

11. A. S. Kondrashov, Deleterious mutations and the evolution of sexual reproduction, *Nature* **336**, 435–40 (1988).

12. M. Ridley, *The Red Queen: Sex and the Evolution of Human Nature* (Viking, London, 1993).

13. S. Jones, *The Language of the Genes* (London, HarperCollins, 1993).

14. This imagined 'Handbook' I do not regard at all as the technical hurdle. Its creation is going ahead speedily and is just a lot of systematic and rather boring work: before long the Human Genome Project will have provided us with the first draft of a manual. Big questions, however, still loom about what is to be treated as normality among the many gene variants that are all the time being discovered, but that is being discussed. For many genes obviously is no single normality and many are of equal status; here a philosophy based on facets of Parasitic Red Queen theory may yet enter again and enable us to extend the concept of normality appropriately. All this is not the great problem.

15. It had better be perfect, of course, for the public, especially the American public, will be likely to sue genetic normalizers for perhaps many lifetimes's worth of damages if something can be proven to have made a mistake—to

have allowed to enter something worse than the condition supposed to be eliminated.

16. T. Murray, 'Shall I be a mother?', *The Independent on Sunday*, 5 only (25 May 1997); J.-A. Goodwin, 'When your husband is a hunk', *The Independent on Sunday*, 4 only (25 May 1997).

17. My own idea, though it is backed as yet no clear insight, is that such a 'Genome Demon' may be going to prove as impossible to put to work in practice as his prototype has proved in physics. In other words, perhaps it can be formally shown that a lone human can never understand the full integration of the processes that create a lone human. Such understanding of a single human in itself would not be impossible but it requires a higher level of being to effect it, of which our wistful understander cannot be more than part. In my scenario the situation is actually worse: if the would-be understander is already degenerate due to mutation, as I suppose, he or she may get less far in understanding him- or herself than do humans today under conditions of more independent life. As to the higher being that understands its 'single cell biology' completely (that is, it understands a human), this being in turn will still be mystified by many aspects of its own construction and activity even if these are wholly based upon human cells.

18. In writing this, a sharp curiosity crossed my mind as to whether descendants of Theo Van Gogh, Vincent's brother, have in fact benefited from the cruelly ironic prices their grand-grand- . . . uncle's paintings fetch in modern millionaires' salesrooms. Given the unflinching solidarity expressed in both directions between Theo and his brother, without which the artist's great work certainly could never have been accomplished or preserved, Theo's descendants certainly deserve to be rewarded (W. H. Auden, *Van Gogh: A Self Portrait. Letters Revealing his Life as a Painter* (Thames and Hudson, London, 1961)).

Very distant from the life of Vincent may seem Pharoah Akhenaten of eighteenth-dynasty Egypt, then the world's richest man and living most of the time, as one writer put it, a palace regime of 'refined sloth'. Yet, following as probably indirect consequence of his deformities (B. Brier, *The Murder of Tutankhamen* (Weidenfeld & Nicolson, London, 1998)), he was to end in disgrace and blindness, a wandering exile (I. Velikovsky, *Oedipus and Akhnaton: Myth and History* (Doubleday, New York, 1960)). His gifts of scientific monotheism (to be dispersed mainly later via Greece and via Judaism) and of realism in art (Greece again) still struggle among us to multiply but slowly do so. Unless Moses was an actual descendant of Akhenaten, it is not clear that either direct or close indirect biological fitness played any part (S. Freud, *Moses and Monotheism* (Vintage Books, New York, 1955)).

19. But perhaps I will find myself surprised here. A science reporter's note (P. Cohen, Not disabled, just different, *New Scientist* **160**, 18 only (24 October 1998)) has indicated that some deaf people prefer deaf offspring and a few even tell that if the technique was legal and available they would seek to abort a non-deaf fetus to re-try for a deaf one. The note indicates that the same may hold for some couples who are both achondroplasiac dwarfs. But both these responses appear to be minority views among the affected at present and I suspect that if the present degree of state help for such people were for some reason to disappear, they would become much rarer still.

20. In all this I am far from saying, of course, that any slight physical handicap that is known to be genetical and that sets a person apart has to be considered a product of deleterious mutations and therefore a downward step, such that the conscientious sufferer would decide not to reproduce. Even quite large handicaps may not be like this. I have expounded elsewhere and will again in the paper in the last chapter of this volume (Chapter 18) how the Red Queen theory of stability of sex through parasitism indicates that there may be many genes passing through the lows of cycles where they act for now like bad mutations but which are due to show their worth as resistors of disease in some future phase. Generally such conditions should be distinguishable from those due to unequivocally bad mutations by their existing at frequencies that such mutations (prior to human interventions) could never have attained. Because all of us have physical defects, this view is generally a message of hope and particularly it should be so for those who have more than average burdens—every one is justified to think that by struggling still to exist and to reproduce against a handicap, he or she may be carrying something through that is worthwhile to the overall great enterprise of our species.

21. P. Singer, *Rethinking Life and Death: The Collapse of Our Traditional Ethics* (Oxford University Press, Oxford, 1994).

22. K. Lorenz, *King Solomon's Ring*, 2nd edn (Methuen, London, 1960).

23. J. Archer, *The Nature of Grief: The Evolution and Psychology of Reactions to Loss* (Routledge, London, 1999).

24. These include (in what I might paraphrase Doughty and call a 'subcynical darkness of human thought') sometimes extremely high numbers of priest offspring—the numbers being far greater (e.g. 65) than almost any layman could attain even when *not* following the strict Church-imposed monogamy. See W. E. H. Lecky, *History of European Morals from Augustus to Charlemagne* (Longmans Green, London, 1884), Vol. 2, p. 331 and surrounding pages. The

onset and the failures of emphasis on clerical celibacy and chastity in the Dark and Middle Ages are fairly well-known historical subjects, but the simultaneously increasing religious emphasis on the sinfulness of abortion and infanticide, as a possibly parallel 'subcynical' theme with an obvious possible sociobiological connection (the promiscuous male should biologically want his many conceptions to be born and to survive), seems not to have been commented or researched.

25. W. D. Hamilton, Haploid dynamical polymorphism in a host with matching parasites: effects of mutation/subdivision, linkage and patterns of selection, *Journal of Heredity* **84**, 328–38 (1993).

26. R. Hoban, *Riddley Walker* (Jonathan Cape, London, 1980).

27. All this is referring obliquely to the evolutionary history of copulation in the Anthocoridae, Cimicidae, and certain other closely related 'true bug' families within the great insect order Heteroptera. Within that superfamily of predaceous and ectoparasitic insects it highlights an event in an ancestral species in which an over-eager male came to pierce the female's abdominal dorsum with his penis during copulation—that is, he pierced the armoured upper hind part of her body, penetrating between the segments instead of entering her through the natural aperture. On this occasion, so far from killing her, he initiated a totally new and successful route to fertilize her eggs and started the trend (probably itself gradual in the sense of involving slow changes to probabilities of the piercing method being used) that now allows males *always* to send their sperm direct through the female's abdominal cavity to the ovaries. Special tissues in the female for handling the trauma within the abdominal wall (the first event is now really perhaps 100 million years back) have evolved to accompany the new style of penetration and the former natural aperture has closed.

28. R. Lynn, Further evidence for the existence of race and sex differences in cranial capacity, *Social Behaviour and Personality* **21**, 89–92 (1993).

29. W. R. Trevathan, *Human Birth: An Evolutionary Perspective* (Aldine de Gruyter, New York, 1987).

30. J. Masefield, *The Collected Poems of John Masefield* (Heinemann, London, 1928).

31. Evolving back from a commitment to Caesareans will be difficult but not impossible and will need to be begun before matters have gone too far. It will involve, however, the very sorts of eugenics measures that are typically most strongly rejected. Caesareans would, of course, continue to be performed where necessary, but each would be a notifiable event entailing negative marks recorded in some government records with such marks being applied to

both of the parents and also to the offspring. The marks would be used in state-organized counselling and reproductive disincentive schemes.

Selection so imposed would, of course, tend to work against the further evolution of capacious human crania at least at birth, which, because of the long-term correlation that has built our species to what it is, would be regrettable. As a counterbalance perhaps positive eugenic marks would be awarded to encourage a trend towards the combination of (1) 'premature' birth and (2) ability of 'premies' to thrive with minimal high-tech intervention after the event (the 'thriving', of course, including demonstration of good brain expansion after birth). As regards how head evolution is headed, so to speak, and how abetted in that path by human mate choice, see, for example, D. I. Perrett, K. J. Lee, I. Penton-Voak, D. Rowland, S. Yoshikawa, D. M. Burt, *et al.*, Effects of sexual dimorphism on facial attractiveness, *Nature* **394**, 884–7 (1998) and also a commenting later letter (D. A. Meyer and M. W. Quong, The bio-logic of facial geometry, *Nature* **397**, 661–2 (1999) plus the reply of D. I. Perret and I. Penton-Voak)—and note particularly the illustration of a projected head shape for future *Homo*.

Perhaps neonate Isaac Newton should be taken as the prototype for the second part of the suggested reproductive style that I believe will indeed slowly arrive (or, rather, will continue—see, for example, J.-L. Arsuaga, C. Lorenzo, J.-M. Carretero, A. Gracia, I. Martinez, and N. García *et al.*, A complete human pelvis from the Middle Pleistocene of Spain, *Nature* **399**, 255–8 (1999)). Newton was described at birth as being of a size almost to be 'fitted into a pint pot'. To my mind, however, Newton's mother's act of abandoning him to the care of her own mother at age three in order to marry another man must offset very largely her eugenic 'good marks' towards her acceptance as our Neo-Eve!—and possibly it was on these lines that Newton likewise, summing all the aspects in his superb and love-famished brain, decided firmly and quickly never to reproduce.

As regards where we are going or might go on male–female differences, see note 7 of Chapter 9.

32. I. Illich, *Limits to Medicine. Medical Nemesis: The Expropriation of Health* (Penguin, London, 1990).

33. J. Bonner, Hooked on drugs, *New Scientist* **153**, 24–7 (18 January 1997).

34. E. D. Parker, Jr, Ecological implications of clonal diversity in parthenogenetic morphospecies, *American Zoologist* **19**, 753–62 (1979); R. C. Vrijenhoek, Factors affecting clonal diversity and coexistence, *American Zoologist* **19**, 787–97 (1979); R. A. Angus, Geographical dispersal and clonal diversity in unisexual fish populations *American Naturalist* **115**, 531–50 (1980).

35. S. J. Gould, *The Mismeasure of Man* (Penguin Books, London, 1996).

36. L. J. Kamin, *The Science and Politics of IQ* (Penguin Books, London, 1974);
    L. J. Kamin, *Intelligence: The Battle for the Mind* (Pan Books, London, 1981);
    S. Rose, R. C. Lewontin, and L. Kamin, *Not in Our Genes* (Pantheon, New
    York, 1989).

37. This accusation is serious and should be given examples. In 1996, S. J. Gould
    published a new and expanded edition of his 1981 book *The Mismeasure of
    Man*. In 1990, S. Rose, R. Lewontin, and L. Kamin reprinted their 1984 book,
    Not in Our Genes. Both books contained numerous posthumous defamatory
    statements about the life and work of the educationalist and psychometrician
    Sir Cyril Burt. None of the authors in their respective republications gave any
    acknowledgement of the fact that, since their first publications, at least two
    books (and in Gould's case clearly three books) (R. B. Joynson, *The Burt Affair*
    (Routledge, London, 1989); R. Fletcher, *Science, Ideology and the Media: The
    Cyril Burt Scandal* (Transaction Publishers, New Brunswick, NJ, 1991); N. J.
    Mackintosh, *Cyril Burt: Fraud or Framed?* (Oxford University Press, Oxford,
    1995)) had been published. In two cases these had exonerated Burt from the
    charges of fraud while the general import of the five-author remaining one was
    of mostly muddles (of an old man, sometimes muddling away from his
    preferred beliefs) and they left the import of his scientific claims, never
    seriously challenged subsequently, almost unchanged. Charges by detractors
    had begun 2 years after Burt's death: those made much of as primary detractors
    in the anti-hereditarian books I mention above were L. S. Hearnshaw,
    O. Gillie, L. Kamin, and Alan and Anne Clarke; later, and equally careless of
    truth in the making of a BBC so-called 'documentary' film in 1994, came
    S. Davis and M. Freeth—see Joynson (1989) loc cit.).

    It is a pity in some ways that defamation of the dead is not legally
    indictable; possibly this recourse would not have helped Burt's name, however,
    because he had neither surviving wife, offspring, or sibling to defend his
    memory (his older unmarried sister, Marion Burt, died, perhaps of the misery
    inflicted, a month after she heard from the man she had deemed her brother's
    friend and had chosen to be his biographer, about the coming charges of
    dishonesty he would make in his book). If posthumous libels were indictable,
    descendants of Henry H. Goddard, an early American follower of Binet in
    testing intelligence and 'mental age', might also consider requiring S. J. Gould
    to defend his distortions of their relative in his *Mismeasure* book too. In his
    account of Goddard's work (pp. 184–204), Gould heads a passage 'Preventing
    the immigration and propagation of morons' and in it he leaves unmentioned
    the conclusion of the paper he is discussing, which is exactly the opposite of

his subtitle. The conclusion was that, despite the very low scores on the pioneer Binet intelligence test Goddard had found for some of the would-be immigrants to the USA, he believed all of the set he had examined could live usefully and procreate good citizens: he stated, in fact, that the scores were likely to be not hereditary but due to poverty of background in Europe and Russia. Noting that similar immigrants had been coming for some time and most descendants of them seemed to be doing well, he wrote of his studied set: 'we may be confident that their children will be of average intelligence and if rightly brought up will be good citizens'. Gould conveyed a sense exactly opposite to this: that Goddard believed their defects were hereditary and was out to exclude them if possible from the USA.

The passage under the misleading title contains also Gould's imputation that Goddard believed his results showed that four-fifths of the inhabitants of countries from which his tested aspiring immigrants came (Jewish, Hungarian, Italian, and Russian) must have intelligences scoring below 'age 12' on the Binet scale—an idea that one would need to be near moronic oneself to believe and which Goddard made clear he did not. There is only one slender implication, quite likely a typographical slip, supporting any such implication about average intelligence in Goddard's paper. Gould himself admits that Goddard made quite plain in opening sentences that he had not tested a random sample but had chosen specifically a below-average one. After that explanation it does seem that Goddard followed with a rather contradictory reference to a group of 'average immigrants'. A reasonable interpretation, however, is that he intended 'below average' or else 'average' for the "third class" of the immigrant ship (see quote on p. 197 of *Mismeasure*): either intention would make much more sense with the rest of the paper and I, for one, would have read his reference to 'average immigrant' in one of these ways. The explicit idea of the study, then, had been to assess a set of 'below-average immigrants'—those that, given the lenient rejection criteria the immigration authorities were applying (they were rejecting only those they thought would be unable to make a living), were going to be admitted. 'Normals' Goddard himself told that he was avoiding for his study: thus his set presumably began at a point somewhat below the mode and extended downwards but in the 'low' tail necessarily excluded those already rejected (both the boundaries to his set, one has to admit, however, were left lamentably vague). On this interpretation—a slip in the text being admitted—the low figures he obtained are not very surprising and there is no need to impute to Goddard conclusions about average immigrant mental ages, which in any case he could not have expected any educated reader to believe.

In part my points here about Gould's first edition of *Mismeasure* have already been made (along with others) by M. Snyderman and R. J. Herrnstein (Intelligence tests and the immigration act of 1924, *American Psychologist* **38**, 986–95 (1983), by B. D. Davis (*Storm over Biology: Essays on Science, Sentiment and Public Policy* (Prometheus Books, Buffalo, 1986)), and by yet others. Gould ignored the first of the above; he attempted to rebut Davis's criticisms but still ignored the more serious—that, for example, concerning his grossly misleading subheading (S. J. Gould, Who has donned Lysenko's mantle?, *The Public Interest*, 148–51 (Spring 1984); see again Davis, loc. cit.). In his introduction to the second edition Gould portrays his exchange with Davis as a triumph against a 'ridiculous personal attack' by a 'dyspeptic colleague' (p. 45) but left unchanged his whole account of Goddard, including the derisive subheading.

As was common early in the twentieth century amongst intellectuals and psychologists of both left or right political stance, Goddard was a believing eugenicist; this fact, however, provides no excuse for publishing gross distortions of his character and work and then never changing a word after the erroneous implications (whether they were made in good faith or not) are made plain. As portrayed in a recent book devoted to his life (L. Zenderland, *Measuring Minds: Henry Herbert Goddard and the Origins of American Intelligence Testing* (Cambridge University Press, Cambridge, 1998)), the most careful evidence is that Goddard, like Burt, was not the racial and social bigot that Gould, despite his opportunity for revision, has persisted in trying to display to us. Goddard, descendant of practising New England Quaker parents, was not a brilliant or imaginative scientist but he was conscientious and there is no evidence that he was racist or perpetrated any fraud.

Many similar examples of non- or slapdash response to criticism exist in Gould's writing on social issues. Many are detailed, for example, including with abundant and recent documentation from the author himself, by J. P. Rushton (Race, intelligence, and the brain: the errors and omissions of the 'revised' edition of S. J. Gould's *The Mismeasure of Man* (1996), *Personality and Individual Differences* **23**(1), 169–80 (1997)).

38. B. D. Davis, *Storm over Biology: Essays on Science, Sentiment and Public Policy* (Prometheus Books, Buffalo, NY, 1986); R. B. Joynson, *The Burt Affair* (Routledge, London, 1989); R. Fletcher, *Science, Ideology and the Media: The Cyril Burt Scandal* (Transaction Publishers, New Brunswick, NJ, 1991); J. P. Rushton, The equalitarian dogma revisited, *Intelligence* **19**, 263–80 (1994); J. P. Rushton, *Race, Evolution and Behavior, a Life History Perspective* (Transaction Publishers, New Brunswick, NJ, 1995); R. Lynn, *Dysgenics: Genetic Deterioration in Modern Populations* (Praeger, Westport, CT, 1996);

P. R. Gross and N. Levitt, *Higher Superstition: The Academic Left and its Quarrels with Science* (John Hopkins University Press, Baltimore, MD, 1998).

39. See Rushton (1997) in note 37.

40. The speed with which connections are discovered between autoimmune diseases and infections can be judged also from popular science articles; for example, that by H. Bower, 'The germ bug', *The Independent on Sunday*, 44–5 (8 June 1997), an article hinting that many mysterious human complaints such as sudden infant death syndrome and schizophrenia (plus other forms of acute depression) may eventually be explained by molecular mimicry. Another article hints at quite subtle personal, not necessarily pathological, idiosynchracies being possibly of this nature (P. Brown, 'Over and over and over . . .', *New Scientist* **155**, 27–31 (2 August 1997)). Because the implicated infectious agent is often a common one that many of us catch without progressing to any severe symptoms and for many will never have any after-effects either, human genetic variation in reaction to the agent seems very probable.

41. C. Bompard-Gilles, P. Rousseau, P. Rouge, and F. Payon, Substrate mimicry in the active center of a mammalian α-amylase: structural analysis of an enzyme-inhibitor complex, *Structure* **4**, 1441–52 (1996).

42. Brown (1997) in note 40.

43. See also S. Jones, *In the Blood: God, Genes and Destiny* (HarperCollins, London, 1996).

44. A. Coghlan, Perfect people's republic, *New Scientist* **160**, 18 only (24 October 1998).

45. D. R. Lincicome, The goodness of parasitism: a new hypothesis, in T. C. Cheng (ed.), *Aspects of the Biology of Symbiosis*, pp. 139–227 (Butterworths, London, 1971).

46. J. Connell, On the role of natural enemies in preventing competitive exclusion in some marine animals and in rain forest trees, in P. J. van der Boer and G. R. Gradwell (ed.), *Dynamics of Populations*, pp. 298–310 (Centre for Agricultural Publication and Documentation, Waageningen, 1971).

47. D. H. Janzen, Herbivores and the number of tree species in tropical forests, *American Naturalist* **104**, 501–28 (1971).

48. J. B. Gillet, Pest pressure, an underestimated factor in evolution, in D. Nichols (ed.), *Taxonomy and Geography: A Symposium*, pp. 37–46 (London Systematics Association, London, 1962).

49. C. Wills, *Plagues: Their Origin, History and Future* (HarperCollins, London,

1996); C. Wills, R. Condit, R. B. Foster, and S. P. Hubbell, Strong density- and diversity-related effects help to maintain tree species diversity in a neotropical forest, *Proceedings of the National Academy of Sciences, USA* **94**, 1252–7 (1997).

50. In spite of recent discussion, multiline planting does not seem to have been much adopted, the apparent pragmatic reason being the difficulty of harvesting a crop of mixed species, or even a single-species crop that is just non-uniform in growth.

In lighter vein, perhaps something else is also subtly wrong with it: according to the Bible, God apparently thought so. Thus, unless accompanied by some ingenious trickery to hide the event from God's eyes, Orthodox Jews don't allow multiline planting because, amongst rules that He gave to Moses (and only second one down from his important one about 'thou shalt love thy neighbour as thyself') came the rule that 'thou shalt not sow thy field with mingled seed' (Leviticus 19:19). Some injunctions in the Bible are so mysterious in this why-would-anyone-ever-have-thought-of-that? way that I find myself wondering whether some of the days of dictation in the desert were either so very hot or subject to such thin mountain air that Moses, with concentration wandering, caught the wrong end of some stick whose true leafy nature we will never know. Another example I would like to see clarified (which I found, I think, during my early youthful researches into whose female nakedness I mightst or mightst not uncover), was the one about not pissing against a wall, which act my prep school at the time was virtually insisting that I commit every few hours; but here at least thoughts like: 'Palestine after all, very dry, ammonia stinks, contagion, wall perhaps not your own, good idea indoors anyway . . .' could drift in my mind. But this of mingling seed—real seed for planting, why? Well, as is usual when a new advantageous technique is in the offing, Orthodox Jews have found ways around: and if a sower doesn't *know* that seed has been 'accidentally' mingled, it seems he commits no sin in planting it. What is impressive in a scientific and practical way, however, is that it appears that this technique was being tried on Israeli kibbutzes (along with the appropriate and ridiculous rabbinical subterfuge, as amusingly told by Israel Shahak in the work referenced in note 35 of Chapter 8) as early as 1945.

51. M. A. Parker, Pathogens and sex in plants, *Evolutionary Ecology* **8**, 560–84 (1994); S. A. Frank, Problems of inferring the specificity of plant-pathogen genetics (a reply to Parker), *Evolutionary Ecology* **10**, 323–5 (1996).

52. R. A. Kaslow, M. Carrington, R. Apple, L. Park, A. Munoz, A. J. Saah, *et al.*, Influence of combinations of human major histocompatibility complex genes on the course of HIV-1 infection, *Nature Medicine* **2**, 405–11 (1996); M.

Westby, F. Manca, and A. G. Dalgleish, The role of host immune responses in determining the outcome of HIV infection, *Immunology Today* **17**, 120–6 (1996).

53. M. Samson, F. Libert, B. J. Doranz, J. Rucker, C. Liesnard, C. M. Farber, *et al.*, Resistance to HIV-1 infection in Caucasian individuals bearing mutant alleles of the CCR-5 chemokine receptor gene, *Nature*, **382**, 722–5 (1996); M. Carrington, T. Kissner, B. Gerrard, S. Ivanov, S. J. Obrien, and M. Dean, Novel alleles of the chemokine-receptor gene CCR5, *American Journal of Human Genetics* **61**, 1261–7 (1997); S. S. Bakshi, L. Q. Zhang, D. Ho, S. Than, and S. G. Pahwa, Distribution of CCR5 Delta 32 in human immunodeficiency virus-infected children and its relationship to disease course, *Clinical and Diagnostic Laboratory Immunology* **5**, 38–40 (1998); F. Libert, P. Cochaux, G. Beckman, M. Samson, M. Aksenova, A. Cao, *et al.*, The Delta CCR5 mutation conferring protection against HIV-1 in Caucasian populations has a single and recent origin in Northeastern Europe, *Human Molecular Genetics* **7**, 399–406 (1998).

54. Subsequently, however, it has been suggested that it may have been via a fortuitous cross-resistance that Europeans had the (very grim) 'luck' of a pre-selection of a different kind. It has been suggested that the *Yersinia pestis* plagues in Europe raised frequency of the propitious gene (J. C. Stephens, D. E. Reich, D. B. Goldstein, H. D. Shin, M. W. Smith, M. Carrington, *et al.*, Dating the origin of the CCR5-Delta 32 AIDS-resistance allele by the coalescence of haplotypes, *American Journal of Human Genetics* **62**, 1507–15 (1998)). This conjecture is partly based on molecular similarities but first came up because the areas of Europe where plague struck hardest were those where the HIV-1 resistance CCR5 gene has its highest frequency.

55. M. J. Hobart, Phenotypic genetics of complement components, *Philosophical Transactions of the Royal Society London B* **306**, 325–31 (1984).

56. K. F. Lindahl, Minor histocompatibility antigens, *Trends in Genetics* **7**, 219–24 (1991).

SEX AND DISEASE[†]

W. D. HAMILTON

Nothing 'gainst time's scythe can make defence
Save breed, to brave him when he takes thee hence.
SHAKESPEARE, *Sonnet XII*

The decade that is witnessing the emergence of a new lethal, sexually trans-mitted disease may seem a bad one in which to be supporting a theory that sex is life's main adaptation to combat disease. But such a theory exists, and it seems to me the best contender in a great puzzle.[1] The theory began before AIDS[2,3] (see also Chapter 2) and my contribution to this symposium will be devoted to showing the arguments and the evidence.

'Can you be claiming we need sex to create immunity to diseases that are spread by having sex—this must be nonsense!' This seems a reasonable reac-tion at first, and certainly venereal diseases are going to have to be reckoned on the debit side of any such general adaptation as I will claim sex to be. But, of course, the number of human diseases that are transmitted sexually is quite low compared with those transmitted in other ways. It just happens that today most of the other deadly diseases (e.g. plague, smallpox, and cholera) have been on the wane for some time due to progress in medicine and public hygiene, so the new, venereal, one looms large. However, many of the others are far from negligible even now. Malaria has long been one of humankind's greatest natural killers, apart from the diseases of old age; it still is and indeed is resurgent. Outside of Europe and America, certain enteric and pneumonic diseases are still serious, especially as infant-killers; and then there are measles, hepatitis, dengue, all the parasites that may not kill but certainly weaken, and so on.[4] Looking back merely two centuries in Europe, a theory that took for granted that disease was one of the most common, per-vasive, and terrible of the threats to human life would have been easy to accept. Two hundred years ago, a theory that claimed in essence to explain why the hand of death seemingly so arbitrarily plucks some members of a family while sparing others equally exposed, and why the hand of death

---

[†]In G. Stevens and R. Bellig (ed.), *Nobel Conference XXIII: The Evolution of Sex*, Ch. 4, pp. 65–95 (Harper and Row, San Francisco, 1988).

plucks some families entirely while it spares others equally entirely (and also why it favoured some races against others), might then at least have had the benefit of appealing to common experience. In old literature, one does not have to be reading Defoe's account of the great plague in London to see this: abundant reference to the ever-present threat can be found in the writing of almost any author from Chaucer to Dickens.

I took up what is really the very peripheral issue of AIDS in humans to emphasize the seriousness of disease in general; but let us just consider for a moment whether the issue of genetic resistance is really so passé, even now and even for AIDS. May we not have a prospect of a common experience, just as in the Middle Ages? Please, for a moment, take the imaginative step of allowing my idea—that is, assume that sexual reproduction might serve to bring together new combinations of genes for offspring that could resist current important *non-sexual* diseases. Surely it is plausible that such a system will also work to alleviate its own inevitable hangers on—the sexually transmitted diseases. Such a possibility seems quite strong in the case of the human immunodeficiency virus (HIV), the cause of AIDS.

Clinical genetic studies, published in 1987,[5] indicate that you may indeed be able to reduce the chance that your offspring will die of AIDS if you select your partner on the basis of certain well-defined, biochemically identifiable, genetic traits. You yourself may already have the lucky genotype; or you may have one of the less-lucky ones, or the very bad one. It seems, if this report is right, that there may well be today some 8 per cent of Europeans who will never die of AIDS, or even develop any signs of HIV infection after being constantly exposed to it.

Can it be just by chance that the human population had a common set of neutral genetical variants for a protein (i.e. a neutral polymorphism) that profoundly affects susceptibility to HIV, and that this polymorphism was drifting under no selection right up to the time when the virus made its first appearance—that is, drifting for millions of years? This seems incredible. The particular immunoglobulins that are coded by the genes that I am talking about were already under investigation before their associations with HIV were noticed. This was because of intrinsic interest—their involvement with the transport of vitamin D and certain aspects of cell communication in the immune system—and also because of some other weak associations with obscure diseases such as rheumatoid arthritis. Thus it seems to me far more likely that the immunoglobulins were already involved with disease defence, most likely with defence against other viruses, perhaps ones that are not currently very pressing or dangerous. This is only a guess, I have to admit, but it can serve to lead to my next point.

It is a necessary part of the theory of sex that I am presenting that there is no feasible complete genetical answer to the disease threats pressing on a species, nor even any moderately satisfactory answer that remains good for any length of time. Disease threats are in process of continual change. Thus if HIV infection were to remain medically uncontrolled (and people were to remain foolish about the obvious behavioural measures), the end of the epidemic might leave the frequencies of the genes just mentioned at completely different values from what they have today—the immune genotype mentioned above, for example, might go from its present 8 per cent to nearer, say, 80 per cent. But maybe a decade or two later the genotype that was the worst for HIV infection might prove the best when a new virus spreads to human beings. The frequencies might swing back, or perhaps they would swing towards some quite new state.

This would be natural selection, but it is nothing like an answer to the problem of sex. A human population that reproduced by parthenogenesis could respond in much the same way and would probably achieve a given adjustment of frequencies faster. (This is only not true if the good trait tends to be dominant, which is claimed not the case in the example given. Basically, under sexual diploidy and random mating, selection works on means of samples of two—the two homologous genes. This is only half as efficient as when selection works on single genotypes, as is the case under parthenogenesis.) For a theory of sex based on disease resistance to have any hope at all, it must discuss not just a single locus but a minimum of several, and the alleles at the different loci must be interacting in some interesting way. It has to be the advantage of special combinations of alleles that favour having sex. New gene combinations are sex's specialty; indeed, in nuclear genetic terms, new combinations are perhaps all that sex can supply, although outside of the nuclear chromosomes there are other interesting cytogene possibilities that still deserve study.[6]

Take the viewpoint of a gene in a very common situation, housed in a chromosome of a female in a species where the male does little or no work for the brood. Imagine this is also a species where there is an option of parthenogenesis—the female doesn't have to reproduce sexually, but she can. Species in which such an option exists are extremely rare, and there are fairly obvious reasons why populations will quickly move off such an evolutionary knife edge if they even come to alight on one; but let us assume that the option exists. (When the option of asexual reproduction is completely open, as is implied here, in any period of environmental constancy the variant wastefully performing sex is likely to be lost completely, and extinction of the species may then follow when the environment again changes; this point will be discussed in more detail later.)

Suppose you are the gene that has the option—you work the switch that determines whether your carrier is going to be parthenogenetic or not. What sort of information would you want to gather to decide which was the best option (i.e. either sex or perfect replication of your bearer's genotype), giving regard to the long-term survival of copies of yourself?

If things seemed to have gone well during the lifetime of the population of cells that is your present bearer, and if your information about the environment suggested it was going to stay the same, then I think you would undoubtedly choose parthenogenesis. After all, if you accept a mate then, subject to the rules of genetics, you incur immediately a 50 per cent chance for any particular offspring that you won't be transmitted at all; whereas with parthenogenesis, you are certain to be in every offspring. This rule of inheritance that so strongly favours the option of parthenogenesis is called the *cost of sex*, although according to circumstance it might more aptly be called 'the cost of tolerating cheap sperm cells', or of 'having lazy males' (strictly, of allowing there to be fathers who don't do their share in child-raising—males are far from lazy in the other things that they have to do).

If, on the contrary, you perceive that your bearer had a bad time, was not well adapted, and only just made it to being able to reproduce at all—or else if, although your own life was good, you perceive the environment to be in the process of very rapid and unpredictable change—then you might opt for sex on the ground that retaining your present dubiously serviceable gene combinations (i.e. retaining your allies within your current genome) may make your offspring quite hopeless. Then you might well reckon that the mere half chance of transmission due to sharing with a mate is more than offset by the chance of making up new random alliances that will turn out well. The object is these better new combinations—the mixture of some of your present allies along with some of someone else's. It doesn't greatly matter which happen to mix with which; or at least it does matter, but you can't predict what will be good, and random chiasma formation will see to it that you can't choose. (Actually this is not quite true; as a gene you can't choose anything, of course, certainly not any particular combination, but in the ensemble that is your bearer there may be the power of deciding what is a better bet. For example, the best may well be nearly the same as was best in the present generation. And the female you are in may have some power of detecting the males more likely to slant your offspring the right way—to carry genes that would be good to join or to replace your current allies: but this sexual-selection story would take too much time to tell.)

For sex itself, arguing that things are so bad so often that union with a mate is the best course to take is not going to be at all easy. It has to be remembered that in the effort to make up better new combinations by sex, worse ones—and more of them—are also bound to be produced. This factor is offset somewhat if some elimination of hopeless offspring can occur very early, when not much biomass is lost and when they can be replaced by siblings. The effect of sibling competition in improving models of sex has long been recognized.[7,8] But not all highly sexual groups have such competition (e.g. fish or crabs having tiny planktonic larvae), and without this factor the situation still looks rather bleak. Many who have tried seemingly hopeful assumptions in models have found how hard is the task. As I came to realize it myself, I became more and more pessimistic that the generally considered factors of evolutionary change could be strong enough or of the right kind. Constant strong pressure for a set course of change certainly will not do; the pressures themselves need to change all the time. They must change rapidly and promote now this and now that gene combination, and ideally it should all be done in a fairly erratic sequence so that no regular sequence of poly-morphism (such as is produced out of a single genotype in a clone of aphids or water fleas throughout the seasons of the year) could succeed. Nor is it enough if, although unpredictable, the changes occur smoothly over a span of many generations. Then it is still going to be advantageous at least in the short term to stay with a known successful genotype. But if the long term is considered, might sex eventually succeed nevertheless? The issues at this point become rather subtle.

It could well be that, if parthenogenetic, a population can't keep up with change for any great length of time, because mutations adapting to the change do not occur fast enough. This will particularly be the case if needed adaptations involve large, carefully constructed molecules doing functional tasks: the loss of the gene for any such molecule because it is temporarily not needed may be the gain of a dormant seed of long-term disaster. Thinking of yourself as a sex-deciding gene again, it may seem that if you could be far-sighted, you would opt for sex even though for some generations you would become increasingly overwhelmed by the numbers of competitors who show parthenogenesis. The question is whether the short-term numerical advan-tage of the clones becomes so great that you as sexual determiner go extinct while still waiting for your farsightedness to pay off. The chance of extinction makes it unclear if you will survive if the pace of change is slow, but equally unclear is the ability of asexual clones to survive in the long term.[9] The rather unexpected implication of the two ideas just stated, taken together, is that a form of group selection may have to play a part in explaining the evolution and maintenance of sex. For the one we can try to find factors in nature that are rapidly changing (diseases are one obvious good bet), and try to construct models that achieve the difficult task of paying the full cost of sex in the short term (see Chapter 2). We might reluctantly accept the argument that there has been a very long history of population and species extinctions and this has borne hard on all units that were able to go asexual too easily, leaving only the units that, while they were sexual, had accumulated by accident (perhaps as side-effects of genes selected for other reasons) various inbuilt barriers to the easy acquisition of parthenogenesis.

Such barriers do seem to exist. For example, entry of sperm seems to be essential for the further development of ova in many groups, leading to a situ-ation where parthenogenesis, if it exists at all, is still disguised in lineaments of sex. Females of so-called 'gynogenetic' asexual strains avoid the dilution involved in getting genes from male contributors, but their ova still have to be penetrated by a sperm before they can develop. An important point about this system is that the clones cannot exterminate the sexual strain that supplies the sperm without immediately going extinct themselves; and, in fact, the system seems always to set up an uneasy equilibrium of the two strains even though this is not an outcome of the most obvious theoretical dynamics.[10]

In so far as barriers to a quick, efficient start to parthenogenesis exist, we can accept progressively longer and longer-term environmental changes as supporting sex. During any phase of a cycle that lasted many generations, there might be rounds of selection towards improving a particular mutation to parthenogenesis because it is linked in a currently excellent and increasing genotype; but if these runs only start work on a process that is very inefficient—one, say, with only little over half of the parthenogenetic embryos surviving—there will not be enough time for progress to reach a point where the clone could drive the sexual strain extinct before, with changing phase of the cycle, the sexual's continued recombination began to pay off. Then it would be the imperfect asexuals that went extinct, and after that there can only be another equally inefficient start repeated during another phase.[11-13]

However, quite a lot of organisms—especially plants, but also lower animals at least up to the level of arthropods—seem to have easily available efficient parthenogenesis. This judgement is based on how often strains lacking males have been established and from the frequency of claims of having found examples of facultative parthenogenesis. Clearly, then, the problem of short-term maintenance remains quite acute. Therefore, let us now continue our search for sources of rapid and severe environmental change that might be able immediately to justify the twofold cost of sex. I aim to put any candidate process through a severe test: if there is one that kills the twofold dragon, it will almost certainly kill the lesser (perhaps more realistic) group-selection dragons quite easily.

It might be thought that, even if the annual climatic cycle (as it applies to very short-lived organisms, such as water fleas) is too regular for sex, there are still plenty of unpredictable longer-period climatic changes. These changes push back and forth, but mostly are not regular enough to be called cycles.[14] At the long extreme, an example would be the ice ages. The trouble with them is that although they apply to both shorter- and longer-lived organisms on the right scale of generations, and although they may have the right degree of autocorrelation (varying from near lack of it to a good periodicity, as in the case of, say, sunspot cycles), the fluctuations are too unresponsive to the states of the populations they affect. Their extreme runs sooner or later drive all polymorphisms to fixation (see Chapter 1)—that is, one by one all the alleles go extinct until only one at each locus is left.

Might polymorphism, in spite of such runs of selection, be maintained by recurrent mutation? It might, but it is not very likely to work as a help for sex for two reasons. I have already suggested that it isn't likely that simple re-mutations to gene states that code elaborate functional molecules will occur. Second, even if such mutations were somehow numerous enough, they assist asexual clones to cope with change just as much as they help the sexuals. Once polymorphism is truly lost, there is, of course, nothing that sex can do to make new combinations, and all possibility of advantage is gone. What seems to be needed is a more relenting set of pressures that ease off when alleles are near to being lost. This will tend to ensure that there is always something with which recombination can work.

It is a well-known principle in ecology that common organisms tend to attract many exploiters. This happens both within the lifetimes of particular populations of the organism, and over long evolutionary times. If you look at the leaves of the commonest species of tree in your neighbourhood, you are likely to see more of them damaged or infected than if you look at the leaves of a rare tree. This is the principle that makes the living world so varied. It implies that biotic pressures ease off when any species becomes rare. This slows its path towards extinction, and perhaps gives it a chance to await the opening of a new niche and a return to abundance. If the same principle can be applied to genetical variation within a species, then we see a possibility of a biotic adversity view of sex:[15] variation will be preserved in the species in the same way that species are preserved in an ecosystem. The crucial ques-tion, then, is whether such variation is largely static, or whether it is kept dynamic by some intrinsic ecological churning of the coadapting systems. If the latter is the case, there is hope that we are on the track of an answer to the puzzle of sex.

Before discussing possible dynamics, however, let us ask what kinds of enemies are most likely to maintain variation in a species. Enemies are basically of two kinds: predators who are big and eat whole individuals at a time, and parasites who are small and nibble or usurp nutrients—who damage but typically do not directly kill. The predators are almost always generalists and feed on many prey species. They are also normally longer-lived than their prey. I can see them at present as being rather less likely to maintain an abundance of variation in prey, or to render the variation dynamic to the degree needed. This view may be mistaken. Some cycles of wildlife in the arctic, for example, have been attributed to a system of over-shoots and crashes by populations of predators focused on a few normally abundant prey. During these cycles, the selection pressures on the prey cer-tainly will change dramatically, and not only with respect to the predation itself. There might in these cases be a tendency to spare rare variants. But at present I cannot easily rationalize these possibilities as support for sex unless the other kind of biotic 'pursuit'—that is, parasite—is involved (see Chapter 8). Parasites have a much more intimate approach to their food supply and are much more likely to differentiate between varieties of species. In fact, it is known that they often do differentiate; and because of the economic importance of parasites, in many cases the pattern of their discrimination has been worked out in detail.

'Parasite' is almost another way of saying 'disease', so now I am back close to my title. A virus, a gut bacterium, a flea, a blood fluke, and a caterpillar are all parasites.[16] Some of them—the flea, the bacterium, the caterpillar—do not normally cause anything described as disease, but that does not mean that they exert negligible selection. The flea can transmit serious diseases; the bacterium can turn aggressive when host defences are low, indicating that it probably has to be under some restraint in our gut all of the time; and caterpillars do sometimes defoliate trees completely and kill them. Parasites, moreover, are ubiquitous; there is probably no sexual species that does not have several, and most have many with a fair proportion of their parasites devoted to the exploitation of that host alone. Parasites must undoubtedly have made their appearance at a very early stage of the evolution of life. I think this is one point on which the view of Lynn Margulis will agree with mine. Such an immense past span of interaction plus such continuing ubiquity means that parasites, unlike most other factors that have been suggested, have a potential to be the universal cause of sex. Other factors might exist or might not: they would not be needed.

What pattern of interactions of parasites with their hosts is most likely to favour sex? Having decided this question, we may then further ask whether we find the specified pattern commonly in sexual species, and whether it is less common in species that have gone asexual.

One pattern of interaction that needs to be dismissed at once is overdominance within loci—that is, patterns where heterozygotes at each locus provide the fittest genotypes. Overdominance is the phenomenon usually illustrated by the well-known case of the sickle-cell anaemia gene, which confers resistance to malaria in humans; and, in fact, that case is so well known that we have to beware not to assume all kinds of disease resistance are like it. Actually the pattern, at least with such strength of effects as is shown by the sickling gene, seems to be very unusual. I cannot afford space to describe the case and my objections to it as a paradigm of resistance, but will note only the following:

1. Such situations have stable population equilibria and do not lead to the *perpetuum mobile* that we require. The existence of such equilibria, however, makes like cases easier to detect and understand; thus perhaps the well-studied examples give an exaggerated impression of their importance.

2. A population would be considerably more fit if the heterozygote could breed true. Under sex, the heterozygote can't breed true; but after a duplication of the locus (which is not rare as an evolutionary event), a situation equiv-

alent to having a true-breeding heterozygote is easily established. Both loci would be homozygous (for different alleles) and there would be no contribution from sex. But parthenogenesis in the heterozygote would also achieve the same benefit and it would have the intrinsic efficiency of this mode of reproduction as a bonus. Thus, heterozygote advantage may be good at keeping genes in play: but, like recurrent mutation, it gives genes its protection only in circumstances that allow even greater advantages for parthenogenesis. (Only half the offspring of a heterozygote through a sexual union are like the parent. Because by definition of overdominance the two homozygous classes are less fit, it cannot be advantageous to produce them. Thus parthenogenesis, which makes all offspring like the parent, is the best option even before its other advantages, accruing through the non-production of males, are taken into account.)

A more promising scheme for sex will have most of the interaction of genes required for resistance not within loci, as in the example just mentioned, but between them. A type of interaction that, like the last one, tends to preserve variability and protect alleles, is one where most genotypes have typically a rather uniform level of fitness, but all the time the environment, by temporally varying criteria, is singling out some genotypes as being extremely bad. This pattern preserves variability much better than a pattern where there are instead a few exceptionally good genotypes, while most are indifferent.

To visualize these two cases, consider the latter case first. Imagining all fitness possibilities for genotypes arrayed before us in some orderly way, we see sharp peaks of fitness standing up out of a uniform or undulating plain. The height of peaks reflects the relative fitness of the genotypes that map to them. During times when there is little change in the environment, there is a rapid concentration of the population on the peak genotype with accompanying rapid loss of alleles that happen to be represented only in the plain. To see how genetic variability is better preserved in the first option of few extremely bad genotypes, imagine deep pits in a flattish, coalfield-like landscape of fitness. There will be no strong tendency to lose alleles in this representation. On the theory already outlined, the parasite will concentrate on the commonest genotypes. Thus it is the common ones that tend to become pits. As a common genotype becomes a pit, it is the property of sex to be able to send many of the offspring of its outcrosses (i.e. of matings other than with those with its own type) scrambling out of the pit. This coalfield-like landscape undulates over time, as did the previous landscape; but although pits may grow swiftly deep, like mine shafts in course of excavation, they are less likely to get anywhere near to engulfing the whole population (thereby driving many alleles to extinction) than is the previous selection we envisioned based on peaks.

A coalfield landscape of genetic fitness interactions can be argued to be a more likely outcome of parasite ecology and evolution than the peak landscape: peaks imply that some multilocus genotypes exist that resist almost all types of parasite attack. Considering the different natures of the various parasites that beset a species, this seems intrinsically unlikely. The coalfield, on the other hand, implies merely that whenever a genotype has grown common, all kinds of parasites evolve to become specialists at attacking it, causing the common genotypes to sink to form a pit.

Such a verbal and pictoral argument is, of course, very far from showing that any given situation can be changing fast and radically enough for a twofold cost of sex to be met. Can it be worth such loss of efficiency just to have the chance of throwing some (certainly not all) of your offspring out onto the shoulders of an engulfing pit? Continuing in such merely heuristic reasoning, there is at least one plausible line from ecology and behaviour that may indicate how, for some species, the landscape may always be changing sufficiently fast to warrant the choice.

Many species have social behaviour or other interactions that make small differences in phenotype magnify into much larger differences in expressed fitness. Two animals may be very close on a scale of size and health; and yet one of the two may achieve its full potential reproduction, while the other, just slightly below it on the scale, may achieve nothing. All animals of still lower ranks would likewise fail to reproduce. This pattern characterizes those higher species that are sometimes said to be more 'K-selected', the term referring to their tendency to maintain set maximum levels of population density. However, a more appropriate term for them in the present context is that they are 'soft-selected',[17,18] this referring to the idea that selection is based on rank ordering, so that the achievement of a particular genotype varies greatly with the state of the population. Individuals in such species tend to compete face to face over limiting resources. In fact, they often seem to make more fuss about gaining control of these resources than one would have thought justified by necessity. The resource could be, for example, a territory or a nesting place within a colony area. Those animals that don't obtain any share in the contested resource become 'floaters'; and as such, even though the resource they lack may not seem very crucial, their present reproduction is usually nil, and their chance of survival also drastically reduced.[19]

If parasites affect health, and if even small differences in health affect the winning of contests, then such a pattern will markedly accentuate both the depth of pits in the fitness landscape and the rate of change of depth and position of those pits. It will produce the rather flat landscape already described, whose main features are the exceptionally deep holes; moreover, the holes will be filling and others opening all to a rather quick beat of the generations. With Robert Axelrod and Reiko Tanese, I have found in computer simulation that such a rank-based 'soft-selection' scheme does make a model that can advance sex very effectively against the twofold cost. The models concerned are largely of a gene-for-gene nature (see below): but I have hopes that the same can be shown in a broader class, including ones where no subtle biochemical interaction has to be assumed to create the interlocus interaction. It is hoped that the interaction needed could arise from the stepwise fitness function implied above. In other words, if you are far enough up a rank order, for whatever reason, you get full fitness; if you are not far enough up, you get nothing. However, it has not yet been shown that this model, based on polygenic traits and this simple kind of interaction, will work.

So-called soft selection is, of course, often anything but 'soft' for the organisms experiencing it. As already implied, soft does not mean weak, although there is no reason why soft selection should not sometimes be weak. It is called soft simply on the ground of no fixed relationship between a geno-type and how many offspring it produces; instead this depends on the social situation—how many better individuals there may happen to be than a given one. There can be little doubt that something approaching this scheme does hold very widely for large long-lived organisms, especially animals. Such organisms are exactly the ones that can be expected to have exceptional problems with parasites, both because they are conspicuous to parasites and because parasites have such an advantage in adapting to them through the great discrepancy of generation times (Chapter 5). Satisfactorily for the the-ory, these large organisms are indeed the ones that are most consistently sex-ual. Small, short-lived organisms tend on the contrary to be opportunistic. Their regime is more 'hard selected'—that is, their reproductive performance is rather smoothly and rigidly linked to how well a particular genotype fits with its environment. Fitness shows no abrupt steps. A majority of these smaller species are actually parasites, or even parasites of parasites. Again corresponding with theory, the smaller and the more short-lived is the species, the more likely the species is to be asexual (although there are many exceptions).

It has been indicated above that the theory will work best if it is impossible to find genotypes resistant to all parasites at once, so in effect preventing the peaky landscape. Resistance instead should be rather a piecemeal affair, with many of the lines of adaptation mutually exclusive. Resistance has been so important to animal and plant breeders that there is actually quite a lot of evidence. Unfortunately I cannot state that, as a whole, it currently leans the way I would wish. Even in theory one can see ways in which whole sets of resistances could be strengthened simultaneously, and for the most part the evidence at present does not indicate mutual exclusivity for lines of adapta-tion.[20] For example, the immune system is a general-purpose adaptation that aids higher animals in coping with the vast range of parasites—the 'inven-tion' of the immune system was, in fact, probably the crucial adaptation that permitted large homeotherms to evolve at all (see Chapter 5). This system is very complex, however, and it is hard to imagine a single mutation that would affect all parts beneficially. Nevertheless, a gene that enlarged allocation to lymph nodes in embryogenesis, or one that enlarged the thymus, or one that enlarged bone marrow, could well have positive consequences for many facets of resistance at once. In so far as cost-free, such sweeping and beneficial modifications are hostile to my current theory.

Likewise, turning to large plants, trees, though they lack an immune system, could give the same sort of broad encouragement to resistance through genes that caused leaves to be more loaded with phenols or tannins, or simply to have more hard fibre or silica inclusions. These attributes could confer a general resistance by deterring insect attack (although one needs caution here when the 'deterrent' is not a poison).[21] For plants in general versus insects and fungi, a thicker cuticle might be of help.[22] Adaptations of such general nature are referred to as 'horizontal' resistance by phytopathologists. What I have to admit is that descriptions of resistances of the horizontal kind seem to be more common, so far, in the literature of plant and animal breeding than are descriptions of highly specific resistance, and they are still more common relative to cases of which I have begun calling 'complementary' resistance—that is, to cases where the genotype resistant to one pathogen proves highly susceptible to another and vice versa. The last pattern does occur,[23,24] and it may be that as between diseases of very different types that are not often compared it is more common than is known. 'Complementary' resistance is close to, although not identical with, the much discussed 'vertical' or 'gene-for-gene' resistances that phytopathologists contrast to the 'horizontal' type discussed above. The latter seems to be particularly common between crop plants and their parasitic fungi, but some other parasites are involved and even some animal systems show it,[25–27] although here it appears to be much more rare. Even in plants it may not be so clear-cut a system as has been claimed.[28] Taken all together, then, the evidence for complementary resistance is scrappy; that for general resistance is much more substantial.

General resistance, however, also must carry a cost of some sort. Otherwise every host species would have evolved to be resistant to every parasite, and it would follow on from this that all parasites would be extinct. This is *reductio ad absurdum* indeed, for the true situation couldn't possibly be more different: in species numbers, at least parasites rule the world. A parallel *reductio* is forced on us when we read the applied plant and animal breeding literature. Heritable resistance seems almost always to be found when it is looked for, and yet almost all the disease problems of crops and domestic animals steadily continue. If anything, they grow worse or at least become a bigger facet of husbandry, because as the variety of the genotypic bases of crops and breeds is reduced in response to the needs of mechanization, and through the possibility of the almost worldwide dissemination of particular successful varieties, the spread of pathogens is greatly facilitated.[29] Such a situation would be a complete puzzle unless there were a cost working in such a way that breeders are finally rejecting the most highly resistant lines because they are of low productivity;[30] or else are duly selecting a seemingly successful resistant strain only to find new problems coming from the complementary adaptations of an unthought of disease, which makes itself known soon after the new variety is planted,[31] rather as the Bad Fairy, forgotten at the party, comes remorselessly for her revenge on Sleeping Beauty.

If the former situation is the more common (i.e. in a given case it was horizontal resistance that came to light and it had a cost), then in nature we would find that whenever a parasite epidemic ended, non-resistant genotypes would outbreed the others and susceptibility would come back. It is certainly possible to imagine a repeated cycle like this producing an advantage for sex, but the possibility has not been pursued. There might be cycles of abundance based on recurrent epidemics, and the epidemics could possibly be of quite different pathogens.[32] Suppose one cycle of epidemics has a different time period of repetition compared to a cycle of some other kind affecting the species—say, that coming from pressure made on the host's resources by a second competing species that has its own cycle. If the genes for resistance to the plagues interact non-additively with the genes that have to do with adaptation to competition, we have the essence of a system that could support sex.

Non-additivity (which as a technical point here needs to be thought of in terms of logarithmic fitness measures) might be very simple. It may not need to be anything more than the natural outcome of the rank-based soft-selection scheme already outlined. A sexual variant in this population might much more effectively track the changing requirements of its environment compared with a clone. Sex genes, hitchhiking along with changing instead of fixed combinations, could do well; and, among sex alleles, those that do best may be the ones that promoted intermediate levels of recombination (Chapters 1 and 5, and ref. 33). The level of recombination that evolved would depend, probably, on the periods and relative strengths of the cycles.[34] It has not been shown yet, but such a model might well pay the substantial cost of having sex. I regard the absence of models demonstrating how sex can be maintained in broad-spectrum and polygenic-resistance systems when treated along with reasonable opposed parasite ecologies, as one of the main current weaknesses of the theory.

In essence, under random mating, sex has one simple genetical effect: to release correlated associations of particular genes. It allows them to move towards more random association, and opens the possibility of quick formation of new correlated blocks. Under natural selection, some of the new blocks may be advantageous and become abundant (in diploids the new blocks also have opportunity to become more often homozygous); but it has to be emphasized that this later part of the process—selection causing changing abundance—has nothing to do with sex: given the same combinations to start with, asexuals would change as much or more. So, looking both at the old combinations and the new after selecting, necessary signs to show that sex is not merely present but achieving something are (1) *non-zero linkage disequilibria* (this means that particular combinations of genes along chromosomes are unusually common—more common than could be explained by chance); and (2) finding that these linkage disequilibria are *changing with time*. Are such non-random gene associations and changes in them known?

Unfortunately they are by no means yet known on the scale that my theory requires. Surveys of electrophoretic variation covering many gene loci do sometimes reveal linkage disequilibria, but they are usually quite small. I know of hardly any evidence of non-random associations that are changing with time. However, this latter fact may reflect more that lack of repeats of the surveys than a real absence of change. More generally, an excuse for the weak and absent effect involving linkage disequilibria can be that most of the enzyme systems studied in the surveys are definitely *not* those expected to be in the front line of defence against parasites. It might be asked here whether it isn't inevitable that some of the many gene loci surveyed will be at least linked to loci that *are* involved in defence. Won't the combinations of such linked genes be changed, if to a lesser degree, along with the breaking and forming of the others that are really used in coadapting to the parasite? But the answer here is that if the frequencies of the marker genes that are seen are static, and if the population is large enough so that the effects of bottleneck combinations can be ignored, then such induced linkage disequilibria in neutral alleles will not occur. I can therefore remain hopeful that when the right genes are surveyed, fairly strong linkage disequilibria in the process of change will be found. On a more positive note, pointers to particular loci that may be connected with defence and disease are accumulating. These loci are not yet known to show any suitable non-random associations with each other or with other polymorphic loci; but then such association has not been sought.

The huge polymorphism of the major histocompatibility complex is slowly revealing its connections with infectious disease. A truly functional role for its molecules in disease defence is at last beginning to appear: it seems that these molecules may actually hold things—pieces of virus or virus products[35] —up to the view of other cells as signs that the cell in whose membrane they are standing has been infected. On this view they do not (as in the earlier view, which was always strange to an evolutionist) simply signal the identity of their bearer. The finding of function for these molecules, taken together with the inevitable process of adaptation of microorganisms to avoid being caught up and revealed by the currently commonest histocompatibility antigens of defending hosts, is certainly a great help towards understanding the diversity at this locus.

The major histocompatibility complex does show some non-random association of alleles at loci within itself, and doubtless sex has played a part to help these to be formed. However, linkages with the complex are too tight for them to be relevant to the short-term sex maintenance we are seeking at the moment. We need to look for other systems with which this one could be interacting.

Other loci with moderate polymorphism but, above all, showing exceptionally high rates of evolution have been revealed and discussed. These loci produce protease inhibitors in vertebrates.[36,37] They are as yet only conjectured to be connected with disease, but the idea is very plausible. The function of the inhibitors is to block the active sites of certain of the enzymes that digest proteins, so rendering those enzymes ineffective. My hopelessly unbiochemical imagination tries to sharpen its image of the interaction by seeing the product molecules as specially shaped logs that are thrown to jam the jaws of alligators as the latter swim towards a frail defending canoe. The canoe, of course, is the Vertebrate Body as it picks its way through a choked Swamp of Disease and unavoidable. In the real world, the conjectured interpretation is that since many pathogens, from bacteria to worms, are known to promote their attack by releasing proteases to digest host proteins, the protease inhibitors are a facet of the host's countermeasures. The 'logs', in other words, have to be of just the right shape to jam the 'teeth' of alligators. Thus again, the Swamp of Disease keeps sending in new alligators already bred to be different. So you, too, must keep changing the shape of your logs. In a kind of strip cartoon vision, I so interpret the very high rates of evolution known to be occurring in loci, coding the contact sites on the inhibitor molecules.

So far, the rapid evolution disclosed for these molecules indicates high rates of gene replacement rather than a current polymorphism. The degree of actual polymorphism in the wild is not known, probably because it has not yet been a point of interest. The situation taken at face value prompts the thought that in general, mixed models having both dynamical polymorphism and rapid succession of complete gene transiencies (0–100 per cent) seem a hopeful and an aesthetically acceptable solution for sex. The idea of numerous novel complete transiencies—new universal tooth patterns for alligators and new universal notches on logs—carries us back towards the old and currently rather discredited biological theory of sexuality, which said that sex existed because it permitted good new mutations to be brought together in single stocks with maximum speed. This idea was discredited partly because it was realized that it could not be imagined that new good mutations were occurring fast enough. Biologists were thinking then of genes like those that, since the Eocene epoch, have been improving the adaptations of a horse's limbs for running, and good new genes in this sense do indeed seem to be very rare. But in the context of coevolution with microorganisms, there appears almost no limit to the potential demand for novelty—almost any new log with new shape may be better than nothing for the Swamp of Disease is changing the teeth of its alligators very fast indeed. Bacteria have the potential to breed—which means to double their numbers—every 20 min in the warm, rich human body, and were it not for the body's defences they might do so. They get along very well with little sex, plenty of mutation, and a system of acquiring plasmids which to me, incidentally, seems rather like villagers hiring somewhat expensive and unruly samurai to help protect their enterprises. (Plasmids in a sense *are* sex for the bacteria, but that, too, is another story.) As for the evidence of rapid change in microbes, it is only too evident in the evolution of resistance to antibiotics by bacteria and in the constant trans-mutations of cold and influenza viruses.

Altogether it seems possible that there can be a fairly happy marriage between the 'combining good genes' theme and the 'dynamic multilocus polymorphism' theme that I favour. A cycle of wide amplitude so that it shows fixations and remutations while essentially still going around the cycle may be a case that can be understood in the light of both the old theory and the new. However, I still very much doubt that such extremely long cycles as are implied in fixation–mutation successions can give short-term maintenance of sex, and so still expect to see true dynamic polymorphisms, that seldom fix and that carry strong non-random associations to and fro, playing their part.

If the idea that sex primarily defends against parasites is right, we would expect sex to be most universal where parasites bear heaviest, and to disappear where their pressure is relaxed. Inbreeding (which reduces the effectiveness of sex) and parthenogenesis are both well known to be unusually common in organisms of extreme environments. This has been a puzzle in the past. It fits badly, for example, with the idea that sex gets together good genes for overcoming vagaries of a harsh physical environment. On the other hand, it may turn out to fit well with the idea that reduced biotic pressures in such environments cause reduced need for sexuality.[38,39] The sheer harshness of the arctic, high mountain slopes, edges of deserts, and disturbed agricultural land, where the self-pollinators and parthenoforms are abundantly found, may by itself keep down parasites. Microbes, for example, may be too cold, too sterilized by sunlight, too desiccated, and so on, to be much transmitted. But, in addition, parasites of all kinds may fail to coevolve along with their hosts into such habitats right at the start;[40] or, when they do establish, they may still fail to cross-infect and contribute to the burden of other hosts of such localities because there are too few closely related potential host species on which they have any chance—a situation utterly different from what obtains, say, in old ecosystems in the wet tropics, where coexisting closely related species are common. However, we should be careful here, because the plants and animals that live in harsh environments may be extremely abundant where they occur, and this may give those parasites that likewise succeed in reaching the habitat very easy opportunities of transmission. The reindeer tolerates a more arctic climate than any other deer; but, to judge from its polygynous system and its huge horns in the male (which are largest in proportion to the body for any deer), it is very insistently an outbreeding sexual animal. However, the reindeer is large and warm-blooded, as well as both common and gregarious, and all of this, of course, makes it a very satisfactory host. Sure enough, parasites and diseases are present in abundance, and are an important factor in reindeer demography.[41]

Still in the arctic habitat, however, let us turn to the grasses that the reindeer eat. These are often those most explicit of parthenogens that produce mitotically green seedlings growing out of their spikelets while the latter are still borne on the parent plant; and this and other more cryptic types of asexual behaviour and inbreeding in plants of the arctic are extremely common. Such perpetuation of unchanged genomes almost certainly would not succeed for the common species in a dense temperate meadow as some other dramatic recent evidence shows. Thanks to studies of grasses by Antonovics and his coworkers,[42] we begin to see not only that it is true that it does not succeed, but also hints that disease is the main reason. In a recent experiment they reared sexual and clonal offspring of sweet vernal grass (*Anthoxanthum*) and replanted them at the stage of rooted tillers in the meadow where the parents still grew. The results showed in various ways a high advantage to the sexually produced offspring in their ability to thrive and produce their own seed. In the latest experiment, the performance of the sexuals is already, after 2 years' observation, 1.55 times that of the asexuals. This is not quite the twofold factor, but since seedling mortality was not estimated due to the use of established tillers for the test, the real seed–parent factor may well exceed two; and in any case, the grass being perennial and the experimental plants still alive, the observed differences are likely to grow. These experiments are the only ones I know that show that sex may pay immediately, or almost immediately, its full cost. Because other influences bearing on success were carefully controlled for, and damage by insects or even fungi was not apparent to a degree that could explain the difference, the investigators have suggested viral diseases to be a likely main cause.

It would be very interesting to know whether the same advantage of sexual offspring would have been found if *Anthoxanthum* had been studied at the very edge of its known range. Taking the life zones of the planet in a broad view, it is already established that tropical species have more of their gene loci polymorphic and also more general heterozygosity per locus.[43] In the case of some of the grasses ancestral to the cultivated cereals in Israel, it is also being found that marginal populations show both less disease incidence and less resistance to disease.[44] Taken together, such facts hint at a correlation of polymorphism with disease incidence. The pattern would no doubt partly be based on the density and type of spatial distribution, and partly on the levels of favourability of the environments to the transmission of disease propagules. It would be expected to hold both across ranges of species and within their ranges.

Obviously, much more needs to be known before we can say whether the ecological distribution of sex accords in detail with expectations of the parasite theory, and many more experiments of the type pioneered by Antonovics are also needed. Ideal for testing experimentally in a similar way would be short-lived, easily reared species that have both sexual and asexual races; or, better still, those species that tread the wobbling tightrope of being able to use sex or not as they please. Such species might be made to evolve towards sexuality or asexuality by manipulating the extent of parasite attack. This has not yet been done, but a recent study by Lively[45] somewhat on these lines is almost equally well chosen. The study was of the distribution of sex in the wild in a New Zealand lake snail, and aimed to differentiate the present theory from several alternatives. Of all the factors reviewed by the author as potentially bearing on the snail's breeding system, only the extent of parasitism significantly predicted whether the snails would have more or fewer males in their populations. The fraction that were male was used as an index of the degree to which a population was behaving sexually. Unfortunately it is not known if the snails have mixtures of sexual and asexual strains or facultative asexuality; but either way, of course, the conclusion still accords well with the theory.

Using the obligately sexual mammals instead of the more undecided snails and grasses, Burt and Bell,[46] through a survey of the known chiasma cytogenetics of primates, found that only generation time was a good predictor of the rates of recombination.[47] If the environment deteriorates at the same absolute rate for all species, and if recombination is a way of offsetting this deterioration as argued in this essay (as also by Burt and Bell), then the positive correlation between generation time and rates of recombination is expected.

If this is a correct line to the explanation of chiasma frequencies, the involvement of age here forces to our notice an implicit point of the theory that has not yet been emphasized. (This point is especially important to an allied version of the theory of dependence of sex on parasites that attributes most force to very local deterioration of environment for each genotype, with sex protecting offspring from the parasite set accumulated by parents and neighbours.[48] The environment is supposed worsening appreciably with respect to disease within the lifetimes of individuals. Theoretical simulations of a population coevolving with its parasites, conducted by Robert Axelrod and myself, have confirmed the effect. They show not only that sexuals can beat asexuals in the short term when the parasites had much shorter lifetimes than the potential of their hosts, but they have also shown increasing mortality rates within a lifetime. This change is due to the evolution of the parasites.

Even though the reproductive and mortality properties of the hosts are set at the outset (i.e. the hosts were potentially immortal), host mortality increases with age. The effect seen is slight, and in nature we would not suggest it to be a major factor in shaping the senescent schedule of mortality; there are other factors more likely to produce this main effect.[49–51] However, perhaps here is one more reason why influenza epidemics seem to be particularly dangerous for the elderly. It is not just that the elderly are weaker due to their senescence; it may also be that, in helping to 'passage' the pathogen—that is, through their participation in epidemics 10, 20, 40, or 80 years ago—the elderly have contributed most to its current adaptations: they may, in effect have customized the virus to hit them.

So, new genotypes are always needed. But are they? Will continuing strides in medical science make ancient sex truly unnecessary? Will we have technological fixes, where for so long we gambled our genes in the way I have outlined? Will we at last be allowed to experience humankind's greatest dream and dread—growing old forever—doing so by a process of patching, pushing in new or replacement genes or their products from the outside?

On other fronts, technology seems to be striding faster still: soon we may be able to clone babies from cells of our own bodies. Once that is possible, will we perhaps—more even than through the (to my eyes) dubious progress in genetic engineering—be finally set free from the crazily inefficient and yet decorative, even risible, protective process that has dogged our line for a thousand million years?

I don't know the answer to any of these questions, but will give my own feeling. I certainly *wish* that the answer to all of them could be 'No'; or at the least that it could be 'Not for a long time'. Speaking of the real danger to the species that might attend abandoning sex, rather than of my own romantic preferences in the matter, I think that we should reflect that whether we name the giver God or chance, we may have been lucky in having AIDS as a first warning of things to come—warning that our own monoculture is already being planted too close, and that our own engines are making the winds that blow our own enemy spores around. Had the new plague bursting on the massed jet-travelling human population of the second half of the twentieth century been a new virus caught in the manner of flu while still as lethal as HIV, we should probably be already well aware of how fragile have been our victories so far over disease. With the factories unmanned, with the last stocks of antibiotics and vaccines failing, with insulin for diabetics used up, with blood and blood factors for haemophiliacs no longer collected, and with blood dialysis machines unpowered, we should be finding little time to throw even consoling words to those *extra* dying—and those words the same barren ones, I suppose, that must be given to haemophiliacs dying with AIDS already: 'Well, science gave you some additional years: we do hope that you have enjoyed them.'

But that is only the beginning. If such lethal flu strikes in the twenty-second century, when clonal man, or perhaps clonal woman, has become the human mode, with all of the even greater necessity for sterilized glassware and reagents and power supply that this implies, let us hope that such dangers of a sudden breakdown of civilization have at least been well forseen and prepared for. For by then all people—*all*—may have within them the saved and 'corrected' genes that are lethal in the absence of continual medical attention. If to be born again, I personally would prefer not to be born into the modish set of that time if there was any way not to be in it. I would prefer to be a remote, unlettered islander who had never seen even a penicillin pill, still less an implanted gene. Certainly I would not wish to be present with the mind and background that I now have, neither in my present body nor in my clonal rebirth, even assuming that a neogerontology (or a technoembryology, or something of the kind) gives me the chance tomorrow. I am a primitive, I think: I was once told that I look like the last of the Neanderthals: perhaps because of this streak, I find I don't even want to be a part of a culture that considers it normal to have babies by Caesarean section. Even that seems to me to be giving too many hostages to sterilized glassware and to chance. To this extent, at least, I am a believer in natural selection. Along with this belief goes a continuing quaint faith in sex—that it may for a long time work for us in ways that human planning can't.

This is not to say that I believe that the changes implicit in the set of questions I asked above won't happen, or that if they do they will necessarily be cut back by disaster, or that humans will go extinct: the currents are already plainly set, and most people seem to see none of the dangers that I do. And perhaps they are right. It may be that the building of the high-technology human modules of the future out of our increasingly (individually) hopeless bodies is as inevitable as was the building of the large long-lived multicellular creatures, once sex had allowed this to be possible. Of that long-past evolutionary building process I am now an example. Had I been an independent protist flowing on the bottom of a pond, a crawler that had not yet even dared to adopt a spirochaete to be a flagellum,[52] would I have hung back on the threshold of that ancient revolution too? I suspect that I would, and doubtless would have looked foolish then as well. This essay is being poured hopefully, respect-fully, as will be seen at the end, to modernize a gap between two sonnets of Shakespeare, to give him arguments for the future if he would need them. (The question is why men should continue to exist, a point sometime that may I fear become as unobvious in the eyes of women as it is already in the eyes of biologists. Shakespeare's almost undivided theme in his first 17 sonnets was to persuade an admired male friend that he should reproduce his beauty in offspring before he no longer can or dies. His verse emphasizes both male beauty and male heredity. The ending of Sonnet XI, for example, says 'She [Nature] carv'd thee for her seal, and meant thereby/ Thou shouldst print more, nor let that copy die.' In the modern future world, for peaceful life as an author and chance of tenure in a good department, Shakespeare had better not assume that male–male parthenogenesis is either the most desirable or the easiest to achieve.

But even Shakespeare, I think, is probably more modern than I on the point I am worried about. At least the trend of unnatural human reproduction, for example, had begun long before his time. He knew about it and, to judge from some of his characters, it intrigued him also. His able, sturdy Macduff killed Macbeth, evading prophesy of the witches, by virtue that he (Macduff) was 'not of woman born', but was 'from his mother's womb untimely ripped'. The play hints that the mother survived this, for Macduff had brothers. Further, as is well known, so ripped was the more-ambiguous Julius Caesar of another play. But in Shakespeare's time Caesarean section had not begun its bid to become a normal method of human childbirth. Caesarean birth, although unrelated to the theme of infectious disease, is a good paradigm of the coming dilemma. As evolutionary background, it needs to be noted that the broadening of the female human pelvis, compared to other primates and the human male, to accommodate increasing fetal head size, clearly shows that substantial mortality of mothers in childbirth and/or a mortality or disability of birth-stressed offspring has been occurring. Accompanying the change, the shape of the pelvis has become a major constraint on women's average athletic ability and was perhaps the main factor underlying other trends of human sex difference occurring in the Palaeolithic. These differences are usually regarded as having been favourable to men, although this seems to me possibly more a fashion of attitude than real. If we do decide that reducing the differences is desirable, it will certainly be easier to do if Caesarean birth becomes universal. But eventual costs of the new birth system will be (1) necessity for medical facilities and personnel to be present at every birth; and (2) as the pelvis reconverges and/or fetal head size further increases, a prospect of never going back. The increasing medical dependence implied in this example already has a momentum in public attitudes and therefore a kind of inevitability. Should we not want this to continue, it is already hard to see what should be done. Mothers' lives obviously must be saved by Caesareans when necessary; but perhaps operations that are done in true necessity could be followed by an offer of a state reward for a pledge by the woman not to bear more children.

What I want is not to reject medicine and the use, sometimes, of its extreme measures, but rather to say that I believe we should go carefully. We should try to understand not just the mechanisms by which our bodies work, but what might be described as the philosophy underlying that mechanism. We should try to understand the process that helped us to become as extraordinary on the intellectual side as it is— in which process sexuality certainly has played a great part. I want us to understand the whole chain of our being, that which connects us to the rest of living nature, not just the alloys and welding of special links (our own) and how to fix these. We need at the least a theory of human psychical evolution that pays attention to the net of kinship behind us and still with us, to the effects of group competition in our makeup, to the extra innovations made by barter—and this is naming but a few of the factors that are already coming to be better understood.[53] We need this before, if ever, trying to grasp inflexible dogmas about what is right. Otherwise our attempts to change the world towards how we believe we wish it to be is almost certain to produce monsters. I feel that such evolutionary understanding will make us less cowardly about death and will deter the cant of claiming it is dogmatically and unequivocally right to cure disease, save lives, or fix infertility. The offspring that sex always dropped deeper into the mineshafts I have described, those that it never expected to save—indeed which its very object was to eliminate efficiently[54]—should be borne in mind. We should reflect whether, in the longer run, technology has really any hope of saving such 'selective deaths' even now. In the most immediate way, as dependent on my theme, I would like to see respect given for the views and rights of those who refuse any combination of medical intervention for themselves and their children. I find myself in this an improbable evolutionist ally to Jehovah's Witnesses against the Establishment of Medical Science. I think I look with eyes very like those of Witnesses at some things—with horror, for example, at the white pall of hospitals that is rearing so high over our concrete and asphalt suburbs all over the world, seeing here the beginning of even greater arrogance and excess than when the churches of the Middle Ages spread (at least much more beautifully) over the then-greener face of Christendom.

Sex had done far more for us in evolution than simply to hold back disease—far more in fact than I here discuss. I would like to see sex kept not only for our recreation but also, for a long while, allowed to retain its old freedom and danger, still used for its old purposes.

> O! none but unthrifts:—Dear my love, you know
> You had a father: Let your son say so.
> SHAKESPEARE, *Sonnet XIII*

References and notes

1. G. Bell, *The Masterpiece of Nature: The Evolution and Genetics of Sexuality* (University of California Press, Berkeley, CA, 1982).

2. J. Jaenike, An hypothesis to account for the maintenance of sex within populations, *Evolutionary Theory* **3**, 191–4 (1978).

3. H. J. Bremermann, Sex and polymorphism as strategies in host–pathogen interactions, *Journal of Theoretical Biology* **87**, 671–702 (1980).

4. R. M. Anderson and R. M. May (ed.), *Population Biology of Infectious Diseases*, Dahlem Konferenzen, 1982 (Springer, Berlin, 1982); in particular, the contribution of M. S. Pereira, pp. 53–64, summarizes the current situation of human disease.

5. L. J. Eales, K. E. Nye, J. M. Parkin, J. N. Weber, S. M. Forster, J. R. W. Harris, and A. J. Pinching, Association of different allelic forms of group specific component with susceptibility to and clinical manifestation of human immunodeficiency virus infection, *The Lancet* (May) **2**, 999–1002 (1987).

6. A. Grafen, A centrosomal theory of the short-term evolutionary maintenance of sexual reproduction, *Journal of Theoretical Biology* **131**, 163–73 (1988).

7. G. C. Williams, *Sex and Evolution* (Princeton University Press, Princeton, NJ, 1975).

8. J. Maynard Smith, *The Evolution of Sex* (Cambridge University Press, Cambridge, 1978).

9. M. Treisman, The evolution of sexual reproduction: a model which assumes individual selection, *Journal of Theoretical Biology* **60**, 421–31 (1976).

10. N. Stenseth, L. R. Kirkendall, and N. A. Moran. On the evolution of pseudogamy, *Evolution* **39**, 294–307 (1985).

11. H. L. Carson, Selection for parthenogenesis in *Drosophila mercatorum*, *Genetics* **55**, 157–61 (1967).

12. A. R. Templeton, H. L. Carson, and C. F. Sing, The population genetics of parthenogenetic strains of *Drosophila mercatorum*. II. The capacity for parthenogenesis in a natural bisexual population, *Genetics* **82**, 527–42 (1976).

13. R. Y. Lamb and R. B. Willey, Are parthenogenetic and related bisexual insects equal in fertility? *Evolution* **33**, 774–5 (1979).

14. Examples of such selection are described in A. J. Van Norwijk, J. H. Van Balen, and W. Scharloo, Heritability of ecologically important traits in the Great Tit, *Ardea* **68**, 193–203 (1980); and in P. R. Grant, *Ecology and Evolution of Darwin's Finches* (Princeton University Press, Princeton, NJ, 1986).

15. R. R. Glesener and D. Tilman, Sexuality and the components of environmental uncertainty: clues from geographical parthenogenesis in terrestrial animals, *American Naturalist* **112**, 659–73 (1978).

16. P. W. Price, *Evolutionary Biology of Parasites* (Princeton University Press, Princeton, NJ, 1980).

17. F. B. Christiansen, Hard and soft selection in a subdivided population, *American Naturalist* **109**, 11–16 (1975).

18. C. Wills, Rank order selection is capable of maintaining all genetic polymorphisms, *Genetics* **89**, 403–17 (1978).

19. See, for example, D. Jenkins, A. Watson, and G. R. Miller, Predation and red grouse populations, *Journal of Applied Ecology* **1**, 183–95 (1964); S. M. Smith, The underworld of a territorial sparrow: adaptive strategies for floaters, *American Naturalist* **112**, 571–82 (1978).

20. As one example of resistance to several taxa of pathogen, although all intracellular, see E. Skamene, P. Gras, A. Forget, P. A. L. Kingshorn, S. St Charles, and B. A. Taylor, Genetic regulation of resistance to intracellular pathogens, *Nature* **297**, 506–9 (1982).

21. W. D. Hamilton and N. A. Moran, Low nutritive quality as defense against herbivores, *Journal of Theoretical Biology* **86**, 247–54 (1980).

22. See, for example, A. W. Farnham, K. A. Lord, and R. M. Sawicki, Study of some of the mechanisms connected with resistance to diasinon and diazoxon in diazinon-resistant houseflies, *Journal of Insect Physiology* **11**, 1475–88 (1965).

23. Some examples are in Chapter 5 and also B. R. Murty, Developmental traits in breeding for disease resistance in some cereals, in *Proceedings of the International Atomic Energy Authority Panel Meeting on Mutation Breeding for Disease Resistance, 12–16 October 1970*, pp. 93–105 (IAEA Publication No. 271, 1971).

24. Both patterns in one system are shown in E. Bruzzese and S. Hasan, The collection and selection in Europe of isolates of *Phragmidium violaceum* (Uredinales) pathogenic to species of European blackberry naturalized in Australia, *Annals of Applied Biology* **108**, 527–33 (1986).

25. R. L. Gallun, Genetic basis of hessian fly epidemics, *Annals of the New York Academy of Sciences* **287**, 223–9 (1977).

26. D. M. Parrott, Evidence of gene-for-gene relationship between resistance gene H1 from *Solanum tuberosum* ssp. *andigena* and a gene in *Globodera rostochiensis*, *Nematologica* **27**, 372–84 (1981).

27. J. M. Rutter, M. R. Burrows, R. Sellwood, and R. A. Gibbons, A genetic basis for resistance to enteric disease caused by *E. coli*, *Nature* **257**, 135–6 (1975).

28. J. A. Barrett, Plant–fungus symbiosis, in D. Futuyma and M. Slakin (ed.), *Coevolution*, pp. 137–60 (Sinauer, New York, 1982)

29. S. Connor, Genes defend plant breeding, *New Scientist* **112**, 33–5 (1986).

30. See B. Murty, Breeding procedures in pearl millet (*Pennisetum typhoides*) (Indian Council of Agricultural Research, New Delhi, 1970); for an effect of the same type, although not in an economic organism, see also D. H. Minchella and P. T. Loverde, Laboratory comparison of the relative success of *Biomphalaria glabrata* stocks which are susceptible and insusceptible to infection with *Schistosoma mansoni*, *Parasitology* **86**, 335–44 (1983).

31. See Murty (1971) in ref. 23.

32. See, for example, H. M. Canter and J. W. G. Lund, The parasitism of planktonic desmids by fungi, *Osterreichische botanische Zeitschrift* **116**, 351–77 (1969). The paper concerns population dynamics without reference to genetics: however, genetic variation in resistance to phytoplankton is known.

33. G. Bell and J. Maynard Smith, Short-term selection for recombination among mutually antagonistic species, *Nature* **328**, 66–8 (1987).

34. A. Sasaki and Y. Iwasa, Optimal recombination rate in fluctuating environments, *Genetics* **115**, 377–88 (1987).

35. See 'Self-examination' in *Scientific American* **256** (5), 70 (1987); also M. S. Pollack and R. R. Rich, The HLA complex and the pathogenesis of infectious disease, *Journal of Infectious Disease* **151**, 1–8 (1985); R. S. Smeraldi, G. Fabio, A. Lazzarin, N. B. Moroni, and C. Zanussi, HLA-associated susceptibility to acquired immunodeficiency syndrome in Italian patients with human-imunodeficiency-virus, *The Lancet* (22 November), 1187–9 (1986).

36. R. E. Hill and N. D. Hastie, Accelerated evolution in the reactive centre regions of serine protease inhibitors, *Nature* **326**, 96–9 (1987).

37. M. Laskowski, Jr, I. Kato, W. Ardelt, J. Cook, A. Denton, M. W. Empie et al., Ovomucoid third domains from 100 avian species: isolation, sequences and hypervariability of enzyme-inhibitor residues, *Biochemistry* **26**, 202–21 (1987).

38. R. R. Glesener and D. Tilman, Sexuality and the components of environmental uncertainty: clues from geographical parthenogenesis in terrestrial animals, *American Naturalist* **112**, 659–73 (1978); D. A. Levin, Pest pressure and recombination systems in plants, ibid. **109**, 437–51 (1975).

39. For the relative inability of asexual and selfing plants to withstand parasites, see J. J. Burdon and D. R. Marshall, Biological control and the reproductive mode of weeds, *Journal of Applied Ecology* **18**, 649–58 (1981).

40. E. Reimers, L. Villmo, E. Gaare, V. Holthe, and T. Skogland, Status of *Rangifer* in Norway including Svalbard, in E. Reimers, E. Gaare, and S. Skjenneberg (ed.), *Proceedings of the 2nd International Reindeer/Caribou Symposium, Røros, Norway* (Direcktorat for Vilt og Ferskvannsfisk, Trondheim, 1979).

41. O. Halvorsen, Epidemiology of reindeer parasites, *Parasitology Today* **2**, 334–9 (1986).

42. S. Kelly, J. Antonovics, and J. Schmitt. The evolution of sexual reproduction: a test of the short-term advantage hypothesis, *Nature* **331**, 714–16 (1988).

43. E. Nevo, Genetic variation in natural populations: patterns and theory, *Theoretical Population Biology* **13**, 121–77 (1978).

44. For evidence of both fewer parasites and reduced resistance in marginal populations, see J. G. Moseman, E. Nevo, M. A. E. Morshidy, and D. Zohary. Resistance of *Triticum dicoccoides* to infection with *Erysiphe graminis tritici*, *Euphytica* **33**, 41–7 (1984); A. Segal, J. Manisterski, G. Fishbeck, and G. Wahl, How populations defend themselves in natural ecosystems, in J. G. Harsfall and E. B. Cowling (ed.), *Plant Disease: An Advanced Treatise*, Vol. 5, pp. 76–102 (Academic Press, London, 1980).

45. C. M. Lively, Evidence from a New Zealand snail for the maintenance of sex by parasitism, *Nature* **328**, 519–21 (1987).

46. A. Burt and G. Bell, Mammalian chiasma frequencies as a test of two theories of recombination, *Nature* **326**, 803–5 (1987).

47. For similar facts for plants, see G. L. Stebbins, Longevity, habitat and release of genetic variability in the higher plants, *Cold Spring Harbor Symposia on Quantitative Biology* **23**, 365–78 (1958).

48. This version is emphasized in: H. J. Bremermann, Sex and polymorphism as strategies in host–pathogen interactions, *Journal of Theoretical Biology* **87**, 671–702 (1980); J. Tooby, Pathogens, polymorphism and the evolution of sex, *Journal of Theoretical Biology* **97**, 557–76 (1975); W. R. Rice, Sexual reproduction: an adaptation reducing parent–offspring contagion, *Evolution* **37**, 1317–20 (1983).

49. W. D. Hamilton, The moulding of senescence by natural selection, *Journal of Theoretical Biology* **12**, 12–45 (1966) [reprinted in *Narrow Roads of Gene Land*, Vol. 1, 94–128].

50. G. Bell, Evolutionary and non-evolutionary theories of senescence, *American Naturalist* **124**, 600–3 (1984).

51. L. D. Mueller, Evolution accelerated senescence in laboratory populations of *Drosophila*, *Proceedings of the National Academy of Sciences, USA* **84**, 1974–7 (1987).

52. L. Margulis, *Symbiosis and Cell Evolution* (Freeman, San Francisco, 1981).

53. R. D. Alexander, *The Biology of Moral Systems* (Aldine de Gruyter, New York, 1987).

54. I owe this stronger view to S. Nee.

CITED BUT LITTLE READ

This Week's Citation Classic

It remains to note that I have written this essay with some sense of urgency, given the current significance of sociobiology, and the good possibility that it will soon disappear as science, . . .

M. SAHLINS[1]

NOW comes a two-chapter break from disease and sex and a return to the topics that largely made up Volume 1 of *Narrow Roads of Gene Land*; but the present one is a very slight chapter. Sometimes it has seemed to me that the challenge to species and group-benefiting evolution issued in my first three papers has brought harpies to follow me ever since. Social behaviour, the topic of this and the next chapter, had been a four-year absolute obsession to me as I wrote the papers in the early 1960s—I was deeply immersed in it, a monomaniac, could do almost nothing else. But in the 1980s the requests for the papers of these two chapters brought a rather opposite feeling to me; they were distractions from the theme I most wanted to work on, which, by then, was sex. And now those distracting works come again as secondary distractions as I write these introductions. The present 'paper' in particular is so trivial it might seem hardly to merit inclusion in this volume at all, leave alone to deserve an introduction; but what I wrote in it is already history of a kind and it does mention a few things I didn't in Volume 1. In all, I'll be extra brief in this introduction but let me take the opportunity it offers to explain how the two coming papers should be considered together.

The 'paper' of this chapter, then, itself is an introduction—a one-page history I wrote in 1988 about the period 30–25 years previously when

I was working on my first papers on kin selection. But in 1989, the year after writing this brief historical sketch, I was actually back with kin selection and related matters of social behaviour again due to a commentary I was writing jointly with a colleague at Oxford, Alasdair Houston. The commentary was being applied to a paper that was due to come out simultaneously with it and also with many others like it—indeed, as it turned out there were to be comments of even some 30 or so others. More of the how and why of what may sound to you—and was—a confusing orgy of commenting, will be explained in the introduction of Chapter 14.

What deserves to be mentioned here, however, is that the target paper of that coming commentary was a modern critical reaction directed at recent trends in the principles of social psychology and economics, trends that had been in part started by the same early paper of mine that is the focus of the present chapter. Thus the critique was also in part directed against 'inclusive fitness' arguments in general, although the particular strong heat the paper indirectly generated, as of so many others around that time and from much earlier, seemed to arise from a whole subset of similar microevolutionary ideas being applied to humans. The authors in effect were targeting several other lines of social reasoning including all the general realist approaches that had begun even centuries earlier—for example, with Adam Smith and his *Wealth of Nations*, with David Ricardo, with Thorstein Veblen, and so on; perhaps even I might add Machiavelli.

On this side it was the developments dependent on Robert Axelrod's *a priori* theory of reciprocation that were in the line of fire rather than kin selection because the criticism and the new study focused most on how normal, unrelated people played the game of classic (unrepeated) Prisoner's Dilemma. But I had been involved a bit with that story too (see Chapter 4 of this volume and also 4 in Volume 1), so, again, some of the criticism ricocheted into my field of interest. Most centrally targeted by the critique to which Alasdair Houston and I were responding, however, was the basic concept of an intrinsically selfish 'economic person', the idea which both kin-oriented and reciprocation theories had postulated as their starting point. The notion of a spontaneous striving egoist has long been a foundation for economic reasoning. Well, will the ghost of Adam Smith stand up and face these harpies on all our behalves—gladly do I cede to him the priority and I suspect Edward Wilson, Richard Dawkins, Robert Axelrod (and all other realists I imagine to be similarly pursued) will do the same . . . Adam's prime point, made right at the start of his two-volume work in the eighteenth century, was that to try to found sociology and economics on any notion of an unconditional altruist has never worked.

Indeed in the context of the ideas of this volume, selfish economic man seems to have come to birth in a way quite relevant to this volume—parthenogenetically: Adam Smith is his/her mother; hitherto unexplained oddities of British national and international markets that Adam Smith thought he could understand seem to have acted as a Holy Spirit stimulating the new conception. (Readers who have accompanied me this far will also follow my metaphor in the above mile-long biological adverb used for 'fatherless birthing'; but those who are dipping or simply don't like my parallel can put 'autochthonously' there instead if they prefer—a trifle shorter but by no means prettier—and they could then eliminate the rest of the last sentence also if they choose.) In brief, many criticisms that have fallen on sociobiology and related lines are seen in proper perspective when we note that in the original incarnation, this *Homo economicus* that we authors have considered to be reasonable as a starting point for our ideas, cannot have had a single selfish *gene* in his original philosophic make up—he couldn't because the Adam Smith (and the closely following Ricardo) prototype for the notion well pre-dated both Mendel and Darwin.

In theory, then, the pair of publications of these two chapters could mark a good place to take stock of what was still happening in the late 1980s to my earlier 'social' themes and especially to the aspects of those that most affect humans. It would seem also a good point to say something about convergences of disciplines that have been and are still going on. As I said, however, my heart isn't in it for a long ramble at this point, even supposing the reader wants one; rather I want a speedy return to my main current obsession, sex, and to what I feel to have been my next worthwhile contribution to that. As a reminder of the old theme and, especially because a lot of the detailed history has been covered in Chapter 2 of Volume 1, the sketch that follows speaks for itself and little remains to say except how it came to be commissioned.

The Institute of Scientific Information (ISI) publishes several of scientists' most useful bibliographic aids and its various lines of pamphlet-like issues of *Current Contents* are among them. The purpose of each issue is to list for subscribers all titles of papers that have recently come out in their field, covering all the most important journals. Several additional sections are also included that seem aimed to give general interest—perhaps aiming to stretch the journal beyond a readership of just anxious priority hunters who are watching for the work of the rival they may be narrowly not beating: the pamphlet you handle is thus almost like a very miniature *New Scientist*. The section called 'This Week's Citation Classic' contains each time a commissioned one-page article by a scientist about one of her or his papers that has achieved a critical count of being cited in other journals (this being known from a count that ISI conducts for another of its publications, *Science Citation Index*). CC comes out in seven series each devoted to a different field of science. Three simultaneous issues among these seemingly chose to highlight my paper and my outline, the series doing so being those of agriculture, biology, and environmental science; of arts and humanities; and of social and behavioural science. (Why did I cite it only once in my report to the Royal Society that year? Hell, I could have put it in three times!) Anyway I presume the triple usage means my paper must have been scoring well in journals of all those varied fields. Finally I am proud to draw attention to an added note in the issue telling that, for the journal it came out in, *Journal of Theoretical Biology*, it was standing as the most cited paper ever. Whether that status continues I don't know.

There is reason to believe that the degree to which the paper was being cited, however, didn't reflect the degree to which it was being read: I have mentioned this story in Volume 1 (Chapter 2, p. 21). In brief, Jon Seger and Paul Harvey detected that most citations contained exactly the same error in the wording of the title. This slip seemed to have first appeared in the reference list of Ed Wilson's famous and massive tome *Sociobiology*. Once that was published, it appears that most who thought to cite something about the general theme of kin selection copied the reference from the much-bought volume. Of course, many who did that may have really read the paper but there was plenty of evidence in erroneous notions of inclusive fitness that were circulating that many had not.[2] Some citations and some ideas, right or wrong, acquire, like buzzwords, a life of their own; it simply becomes fashionable to put them in.

When the pre-set level for a paper is reached, the editor asks the scientist to provide the 'Citation Classic' outline: it should contain how the ideas originated, difficulties of the research, how the paper was written, why the author thinks it is so much cited, and what she/he sees as the most important growth areas today that the paper stimulated. So far as I know two of my papers have achieved the distinction of drawing this request from CC. First was this one—my 1964 paper on kin selection—and then a few years later followed my paper 'Geometry for the selfish herd' (reprinted as Chapter 7 in Volume 1). As to the latter, I didn't have time to write the article and after one repeat request had come I heard no more. The honour of the second invitation has thus lapsed but it has brought a dubious consolation of my kidding myself that, by now, it could be, I have yet other papers that have crept up to the critical level and I am not told: the editor, his offer once declined, has marked me in default and doesn't invite again.

References

1. M. Sahlins, *The Use and Abuse of Biology* (University of Michigan Press, Ann Arbor, 1976).
2. R. Dawkins, Twelve misunderstandings of kin selection, *Zeitschrift für Tierpsychologie* **51**, 184–200 (1979).

THIS WEEK'S CITATION CLASSIC[†]

W. D. HAMILTON

———

An approximate criterion for increase of a gene for a social action is br−c>0, *where* b *is conferred benefit,* c *is cost of the action, and* r *is Sewall Wright's coefficient of relationship of interactants.* Inclusive fitness, *based on the criterion, is proposed as a guide in social reasoning. [The* SCI® *and* SSC® *indicate that this paper has been cited in over 1335 publications, making it* Current Contents' *most-cited paper.]*

This paper[1] (part one of a consecutive pair) was the first that I published apart from one short note embodying the same idea.[2] The note came out the year before but was written after the main work; in spite of one rejection, it had, through its shortness, a swifter editorial passage.

The theme of all three papers—the condition for the evolution of genetical altruism—began for me while I was an undergraduate reading natural sciences at the University of Cambridge in 1958. I discovered R. A. Fisher's *The Genetical Theory of Natural Selection* in the St John's College Library and immediately realized that this was the key to the understanding of evolution that I had long wanted. I became a Fisher freak and neglected whole courses in my efforts to grasp the book's extremely compressed style and reasoning. I quickly noticed, however, that Fisher's arguments implied a basically different interpretation of adaptation from what I was hearing from most of my lecturers and reading in other books. Was adaptation mainly for the benefit of species (the lecturers' view) or for the benefit of individuals (Fisher's view)? Clearly it was Fisher who had thought out his Darwinism properly; where interpretations differed, therefore, he must be right—but were the others *always* wrong? I started on what seemed the key theme in this puzzle—*altruism*. Did it exist? Could one evolve it in a model?

What had been a distraction in undergraduate days became, in 1960, a problem of funds and survival once I had my BA. Most whom I consulted could not see that a problem existed; those that could see something averred that what little was worth saying about it had certainly all been said by

---

[†]*Current Contents* No. 40, 16 (3 October 1988).

J. B. S. Haldane, although none could tell me where. That both Haldane and Fisher had said things, albeit few and special, was true, as I saw later after a lot of reading.

I found some interest in my ideas, surprisingly, at the London School of Economics (LSE) and an opportunity there for graduate study. Isolation and disinterest in my theme continued, however, in spite of additional enrolment at University College London, after I had begun to be too genetical for LSE.

And as I look back, the attack that I began on deriving a general measure of relatedness was indeed extremely circuitous and ill-conceived. I even delved into anthropological literature in the hope of seeing from the way people actually behaved some hint of the quantitative measure that I needed. All this led nowhere. It would have been sensible, at some point, to have asked Fisher (whose department I had trained in) for rays of guidance, but I hated to expose my evident naivety in writing. Fisher had retired a few years back and was working (and, in 1962, dying) in Australia; Haldane was in India and also was soon to die.

All I wanted was a measure of relationship that would enable me to generalize from the case of altruism to sibs that I had already worked out. Invariance that had appeared in the criterion for altruism with respect to gene frequency in the case of sibs had seemed a gift from God, and I did not expect to see it repeated in the more complex trial cases I had moved on to. So it was with joy and almost with incredulity that I at last found emerging out of acres of my tedious and usually wrong algebra for the case of uncles, and then for the case of cousins, the same invariance as I had found before. Shortly after this I saw how I could generalize still further and could invoke (slightly incorrectly, as I and others saw later[3]) Sewall Wright's coefficient of relationship for my measure. Finally, I saw how I could formulate a new concept of biological fitness—*inclusive fitness*—that would serve as a guide to reasoning in social situations.

My manuscript had a rather slow passage with the *Journal of Theoretical Biology* (*JTB*), largely because of the many biological applications that I wanted to include to illustrate the idea's usefulness. Realizing that it was going to take a long time to get it through, I wrote the shorter paper.[2] I had an urgent need to get something published because I had given up on the idea of getting a PhD out of my work and needed some published achievement to back my hunt for further research opportunities. An editor's acceptance would also encourage me personally because of my main fear about my work: that I was simply a crank.

At its first submission, to *Nature*, my short paper was rejected by return of post (possibly my address, 'Department of Sociology, LSE', weighed against it), but then, on the next attempt, it was accepted by *American Naturalist*. I learned that the long manuscript would be accepted by *JTB* subject to rearranging and writing as two; this slowed it because the rewriting had to be done in the midst of a trip to Brazil that I had arranged to try to check some of the predictions arising from the ideas. In the end, the first part had the maths and the second the biological discussion,[4] including that of the evolution of kin recognition, which is one of the growth areas for citations today.[5]

References

1. W. D. Hamilton, The genetical evolution of social behaviour. I, *Journal of Theoretical Biology* **7**, 1–16 (1964) [reprinted in *Narrow Roads of Gene Land*, Vol. 1, pp. 31–46].

2. W. D. Hamilton, The evolution of altruistic behavior, *American Naturalist* **7**, 354–6 (1963) [reprinted in *Narrow Roads of Gene Land*, Vol. 1, pp. 6–8].

3. A. Grafen, A geometric view of relatedness, *Oxford Surveys in Evolutionary Biology* **2**, 28–89 (1985).

4. W. D. Hamilton, The genetical evolution of social behaviour. II, *Journal of Theoretical Biology* **7**, 17–52 (1964) [reprinted in *Narrow Roads of Gene Land*, Vol. 1. pp. 47–82].

5. D. J. C. Fletcher and C. D. Michener (ed.), *Kin Recognition in Animals* (Wiley, Chichester 1987).

WIND IN THE BAOBABS

Selfishness Re-examined: No Man is an Island

When living beings first crawled on earth's surface, a dumb and filthy herd, they fought for acorns and lurking places with their nails and fists, then with clubs, and so from stage to stage with the weapons which need thereafter fashioned for them, until they discovered language, by which to make sounds and express feelings. From that moment they began to give up war, to build cities and to frame laws that none should thieve or rob or commit adultery.

HORACE *Satires* I, 3[1]

THE Zoology and Psychology departments in Oxford are housed in one concrete block consisting of two wide tower buildings connected at ground level and then by two bridges on the third floor. From the air the two must look like rather rectangular ciliate protozoans in syzygy. This beautiful word, which was among the more mystical factors attracting me to biology when I was at school, has a meaning that is properly a secret amongst us biologists and no one who hasn't suffered a three-year degree course is supposed either to understand or use it. I break the rules of secrecy briefly here, however, to tell you that it refers simply to a state of symmetrical copulation—each individual acting as donor and recipient at once (an enviable hedonic situation one would think, although our textbooks don't emphasize this). Usually tubes are formed through which the partners exchange hereditary material. The process is possibly akin to the attempted mutual cannibalism from which Lynn Margulis and Dorion Sagan originate sexuality (see Chapter 12); but over the thousands of millions of its repetitions the exchange has ended with an orderly neatness and symmetry. It has come to bear a hallmark of perfection by natural selection—a selection, according to my thinking, likely to be spurred by some as yet little-known struggles that the unicells have, at other times, with their parasites.

I work on the Psychology side of the syzygized building but am part of what I may call a certain string of 'code' and associated 'cytoplasm' known as the Animal Behaviour Research Group. Originating out of Zoology, we are in, and thus are presumably supposed to be fertilizing, Psychology; but during my time of many years not having been either digested or incorporated, I suppose we are somehow judged incompatible and will sometime be retracted into our home tower—perhaps, however, we'll then come back, dragging some experimental psychology along with us, just as in the Margulis–Sagan model of the origin of sex. Zoology will then become more intelligent (or 'cognitive' as seems the more modern word) or else (by the Parasite Red Queen thinking) we will just be more resistant to infection by wrong ideas than we were previously. The act of syzygy initiated by Zoology's former Professor J. W. S. Pringle, who had the dual building designed, will have borne its important fruit.

This all leads me to the point that perhaps because of its state of copulation the Psychology–Zoology complex has a very confusing architecture, which, in turn, has some odd consequences for our social life. One is that people you meet on stairways are usually people you know or ought to know because the stairs are very hard to find, much harder than the lifts. People using stairs, therefore, are mostly 'insiders' whereas the people you meet in either corridors or the lifts are mostly lost strangers and are asking you the way. A probabilist mathematician, George Pólya, proved that in an infinite tower block if one searches for a colleague's room known to be near one's own and uses the simple though extremely amnesiac method of a series of random door-to-door moves looking for his name and moving on in all six available directions—N, S, E, W, Up, and Down with equal probability—one has a chance somewhat greater than 0.37 of wandering off indefinitely and never finding your colleague. Meanwhile, as to finding yourself ever back at your own room, that figure of 0.37 . . ., etc. is then exactly the right one. If, however, you work in a single-storey building instead of in a multi-storey one, it can be proved that you would *always* find your colleague in the end and, likewise, that you would get back to your office although you would be well advised not to rely on the procedure as your usual course of action; both these 'random walks' have appreciable chances of occupying an extremely long time.

In some ways our Zoology Department provides opportunities a little less bad than Pólya's amnesiac's walk in the infinite 3D block—that is, our department is finite—but in others it's worse. For example, the Zoology Department has the added impediment of fire doors in the corridors every few yards. All these swing stiffly and are coloured red, thus wearing you down and abetting the amnesia. The stairs, too, have their problems. If, as I said, the stairs are the best places to meet colleagues they remain far from good places where to talk to them due to their narrowness—I refer, of course, not to the colleagues but the stairs and to how having conversations on them gets in people's way . . . If you want to talk to colleagues and are likely to block them for more than, say, a half hour (once again 'blocking' means, of course, the stairs, not the colleagues, even if narrow), it is far better to go instead to one of the truly spacious and well-known sites in our building, such as the coffee areas (Level C) or the basement car park (A) . . .

It was on the stairs of the Zoology side one day in 1988 that I met Alasdair Houston, and he asked me in passing if I had been invited to comment on a paper due for *Behavioral and Brain Sciences* (BBS) about co-operation. Such invitations used at that time to come to me frequently from the editor of the journal, Stevan Harnad. The only time I responded positively was in the case of the article of this chapter; the idea of the paper was initiated on this occasion by Alasdair Houston right there on the stairs.

Probably because I respond so seldom, I now seem to have been dropped from *BBS*'s mailing lists. Stevan Harnad evidently found in sociobiology a rich energy source for the storms he liked to stir up in the teacup of his journal. 'Journal of Instant Controversy' wasn't the name he chose for it but I wonder if that crossed his mind as he started his plans. Its principle, which I admit to have been very ingenious, is to persuade people to write sharp rejoinders to target pieces that have been either spontaneously submitted to the journal, or which Harnad has encouraged, in controversial areas, and to generate all these comments before the submitted piece had been published; Harnad would then arrange rejoinders by the main authors to those comments, and finally would package the whole to come out as one big block in the journal. I didn't care for the idea greatly; it seemed a bit artificial and too apt to give a lot of space to very half-baked thoughts: in short, a decision about a controversial idea needs longer after the first encounter with it than the time allowed for writing the replies. There was needed usually a literature search, for example, and a checking over of evidence, as well as time for reflection. It is true that there is a common context in the procedures of science where one has to decide on something at least as quickly. This is over the manuscripts sent for review by editors of more ordinary journals; but that is a kind of duty to a topic in which you are supposed an expert and often it is a field whose advance you care about, so that the effort is worthwhile. But even from doing such reviews you quite often realize later that your haste has caused a mistake. As regards *BBS* I had to admit that the issues I happened to hear about through Harnad's preliminary postings often gave me intriguing reading; but by the day when I met Alasdair on the stairs my colours on the matter of contributing to *BBS* had already been hoisted to my mast head. Abstention from the series was to be my contribution to conservation: buzz saws, paper pulp, then more journals, then the buzzes of adrenalin these in turn were designed to generate; old forests, their wonderful insects, and their nest sites for great black woodpeckers seemed to me better than that last product.

On the stairs I told Alasdair I had received a vaguely remembered abstract for the article he mentioned and I explained some of the above about my policy. He frowned (much as I expect he will again at these sketchy remembrances) and after a moment said he had thought of sending something himself. The paper sounded to him interesting and just from the abstract there were obviously matters the authors didn't understand about current behavioural theory—for example, about 'selfish genes' not implying 'selfish people'; and then there was also the old chestnut of 'group selection' versus 'kin selection'. To the detriment, as usual, of free passage on the stairs Alasdair's phrases churned rather slowly in a mind that had become, over many years now, much more attuned to ideas about polymorphisms and sex than about altruism and selfishness. But I thought of how I was involved in all this on at least two fronts: on the one hand the kin selection and all that 'selfish gene' stuff that is so often thoroughly misunderstood, while on the other hand there was the reciprocation theme on which I had worked with Bob Axelrod (Chapter 4). Perhaps I ought to do something. I suggested to Alasdair that he write that he would like to receive the full article: then we should both take a look and, if it still seemed a good idea, perhaps we could write something jointly. Alasdair agreed. He reminded me that these people had some experimental results out of what sounded to be once-played 'Tragedy of the Commons' games. This made me the more curious, but to cease blocking the stairway we now had to move on, I with my flag halyard loosened a little from its cleat.

Very different from the task, say, of detecting a fire door leading to a stairway in the Psy–Zoo block, Alasdair himself is not difficult to spot from a distance. Tall and with a big head he seems exapted (if that's the modern word) to the low lintels of the world by wearing a cushion of curly black hair. Even so he often stoops a little as he comes through your doorway and is apt to frown as if some other lintel had scored on him recently, though to be honest I think his standard frown is more likely to reflect a current problem—a tricky backward integration or something like that. He was a postdoc in Zoology when I arrived and he stayed that way for a long time. Chekhov used to cast special parts in his plays for 'eternal students' but nowadays to be with it in Western Academia he would need parts for 'eternal postdocs' instead. Sometimes being such an 'eternal' is a bad sign but I soon realized, from the way he seemed to be turning down lectureships and alternative postdoc positions, and from his eventual acceptance of one he was offered at Bristol where he at once became professor, this was not so in Alasdair's case. As I realized later he was simply applying the technical wisdom he had gained (or had created) in the Zoology subgroup he worked in (John Krebs's). He was perfecting his own 'optimal foraging strategy' and applying it in the patchy landscape of Euro-American academia. Before he left us he must have passed tips or given a talk on his technique because others in the group soon began to imitate him to an extent that became quite destructive: even the group's two research professors (Krebs himself and Robert May) soon foraged away towards loftier posts.

I hadn't remembered, in the *BBS* abstract I received, the points Alasdair had told me about on the stairs but had a high opinion of his judgement on evolutionary biology and I was very inclined to the idea of writing with him. Indeed, if I hadn't been so tied up with the sex problem, which was outside Alasdair's sphere of interest, we might possibly have collaborated more. We both appreciated the computer as a magic carpet to reach places we couldn't reach by analysis and we not only appreciated but slightly competed for assistance to be so carpet-carried by our department's best and most footloose computer genie, Brian Sumida. Brian always had a kindly outgoing approach to science (a Hawaiian's approach you might say, and Hawaii was where he was born) and was always willing to dash off a bit of C programming, or of UNIX machine management, to apply on user-unfriendly work stations, all in the interest of this or that evolutionary problem that we had managed to make appeal to him. But for the present project we needed no programmer and were not in competition.

Alasdair, I think, wrote our first draft—always the hardest part. The points in our response that most affect me and that I added to are really just two. First, there is the point that the ambit of 'selfish-gene theory' or 'kin selection', or 'sociobiology' or whatever you choose to call it, was being, as so often, misappreciated. None of the more recent theory stood outside the scope of Neodarwinism in the way that the authors supposed; there was, therefore, no point in talking of a return to a supposedly more pure Darwinism. Second, the authors weren't taking seriously enough the possibility that the results they were getting might reflect currently maladaptive behaviour. They might be watching reactions from anthropoid brain levels, and listening to discussions among their volunteers, all of which were of a spirit more appropriate to a music of winds stirring leaves of a baobab and to the buzz of flies over meat laid on the savannah grass under the shady tree, then to the sigh and rattle of air conditioners in the university departments of Oregon, Utah, and Arizona. If not meat, stone handaxes would be better imagined in the hands of the participants as the items of trade and the backup of trade that might fail than the dollar promissory notes the researchers had been passing out . . . But because our paper and such comments are hardly going to be comprehensible to the reader without more detail about the experiments that are the target for what we wrote, I had better explain more.

The four authors of the *BBS* paper, led by Linda Caporael from the Rensselaer Polytechnic Institute in New York State, were claiming to show by new experimental evidence that currently accepted theories of human behaviour were inadequate. Their experiments were done with 'students and townspeople' they had induced to participate in 'social dilemmas'. You might call them social 'games' if you were a games theorist, but if a more ordinary person who thinks games are 'for fun' you should realize, as I have emphasized previously in these essays, that games theorists have sometimes a different and pretty grim idea of them—as, for example, when they discuss 'games' of nuclear warfare. The authors were thus right to emphasize in their paper that the situations they set up were not 'games' at all to those who agreed to participate but instead real social dilemmas. If you become a volunteer for the experiment, the money rewards you received for being one depended on what you did. The amount of money was not very serious, a point that several commentators were to make (thus to richer participants the study might really have seemed just a game); however, the shape of the situation was always cast so as to resemble a 'game' only too well known to us all in real life, whether we live among groves of savannah trees or groves of industrial steel. The dilemma of the game is whether to contribute resources that we could keep for ourselves to a communal project, hoping to win eventually a benefit that all in our group would share equally, whether they contribute or not.

In practice the participants were admitted to a room in small groups and, as they entered, each was given a note promising them a (small) sum of money. If they retained the note to the end of the session they would certainly be paid its value. Effectively this was a priming gift, such that every participant had a 'resource' he or she could decide what to do with. As was next explained to them, the promissory note could then optionally be gifted onwards, in a concealed fashion, into a group pool; that pool might (or might not, if insufficient) then bring back a much- amplified group benefit later. The nub was that non-contributors to the group pool would receive benefits that came back along with, and just as much as, those who had contributed to the pool. The group benefit would only flow, however, if enough had been pooled for a threshold to be reached. (If the threshold isn't reached it has to be imagined that, in real life, for want of crucial effort or investment, the dam project is washed away, the whale hunt fails, the village fortification is incomplete when the raiders come, the church roof falls in, or whatever.) In every case non-contributors would receive as much of these returns, be they zero or a lot, as the contributors received. There were many variations (with rules always painstakingly explained to the participants) and these were progressively designed to elicit why people acted as they did. Basically all variations fanned out around what the authors call the 'standard dilemma'. This had $5 given to each of seven participants initially, and, then, if x or more of the $5 notes were contributed to the pool, $10 were given to everyone in the group: thus contributors in effect doubled their investments and non-contributors got for nothing a bonus additional to what they had already selfishly kept ($5), thus 50 per cent more than contributors. Moreover, if the set threshold wasn't reached, everyone except for the non-contributors lost everything. The threshold x was set somewhere in the mid-range, such that just under or just over half would be enough to win the group reward, or 'public good' as economists call it.

Think of this social situation and you will see that, if the amounts are substantial enough to matter to you, it is a severe dilemma indeed. You will also see it cast in exactly the same mould as the Prisoner's Dilemma we examined in Chapter 4—at least, as the basic one-off situation. And the dilemma is one that all of us have faced many times. Why not hope that enough others will contribute enough to win the public good and keep your personal \$5 as a private bonus? The results of the experiments were very diverse. Some leaned more to the selfish side than had been expected, sometimes leaning even more than could be argued to be rational. Some leaned towards an equally puzzling and irrational overenthusiasm to subscribe for a public good that is not likely to materialize. On the whole, however, participants were more generous than arguments from game theory, microeconomics, or 'optimal behaviour' evolutionary biology says they should be. Most consistent and dramatic of the findings was that if the participants were given an opportunity to discuss the situation before making their (still secret) final decisions, the rates of contribution invariably increased. This the authors interpreted, rightly I think, as meaning that if participants are given a chance to feel, through discussion, that they really do constitute *a group*, they eagerly promote that idea and start to act as one—at least, the majority do. There was a slightly sinister feature that I think in retrospect was quite significant although it was little commented on in the paper and not at all, so far as I remember, by the respondents to it. This was that in those of the groups where discussion was allowed, those who contributed gave often irrational explanations for what they were doing, whereas those who, frequently with considerable profit to themselves, decided not to contribute, gave more rational explanations.

Had this point among the results struck me at the time I was first reading the target paper I would have commented on it in our piece because it bears out one theme Alasdair and I were already emphasizing in our response—the difference between the ambiences where our social reactions were evolved and those where they are now played out. On the one hand is what I may call the 'baobab' environment where you might meet under the same tree for most of the weeks of your whole life, rarely seeing even a distant relative, still more rarely encountering a complete stranger and virtually never coming face to face with a person of a different race unless that person is a hunted quarry turning at bay. On the other hand is the world of the psychologist's reception room where, suddenly, as you come through the door, you see everyone you are going to interact with around you for the first time—or you ought so to see them if the methods are correct. Recall here that humans spent millions of years in Africa plus a quarter more subsequently still being hunter-gatherers as we spread out around the globe; then recall that following this we have spent only a period of at most 10 000 years living in conditions that could be called anything like civilized. Undoubtedly most of our social psyche was built up during the tens of thousands of years before substantial division of labour brought us to mix with numerous and unknown strangers, leave alone before it contrived for us sets of interactants who changed every day (Chapter 9).

That I didn't comment on that point at the time we wrote almost certainly is because I hadn't read the target paper well enough, my mind probably having been just too full of sex. Now that I have read it for the purpose of this introduction, I have to say that the article seems to me a better effort than it seemed at the time: it has both the neat and revealing experiments and a very useful review of the varied previous theories of human co-operation. This is not to say that I feel a need to retract anything that Alasdair and I wrote in our commentary. The misunderstandings of points of previous and current genetic theory, the apples we object to being heaped in among oranges by the writers, and the lack of attention to what I image as that windy and aeon-long space under the trees of Africa, are all still there, but the authors were undoubtedly raising serious points and for some of these the mystery that they outline, even on the side of the theory, continues to this day. What actually is a human group, for example; can group minds stably evolve in humans independent of both blood relatedness and group permanence (see the introduction to Chapter 4)? Is David Wilson right that groups by ties of blood can evolve striking properties even when they are also always ephemeral?[2] What is the least kin-related route to useful reciprocatory activity?[3,4]

To defend my longstanding views a little, contrary to the picture of inclusive fitness theory applied to humans that was presented in our target paper, I certainly have never neglected the importance of pre-human groupings in anthropoids. It wasn't so even when I was in what I may call my pre-Pricean frame of thinking (see Volume 1, Chapters 5 and 9) and was still hostile to the phrase 'group selection': for example, even in my first 'wild oat' touching upon humans published in 1971 I finish a comment that speculatively (and sadly) explained the evolution of the group activity of cruelty, as follows: 'considering inclusive fitness, or group selection (which may be a really appropriate term for many human situations), there is no theoretical difficulty'. I was referring here of my own impression of how humans defined themselves into groups and I was speculating exactly on what Caporael and her colleagues were later to research and prove to be more than just my impression. As for imagining people evolving out of a solitary primate background—well, that's certainly from no paper of mine! Even the somewhat unsocial orang-utan, the anthropoid least cited in discussions of human antecedents, is not asocial as an adult. Again in 1975 (Volume 1, Chapter 9, p. 341) I had written, for example, that: 'Roughly, as we currently see it, a cunning ape-like creature once pushed boldly out from near niches now held by baboons and chimpanzees'. Neither of these animals could remotely be said to be a solitary species nor I to be referring to them in that way.

That the theoretical side of kin selection had to start from an assumption of a species that was lacking in at least one social trait, the trait the evolution of which was about to be discussed, was inevitable—you mustn't assume that which you are going to prove. I believe that the target paper would have been greatly improved if it, too, had had a model showing how its imagined evolution was supposed to carry us from a state where no minds were willing to join another in a group action to a state where all are so willing. Looking through the commentary articles likewise, as I have now done, I found two attitudes that could make a commentary much more interesting to me. One was willingness to use models (though I admit that our own didn't). The other was specific examples of groups and of cheating taken from real life. The target article certainly drew much interest and perhaps even scored a record for *BBS* (28 responses) but as objective commentaries some of those included were very weird. One commentary saw in the coincidence of several experiments finding co-operators with a frequency near to 0.62 a possible expression of the 'golden ratio' $[\sqrt{5-1})/2 : 1]$; but no explanation why this should appear was suggested. Psychology through a Renaissance window! Amid such nonsense how one woke up when encountering precise statements and a model!

A little sadly I still have to maintain a position that I don't think that in the long run group minds lacking relatedness or long-term interaction can be stable from within. They may try to be in the modern world and all sorts of aids and persuasive religious devices have been invented to help them to do it. But basically all our attempts are still infused assumptions made under the acacias and baobabs and the wind that made those situations live has long since died away. Meanwhile amongst us move those who have noticed and know only too well how to exploit the naive group reactions that the Caporael experiments revealed. Through their success these new 'free riders' must be becoming slowly more common. Schooled himself in a rough country life, famed and popular author Mark Twain knew this and through a series of humorous warnings he gave us the characters The Duke, The King, and others relevant to the matter in hand in his *Huckleberry Finn*. He made these men pathetic exaggerations but many such 'knowing' men have indeed become professionals, including leaders in politics and in religions new and old. Thence, using proven techniques, they prolong bemusement for the rest of us—quite commonly, be it said, greatly to our advantage, but far more consistently to their own. Possibly I overstate; certainly there is still room for much argument and equally certainly I am not saying that all group beliefs and utopias or all showmen have been bad for us. Helping themselves, they may help us to tame a huge energy source, a potential for group achievement, that lies locked away and wasted in Prisoner's Dilemmas of various kinds that have gotten themselves resolved at the 'DD' (defect–defect) vertex. A lot could be written on religions and priests, in the course of much selfish benefit to themselves, lifting people, by guile and persuasion, out of deep and long-established ruts of 'DD' behaviour.

Anyway, in open societies today plenty of argument over these matters continues.[5] Quite possibly new memes that we are creating and are themselves vigorously 'Darwinizing' and thus self-improving may acquire strength to take us over almost completely, much in the way a 'brain worm' takes over behaviour in an ant. Whenever a weakness in the meme-acceptance system has been discovered there still seems little to stop this happening, at least within an individual lifetime. Whether memes can take over our progeny also for many generations through the same unchanging given property is less clear. The genetic system is bound to strike back and may well do so in some almost paradoxical ways. Consider a meme for respecting the environment, for example, and for in its interest not having too many children. Some progeny (perhaps mutated) may be simply too stupid to understand this idea that their parents try to pass to them; thus the success of the meme to establish among the intelligent and its failure to do so among the stupid may lead to genes for low intelligence coming to preponderate, all contrary to the long trend of our history.

Whether any of the increasing importance of intellectual ideas in human affairs can be said to be wholly for 'our' good, seems to me, yet again, open to question and with answer that must depend on what we believe 'we' are—what do we imagine these 'selves' we refer to, to be made of, our genes or our memes—and, if we are mixed, how much of each? We are brought back in part to such specific questioning as, for example, I raised in Chapter 12 (and also earlier still) with respect to the kinds of memes and the kinds of leaders that are offering us our Big Mother plus her home—the coming Planetary Hospital.

References

1. Horace in E. C. Wickham, *Horace for English Readers* (Oxford University Press, Oxford, 1903).

2. D. S. Wilson, Incorporating group selection into the adaptationist program: a case study involving human decision making; in *Evolutionary Social Psychology* (ed. J. A. Simpson and D. T. Kenrick), pp. 345–86 (Erlbaum Press, Hillsdale, NJ, 1997).

3. M. Doebeli and N. Knowlton, The evolution of interspecific mutualisms, *Proceedings of the National Academy of Sciences, USA*, **95**, 8676–80 (1998).

4. G. Roberts and T. N. Sherratt, Development of cooperative relationships through increasing investment, *Nature* **394**, 175–9 (1998).

5. E. Sober and D. S. Wilson, *Unto Others: The Evolution of Altruism* (Harvard University Press, Cambridge, MA, 1997).

SELFISHNESS RE-EXAMINED: NO MAN IS AN ISLAND[†]

ALASDAIR I. HOUSTON and WILLIAM D. HAMILTON

The target article[1] combines experiments on human choice in a social setting with some general discussions on selfishness and co-operation in an evolutionary context. We find the experiments interesting and agree with the authors that they provide a challenge to sociobiological views of human behaviour. They do not seem, however, to imply such a radical difficulty for current sociobiological and evolutionary theory as Caporael *et al.* suppose. In the first place, the extension of theory that they themselves suggest is needed is not really outside the existing framework. This is the major point discussed below. In the second place, the experiments, although throwing light on a socially welcome aspect of human nature, may be no more puzzling, ultimately, than the maladaptive behaviour of a butterfly beating on a window-pane. In other words, psychologists' experiments, like other single-encounter social situations of civilized life, may be just as evolutionarily novel for us, and therefore just as confusing, as glass panes are to butterflies. This idea does not detract from the value of the experiments, which, if the idea is right, warn us that our behaviour is not stable at its present level of co-operativeness: cheats, though uncommon, may be doing uncommonly well, and the proclivity to co-operate may be slowly declining.[2,3] Thus, the target article and its discussion may encourage us to consider remedial measures.

Turning now to Caporael *et al.*'s own interpretations of their findings, one of our major objections concerns the characterization of egoistic incentive (EI) theories and an apparently alternative Darwinian framework. Selfish gene theory is of much broader scope than Caporael *et al.* suppose. It is in fact the basis of the current Darwinian approach. Although Darwin himself couldn't refer to genes, because they weren't known, the present theory has the same definite and individualistic spirit of interpretation that he used.

As follows from this, the authors' view of EI theory as part of sociobiology is open to objection. We clearly cannot speak for everyone who has used evolutionary considerations as a basis for speculation about human nature. To the extent that some people see selfish-gene theory as a basis for only selfish

[†]*Behavioral and Brain Sciences* **12**, 709–10 (1989).

behaviour, Caporeal *et al.* may be justified in the line that they take. We would argue, however, that their view of selfish-gene theory and EI theories is misleading. They say: 'According to EI theory, people will always choose the selfish strategy in social dilemmas' (sect. 1, para. 5). This is not true if kinship is involved or if there are repeated interactions. The authors go on to consider a variety of ways in which an individual's incentives may change in such a way that co-operation emerges. These ways are coercion, conscience, reciprocity, and inclusive fitness (IF) maximization. We feel that the last two categories do not really belong with the first two. The authors give the impression that IF theory attributes explicit incentives to individuals such that each rationally computes the IF associated with various options ('behavior . . . can be explained in terms of people's attempts to maximize their inclusive fitness', sect. 3, para. 12). Rationality is actually seen as a corollary of EI theory (sect. 1.4, para. 2). But IF says nothing about rationality, or the exact nature of the mechanistic and psychological processes involved.

Coercion and conscience may indeed change an individual's incentives so as to remove dilemmas, but reciprocity and IF theory do not rest on this basis. IF theory[4] is a way of looking at the spread of genes—that is, a way of implementing the Darwinian approach that the authors themselves favour. Similar remarks can be made about reciprocity ('Darwin's emphasis on individual advantage has been formalized in terms of game theory. This establishes conditions under which co-operation based on reciprocity can evolve'[5]). The fact that such evolutionary conditions may underlie human behaviour does not necessarily remove a dilemma. We may in fact be conscious of a conflict between 'our' interests and the maximization of IF.

Although we have argued that conscience differs from reciprocation plus IF as a potential explanation for co-operative behaviour, we are not barred from believing that conscience has been shaped by our evolutionary past so that certain sorts of behaviour are encouraged. (This form of evolutionary explanation may be an example of what Caporael *et al.* refer to in sect. 1.4, para. 3 as 'positing intervening selective processes'. The use of such a two-level approach is by no means unique to cultural evolutionists: in discussions of the evolution of behaviour it is common to distinguish the evolutionary advantage and the behavioural mechanisms—for further discussion and references see refs 6 and 7.) In their treatment of conscience (sect. 2.7), Caporael *et al.* ignore the possibility that conscience may be group-biased. People may be automatically applying rules for co-operating in a way appropriate to outdated circumstances whenever they find themselves in any face-to-face or generally 'believed-in' group, treating this group as if it were not so ephemeral as the conditions of the experiment say—as if they couldn't believe that it could be so ephemeral because, in the Palaeolithic, for which their reactions are evolved, groups never were.

And in reality, when participants are allowed to discuss face to face they may also not want to be branded as even possible or probable cheaters, in case the same people are met again in real life outside. If the experimenters want to get as close as possible to true rational motivations, maybe they should do everything by mail: send out the dollar bills with a description of what the game is, how things will be arranged and grouped when and if the bills are sent back, and so on. The improved generosity in face-to-face discussion groups suggests to us that either the rationality of the subjects refuses to unlink the experiments from reputations valued in real life or else it reflects a quasi-innate reaction to the perceived formation of a group, a reaction that was adaptive long ago when groups were hardly ever ephemeral, and is now no longer adaptive. We suspect that the authors favour the second alternative, but they somehow seem to see it as contrary to sociobiology or selfish-gene theory. This fundamental issue recurs in the context of revealed preference.

Caporael *et al.* says that the sociobiology of choice rests on the revealed preference approach. We do not agree with this claim. Revealed preference amounts to seeing a concept such as utility as inferred from choices; as long as choice is consistent, such a function can be constructed to describe choice. A sociobiological approach, whether it is called selfish-gene theory, IF maximization, or a Darwinian framework, tries to find costs and benefits that would render the observed behaviour favoured by natural selection. Even if behaviour is consistent, no such explanation may be possible (see also refs 7 and 8). In other words, behaviour can be consistent without making evolutionary sense. (We note in passing that a male elephant seal has been recorded as showing consistent and indeed seemingly thoughtful 'psychopathic' behaviour towards members of another species[9] and that dolphins sometimes extend altruistic behaviour to members of other species.[10])

In the context of Caporael *et al.*'s emphasis on the importance of groups in human evolution, it is interesting that in his discussion of the prisoner's dilemma, Sen[11] writes: 'I would argue that the philosophy of revealed preference approach essentially underestimates the fact that man is a social animal, and his choices are not rigidly bound to his own preferences only.' We feel that Caporael *et al.* have underestimated the extent to which previous work has considered this idea. Wilson[12] writes: 'It is likely that the early hominids foraged in groups', and he goes on to mention that this may have given some protection from predators. Hamilton[2,3] emphasizes group-living and attendant group selection in early humans, pointing out that because group members would have been almost always related, group selection and IF theory offer equivalent interpretations.

References

1. L. R. Caporael, R. M. Dawes, J. M. Orbell, and A. J. C. van de Kragt, Selfishness examined: cooperation in the absence of egoistic incentives, *Behavioral and Brain Sciences* **12**, 683–99 (1989).

2. W. D. Hamilton, Selection of selfish and altruistic behavior in some extreme models, in J. F. Eisenberg and W. S. Dillon (ed.), *Man and Beast: Comparative Social Behavior*, pp. 57–91 (Smithsonian Press, Washington DC, 1971) [reprinted in *Narrow Roads of Gene Land*, Vol. 1, pp. 198–227].

3. W. D. Hamilton, Innate social aptitudes of man: an approach from evolutionary genetics, in R. Fox (ed.), *ASA Studies 4: Biosocial Anthropology*, pp. 133–53 (Malaby Press, London, 1975) [reprinted in *Narrow Roads of Gene Land*, Vol. 1, pp. 329–51].

4. W. D. Hamilton, The genetical evolution of social behavior, I and II, *Journal of Theoretical Biology* **7**, 1–52 (1964) [reprinted in *Narrow Roads of Gene Land*, Vol. 1, pp. 31–82].

5. See Chapter 4 (this volume), p. 156.

6. A. I. Houston, Godzilla v. the creature from the black lagoon, in F. M. Toates and T. R. Halliday (ed.), *Analysis of Motivational Processes*, pp. 297–318 (Academic Press, London, 1980).

7. A. I. Houston and J. M. McNamara, A framework for the functional analysis of behaviour, *Behavioral and Brain Sciences* **11**, 117–54 (1988).

8. A. I. Houston and J. E. R. Staddon, Optimality principals and behavior: it's all for the best, *Behavioral and Brain Sciences* **4**, 395–6 (1981).

9. P. B. Best, M. A. Meijer, and R. W. Weeks, Interactions between a male elephant seal *Mirounga leonina* and Cape fur seals *Aretociphalus pusillus*, *South African Journal of Zoology* **16**, 59–66 (1981).

10. R. C. Connor and K. S. Norris, Are dolphins and whales reciprocal altruists? *American Naturalist* **119**, 358–74 (1982).

11. A. Sen, Behaviour and the concept of preference, *Economica* **40**, 241–59 (1973); quote on pp. 252–3.

12. E. O. Wilson, *Sociobiology: The New Synthesis*, p. 567 (Harvard University Press, Cambridge, MA, 1975).

TIME LIKE
A DRIPPING TAP

Memes of Haldane and Jayakar in a Theory of Sex

Out of the earth to rest or range
Perpetual in perpetual change,
The unknown passing through the strange.
MASEFIELD[1]

Manuscripts stacked and waiting, tired editors, slow contributors to multiauthor volumes, finally, and not least, obtuse and sometimes malicious referees—all these combine to confuse the sequence of a scientist's ideas as they proceed, via pencil or keyboard, to their printed form. The paper of this chapter is another example that is out of order not only in my sequence of thinking but also of writing. It was started after several versions of the chapter that follows it in this volume had been completed and had been rejected by more than one journal. It expresses some of the estrangement I was feeling through such unsuccess. That paper was one I considered among the most important I had completed, perhaps my second most important ever. Certainly it was one that represented a lot of invested effort since I had worked on it for about 3 years. This was as long as I spent, for example, on my first long paper on kin selection. Bob Axelrod and I believed we had found a respectable solution for a problem central to all biology plus much evidence in support of it: yet no one seemed to understand what we had done or were saying.

At the time I wrote the present 'Memes of' in the summer of 1989 I didn't even know whether our HAMAX model would ever be published and more than a year of work on the writing up alone seemed to have been wasted. It had been a model laying out a complex simulation and as such inevitably, both on its construction side and concerning why our runs had proved so successful, had left a lot of choices and details not fully explained. Yet, for me, the notion I had started with, that even against sex's full halving inefficiency the problem could be solved by looking at the need of a population to manoeuvre against its many rapidly evolving parasites, with these differentiating resistance tendencies at many host loci (the more the better), had been vindicated.[2] Yet while I felt quite estranged from fellow Darwinians and dynamists in my seeming failure to be understood (which applied not just to the anonymous referees but to the attitudes of many immediately around me at the time), I was also happy. Whatever my failure in explaining myself, at last I knew I had a robust answer, an answer that seemed to be fitting with a vast range of the facts of nature. For me, the ache of a 20-year problem was over. The present post write-up, I think, expresses that split spirit of estrangement and triumph.

When I received an invitation from Carlo Matessi to contribute to a memorial symposium in Pavia in 1989 for a recently deceased scientist I had known, my efforts to remould HAMAX to appease the latest whims of referees were still in progress. They never worked: the referees always had new objections; dislike for our solution seemed to be unbounded. But in Italy and in India things seemed friendlier. The background of the meeting in Italy was that Suresh Jayakar, an Indian evolutionist who had started with J. B. S. Haldane and later adopted Italy as his country of work, had died unusually young in Pavia in January 1989. It at once became the idea of some of his colleagues, notably Carlo Matessi and Francesco Scudo, to honour his memory by convening the meeting.

Jayakar had been only a few years older than me. He was one of the first graduate students of Haldane and Helen Spurway (who was Haldane's companion of his later years) when they both migrated to India in anger over the Suez Canal Crisis of 1956. Later, after both Haldane and Spurway had died, Jayakar had gone to a post in the University of Pavia, Italy, and worked there for the remainder of his life. I told Carlo that due to my efforts with another paper I would have no time, if I came, to write the publishable version of my talk that he wanted for the memorial volume. Carlo replied that in that case they would like to tape-record and then print a transcript of my talk. This idea didn't appeal to me because my lecturing is not very good and certainly far from directly transcribable;

however, I accepted the idea with the proviso that I be given time to look over and revise the transcript.

By the time of the symposium, at the end of September of 1989, the manuscript of the HAMAX paper had gone off to *Science*. *Nature* had rejected it a second time about a month previously. I had sufficiently re-skimmed Jayakar's papers and others connected to them to be reminded that much of his work was very relevant background to our HAMAX project. Thus it seemed good not only to support the idea of the memorial but also to improve my understanding of evolution history and of my own topic to expand my spate of reading still further. So I began to shower the subterranean stacks at the Radcliffe Science Library in Oxford with blue journal-request slips, targeting as I did so ever dustier basement shelves. In September 1989 I told Carlo that if he could give me a month or so extra I would deliver to Sharat Chandra, the editor of *Journal of Genetics* who had accepted the idea to publish a special Jayakar memorial issue, a fully worked manuscript, just as Carlo had originally requested. By a coincidence I was due to visit Bangalore in Karnataka just at the end of the promised months. Since Bangalore was Dr Chandra's base, I was able, in an act probably now about as outmoded in editorial practice as phlogiston is in chemistry (and particularly rare, I imagine, for a Westerner publishing in an India-based journal), to give the manuscript directly into Sharat's hands. By the time I received the proofs, the struggle with HAMAX had at last taken a more optimistic turn: two population geneticists had seen a point in our paper and one was even enthusiastic. Subject to only minor adjustments it was accepted by *Proceedings of the National Academy of Sciences, USA*. This was the fourth journal we had sent it to. I was thus pleased to be able to put into the 'Memes-of' proofs a reference to HAMAX as 'in press'. Fast as *PNAS* is generally in getting manuscripts into print, *Journal of Genetics* beat it and the present late starter came out first by a few months.

Unlike most of my 'wild oats' (for example, the papers of Chapters 9 and 10) this one isn't a paper written for the general reader: it is aimed mostly at ecological and evolutionary theorists—at my own kind. To some extent it is aimed especially at the purists who hadn't understood what we were trying to do. But in spite of this aim the paper is formula-free and I hope it will give almost anyone a feeling for how a theoretical front in science progresses, including, perhaps, for the delays intrinsic to that progress. Perhaps also it shows why a model like ours has difficulty in being accepted. To be truthful, this is still largely a mystery to me in the present case, but I try to explain what may be some of the causes in my remarks about applied mathematics and standards of progress on pp. 595–6. I put more effort into the writing of this 'Memes of' paper than I did with most other symposium efforts both because of my genuine admiration for both Haldane and Jayakar, and because Carlo Matessi and Sharat Chandra had actually wanted me to write something and had not stipulated that there should be an avenue of a dozen more hostile referees.

Haldane had founded *Journal of Genetics* while he was in England and had taken it with him to India when he departed there from Britain in his annoyance over the Suez affair. Sharat Chandra told us the journal was currently struggling a bit with its circulation. His attitude in believing that what I wrote might be beneficial to the journal, an outlook so different from the one telling me that I was over the hill and my work on sex was barely publishable (roughly what Bob and I had gleaned from the referee reports we had been reading), was very refreshing. He hadn't even asked me if I was going to write up for Jayakar the ideas for which I was well known (kin selection, sex ratio, and such like) and he in no way shied from the ideas for which, in that decade, I was infamous—my supposed overemphasis on parasites in sex and sexual selection. Jayakar, I may mention, has long been in my reading list on both the main sides of my life's work: one of his first papers with Haldane had been on measuring genetical relatedness and then there are the ones more discussed in the present paper on temporally varying environments and on antagonist cycling. Anyway, in a spirit of gratitude I decided I would write my best because it was Haldane's journal. And for Jayakar's spirit's sake I would try to make one of my 'wild oats' transform itself into at least a fairly well-cultivated Indian millet—that whose progenitor the two men had planted.

The actual meeting tends to confuse in my memory with others occurring in Europe in the late 1980s. Perhaps also it confuses a little with the visit I made to Pavia at Suresh Jayakar's invitation just over a dozen years earlier, a visit that had been one of the first signs of international recognition for my work on kin selection. Apart from that, one of the more recent European meetings had been in Switzerland and had launched a new Europe-based international evolutionary journal (*Journal of Evolutionary Biology*—*JEB*) while another was again in Italy and had launched yet another journal (*Ecology, Ethology and Evolution—EEE*). Why were these two journals launched almost at once? Both were definitely good news regarding the growth of evolution study in Europe, because in the continent formerly the activity had tended to be centred very much in the north and west. I pushed aside my thoughts of the new journals pulping potential homes of sad woodpeckers in the Alps and Apennines, reminded myself instead of how far more popular and pulp-hungry journals would sadden many more of the birds than we Darwinians ever would. Meanwhile surely Suresh, scion of Haldane and now gone from me beyond a greater divide even than the Alps, would be delighted to know of the new outlets his move to Italy may have helped to begin. Much of the new evolutionary activity, in fact, seemed to be happening in a 300-kilometre radius around Pavia. This included Basel (home of *JEB*) to the north of the Alps and most of northern Italy to the south of them. As to my memories of the two later meetings, the Italian conference in Florence in 1987 brings to me especially Bob Trivers, late arrived and deeply asleep on the crowded floor of the meeting's foyer. He had complained as he lay down there of a combination of jet lag and an unkind assignation of a rather distant hotel. Whether stumbled over or tiptoed around, his massive and torpid form seemed still, radiating into the old rooms of Florence's Società della Storia e Filosofia della Scienza something of a re-Renaissance spirit in which sound evolutionary insight and genetic reasoning stood out as the new components, like the mechanics of Galileo of old. Sages of the Renaissance were often eccentric, too, but at that time the punishment for going too far (as in the case of the Italian philosopher Bruno) was burning, surely a worse fate than failure to get tenure. Inviting Trivers to a meeting is rather like inviting Professor Challenger to step from the pages of Conan Doyle's *Lost World* into real life; it enlivens the proceedings but brings a big headache for the meeting's organizer as well.

In contrast to this memory, trying to recapture the other conference that was in Basel about a year earlier brings me only a dim front-row of faces listening reluctantly to an ill-prepared tirade on parasites, sex, and sexual selection. I have also a memory of seeing, in Basel, the other rooms where Erasmus had taught and where, much later, Nietzsche had taught also. Nietzsche I had always thought of as a man as monomaniacal in his themes of moral philosophy as I was in those of evolution, plus I thought of him as a genuine crank and flawed as a man. His philosophy seemed to me

pseudo-Darwinian. One poem of Nietzsche's that focused the loneliness of
independent thought (something he undoubtedly practised and for which
he was at least as guilty and deserving flames as Bruno—see Chapter 9), a
poem, moreover, giving expression through the image of flight not this
time to the ecstatic side of discovery but rather to the outsider's gloom, has
stayed with me through many difficult periods of my life and in rather the
same role as Bruno's.[3] When I was shown the theatre in Basel where both
taught, Erasmus had seemed in a different league from Nietzsche, but
recently I have been wondering. What, for a start, of the coincidence of the
two putting out their seeds of philosophical revolution from central and
conservative Switzerland? And what of a possibility I had read that
Erasmus, upright prophet of rationality and religious reform as I had always
thought, observing through the reversed telescope of four centuries plus
school history books doubtless omitting all gritty details, had had a wilder
and more Nietzschean alter-ego himself as well? For I recently read that
Erasmus—for a start an illegitimate son of a priest, which I had not
known—had many lovers himself and died of an illness hauntingly
resembling AIDS.[4]

　　Anyway, backing away from that historic theatre and carrying myself
in imagination to others I spoke in around this time, in Basel and
elsewhere, always I see opposite to me the same front-row of disbelieving
faces. And always among them or behind them hovers in this imagining
Steve Stearns—avuncular and dubious, and a man who, by the way, has
been still more of a foreign shaker and mover in the recent European
renaissance of evolution theory than Jayakar was, this through, for
example, his starting the journal *JEB*. It was during another visit and
another talk at Basel that he asked me, with what I thought must be a
touch of sarcasm, how, if the victory of sex in the midst of such a world of
parasites as I described seemed so certain, the parthenogens ever got as
many breaks as they did. Perhaps it was in the summarizing of the same
talk, which I had had to cut off only halfway to my conclusion (another
class, I think, had been already stamping outside the door), that he
remarked with gentile acidity that my theme had perhaps relied on too
much darkness and too many slides—shouldn't somebody steal some to my
advantage? I had known of these slide-excess and time-overrun traits for a
long time and they were why I had had those doubts about Carlo's plan for
a transcription. First, though, the slides are like steadying notes for me—

the visual cigarettes if you like, making them hard to give up. Second, I feel I am invited to places to interest people and that some who cannot fail to find my talk boring may still enjoy seeing a landscape or a beautiful bird, rather as I, when bored with a poor Western film on a transatlantic jet, discard the head set and peer past the tough talk and heroism of the story in hope of seeing at least beautiful desert flowers, the items of the set that the photographer could not fail to include. What if I could spot a *Haplopappus*, for example, which to my knowledge I have never seen? Uniquely amongst all flowering plants of the world, *Haplopappus*, a real desert hero if you like, can make and remake all it has—including its yellow flower heads—out of just four chromosomes.

Of the Jayakar symposium itself my direct memories are elusive. I recall how I came by air and in Milan airport met Ilan Eshel, arrived simultaneously from Israel; then how we were driven by Carlo Matessi to the dormitory where we were to stay in Pavia. It was evening rush hour and at one point Carlo hesitated for a pedestrian at a crossing but then drove on, muttering that one killed more in Italy if one stopped for pedestrians at crossings: confused and amazed by your action they then stepped off and the next car ran them over. It seemed likely enough: the street was like a cavalry charge and no one else had even paused. Ilan remarked that it was the same in Tel Aviv but there was this difference: you could assume a pedestrian who set foot on a crossing on a busy road in Israel must want to be killed. You felt a kind of duty—it was distasteful, of course, but was it fair to leave it to others? I thought how dull our London driving was by comparison, and couldn't cap these stories, instead I sat in silence nursing my suspicion that it wasn't only for the pedestrians that cars were dangerous in the Italian rush hour. I thought of the 'loose scrum' of my school rugby days and of how here I saw motorized equivalents to our much-loved 'scissor' movement, as also equivalents to those one-foot feints that carry you off past the tackler in a new, unexpected direction. Fortunately the more physical moves of rugby, like hand-offs and the outright falling tackles, seemed rare, though, judging from car body work around me, they were not unknown. Eventually Carlo announced, looming ahead, the ancient and calm building where the conference was to be held (sorry, Carlo, to remember best the black humour and the scrimmage of that drive instead of the Roman stone work and medieval towers that you, from time to time, told us we were passing).

Just as I remember coming to the conference, so also I remember well how I left—by rail from Pavia direct towards Genoa whence next day I was to fly back to Britain, but then breaking my journey to walk in the hills between Prarolo and Isole del Cantone in the Ligurian Apennines. I recall how late that evening I unrolled my sleeping bag amidst grassy scrub on a hillside that was prickly with genistas and pungent with thyme and then how, after I had laid down and was composing myself for sleep, I listened to a succession of shotgun reports emitted uncomfortably close by among the bushes. After the shots there floated in the breeze what sounded like muttered curses—pheasant, hare, or whatever seemed not to have been hit. When the voice and the shots had ceased I spent a long time quiet but still awake. The hunting season in Italy has a reputation similar to that in the US where farmers in desperation are known to paint 'COW' in large white letters on the sides of their animals: had I had some paint perhaps I'd have sat up and painted a white 'UOMO' on the side of my bag. But then again, in a country where the news is more of shotguns and fathers mistaken for pheasants than of repeating rifles levelling innocent cows, it seemed to me that silence, the long grass, and scabious heads nodding among the stars above me might be protection just as good.

Hunting and being hunted—those are so often one's bedtime thoughts whenever one sleeps tentless in the grass, as also it is a bit even if one sleeps inside the flimsy walls of a tent. Slowly a sense of security returned. The man had probably gone. I thought more about the conference. Think yourself, therefore, to be with me. Think how human hunters with guns kill more indiscriminately than other predators do: they don't have to run or fly in pursuit, so in the limit, as in these darkening Apennines, hunters are probably still more indiscriminate and blaze off, just like I had used to blaze off when poaching as a boy, at only vaguely plausible prey. My justification then was that I was shooting in the aftermath of World War II, during which my father had had actual permission to cull game for his family in our local woods. I, too, could claim I was shooting because my ancient hunting soul needed to shoot. I never had the permit myself but in a sense I deputized for my father. As to firing at dusk (also illegal, I learned later), pheasants roost, so I shot only upwards and there was no chance whatever to hit a human except by the spent pellets. Normally, anyway, in those lonely woods there was only me, and my dusk 'pheasants' commonly turned out to be leaves or even bare twigs chance sculpted by silhouette into an unusual pattern. Occasionally, however, I hit a real one, and such success kept me addicted . . .

Anyway, the point for now, to think of as you lie with me amidst the grass and stars, is that real predators are usually better assessors of bird health than parasitologists are, and certainly they are vastly better than shooters—our expert distant-acting weapons kill even the good ones. Predators, in fact, have been shown to kill more parasitized animals even in those cases where human experts have told us the parasites in question are harmless.[5] Clearly, therefore, the parasites aren't harmless: even mildly parasitized prey are showing themselves to be less agile and wary. This common pattern highlights three points that are important for the theme of both this chapter and the next. First, in discriminating between individuals, predation is only weakly differentiating the genes for resistance. Second, however, the selection is common and widespread and thus may have serious effects. Third, the selection is 'soft', a strange euphemism, meaning in this case that once the predator has caught its prey (which, commonly, as already hinted, will mean a near to bottom ranker in health), it goes off to digest the victim and other local prey are then spared. The predator tends to truncate the bottom end of a distribution of prey qualities. The technical point for modellers is that under such soft selection the fitnesses are relative, with their values staying undefined until the genetic condition in the entire rest of the population has been surveyed. In contrast, in 'hard' selection models the chances of survival are set from the start; given a genotype you know its reproductive potential without having to refer to any of the rest of the population. More precisely, the 'soft' selection against the squirrels and in my models is of the special kind I already indicated: 'truncation selection'. Our predator, again, is imagined chipping away—truncating—from the bottom of a fitness distribution. All these ideas had been worked into our HAMAX model and our manuscript. Since I thought them important, inevitably I re-emphasized them in the present paper; inevitably, also, I lay there reviewing them and how well I had expressed the idea that night in the Italian hills.

Unfortunately for steady progress of understanding of dynamic polymorphisms by using classical population genetic methods—through, as I will put it in the paper, 'worthwhile, proven results, the stuff of applied mathematics'—truncation selection is notoriously difficult to analyse mathematically. As has recently been realized, the difficulty is due to the

strong tendency for truncation to produce 'chaotic' dynamics. It is, of
course, possible still to reason logically about truncation effects and chaotic
dynamics and as always mathematics provides the best training and the
best routes known. But if technical 'chaos' indeed commonly follows from
truncation, then, even with the best of preparation, the only precise
prediction is paradoxical: precise prediction is impossible. The fact is that
the outcome is bound to be vague compared with what can be predicted if
more traditional hard-selection assumptions could be applied. This is
because trajectories of chaos, even when started very close together, show a
worsening divergence as they proceed; they are divergent by 'chaos'
definition. In short, they bring in unpredictability at a deep level. But the
reason I didn't proceed to the proper diagnostic tests of chaos is not just
that I am too old to acquire easily more than a smattering of the new
methods (and jargon) of chaos studies, but rather that for the time being I
can't see much incentive. It was obvious enough that no simple formulas
were going to summarize the extremely complex trajectories I was seeing.
So even if the paths were completely determined—nothing was due to
chance—I might as well simulate them in powerful computers and then
treat 'results' so obtained statistically, as if they were part chance-
controlled and from nature.

Many theorists were assuming that the newly defined chaos was
unlikely to apply to real population processes. It is also possible that a lot of
my seeming chaos is not, actually, the official kind. Even though the
trajectories on my computer monitors look so very like wool messed by a
kitten that they fit well with an everyday notion of chaos, wool messed by a
kitten is unreferred to in the technical definition. The point for me is that
for sex it probably doesn't matter whether at the ends of my runs I am in a
state of 'proper' mathematical chaos or not. Complicated confinement to a
torus (doughnut-shaped) surface, or to that beautiful folded Mobius strip
known as Ruelle's attractor that I sometimes seemed to see, or confinement,
if that is the word, to the kitten-tangle attractors that I so commonly found
and ought to study and define more carefully—possibly any of these can do
equally well for sex. But perhaps not quite equally. Really there need to be
about four attributes: (1) the situation has to be restless; (2) it has to
involve, again a necessity, changes in gene associations across loci; (3) these
changes should have a suitable rhythm, such that the uncoupling caused by
recombination turns out useful; and (4), and now more an ideal than a

necessity, the changes should be irregular so that they are hard to predict. This last, of course, is where chaos might be desirable; chaos would be a state where an opposed species partner would face impossibility to make any long-term prediction—no constant policy of preparation for coming distant changes in its exploited (or fled-from) partner would pay off.

All the above tells us that the one kind of 'attractor' of the dynamics of the system that has to be totally banned, if sex is to have a chance, is a completely steady state in the genotype frequencies—mathematically the point outcome. A point outcome in the state space implies a genuine steadiness of all frequencies: each genotype takes on a single unchanging frequency, which is zero or otherwise. Steadiness—stable equilibrium to this degree—cannot help sex to survive because recombination creates nothing of advantage, only what is already there. In contrast, any system under selection that at least changes the statistical associations of the alleles has a chance of enabling sex to prevail among competing clones: clones can follow sex only by their slower selection of mutations. If sex lacked its usual cost (that is, if it costs nothing to be a 'single mother', as Olivia Judson chose to put the nub of the problem), it would be very easy to find conditions where sex would prevail in a competition (see note 22 in the introduction to Chapter 2 on linkage disequibrium, noting further that for sex to succeed we must have that recombination tends, more often than not, to fill up a scarcity of what is, or is about to become, the newly required type of the Red Queen race). If, on the other hand, sex has the usual single-mother's disadvantage, then the outcome is uncertain but the chance for sex is often good—as our HAMAX simulation will show in Chapter 16. Repetitive cycles (as in Chapter 1) or the non-repetitive near-cycles in Ruelle's attractor or our balls of wool could equally serve, as also could any quick-fire incidence of interacting mutations—these beneficial or/and detrimental. Furthermore, and quite unlike those dreaded 'impotent' horizontal two-point cycles I mentioned in Chapter 1, the 'tangled-wool' (therefore probably chaotic) cycles that we found as the most frequent outcomes under our most natural assumptions are quite guaranteed to be altering the gene associations. After all this, however, it has to be admitted that how often they occur and how great the benefits are that can flow from sex and recombination—making or unmaking the gene associations that selection previously had formed (how often does it get things right?)—still need much more investigation.

Obviously not much of the above can have been passing in my mind as I lay in the grass on the Italian mountain and listened to the dusk sportsman and then later to the bush crickets, their songs fading in the night's cold. Something in the evolution line certainly was passing, however, as things quietened down. Evolution generally provides my bedtime thoughts wherever I am and only a bare touch of something connected to my work is enough to set the thoughts going. I have even learned that it is worth trying to exploit the more special places; my nightly thinking is apt to be most creative exactly when my bed is unusual. For that particular night I don't remember if it was sex but, given the preceding meeting, probably that came in and so I realistically reconstruct. On the other hand, I do remember crushed thyme reminding me of other nights when, following butterflies and orchids in the hills of Kent in my boyhood, I had slept in similar lonely places, where also flowering scabious was over me and certainly there was a fantasy of them becoming 'other' or 'again' daisies before my eyes even as they nodded through drifting stars. I suspect I hadn't, at Cantone del Isole, yet read Stanley Temple's paper about his tame hawk and those more and less parasitized (that is, naturally variable) squirrels he loosed it against; however, ancient fears of predation come to us with night winds everywhere and not even an Eskimo, I suspect, fails to be haunted by ghostly leopards from Africa as well as by his present polar bears. What else might I have been thinking that night?

Of Nietzsche and his poem, perhaps, and how he finally took sick and mad in Turin, another town in the broad flat valley I had come from, west of Pavia, under that Great Bear sparkling over. And there, too, beyond the Alps, in Basel, had died Erasmus . . .

Of starlings, perhaps, and Italian cooking and jokes from either I might have used to enliven my front-row faces . . . A starling puts aromatic herbs among the materials of its nest, and it has a plausible and effective evolutionary intent to deter lice and fleas that could otherwise plague both the adult sitting bird and its young. And way to the south chimpanzees eat the rough aromatic leaves of yellow daisies. In this case the chimps probably have an unconscious evolved tendency to terrify and abrade as well as intoxicate their intestinal worms . . . Surely, then, it must be that the aroma of this Italian thyme around me was pleasant for the same reason; whether culturally or genetically we have learned to like its protection. Perhaps it still is protective, though probably was more so in times when, here in Italy and places similar, mosquitoes, malaria, and worms were more threatening. Here was the joke I'd missed connecting Italian cooking and Italian longevity, a serious joke about a 'taste' perhaps acquired culturally first in Italy or all around the Med and then becoming a first step towards a new 'exosomatic' resistance for us all; finally, due to antibiotics, it slipped back to be a 'toenail' of our past. On this view it is fools wanting to die of cancer who have their curries too hot.

Astringencies, tastes, pleasurable and harmless scents in some but subtle and tasteless poisons in others, all these are examples of effects due to what we call plant secondary substances—that is, substances that appear as gratuitous additions to what a plant needs for its growth, energy supply, and the differentiation of its cells. Scents like thymol in thyme have been something of a puzzle to an evolutionist—why they should be there. Well, same thing but more definitely genetical; they are now believed, along with most of the other secondary substances, to be serving the plant's defence against insects, fungi, and even vertebrate grazers.[6] Sometimes, then, the defence is such a success that we vertebrates learn to adopt and exploit it; we pick up the rifle that was partly intended for shooting us. Fleabane, bugbane, cowbane, wormwood, feverfew—the essences of these plants whose uses the names in the old herbals emphasize were undoubtedly all meant originally as deterrents to creatures that eat the plants. Absinthe may give us visions, as in the great vogue at the turn of the century, but the visions are so strong as to be stunning to worms within us and cause them to lose hold and be expelled like LSD in the more modern vogue, I suppose, causing you to jump out of an upper-floor window thinking you can fly. I could have mentioned, too, at that symposium the penicillium out of Gorgonzola, although probably not much of its essential antibiotic makes it through the intestine wall. (Why do I always think of these enlivening additions just after I have finished a talk?) Perhaps cattle raisers of the world have evolved, genetically, fortification against the decay microbes of cheese and milk they must continually encounter, and this might explain why Rosanna, my cook in Amazonia, was so utterly disgusted at the cheeses I brought to the flutuante—her aversion beginning even when I arrived with my oily, sweating parcel at the end of one day of travel, leave alone a week later. The Japanese are disgusted too, and the real reason the crow dropped that cheese in the fable may have had less to do with the fox's flattery and more with its foreboding of danger—the crow might have loved carrion but that of a cheese was the wrong kind for it, just as their kind is wrong for us.

My particular ancestors, my many former incarnations I might put it, surely were cattle raisers but, even so, one sibling of mine hated milk even in childhood. But then another in our family was a hater of spinach, and my own dread was for turnips. My mother (at the time when I remember the pressure most intense) had tried to make us all to like all, on a ground basically the same as that which made her not ask my father (or later me) where a shot pheasant came from, from our own garden or outside, and made her tell us, too, of the refugees suffering more than we from the war and they longing for all that we rejected (most of which, I may add, she grew by her own hands in our garden). She said that what one of us could thrive on, so could we all. But was it true? Theory has led me to overturn these long-respected injunctions of my mother and to think that the 'whims' my parents tried to anull might be justified by our enzymes and our genes if only we knew, the differences within being like all the others that are so visible on the exterior—the blue eyes and the brown, hair curly or straight, and thus all our tastes all foretold, in fact, ever since these very two, our parents, shuffled their decks and dealt us the Mendelian cards they left us to play.

On such lines, even before I had brought my two small daughters to Michigan with me, which corresponded roughly to the time when I became the variation-by-parasitism freak that I now am, I became myself far more lenient in this matter of dislikes of food than my parents had been with me. When I was an undergraduate at Cambridge I tested all my own family members, including my parents, for the 'taster' gene for phenylthiourea, using methods that had been laid out by Hans Kalmus (shortly to be my colleague at University College London—and, remarkably, the only one who did occasionally seek to talk to me). I found that we six (as we were then) expressed the 1:1 ratio Kalmus predicted to perfection and that the differences neatly explained all our sharply defined dislikes or indifferences to cabbage and turnip. So mightn't those likes and dislikes index differing abilities to catabolize substances dangerous to our development (Kalmus mentioned, I think, effects on the thyroid), while at the same time the identical drugs could potentially be hostile within the body to certain once-serious parasites? The parasites had nudged us—that is, had evolved us, to coin a new active verb—into a hard place between Devil and the sea, a situation where we resisted worms, say, only at cost of some injury to our own development. It was an evolution that came to seem to me very likely and potentially very important for the immense sweep of organismic variation I wanted to resurvey.

Climbing from the station to where I camped by perhaps a thousand feet, I had found the vegetation ever more 'English', even for certain none of it had been introduced from Britain into Liguria, if anything the other way. This similarity and its implication of an old flora and the contrast of that to the 'weedy' situations I passed in the fields on the way up, for sure would have engaged that day one of the lines of speculation that has become almost a constant companion as I travel anywhere in the world. How different are the native, mixed floras of grasslands and mountains from those I see in more human-disturbed places! That day, for example, alongside all the roads up the rich, slanted agricultural countryside the trees had been robinia (locust tree), and then more and more of the same—the slender, thorny-barked, crazily sideways-suckering trees of the east Alleghenies, making thickets here in Europe wherever it could stretch out its yellow, far-reaching roots; and all of this 'monoculture' of robinia had come here, ultimately, from that other continent! Robinia I found, usually in woodlands or else near to houses, had grown to be bigger trees in Italy and Poland than I had ever seen in its American homeland, where the lead stems were so often killed back by the locust borer while a dozen other common insects made feast of the leaves and imbibe the protein-rich 'refrigerante' of its leaf sap.[7] There was, for example, the beautiful hard-backed, long-antennaed, magnificently yellow-striped and waspish, shoot-destroying beetle, the locust borer. This hasn't succeeded—yet—to accompany the plant to Europe and nor had all the others, the bagworms, leaf miners, scales, bugs and treehoppers . . . Their absence became, doubtless, a main factor in the tree's triumphal rampage across Europe.

What has been happening during the two centuries or more since the black locust has been in Liguria and the black locust borer not? Are local insects and fungi switching to it, taking advantage of an initial uniformity, as has begun happening with some other 'wonder' plantation tree crops in other parts of the world? If so, with what effect? And how, in contrast, are fading the old guard of the plants and animals (in Europe say) that such newcomers oppress by their competition; how do these sets undiversify as they die out—the red squirrel, for example, facing the grey, and big woodpeckers needing the big, old trees and being denied them (as I speculated earlier) by the creation of new journals? Such general questions could well be applied to some of those plants more rare than the thyme under my sleeping bag. For, along with, and more serious than, tramps such as me coming to lie in the grass, the fields and artificially seeded pastures were themselves pushing ever higher, squeezing out as they did so the plants of the natural grasslands that once, doubtless, had been encouraged (not prevented) to exist there by the combined efforts of deer and hares and chamois.

The world today is full of discussions of species biodiversity and how it is being affected by human pressures. Genetics sometimes enters these discussions because of the evidence that genetic impoverishment may be more crucial than actual numbers in determining the point of no return for declining species. Doubtless there are other models besides mine for habitat diversity and its change and ones that predict similar criticality, but one thing I could have pointed out in my lecture and possibly did (the written paper often misses things I say and still more vice versa), is that a rough critical population size is an easily seen prediction from the Red Queen's viewpoint. In fact, its effects were often watched by Bob Axelrod, Reiko Tanese, and me during our HAMAX runs in Ann Arbor: on my visits I would sometimes sit late some nights in Bob's office watching the displays subprogrammed by Reiko to give me 'live' synopses of the progress of genetic variability in both sexuals and asexuals as I watched. I would see that as the Queen's tool-kit—her gene variation—became impoverished, as has to happen in any dwindling species and as was represented in our simulation by a dwindling sexual subset of hosts, the Queen in a sense went progressively down in her rank: Rook, Bishop, Pawn . . . down, down, until at last, unable to resist anything, she dies and the population flips to monomorphic. Monomorphism has brought her to a standstill, to her knees, her parasites adapt perfectly; in real life, extinction follows. In the successful runs, of course, it would all go the other way, with the set of clones of the host being that which lost variability and died out: after we brought in soft selection this became the more common outcome under realistic settings. Maybe a still better image for those events I watched would be a *Scolopendrium* centipede progressively losing more and more of that seeming endless excess of legs it starts with, gliding the whole grotesque beast forward on the walls of a wine cellar. This image might

simulate better in our failing populations the rather sudden onset of ability
to progress against the parasites at all. The reality of all this, of course, is
one gene locus after another going to 'fixation'—that is, having only one
morph, or allele, of the gene available.

Treating the interplay of genetic with species biodiversity as
affecting no more than the terminal conditions of dying species is,
however, over the whole arena, a much too limited a view of introduction
species. Think of the creeping accumulation of parasite switches currently
being recorded for eucalyptus as a plantation tree all over the world[8] and
also of that switching towards superabundant *Homo sapiens* that is
happening similarly; and think of that which, for robinia in Europe, is just
beginning. Basically there are two ways in which species, becoming
common by some new adaptive success, may defeat the accumulation of
parasitism that naturally pursues such success. One is the Red Queen route
to ongoing defence as I define it in the sex theory. First, this one aims to
increase the diversity of your anti-parasite technology (patches for the tyre
punctures you experience) so that you may keep running in local gene
pools both of your own kind (against competition) and in those of your
parasites. Second, it aims to stay all the time outbreeding so that your
changes of frequency, your passing on of a 'patch' technology, can be made
swiftly. Third, it tries selectively, by mate choice, to outbreed so that your
descendants' reactions to each new parasite threat may come more swiftly
still. The other more simple route is to speciate—to divide your population
perhaps many times over, and thus draw a little ahead of your parasites.
They, however, may well accompany you by doing the same; but when
again they are cospecialized (as the great majority seem to become
eventually, especially those that are most serious), there still remains the
recourse of hiding from your own kind by moving away in space: you hide
amidst what you have caused to become an ambient multitude of different
species, as I discussed in Chapter 8. Hatched no longer out of mere
varieties of your kind but from what have become much more different host
environments, full species though still related, most of the parasites you
encounter can no longer attack. This second course had to be in itself, I
believed, the ultimate explanation of the floristic richness of the limestone
grassland such as that where I lay—the thyme here and scabious there and
also that rest harrow, more thorny than ours in Britain, I had so soon rolled
away from, preferring even the low *Cirsium* and *Carlina* thistles whose

shorter prickles more seldom penetrated . . . And so it went on, while through all the gaps of those deep indented trifoliate and pinnatifid leaves, down leaned the black grasses to tickle my face—delicate quaking grass here as in Kent and even already here, too, those old fiends of mine that had so spoiled our hillsides after the myxomatosis and under our new nitrated rain, nitrated soil, the upright brome grass (so called, but really it droops) and the bitter-stemmed false oat that no farmer chews, leaning on his gate . . .

At last I felt comfortable. Under such cataracts of bedtime speculation as I try to outline here, speculation snowed upon me only the more vividly, and sometimes usefully, in such strange sleeping places, it came hardly to matter that I had only sparse cover if it should, in fact, come to rain as the now-vanishing stars suggested, or that I still felt urgently my need not to snore in case I sounded too like a foraging *cinghiale* (I hadn't actually, I recalled, heard the hunter go) . . . So Jayakar's successors, I thought as I began really to doze, if they ever caught up to where he had left off,[9] they could have pleasure in all that precise and proper mathematical theory that was yet to come . . . later, too, in all that proven Popperian exorcism of devils in my papers that likewise must surely come. But I instead would have this—this beauty around me and a sense out of the very earth of where I was going.

References and notes

1. J. Masefield *The Collected Poems of John Masefield* (Heinemann, London, 1928).

2. Papers by J. J. Burdon and his co-workers on wild plants and their parasitic fungi can be suggested as a set to illustrate this statement. The following subset, for example, illustrate various aspects of a particular host–parasite pair: J. J. Burdon and A. M. Jarosz, Host–pathogen interactions in natural populations of *Linum marginale* and *Melampsora lini*. 1. Patterns of resistance and racial variation in a large host population, *Evolution* **45**, 205–17 (1991); A. M. Jarosz and J. J. Burdon, Host–pathogen interactions in natural populations of *L. marginale* and *M. lini*. 2. Local and regional variation in patterns of resistance and racial structure, *Evolution* **45**, 1618–27 (1991); J. J. Burdon and A. M. Jarosz, Temporal variation in the racial structure of flax rust (*M. lini*) populations growing on natural stands of wild flax (*L. marginale*)— local versus metapopulation dynamics, *Plant Pathology* **41**, 165–79 (1992);

A. M. Jarosz and J. J. Burdon, Host–pathogen interactions in natural populations of *L. marginale* and *M. lini*. 3. Influence of pathogen epidemics on host survivorship and flower production, *Oecologia* **89**, 53–61 (1992); J. J. Burdon and T. Elmqvist, Selective sieves in the epidemiology of *M. lini*, *Plant Pathology* **45**, 933–43 (1996). Other work by Burdon on other pathosystems is also relevant.

Rather differently from me, Burdon himself emphasizes in his work the unavoidable metapopulation structure most plant pathosystems have and following from this the 'escape in space' aspect of defence rather than the 'escape by recombination' (for example, see P. H. Thrall and J. J. Burdon, Host–pathogen dynamics in a metapopulation context: the ecological and evolutionary consequences of being spatial, *Journal of Ecology* **85**, 743–53 (1997)). As I discuss in Chapter 16 and its Appendix, however, these two modes can be much entwined and it hard for me to believe that advantageous recombinations are not playing a substantial part in the situations he details.

Overall, in fact, the ensemble of Burdon's findings seems to me in accord with my own line on dynamic polymorphism maintaining sex. Three further aspects are:

(a) *Whole genome involvement.* Many resistance and virulence genes are shown widely scattered in genomes of both hosts and parasites (D. C. Abbott, A. H. D. Brown, and J. J. Burdon, Genes for scald resistance from wild barley (*Hordeum vulgare* ssp. *spontaneum*) and their linkage to isozyme markers, *Euphytica* **61**, 225–1 (1992); D. F. Garvin, A. H. D. Brown, and J. J. Burdon, Inheritance and chromosome locations of scald-resistance genes derived from Iranian and Turkish wild barleys, *Theoretical and Applied Genetics* **94**, 1086–91 (1997)).

(b) *Varied interactions.* Instead of pure 'vertical' or 'gene-for gene' interaction, some sets show partial additivity for resistance (A. H. D. Brown, D. F. Garvin, J. J. Burdon, D. C. Abbott, and B. J. Read, The effect of combining scald resistant genes on disease levels, yield and quality traits in barley, *Theoretical and Applied Genetics* **93**, 361–6 (1996).

(c) *Dynamic interaction.* Rapid changes of frequency of some genes have been shown, these sometimes following an observed epidemic (J. J. Burdon and A. M. Jarosz, Temporal variation in the racial structure of flax rust (*M. lini*) populations growing on natural stands of wild flax (*L. marginale*)— local versus metapopulation dynamics, *Plant Pathology* **41**, 165–79 (1992); J. J. Burdon and J. N. Thompson, Changed patterns of resistance in a population of *L. marginale* attacked by the rust pathogen *M. lini*, *Journal of Ecology* **83**, 199–206 (1995); J. J. Burdon, L. Ericson, and W. J. Muller,

Temporal and spatial changes in a metapopulation of the rust pathogen *Triphragmium ulmariae* and its host, *Filipendula ulmaria, Journal of Ecology* **83**, 979–89. (1995)).

3. *The Solitary*

> Harsh cry the crows
> And townward take their whirring flight;
> Soon come the snows—
> Happy who has a home this night.
>
> With glances dead
> Thou gazest backward as of old!
> Why hadst thou fled
> Unto the world from winter's cold?
>
> The world—a gate
> To freezing deserts dumb and bare!
> Who lost what late
> Thou lost is homeless everywhere.
>
> Pale one, to bleak
> And wintry pilgrimages driven,
> Smoke-like to seek
> The ever colder heights of heaven.
>
> Soar, fling wide
> That song of birds in deserts born!
> O madman hide
> They bleeding heart in ice and scorn.
>
> Harsh cry the crows
> And townward take their whirring flight;
> Soon come the snows—
> Woe unto him who has no home this night.

Translated by Ludwig Lewishon in M. Van Doren (ed.), *An Anthology of World Poetry*, p. 843 (Cassell, London, 1929).

4. T. Appelboom, *et al.*, The historical autopsy of Erasmus Roterodamus (c. 1466–1536), in T. Appelboom, *et al.* (ed.), *Art, History and Antiquity of Rheumatic Diseases*, pp. 76–7 (Elsevier, Brussels, 1987); M. D. Grmek, *History of AIDS: Emergence of a Modern Pandemic* (Princeton University Press, Princeton, NJ, 1990).

5. D. Jenkins, A. Watson, and G. R. Miller, Predation and red grouse populations, *Journal of Applied Ecology* **1**, 183–95 (1964); S. A. Temple, Do

predators always capture substandard individuals disproportionately from prey populations?, *Ecology*, **68**, 669–74 (1987); A. P. Møller and J. T. Nielsen, Differential predation cost of a secondary sexual character: sparrowhawk predation on barn swallows, *Animal Behaviour* **54**, 1545–51 (1997).

6. It was once believed likely that most secondary substances had no adaptive significance; they were regarded as waste or accidental byproducts of the important 'primary' processes. But this view seems abandoned and for many such substances it is known that their manufacture entails considerable cost to the plant in slower growth (K. Han and D. E. Lincoln, The evolution of carbon allocation to plant secondary metabolites: a genetic analysis of cost in *Diplacus aurantiacus*, *Evolution* **48**, 1550–63 (1994)).

7. W. W. Hargrove, An annotated species list of insect herbivores commonly associated with black locust, *Robinia pseudoacacia*, in the Southern Appalachians, *Entomological News* **97**, 36–40 (1986).

8. In the case of *Eucalyptus* it seems that where this immensely useful Australian tree has been planted in the temperate zones it has been mainly pests 'catching it up' from Australia that have become its problem; in the tropics, in contrast, such 'catching up' has also happened but far more of the present pests are cases of indigenous insects switching their attentions to the tree (C. P. Ohmart and P. B. Edwards, Insect herbivory on eucalyptus, *Annual Review of Entomology* **36**, 637–57 (1991)). Possibly this reflects the greater diversity of the local tropical parasites and also possibly a flora with more plants of the Myrtaceae to provide likely opportunists.

9. Population genetics catches up but has taken time: L. A. Zonta and S. D. Jayakar, Models of fluctuating selection for a quantitative trait, in G. de Jong (ed.), *Population Genetics and Evolution*, pp. 102–8 (Springer, Berlin, 1988); A. B. Korol, V. M. Kirzhner, Y. I. Ronin, and E. Nevo, Cyclical environmental changes as a factor maintaining genetic polymorphism. 2. Diploid selection for an additive trait, *Evolution* **50**, 1432–41 (1996).

MEMES OF HALDANE AND JAYAKAR IN A THEORY OF SEX[†]

W. D. HAMILTON

The history of mathematical modelling of communities and polymorphisms under intrinsic fluctuating selection is outlined. Authors have usually encountered difficulty in obtaining stability in cases involving host–parasite relations. Stability may, in fact, be uncommon. On the other hand protection of diversity (non-extinction of rare species or variants) may instead be common and important.

Within multilocus systems, mild truncation selection on a host–parasite system both protects variation and is supportive of sexuality against parthenogenesis even when sex pays a full cost—that is, even if sex has halved the efficiency of reproduction due to production of males. Truncation based on heritable health, which is itself based on polygenic resistance to parasites, provides the most robust and universal model supporting sex yet presented.

The separate and joint roles of J. B. S. Haldane and S. D. Jayakar in originating ideas now incorporated in the model are discussed.

For the past 10 years I have developed an idea that the ultimate evolutionary cause of sex is disease. The idea is increasingly credible. This is not so much because favourable evidence is accumulating, although the scraps that come to light from time to time (usually having been published incidental to some other objective) are mostly favourable. It is rather the model itself, its engineering and appropriateness for its task, that has improved. A version now evolved with R. Axelrod and R. Tanese (see Chapter 16) requires only as a workable and realistic minimum to make sex secure, even when carrying the full twofold cost, the following three assumptions: (1) that diseases are many; (2) that resistances are polymorphic; and (3) that selection has 'softness' or ideally a pattern of truncation in its operation. The last assumption indicates that if the theory is right sex ought to be most universal in species whose members compete for places in hierarchies and live in saturated habitats; parthenogenesis, in contrast, should be most common in colonizing species living in unstable, physically severe habitats. This is exactly what is found, a

[†]*Journal of Genetics* **69**, 17–32 (1990).

point that seems to favour our theory over others that for the most part leave the ecological correlates of asexuality unexplained (see reviews in refs 1–3).

Feeling at last in sight of a theory that can respect if not yet quite reverence Mother Nature's bizarre invention of maleness, wherever it occurs, and thus also slightly more at ease about the future continuation of my own kind within the human species (I can now tell people that we males are necessary for health), I find myself more interested in the history of the ideas that have become the ingredients of what I see as the successful theory. Ingredient is the word. Evolving the theory has been like working out a recipe for a cake, or, rather, as I would hope, trying to reproduce someone else's unknown recipe—Mother Nature's. It has been a matter of experimenting with materials, putting jars back on the shelf, taking others—in fact, trying to use various insights of population genetics and ecological theory that have been evolved over the past 35 years (and some from even farther back as will be seen). These insights and their origins are the topic of this essay. Its central theme will be the insights that foreshadow the demise of the ecological and genetical expectation of *stasis* or, in other words, that tend to reject a supposed abundance of stable equilibria in the genetical living world (see Chapters 5 and 8).

EVOLUTION AND DISEASE

In writing this historical sketch I have not had the time to search the literature as thoroughly as I should, and even out of work known to me only a skeleton is cited. A good fairly recent review of much of the rest is in Levin.[4] How thorough should one try to be when writing history of the future—and of an unlikely future at that, as most see it? In the eyes of many, irrespective of coherence of theory, I will seem way ahead of my facts already and to be as if writing the history of how world population growth was halted long before any such event has happened. Why write history of what may prove to be false? Hardly anyone seems to agree with me that disease is particularly important for sex, leave alone endorses my particular model.

My excuses for an admittedly presumptuous approach are twofold. First, I was asked to write a paper relevant to the work of Suresh Jayakar and, for sure, Suresh is deep in the story I want to tell. Second, irrespective of what others may feel, I am now much happier with my model than at any time since the 1960s 'reformation' did away with an old faith.[5] The latest version using many loci and truncation selection has greatly improved in both robustness and realism, giving us an idea that at last makes sense on all fronts. It has been said that any new and true theory has to go through three stages as it becomes accepted. In the first, it is described as obviously wrong; in the second, as contrary to religion; in the third, as obviously right—as was known long ago. Perhaps here is a justification for pushing my idea directly towards the third stage because, as my history is about to show now, part by part, all of the idea of sex and parasites was known long ago.

If the Mendelian system is for the 'purpose' that the theory of sex from parasitism suggests, then it is no surprise to find that suitable theoretical properties exist in the sexual process. Nor is it surprising that, in making a striking product, the properties have often been striking themselves and therefore noticed. What is surprising is that the whole compound of diploid recombining Mendelian genetics, source of all such properties, should be accepted for so long accompanied by only very weak theories of why it should exist.

Usually when one looks for the earliest hint of any new idea in evolutionary biology it turns out that a careful search of the writings of Charles Darwin reveals that he wrote and understood more about it than anyone has yet noticed. Alfred Russel Wallace, too, is worth searching for he also ranged deep, although on a narrower front. If nothing is found in Darwin or Wallace the next likelihood is to try the contributions of August Weismann, another thinker as opposed to mere documenter of evolution; then failing Weismann one turns with high expectations to the writings of what may be called the Founding Peers of the genetical unification—Sewall Wright, R. A. Fisher, and J. B. S. Haldane. Born over the 3 years 1889 to 1892, these three made surprisingly equal and independent contributions. At their hands Darwinian theory was revised into line with Mendelism and many further consequences were described.

Following the above protocol with respect to a possible understanding of sex as a response to parasitism, it turns out that Darwin's contribution was, first, not to fail to point out disease as an important agent of natural selection, and, second, and quite separately, to investigate a related and ultimately convergent theme that arose on the side of sex. He gave plentiful discussion to the adaptations by which sexual organisms secure outbreeding while admitting to remain puzzled why they were there. For a long time he saw outbreeding as somehow vaguely helping to cause variation. Later, however, he was led by the 'blending' arguments of Fleeming Jenkin to a quite opposite view, that sexuality might be hindering variability. Forced to look for new abundant sources to replenish what 'blending' seemed to be draining away, he fell back in later editions of *The Origin* on 'use and disuse' and Lamarckian concessions. As became clear with the advent of particulate Mendelism blending is not really a problem, but the continued existence of so much variability and how outbreeding affects it in the long run are still unsolved issues. Possibly it is only with understanding of the role of parasites on lines that I outline below that Darwin's intuition that outcrossing would help macroevolution by creating variability begins to be justified. This help may be coming, however, by more devious pathways than anyone from Darwin to Lysenko, even including the intervening founders of Neodarwinism, ever suggested (see ref. 6 and Chapter 16). Even while such new themes arise, however, proponents of sex theories that do not attempt to account for outbreeding (e.g. refs 7 and 8) continue unwisely to neglect Darwin's early concern and findings.

On the front of disease, after Darwin there followed a gap during which information accumulated on the heterogeneity and detailed genetics of disease resistance in natural and domestic populations but little was said about any general implication, even though, as it seems to us now, the high heritability, so contrary to expectations for a fitness-enhancing trait, might have warned of an odd and interesting situation. The failure to relate such findings to sex, however, is not surprising because nearly every one was satisfied that they already understood what sex was for. The spectre of blending inheritance had been exorcised but a new, cheerful, yet almost equally light-weight opinion had arisen in its place. For the time being it seemed to exemplify nicely what Darwin of the earlier period had been grasping for. The idea was that sexuality and Mendelian genetics helped to speed up evolution by allowing the occasional beneficial new mutations to be combined quickly in a single evolving stock. It rested on the kind of broad group- or species-selectionist justification that was quite usual for biologists at the time. Slightly more surprisingly the idea was accepted even by the Founding Peers, contrary to their usual insistence on detailed individual-selection arguments. False security in sex's supposed assistance to a long-term 'onward and upward' aspect of evolution progress was to last approximately until the mid-1960s. The story of its eventual collapse has been told several times (see refs 1, 9, and 10) and will not be repeated. Both before and after, the theoretical stirrings that I believe were quietly laying the foundations for a better theory were not being generated out of any perplexity about sex itself but appearing as comments on neglected but interesting properties discovered in various subdisciplines—theoretical ecology, demography, and genetics.

'HALDANE (1949)'

In 1948, J. B. S. Haldane attended a symposium on ecological and genetic factors in animal speciation at Pallanza in Italy. The paper he read was entitled 'Disease and evolution' and it was published in 1949 in a symposium volume that appeared as a supplement to *La Ricerca Scientifica*.[11] This is a particular supplement of the nineteenth year (19° anno) of the journal. Many (myself included until recently) have not understood this and consequently have cited the work as 'Supplement 19'. Actually *La Ricerca* often has supplements on several different topics in a year; hence, in the case of 1949, a reader pursuing the usually given reference is as likely to find him- or herself reading a bibliography of polarographic analysis as the desired volume that contains Haldane's article. From the almost universal yet slightly inappropriate form of the citation I judge that not many people have actually read it: I therefore summarize what it contains.

It is a forceful essay on selection by microbial disease and a plea to end its neglect as an evolutionary factor. It details the special properties selection by disease has. Haldane emphasizes the *specificity of parasite attack* and discusses how different this is from selection by larger enemies such as large herbivores and predators. The properties of parasites, he suggests, create a strong frequency dependence and this can make disease and parasitism a major factor maintaining variation. As to the kinds that are maintained he draws particular attention to variation at the level of proteins, first signs of whose abundance he had been quick to notice. He cites papers showing such abundance more than a decade and a half before the key paper that later expanded our interest in the matter.[12]

All of these features—the ubiquity of parasites, their particularity and frequency dependence, and the consequence for hosts in abundant protein variation—Haldane saw clearly, and they are indeed important initial ingredients in the theory I now support. However, equally striking in Haldane's paper are the topics he does not discuss. There is no single mention of sex, or recombination, or linkage. There are many mentions of mutation and it becomes apparent that what he had mainly in mind for variability was a polymorphism of host resistance somehow statically balanced against virulence variants that are present within one or many species of parasite. Both sets are from time to time increased in their variety by mutation. There is just one mention of heterosis in diploids (the sickling and malaria story) but otherwise his hosts and parasites might be all, from his discussion, not only asexual but haploid. Thus, as regards why sex exists, the paper provides only very raw ingredients for an argument: Haldane had nothing to say about how disease interacts with multilocus genetics or how a great inefficiency of possessing sex might be offset.

FLUCTUATING SELECTION

Searching forward in Haldane's work for later possibly relevant contributions, the next I find is in 1963 and it is here that Suresh Jayakar first enters the story as a coauthor. 'Haldane and Jayakar (1963)' is, in fact, probably the best-known joint paper of their collaboration.[13] Their subject is temporally varying selection and this is another crucially important ingredient. Once again, however, the topic is presented quite independently: nothing connects it either with Haldane's earlier disease paper (even though disease has a passing mention), or with multilocus genetics, or with sex. During the writing of it the two were in India and the paper came out in Haldane's *Journal of Genetics*. The main idea was the importance of looking at geometric means of fitness over time to determine the permanence of a dynamical polymorphism.

As it turns out Haldane and Jayakar were not quite the first to investigate this idea although that the earlier more limited results should have escaped their notice (as it did mine and seemingly most other writers') is not surprising. The forerunner was Dempster[14] contributing at the Cold Spring Harbor Symposium of 1954. His paper discusses theoretical forces that maintain polymorphism. Dempster starts with Levene's then recent model for stable polymorphism in a population spread over two niches,[15] a part of his purpose being to point out that Levene's model only works under what now might be called 'soft' selection, and not under 'hard'. This itself is an interesting hint of an ingredient of my model that I will return to later. The second and main part of his purpose seems to have been to emphasize the potentially diversifying effect of frequency dependence and, in course of discussing this, to demonstrate what was probably a quite new idea for the time: that for a single locus with diploidy, randomly fluctuating selection also can establish permanent polymorphism. He showed that this may happen when there is neither explicit frequency dependence of selection nor any necessary mean-fitness overdominance arithmetically either within or across generations. He discusses the importance of geometric means and refers to 'a kind of cumulative overdominance'. Thus he showed awareness of the principle that Haldane and Jayakar were later to annunciate more clearly. In other respects he may even have seen farther than they did—again only tentatively but certainly way ahead of his time. Thus he stated that 'Much more complex interaction systems, involving epistasis in addition to dominance and variable selection, are likely to be common in real populations and might be more effective [than single-locus effects] in maintaining alleles at intermediate frequencies'. Later, likewise, he remarked that 'selection pressures varying in space or time could act to maintain genetic variance of fitness with a substantial additive component'. All this foreshadows themes whose reality can now be confirmed.

Haldane did not attend the Cold Spring Harbor symposium where Dempster gave his paper. Many of the ideas about maintaining variation that Dempster had raised were mentioned again in the introduction of the joint paper with Jayakar. As the main matter of the paper the two gave various demonstrations of one-locus varying selection and then synthesized an important verbal conclusion that polymorphisms can persist if the geometric mean fitness of each homozygote relative to the heterozygote is less than unity—in other words, if the heterozygote is *geometrically*, over time, the most fit. They add: 'Once again this may occur, even if the mean fitness of homozygotes relative to heterozygotes exceeds unity provided the relative fitness of homozygotes occasionally falls very low.' The last remark gives an intuitive lead because a low fitness has, of course, far more effect on a geometric mean than it has on an arithmetic mean. Low points in fitness of a homo- or hemizygote can drop a gene frequency from abundance into the midrange in a single step, and in the hemizygous case even to zero: however, with diploidy and with the assumed relative stability of the heterozygote fitness, such events have little effect once the gene is rare and cannot send it to extinction. We see at once in all of this a possibility of a kind of *one-locus* protection of sex against parthenogenesis foreshadowing certain later models. However, for me at least, such one-locus models still remain completely[16] (see also Chapter 7) or else largely[17] unrealistic unless other recombining loci are an additional factor. More importantly, I think, for the important case of non-absolute inviability, Haldane and Jayakar's remark foreshadows Treisman's point about clone extinction under fluctuating selection.[18] All of this is reinforced later in their discussion where they say: 'one result of Mendelian inheritance is that polymorphism may be permanent, if not exactly stable, under fluctuating selection when it is not so for a difference inherited clonally, cytoplasmically, or in a haploid'.

In spite of such insights Haldane and Jayakar's paper fails to connect matters that might, from the present perspective, have well been connected. It has nothing, for example, on events at more than one locus and little discussion of possible sources of the fluctuating selection that is assumed. The only source dealt with in any detail is selection by alternating seasons on a bivoltine insect. However, Haldane and Jayakar did not quite omit Haldane's 1949 topic: one sentence states how occasional severe epidemics could keep the gene for haemoglobin S in a population in spite of its constant disadvantage in the absence of such disease. The connecting theme is present but it seems to me a handicap for them that they were thinking of that example: the exceptional strong overdominance in fitness and consequent stability in the sickling polymorphism may have often been an unfortunate lead distracting from far more common patterns of disease resistance (Chapter 5).

JAYAKAR AND CYCLES

So far, referring back to my title, I have shown relevant 'memes' of a certain HH homozygote (Haldane 1949)[11] and of an HJ heterozygote (Haldane and Jayakar 1963):[13] it remains therefore to describe contributions of a JJ homozygote. These are in Jayakar (1970).[19]

Evolutionists frequently point out that macroevolution is opportunistic. It appears so on the present 'memetic' evolutionary road of discovery also: the microevolutionary steps described so far give little clue of what the next bend is about to reveal. However, even Jayakar's title of the next paper, 'A mathematical model for interaction of gene frequencies in a parasite and its host' taken with his previous work and that of Haldane suggests converging paths. And his argument for the first time gives us explicit coevolutionary cycling at the genetic level. Equally important, such cycling arises out of the simplest imaginable form of interaction: single-locus costly virulence in a parasite opposed by single-locus costly resistance in a host. Looking ahead to our own recent model on this theme (see Chapter 16) for a moment, while complex multilocus resistance applied to each parasite or even to a variety of parasites may fuel the model best, an encouraging feature has been that numerous one-locus resistances have also proved to be perfectly adequate. It is a one-locus resistance and one-locus virulence, as interacting in isolation, that Jayakar models in this paper following his customary aim for analytical completeness. While it seems obvious (at least to me) that interaction at more than one locus is going to have to come in somewhere before a theory of sex can really work, it was certainly not at all clear until recently (at least wasn't to me) that the required interaction might just as well emerge from external assumptions of ecology and sociality as out of internal complexity of the genetics and biochemistry of the infection—that is, that there are other ways of getting epistasis than out of the codes, tactics, and armaments involved in the host–parasite battle. But these remarks are jumping half as far ahead of Jayakar's important paper of 1970, as now, to connect with his ancient possible antecedents, I first have to jump back.

An audience in Italy is especially qualified to view Jayakar's paper critically, and to question if there is anything novel in writing about a theoretical host–parasite cycling in 1970. Most people who know little else about the work of Vito Volterra at least associate his name with cycles or, phrasing it more biologically, with 'predator–prey cycling'. His work and that of his near contemporary Alfred Lotka, who covered similar themes, long pre-dates Jayakar. The two had important papers in the 1920s and even earlier. Even if both connected their cycling more (and perhaps less appropriately) with predators than with parasites, there is little conceptual difference if the interaction in both cases is assumed one-to-one; moreover, certainly neither of these pioneers failed to mention parasites as an alternative. Several other of the so-called 'ingredients' of the present essay, moreover, have also been waiting to be picked up either from works of these two or from those of their close theoretical followers such as Kolmogorov and Kostitzin. However, offsetting all the undoubted claims to priority, which have been usefully made visible in the republication by Scudo and Ziegler,[20] it has to be said that as regards starting a theoretical ecology or theoretical evolution that would catch the attention of biologists, all the pioneer theorists mentioned suffered from a certain dry-as-dust and almost example-free presentation as well as from a degree of detail and rigour in their mathematics which biological assumptions would hardly justify and which most readers still, I think, find stupefying. Moreover, they made almost no effort to link their work with other developments in theoretical biology such as those going on with Wright, Fisher, and Haldane.

My complaint of dryness implies that unlike the Westerners, the more Eastern three, Lotka, Volterra, and Kostitzin (I now omit Kolmogorov because he seems to have had only one paper in the line in question, although one of characteristic elegance), in spite of their reasonable claim that they were reviving Darwin's own emphasis on demography and ecology, made little effort to test their ideas against data, or even to suggest what data would be needed to confirm any outcome of their models against any another. Wright, Fisher, and Haldane, on the contrary, constantly related their models to facts, and sometimes worked closely (even to the extent of camping in tents on islands) with field ecological geneticists, a branch of study they may be said to have directly inspired. Perhaps what isolated the other three still more, however, was their very different view of the process of natural selection. True to a particular view of competition and coevolution, they saw this being carried on at the level of competing species. We understand now that this is a very weak form (e.g. ref. 21), and as much indeed had been commented with typical terseness by Fisher long ago.[22] Species selection is certainly not non-existent but a little thought shows that it is a bad track for generating short-term testable predictions for field or lab—it fares better perhaps across the layers of the cliffs of palaeontology. Intimidating to biologists by their extreme formalism and failing to appeal to them by predictions or data, work of the pioneer coevolutionists tended to submerge. Altogether there seems little reason to be surprised that Jayakar failed to cite, for example, Kostitzin, the one of the three who had both shown cycles and toyed independently with evolutionary genetical theory. Kostitzin certainly did not bring his two interests together: even while he theorized natural selection he appeared to remain very nervous of its intrinsic individualism. Jayakar did bring the themes together.

As with Haldane and Jayakar,[13] however, Jayakar also had a recent forerunner, and this time he does not miss him. C. J. Mode[23] may have gained his 1958 citation more by being close in time than by being different in spirit from the very early forerunners (whom Mode, like Jayakar, further to illustrate the older submergence, himself did not cite). Mode perhaps also gained from attaching his discussion to a real-life example in the problem of the gene-for-gene pathosystem of host plant and fungus that had by then been repeatedly found in agricultural phytopathology and plant breeding. Apart from this, and perhaps because again a mathematician by training, his work still had much of the same extreme formalism and biological inconsequentiality as the early forerunners. Of the two relevant papers by him that I know, Mode (1958), the first and the one referred to by Jayakar, outlines the genetics of the known flax–rust interaction and then greatly simplifies it in order to effect an analysis. Even with this simplification the work looks formidable. He concludes that the system readily fulfils conditions for a stable interior equilibrium. He comments that, at least in a vague species-benefiting way, which he neither justifies nor criticizes, this is a satisfying outcome. He doesn't say how widely in the reasonable range of the parameters such stability occurs; moreover, his parameterization of selection coefficients has several features that I find unrealistic: for example, certain combinations of host and parasite genotypes one would expect to manifest less-effective resistance or virulence manifest more. Mode's next paper (1960)[24] also does little to allay my doubts about the stability found in the first paper being realistic. He discusses a more general host–parasite interaction, this time without any genetics and finds that under what he presumably considers reasonable assumptions the system will never have a stable interior equilibrium. The lack of diploidy and outbreeding in either host or parasite may well underlie some of the reduction of stability, but the global instability found (except for at fixations) hints that reintroducing these factors is unlikely to conjure back more than a modicum.

Curiously (or should I say understandably, in view of opposite conclusions?), Mode's second paper does not refer to his first even though it was published only 2 years earlier. Altogether he seems as if embarrassed by his new finding and tries to make the best of a possibility, not sustained by much argument, that departure from equilibrium is probably slow, and therefore might be tracked and prevented by an agronomist (who would here be serving, I suppose, in something like the role of the autopilot of a modern unstable aircraft). Unluckily for me he doesn't indicate when departures from his interior equilibrium are likely to be fluctuating as well as slow nor whether limit cycles are possible. I could find this out, of course, by closer attention to his equations, but I am slow at such work and have not had the time to try even though I ought to because the differences between a stable and an unstable equilibrium, and, in the latter case, between a fluctuating and a direct departure, are important to me: where others seem to want stability I always hope to find, for the benefit of my idea of sex, as much change and motion within reasonable non-fixating limits as I can get. A final minor point about Mode's papers is that both his analyses are by differential equations: with discrete timing such as that imposed by the annual cycle the tendency to instability would be stronger.

Altogether I think that Jayakar[19] did little injustice to Mode by missing his 1960 paper in a symposium book, or by citing only his 1958 paper with just the brief comment that he had covered 'a very particular and complex situation occurring in flax and flax rust, and [had found] a stable polymorphism for the system', even though it did slightly disappoint me that Jayakar gave no critical discussion of Mode's conclusion, which differed so markedly in emphasis from his own. It appears as if every one around that time, Jayakar included, expected and wanted stability: oscillatory departures were seen as perhaps intriguing and ornamental but unlikely to be more relevant to biology than outright monotonic departures were, even though the latter had to be more automatically disastrous or at least simplifying to the systems where they occurred.

Instability phobia continues in papers that follow Jayakar's. Bypassing many, we find Lewis[25] in 1981, for example, clear about the limitation and fragility of stabilities in a more general genetical coevolution model and emphatic on the sheer non-existence of stability in his Mode-like remodelling of another reasonable prototype of a gene-for-gene agricultural pathosystem, nevertheless remarking, like Mode, that it would actually be beneficial for both host and parasite if they could attain the potential equilibrium because this would maximize average populations of both interactants. Similarly, 2 years later Levin,[4] in an important review of efforts to model host–parasite coevolution and while discussing and often criticizing all the unlikely, if expert, balancing acts of his predecessors, remains himself inscrutable about what he thinks is likely in nature. His discussion certainly does not make stability seem easy. He clarifies, for example, that 'mass action' diploid models are usually going to need marginal overdominance to attain it and points out that even fulfilment of this condition may still be far from enough. His paper and several cited in it all refer back to May's now classic account of basic principles of community dynamics;[26] yet all of these papers seem to me to dwell surprisingly little on concepts by which May demonstrates the source of everyone's difficulty about stability of hosts with parasites or indeed inter-species relations generally. In essence, May showed that species (or varieties) that do not look after their own stability through self-directed negative density dependence are unlikely to find that stability taken care of for them through interactions with the rest of their community. More technically this is equivalent to saying that the sum of diagonal elements, or 'trace', of the community matrix determines a kind of total instability tendency associated with a community. Individually the diagonal elements are just the self density dependences of the 'species' of the community. Using the mathematical fact that the trace is also the sum of the eigenvalues, we see that stability involves keeping all of the eigenvalues small. The trace can, of course, be allocated to eigenvalues in a great variety of ways, but to keep stability constraints quickly appear.

1. If the trace $\geqslant 0$, no pattern of suballocation can prevent at least one eigenvalue having positive real part, implying instability. Paying attention to off-diagonal elements as well:

2. If a majority of off-diagonal elements become large when of *like sign* (mutualism and/or competition), it becomes increasingly unlikely (impossible?) that diverging conjugate pairs fail to have one member cross the imaginary axis, so causing instability. Departure here tends to be direct and consequently drastic.

3. If a majority of off-diagonal elements become large when of *opposite sign* (the 'pursuit' cases—host/parasite and predator/prey), the final divergence of the conjugate pairs of eigenvalues tends to be in the imaginary direction.

Such dispersal of roots implies oscillatory components in the motion of departure or approach near to the fixed point rather than a forcing instability *per se*. However, when linkages in the community are made other than simple and/or when values of coefficients are increased, the roots of the characteristic equation spread out like sparks of a firework with a consequent high probability that a pair eventually crosses the imaginary axis (or in the discrete analysis case and now with inevitability, passes out of the unit circle). Again instability supervenes, the difference from (2) being in an *oscillatory departure*.

In spite of laying out the basis of such a demonstration May himself, in his early biological writings, did not emphasize permanent demographic turmoil being the almost inevitable outcome of certain types of species relations and even at times appeared to join those seeking the nirvana of universal stability. With theorists so unenthusiastic for anything except stability it is not surprising that the slow growth of an acceptance of turmoil as almost inevitable in host–parasite genetical coevolution seems to have sprung up from grass-root and quite distant sources. 'Grass-roots' are indeed almost literal in the case of Levin,[27] who based his ideas on his experience of communities of low vegetation. Shortly following Levin, papers of the same slant by Clarke[28] and Eshel,[29] though progressively more theoretical, also do not draw inspiration from the community matrix approach or from questions about stability of ecosystems. Still more widely in parallel to this, and in coevolutionary contexts allied more to game theory than to coevolution, Selten[30] shortly proved certain related inevitable instabilities. Finally, independently yet again and overlapping three disciplines—game theory, ESS (evolutionarily stable strategy) theory, and coevolutionary dynamics—Eshel and Akin[31] gave similar proofs to Selten's. In their paper, the importance for biology of the particular combination they were discussing is at last made clear: in effect, they prove from reasonable assumptions the eternal motion that has to result when central instability combines with extinction protection near to boundaries and they embrace the result as a robust, expectable characteristic of nature, unlikely to rationalize away. The nettle is grasped. Almost simultaneously with this came May's demonstration of eternal chaotic motion in a simplest imaginable host–parasite model.[32]

Probably being published in the first volume of a new journal does not help a paper to be noticed. Jayakar's 1970 paper was in the first volume of *Theoretical Population Biology*. What the paper's influence was is hard to know; however, as to showing the likelihood of genetical coevolutionary cycling, it certainly should have opened a door. Jayakar gave diagrams to illustrate the cyclical departures that arose as soon as he brought costs of virulence or of resistance into his model. He opposed costly resistance treated on the host side as alternative susceptibility, to unconditional costly virulence made alternative on the parasite side to partial virulence (the latter over-coming only susceptible hosts). As I have already stated, such a pattern seems both simple and realistic, and there are a multitude of suggestive examples in the literature, some of which are well documented (see Chapters 5 and 16). Searching the same resistance literature, however, had forced me to doubt the generality of another pattern that I would have liked to have believed common because it too seemed a reasonable expectation while appearing even more certain to give coevolutionary cycling. This other pattern, which I called 'complementary resistance' (Chapter 5), is of the following form: a parasite of type a_1 successfully attacks a host variety A_1 but not A_2, while parasite a_2 attacks A_2 but not A_1. Rather surprisingly this seems much less common than the alternative simple pattern based on a cost modelled by Jayakar (a_v attacks both A_+ and A_r; a_+ attacks only A_+; a_+ and A_+ grow faster than a_v and A_r). Thus it has been a particular pleasure and reassurance to find from Jayakar's paper, in the re-reading necessitated for this essay, that the system I had thought less promising does also cycle; moreover, although Jayakar himself never produces a limit cycle, it is easy to imagine added reasonable considerations of frequency dependence, mutation, or (less obvious but much more exciting) further loci that can create this. At the same time, his formal arguments reveal how dauntingly complex proper analyses of coevolutionary models are when diploidy, multiple alleles, and other mere one-locus assumptions are brought in, to say nothing of the more realistic horrors of variation at multiple loci (interestingly, however, as a first indica-tor of what may be expected, the old genetic-load arguments showing how soft selection can allow more variation than hard might have been invoked as a first step towards showing protection of general multilocus polymorphism (see refs 33 and 34)). As already pointed out, multiple loci almost have to enter when sex is in question, so it was exactly a feeling of helplessness in the face of such complexity that already spurred our recourse to simulation.

One spiral put in a diagram is more likely to impress me than a dozen complex eigenvalues mentioned in text. It is therefore likely that Jayakar's diagrams must have caught my attention when I first saw them. I have tried to remember how much the diagrams or conversations with Jayakar may have hinted the possibility of permanent cycling of hosts and parasites, or hinted at that more complex and permanent unrest of many loci that I now believe to be the key of sex. I certainly did not remember that he had such diagrams or remember anything much about his conclusions until I began work for my talk at this memorial symposium and reread his paper. I possess a reprint of the work, however, and almost for sure Suresh must have given it to me when I visited Pavia as an invited speaker in 1973. On that occasion I would certainly have mentioned my admiration for his fluctuation paper with Haldane. I recall Suresh talking about extensions of some work with Haldane, and this, given the time, must have meant his host–parasite modelling. But his ideas cannot have impressed me consciously; perhaps they did so in the same way as other relevant conversations I was hearing from parasitologists at Imperial College Field Station, which means, for the time being, almost not at all. At lunch time in the Silwood Park refectory Professor Garnham would talk of the malarias of chamaeleons and South American monkeys, of dying birds he or others had collected and had found to be infected with blood parasites, and of other intriguing but, for me, then extremely unmainstream matters of parasite evolution and ecology. I can still almost feel my attitude of those days that the evolution of parasites must be so easy once they are in the rut that it might hardly be considered true evolution at all. I was thinking, of course, of morphology and behaviour, not of biochemical attachment and the like, or in other words I was thinking of evolution on a macro- rather than a microscale.

In any case, without doubt conversations both in Pavia and at Silwood must have left a deeper record than I was aware of at the time; fermenting with others they must have helped create the parasite–coevolution freak that I now am. But this is of little importance: quite independent of how I reacted, Jayakar's paper seems to have its priority in population genetics theory. I have so far found no evidence against his hesitant claim in the paper to have made 'the first truly genetic model where cycling behaviour has been shown to be possible'. Mode had come near, but the model that he geneticized at least in cases that he chose to discuss was stable. Jayakar not only recognized but thought worthwhile to illustrate his model's cyclicity, and, as I have emphasized, I believe his conditions for his model were realistic. His results therefore provide another important element for the theory of sex I now support. In fact, after his paper, the elements I see still lacking to make a workable hypothesis are merely two. First and most important we still need the many loci in simultaneous coevolution already alluded to. These loci must work through different modes of resistance, one or more loci applied to each parasite, and their interaction must be not such as either to dampen or synchronize the cycling. Second and slightly less essential, it will help to have an additional environmental and/or social factor present: 'soft' or 'truncation' selection (refs 33–36 and Chapter 16). As discussed below such a selection pattern serves as an aid both to instability and to the long-term conservation of variation.

SEETHING AND MUTATION

The effect of *multiple coevolving loci* in a host has been almost totally ignored by all the workers on disease and temporally varying election I have mentioned. The reason is perhaps that as trained theorists all followed a set code of practice. They found already enough difficulty and interest with the mathematics of the limited cases they treated. It was a pattern of careful advance and of worthwhile, proven results, the stuff of applied mathematics. To add loci and parasites multiply and, it must be said, brainlessly (i.e. by simulation) and so to bring all the previous work to bear on recombination and the cost of sex, has been left for me. It has, of course, led me only to the typical position a simulator has, where as such I cannot really explain exactly how my model works or how far any of its results can be extrapolated. However, a simulator can test all the reasonable conditions quite thoroughly. Given the final ingredient about to be mentioned, first added to the present theme in my work with Robert Axelrod and Reiko Tanese, the results of the model now appear to be extremely robust. Interpreting our results by means of a physical image that applies genetical detail to an insight of Treisman,[18] I now see inevitability to the success of sex given a set of reasonable and natural conditions with as much certainty as if I had proved it.

This brings me to my last ingredient. The *soft and/or truncation selection* already mentioned has been brought into the model also by simulation. But with this addition it was easier to gain a little preliminary understanding of its probable usefulness from literature on its power[37,38] and potential protective function[33,34] and from previous simpler unpublished simulations of my own that had confirmed the permanence of polymorphism under 'coal pit' hard selection, which mode at least imitates some properties of low-level truncation (see Chapters 1, 2, and 12). The last ingredient was thus partly 'off the shelf' and partly made for the present purpose. The making side had, in fact, begun more than 10 years ago with a model first programmed by P. A. Henderson (Chapter 1).

The strength of the synergism of truncation with multilocus cycling, with its creation of high instability near to an interior equilibrium and then protection of variation once far away from it, all in spite of an incessant and irregular motion, came as a surprise. Looking back I doubt whether the theorists who had noticed the maximal power of truncation selection in effecting various simultaneous changes in population genotype guessed a possible importance for the very fact of Mendelism. If any theorists did foresee such possibility, the most likely would seem to be Dempster or more recently A. S. Kondrashov. Kondrashov's rationale for the support of sex so nearly approximates in certain soft-selection-like aspects to my own that I have wondered whether he may not have entertained exactly the same hopes of sex out of pathogens. If so, perhaps he had already dismissed the idea on account of the apparent lack of cyclicity in nature. This is a worrying thought. Such a lack remains the main weakness of the theory supported here, against which for the present I am only able to argue (1) that electrophoretic surveys of variation have seldom been repeated over requisite spans of several generations, so giving little chance of looking for the changes predicted, and (2) that because our work shows that the final amplitudes of changes of both gene frequencies and linkage disequilibria diminish as the coevolving loci are made more numerous, the expected changes may be intrinsically rather faint and indefinite in nature (although, if so, they must be numerous), and thus may easily pass as mere stochastic variation unless specially tested.

Kondrashov's alternative to parasites is based on sex's power to clear harmful mutations efficiently.[39] The striking convergences of his theory and ours arise specifically through his assumption of superadditive harm for individuals carrying multiple mutations. Although he has hard selection where we have soft, this is probably mainly because hard allows him more nearly analytical results: the forms of interaction he invokes certainly approximate to the forms that would arise out of softness and truncation and, moreover, he has expressed his awareness of the still greater power to support sex that truncation could have if it was present.[40] In my view his theory undoubtedly covers an important function that sex has for its possessors but does not provide a likely reason why sex exists: the question is whether mutation clearance has power to originate sex, or is merely auxiliary. In my opinion the latter is more likely. Our arguments on this, based partly on ecological evidence and partly on genetics, are set out more fully elsewhere (see Chapter 16). However, the biggest cause for doubt is simple: if mutation was the only trouble, surely there would to be more efficient ways to handle it. Surely more careful management and correction procedures could be used for the genome without reaching outside of a germ cell line, leave alone reaching far across habitats and populations in the way Darwin and his contemporaries had shown to be almost an obsession of wild nature. There are indications that when fidelity of inheritance is made crucial it can be achieved internally (ref. 41 and Chapter 16) and so could be achieved also in a parthenogen. Sex, in other words, is a cumbersome strange tool to have evolved for a housekeeping role.

The important difference in what soft multilocus selection achieves in our model compared with what it would achieve in the equivalent 'soft' version of Kondrashov's is that in ours it conserves, in his it merely eliminates. The full fitness of every non-truncated genotype combined with the dispersion of alleles by sex into an extremely numerous and diverse set of genotypic combinations ensures that always some instances of rare and only currently 'bad' alleles are being preserved safely. In Kondrashov's theory, this conservation effect is a tolerated detriment, one that is only less bad when a species is sexual rather than asexual. In ours, once their turn comes to be advantageous selection reverses the slow terminal decline of 'bad' alleles and recombination spreads them into new unique combinations—new homes for present advantage and future survival. Thus, like good ideas, and like my own feelings about my generous and modest host of Pavia of 1973, about whom it has been so difficult to write in the past tense, such alleles have become almost immortal,[42–46] and, beyond mere persistence, they can grow in complexity as their modifiers accumulate.

References

1. G. Bell, *The Masterpiece of Nature: The Evolution and Genetics of Sexuality* (University of California Press, Berkeley, 1982).

2. S. C. Stearns (ed.), *The Evolution of Sex and its Consequences* (Birkhauser, Basel, 1987).

3. R. E. Michod and B. Levin (ed.), *The Evolution of Sex* (Sinauer, Sunderland, MA, 1988).

4. S. Levin, Some approaches to the modelling of coevolutionary interactions, in M. Nitecki (ed.), *Coevolution*, pp. 21–65 (Chicago University Press, Chicago, 1983).

5. W. D. Hamilton, Gamblers since life began: barnacles aphids elms [a review], *Quarterly Review of Biology* **50**, 175–80 (1975) [reprinted in *Narrow Roads of Gene Land*, Vol. 1, pp. 357–67].

6. D. Hillis, Co-evolving parasites improve simulated evolution as an optimisation procedure, in C. G. Langton, C. Taylor, J. D. Farmer, and S. Rasmussen (ed.), *Artificial Life II*, pp. 313–24 (Addison-Wesley, Reading, MA, 1991).

7. I. Walker, The evolution of sexual reproduction as a repair mechanism. Part I. A model for self-repair and its biological implications, *Acta Biotheoretica* **27**, 133–58 (1978).

8. H. Bernstein, F. A. Hopf, and R. Michod, Is meiotic recombination an adaptation for repairing DNA, producing genetic variation, or both?, in R. E. Michod and B. Levin (ed.), *The Evolution of Sex*, pp. 139–60 (Sinauer, Sunderland, MA, 1988).

9. G. C. Williams, *Sex in Evolution* (Princeton University Press, Princeton, NJ, 1975).

10. J. Maynard Smith, *The Evolution of Sex* (Cambridge University Press, Cambridge, 1978).

11. J. B. S. Haldane, Disease and evolution, in *Symposium sui Fattori Ecologici e Genetici della Speciazione negli Animali, Supplemento a La Ricerca Scientifica*, Anno 19°, pp. 68–75 (1949).

12. R. C. Lewontin and J. L. Hubby, A molecular approach to the study of genic heterozygosity in natural populations. II. Amount of variation and degree of heterozygosity in natural populations of *Drosophila pseudoobscura*, *Genetics* **54**, 595–609 (1966).

13. J. B. S. Haldane and S. D. Jayakar, Polymorphism due to selection of varying direction, *Journal of Genetics* **58**, 318–23 (1963).

14. E. R. Dempster, Maintenance of genetic heterogeneity, *Cold Spring Harbor Symposia on Quantitative Biology* **20**, 25–32 (1955).

15. H. Levene, Genetic equilibrium when more than one ecological niche is available, *American Naturalist* **87**, 331–3 (1953).

16. M. Kirkpatrick and C. D. Jenkins, Genetic segregation and the maintenance of sexual reproduction, *Nature* **339**, 300–1 (1989).

17. D. Weinshall, Why is a two-environment system not rich enough to explain the evolution of sex?, *American Naturalist* **128**, 736–50 (1986).

18. M. Treisman, The evolution of sexual reproduction: a model which assumes individual selection, *Journal of Theoretical Biology* **60**, 421–31 (1976).

19. S. Jayakar, A mathematical model for interaction of gene frequencies in a parasite and its host, *Theoretical Population Biology* **1**, 140–64 (1970).

20. F. M. Scudo and J. R. Ziegler, *The Golden Age of Theoretical Ecology*, Lecture Notes in Biomathematics 22 (Springer, Berlin, 1978).

21. W. D. Hamilton, Innate social aptitudes of man: an approach from evolutionary genetics, in R. Fox (ed.), *ASA Studies 4: Biosocial Anthropology*, pp. 133–53 (Malaby Press, London, 1975) [reprinted in *Narrow Roads of Gene Land*, Vol. 1, pp. 329–51

22. R. A. Fisher, *The Genetical Theory of Natural Selection*, 2nd edn, p. 50 (Dover, New York, 1958).

23. C. J. Mode, A mathematical model for the coevolution of obligate parasites and their hosts, *Evolution* **12**, 158–65 (1958).

24. C. J. Mode, A model of a host-pathogen system with particular reference to the rusts of cereals, in O. Kempthorne (ed.), *Biometrical Genetics*, pp. 84–94 (Pergamon Press, Oxford, 1960).

25. J. W. Lewis, On the coevolution of pathogen and host (Parts I and II), *Journal of Theoretical Biology* **93**, 927–85 (1981).

26. R. M. May, *Stability and Complexity in Model Ecosystems* (Princeton University Press, Princeton, NJ, 1973).

27. D. A. Levin, Pest pressure and recombination systems in plants, *American Naturalist* **109**, 437–51 (1975).

28. B. Clarke, The ecological genetics of host–parasite relationships, in A. E. R. Taylor and R. Muller (ed.), *The Ecological Genetics of Host–Parasite Relationships*, pp. 87–103 (Blackwell Scientific, Oxford, 1976).

29. I. Eshel, On the founder effect and the evolution of altruistic traits: an ecogenetical approach, *Theoretical Population Biology* **11**, 410–24 (1977).

30. R. Selten, A note on evolutionarily stable strategies in asymmetric animal conflicts, *Journal of Theoretical Biology* **84**, 93–101 (1980).

31. I. Eshel and E. Akin, Evolutionary instability of mixed Nash solutions, *Journal of Mathematical Biology* **18**, 123–33 (1983).

32. R. M. May, Regulation of populations with non-overlapping generations by microparasites: a purely chaotic system, *American Naturalist* **125**, 573–84 (1983).

33. C. Wills, Rank order selection is capable of maintaining all genetic polymorphisms, *Genetics* **89**, 403–17 (1978).

34. S. Karlin and R. B. Campbell, The existence of a protected polymorphism under conditions of soft as opposed to hard selection in a multi-deme population system, *American Naturalist* **117**, 262–75 (1981).

35. B. Wallace, Hard and soft selection revisited, *Evolution* **29**, 465–73 (1975).

36. A. Lomnicki, *Population Ecology of Individuals* (Princeton University Press, Princeton, NJ, 1988).

37. J. A. Sved, Possible rates of gene substitution in evolution, *American Naturalist* **102**, 283 (1968).

38. M. Kimura and J. Crow, Efficiency of truncation selection, *Proceedings of the National Academy of Sciences, USA* **76**, 396–9 (1979).

39. A. S. Kondrashov, Deleterious mutations and the evolution of sexual reproduction, *Nature* **336**, 435–40 (1988).

40. A. S. Kondrashov, Selection against harmful mutations in large sexual and asexual populations, *Genetical Research* **40**, 325–32 (1982).

41. H. Nothel, Adaptation of *Drosophila melanogaster* populations to high mutation pressure: evolutionary adjustment of mutation rates, *Proceedings of the National Academy of Sciences, USA* **84**, 1045–9 (1987).

42. P. A. Cazenave, A. Benammar, J. A. Sogn, and T. J. Kindt, Immunoglobulin genes in the feral rabbit, in S. Dubiski (ed.), *The Rabbit in Contemporary Immunological Research*, pp. 148–63 (Longman, Harlow, Essex, 1987).

43. W. van der Loo, Studies on the adaptive significance of the immunoglobulin alleles (Ig allotypes) in the wild rabbit, in S. Dubiski (ed.), *The Rabbit in Contemporary Immunological Research*, pp. 164–90 (Longman, Harlow, Essex, 1987).

44. F. Figueroa, E. Günther, and J. Klein, J., MHC polymorphism pre-dating speciation, *Nature* **335**, 265–7 (1988).

45. D. Lawlor, A. F. E. Ward, P. D. Ennis, A. P. Jackson, and P. Parham, HLA-A and B polymorphisms predate the divergence of humans and chimpanzees, *Nature* **335**, 268–71 (1988).

46. T. Sagai, M. Sakaizumi, N. Miyashita, F. Bonhomme, M. L. Petras, J. T. Nielsen, T. Shiroishi, and K. Moriwaki, New evidence for trans-species evolution in the H-2 class I polymorphism, *Immunogenetics* **30**, 89–98 (1989).

THE THREE QUEENS

Sexual Reproduction as an Adaptation To Resist Parasites (A Review)

Between red death and radiant desire
With not one sound of triumph or of warning,
Stands the great sentry on the Bridge of Fire.
                                        J. E. FLECKER[1]

THE protracted states of submission and revision undergone by the paper of this chapter have been mentioned several times already. I told how the paper eventually was accepted and that, in spite of what seemed the contrary opinions of so many, I rank it as the most important of my papers in this volume. After this, what remains? Still a lot. More needs to be said of the crucial roles of my coauthors and more also of the mistier roles of the various unknown opponents; then there is the load that slowly piled on my mind about why parasite coevolution in general should be so underestimated and the sex idea based on them so resisted; also how I think the notions of this paper and the opposition they generated have all fared since it was published. Lastly I would like to explain, though most of this will come in the next chapter, how the biological fuss about sex has affected me—the way the light shed by my solution has changed my view of life and of our human future.

One of the puzzles about the dislike, even contempt, the work of this chapter seemed to arouse in my evolutionary peers is that it was as if we had been unable to explain what we were thinking. I could understand a failure of communication for some of my papers, those more purely my own. There were, for example, all those 'oats' that a patient reader may have sighed through in both this volume and the last. In contrast I believe that the work in the present paper is unusually well explained. In so far as it is true, it is due to Bob Axelrod who has always been a restraining influence on my flights of both prose and fancy. He is insistent that we should say just what would be helpful to our reader on our theme and no more. And yet while one referee praised our style, another described the paper as written very badly; because neither said anything good about the ideas or content I presume that even the one that liked the writing found it a kind of eloquent twittering.

From my personal viewpoint, years lived under a cascade of comments from people who seemed not to see at all the major themes of sex and sexual selection that I was addressing were difficult and frustrating for me. I had expected some discussion of the implications if our methods and claims were valid, also about how others imagined, if we weren't right, the manifest continual advances of parasites (so increasingly evident with every passing year for the human case with AIDS and, on the scales of decades, also through the antibiotic and pesticide resistances of many kinds that were sweeping the world) were being handled by species—if not in the way we were suggesting, then how? It was clear enough that the 'facultative' or 'acquired' side of the immune system was not doing everything because even a few years into the AIDS epidemic, for example, there were signs of a huge variability in the swiftness of progress to the full disease and, recently, as I had always suspected (as it is so with every other disease we know) completely resistant (or at least non-progressing) types were proven. Why could people not see the unexplained variability with respect to resistance to every disease that I was seeing? Had I become at last the crank I once feared and decided I was not, and why was my world once again seemingly so differently shaped from everyone else's? Worse, had I dragged an independent thinker from his own far discipline into a rut of my making—a person perhaps unluckily not enough versed in biology to realize the improbability of assumptions I was supporting? Lastly, had I hijacked also a superb programmer into, similarly, a waste of her time and perhaps even a disillusionment with academia and science? About the time we finished the main experiment of our paper, Reiko Tanese abandoned not only our project but academia altogether: I hope and believe this project wasn't the main cause.

Our model had achieved results that others had stated impossible with the tools we were allowing ourselves. Many of the dragons that had oppressed individual-advantage models in the past seemed to us to be slain. Part by part, perhaps, not much in our work was new although our explicit modelling of a large number of loci in a Red Queen situation certainly was and the increase of stability of sex that came with the growth of numbers of loci made the most dramatic feature in our results. Simulation in itself admittedly isn't understanding and various previous papers, including some of my own (Chapters 1 and 2 and see also 15 in this volume), had already drawn attention to the kinds of possibilities we were now testing. The simpler analytical discussions and models, however, including again my own, all had had severe snags and none showed any chance to be general. Besides treating many loci and many parasites at once—obviously much closer to the real situation (and the importance of our studying truly many loci, not just three or four, cannot be overstated)—we had brought in a variable life history that I consider to be much more realistic than is typical in most evolutionary modelling. A major artificiality remained in its lack of intrinsic senescence[2] but this seemed unimportant and no one has explicitly criticized it. The life history's combination of births and deaths depended on a supposed habitat saturation: this is not always the case, of course, but it seemed to me quite realistic for the ecological situations where sex is most universal.

The combination chosen implies a population balanced by its resources and thus constantly at an upper limit. Imagine birds in a wood with an absolute dependence on nestholes and the number of these remaining constant. The birds in such an example actually never do have alternative asexuals lurking and competing but certain sister groups, as among the lizards and notably in rather different open situations where their numbers can swell and diminish dramatically, sometimes are asexual. Excluding erratically exploding and contracting populations in both plants and animals—the realm of 'weedy' organisms if you like—we can relax a bit on the stated population constancy above and say that it is not just 'many' but 'most' populations that fulfil the conditions. And almost all of them are sexual. Via territoriality and similar traits, a great many animals and plants tend to create their own population restrictions and thus to stabilize their numbers.[3]

In short, apart from some questionable details in our assumed genetics of resistance, to which I return below,[4] we felt we had put together a very plausible model. When the model was run the need for recombination proved so continuous and strong that even a few generations of the parthenogenesis of a successful clone—a period, that is, of failing to change—brought the clone to a state where, by comparison to the offspring of sexuals, it had become desperately infectible by parasites and therefore unfit. Despite the reproductive efficiency of the clone—or, as it might be put, *because* of that efficiency when the inevitable fate of the seductively uniform product is to be adapted to and eaten by the much faster evolving parasites—it was rapidly beaten back. Neither were high barriers to a sudden onset of parthenogenesis needed for sex in our model to persist: we were all the time allowing the best sexual to mutate into an efficient parthenogenetic strain with an approximately one-in-thousand chance.[5] Therefore except in this unavoidable and minor sense of an endless succession of clones all being given their chance to invade, the model was successful without needing to use any group selection.

Attempts to enliven faltering human marriages by experiments with group sex don't usually save them: in our modelling it is the same if groups in the present sense are brought in to save a theory of sex. Using group sex makes sense only in special situations and, also the step has to be a desperate last resort. Probably, gentle reader, you haven't worried yet about parthenogenetic prostitutes looming to take over the world. But should you happen to have worried, even if only like me in a slightly humorous spirit, then I hope you will cease once you have read our paper that follows. Groups, other than the standard challenge clones that I had already admitted in order to check out our idea as above, have been shown to be un-needed by our HAMAX model. As for dreaded human clones—the gynogenetic prostitutes, the swarms of stem-cell 'single mothers', human versions of Dolly the sheep, etc., which either will already exist when this is published or are to come—I predict all these will become too unhealthy too quickly for them ever to matter much in world affairs (unless and until we go very far down the technological superorganism road, as in Volume 1, Chapter 6). And note that all of this prediction has nothing to do with sexual diseases, or only indirectly.

I write this in a light spirit but am not joking about what I see as the achievements of the model. I was sure at the time that they must be surprising even to those who were not already primed to feminine take-overs. Nor could anyone pretend that this theme of evolution of sex was a narrow one nor of specialist interest only: from Erasmus Darwin to the present time, sex has repeatedly been saluted as one of biology's supreme problems, perhaps its very greatest.[6] Hence Bob Axelrod and I at first believed that our model, with its realism and its dramatic success under conditions others had deemed impossible for it, was virtually sure to be acceptable to one of the major general scientific journals such as *Science* or *Nature*.

Given the basic twofold advantage of asex, it was inevitable that when an especially favoured sexual genotype happened to mutate to asex we would see in our model an initial explosive increase in the new mutant clone's frequency. Equally inevitably, however, the nemesis due to the uniform abundance would rapidly follow: as the clone increased, we would watch parasites 'find it' and beat it back. Intrinsic to this process, defence components possessed by the 'very fit sexual' genotype that the parthenogen had come from, would be moving into new alliances that the parasites weren't immediately able to match—they'd find them again eventually, of course, but only by mutation. Our model showed that we needed no special arrangement of populations on the ground for this to happen: sex was being proven stable in a single and thoroughly outbred population. Again, this was a version of the problem that most theorists had admitted especially difficult to conquer. Indeed ours is still, so far as I know, the only model that has 'solved sex' with anything like realism in the situation of a single and thoroughly mixed population. Again I had supposed my peers would have to be impressed by this evidence of a success that probably would become stronger still if we extended our setting to be the more usual and realistic archipelago of populations linked by migration (the 'metapopulation' as evolutionists call this kind of ensemble), or if we scattered our individuals spatially and gave them (we could call this a 'snails-in-the-meadow' version of the metapopulation concept) only very limited movement.[7]

Some theorists working on the sex problem, especially George Williams,[8] had shown good reasons why infecund species should have particular difficulty containing the spread of mutations causing efficient parthenogenesis. Our own species, *Homo sapiens*, is one that particularly prides itself on low fecundity, having made the route possible by its unusually careful nurturing of each offspring born. The idea that my kind of modelling would need such strong selection that it could never account for non-parthenogenesis in a species like our own—which in effect amounts to

a claim that the stem-cell feminists and/or parthenogenetic prostitutes that
I have mentioned above must always loom over us—was one I often heard.
I had decided early on, therefore, that we would arrange the life-history
pattern and the potential for a wide range of overall intensity of selection
first; then, later, and as far as possible separately, we would arrange, within
the wide specification, how the natural selection would be handling the
genes themselves, the host–parasite interactions, and the resulting fitnesses.
Then I would be in a position to change the life-history pattern in various
ways and then, for sure, because it seemed to be regarded as a particular
challenge, I would include taking a hard look at cases of truly low
fecundities. Indeed, there are some very sexual animals such as orang-utans
and still more some dung beetles that are much less fecund than we are.
Has it crossed your mind that you are more weed-like and care for your
offspring less well than does many a dung beetle living in the soil under a
cow-pat? Perhaps it's true, in modern times, that we are at last coming to
narrowly beat the dung beetles in terms of our effectiveness in infant care,
but it is still a narrow thing and has been achieved only at the cost (and in
consequence of) our extreme recent expansion and energy depredations
with all their resulting detriments to our environment. Dung beetles,
having been strictly limited from above by the availability of cow-pats and
similar have managed to avoid that.

Returning to the story of the work, once Reiko under Bob's guidance
had done the program, I experimented with it by e-mailing her or Bob with
requests for chosen runs. At one point I visited the University of Michigan
at Ann Arbor and worked for a fortnight intensively on modifications to
the program with Reiko—this came after a bad patch of misunderstandings
and unpromising runs that had caused us all to become somewhat
pessimistic. By Bob's arrangement I was accommodated at Martha Cook
Hall, a residential college for female students at the university. I could
almost say, given the character of families and daughters I met there, it was
a college for lady students. I may have been at the time the only man living
in the building. Variously at breakfast or on social occasions (there was a
weekly formal dinner that I remember), I was asked by the inmates what I
was researching and I recall many very wide eyes in beautiful young faces at
what I explained, even though I was taking care for the most part to keep
my stem-cell feminist and prostitute theme fairly far in the background.
Sex?—isn't it obvious, how else babies?—the surprised eyes and sometimes

the voices said to me. Reiko worked mainly on her personal computer at her apartment but sometimes in Bob's office in the Institute of Public Policies Studies, a tall room in a stately building, Lorch Hall, where I would run up four lofty flights of stairs to keep fit each day. Reiko instead took the lift and jogged for her fitness at other times, a neglect of stairs and of conservation about which I chaffed her a bit, but perhaps socially she was right for I have noticed in other contexts that when you enter a room still panting after a flight of stairs people react as if you are frightened or angry. Daily Reiko sprinkled me and Bob, like tender house plants, with her floppy disks bearing her updated codes. Some that were intended to produce substantial result sets were put to work in Bob's office overnight and I think Reiko's fiancé's computer was also brought in at times. Other copies I took away to work on by day in the NUBS computer basement to the university's herbarium (see Chapter 1) or else, late in the evening, took them to Bob's office. Finally, after I had returned to Britain, I ran copies that Reiko sent me on PCs in my department.

At some point, with the debugging finally finished and while I still experimented with effects under varied life patterns, I noticed that I was observing stable success of sex in what was, accidentally at first, a rather humanoid life history that was also experiencing quite a low selection severity. It was a pattern, in fact, which could be quite appropriate for chimpanzees or even for Palaeolithic humans. In preparing to report our results we needed to centralize some state of the variables as a starting point within the rather large and multiple parameter range we wished to cover, so I decided to perfect those primatoid features that had initially arrived almost by accident. I found that sex stayed comfortably stable under the changes. We therefore built our large final experiment varying everything around this case, aiming thereby to reply frontally to a challenge I had often experienced—that whatever our gimmick might be with parasites, it was bound to fail for the cases of low fecundity. As regards getting our work into print, however, this turned out to be one of our mistakes.

Somehow the choice and mention of *Homo* appeared to annoy several of our reviewers almost as if they had been creationists. They wrote that they couldn't see why we should centralize such a case and hinted that our success in it must collapse anyway if any dilution of the selection was brought in—that is, would do so if some of the factors that determined position in the rank order we set up ready for truncation were random, unconnected with genotype. The last was a fair objection and needed to be tested: I therefore added random influences in determining the rank order to an extent equal at least to the effect of the genetic variation and found that the support for sex still held. When our number of gene loci was set to its maximum (about 14) I found I could even add a random component three times greater than the genetic variation and still sex would be successful most of the time. It followed that hot and cold periods—fickle winds off deserts or ice-sheets—weren't too important: our 'humanoid', be she tropical 'Lucy' or an arctic Neanderthal, stayed as sexy as the ibex or the reindeer that she, he, or it may have hunted. Nevertheless when the revised paper went back to the referees with these new experiments included, but with no change to our centralizing of the *Homo*-like life history, we found all our new points left uncommented and the manuscript rejected by the referees even more curtly than before. Two of them indeed dug out new objections they hadn't thought of first time and claimed to see no substantial changes in the rest. Earlier, under my half-humorous view of the problem, I had been pleased with how my model was showing that the 'man' side of Man now had back-up security against any cloning armies of feminists; I'd also supposed that at the least my centralizing of *Homo*, and thus my implied defence of Man, would have brought smiles like my own to any readers who weren't really creationists. Instead the idea fell as flat as a joke in a customs hall.

My serious advice out of all this to a young evolutionist is to be like Darwin and like Wallace, not as we tried to be. Leave the *Homo* side of your thinking to your Volume Two; safest of all leave this side to your followers in the next generation. Perhaps not even that. To be popular, be like the cave painters (or a modern popular evolutionary writer): make of your own species a rare and a match-stick symbol in all you do, postpone analysis of what he/she is really like for about 25 000 years as, after showing they were perfectly capable of realistic painting of animals, the cave painters did.[9] In brief, never expect honour for realism about humans except from the fewest of few—this even among evolutionists.

Although I ought to have known this, at the time I felt optimistic and dealt quite oppositely. The danger of seeming irreverent to humanity—*Homo sapiens*—didn't cross my mind although that I believe is what bothered the referees. We were the first to be able to account for all sex's major manifestations in one model. We accounted not just for the patterns of selection sex needed across the genome but for its geography, its ecology, its life histories, and its size preferences among species . . . Surely all of this, but perhaps especially the light the theory sheds upon human sex, does have a popular appeal: you may smile at what our results imply or you may turn away from them, or you may weep but the consequences, if the idea is right, are certainly there. Even though most scientists seemed hardly to have realized the severity of the evolutionary puzzle or perhaps were still refusing to see it, we felt we should be able to summarize our work vividly enough to catch any scientific reader's attention and thought we had done so. Undoubtedly it should have been possible, as excellent sales of books that followed us, like *The Red Queen*,[10] have shown, as also have popular articles on the same topic.[11] But somehow we seemed to lack skill or luck or were forcing our results into too narrow a mould. We sent our manuscript to *Nature* twice (in 1988 and 1989) and then once to *Science* (in 1989). Failing with these I sent it in preliminary way to an editor of the Royal Society journals to see if they would be interested, but the comments I received were as discouraging as the rest. It particularly shamed me to have to tell Bob that even the society that supported me in general believed me to be over the hill on this topic. I think Bob was more nervous than I on account of the hostile referee reports because he had to trust me that we were picking reasonable biological assumptions. Finally Bob himself submitted the manuscript to *Proceedings of the National Academy of Sciences of the USA*, which as a member of the Academy it was his privilege to do unsupported. There, early in 1990, it found referees who at last seemed to understand it, we had found some of those 'fewest of few' and one at least seemed even enthusiastic.

The above record of rejections probably actually isn't long compared with some that much more revolutionary yet valid papers have received from journals. What, for example, about the attempts of Alfred Wegener to publish on continental drift, or Ignaz Semmelweiss to publish on puerperal fever, or Richard Altmann on the symbiotic origin of the mitochondrion? On the other hand, at the time we were submitting neither Bob nor I was an unknown scientist and neither of us had a reputation for mistaken or trivial ideas. The number of suspicious and hostile referees we found had come, therefore, as a considerable surprise. It puzzles me still that I had more trouble with this paper than I had had with those papers in which I started the theme of inclusive fitness, to which this one was so closely connected.

The main potentially serious implication in the reviews we received was that our (admittedly) complexly specified model had to be a ship in which we were somehow contriving to balance ourselves delicately in the mid-deck. With clever footwork we were keeping our vessel afloat for now but as soon as any wind arose a capsize would become inevitable. Put another way, when our assumed ecology and life history were changed, environmental variation was brought in, or any other changes, we would be sure to lose variation and asex would triumph. So the referees seemed to imply anyway, and such a view had been made explicit in the challenge to put in more random variation as I mentioned above.

Our statement that we had tested the model much more widely than we covered in the states we reported evidently wasn't believed, as also was the case with our description of the model. Several referees said this wasn't adequate; and yet it was quite as thoroughly described as models usually are in papers whose results rely on simulation. Moreover, our printed description was backed up by our offer to any one writing for it of a copy of the computer program plus the detailed annotation and description written by Reiko. But for details of what the model did in a few very unusual states—states that never actually arose in the simulations whose results we presented—the model was completely described in the paper, although we were admittedly terse. A subsequent team (Richard Ladle and Rufus Johnstone, later joined by Olivia Judson) reproduced and extended our model[12] purely for the paper's specification. Ladle and Johnstone did not even tell me they were working on this until our major results had been verified. In contrast, another paper on sex by two authors whom I will just call W and X, presenting a model of similar design, had been accepted recently by *Nature* in spite of a justifiable suspicion a referee might have had that the description had omitted details, notably a mention of a non-zero mutation rate in its resistance genes. This seemed likely to be necessary to prevent gene frequencies wandering to fixation, the problem I have alluded to several times in this volume. Mutation actually had been used in their model as W and X admitted later to my colleague Sean Nee when he wrote to ask. At the same time this model came not within a league of being able to cope with a twofold cost of sex. Thus the ready acceptance of a paper that couldn't solve the problem, established little not already known, and finally gave (to be a little more technical) a misleading impression based on two gene loci that quantitative genotype matching would be more effective in maintaining sex than would a system of gene-for-gene matching,[13] reinforces, it seems to me, a general truth that most scientists don't feel challenged by small steps in their field—matters almost already known—and are happy to see them published, but are scared to hostility by larger ones.

    Nature also published another paper on sex at about this time by two other authors whom I will call Y and Z. This paper was more purely algebraic in foundation and certainly can't be criticized for its technical description; but it, too, skated by leaving important biological countertopics unmentioned. Y and Z gave no mention to all the great groups and the whole regions of natural history in which the model could not possibly account for sexuality. For me this made the idea a non-starter for the general problem, even more so than the model of W and X had been. Instead of looking for a benefit to sex emerging from the interactions of genes replicating at different loci on chromosomes as most theories, including all mine, have done, Y and Z's paper suggested that interactions simply within the pairs of homologous genes (mother's gene/father's gene) reassorting within loci—that is, Mendelian single-locus dynamics—might be enough. The authors showed that in some circumstances new beneficial genes could be more efficiently incorporated into the genome in diploid organisms—that is, in organisms with such parental pairs of chromosomes in every cell—than they could be in haploids. The argument was clear and the point was relatively new, but what wasn't mentioned was the large class of organisms in which sex is still abundant but the major phases of life are haploid. In these the described support for sex couldn't occur. Many organisms that maintain both sex and large pools of variation are haploid for virtually all of their lives.[14] I have already addressed this at some length in Chapter 2 and won't repeat the arguments here. The main point is that in a broader vision the theory had to be seen as yet another partial one, a tacit admission that if this was indeed the correct explanation of diploid sex, we were going to have to search all over again for a cause that could apply to all the extremely similar effects that are seen, for example, in all mosses and in the haploid seaweeds: nave and transept of our cathedral had to have been designed by different architects; or, reverting to another image I have used, we seemed to have found a gap in the forest canopy on the hillcrest and to be looking out only to see not some unified natural vista but a patchwork of suburbs . . . Provided one kept to the authors' narrow focus, however, nothing was wrong: again, this *Nature* paper was a valid and safely small elucidation of one extra factor that could work in diploids.

I feel ashamed to reveal my pettiness in such comments as the above but I think many scientists will recognize my feelings—perhaps general readers are helped to see what we scientists are like. Unmistakably I harbour a resentment towards some particular papers because they were readily accepted in a prestigious journal when one of mine wasn't. The acceptance there probably even colours my estimation of their content. But the issue of what steps in science are acceptable and favoured is worth discussing even if it has to include this boring display of a person's egotism.

Probably I was particularly sour towards *Nature* after the final rejection because, shortly before I submitted our manuscript of the present chapter, I had been trying to help to be published in the same journal yet another manuscript that I thought to be likewise an important paper. Two biologists in Riverside, California, Richard Stouthamer and Robert Luck, had shown that a state of asex in a certain tiny parasitic wasp could be 'cured'—that is, in its hitherto all-female line, males be caused to reappear—simply by feeding a female an antibiotic to wasps in their honey diet. The finding had the obvious implication that the parthenogenesis here had been caused by some microbial agent and that this agent was still present and alive. I feel I barely justify my place as third author along with the Riverside two because I did so little in the work except to help it to be published; anyway, the point for the present is that the claim that even one case of parthenogenesis could resemble a curable disease seemed at a first glance extremely surprising and unlikely; however, it was also much more directly supported by experiment than were our biological claims arising from our HAMAX model. Therefore whether the matter was experiment or theory seemed to make little difference: again the trouble was that the potential change to current views of parthenogenesis was a big one and the referees again recoiled.

You don't have to say much that is bad to ensure rejection of a paper by a prestigious journal. In the case of Stouthamer and Luck's paper, one of the main weapons I recall being wheeled out by the two was that old and heavy howitzer of medical microbiology called 'Koch's Postulates'.[15] In its full form, K's Ps are often not insisted on nowadays before accepting a disease symptom as being caused by a microbe—they never were insisted on, for example, with HIV-1 (a fact that was made much play of by Peter Duesberg)—but in Stouthamer and Luck's case the full barrage was applied. For every other detail of the text, too, scalpels and fine combs had been unwrapped and also some much blunter instruments. As with our HAMAX paper I found it hard to believe the quality of some of the referees' comments that proved enough to persuade the editors first of *Nature* and then of *Science* not to accept. One has to sympathize with staff of journals receiving hundreds of manuscripts in a week but I would have thought statistical linguistic tests on referee reports could by now have been devised to weed out those using some of the more common techniques of prejudice.

Coming back to our HAMAX paper, surely not even a statistical test is needed to tell an editor that a 79-word review (61 words if we except the first, which was just a restatement of our main finding) and which included one 'sentence' ('Also there is data showing the goodness of fit in existing.'—full stop is as in the original) that made absolutely no sense at all, deserves a straight passage to the bin.[16] As can be seen in note 16 where I give the review in full, really the only intelligible claim in the piece was that we had not reported on any simulations outside the range we had studied in detail. Scientists, however, can surely be mostly trusted to have probed into the general surround of the particular set of conditions of the model they choose to present. But even if we hadn't so probed (which we had abundantly), our results should still be regarded as remarkable for their breadth, not their narrowness. There was certainly nothing equally broad already in the literature. If you are worried about what happens over the edges of the marginal 'cliffs' of our Figure 2 (16.2 in this volume), just look at the secure and evenly sloping hillsides that are shown. The diagram covers in effect four dimensions of parameters, two of them are over binary ranges, one (recombination) covers its entire possible range, and the remaining dimension (number of loci) covers as broad a range of specific numbers of loci as I am aware of ever being covered before. Thus the only intelligible and only conceivably worthwhile sentence in this four-sentence report—that we had not tested things widely enough—seems extra-ordinarily offbeam. If one criticized every paper studying some feature of one-locus population genetics, for example, on grounds that it hadn't yet probed into even just possible two-locus complications (or hadn't reported

having done so), a substantial fraction of the literature of population
genetics would have stayed unpublished. It is, surely, pointless to heed such
a review; equally pointless to send it on to the authors.

Others almost as bad as this may be paraded when I write more about
the Stouthamer and Luck paper. During the time I am describing my main
consolation was to recall the efforts I had made to support other works,
quite independent of my own, that had seemed to me clearly important and
that had been also disliked and denied by my contemporaries, as expressed
through the referee reports, just as mine was disliked and denied now. I
describe some of those efforts I made in Chapter 9. I knew those papers
pioneered new fields, proving me right to have given the support; I hoped it
would prove the same now with both HAMAX and curable parthenogenesis
once the topics managed to see true printer's ink as opposed to just the
black dust of photocopiers. Memory of those past cases gave me a wry pride:
I still to myself seemed on the right side—young enough, it appeared, to be
creating and promoting ideas that could sting referees into writing
indignant reports; I could still seriously annoy people; had not yet switched
over to the establishment nor taken up the role of trying to suppress them.
Now, seeing the neat experimental work of Richard Stouthamer and Bob
Luck resisted by insinuations of unreliability so extremely similar to those
being directed at my work with Bob and Reiko, or that had formerly been
directed at Warren Holmes and Bruce Waldman, separately trying to get
their findings on kin recognition published in the 1970s and early 80s (see
Chapter 9), tended to reassure me that our rejection didn't necessarily mean
that, in the years between, I'd completely lost my marbles. Put-a-stop-to-it
reports are, indeed, recognizable whenever you see them. All the same, in
the case of the curable parthenogenesis I had clearly made a tactical
mistake in trying to help the Riversiders while letting myself go onto their
paper as third author. Once added to an author list obviously you lose your
halo of supposed unbias.

To wrap up this rather dreary subject and to lead into a more
substantial one about the theory I consider runner-up to the Parasite Red
Queen (another topic due to be a bit dreary—or rather tragic—for a quite
different reason), I present below citation scores for all the recent papers
just mentioned up to mid-1998. The undiscussed one in the list,
'Kondrashov (1998)', is that which I will focus on shortly; it is a
representative of the Parasite Red Queen's main rival theory. The two

figures I give for each paper are of kinds that play increasingly delphic roles for persons, journals, and even institutions in modern science. There are many objections to such figures and I doubt that anyone supposes them the best to establish the worth of a paper. Eventually, of course, one will do far better to look in review papers, textbooks, to ask leading researchers, and so on. Nevertheless the figures are easily understood and can hardly be denied to give rough measures of the interest that up to a given time a set of papers have generated.

	Year published	Journal	Citations since publication	
			Total	Mean per year
W and X	1987	*Nature*	29	2.9
Kondrashov	1988	*Nature*	135	15.0
Y and Z	1989	*Nature*	30	3.8
H, Axelrod, and Tanese	1990	PNAS	123	17.6
Stouthamer, Luck, and H	1990	PNAS	68	9.7

Data in mid-1998 from *Science Index*, Institute of Scientific Information.

To save space I have put H for me, and *PNAS* for *Proceedings of the National Academy of Sciences, USA*. All five works were being gestated in the mid- to late 1980s with 'W&X (1987)' and 'Kondrashov (1988)' a little in the lead, having had shorter 'pregnancies'. 'H,A&T (1990)' would have come out at least a year and 'S,L&H (1990)' at least 2 years, earlier than the others, and, had they not had the difficulties I have described, they would have been in *Nature*. Enough time has now elapsed to show that none of the works was seriously flawed and none, as proved by the citations they have received, could be called a flop. Three could be said to have proved considerable successes; two are the *Nature* and *Science* rejects that ended up in *PNAS* and the other the *Nature*-accepted paper by Alexei Kondrashov.[17] A Russian geneticist, and about this time coming to work in the US, Kondrashov was establishing himself as leading proponent of what I am going to call (following roughly his own words) the Deterministic Mutation Hypothesis (DMH) of sex. I had been thinking of a name for it myself—'Mutation Black Queen' theory. So mine was again ex Lewis Carroll but coming a little back to normal from Carroll's own odd choice of chess colours and with the new one seeming to me quite fitting its implications, as you will see; but I come to admit DMH to be on the whole more explicit and better.

Ever since through Kondrashov's writings DMH theory came fully to my attention it seemed to me clearly the main rival to the Parasite Red Queen (PRQ) theory. But I myself at all times have continued to prefer the latter. Just as the big paper by Kondrashov in *Nature* gave DMH its major review support for the time, so, I like to think, our HAMAX paper of this chapter and 2 years later outlined the major support for PRQ.[18] Right now the table above might be likened to a glimpse of our offspring and Kondrashov's running neck and neck in a school sports of scientific competition. But, of course, these are just single papers and which theory, as a whole, if either, will ultimately go on to win 'gold' in the Olympics, it will need far more time to see. Extending figures of the kind I show here by future round ups from *Science Citation Index* probably won't tell us because each time someone brings out a new good overview of DMH (say) that may reduce growth of 'K88's citations, because scientists often just cite the latest approved review that covers a topic. Ours, 'HAT90', may suffer similarly; indeed, it is my high hope that this very book may bring this about. DMH has had an advantage of being published in *Nature* but PRQ may have its counteradvantage from publicity in a popular book by Matt Ridley that shortly followed it.[10] Well, let us wait and see. As we point out in the paper itself, the final decision between the two theories won't be quick or easy: (1) the two are far from mutually exclusive—indeed each is almost guaranteed to be right in part; (2) many of the predictions are similar and thus serve only to distinguish the pair jointly from the rest in the field. Nevertheless, several points give me encouragement about the future of PRQ.

First, DMH at its present stage says almost nothing about the ecology of sex. All populations are subject to mutations and the bad ones are inevitably by far the commonest. As we all know, tinkering at random with the settings of any complex machine—think of, say, the combined injection and ignition system of a motor car—rarely makes its performance better. Therefore, at a first glance at least, all populations should benefit from having sex helping to annul random prods with hammer and screwdriver executed by an angry, untamed environment. But on this view all populations should benefit about the same from sex and the fact is they don't; sex has a great deal of geography in its distribution, and it also has much of even narrowly local ecological variation. Some differences for different places can be extracted from DMH[19] but at present information of this kind is short on convincing detail. For example, DMH predicts high levels of sex on mountain tops where mutations from ionizing solar radiation should be common.[20] Actually it is parthenogenesis that increases with altitude on mountains[21] (although I have to admit that many high plant species remain large-flowered and therefore presumably highly sexual much as reindeer stay highly sexual in the extreme north). As I will explain later (in the Appendix to this chapter and Chapter 18), the predictions from PRQ are likewise ambiguous about the expectations for very harsh places but at least in my case there is a wealth of more detailed prediction that can be investigated. Such predictions might be said to weave like a fungus through all the tissues of what ecology and biogeography have long been telling us. An important review paper of the 1970s was the first to point this out[22] and in 1982 an important book by Graham Bell followed it, highlighting the issues still more vividly.[6] Where such detailed predictions have been investigated since that time, it seems to me they are mostly coming out very well for PRQ.[23]

DMH does have some things to say about behaviour—for example, mate choice should aim to recognize and should avoid highly mutated individuals—but, as we point out in the paper, the predictions here are mostly identical to those of PRQ. To illustrate this difficulty consider the case of using age as a criterion of mate choice. It is known that mutation rates increase with age in individuals. This includes mutation in the germ line and I emphasize that it is rate, not just accumulation. At first sight it would seem that the mutation clearance theory should postulate that young mates should be preferred by choosers in order to avoid mutations. The mate must be adult, of course, and some concession might be made to finding a mate large enough to be a powerful protector in the cases with paternal care: otherwise the best chance under a supposed copious and increasing rain of mutations is to choose the youngest. But young males are untested. Choosing from among them you might by mistake pick the one who actually has many minor bad mutations but who has not had time to die of them; if you had waited a year or two he would have died of them and although a less-loaded peer would have survived . . . the field would be narrowed and more tested. So here we are arriving at the same argument as we use more obviously to say that picking oldies is the best course under long-cycling PRQ: old animals have proven themselves to be good survivors in all ways, including in their survival of disease. And it turns out that moderate gerontophilia is indeed common in species where sexual selection is highly developed.[24]

Thus the case has gridlocked in exactly the sort of situation where the biologist should seek a mathematical model to try to refine principles by which to discriminate the theories. The only model I know of here (Kokko and Lindstrom's) comes to the surprising finding that DMH actually predicts moderately *more* emphasis on preferring age than PRQ does.[25] Of course we've all known intuitively—or at least those of us have known who are old enough to be rightly forming mature feelings and judgements about the situation—that choosing age and white hair in human pairing is the wise course; but the evolutionary arguments underlying our intuition must be admitted to be on a high plane of subtlety, akin, say, to perceiving intuitively that $\exp(i\pi) = -1$. So at present I feel I need more than my first hasty reading of Kokko and Lindstrom's paper and its assumptions before fully accepting the surprising conclusion. If its result is realistic, it is clear that the difficulty of find discriminators of the two theories, even where we might have thought we might have one, hasn't gone away. Rosemary Redfield[26] is among those who made the gerontophilia point in favour of PRQ, but after Kokko and Lindstrom perhaps we should take not only this point but a second one as equivocal. Redfield drew attention to the fact of mutation rate being usually much higher in males than females. This seemed to imply mate choice being directed mainly the wrong way in the first place (females selecting males; the other way the selector might often see progeny tests) and then still more perversely compounding trouble by selecting older males. A recent review of mutation rates, however, seems to indicate an inevitable proportionality of genome mutation rates to the number of cell divisions preceding gamete formation in germ lines in the two sexes and also to unreducable rates of replication errors for every division.[27] This would make the high rates of mutation in males apparently inevitable; but mutations present in male gametes rise more than proportional to a male's age. Is it constant damage from ionizing radiation or from metabolism-produced free radicals? More research seems needed before these matters can be resolved.

Moving to yet another point, Alexei Kondrashov, in bravest and best Popperian tradition and more quantitatively than we on the PRQ side have ever dared or found ways to be, has stated a threshold on the proof of which survival of his version of the DMH depends.[17,28] He explained that

DMH sex needs a rate of at least one mutation of about 10 per cent defect happening to each individual in each generation. Unfortunately for deciding, it looks as though the rates currently appearing for various different organisms and genes studied may hover just around exactly this 'Kondrashov' limit: some studies indicate a level a bit higher, others considerably lower.[27,29] One study on yeast looking at fitness directly seems to favour the DMH story, but I remain doubtful because little effort was made to convince the reader that the same results disfavour PRQ.[30]

Does all this mean that the evidence is going to be ever indecisive about whether mutation clearance is an adequate evolutionary cause for sex? I'm not sure; but what the present state of indecision *per se* inevitably suggests is that if the genome mutation rate continues in the critical region with regard to Kondrashov's prediction, I will claim this very fact to favour the Red Queen. This is because if the PRQ alone is really powerful enough to cause sex to be stable, as I believe, sex's retention means that individuals have the opportunity to cream extra advantages from higher genome mutation rates than they would otherwise be expected to tolerate—that is, given so inevitably restless a population state as PRQ describes, such rates may exist at little extra cost to the organisms. I have not worked out this idea in detail but it seems plausible that if a species is changing its defences all the time to keep pace with rapid parasite coevolution, then the same selection that is facilitating escape from current parasites can eliminate bad mutations simultaneously at little extra cost. Bad mutations, obsolete gene combinations can all go out in the same handfuls of death—twos and threes or mores at a time—while, as the bonus, the rare good mutations produced by the higher mutation rate can be promoted. It is obvious that the PRQ and DMH processes cannot avoid interweaving. Natural Selection, grim sentry on the Bridge of Life, attends to the Geneva Convention less even than the sentry of the army joke: she shoots first, doesn't even pretend to ask questions later. All causes of ill health—the new bad genes and the formerly good ones that are now bad—are treated by her completely indifferently.[31] Among the mutations she tests have to be, as usual, the rare good ones mixed with the bad. The good are the potential for real onward and upward progress as opposed to all the mere rushing around while changing resistances due to the Red Queen. The opening of opportunity for more of these good mutations to occur and to come through is the incentive to let the mutation rate rise if the overall

cost can be low. If, on the other hand, the Red Queen isn't the effective
storage agency that we portray in our paper (which our critics clearly
believed), then currently I can't imagine what other selection pressure can
prevent the genome mutation rate being selected to levels far below those
which we see in most sexual organisms. We know that higher standards in
repair and replication are possible. We know this through lab studies of
mutation rates showing them to be fairly responsive to selection,[26,32,33] and
we know it also through prokaryotes having their mean rates set at a mere
hundredth or less of the values characteristic of sexual larger organisms.
Unless PRQ dynamics are indeed powerful and sufficient selectors for sex
and pave the way for DMH, it seems inexplicable to me that genome
mutation rates continue high enough to be near to the Kondrashov limit—
that is, at the rate of one or more mutation per individual per generation.

There is also the question of whether DMH can really cope with the
twofold cost of sex on a basis of a short-term advantage. The need for the
kinky escape—the group-sex argument—may be again rearing its head. I
doubt, in fact, if the twofold challenge can be met and have yet to see a
convincingly run model; certainly I have seen nothing published to match
the demonstration our paper gives for PRQ even though we have no
mathematical demonstration. If we put into either kind of model freedom
of the sex locus to mutate to parthenogenesis we have to remember that
the mutation to asexuality can occur in any genotype of the population,
including in the genotype that is most mutation-free and thus most
successful.[34] Patents don't apply here: Sex Inc.'s expensive and ingenious
new model, brought triumphantly to the market, is immediately wide open
to copying by Sex Inc.'s rival, Asex & Co.; merely mutation suffices to set
up the perfect copy of the product. What is to stop this imitation, stream-
lined and cheap through its parthenogenesis, taking over the market?
Against the imitation, of course, runs the steady click of bad mutation once
parthenogenesis has begun—no further research and development by Asex
is finding out how to fix this, there is no further interweaving of ideas, it
was pure parasite, a sham constructive company. But is this enough for Sex
Inc. to beat the cheapness of the Asex & Co. imitation? It certainly wasn't
enough, for example, in the recent and unmetaphorical case of the clones
of the Macintosh computer: they quickly beat their parent and to save itself
the Apple Company had to refuse to renew their licences. The biological
case for DMH in short, seems highly questionable.[35]

Quite differently from the way adversity mounts against any abundant type in the PRQ theory, under DMH no rapid change of the environment counters any genotype just because it is common. The beauty of what sex, given a chance, could do against mutation is very apparent. Parthenogenetic lines diverge, they become burdened with new bad mutations in different ways; but where sex could be assembling re-unburdened genotypes from among these, asex is unable to. Though the asexual lines may have to be less fit in the end than the average sexual genotype might have been, the problem is whether any sexual genotypes are by that time around to take advantage: why are they not all swept to extinction during the first expansion of some successful clone? We need explicit models here to convince us.[36]

If we could be not worried by such points, however, a strong and even biblical one can be added in favour of DMH. Sexuals can go lower in numbers without disaster under DMH than they can under PRQ. In fact, old Noah's system of conservation, just a pair for each kind to escape the Flood, only really made sense under DMH. After the Flood, selection could begin immediately on the offspring of the single surviving pair Noah had brought through. It would be maintaining standards from that very point onwards. In contrast, had Noah believed in PRQ, he should have fetched not just two but a few hundred of each species to his Ark because he would be needing to preserve many alleles at many varying loci. Even with his most careful selection of individuals, a pair would preserve for him only four gene variants at each locus, and, even had he the techniques, his chance of finding a pair with such a condition at more than a few loci is extremely small. To judge by MHC, to have four at each mightn't be enough anyway and this is yet saying nothing of his getting enough variation at other loci. When the population itself is small, the breeding drift to uniformity quickly dominates; variation dwindles and then recombination can do nothing. For PRQ with its short-term selection an adequate-sized sexual population needs to continue somewhere all the time, whence sexuals can spread to replace the failing clones.[35] But although DMH needed but a pair to be saved in the Ark, where in the world, since that event, can at least one pair have been hiding all the time from all the deadly competitive clones that will have sprung from excellent mutation-free stocks? Sooner or later we need a multitude of Ark events to save each species from the repeating flood of its own clones. No escape seems possible; the setting up of metapopulations helps only marginally.[37] In contrast, in the PRQ theory, as the paper I introduce proves, islands of preservation may be advantageous but are not essential;[38] in any case metapopulations set them up almost automatically.

Altogether, DMH seems to be founded in part on the assumptions that the 1960s decreed that Hermann Muller and Ronald Fisher shouldn't have made. Could those be returned to, Weismann–Muller–Fisher theory itself could still stand. Long before Muller and Fisher, before ghosts of Darwin and Mendel nodded to one another in the evolutionary writings of the first quarter of the twentieth century, and really ever since vastly earlier in the sense that the potential of crossing to combine good traits has probably been appreciated by animal and plant breeders since farming began,[39] everyone has understood, roughly, that recombination through sexual outcrossing does yield a big long-term advantage—that is, if you like, it obviously aids long-term and 'macro' evolution. The key problem is the survival of short-term, very powerful competition from asex. Some have tried to argue the short-term problem away because new clones (and perhaps old ones as well) are too imperfect in executing their new way of reproduction.[8,40] It may be true of many mutations to asex but some mutant parthenogens have been seen arising with high reproductive efficiency and, as long as any such occur, sex is under threat.[41] Yet in most environments the spread of asexual morphs seems limited much too quickly[42] for the accumulation of mutations to be responsible.

Under PRQ thinking the more realistic danger than this is that sex during its eventual rebound, forces each defeated successful clone to low frequency but doesn't extinguish it: the clone persists. A succession of such events could lead eventually to a sufficient set able to accomplish much of what we claim to be the achievement of sex.[43] Why should clones not run with the Red Queen too—and, it seems, on checkerboard patterns? In the metapopulation, a rich pattern can be expected[44] and, albeit with only few signs of the expected flickering dynamic so far,[45] a 'checkerboard' image for the clone distributions has been indeed independently invoked for the zooplankton of ponds.[46] Importing another metaphor that I will re-use later, a seething of clone changes, as clones (or, on a longer scale, pairs of them) take turns to be predominant, may, in some sense not yet very clearly defined but basically frequency dependent, protect the overall diversity. The point to grasp is that in many situations of few but damaging parasite species the sexual strain is likely to find itself ousted completely by such an asexual set and this may be just what has happened in many of the 'species' that seem able to exist for long periods under pure parthenogenesis. Thus long-term stable asexuality may also be patterned by PRQ processes.

Polyclonal metapopulations of this kind have been studied especially in *Daphnia* and related waterfleas, but are also recorded in many other groups: rotifers, ostracods, snails, earthworms, a fly, a moth, a woodlouse, geckos and fish, bacteria, arctic apomictic grasses, even dandelions. Surveys of local genotypes in such species generally show the pattern of mixed clones.[23] In ever more cases there comes evidence of differential biotic susceptibilities of clones to parasites (and in one case to a predator).[47] In at least one case[48] further evidence show parasites more affecting the common genotypes; however, for most such cases it must be admitted we still don't know who or what maintains the mixtures. My suspicions, of course, definitely focus antagonist species as the agents—'whos' rather than 'whats'—with Schmid's study[48] picked out as showing the trend. The literature more traditionally, however, has tended to favour 'whats', with explanations in terms of neutrality, history, and tiny local variations of conditions frequently mooted—or, as Little and Ebert recently summarized the less-specific faiths in general, by simple appeal to 'other forces and ecological complexity'.[49] One example is a notion that the mixture is a record of a remote population founding event.[50] For several reasons, including the rapid changes in clone frequencies within a year or over several that are being increasingly observed,[45,49] none of this seems very likely.

Mixtures of old clones not derived from each other but rather separately arisen from a sexual strain that has continued to survive somewhere, are much harder to explain by DMH than they are by PRQ. Under the latter they can appear and be understood quite naturally;[38] but the DMH theory dictates that the old clones should always be worse so that in a reverted asexual population or part population there should be only minor variants of relatively young clones, all of them quite closely connected to a single origin: there should be no force bringing together diverse mutant clones in any organized way. Such minor satellite clones undoubtedly do exist[51] but mixtures of much older ones are still more abundantly proven.[52] Although in a general way the existence of mixtures of what we may call 'Methuselah' clones can be said to be favourable to ideas of PRQ and to make the most likely rival variation system to it in the face of ever-changing parasites, we need further theoretical studies to show clearly what the limitations of asexual mixtures are and also to show to what extent local niche adaptations might rationalize deeper diversities under DMH (though, I admit, I will watch this one askance—it sounds like a weak government saving itself by coalition). Anyway, what life histories, ecologies, genetics, and parasites are likely to let such a pattern work and what others instead may give premium to sex's ability to provide its almost unlimited array of genotypes, an array far greater than that of any mixture of clones yet found?

Methuselah organisms rear grey heads and mouth off their boring opinions, irritant to the young, at many levels in biology. Perhaps PRQ forces always have a part in creating such truculent oldies, an idea that might be true whether we speak of relict taxa of plants or animals or whether we look at a level even below that of clones and consider supercentenarians individually. The British yew tree, *Taxus baccata*, is an example both ways. Barely touched by insects[53] and seemingly also very little attacked by fungi and other parasites, yews stand ageless, perhaps even stand for a thousand years, in Oxfordshire churchyards. Below them their roots reach past the human bones into Oxford Clay, a formation that has lain undisturbed since it settled off coast in tropical seas of 150 million years in the past. In the clay, yew roots touch bones astonishingly similar to bones of the living tree itself. They are fragments of 'taxad' wood that sank not far from coral coasts where they originated, there having been torn, perhaps, from living trees by dinosaurs—trees pulled down perhaps by that fearsome 'thumb' of *Iguanodon*. Some plant forms, like some animal ones, most mysteriously don't change. Often, like the yew, these Methuselah species are taxonomic loners. Were there once thousands of insects plaguing the yews of the Oxfordshire forests of the dinosaurs, like insects plague oaks today—plague them in hundreds, as was shown in T. R. E. Southwood's paper, when for yew he found only one?[53] And did dinosaurs trim yews with impunity, the way cows and horses trim oaks in our parks today? When cows and horses try to eat yews they die. Their detoxicant systems have moved far off from, or perhaps never appreciably encountered, the forgotten poisons of rare and relict species. Is this why relict taxon *Taxus* can produce such special drugs as are now being used to fight cancer?

Descending to the individual rather than the species scale, HAMAX itself sometimes produces individuals of great longevity both in its sexuals and its asexuals. These are monentary islands in a sea of change, which, even if never repeating itself, still shows no progressive evolution—the 'Red Queen' is equally appropriate on the two different scales. Easily I imagine some of our HAMAX Methuselah genotypes as 'churchyard yew trees' of our program; outliers that were born far from the current hub of change in the genotypic space of our model, just as the real trees are often far from, in Britain at least, anything that could be called a 'wood' of yew trees—that sort of place where yew parasites might congregate.

As I already hinted, the interesting parallel fact at the other scale is that there are extraordinarily few yew parasites living anywhere, either in yew woods (there are such on the South and North Downs of England) or in churchyards; the tree species had the luck to survive as a taxonomic outlier as well.

Of course we have no senescence process either built in or open to evolution in HAMAX other than the steadily increasing susceptibility to parasites occurring for any type that tends to abundance. But what if a type, during all the multidimensional cycling, manages to stay uncommon? If you were born lucky enough to be an outlier of the current genotype cloud, continually on the edge of your population's cloud in that vast space that multilocus genetics always provides, then it seems there is a moderate chance for you personally to become Methuselah too.

So might some of your linked gene combinations with you. But leaving out the combinations, what about Methuselah *genes*? This question, as it turns out, is an extremely important topic for the survival of PRQ as a theory—yet another level if you like—and this leads me on to some first comments on degrees of gloom within the spectrum of what can all be called 'Gloomy Sciences'. Let us look at the Red and the Black Queen theories for what they foretell about the future of humans. In Lewis Carroll's original, it will be remembered that the illogical, forgetful, affectionate White Queen was the one really likeable character: we must take her to represent, sadly, only the comfortable fairy-tale views of sex and selection and their wishful thinking. The Red Queen, in contrast, you will remember in 'Alice' as pushy and domineering. But what about the new character I have brought in, the Black Queen? Alas, this lady is gloom itself.

It needs to be emphasized that the digression I begin here has not a scrap of scientific bearing on the relative merits of the DMH and the PRQ views, nor (and which is to say almost the same thing) how the theories would appear from an extraterrestrial's point of view. Readers purely interested in the scientific questions are advised to skip the rest of this essay. If, however, you happen to have a strong faith in a benign Creator, or are a White-Queenist vaguely hoping for the best and counselling inaction, then the thoughts I give below may lead you to attach more likelihood, it seems to me, to PRQ as the more humane in a sorry set of unavoidable rational options. Both the Red and the Black tell us that our genotypes over the generations are degrading. In the Black option they degrade by accumulating bad mutations and in the Red through parasites bounding continually ahead of us and reducing the positive health effects of common genes. From this pressure of parasites still under PRQ only some lucky new genotypes thrown up by recombination—outliers as mentioned above but in general not such extreme or rare ones as to manage to continue outlying for long—are able to escape. The difference is that for anyone finding his health below average and, moreover, on a descending side of my treadmill (or my floating ball of fluctuation[54]) see, in the DMH world, no light at the end of his tunnel unless it is the shining white wall of a hospital; on the other hand an individual of the same degree of unhealthiness in the PRQ world can see both light and open country. For many, this light may again prove illusory but for some it is real and will be reached by the weakly individual's descendants. As in the other Alice story of Lewis Carroll, the combination of having the key and being the right size are, in the end, not going to be impossible.

The idea of gene storage through 'low' phases of cycles in our paper is a crucial difference of our model from DMH, as we explain. If the theory should be right this is important. We present a quite new way in which ancient alleles—that is, alleles preserved in polymorphisms for longer than the lives of the species they inhabit, sometimes for much longer even than the spans of genera[55]—may be under preservation. As yet the experts of population genetics, although coming near this topic when they have treated gene preservation under the topic of *random* variation of environment, have paid selective maintenance by determinately dynamic processes little attention.[56] After noting the uninformative spirit of the reviews of our paper, it almost seems to me that the experts draw aside their cloaks from any such possibility of contact or serious thought as they pass, pointedly disdaining ideas that no one of them has yet seen how to anoint with his alphabet soup. Having myself played with the dynamics of not only HAMAX but other dynamical models—dynamics that, as seen from a distance, one might call states of stable preservation even if they always include a jittering of frequencies of some kind (beautiful, too, sometimes, if figures I obtained later are relevant)—I have come to believe that theorists ought to pay this kind of preservation much more attention. Increasing numbers indeed begin to[57] but many still pursue only the older focus on full stabilities. All full stabilities have to be, of course, uncompromisingly hostile to PRQ sex.

To put my point of hope, my anti-gloom, more vividly, let us imagine an individual, say a male, living in my PRQ world who has started life in a terrible but not fatal ill health—the snuffliest, most adenoidal, and most spotty child within his sibship; see him later a schoolboy who almost permanently is 'off-games' because of his various ailments. From weakling there, later he is to be a boon patient to his homeopathist . . . surviving whose treatment (usually not dangerous), watch him, of all surprises, have the eventual luck to find a kind (and perhaps PRQ-clairvoyant) wife by whom he produces children. The point now is that even when those children begin to reveal the same debilitating traits their parent suffered from, this man may still rationally console himself, if he believes in PRQ, that his efforts to survive were and are worthwhile, and that the line he is founding may yet, even in positive aspects of health, be significant for the whole of the human species. Believing in PRQ he should not think, as might a batchelor trying for perfect altruism in a pure DMH world, that it were better for humanity had he committed suicide in his early years and that at the least he should ever avoid committing serious matrimony. Through his descendants our PRQ weakling may prove himself a crucial preserver of genes, which, a few hundred years hence, the human population may once again need—not necessarily as novelty but, I emphasize, *again*.[58] Genes and combinations, on the view in our paper, can come into favour, go out, any number of times, all this over many generations. In short, each weakling may realistically believe himself guardian of immortal and valuable alleles.

This topic will be the main subject of a paper in Volume 3, although that is a paper written, I regret to say, in yet more of my free and 'wildoaten' style and similar to the present essay but, unlike its paper, lacks all benefit of Axelrod. For the present I simply ask you to contrast the 'gene-guardian' picture of my weakling as above with the funereal gloom, the irrevocable ill health, that has to envelope the weakling of a DMH world. The man who is unhealthy to the same degree in this other world must think of himself as subject to unchanging bad mutations; his only hope for himself and for his family lies in a series of medical patches that he must envisage continuing for ever. He must trust that invention and production of such patches will keep pace not just with present ailments but with all those continually being added in his descendants at the rate (*fide* Kondrashov) of at least one new (perhaps 10 per cent?) defect each generation. If that theory about sex is right no wonder the John Radcliffe Hospital seems to rush towards you, bounding as it were over the heads of we unthinking academics in Oxford beneath. No wonder on the other side of the Atlantic I saw the University of Michigan Hospital raising her motherly eyebrows at me above the Nichols Arboretum as I thought on my various ways to work: 'Why are you upset?', I still hear her say, 'don't you see what you must accept? You think me an unbeautiful, barren architecture. Maybe, but so in proportion you should see your ailing self. How are your xxxs today—aren't they worse each year just like your father's at your age, didn't they begin earlier in life even than his? A few more such troubles added in your descendants and you are going to need me from birth to grave, so don't be proud. Am I not going to be your necessary and best angel for all the centuries of your family? And why, by the way, are you not watching more television where I explain all this by dramas among other ways—all about my new utopias, loved by all but you.'

Only a century and half ago we had to use—we couldn't avoid—sex and death as our exclusive help to escape the ditches of heredity. The beautiful and the healthy worked, married, and had children in large numbers; the ugly and sickly helped them or else they died. The sickly infants also died; the best of medicines for them in those days were of little use. But this has been rapidly changing, and now almost any breathing human matter can be perfused and kept alive if given an adequate effort of technology. At our present point of acceptance of the routine nature of Caesarean childbirth we have pledged ourselves to a certain course, opened our arms to the hospital and to all of its satellites, and they are rushing to enfold us. Home medicine cabinets and pharmacies march in phalanx with the big buildings; it's just that you can't see them in my photos because they are out of sight among the trees.

To be accurate there is, or was, a third alternative to these horrors that I didn't yet discuss. Until recently I wondered if this idea might be serving as a kind of mental insurance, a fall-back position, in the minds of those seemingly most optimistic about the coming medical utopia—a possibility the optimists realize rationally but don't care to mention because of the modern moratorium on any sort of discussion of eugenics. But only a few months ago my belief that this, to me, somewhat more rational idea was being entertained by gene-therapy enthusiasts was rudely shaken and the gate I thought I had seen open closed with a bang. The idea would have been to advance with genetic-code reading and with really fine 'engineering' in this field so far and fast that the whole coming swarm of 'cured' mutations might be really cured—cured in the germ line, not just in the ordinary cells of our bodies, 'Germ line' means cured in those cells of you or of your offspring (such offspring presumably being treated at extremely few-celled stages) that are headed to form your sperms or eggs. By this course we could in theory fix all the accumulation of bad mutations permanently. But to do it we would need to check all genes of every offspring against a standard and put the divergent ones back to their unmutated forms. The standard presumably would be some chosen best (with many polymorphisms, doubtless, still to be allowed) out of that great coming 'Handbook' of the human genome, which, we understand, needs merely a few more years to be published. By this we will soon know, potentially, all our defects.

Personally I doubt for technical reasons that this third course ever could be taken, and I doubt even more that it will be, although I can imagine it as an auxiliary treatment in hospital—an expensive privilege perhaps (the dread of this kind of eugenics I would suppose forgotten but not its immense cost) of the most important people. But my main doubt is because of just too many ways for the checking, correcting, and reinsertion of DNA to go wrong. Nevertheless, let me go back to expressing my astonishment at the fervour of abhorrence that the idea that such a radical and potentially vastly more hospital-avoidant course seem to cause in people: it really seems that life-long series of visits to doctor and hospital are much preferred over the service of a super-high-tech and once-visited clinic. Given the optimism with which we rush for the doctor's waiting rooms and the wards, I would have thought that similar optimism could spill towards this kind of reproductive clinic as well and to what—far in the future—it might be possible to do in un-mutating DNA back into the shapes we prefer. Maybe this course will come back into focus as it begins to be seen that the hospital/medical-fix approach is itself not working so well as was anticipated (too many break downs, too few healthy doctors and nurses to run things). For the moment, though, even talking about this third course, leave alone doing anything (currently, of course, illegal in many countries) makes one even more a mad dog in the eyes of friends than does attempting to explain honestly and without White Queen illusions the human implications of the DMH-hospital symbiosis—which implications, I may comment, were quite well and bravely pointed out, if to my mind a little over optimistically, by Alexei Kondrashov in what I have called his 'K88' paper.

I am personally lucky not to be an unhealthy person though like everyone I have deformities and weaknesses (not mentioning, of course, the oddities of my mind easily perceived in these writings). Some of these defects I strongly suspect to be mutant and hereditary, though whether this is by mutations having occurred in lines of my cells, or in my constituent sperm and egg, or whether it's from mutations that happened some generations back, generally I don't know. Mentioning some of the defects, I may begin with one that seems both trivial and serious at once—a bad musical ear, an un-faculty so hampering to me and so setting me apart socially that I sometimes wonder if I've had from birth and unrecognized a never-ceasing tinnitus that turns every harmony ever directed to me into a discord. Besides this, a dreary list begins: stiff joints and heavy limbs (such that I may hardly kneel or squat), an unruly anus, ingrown toenails, a skin all my life busy in a hot-and-cold love affair with *Staphylococcus aureus*—to say nothing of my father's heart attack at 74 and a grandmother's breast cancer that caused her death at 72. And so on. Still, my list of ailments seems middling compared with others I see around me. On the one hand, and far above me, are both athletes and effortless mental geniuses; on the other and below, are the sickly, the inane, the lazy, and below these again, and envied only by me on account of car-park places in crowded streets, are the innately disabled. Qualities of inanity and laziness I regard as commonly strongly linked to ill health even when the health *per se* of the victim may seem to be perfect. It is a matter of stress: force some to work or to think and the stress of extra activity brings out the weakness. The sloth seemingly learned in childhood and youth may indeed be learned and yet may still be the individual's proper evolutionary adaptation. Such ideas make up why I like to think of my children having lives at least as independent physically as mine has been and of their being capable for the same activities. I would like them to be at least as ready, say, for the next Earth-impacting comet or asteroid that is even now on course for us. The toughest of our present-day physiques and races might survive the rigours of the planet after a medium asteroid impact but all people who have become permanently tethered to hospitals are probably doomed. Like crocodiles and birds surviving after the end-Cretaceous event that finished the dinosaurs, Inuit people, for example, might become our representative survivors. Spreading south in the many-years of post-impact winter I doubt that they'll exactly dine on us, as crocodiles may have done on their monstrous collapsing cousins. But in place of wet igloos, what could be more enticing than all the huge, empty, and collapsing hospitals dotting the countryside to the south, in which most of us had been living? They are welcome to them; but, in starting the next cycle, let them, in turn, take care.

Lastly, it seems worthwhile to persuade to be seen in these writings a likeable and even 'politically correct' pan-humanism, a levelling of pride underlying the practical human implications of my own PRQ version. These are points on which DMH is quite neutral. The suggestion above of a golden future for the Inuit can be our start, but to follow it I ask you to turn back to the passage in the introduction to Chapter 8 where I dealt with the operations of my Cloud Cuckoo International Marriage Bureau (p. 275). There you will find my advice concerning three potential ideal mates for the modern world: the Nairobi prostitute, the mildly Tays-Sachs-affected descendant of the East European Jewry, and select West Africans from The Gambia. You know the one about the black prostitute who fell asleep on the Temple steps? No? Well, the punch line is that she woke up in the night to find herself covered by a heavy Jew. I take this joke seriously, as you can see, suggesting that it only needs adding that the girl was of mixed origin (not surprising for the profession) from West and East Africa, and then predicting that she is due to have a very large descendance. The diseases in question against which these ideal mates were suggested as protection were, respectively, AIDS, TB, and malaria.

The point to add here is that if you believe all three disasters are already on our side of the horizon (it may be so) and you desire triple resistance then the goal is set back merely by about 60 years because a minimum of one whole generation (about 30 years if you are a hunter-gatherer or an intellectual) is needed for some of your descendants to incorporate each new allele. But the concatenation doesn't need to be complete for it to begin to be valuable for you and for humankind. Through the extra spacing that your very first fertile marriage starts, the intermingling of your resistant offspring among the susceptible (just as does a multicrop or multiline planting in a modern field, and resembling, too, the fallows and the rotations in more traditional practices), a further benefit will have begun. Even the first resistant offspring can contribute their mite to world health in two clearcut ways: at the same time that they improve personal health chances, they will also, in the next generation, reduce the incidence for others by lowering transmission rates (the spacing effect already mentioned). In addition, and via these lowered rates, they will slow down or even reverse the *evolution* of the disease organisms towards greater virulence.[59]

More generally, and independently of whether DMH or PRQ wins the Olympics as the explanation for sex, far-reaching changes to the norms of human ethics are going to be needed if *H. sapiens* is to survive the next asteroid collision in spite of our dysgenic practices that will undoubtedly continue between now and then. It will only become untrue if, by the time of the impact, the supertechnology we have by then deployed (and been, probably, submerged by—*qua* individuals[60]) has neutralized the threat. This might have happened, for example, by invention of means to deflect the asteroid in time, or by starting dispersal from our present planet to others, or, while still remaining on our own, building and provisioning secure bunkers. But, as I already explained, if it is DMH that rules sex then the needed changes to our norms have to be similar but even more drastic. Integrity of the genome through medical patches is going to require that eventually all ideas of the sanctity of intra-uterine and *in vitro* human zygotes will be blown away. They will come to be abandoned as thoroughly as the idea always was, physiologically, within that necessarily more limited surveillance system used naturally by the female mammalian body to check over its fetus. Natural postnatal surveillance commonly went even farther than that natural internal one, for there are signs that maternal psychology about the time of childbirth is pre-set for a further screening for gross detectable defects in the newborn. Across anthropology and throughout history, and extending certainly to the civilized tribal cultures, abandonment under some combinations of resource lack and degree of defect seems to have been a socially condoned practice. Setting aside some religious ideas, for which I see neither evidence nor evolutionary rationale (nor even much in the way of deep scriptural precedent), I cannot see why such natural infanticide would not be expected to be a norm. Under either of the 'Queen' views treated above, or indeed under both at once, or again under any view that squares what we know about evolution, this natural and decidedly not cruel attitude deserves, in the interests of everyone, to be returned to. Obviously the avoidance of cruelty is a substantial issue, and having the welfare and psychology of both parents and offspring in mind, the time cut-off for humane infanticide should be set by some reasonable combination of the onset of bonding by parents and ability in a neonate to notice changes of parental attitude that could be distressing to it. In contrast to this morality, I believe that the 'pro-life' lobby throughout the ages has been largely a cultural artefact, neither rational for the parents nor deeply felt by them. Indeed possibly it is not merely cultural: it seems to me primarily a male-inspired idea and thus could be suspected of having an innate adaptive underpinning.

To explain the last idea a little more, it should prove the case if it is right that the ratio of men to women is significantly higher among strongest proponents of the 'pro-life' lobby than among those of the opposed 'freedom of choice' lobby. Only a moment of evolutionary thought is needed to show that two parents of a paternally unacknowledged fetus or baby are expected often to have quite different attitudes to its survival: a woman foreseeing the 15 or so years of lone child-raising or/and the possibility of a break up with her husband, should, compared with the biological father, set a much lower priority on maintaining a fetus or baby. She is expected correspondingly to set a higher priority on having at least a right to terminate. The man in the same situation may not have to think about costs at all. Admittedly if he doesn't acknowledge it the child is unlikely to become his legal inheritor; still, once created and whether or not it emerges as a healthy baby, it carries onward, if nurtured, half of his genes towards at least some chance of further propagation . . . By this evolutionary reasoning, it is expected that in so far as casuistry and hypocrisy allow the man will represent abortion to the woman as an undesirable or an evil course. In so far as his casuistry and hypocrisy don't allow—he's too honest, he doesn't dare, wants to seem more objective—he is likely to pay his dues and to encourage his own and his lover's adherence to a religion that tells the woman the same thing—that abortion and infanticide are sins.

Gloomy thoughts, but as I warned I wanted to unload one or two ideas that will now seem strange to modern civilized ethics but would not I fancy have sounded strange at all in the world of Abraham and Hagar or in that holding in many distant places in the world today. I would like to sow useful seeds, as I see them, of a little honesty in a field I perceive as choked with the tangled briars of cant and chauvinism. For the moment let it be only a first sprinkling of seed on what I predict to be an intensely dry and hostile soil where few will grow, and let us now rush back to the main path, leaving open only a possibility of a return to more of this 'moral genetics' in Volume 3, in which among the papers themselves more than one essay by me touching themes of the same kind will be republished. For now I will be happy if a single seed can grow in a single mind: perhaps I have persuaded a humane reader somewhere to think a little more about the newly revealed countryside through which our theories plod or run. No slur in all this discussion, I would like to emphasize, is reflected towards Africans or Melanesians through my name for the alternative theory as Mutation Black Queen. Whether the 'Red' or 'Black' theme is due to win the Olympics, the darker races are clearly my heroes at the present point, simply by not being so much affected by 'modern' ethics and 'modern' medicine. I believe the reasons they do already disproportionately win the actual Olympics are exactly those I am giving. Many a very black race could be substituted for the Inuit as my expected survivors of the asteroid. Moreover, I suspect that Inuit, Melanesians, and Africans may have a better intuitive grasp of some ideas of this chapter than have most readers who have become acculturated by the male-dominated mega-religions of the past 2000 years.

References and notes

1. H. Monro (ed.), *Twentieth Century Poetry* (Chatto & Windus, London, 1929).

2. A form of *extrinsic* senescence is present in the sense that each genotype favours parasite genotypes that can attack it, so that the longer an individual has lived the more chance that the annual round of parasite attack will lower its fitness and it will die. Chronic internally evolving infections are not explicitly involved in this form of senescence but the effect is similar.

3. V. C. Wynne-Edwards, *Animal Dispersion in Relation to Social Behaviour* (Oliver and Boyd, Edinburgh, 1962).

4. M. A. Parker, Constraints on the evolution of resistance to pests and pathogens, in P. G. Ayres (ed.), *Pests and Pathogens: Plant Responses to Foliar Attack*, pp. 181–97 (BIOS, Oxford, 1992); M. A. Parker, Pathogens and sex in plants, *Evolutionary Ecology* **8**, 560–84 (1994).

5. A test of this severity, allowing *any* genotype, including the best, to mutate at this rate (or faster—see the end of the paper) has not yet been applied to the mutation clearance theory of Alexei Kondrashov (A. S. Kondrashov, Deleterious mutations and the evolution of sexual reproduction, *Nature* **336**, 435–40 (1988)).

6. See note 1 in Chapter 2; also G. Bell, *The Masterpiece of Nature: The Evolution and Genetics of Sexuality* (University of California Press, Berkeley, 1982).

7. The only exception known to me of a model that is fully proven to beat the twofold advantage of asex is another parasite coevolution model presented by M. J. Keeling and D. A. Rand (A spatial mechanism for the evolution and maintenance of sexual reproduction, *Oikos* **74**, 414–24) in 1995. It uses simple but realistic assumptions of limited mobility for both hosts and parasites. As with ours, eternal unrest of local genotype frequencies is the outcome that mediates the success of the sexuals, but the authors claim that this unrest could not be described as cyclical—that is, it is essentially different from the (highly irregular) cycling that is the principle factor in ours.

8. G. C. Williams, *Sex and Evolution* (Princeton University Press, Princeton, NJ, 1975).

9. Our species virtually remained unrepresented in graphic art that seriously tried to maximize human realism until some works of Ancient Egypt's Middle Kingdom (e.g. art of el-Amarna); another hiatus then intervened until classical Athens. My estimate of 25 000 years is based on the Chauvet Cave paintings in France, in which, despite superb draughtsmanship of animals, not a singe human appears.

10. M. Ridley, *The Red Queen: Sex and the Evolution of Human Nature* (Viking, London, 1993).

11. See, for example, R. Cowen, Parasite power, *Science News* **138**, 200–2 (1990).

12. R. J. Ladle, R. A. Johnstone, and O. P. Judson, Coevolutionary dynamics of sex in a metapopulation: escaping the Red Queen, *Proceedings of the Royal Society of London B* **253**, 155–60 (1993); O. P. Judson, Preserving genes: a model of the maintenance of genetic variation in a metapopulation under frequency dependent selection, *Genetical Research, Cambridge* **65**, 175–91 (1995).

13. See the caption to Fig. 16.1 in the paper (p. 653). For further demonstration of how quantitative antagonistic species interaction may indeed cycle, see L. A. Zonta and S. D. Jayakar, Models of fluctuating selection for a quantitative trait, in G. de Jong (ed.), *Population Genetics and Evolution*, pp. 102–8 (Springer, Berlin, 1988); A. B. Korol, V. M. Kirzhner, Y. I. Ronin, and E. Nevo, Cyclical environmental changes as a factor maintaining genetic polymorphism. 2. Diploid selection for an additive trait, *Evolution* **50**, 1432–41 (1996); V. M. Kirzhner, A. B. Korol, and E. Nevo, Abundant multilocus polymorphisms caused by genetic interaction between species on a trait-for-trait basis, *Journal of Theoretical Biology*, **198**, 61–71 (1999). None of these papers addresses the difficulty for sex of the two-cost. Thus, although the robustness of cyclicity is certainly relevant to sex, I still believe that 'sex itself' will be found more easily stabilized under more 'gene-for-gene' patterns of interaction; these likewise generate robust and commonly allele-protective cyclic polymorphisms (W. D. Hamilton, Haploid dynamical polymorphism in a host with matching parasites: effects of mutation/subdivision, linkage and patterns of selection, *Journal of Heredity* **84**, 328–38 (1993)).

14. D. Mishler, Reproductive ecology of bryophytes, in J. Lovett Doust and L. Lovett Doust (ed.), *Plant Reproductive Ecology: Patterns and Strategies*, pp. 285–306 (Oxford University Press, Oxford, 1988); R. Wyatt, A. Stoneburner, and I. J. Odrzykoski, Bryophyte isozymes: systematic and evolutionary implications, in D. E. Soltis and P. S. Soltis (ed.), *Isozymes in Plant Biology*, pp. 221–40 (Chapman and Hall, London, 1989).

15. 'Koch's Postulates' (Robert Koch, 1843–1910) state that to establish that a suspect microbe is indeed the cause of a particular infectious disease it is necessary to show all of the following: (1) that the microbe can be found in every case of the disease; (2) that the microbe can be obtained in pure culture; (3) organisms from a pure culture reproduce the disease in healthy hosts when inoculated; and (4) the organism must be recoverable again from such infected hosts. A susceptible host other than a human for (3) and (4) seems allowed in cases of human disease.

16. In full, this referee's report was as follows:

> The paper presents a model which demonstrates that sexual reproduction is an advantage in the defence against parasites. The

paper lacks a sensitivity analysis—not only as to how dependent the results are on the values used for the parameters but also how sensitive the results are to changes in the model itself. Also there is data showing the goodness of fit in existing. Wouldn't a better title be 'Model showing sexual reproduction as an adaptation to resist parasites'?

Omitting the restatement of our paper's claims, this musters 61 words and it includes one uninterpretable sentence.

Our paper was not the only one on the parasite–sex theme to receive referee reports of such uninformative brevity during this period. For me, one of the worst experiences of the whole difficult period came a year or two later when I realized, through an editorial slip, how a referee's report that one of my associates had received for a manuscript on the sex–parasite theme, as brief and uninformative as the above but more cutting as well, had been written by one of my supposedly friendly associates in Oxford—and this a person whom I myself had recommended to the journal as a reviewer, believing he would be at least conscientious and probably favourable. Hearty and friendly discussants, even just around our coffee area, I had to decide, were not always what they seemed.

17. Kondrashov (1988) in note 5.

18. Kondrashov had a predecessor in DMH theory in J. T. Manning (The consequences of mutation in multiclonal asexual species, *Heredity* **36**, 351–7 (1976)); rather as I had one in PRQ in J. Jaenike (An hypothesis to account for the maintenance of sex within populations (*Evolutionary Theory* **3**, 191–4 (1978)). But Manning, like Jaenike in my metaphor of Chapter 1 (p. 15), followed what seems a pre-mammalian pattern of leaving his eggs to develop by themselves compared with Kondrashov's approach (and mine, I hope), which was to give weight to the idea by a detailed setting out of the conditions, consequences, predictions, and evidence.

19. J. R. Peck, J. M. Yearsley, and D. Waxman, Explaining the geographic distributions of sexual and asexual populations, *Nature* **391**, 889–92 (1998).

20. M. R. Berenbaum, Effects of electromagnetic radiation on insect–plant interactions, in E. A. Heinrichs (ed.), *Plant Stress–Insect Interactions*, pp. 167–86 (Wiley, New York, 1988); J. H. Sullivan, A. H. Teramura, and L. H. Ziska, Variation in UV-B sensitivity in plants from a 3000m elevational gradient in Hawaii, *American Journal of Botany* **79**, 737–43 (1992); L. O. Björn, T. V. Callaghan, C. Gehrke, D. GynnJones, J. A. Lee, U. Johanson, *et al.*, Effects of ozone depletion and increased ultraviolet-B radiation on northern vegetation, *Polar Research* **18**, 331–7 (1999).

21. P. Bierzychudek, Patterns in plant parthenogenesis, *Experientia* **41**, 1255–64 (1985); but see also R. A. Bingham, Efficient pollination of alpine plants,

Nature **391**, 238–9 (1998); D. B. O. Savile, Arctic adaptations in plants, *Canada Department of Agriculture, Research Branch Monographs* **6**, 1–81 (1972). For mosses see P. Convey and R. I. L. Smith, Investment in sexual reproduction by Antarctic mosses, *Oikes* **68**, 293–302 (1993). For arthropods (crustaceans, insects, and mites) see J. A. Downes, What is an arctic insect?, *Canadian Entomologist* **94**, 143–62 (1964); J. A. Downes, Arctic insects and their environment, *Canadian Entomologist* **96**, 279–307 (1964). P. Bierzychudek, Pollinators increase the cost of sex by avoiding female flowers, *Ecology* **68**, 444–7 (1987).

22. R. R. Glesener and D. Tilman, Sexuality and the components of environmental uncertainty: clues from geographical parthenogenesis in terrestrial animals, *American Naturalist* **112**, 659–73 (1978).

23. This is shown in the following group of citations amongst which note especially those involving Curtis Lively and Mark Dybdahl and their New Zealand snails. P. D. N. Hebert, Genotypic characteristics of cyclic parthenogens and their obligately asexual derivatives, in S. C. Stearns (ed.), *The Evolution of Sex and its Consequences*, pp. 175–95 (Birkhauser, Basel, 1987); O. P. Judson, Preserving genes: a model of the maintenance of genetic variation in a metapopulation under frequency dependent selection, *Genetical Research, Cambridge* **65**, 175–91 (1995). For examples (plus more theory), see: E. D. Parker, Jr, Ecological implications of clonal diversity in parthenogenetic morphospecies, *American Zoologist* **19**, 753–62 (1979); R. C. Vrijenhoek, Factors affecting clonal diversity and coexistence, *American Zoologist* **19**, 787–97 (1979); R. A. Angus, Geographical dispersal and clonal diversity in unisexual fish populations, *American Naturalist* **115**, 531–50 (1980); H. Ochmann, B. Stille, M. Niklasson, and R. K. Selander, Evolution of clonal diversity in the parthenogenetic fly *Lonchoptera dubia, Evolution* **34**, 539–47 (1980); R. J. Jeffries and L. D. Gottlieb, Genetic variation within and between populations of the sexual plant *Pucinellia* × *phryganodes, Canadian Journal of Botany* **61**, 774–9 (1983); Bierzychudek (1985) in note 21; L. G. Harshman and D. J. Futuyma, The origin and distribution of clonal diversity in *Alsophila pometaria* (Lepidoptera: Geometridae), *Evolution* **39**, 315–24 (1985); N. C. Ellstrand and M. L. Roose, Patterns of genotypic diversity in clonal plant species, *American Journal of Botany* **74**, 123–31 (1987); G. Pasteur, J. F. Agnese, C. P. Blanc, and N. Pasteur, Polyclony and low relative heterozygosity in a widespread unisexual vertebrate, *Lepidodactylus lugubris* (Sauria), *Genetica* **75**, 71–9 (1987); L. J. Weider, M. J. Beaton, and P. D. N. Hebert, Clonal diversity in high-arctic populations of *Daphnia pulex*, a polyploid apomictic complex, *Evolution* **41**, 1335–46 (1987); B. Christensen, M. Hvilsom, and

B. V. Pedersen, On the origin of clonal diversity in parthenogenetic *Fredericia striata* (Enchytraeidae, Oligochaeta), *Hereditas* **110**, 89–91 (1989); J. M. Smith, C. G. Dowson, and B. G. Spratt, Localized sex in bacteria, *Nature* **349**, 29–31 (1991); C Moritz, T. J. Case, D. T. Bolger, and S. Donnellau, Genetic diversity and the history of Pacific island house geckos (*Hemidactylus* and *Lepidodactylus*), *Biological Journal of the Linnean Society* **48**, 113–33 (1993); D. T. Bolger and T. J. Case, Divergent ecology of sympatric clones of the asexual gecko, *Lepidodactylus lugubris*, *Oecologia* **100**, 397–405 (1994); B. Schmid, Effects of genetic diversity in experimental stands of *Solidago altissima*: evidence for the potential role of pathogens as selective agents in plant-populations, *Journal of Ecology* **82**, 165–75 (1994); M. F. Dybdahl and C. M. Lively, Host–parasite interactions: infection of common clones in natural populations of a fresh-water snail (*Potamopyrgus antipodarum*), *Proceedings of the Royal Society of London B* **260**, 99–103 (1995); M. F. Dybdahl and C. M. Lively, Diverse, endemic and polyphyletic clones in mixed populations of a fresh-water snail (*Potamopyrgus antipodarum*), *Journal of Evolutionary Biology* **8**, 385–98 (1995); K. A. Hanley, A. N. Fisher, and T. J. Case, Lower mite infections in an asexual gecko compared with its sexual ancestors, *Evolution* **49**, 418–26 (1995); B. F. Theisen, B. Christensen, and P. Arctander, Origin of polyclonal diversity in triploid parthenogenetic *Trichoniscus pusillus pusillus* (Isopoda, Crustacea) based on allozyme and nucleotide sequence data, *Journal of Evolutionary Biology* **8**, 71–80 (1995); J. G. Vernon, B. Okamura, C. S. Jones, and L. R. Noble, Temporal patterns of clonality and parasitism in a population of fresh-water bryozoans, *Proceedings of the Royal Society of London B* **263**, 1313–18 (1996); T. J. Little, R. Demelo, D. J. Taylor, and P. D. N. Hebert, Genetic characterization of an arctic zooplankter: insights into geographic polyploidy, *Proceedings of the Royal Society of London* **264**, 1363–70 (1997); T. J. Little and P. D. N. Hebert, Clonal diversity in high arctic ostracodes, *Journal of Evolutionary Biology* **10**, 233–52 (1997); J. Jokela, C. M. Lively, J. A. Fox and M. F. Dybdahl, Flat reaction norms and 'frozen' phenotypic variation in clonal snails (*Potamopyrgus antipodarum*), *Evolution* **51**, 1120–9 (1997); R. D. Semlitsch, H. Hotz, and G.-D. Guex, Competition among tadpoles of coexisting hemiclones of hybridogenetic *Rana esculenta*: support for the frozen niche variation model, *Evolution* **51**, 1249–61 (1997); R. C. Vrijenhoek and E. Pfeiler, Differential survival of sexual and asexual *Poeciliopsis* during environmental stress, *Evolution* **51**, 1593–600 (1997); M. F. Dybdahl and C. M. Lively, Host–parasite coevolution: evidence for a rare advantage and time lagged selection in a natural population, *Evolution* **52**, 1057–66 (1998).

24. An example from Marlene Zuk's crickets is discussed in the Appendix to Chapter 6 (p. 800), and as one of many recent cases see S. Perreault, R. E. Lemon, and U. Kuhnlein, Patterns and correlates of extrapair paternity in American redstarts (*Setophaga ruticilla*), *Behavioral Ecology* **8**, 612–21 (1997).

25. H. Kokko and J. Lindstrom, Evolution of female preference for old mates, *Proceedings of the Royal Society of London B* **263**, 1533–8 (1996).

26. R. J. Redfield, Male mutation rates and the cost of sex for females, *Nature*, **369**, 145–6 (1994).

27. J. W. Drake, B. Charlesworth, D. Charlesworth, and J. F. Crow, Rates of spontaneous mutation, *Genetics* **148**, 1667–86 (1998).

28. A. S. Kondrashov, Selection against harmful mutations in large sexual and asexual populations, *Genetical Research* **40**, 325–32 (1982).

29. P. D. Keightley, Nature of deleterious mutation load in *Drosophila*, *Genetics* **144**, 1993–9 (1996); P. D. Keightley and A. Caballero, Genomic mutation rates for lifetime reproductive output and lifespan in *Caenorhabditis elegans*, *Proceedings of the National Academy of Sciences*, *USA* **94**, 3823–7 (1997).

30. C. Zeyl and G. Bell, The advantage of sex in evolving yeast populations, *Nature* **388**, 456–8 (1997).

31. R. S. Howard, Selection against deleterious mutations and the maintenance of biparental sex, *Theoretical Population Biology* **45**, 313–23 (1994); R. S. Howard and C. M. Lively, Parasitism, mutation accumulation and the maintenance of sex, *Nature* **368**, 358 only (1994); R. S. Howard and C. M. Lively, The maintenance of sex by parasitism and mutation accumulation under epistatic fitness functions, *Evolution* **52**, 604–10 (1998).

32. H. Nothel, Adaptation of *Drosophila melanogaster* populations to high mutation pressure: evolutionary adjustment of mutation rates, *Proceedings of the National Academy of Sciences*, *USA* **84**, 1045–9 (1987).

33. But see also A. S. Kondrashov, Sex and deleterious mutation, *Proceedings of the Royal Society of London B* **369**, 99–100 (1994).

34. Alexei Kondrashov in a paper of 1982 (Selection against harmful mutations in large sexual and asexual populations, *Genetical Research* **40**, 325–32) illustrates a case of a mutation to parthenogenesis first increasing but later being beaten back and extinguished by competition from the sexuals, but it makes the assumption that the mutant carries the number of mutations that is the average of the sexual population. Thus he does not test whether an asex mutation in a lightly loaded sexual genotype could be beaten back. HAMAX has asex mutations possible in all genotypes. But see also B. Charlesworth, Mutation-selection balance and the evolutionary advantage of sex and recombination, *Genetical Research* **55**, 199–221 (1990).

35. M. Bulmer, *Theoretical Evolutionary Ecology* (Sinauer, Sunderland, MA, 1994).

36. At the time I wrote this paragraph I had not noticed that a paper already exists (R. S. Howard, Selection against deleterious mutations and the maintenance of biparental sex, *Theoretical Population Biology* **45**, 313–23 (1994)), which raises some of the same issues about whether a genome-wide rate of one bad mutation per generation is enough for DMH to work under realistic conditions. In this paper I find that, based on extensive simulations of finite populations but sometimes as large as 10 000, Howard concludes that the per-generation genome mutation rate for DMH to work needs to be >2, not >1. With currently known mutation rates, this confirms that adequacy of deleterious mutation as a sole factor in the preservation of sex is highly questionable.

37. But see Peck *et al.* (1998), note 19.

38. H. N. Comins, Prey–predator models in spatially heterogeneous environments, *Journal of Theoretical Biology* **48**, 75–83 (1974).

39. D. Zohary and M. Hopf, *Domestication of Plants in the Old World* (Clarendon Press, Oxford, 1993).

40. R. Y. Lamb and R. B. Willey, Are parthenogenetic and related bisexual insects equal in fertility?, **313** 774–5 (1979); A. R. Templeton, The prophecies of parthenogenesis, in H. Dingle and J. P. Hegman (ed.), *Evolution and Genetics of Life Histories*, pp. 76–85 (Springer, New York, 1982).

41. J. Jokela, C. M. Lively, M. F. Dybdahl, and J. A. Fox, Evidence for a cost of sex in the freshwater snail *Potamopyrgus antipodarum*, *Ecology* **78**, 452–62 (1997).

42. M. F. Dybdahl and C. M. Lively, Host–parasite coevolution: evidence for a rare advantage and time lagged selection in a natural population, *Evolution* **52**, 1057–66 (1998).

43. Dybdahl and Lively (1995) in note 23.

44. A. Sasaki and W. D. Hamilton, unpublished work.

45. H. Ochmann, B. Stille, M. Niklasson, and R. K. Selander, Evolution of clonal diversity in the parthenogenetic fly *Lonchoptera dubia*, *Evolution* **34**, 539–47 (1980); P. D. N. Hebert and T. Crease, Clonal diversity in populations of *Daphnia pulex* reproducing by obligate parthenogenesis, *Heredity* **51**, 353–69 (1983).

46. C. C. Wilson and P. D. N. Hebert, The maintenance of taxon diversity in an asexual assemblage—an experimental analysis, *Ecology* **73**, 1462–72 (1992); C. C. Wilson and P. D. N. Hebert, Impact of copepod predation on distribution patterns of *Daphnia pulex* clones, *Limnology and Oceanography* **38**, 1304–10 (1993).

47. Wilson and Hebert (1993) in note 46; also Weider *et al.*, (1987) in note 23;
    J. Bengtsson and D. Ebert, Distributions and impacts of microparasites on
    Daphnia in a rockpool metapopulation, *Oecologia* **115**, 213–21 (1998); D.
    Ebert, C. D. Zschokke-Rohringer, and H. J. Carius, Within and between
    population variation for resistance of *Daphnia magna* to the bacterial
    endoparasite *Pasteuria ramosa*, *Proceedings of the Royal Society of London B* **265**,
    2127–34 (1998); T. J. Little and D. Ebert, Associations between parasitism
    and host genotype in natural populations of *Daphnia* (Crustacea: Cladocera),
    Journal of Animal Ecology **68**, 134–49 (1999).

48. Schmid (1994) in note 23.

49. Little and Ebert (1999) in note 47.

50. M. G. Boileau, P. D. N. Hebert, and S. S. Schwartz, Nonequilibrium gene-
    frequency divergence—persistant founder effects in natural populations,
    Journal of Evolutionary Biology **5**, 25–39 (1992).

51. B. F. Theisen, B. Christensen, and P. Arctander, Origin of polyclonal diversity
    in triploid parthenogenetic *Trichoniscus pusillus pusillus* (Isopoda, Crustacea)
    based on allozyme and nucleotide sequence data, *Journal of Evolutionary
    Biology* **8**, 71–80 (1995).

52. As shown in the references of note 23.

53. T. R. E. Southwood, The number of species of insects associated with various
    trees, *Journal of Animal Ecology* **30**, 1–8 (1961).

54. Hardly any evolutionist seems to like my 'floating ball' metaphor for how
    recombination aids defence against parasites, as described in the paper of this
    chapter. But it is intended not so much for evolutionists and mathematicians
    as for outsiders who (like me) never had a course in population genetics. I
    admit that, like all metaphors, it has limitations. Perhaps the most serious
    three may be as follows:

    (a) The image annoys precise minds because the gene-frequency space notion
        that I use—the weighted cube or hypercube that has been inflated into a
        ball—is easily seen by any mathematician or population geneticist as
        hopelessly inadequate to represent the full dynamics of the system. Where
        I use a cube as the primary idea, there are really needed 8 dimensions for a
        truthful image, or 16 when the parasites are represented numerically, or 24
        if the asexual hosts are represented too . . . And all this for just three loci,
        still far short of the number needed for firm support of sex. For me,
        however, it is exactly this enormity of the full problem that make the
        simple shadows in my three-dimensional representation so much better
        than nothing; these shadows cannot be denied to gives clues to the vast
        mobile hyperballs of the full process.

(b) Without additional and limiting assumptions that are not given in my description, it isn't true that parasitism necessarily imposes on all genotypes (cube vertices) that are 'out of the water' and accumulates on them the longer they are out: populations are in no sense localized and growing on the vertices. Instead (somewhat less roughly speaking though still vaguely) parasites either have the centre of gravity (centroid) of their mass in the cube move towards that of all extant hosts (hard selection) or move their mass directly away from the erstwhile centroid of most of their recently vanished genotypes (the soft truncation selection as used). The effect of such changes in the proper image (as near as we can get to it), however, leaves intact the notion of the source of the permanent instability that seemed so well implied by the simpler image—the continual extra weighting, or parasite accumulation, of the 'higher' vertices of the floating ball, making inevitable its instability. The main benefit from trying to take a more accurate view is that one can then see how vertices 'out of the water' may not become parasite-burdened even if sometimes they are out for a long time. Providing vertices keep low, the parasites evolve and achieve their high matching elsewhere. This can be imagined the case for pairs of host genotypes that are on vertices at or near to a (temporary, spontaneous) 'axis of rotation' of the cube or hypercube. Escaping parasitism by keeping relatively rare is a realistic theme, as is discussed in the text of the introduction for this chapter shortly before this note, where I refer to 'Methusaleh' species and genotypes.

(c) More peculiar 'motions' of the ball in the water than simple rotation are sometimes needed and for these the ball image is admittedly not very apt. Implication of one striking contortion was first seen as I watched hundreds of trajectories of 'centres of gravity' of hosts and parasites within a computer-depicted cube during the early days of work leading to a paper of 1993 (W. D. Hamilton, Haploid dynamical polymorphism in a host with matching parasites: effects of mutation/subdivision, linkage and patterns of selection, *Journal of Heredity* **84**, 328–38 (1993)). It was work producing a 'zoo', as I termed it, of beautiful and strange cycles, but one of the simplest and most common deserves special note and has been re-drawn to my attention recently by work by Akira Sasaki. It is a trajectory that commutes rapidly from one vertex to an opposite one, hardly being affected by the six other vertices between, thus suggesting a 'ball', which, instead of spinning, flicks back and forth between mirror images as if exchanging two 'poles' by interpenetration. The recent excellent visualizations of HAMAX-type simulations that Akira created for me have emphasized that this may be an important type of trajectory when recombination is very low, providing for hosts (especially for pure asexuals) their best makeshift against parasites, the alternative of abundant recombination being unavailable. Asexuals in our models often display such pole-to-pole oscillation, and they play a regular part in the onset of asexual takeover when sexuals, as so easily happens in few-locus cases, are chancing to lose too much of their variation. Such alternating opposites (or at least their probable approximation in nature, pairs of super-randomly distant genotypes) deserve to be looked for in nature where we would expect them to commonly constitute a kind of 'makeshift sex' among mixed asexual strains if the PRQ idea is right (for possible cases, see some 'Hebert' and 'Little' references on *Daphnia* and ostracods in notes 23 and 47). The same diagrams that show these opposite pair alternations also re-emphasize the floating ball image for sexuals that are securely maintaining variation: vertices are emptied in an irregular series, which, however, tends to empty all before returning to the first. This is the 'tangled ball of wool' description for the 'cube' of the 1993 paper (the 'wool' is inside the cube but can be imagined projected from the centre onto the ball to draw the approximate path of the ball's highest point).

A light skimming path could possibly roll my ball from one pole to the other with none of the intervening vertices going 'under the water', but this seems too delicate to be realistic. Nevertheless, evidence from my cube diagrams show not uncommon states in hard-selection dynamics where pole-to-pole oscillation occurs but the six non-opposite vertices manifestly affect the course by which the centroid of hosts traces its paths through the gene-frequency space, bowing these courses in the mid-course of the transition elegantly outwards.

55. Anon. Diverse genes bring humans and chimps closer, *New Scientist*, 35 only (22 September 1988); J. Klein and N. Takahata, The major histocompatibility complex and the quest for origins, *Immunological Reviews* **113**, 5–25 (1990); M. K. Uyenoyama, A generalised least-squares estimate for the origin of sporophytic self-incompatibility, *Genetics* **139**, 975–92 (1995); A. D. Richman, M. K. Uyenoyama, and J. R. Kohn, Alleleic diversity and gene genealogy at the self-incompatibility locus in the Solanaceae, *Science* **273**, 1212–16 (1996); M. Kusaba, T. Nishio, Y. Satta, K. Hinata, and D. Ockendon, Striking similarity in inter- and intra-specific comparisons of class I *SLG* alleles from *Brassica oleracea* and *B. campestris*: implications for the evolution and recognition mechanism, *Proceedings of the National Academy of Sciences, USA* **94**, 7673–8 (1997); S.-S. Woo, D. Sicard, R. Arroyo-Garcia,

O. Ochoa, E. Nevo, A. Korol, *et al.*, Many diverged resistance genes of ancient origin exist in lettuce (Paper presented at Plant & Animal Genome Meeting VI, San Diego, 18–22 January (1998)).

56. See N. Takahata and M. Nei, Allelic genealogy under over-dominance and frequency-dependent selection and polymorphism of major histocompatibility complex loci, *Genetics* **124**, 967–78 (1990); also U. Dieckmann, P. Marrow, and R. Law, Evolutionary cycling in predator–prey interactions: population dynamics and the Red Queen, *Journal of Theoretical Biology* **176**, 91–102 (1995).

57. S. A. Levin, L. A. Segel, and F. R. Adler, Diffuse coevolution in plant–herbivore communities, *Theoretical Population Biology* **13**, 171–91 (1990); S. A. Frank, Evolution of host–parasite diversity, *Evolution* **47**, 1721–32 (1993); S. A. Frank, Coevolutionary genetics of plants and pathogens, *Evolutionary Ecology* **7**, 45–75 (1993); G. D. Ruxton, Low levels of immigration between chaotic populations can reduce system extinctions by inducing asynchronous regular cycles, *Proceedings of the Royal Society of London B* **256**, 189–93 (1994); A. B. Korol, I. A. Preygel, and S. I. Preygel, *Recombination, Variability and Evolution* (Chapman and Hall, London, 1994); V. Andreasen and F. B. Christiansen, Slow coevolution of a viral pathogen and its diploid host, *Philosphical Transactions of the Royal Society of London B* **348**, 341–54 (1995); J. D. Van de Laan and P. Hogeweg, Predator–prey coevolution: interactions across different timescales, *Proceedings of the Royal Society of London B* **259**, 35–42 (1995); V. M. Kirzhner, A. B. Korol, and E. Nevo, Complex dynamics of multilocus systems subjected to cyclical selection, *Proceedings of the National Academy of Sciences, USA* **93**, 6532–5 (1996); P. Marrow, U. Dieckmann, and R. Law, Evolutionary dynamics in a predator–prey system: an ecological perspective, *Journal of Mathematical Biology* **34**, 556–78 (1996); S. Gavrilets, Coevolutionary chase in exploiter–victim systems with polygenic characters, *Journal of Theoretical Biology* **186**, 527–34 (1997); A. I. Khibnik and A. S. Kondrashov, Three mechanisms of Red Queen dynamics, *Proceedings of the Royal Society of London B* **264**, 1049–56 (1997); V. Kirzhner and Y. Lyubich, Multilocus dynamics under haploid selection, *Journal of Mathematical Biology* **35**, 391–408 (1997); Korol *et al.* (1996) in note 13; K. McCann, A. Hastings, and G. R. Huxel, Weak trophic interactions and the balance of nature, *Nature* **395**, 794–8 (1998); G. A. Polis, Stability is woven by complex webs, *Nature* **395**, 744–5 (1998); A. B. Korol, V. M. Kirzhner, and E. Nevo, Dynamics of recombination modifiers caused by cyclical selection: interaction of forced and auto-oscillations, *Genetical Research, Cambridge* **72**, 135–47 (1998); Kirzhner *et al.* (1999) in note 13.

58. E. A. Stahl, G. Dwyer, R. Mauricio, M. Kreitman, and J. Bergelson, Dynamics of disease resistance polymorphism at the *Rpm1* locus of *Arabidopsis*, *Nature* **400**, 667–71 (1999).

59. P. W. Ewald, Host–parasite relations, vectors and the evolution of disease severity, *Annual Reviews of Ecology and Systematics* **14**, 465–85 (1983).

60. See Vol. 1 of *Narrow Roads of Gene Land*, Ch. 6, p. 193.

SEXUAL REPRODUCTION AS AN ADAPTATION TO RESIST PARASITES (A REVIEW)[†]

W. D. HAMILTON, ROBERT AXELROD, and REIKO TANESE

———

Darwinian theory has yet to explain adequately the fact of sex. If males provide little or no aid to offspring, a high (up to twofold) extra average fitness has to emerge as a property of a sexual parentage if sex is to be stable. The advantage must presumably come from recombination but has been hard to identify. It may well lie in the necessity to recombine defences to defeat numerous parasites. A model demonstrating this works best for contesting hosts whose defence polymorphisms are constrained to low mutation rates. A review of the literature shows that the predictions of parasite coevolution fit well with the known ecology of sex. Moreover, parasite coevolution is superior to previous models of the evolution of sex by supporting the stability of sex under the following challenging conditions: very low fecundity, realistic patterns of genotype fitness and changing environment, and frequent mutation to parthenogenesis, even while sex pays the full twofold cost.

Parasite coevolution will be shown to be superior to previous models of the evolution of sex by supporting the stability of sex under the following challenging conditions.

1. *Very low fecundity as in scarabs and humans.* In previous modelling,[1,2] it has been found easy to show an advantage to sex when fecundities and mortalities are high, as, for example, in reproduction of trees, fungi, and marine invertebrates. Difficulties with low fecundities (as in humans) have been much greater. One author has suggested that sex is inherently unstable in such groups and that its retention is due to the difficulty of mutation to parthenogenesis.[1]

2. *Realistic patterns of genotype fitness and changing environment.* Attempts to cope with low fecundities have had to use high fitness differences entailing dramatic changes per generation (refs 2 and 3 and Chapters 1 and 2). When more realistic assumptions and values were applied to a model that used two recombining loci (Chapter 2), the model failed.[4]

[†]*Proceedings of the National Academy of Sciences, USA* **87**, 3566–73 (1990).

---

3. *Frequent mutation to parthenogenesis, even while sex pays the full twofold cost.* Authors have pointed out that mutation to efficient parthenogenesis is unlikely and seldom seen and have claimed that this reduces the problem.[1,5,6] Situations where parthenogenesis would be an advantage are extremely numerous, however, and routine successful parthenogenesis, including facultative use of the mode, has evolved hundreds of times in both plants and animals.[3]

4. *Very broadly overlapping generations as in trees and humans.* Coevolutionary models of sex (see Chapters 1 and 2) depend on intrinsic oscillation and are sensitive to factors creating or reducing lags in feedback. Sequential (iteroparous) reproduction dissipates fluctuation. No models have hitherto met the challenge of this pattern, which is common in fully sexual organisms.

PARASITES AND SEX

Parasites are ubiquitous. There are almost no organisms too small to have parasites. They are usually short-lived compared with their hosts, and this gives them a great advantage in rate of evolution. Thus antiparasite adaptations are in constant obsolescence. To resist numerous parasites, hosts must continually change gene combinations (refs 7–17 and Chapters 2, 5, and 11). Contrary to the assumption of the mutation theory,[18,19] the host species needs to preserve not one ideal genotype but rather an array. In the course of this preservation, selective changes in the midrange of gene frequency must be common. Because single and multiple heterozygosity is maximal for genes in the midrange, the effectiveness of events of recombination in uncoupling these changes is great, thus providing power to the model that we now describe.

We simulated a host population of 200 individuals that are either sexual hermaphroditic or else all-female and parthenogenetic; the difference is controlled by a single gene.[20] All are of equal fertility in the sense that chances to be a parent in each year are assigned fairly to all living mature individuals. However, sexual parents are paired and share offspring. Therefore they reproduce their genes with only half of the efficiency of asexual parents and 'fair assignment', as above, embodies the twofold advantage of parthenogenesis. In fact, in the absence of selection through differential mortality, the population extremely rapidly eliminates the allele for sex. It should be noted, however, that we are assuming males or male functions do not contribute to the number or biomass of offspring; otherwise, the mated pairs would have more nearly an average of two offspring born when the lone parthenogen has one.

The population is assumed to be at a stable density with an average death rate $d = 1/h = 1/14 = 0.0714$ per individual per year applying at all ages. Reproduction is assumed to occur on birthdays, but there is a juvenile period of $j = 13$ years during which individuals do not reproduce; thereafter, at exactly age 14 the 'fair chance' to participate in reproduction begins and continues indefinitely. For a population of stable age distribution under no selection, the mean level of fecundity needed to replace deaths is easily found by summing geometric series as in standard theoretical demography to be $f = d/(1 - d)^{j+1} = 0.2016$ per individual per year (see also ref. 21); the mean age of parenthood (a measure of generation length) is $G = j + h = 27$. The particular parameter values above are chosen to imitate primitive hominid reproduction.[22] Among the parameters so far defined, d (accompanied by its h) is different from all others in that, although it is fixed as a mean for the population as a whole, it varies over individuals according to genotype because of the effects of parasites: this is the sole kind of selection on hosts in the model. How the variation in mortality comes about in the model will now be described in full detail.

The host population is subject to n (ranging from 2 to 12) species of asexual parasites; each species also has a population of 200. All parasites have $j_p = 0$ and $G_p = 0 + 1/d_p = 1.1$, so that $d_p = 0.909$. Many other combinations of the host and parasite demographic and selection severity parameters have been tried. The results, not reported here, were generally similar to those that will be described, assuming parasite generations remain short compared to those of the host. As expected, changes that simulated less broadly iteroparous reproduction by means of a menopause[22] placed close to the age of first reproduction made sex succeed better; thus, the schedule having a wide overlap tests a fairly difficult case.

Both hosts and parasites are haploid. Each parasite species is assigned k loci for its own chromosome. The host's chromosome has a defence sector of k loci for each of the n parasite species. Hosts also have a sex-determining locus so that total host chromosome length is $1 + nk$. Alleles are 0 and 1. Each year, every host randomly acquires one parasite of every species, n in all. A parasite's fitness is evaluated by matching its chromosome with the assigned section of its host's chromosome and counting matched alleles. Thus if the host's section reads 00110 and the parasite's section reads 01111, the sum is $S_p = 3$. The maximum here, obtained with identical strings, would be 5. To fix ideas further by a potentially realistic example of only three loci, imagine that humans are haploid but still manifest the Lewis/secretor/ABO blood group system[23,24] except that we also reduce the ABO locus to only two alleles (say A and O). Suppose that, corresponding to the $2^3 = 8$ now possible genotypes and phenotypes of this system (of which none, on the known gene frequencies, would be very rare), there exist eight strains or species of a pathogen, each maximizing reproductive success when the states of alleles at three of its own loci match those of the host and that one unit of 'match score' is subtracted from a maximum score of 3 for each mismatched allele, then this would become an example of a single 'parasite defence sector' in the genome of our '*Haplohomo*', relevant to the model. Quite apart from the haploidy, the example remains fanciful in that no such complete diversity of differing microbial enemies involved with the ABO system is yet known. However, there is little doubt that infectious microbes are involved in maintaining the polymorphism; several very distinct groups of parasites, as well as strains within particular species, have already been implicated. Genera of microorganisms shown to correlate with various phenotypes of the human highly polymorphic Lewis/secretor/ABO system include *Schistosoma*,[25] *Giardia*,[26] *Leishmania*,[27,28] *Mycobacterium*,[29] *Neisseria*,[30,31] *Streptococcus*,[30] *Haemophilus*,[32] *Escherichia*, and *Vibrio*,[33] albeit some of the correlations may still need cautious interpretation (e.g. refs 31 and 34). Given that blood groups—glycoproteins on blood cell surfaces or else related moieties in serum and secretions—seem at first unlikely agents of resistance, it is not surprising that other resistance polymorphisms and heritabilities, often independent also of other immune system loci, prove very numerous (e.g. Chapter 5 and refs 35–41).

In the model, match scores as described are used to determine fitnesses as follows. The 'parasite score' determines both the parasite's rank among conspecifics after it has detached and is competing to survive into the next year and also the parasite's proportionate relative fecundity for the year. At the same time, the parasite's match makes a contribution to the 'host score' of $k - S_p$; when all parasites are considered, the host's total score therefore becomes $S = nk - \Sigma S_p$. Unlike parasite scores, host scores have no effect on fertilities, but they are used to rank all 200 hosts; then the 14 lowest ranked (because $200 \times d$ rounds off to 14) are killed. For parasites, $200 \times d_p$ rounds off to 182, so this number are killed. The system may be described as rank-order truncating 'soft' selection through mortality.[2,42]

Mating of the sexuals is at random and their reproduction is with recombination at a rate r between all adjacent loci. (For example, for a case with $n = 1$ and $k = 3$, chosen to simulate eight parasites' genotypes bearing on the haploid Lewis/secretor/AO blood group system as outlined above, $r = 0.5$ could be chosen realistically because the three loci in *Homo sapiens* appear to be unlinked.) Reproduction of asexuals is faithful to the parent except for mutation. All genes in hosts, both sexual and asexual, including the sex locus, have a mutation rate $m = 0.0001$ or else, in half the runs, $m = 0.01$. All loci in parasites have a mutation rate of 0.01. This rather high rate is chosen to ensure parasite ability to relocate recombinant genotypes within reasonable time despite lack of sex—the parasite threat must be acute. (In earlier versions, parasites also had a sex locus, but if parasite generation was short, they usually, although not under all conditions, lost the sex allele almost immediately. We therefore simplified the model by dropping sex in parasites. In nature, parasites commonly retain sex at least at some stage of cycle, but this may be due to (1) hyperparasitism (especially likely for large and external parasites) or (2) the need to combat facultative defences of their hosts (e.g. by an immune system) such counterthreats being generated on a timescale nearer to the parasite's own life cycle.)

The dynamics of this model can be thought of as a group pursuit around the vertices of an *nk* dimensional hypercube. The parasites evolve toward maximizing and the hosts evolve toward minimizing the matching of chromosomes. Such a process both tends to protect polymorphism and gives an endlessly unstable coevolution.[43–45] Trajectories of gene frequencies (and of linkage disequilibria) in our model are irregular. Gene frequency has the widest range when *nk* is low; as *nk* is raised, the amplitude decreases but the support for sex nevertheless improves. It is worth noting, however, that even our highest value of *nk* ($2 \times 7 = 14$) falls far short of the number of loci known to affect resistance in well-studied higher organisms. In the mouse, for example, 50 loci distributed over 17 chromosomes are known to affect retroviruses alone.[46] Hence our success achieved with relatively few loci augurs very robust success as numbers are increased to realistic levels.

The most acute problem of sex is not to explain how it arose but why it does not currently disappear in view of the much greater efficiency of parthenogenesis. Therefore, except briefly in the final section of this paper where we describe one possible origin for sex, we treat only the potential for sex to resist invasion by asex. To give asex a good opportunity to invade, if it can, we assessed success by initiating the model with equal numbers of sexuals and asexuals and then finding the mean percentage of sexuals present during the last 50 years of 400-year runs. To begin the model in as natural and structured a state as possible, we ran a 70-year 'grace period' before the start. During this, the coevolution proceeded, but all individuals were asexual; then at year 0, half had their sex gene changed from 0 to 1. In the runs, sex tends to succeed or fail completely. Probably this is due to the disadvantage that both sexuals and asexuals suffer when their variation is depleted at low numbers. The internal divide implied, however, is evidently not very high: transitions occasionally occur within runs from at or near fixation for asexuals to at or near fixation for sexuals, and vice versa.

The results for various combinations of *n* and *k* are shown in Fig. 16.1. It can be seen that success of sex increases with the number of loci involved in defence against parasites. In Fig. 16.1(c), it is already asymptotic to 100 per cent, and in Fig. 16.1(d), it is everywhere already near to this except when $r = 0.0$. The high level in Fig. 16.1(d) for $nk = 2 \times 2 = 4$ (i.e. where two parasites are opposed by a total of only four loci) is particularly impressive. Equally so is the fact that success also rises steadily in almost all transects in Fig. 16.1(a) and (c) as loci increase, even if it rises somewhat behind what is being achieved where, as in Fig. 1(b) and (d), fewer parasites have been assigned more loci. The last difference is understandable on the grounds that the sexual host can change sequences in the assigned sets of loci by recombination, whereas the

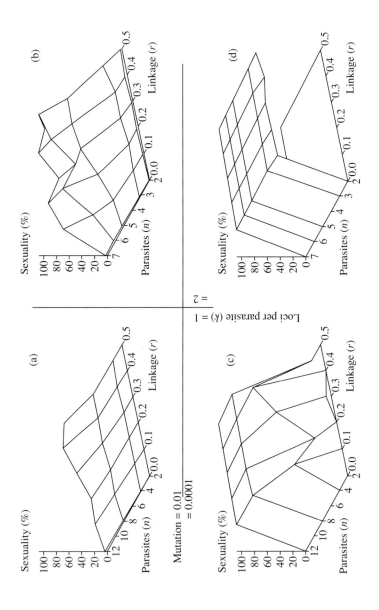

parasite can only change them by mutation. In going from one parasite to a pair of loci to one parasite to each locus, it is as if the extra parasite is giving the pre-existing one the ability to recombine, thus blunting the edge that a sexually recombining host has over both its enemy and its competitor. However, as Fig. 16.1(a) and (c) shows, this detrimental feature of multiple single-locus resistances, which would completely nullify selection for sex under plausible hard selection, is far from doing so when selection is soft. Soft selection automatically lends, even to unifactorial resistances, an epistasis in fitness, which, although changeable, is always of the 'plateau-with-coal pits' form that is already known to protect variation;[47] here, along with variation, it also protects sex (Chapter 12).

Figure 16.1 also shows that the recombination rate is not critical provided it is not extremely low. The runs described and other numerical experiments we have conducted have so far failed to show significant maxima of success for r other than 0.5. We suspect that such maxima may be demonstrated by increasing the lifespan of parasites nearer to that of their hosts so that co-evolutionary cycles are longer.[16,48]

To suggest how the model achieves its results at least for the simple case of n parasites resisted each by one locus (Fig. 16.1(a) and (c)), the following visualization seems helpful. A light ball floats in water; its surface is marked with the outline of a cube. At the eight vertices, single-genotype colonies of our host live and grow. The weight of the largest colonies makes the ball turn toward the side where they are. Eventually the vertex that is heaviest goes under water; its colony is killed by the water and drops off. When the rotation caused by the previous heaviest colony has ceased, another side—the side bearing the remaining heaviest vertices, which will be, probably, those longest out of the water—begins to roll down.

Figure 16.1 Per cent success of the allele for sexuality in the model. Each point is averaged over 10 runs with a population of 200. Total loci per host ranges from 2 to 12 in (a) and (c) and from 4 to 14 in (b) and (d).

Bell and Maynard Smith[17] ran a model similar to a case of ours as at the rightmost corner of (c), referring to it as their 'gene-for-gene' model. They found correspondingly low success in advancing an allele for free recombination as against none (frequency rose from 10^{-5} to 250×10^{-5} in 2000 generations and then showed approximate stability). However, use of 'hard' selection in their model would have contributed to low success, to judge from early hard runs in ours. Our trials of a quantitative matching model similar to their claimed more-successful version showed that this system rapidly dropped back in its support for sex, compared with the model we present, as the number of loci was increased by means of n and/or k; such reversal of success would almost certainly apply to their hard-selection pair of models also.

With a twofold cost, it is unlikely that merely two resistance loci and moderate fitness differences, as in their tables, could advance a gene for recombination. This claim is based on my findings reported in Chapter 2, and in this figure which shows that with four loci, especially if paired and opposed by two parasites (right forward edge of (d)), and with soft selection, the chance of sex succeeding is already very much better.

If the vertices of the cube are labelled with three-locus two-allele host genotypes, 000, 001, 010, etc., in the obvious way, and if weight in the model has not to do directly with biomass of colonies but rather with an accumulation of ill health that follows upon abundance, due to coevolution, then the moving image is very closely analogous to the working of the model. We can see at once, for example, that it is important how heavily the ball rests in the water—the water line is analogous to the mortality rate. Thus, heavy mortality is likely to eliminate alleles even of the sexual subpopulation: such loss becomes inevitable, for example, if a whole face of the cube goes under at once. This emphasizes the difference of the present model from previous ones in which high mortalities were essential[1,2] or merely highly favourable (Chapters 1 and 2). The previous models, however, failed to say how variability is maintained. In our model, rotation about more cornerwise axes of the unweighted cube is good for preserving polymorphism and for sex because it minimizes submergence of faces. However, the most important point that the visualization makes is that no one vertex carrying asexuals is going to escape nemesis for long, because the longer a vertex is out of the water the more weight (i.e. numbers, followed by parasite matching) gathers on it.

If mutation is rare, eliminated asexual genotypes are slow to come back. Of course, sexual genotypes at the same vertex die too, but they are by comparison very easily recreated.[49] Specifically, they are brought back by matings between individuals at appropriate opposite corners. Most important, as the number of polymorphic loci (dimensions) of the model is increased, the safety of all the alleles of the sexual population increases also: each new locus with two alleles diversifies each previous genotype into two new ones, thus doubling the number of places (vertices) where carriers of an allele can remain out of the water. Alleles that survive extinction in this way always have a chance to recolonize empty vertices. By spreading from the genotypes in which they survived, the alleles again enter all combinations: sex has been their Noah's Ark.

While not essential for successful models for sex through parasitism, the soft selection by means of truncation is extremely favourable to it (Chapter 12) as has been verified by comparing hard[2,50,51] runs of the model. Soft selection helps sex simply because, while it readily extinguishes clones,[49] it tends not to extinguish temporarily 'bad' alleles in sexual hosts.

SOFT TRUNCATION SELECTION

The power of truncation to change gene frequencies at many loci simultaneously is well known.[42,52] Under the soft truncation selection of the present model, permanent polymorphism instead of fixation arises plausibly and strongly out of the inevitable frequency dependence of the host–parasite interaction. Multilocus interaction with multiple parasites reduces amplitudes and creates situations more and more favourable to sex. This proceeds in several ways. First, multilocus polymorphism creates numerous genotypes and the possibility for every individual to be genetically unique. Every event of truncation then eliminates members of more than one genotype. It follows that currently bad alleles must often be in 'good' company. Such alleles are both protected from immediate selection and unlikely to be recombined into the very low ranking classes that are subject to truncation in fewer generations than it takes for their disadvantage to be reversed. Second, frequency dependence ensures the removal of genotypes that are common and that carry abundant alleles. Third, inherent lags due to generation turnover and to linkage disequilibria emphasize recent abundance over present or future abundance, inducing overshoot and a consequent tendency to cycle. The combined result of these factors is that, with increasing loci, the danger for sex ceases to be of allele extinction and lost variability and becomes rather that the model may reach multilocus equilibrium and not cycle at all. In practice, however, equilibrium is not observed: the destabilization implied by the third factor and by the truncation itself appears always adequate to keep the system mobile. Figure 16.2 shows that at least under the assumed symmetrical conditions of parasite interaction, after asexuals are eliminated, gene frequency remains mostly in the midrange for multilocus runs, and linkage disequilibria are low. The conditions chosen for Fig. 16.2 are those where sex is maximally stable, similar to conditions on the rearmost corner of Fig. 16. 1(d), except for lower mutation.

To check that polymorphism later continues as well protected as appears in Fig. 16.2, a run with the same parameters was continued for 7000 generations. This ended fixed for sex and with all resistance loci still variable. To challenge the model further, a second run of 7000 generations was conducted in which mutation from asex to sex was not allowed and environmental variation was added. Mutation rates at the sex and resistance loci were made 0.001 (sex to asex only) and 0.0, respectively. Gaussian environmental variation was added to the match scores to an extent that the added variance was four times the variance expected from matching in a neutral model. The result of the test was as before except that incursions of asex were more frequent and of greater magnitude. Still, the incursions hardly ever rose to half the population before declining, and still no fixations of resistance loci were observed. (Runs have shown that without the addition of environmental variation even a one-way mutation rate of 0.01, when combined with other parameters as in back left planes of Fig. 16.1, still does not result in loss of sex: some asexuals are then present at all times but generally remain at very low frequencies.)

The symmetry of the model that leads to gene frequencies distributed around 0.5 is admittedly artificial. However, much more generally significant and more potentially testable is the implication of episodic upward selection on minority alleles. The mutation theory,[19] in contrast, lacks such selection: if

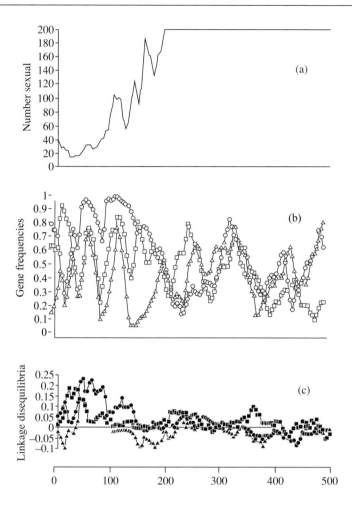

Figure 16.2 Example set of trajectories in one run for gene frequencies of the allele for sex (a), and 3 of 14 resistance loci (b), and linkage disequilibrium among these three resistance loci (c), shown every fourth generation. Model parameters are those of the most distant corner of Fig. 16.1(d) and are as described in the text except that mutation rates are lower: 0.00001 for hosts (all loci) and 0.005 for parasites. The run is started with 25 per cent sexuals.

at all, mutant alleles reach substantial frequencies only by genetic drift. It should be possible, given some identical repeats of the by now numerous surveys of electrophoretic variation in wild populations (reviewed in ref. 53), to test for the intervening occurrence of upward and downward pressures and also reversals.

The essence of sex in our theory is that it stores genes that are currently bad but have promise for reuse. It continually tries them in combination, waiting for the time when the focus of disadvantage has moved elsewhere. When this has happened, the genotypes carrying such genes spread by successful reproduction, becoming simultaneously stores for other bad genes and thus onward in continuous succession. In contrast, asexual genotypes lost in the same phases of truncation come back much more slowly by mutation only. A lost allele that specifies a complex task-specific molecule may not come back at all.

PARASITES VERSUS MUTATION: THE EVIDENCE OF ECOLOGY

Unfortunately for checking the idea against nature and discriminating it from other views, the main contender idea, that of mutation clearance, also works best under soft truncation selection.[54] Hence many of its predictions are the same. Both theories, for example, predict that species should be most sexual when they occupy saturated habitats, compete with members of their own species, and compete by contest rather than fecund breeding. 'By contest' here is meant that, while most individuals easily could breed almost as effectively as those that do, many do not breed because of lack of access to some essential prerequisite. This may be a nest hole (if a bird), territory, membership of a group, position in the canopy of the forest (if a tree), or the like. If success in contests for limiting resources is dependent on health and health depends on either mutation load or parasite resistance, both theories predict that sex will win. Such a prediction does seem to be fulfilled.[3,55] Climax or '*K*-selected' kinds of species are usually sexual, and it is generally the '*r*-selected', colonizing life forms, for which contest competition is minimal, that have the wasteful male aspects of reproductive function reduced or, as in parthenogens, completely abandoned. The basis of such a prediction is, we believe, unique to these theories; however, the facts themselves have often been presented,[3,56–58] usually with different correlates such as harshness or physical fluctuation of environment suggested as the cause.

It may be hoped that further details of the ecology of parthenogenesis and inbreeding may help to evaluate the two theories. However, this may not be easy, even by experiments. Both deleterious mutations and parasites make organisms sick, and both theories suggest that self-assessment of being sick, when there is a choice of sexual or asexual reproduction, ought to direct use of sex or, within sex, direct a higher rate of recombination. Thus the results of an experiment in which a population of *Daphnia*, say, was exposed on the one hand to mutagens and on the other to parasites and in the control to optimal conditions might well come out favouring or disfavouring the theories jointly but fail to discriminate. Both mutation clearance and parasite resistance must have outbreeding to work. Since Darwin and earlier, numerous examples of adaptation to secure outbreeding have been noted. Recently mooted theories such as DNA repair[59,60] and rapid fixation of advantageous mutations under diploidy[61] should work as well or better under inbreeding. Even inbreeding is unnecessary and not ideal: rapid fixation, for example, could come from automictic parthenogenesis, which is potentially much more efficient.[62]

Some circumstantial evidence from nature, however, favours parasites. For example, organisms of environments expected to be mutagenic but not parasite-infested do not seem to be highly sexual, but rather the converse. High levels of ionizing solar radiation in alpine habitats might be supposed mutagenic, and yet parthenogenetic and inbreeding plants are common there.[3,63] In plants more generally, the association of diminished or abandoned sexuality with the colonizing habit is particularly obvious.[15,64–72] This, as already mentioned, favours both theories against many of the alternatives. More supportive of our particular view, however, is that the association has quite often shown the expected lack of resistance to parasites when host densities rise and parasites come to the fore. A particular illustration, out of many,[69–73] is the rust fungus *Phragmidium violaceum* released in Chile for the control of alien brambles. One of the two bramble species, *Rubus constrictus*, is parthenogenetic, while the other, *Rubus ulmifolius*, is fully sexual and self-incompatible. The rust has severely checked the spread of the parthenogen, but it has affected the sexual very little.[74]

PRESERVE OR ELIMINATE

The essence of the mutation theory is that sex facilitates the elimination of unequivocally bad alleles; the essence of the parasite theory is that sex stores temporarily bad alleles and does not eliminate them. In our model the mutation process translates from one potentially useful allele to another. This is admittedly artificial. However, as presented in Fig. 16.1, it serves indirectly to emphasize the conservation aspect. When (a) and (b) are contrasted with (c) and (d) in Fig. 16.1, it is shown that a high mutation rate between alternative alleles for resistance hinders the success of sex. Recalling the image of the floating ball or, better still, thinking of an analogous floating hypercube, note that a mutation moves an individual along an edge of the hypercube. Thus this kind of mutation gives mobility to asexuals approaching that which sexuals can have through recombination and reduces the advantage of the sexuals in replacing lost types. The point also highlights the following argument and recent evidence.

If any change is good, evolving an increased mutation rate must be easy. It follows that if any new variant of a protein is effective as a tool for the host while increasing resistance to a parasite, then sex ought to die out. That this does not happen indicates, in our view, that mutations creating effective new versions of a defence molecule or physiological tactic are uncommon. Such being the case, sex has the useful function of continually recombining a set of preserved defensive elements (Chapter 5). Elements here can mean either parts of a complex molecular structure or parts in some more abstract sense—components of an individual's stance against all its currently pressing parasites. Current understanding of disease resistance does indeed favour the existence of molecular parts that would be hard to revolve once lost.[23,24,75] Molecular structure coming to light in the well-studied immune system of vertebrates, for example, favours such interpretation. So, too, does the growing emphasis on mimicry as a factor in parasitism.[27,28,76–85] Presumably almost any common, necessary, and exposed molecular structure on or in cells of a host can be potentially mimicked, so bringing the genetics of that structure into play.

Mutation rates of genes, as well as rates of evolution, have been suggested to be high for resistance,[86–88] but there is growing evidence that this is not the only reason why defence systems are genetically so diverse. Many polymorphisms that are either known[33] or likely[89,90] to be connected with disease resistance are ancient, spanning many millions of years and the divergence of genera.[75,91–95] This begins to be confirmed even by detailed homologies of DNA.[96–98] The polymorphisms include many of the human blood groups (ABO, Lewis, MN, and Rhesus), the secretor polymorphs associated with ABO, taster for phenylthiourea, lactate dehydrogenase, phosphoglucomutase, the Gm component of polymorphism of immunoglobulin, carbonic anhydrase, and others. Esterases, by being also proteases,[99] may be more connected with microbial control than their name suggests[92,100] They are often multiply varied throughout whole sets of related species, although the variants generally remain of unknown homology. The vast and much-researched polymorphism of the major histocompatibility complex system presents a similar case and has unquestionable relevance to disease. It is now known that, far from being arbitrary badges of identity as was once thought[101,102] (for which purpose any mutant might be satisfactory), the histocompatibility proteins are more in the nature of molecular tongs designed to hold up processed fragments derived from parasite attack in order to alert and inform other cells.[84,102] Random change in such functional molecules is much less likely to be successful than recombinations of components that have already built effective structures in the past. The last claim is also likely to be true of the combinations in the Lewis/secretor/ABO system,[23,24] which contrasts with the major histocompatibility complex system in that the components are not linked. Parasites whose epistatic selection pattern is imposed through direct effect on fitness potential, rather than extrinsically, as by our truncation selection, are not yet known for the Lewis/secretor/ABO system. However, they are not unlikely because parasites (as well as pesticides) clearly sometimes impose direct epistatic patterns in other species.[103,104] In wild rabbits, polymorphism of IgG based on two unlinked genetic loci determining constant regions of the heavy and light chains reveals strong epistatic selection. This is almost certainly due to parasites and is quite likely due to the myxoma virus,[75] but it is not yet possible to apportion the selection to intrinsic complex selection or/and an extrinsic soft process like truncation. If present, intrinsic epistasis probably further favours sex. Runs of our model with such epistasis added through the method of scoring matches gave even stronger support than the version we report.)

Although our model works best with multilocus defences against each parasite (compare success in (b) and (d) to that in (a) and (c) in Fig. 16.1), our findings do not compel us to interpret the model as implying a kind of molecular Lego in nature (success is still near complete in Fig. 16.1(c) at higher n). For a sapling tree racing with conspecifics to be first to occupy a light gap in a woodland canopy, one parasite might be a new virus strain, another might be a stem-mining caterpillar with a new detoxifying enzyme, another a deer (its population rising, perhaps, because it had found a solution to a disease problem of its own), and another an exploitative instead of protective variant of a mycorrhizal fungus. The necessary correctives for these threats are quite different in character. The first, for example, may need the recombination of molecular parts (e.g. ref. 75), whereas the second and third may need change of a more quantitative kind—a thickened cuticle to prevent hatchling larvae boring in or, for the deer, more tannin in the bark. Altogether, assembling adaptations for any such upcoming set may have the same urgent requirement for recombination as the more purely molecular kind of re-arrangement in defence previously outlined. Moreover, although it helps if parasites are capable of causing severe damage, even this is not necessary. Enough parasites with small effects, each still contributing to ranking,[69] can make sex secure, to judge from the trend we have shown. For example, in a gorilla or hominid (to which our life schedule gives a rough approximation), a contribution could come from resistance to the common cold.

LOW FERTILITY AND PENETRANCE

Extremely low fecundities in the animal kingdom are consistent with sexuality in the parasite model but not with the mutation model as it is currently specified.[19] Humans are usually regarded as having low fecundity, but our species does not have the lowest fecundity, by any means. Some scarabs have mean female lifetime fecundities as low as five but are fully sexual and indeed show exaggerated characters from sexual selection.[105–109]

To check that our model can account for *H. sapiens* or an ancestral hominid (which, by arguments based on sexual size dimorphism, would probably lack paternal care and therefore be more directly relevant to the model), we ran a menopause age of 35 and a lower mortality of 1/16 per year (see also note 22). By combining this with 12 parasites being resisted by one locus each, with host mutation at 0.0001 and recombination at 0.3 (a midpoint on the back wall of Fig. 16.1(c) but with a lower, more realistic mutation rate), we found that the model still gave almost full success. The model has hermaphrodite individuals, but there is no essential difference from a version where sexual individuals are born male or female while parthenogenetic progeny, equally numerous, are all female. Hence we confidently expect the same result for a two-sexed case. Calculation shows that when sexuals dominate in this parallel version, females who live right through the reproductive period would seldom bear more than 10 offspring, while the expectation for those just entering the period is (because of mortality) about five. This schedule is almost too infecund to be plausible for a human hunter-gatherer. When we brought in the model maximum of 15 loci (triplets of loci resisting five parasites), lowered at the same time the mutation rate of hosts and parasites by factors of 100 and 5, respectively (i.e. going again to rates more realistic than those in Fig. 16.1), and set mortality to 1/35 while keeping the menopause age 35, all of which gives mean gross fertility of four, as might be appropriate to extreme scarabs,[105-107] we found sex is only occasionally replaced by asex in runs. It would probably be replaced less often still if populations were made larger.

Even with genetic resistance diluted by large amounts of random variation, sex is still uninvadable in our model. This is confirmed by runs with environmental variance approximately equal to the parasite-induced variance (with mean gross fertility as low as three) and, as mentioned in an earlier section, even with runs with four times the environmental variance (for $n = 7$ and $k = 2$).

MUTATION RATE

Keeping germinal tissues deep within a body plus using internal fertilization and viviparity might be expected to buffer germ-line chromosomes especially well. Therefore, in the mutation theory, species with such characteristics might be expected to be experiencing relaxed mutation problems and to be frequently experimenting with gynogenesis and parthenogenesis. The opposite is the case: it is prokaryotes, those organisms that do not buffer the environment at all by a surround of cell layers, or even by a layer of cytoplasm, that reject sex most completely. In spite of their exposure, their rates of mutation per generation are actually lower than those of large eukaryotes by several orders of magnitude. This together with other data[108] suggest that the mutation problem can be solved when sex is not present to help; when, on the other hand, sex is present for other reasons, mutation control can be relaxed, perhaps with long-term advantage for evolutionary flexibility.

ORIGINS

The parasite model of sexual reproduction suggests how sex might have arisen in a natural and continuous manner, as Darwinian theory generally requires. The origin of sex as suggested by Margulis and Sagan[109] has cannibalism by primitive unicells in times of starvation that evolves to a stalemate (at least in some encounters), with would-be cannibals becoming fused but eventually separating when the conditions ameliorate. While fused, some genetic material might, accidentally at first, become exchanged. Our contribution is to suggest the important way that genes encouraging regular chromosome breakage and exchange might confer advantages to their possessors, by means of the newly created gene combinations. We suggest that eventually the new combinations proved their worth mainly during unwanted contacts with smaller unicells that were specialists in exploitation—incipient parasites.

Acknowledgements

We thank B. Sumida for help and advice and A. Burks, M. Cohen, A. Grafen, J. Holland, Y. Iwasa, S. Nee, A. Pomiankowski, R. Riolo, M. Savageau, and C. Simon for helpful discussion. This work was supported by the Division of Biological Sciences and the Museum of Zoology at the University of Michigan (W.D.H.) as well as the National Science Foundation and the Kellogg Foundation (R.A.).

References and notes

1. G. C. Williams, *Sex in Evolution* (Princeton University Press, Princeton, NJ, 1975)

2. J. Maynard Smith, *The Evolution of Sex* (Cambridge University Press, Cambridge, 1978).

3. G. Bell, *The Masterpiece of Nature: The Evolution and Genetics of Sexuality* (University of California Press, Berkeley, 1982).

4. R. M. May and R. M. Anderson, *Proceedings of the Royal Society of London B* **219**, 281–313 (1983).

5. R. Y. Lamb and R. B. Willey, *Evolution* **33**, 774–5 (1979).

6. A. R. Templeton, in H. Dingle and J. P. Hegman (ed.), *Evolution and Genetics of Life Histories*, pp. 76–85 (Springer, New York, 1982).

7. B. Clark, in A. E. R. Taylor and R. Muller (ed.), *The Ecological Genetics of Host–Parasite Relationships*, pp. 87–103 (Blackwell Scientific, Oxford, 1976).

8. J. Jaenike, *Journal of Evolutionary Theory* **3**, 191–4 (1978).

9. H. J. Bremermann, in S. C. Stearns (ed.), *The Evolution of Sex and Its Consequences*, pp. 135–94 (Birkhauser, Basel, 1987).

10. J. J. Tooby, *Journal of Theoretical Biology* **97**, 557–76 (1982).

11. W. R. Rice, *American Naturalist* **121**, 187–203 (1983).

12. H. J. Bremermann, *Experientia* **41**, 1245–54 (1985).

13. D. Weinshall, *American Naturalist* **128**, 736–50 (1986).

14. C. M. Lively, *Nature* **328**, 519–21 (1987).

15. D. A. Levin, *American Naturalist* **109**, 437–51 (1975).

16. V. Hutson and R. Law, *Proceedings of the Royal Society of London B* **213**, 345–451 (1982).

17. G. Bell and J. Maynard Smith, *Nature* **328**, 66–8 (1987).

18. J. T. Manning, *Journal of Theoretical Biology* **108**, 215–20 (1984).

19. A. S. Kondrashov, *Nature* **336**, 435–40 (1988).

20. The complete program documentation is available from W. D. H.

21. W. D. Hamilton, *Journal of Theoretical Biology* **12**, 12–45 (1966) [reprinted in *Narrow Roads of Gene Land*, Vol. 1, pp. 94–128].

22. Humans have paternal care whereas, to judge from skeletal sexual size differences comparable to those of baboons and gorillas, male australopithecines probably did not. A twofold cost of sex thus might be appropriate to primitive hominids; in any case, it applies a conservative test of the model. The more general motivation for using sequential (iteroparous) reproduction is to simulate varied realistic life histories, including, if required, the extreme of once-for-all reproduction (semelparity). Another reason is to forestall the criticism that, if only semelparous reproduction had been used, then introducing overlap of host generations might have so dampered cyclicity that sex would lose support in many realistic life histories. The demography given is easily extended to include a menopause, thus providing a parameter to control a gradual approach to semelparity. If m fertile years start at age $j + 1$, the first menopause birthday is $j + m + 1$. If the mean annual (interbirthday) survival is $v = 1 - d$, independent of age, the formulas of the text become modified to $f = d/\{v^{j+1} (1 - v^m)\}$ and $G = j + h - [mv^m/(1 - v^m)]$. Such formulas (found by summing appropriate finite geometric series) merely display necessary average consequences of the parameters j, m, and d that arise under the given demography and are never used by the model itself. Results of runs in which m is used to set a hominid menopause are mentioned in the section 'Low fertility and penetrance' on page 660.

23. W. F. Bodmer and L. L. Cavalli-Sforza, *The Genetics of Human Populations* (Freeman, San Francisco, 1971).

24. H. Clausen and S. Hokomori, *Vox Sanguinis* **56**, 1–20 (1989).

25. F. E. L. Pereira, E. R. Bortolini, J. L. A. Carneiro, C. R. M. Da Silva, and R. C. Neves, *Transactions of the Royal Society of Tropical Medicine and Hygiene* **73**, 238 (1979).

26. G. L. Barnes and R. Kay, *The Lancet* **i**, 808 (1977).

27. J. E. Decker-Jackson and B. M. J. Honigberg, *Journal of Protozoology* **25**, 514–25 (1978).

28. C. L. Greenblatt, J. D. Kark, L. F. Schnur, and G. M. Slutzky, *Lancet* **i**, 505–6 (1981).

29. R. Overfield and M. R. V. Klauber, *Human Biology* **52**, 87–92 (1980).

30. C. C. Blackwell, K. Jonsdottir, M. Hanson, W. T. A. Todd, A. K. R. Chaudhuri, B. Mathew, R. P. Brettle, and D. M. Weir, *The Lancet* **ii**, 284–5 (1986).

31. C. C. Blackwell, D. M. Jones, D. M. Weir, J. M. Stuart, K. A. V. Cartwright, and V. S. James, *Epidemiology and Infection* **102**, 1–10 (1959).

32. C. C. Blackwell, K. Jonsdottir, M. Hanson, and D. M. Weir, *The Lancet* **ii**, 687 (1986).

33. J. D. Clemens, J. Svennerholm, D. A. Sack, M. R. Rao, M. R. Khan, F. Ahmed, J. Gomes, S. Huda, J. R. Harris, and J. Chakraborty, *Journal of Infectious Diseases* **159**, 770–3 (1989).

34. P. Esterre and J. P. Dedet, *Annals of Tropical Medicine and Parasitology* **83**, 345–8 (1989).

35. J. F. Albright and J. W. Albright, *Contemporary Topics in Immunobiology* **12**, 1–52 (1984).

36. J. F. Boyle, D. G. Weismuller, and K. V. J. Holmes, *Journal of Virology* **61**, 185–9 (1987).

37. T. L. W. Rothwell, S. E. Pope, and G. H. Collins, *International Journal of Parasitology* **19**, 347–8 (1989).

38. E. Skamene, *Review of Infectious Diseases* **11**, Suppl. 2, S394–9 (1989).

39. N. Bumstead, M. B. Huggins, and J. K. A. Cook, *British Poultry Science* **30**, 39–48 (1989).

40. J. A. White, A. Herman, A. M. Pullen, R. Kubo, J. W. Kappler, and P. Marrack, *Cell* **56**, 27–35 (1989).

41. D. T. Briese, in E. W. Davidson (ed.), *Pathogens of Invertebrate Microbial Diseases*, pp. 511–45 (Allenheld/Osmun, Totowa, NJ, 1981).

42. J. A. Sved, *American Naturalist* **102**, 283–93 (1968).

43. I. Eshel and E. J. Akin, *Journal of Mathematical Biology* **18**, 123–33 (1983).

44. J. W. Lewis, *Journal of Theoretical Biology* **93**, 927–51 (1981).

45. R. M. Anderson and R. M. May, *Parasitology* **85**, 411–26 (1983).

46. S. J. O'Brien and J. F. Evermann, *Trends in Ecology and Evolution* **3**, 254–9 (1988).

47. S. Karlin and R. P. Campbell, *American Naturalist* **117**, 262–75 (1981).

48. A. Sasaki and Y. Iwasa, *Genetics* **115**, 377–88 (1987).

49. M. Treisman, *Journal of Theoretical Biology* **60**, 421–31 (1976).

50. B. Wallace, *Evolution* **29**, 465–73 (1975).

51. A. Lomnicki, *Population Ecology of Individuals* (Princeton University Press, Princeton, NJ, 1988).

52. J. F. Crow and M. Kimura, *Proceedings of the National Academy of Sciences, USA* **76**, 396–9 (1979).

53. E. Nevo, *Theoretical Population Biology* **13**, 121–77 (1978).

54. A. S. Kondrashov, *Genetical Research* **40**, 325–32 (1982).

55. R. Trivers, *Social Evolution* (Benjamin/Cummings, Menlo Park, CA, 1985).

56. E. Suomalainen, A. Saura, and J. Lokki, *Evolutionary Biology* **9**, 209–57 (1976).

57. R. R. Glesener and D. Tilman, *American Naturalist* **112**, 659–73 (1978).

58. H. J. Michaels and F. A. Bazzaz, *American Naturalist* **134**, 190–207 (1989).

59. I. Walker, *Acta Biotheoretica* **27**, 133–58 (1978).

60. H. Bernstein, F. A. Hopf, and R. E. Michod, in R. E. Michod and B. R. Levin (ed.), *Evolution of Sex*, pp. 139–60 (Sinauer, Sunderland, MA, 1988).

61. M. Kirkpatrick and C. D. Jenkins, *Nature* **339**, 300–1 (1989).

62. J. J. Bull and P. H. Harvey, *Nature* **339**, 260–1 (1989).
63. P. Bierzychudek, *Experientia* **41**, 1255–64 (1989).
64. G. L. Stebbins, *Cold Spring Harbor Symposia on Quantitative Biology* **41**, 365–78 (1958).
65. J. A. Cullen, P. F. Kable, and M. Catt, *Nature* **244**, 462–4 (1973).
66. A. J. Richards, *Plant Breeding Systems* (Allen & Unwin, London, 1986).
67. H. J. Michaels and F. A. Bazzaz, *American Naturalist* **134**, 190–207 (1989).
68. J. J. Burdon, *Diseases in Plant Population Biology* (Cambridge University Press, Cambridge, 1987).
69. J. J. Burdon and D. R. Marshall, *Journal of Applied Ecology* **18**, 649–58 (1981).
70. J. G. Moseman, E. Nevo, M. A. El Morshidy, and D. Zohary, *Euphytica* **33**, 41–7 (1988).
71. A. Segal, J. Manisterski, G. Fishbeck, and G. Wahl, in J. G. Horsfall and E. B. Cowling (ed.), *Plant Disease: An Advanced Treatise*, Vol. 5, pp. 76–102 (Academic Press, London, 1980).
72. R. K. Koehn, in M. E. Feder, A. F. Bennett, W. W. Burggren, and R. B. Huey (ed.), *New Directions in Ecological Physiology*, pp. 170–88 (Cambridge University Press, Cambridge, 1987).
73. E. Bruzzese and S. Hasan, *Annals of Applied Biology* **108**, 527–33 (1986).
74. E. B. Oehrens and S. M. Gonsalez, *Agro Sur* **5**, 73–85 (1977).
75. W. van der Loo, in S. Dubiski (ed.), *The Rabbit in Contemporary Immunological Research*, pp. 165–90 (Longman, Harlow, Essex, 1987).
76. I. R. Cohen, *Scientific American* **258** (4), 34–42 (1988).
77. R. T. Damian, in B. B. Nickol (ed.), *Host–Parasite Interfaces*, pp. 104–26 (Academic Press, New York, 1979).
78. D. P. Lane and W. K. Hoeffler, *Nature* **288**, 167–70 (1980).
79. D. P. Lane and H. Koprowski, *Nature* **296**, 200–2 (1982).
80. M. V. Haspel, T. Onodera, B. S. Babhakar, M. Hovita, H. Suzuki, and A. L. Notkins, *Science* **220**, 304–6 (1983).
81. D. Vidovic and P. Matzinger, *Nature* **336**, 222–5 (1988).
82. T. Leist, A. Althage, E. Haenseler, H. Hengartner, and R. M. Zinkernagel, *Journal of Experimental Medicine* **170**, 269–77 (1989).
83. V. K. Singh, K. Yamasaki, T. Abe, and T. Shinohara, *Cellular Immunology* **122**, 262–73 (1989).
84. J. G. Guillet, M. Z. Lai, T. J. Briner, S. Buss, A. Sette, H. M. Grey, J. A. Smith, and M. L.Gefter, *Science* **235**, 865–70 (1987).
85. Å. Lernmark, T. Dyrberg, L. Terenius, and B. Hokfelt (ed.), *Molecular Mimicry in Health and Disease* (Elsevier, Amsterdam, 1988).
86. R. E. Hill and N. D. Hastie, *Nature* **326**, 96–9 (1987).
87. M. Laskowski, Jr, I. Kato, W. Ardelt, J. Cook, A. Denton, M. W. Empie, W. J. Kohr, S. J. Park, K. Parks, B. L. Schatzley, L. S. Oeyvind, M. Tashiro, G. Vichot, H. E. Whatley, A. Wieczorek, and M. Wieczorek, *Biochemistry* **26**, 202–21 (1987).
88. J. Klein, *Biology of the Mouse Histocompatibility-2 Complex* (Springer, New York, 1975).
89. E. R. Giblett, *Genetic Markers in Human Blood* (Blackwell, Oxford, 1969).

90. W. J. Miller, *Bioscience* **26**, 557–62 (1976).

91. A. B. Chiarelli (ed.), *Comparative Genetics in Monkeys, Apes and Man* (Academic Press, London, 1971).

92. P. R. Anderson and J. G. Oakshott, *Nature* **308**, 729–31 (1984).

93. N. A. Barnicot, *Science Progress* **57**, 459–93 (1969).

94. R. E. Tashian, D. C. Schreffler, and T. B. Shows, *Annals of the NY Academy of Sciences* **151**, 64–77 (1969).

95. R. D. Sage, J. B. Whitney, III, and A. C. Wilson, *Current Topics in Microbiology and Immunology* **127**, 75–85 (1986).

96. F. Figueroa, E. Gunther, and J. Klein, *Nature* **335**, 265–7 (1988).

97. D. A. Lawlor, F. E. Ward, P. D. Ennis, A. P. Jackson, and P. Parham, *Nature* **335**, 268–71 (1988).

98. T. Sagai, M. Sakaizumi, N. Miyashita, F. Bonhomme, M. L. Petras, J. T. Nielsen, T. Shiroishi, and K. Moriwaki, *Immunogenetics* **30**, 89–98 (1989).

99. J. G. Oakeshott, C. Collet, R. W. Phillis, K. M. Nielsen, R. J. Russell, G. K. Chambers, V. Ross, and R. C. Richmond, *Proceedings of the National Academy of Sciences, USA* **84**, 3359–63 (1987).

100. K. D. Vernick, and F. H. Collins, *American Journal of Tropical Medicine and Hygiene* **40**, 593–7 (1989).

101. L. Andersson, S. Paabo, and L. Rask, *Immunology Today* **8**, 206–9 (1987).

102. H. M. Grey, A. Sette, and S. Buus, *Scientific American* **261** (5), 38–46 (1989).

103. R. M. Sawicki, *Pesticide Science* **4**, 171–80 (1973).

104. C. N. Law, P. R. Scott, A. J. Worland, and T. W. Hollins, *Genetical Research* **25**, 73–9 (1976).

105. H. G. Klemperer and R. Boulton, *Ecological Entomology* **1**, 19–29 (1976).

106. G. Halffter and Y. G. Lopez, *Annals of the Entomological Society of America* **70**, 203–13 (1977).

107. P. B. Edwards, *Oecologia* **75**, 527–34 (1988).

108. H. Nothel, *Proceedings of the National Academy of Sciences, USA* **84**, 1045–9 (1987).

109. L. Margulis and D. Sagan, in G. Stevens and R. Bellig, *Nobel Conference XXIII: The Evolution of Sex*, pp. 23–40 (Harper and Row, San Francisco, 1988).

UCCELLO/OTHELLO

Mate Choice Near Or Far

What men call gallantry, and Gods adultery
Is much more common where the climate's sultry.
BYRON[1]

I T W A S not long after Marlene Zuk and I published our paper in
Science about parasites and sexual selection (Chapter 6) that my idea about
the cardinal in the snowy Nichols Arboretum began to change. Originally I
thought of any male cardinal's song in the way, I suppose, everyone else
did—or at least I thought the way other biologists did. Non-biologists
usually think that birds sing simply for joy. Well, such a popular view may
indeed be part of the truth, but for biologists song has now a long history of
more serious assessment as advertisement for a mate and as a challenge. In
the wintry conditions of Michigan where I heard it, ages before any bird
was nesting, very likely every one of these varied aspects, popular and
professional, were true; but, again, looking more prosaically, could there be
something more to it than the 'health for monogamy' issue I discussed in
Chapter 6?

Papers I was reading and bird incidents I was seeing indeed suggested
something else. Perhaps the biggest change in my old idea came when I
watched two chaffinches through the glass panes of our kitchen door here
in Wytham, my Oxford home (thus later than 1984); but perhaps equally it
came long before that, before 'H&Z' (Chapter 6) was written and again, as
with the cardinal, came in the arboretum, as I walked to work. In this case
it was three American goldfinches engaged in some complicated pursuit
among the willows near the Huron River and the railway.

I witnessed one of the two males suddenly simply give up the chase.

As the other two flew on, he sat on a twig, closed his eyes, and shivered. After observing him for a few moments, I shivered, too, and not just in sympathy for his plight: it was for my own. All over the world, Romanticism, all of that broad path of idealism and artistry that I had grown up in, I was coming to see as dead. Here by river, in the bird world, in which we see ourselves so often and so well reflected, I felt myself watching Romanticism dying once again. A wishful enthusiasm had lasted us a century and a half; it had been too long, and people were persuaded that it did not work. As with our similar fad for chivalry in the slightly deeper past, we'd pulled too far away from real human and animal nature, and now inexorable realism was jerking our lead and we must come back. A better vision—that is, more intermediate and more truthful—had been that more mixed and antecedent one of the poets earlier than Byron and Keats. It was that of, say, Erasmus Darwin with his poems. Now the birds were turning out not so monogamous for us as they had been; they had their love triangles, too, their one nights in other beds. The chase I had just watched needn't be about 'honourable intentions' of pairing and bonding as I would once have assumed; it was far more likely to concern either egg-dumping or fertilization. There were fears by one party of raw acts of either parasitism or sex. I knew in this case that the nesting of the birds I was watching was likely to have been long in progress. Altogether my disillusion about the romantic ideal had been progressive and dreary but it was punctuated by acute moments like this one. As to the bird world as it had stood in the literature, I believe disillusion started much longer ago still—for me maybe about 1960. Along with other romantics I had once believed a pure *joie de vivre* theory of birdsong; a different view began with a verse I read at Cambridge in David Lack's little book about the robin, which he used to head one of his chapters:

> O blind to Nature's all accordant plan,
> Think not the war song is confined to man;
> In shrill defiance ere they join the fray,
> Robin to robin chaunts the martial lay.[2]

It sounds like Erasmus Darwin himself—something from that period anyway—but the authorship is given only as 'Anonymous'. Lack himself had been a pioneer realist over evolutionary competition and I much admired, as honest evolutionary thinking, his approach to understanding brood size and similar matters. Even as an undergraduate I had been cheered by his trenchant and rational opposition to 'Benefit-of-Species' reasoning, which, in the way I now was inclined to see it, was just one of the many sectors of the vast romantic delusions 'Anon.' was criticizing. Blind indeed we had been; but still it was taking me a long time to accept the accumulating evidence that insight of the same kind had to apply to humans.

Yellow vibrant streaks among the willows, beautiful finches: it was that which, like anyone else, I had noticed at first during my walk by the Huron, not caring a bit what they might be busy with. It was the odd action of the one male, that conspicuous giving up, that had made me think harder. Later, as I watched the chaffinches in Wytham I expect that that former ruffled bird on the willow twig shivering and closing his eyes still perched somewhere in the back of my mind. At first I'd have been more aware of—was more 'floating amidst' is perhaps the better expression for those time-out moments when one turns away from what one is doing—all the other merely beautiful bird glimpses, those moments and poses the Chinese catch so well in their pictures. One notices in spring, for example, all the new bright colours, the lively activity. There by the Huron I remember spring optimisms of yellow of many kinds, be they of newly arrived highwaymen yellowthroats in their black masks or goldfinch 'dons' wearing black caps of academe, or simply the yellow greens of the long twigs and leaves of the sprouting willows . . . and then in that far-off case, as background, I can see the silvery Huron gliding in its straight stretch down to the Gallup Park and, alongside, it, those parallel straight steel lines I stepped over every day. I would have hoped for all that kind of spring vividness and no complications again here at Wytham, outside my glass-paned door but, in fact, received both the same vividness and the same disillusion. Spring is always wonderful whether it contains the quarrels and suspicions of wildlife and/or of humanity, or not. Here, to come to the point, it was a pair of chaffinches that I saw hopping in parallel and pecking at something on the ground just beyond the glass panes. As I hadn't thrown out any crumbs recently I looked more closely and on the grey-greenish but smart female saw that an astonishing ornament was appearing. They came very close. She was growing a pale, long moustache, was whiskered like a cat or an emperor tamarin and I soon saw why. Our dog spent a lot of time lying outside that door and she was now collecting its hairs and arranging the bunch neatly in her bill to carry to her nest. When I switched my attention to the male, who came a few feet behind, resplendent in a pink waistcoat where she was grey and green, I had another surprise. I saw him equally busy but there was something amiss. He, too, pecked at the ground from time to time, but he had grown no whiskers at all: not one hair for the nest, seemingly, had he yet managed to find. A slow worker, meaning well—would I see him continue and finish his task after his mate had gone with her load? No such thing: when she at last flew, he was off behind her in a flash and they disappeared over our wall no more than a metre apart.

As if borne to me on the same sharp wings, understanding and a certain sad sense of disillusion once again swooped: all the papers that I had been reading had to be right; I had to accept. Mate guarding, not affection, nor work for the brood, had brought this male to accompany his mate. A new storm, it now seemed to me, was really in the wind of sex and love. Amongst other things it was providing a kind of answer of men to feminists (although actually I still see this issue mentioned in discussions surprisingly little; still we seem to allow almost all high moral ground to them): see, we could say, these new facts are why we males have always been as we have to be about you.

The most basic evidence blowing in this new wind was coming out of the immunological and DNA labs, not out of musings in the field (or in this case, the kitchen) like mine. A 10 per cent impaternity of offspring was being rumoured in 'typical' Western human families; a first survey on this was supposed to have been on householders in undistinguished but beautifully named Kalamazoo, a town not far from Ann Arbor. The new phrases were 'impaternity' and 'extra-pair copulations' (EPCs). In the bird world it was the same, of course, but sometimes the rates of infidelity ran amazingly high and hardly any showed zero. Uccello had become Othello and often was doubting his spouse, it seemed, for much better reason than had Shakespeare's Moorish lord. With humans, as with birds, females were not always the faithful Desdemonas that Romanticism had depicted and, by end of the 1980s—that is, right within the period that ran from my American goldfinches in the early years to my English chaffinches in the late—the great DNA unveiling of the world's intimate secrets and lies that revealed this was in full swing.

If there wasn't some frequent need for it—and often—the male chaffinch wouldn't need to follow his lady so closely, nor pretend to pick up hairs behind her when really just keeping an eye. As I watched my mind had flicked back to those golden cousins in the rough fields by the Huron. Had that male who 'gave up' then, who seemed not entirely well and probably was feverish, had he been the monogamous mate while the one who won the female, at least for the crucial brief moment, had been a particularly resplendent intruder—a male noticed by her perhaps for a week or more past, singing from the top of some far-off bush? Otherwise (and this had been a question bothering me for a long time), why do most songbird males still sing even after nesting has begun and all eggs in the nest he is supposed to care about are laid? Even think of my old winter cardinal: was he, too, perhaps, setting up to have a role in other pairs' lives as well as singing for pairing and for bonding for himself—launching as distantly as he could a particular macho and cold-hardy image for all females to see, hear, and remember far into the months to come?

This theme and ideas related to it have, as I say, come to be abundantly documented in the past 20 years[3] and the idea I raise only as a likelihood in the paper of this chapter (possibly for the first time in the biological literature; see p. 708)—namely, that bright-male species might have higher than usual differences of male quality and that from this may come another correlation, a high rate of extra-pair copulation—can by now be considered proven.[4] Anticipating the snowstorm of EPC studies that had dropped its first desultory flakes into the early 1980s, it became shortly Randy Thornhill's and Glenn Hausfater's initiative to seed a particular shake out (Hausfater likes to fly light planes, I recall, which is perhaps why these atmospheric images float to my mind). Randy Thornhill was a pioneer sociobiologist whose work at the time of the meeting was in mid-flight (in New Mexico) between the scorpion flies on which he had started (showing them almost as distressingly human in their sexual ways as the birds I have discussed above), and jungle fowl; he now works mostly with humans. Glenn Hausfater is best known for work on the sociality of baboons; I mention some later work on parasites below. The papers I am thinking of from the meeting, all of them concerned with the testing of notions of how chronic disease and parasites might underpin sexuality, were collected in a 'symposium' issue of *American Zoologist*. In essence, the symposium was based on all that had happened in the half dozen years since my paper with Marlene Zuk came out—that is, since the paper that I hope you have already read about in Chapter 6 and its Appendix. The occasion for the symposium was an annual meeting in the winter of 1988–89 of the American Society of Zoologists (this fixed jointly with some other scientific societies) in the Hilton Hotel of San Francisco. My paper of this chapter was one in that issue. Glenn had known well how sexual-selection studies were going at Oxford because he had been working with us during a sabbatical, finishing a previous study on primates in Africa (that one secondarily focusing on their parasites) and also managing a study of the parasites affecting a particular tree frog in northeastern America.

Glenn's last work, which actually found rather negative results from my point of view, were also read in the meeting: frogs and toads, it appears, sing, jump, swim, and shamble astonishing well even when carrying quite a zoo of parasites on and within them. Glenn, in short, found surprisingly little effect on the bearers' success in life except when the loads were extreme.[5]

Two others from my Oxford subdepartment attended the San Francisco meeting and brought to it more positive results than did Glenn. They were Howard McMinn, who had been working on parasites as they affect guppy sexual selection, and Nigella Hillgarth whom Glenn knew to be finishing her thesis study on coccidia in pheasants.[6] Marlene Zuk, now working on jungle fowl with Randy Thornhill and David Ligon at Albuquerque, brought to San Francisco yet another galliform perspective that happily for us pointed the same way as Nigella's.[7] Both Nigella's coccidia and Marlene's worms indeed affect what may be called the 'cockerel-type' ornaments of their respective species and thereby, they showed, affect their male host's attractiveness.[8] All this can be read in the appropriate issue of *American Zoologist*: Volume 30, Number 2 of 1990. The title of the symposium was 'Parasites and Sexual Selection: Theory and Data'.

I think Randy and Glenn had determined that, as regards new 'data' they would select for the meeting, it would be slanted to the direct experimentation side rather than to still further attempts to assess parasite effects through interspecies comparisons, such as Marlene's and my first attempt (Chapter 6). Direct experimentation had lagged behind comparative studies. Otherwise it is hard to understand that, except for me and one other 'human-paper' that I mention below, no one working on comparative studies was invited to speak. Possibly this omission (though this was unknown to us as we prepared for the meeting) sowed seeds of controversy over the 'H&Z' idea, perhaps starting those rescorings and retests being planned at Oxford (Chapter 6 and its Appendix) and which were actually occurring simultaneously with the meeting. From the way they were presented to me soon after I had returned from the Americas (I had gone on to Amazonia immediately after the meeting) these studies appeared to me designed with intent to confound rather than simply to retest our original findings. Controversy is the proper medicine for complacency in science, of course, and what was generated in this case undoubtedly led in the long run to better understanding of how to control and interpret all kinds of comparative data, both on this 'H&Z' issue and on many others. The substantial consequences all this had for our particular topic I have tried to lay out more fully in the Appendix to Chapter 6.[9]

At the symposium I was, I suppose, meant to present the first part in the 'theory and data' of the title but I got by, as perusal of the coming paper will show, by presenting only vague thoughts on that and then new aspects of evidence. A colleague to whom I had showed my manuscript before the meeting told me it was a good read but 'wacky' in its content. As from one who would be likely to focus on the fact-to-speculation ratio and happened not to share the aesthetic that had directed my guesses—and no doubt who had not watched American goldfinches in the Nichols Arboretum nor male chaffinches collecting nothing at his door in Wytham—the comment seemed fair and frank. But I am pleased in retrospect to find that my paper seems to have hit several nails-to-come quite squarely—that is, to have noted or predicted several now rapidly expanding themes of the field. Some matters, of course, were not well predicted, at least, as it seems, so far. My idea of an extensive male cheating on monogamy through encouraged egg-dumping, for example, hasn't turned out as I thought it might although the evidence trickles on and it remains not negligible.[10] Possibly there are already a few instances of the particular sire-encouraged egg intrusions I suggested.[11] I have also in the 10 years not witnessed any dramatic increase of evidence in support of my idea that frequent venereal infections might push the master switch in an evolutionary course, potentiating the reversal of the usual sex roles. But some papers moderately positive for the theme have come to notice,[12] including some that I hadn't seen when I wrote the paper,[13] and nothing has been particularly negative: for me, the idea still hangs in limbo.

Counterbalancing such negatives and irresolution, my faith that 'good genes', rather than immediate benefits to mate or young, were to prove the dominant force in mate choice, as also in motivating EPCs, has seemed to march forward very well. The way I wrote this idea makes it apparent that the clues I used were in the main personal and introspective, a set of uninhibited human comparisons, inspired no doubt partly by lust and dreams—ideas that many might think ridiculously unlikely guesses as applied to the far more stereotypical behaviour of birds. I included, too, how these ideas fitted as warp to my varied wefts—that is, other ideas that at first glance might likewise seem rather distantly related. For example, I was trying to unify these mate-choice ideas with the generality of polymorphic variability, the long-pounded but still mysterious mass that seemed to me waiting, despite its huge literature, for some leaven of explanation.

As we see it now, long before we humans thought or spoke in our half joking ways about our 'oldest profession', birds at various points of their tree invented something very like the intersex economics of sexual frustration that we eventually came to. Sex for nectar,[14] sex for nest material (stones, astonishingly enough),[15] sex doubtless for many gifts and allowances that have yet to be identified: the penguins proceed in silence, words and calculations seem not to be needed to understand the exchanges. Notwithstanding the ever more obvious primacy of language in human affairs, each year ethology and related fields bring us further disproof of the idea that we are psychically set apart from animals. For me, the success of my 'wacky' themes in an essay that mainly attended to birds with such different modes of behaviour is only further proof to me of the non-blankness of any slates of human psychic development we have: if I am found to guess birds correctly from my dreams concerning human lust, isn't it likely we can all also guess a lot about humans from watching birds? Our roots reach far down indeed and, so far, we hardly do more than give them a shinier bark.

The exception to the absence of comparative studies at the symposium was Bobbi Low's contribution on the correlation of human marriage patterns with patterns of chronic parasitism across the globe.[16] Bobbi is a biologist and evolutionary psychologist in the School of Natural Resources at the University of Michigan, who has worked in what I think I might rightly call evolutionary sociology, a rare combination and more daring than what might sound the same—plain old sociobiology, which, Ed Wilson's famous chapters notwithstanding, is mostly not focused on humans. It is true that all her study in this case was confined to the one species, us, but the underlying comparative idea was exactly that which Marlene and I had floated. Our own species, in fact, makes a good pretence of being many. No one can deny that there are areas on the slate of development that are blank in humans: these are the areas that can be quickly written upon both by individual idiosyncrasy and by local cultures.[17] In fact, through the implied human virtuosity in learned development it is almost as if we humans are re-attempting everything the tree of life has ever tried before in our or in any branch: meme-mutant experiments of our species fire off continuously in every direction like catherine-wheel fireworks, they attempt every conceivable variety of experience. What is it like for a lately tropical animal to live on ice, seal blubber, and fish like a polar bear? Ask an Eskimo. What is it like to cross waterless deserts like a camel, carrying water in a gourd or an animals' skin? Ask a Taureg. To fly like a bird or climb walls of mountains the way a gecko goes up those of a house? Ask an aviator or a mountaineer. Be a harem master of a hundred like a great stag, or else a weak ally of such, clinging near that kind of master, in his shadow, perhaps charming him with jokes or song, or with novelties of homosexual dalliance. Ask a sultan or, for the other, a troubadour or a court jester. Yet otherwise, what is it like to bed with your sister or even your mother, like a button beetle (and somewhere I'll bet that even bedding a grandmother, like a *Scleroderma* wasplet, has been out tried by humans too)?[18] Or let it be an explorer, inventor, revolutionary, tyrant, artist, paedophile . . . All these courses are being tried as they have been before and elsewhere amid the mazy branches of the tree of life. Some are now re-tried in humans quite hopelessly from fitness, even 'inclusive fitness', points of view, but some are certainly not, and which are which we mostly can hardly yet guess.

Thus for the immensely diverse cultures of *H. sapiens* Bobbi Low's contribution came as a daring comparison to what Marlene and I had started for birds. She found that the prevalence of overt polygyny in cultures around the world correlates, just as the 'H&Z' idea would predict, with the prevalence of parasites in the geographical environment. She found that her correlation persisted even when she restricted her view to cultures wholly within the tropics, a refinement that made it much less likely that the correlation could be due to some broad other factor of climate or resource severity such as, for example, that which might make it

hard for an Eskimo man to support numerous wives. The strength of
Bobbi's correlation surprised me even though I had vaguely imagined and
was glad to see it. Later, her result stood out even more surprisingly when
Andrew Read (of my own department) reported that the 'H&Z' correlation
of showiness with parasitism worked worst (indeed appeared to be weakly
reversed) for the explicitly polygynous species among songbirds.[18] Bobbi's
finding for humans has not been challenged: indeed obliquely, since its
publication in *American Zoologist*, it has been supported. Similar cross-
cultural studies have appeared showing similar effects at a psychological
level[19] and these even have shown morphological oddities linking
statistically to polygyny, to local level of parasitism, and to latitude.[20]

There are parts of the Middle East where an unpleasant and
sometime fatal *Leishmania*, an intracellular protist parasite, can infect you
from the bite of a tiny sandfly. The initial stage of the disease is always
mild—a slow-to-heal sore resembling a boil comes at the site of the bite.
The faster the sore heals the better your prospect for the systemic stages of
the disease that then often follow: rapid healing probably means you had
ability to detect and destroy infected cells and are winning a battle in
which the parasite tries to break into your system more deeply from the
beach-head position it has established under your skin. Fast healing of the
sore might mean you have good nutrition and/or good genotype, and/or
finally, perhaps, a good chunk of luck. Of these outcomes the middle
possibility most concerns us here. The sores always leave a scar, variable in
its extent like a smallpox vaccination scar. It is said to sometimes look as if
a date stone had been pressed into the skin. The point of interest in all this
is that a neat, small visible scar, far from being seen as a blemish to the skin
in local cultures, is treated as a beauty mark. Marlene and I interpret this as
a good fit to our thesis that beauty is, in part, a beholder's unconscious
estimate of health and resistance. The person with the neat scar from a
Baghdad Boil, which is a particular regional leishmaniasis, has proven him-
or herself a resister. 'Marry me and here is one disease you can forget for
your next life,' is the message the healed wound calls—just one extra small
factor for a suitor to notice, doubtless, but for Marlene and me a small joy
for our theory if true, especially in the light of Bobbi Low's dramatic result.
From the example thoughts may spread to all the other scarifications,
tatooings, and cultural inflictions that likewise test and reveal health . . .[21]

Now here is the more interesting part. If the example is right, where is the idea of the beauty of the date-stone scar being developed and memorized? In the genes? I don't discount this but I doubt it. Long indeed may the Kurdish mountaineer have lived with sandflies—their case is perhaps particularly remarkable, it could be perhaps 8000 years with little change—but is even this enough for a specific, gene-controlled image of a point of beauty to have evolved? It seems very unlikely. Instead, and far more likely, the memory and the 'beauty' may be in the culture: marriages arranged to neatly scarred persons may have been not only more prolific than other marriages but may also have been seen to be so. Feeling good about a particular marriage, their own or that of an offspring, parents must have praised scars and thus caused children to see them as comely; and the cultures in which these parents did so may have consequently increased relative to others.[22] But in accepting this, one may still ask whether any conscious idea of healthiness signalled by the scar item has found its place in the notion of beauty. My first thought is that if this idea is present, it is probably only present in the old: for my own memory tells me that beauty for the young is just an overwhelming force seeming to leave no space in one's brain for any conscious analysis. As noted above, however, how easily I imagine gander or goose, peacock or peahen facing potential partners feeling as I do: either this being before me is supremely right for me to draw close to and to copulate with or it is not, and all is known within the time it takes for a very cursory inspection. Such was my first thought, but recently a man who was raised in Rio, Brazil gave me my second one, a phrase suggesting a different view. '*Que saude!*', he told me, would be the exclamation of a group of youths watching an outstandingly beautiful girl. This means simply: 'What health!' A more explicitly healthwise view of beauty than comes in this phrase is hard to imagine—hard also to think of a place more appropriate for its origin than the packed city of Rio within a parasite-oppressed tropic.[20,23] Could the phrase have arisen spontaneously among young people? My third thought on the matter, however, has been that it's also possible the Cariocas were repeating phrases they first hear from their more thoughtful parents and, therefore, that their speech just reflects mature judgements about targets of attraction, even perhaps embodying very rational understanding, almost on the lines of Bobbi's Low's paper,[16] that health in the tropics is more than usually important. As you might guess Carlos denied that the boys themselves were thinking of any opposites to health when uttering the remark. There was no specific fear of what might be caught from an alternative prospect; it was not that dating a merely mediocre beauty was seen as a risk of grippe, ringworm, or TB.

While I remain tantalizingly in doubt about the foundations of human beauty, also thinking that perhaps we are all doomed always to remain in doubt because, in its highest sense, beauty has to retreat beyond what we understand and thus can begin to create, still surely, in spite of this, we are bound to admit that geese and peahens must be unconscious of the health incentive that, as there begins to be proof, actually underlies their choices. With birds it can hardly not be natural selection that directs them to signs and to places that will get them the best offspring, but generally it is simply impossible that they could be imagining distant consequences of a sign of beauty. How could a bird, for example, less than a year out of the nest, understand the consequences of spring courtship in terms of help it will get during a phase of nesting never yet experienced or predict failures of eggs and hatchlings? The same must apply to all those other signs that, with increasing confidence, I connect with primacy of health-oriented prediction. But none of this says anything against natural/sexual selection being the process that creates the preference.

Domestic cocks and male howler monkeys are both polygynous; both species 'crow' in the dawn. Neck anatomy is necessarily involved; but a cock's actions, as he crows, is almost as distinctive as the sound he creates. As he crows he throws into motion all of that sliding and gleaming mantle of long neck feathers that covers also his shoulders. The male howler's throat region, likewise, is a dramatic sight even at rest. And, again, at least in the case of the 'red' subspecies, it shows a golden mantle, this time of fur. When he howls the whole region of his head and neck grows monstrous, and such a sound comes out that when first I heard a lone male howling at a hundred yards in the forest of the Tocantins, I thought it was two jaguars fighting and I tiptoed towards the expected clutching my machete (nervous, of course, that if I came too much into the view of these supposed quarrelsome supertoms they might begin to think about dinner instead of war). Be that noise, excitement, and embarassed amazement as it may, the fact remains that we have on two opposite sides of the planet (your chicken hales from India or Southeast Asia) dawn calls and showy necks. Can there be a connection? There can, I believe, but please wait and hear me tell another story first.

The human warble fly is for humans a dramatic but fairly minor nuisance of the 'Green Hell' of neotropical forest. Only very occasionally is it dangerous; its maggot, however, concentrates the human mind wonderfully when you happen to catch one. I well remember my first: the seeming acne spot that started to weep and twinge and not to go away during my trip from São Paulo to Belem in 1964. Apart from the twinges, which occur when the occupant decides to reopen its breathing hole like a seal under the ice, or just to turn round, I guess it is rather like a Baghdad Boil or else the local Brazilian variety of *Leishmania*, although I have never had that. I will leave to a footnote how one deals with these animals before they get unpleasantly large, merely remarking that persuasion rather than force is the better policy and that rashers of bacon, when available, play a part in one of the best-known remedies.[24] Here I want to concentrate instead on how the bot fly larva reaches you, which to my mind is one of the very tallest of all true stories of biology, this because the story will turn out to have a surprising and sosigonic connection with both cock crows and howler monkeys.

To summarize the theme, the robust and agile female fly (similar to a common bluebottle in her physique though possibly in general style more mimetic to a type of neotropical bee) begins by chasing and capturing a mosquito in flight. Still flying, she attaches a batch of her eggs to the mosquito's body and then releases it, and later the mosquito, unharmed, flies to the human for its meal of blood in the normal way. As she lands (biters are always females) heat radiated from the human (possibly also a raised local level of carbon dioxide gas in your vicinity due to your breathing) awakens well-formed larvae within the attached eggs. In the space of seconds the larvae break their eggshells and jump for the skin of the mammalian host and, with luck, some land on it. Immediately they start to bore in. The hatching and the jump are very hazardous, of course, but occasionally from a visit of a mosquito even several larvae may successfully land on you from the attached batch of eggs and thus create a multiple infestation—in short, two or more 'warbles' or boils from the same parent fly establish themselves not far apart (I had two in the situation described in the footnote). To sketch the completion of the life history, the larvae if not extracted would grow in the course of several weeks to full size, living in the subdermal layer of your skin, actively resisting all painful attempts to dislodge them. When full grown the larvae emerge, drop to the ground, and form pupae in the soil, whence eclose the adults.

It is surprising not to see this egg-deposition trick by the fly of using a less-conspicuous agent to transfer its larvae to its host made more of by creationists, punctuationists, and their kin as showing an adaptation eminently inexplicable by gradualist natural selection, this especially when so much is often made of much weaker claimed examples, such as the evolution of the eye.[25] All the same, I can also see two or possibly three reasons why the example might not be seized on by critics. The first is a doubt that such a fantastic story could actually be true: objecting to it might turn out to be like objecting to the adaptiveness in a unicorn's horn. But, in fact, the human warble fly's story is very well documented. The second reason has to do with the frame of belief out of which most critics may approach the whole matter: the fly is never going to be a particularly nice and cuddly example of whatever vision of evolution or lack of it you prefer—it doesn't seem like the holy harmony of nature was supposed to be. Whether you are a Marxist or a theist, perhaps after all the marvellous eye will serve better for your illustration. Indeed it's still worse when you think of it; for a creationist the human warble fly immediately sets buzzing questions about the real good will of a Creator towards his supposedly favourite species, as well as some more flippant old ones like about where, and in what stage, the 'saved' warble pair existed during their sojourn in the Ark, which is parallel to the doubtless older question about where Noah parked that other marvellous and purely human parasite, the pubic louse . . . The third reason for not using the example, however, is the one that is most important here because it gets to the real snag: it is actually not really difficult to produce a gradualist explanation.

Both blood-sucking and subdermal carnivorous flies have been pests for vertebrate hosts throughout our history and they can be severe agents of selection. As regards the effects of being visited by the individual maggots, as I shall mention below, the oestrids and cuterebrids are perhaps particularly severe. It is therefore not surprising that a dislike of flies and strong reactions to them should have evolved in humans or that similar reactions are present also among our close relatives the monkeys. Slappers and slippers away, humans and flies—it has been an arms race that no one has won. But one can imagine that here is a case where the skill-learning of a vertebrate may sometimes outreach the purely genetic selection for dodging by the flies. A fly is usually either hit and is dead or it is missed, and this isn't so with us: each miss teaches us how the escape occurred and we strike more surely next time, and this is particularly true when the fly is big like a warble fly (most people will know how much more easily and often a mosquito gets to you than a horsefly does). Because the final stage of attraction to the host by parasitic insects is often by smell, the big flies must sometimes get a whiff of their target smell from objects that have recently been in contact with the host. Such objects will include other parasites that have recently visited a host. Occasionally the combination of a strong need to lay eggs with a strong stray whiff of host may trigger the fly to the stage of behaviour that is normally attained only after they had reached the surface of the real host: in short, they will lay eggs on the wrong thing, another fly that 'smelt right' and that they happened to land on. From here on surely the course of the gradualist's argument is plain, even eventually to the catching of the fly in mid-air. Long before this late refinement arrives, however, the more crucial first event among the millions of such chance 'potential proxy' egg layings will have occurred: the mis-laid eggs are brought to a real host, larvae tumble to the skin from the blood-sucker and manage to continue their normal course of colonization. This chance event, followed or paralleled by others that are at first equally haphazard, form the beginning of a 'discovery' by natural selection (in guise of a line of unusually successful, unslapped warble flies) that there is actually a safer way than direct application to deposit eggs on a host. Eventually the previous direct mode of egg-laying in daylight while dodging a rain of slaps[26] is dropped.

If you are the 'stone-on-stone' builders of the castle of evolution facts, as a great student of evolution of symbiosis once characterized the caste of scientists to which he obviously included himself[27] and you are requesting proofs of my every stage of argument, all this may still sound a bit far-fetched. If, however, you are just my own favoured type, not a stone-on-stone person at all, an evolutionary story teller like I am who nevertheless sometimes is gifted to lay the stone that 'becomes the head of the corner', a chain-of-plausibility man content with almost anything rational until he is proved wrong, the story I am involved in now will not upset you. We storytellers, 'we Vulgar Darwinists' as we were once called, are generally content with almost any workable chain that fits with our aesthetic conception and with the facts we know: we are not really concerned with details or with proof until our chain comes to a step that is highly unlikely or is fraught with contradiction—that is, until, in a fairly apt metaphor, a sequence in the fitting together of our jigsaw of plausibilities reaches a difficulty with some part of a larger design. In the same metaphor we may find, for example, that we have for a while been fitting 'sky' pieces into the sea or a chicken has turned up in the picture far separated in the farmyard from her chicks. When such a difficulty comes then, of course, we look back and carefully check, with a hand-lens if need be, every apparent fit in the puzzle that we accepted before. As puzzle-solvers at this point we are losing speed where the stone-on-stone people are steadily forging ahead; but my claim is that we more than make it up when our conjecture that what seemed to be sea was indeed sea and that what seemed to be farmyard dust was indeed farmyard dust, proves to be correct.

Even if you have accepted my demystification of one side of the *Dermatobia hominis* story up to this point you may have noticed, however, that there is another way in which a creationist might come to use the fly in his foray against Neodarwinian Gradualism; for a long time this alternative attack worried me a lot. It was almost as bad as if I was letting my hen stay in the puzzle far separated from her chicks, pretending not to notice that something must be wrong. Dishonesty of this kind in science is a vastly worse error than a tactical stance like adoption of Vulgar Darwinism.

In fact in vulgar parlance, in all the books, *Dermatobia hominis* is called the *human* warble fly. Now, given that the arrival of *Homo* in the neotropics is a mere matter of, at most, tens of thousands of years, and given the extraordinary complexity of its mode of egg-laying as above, and given even more the absence of primate-molesting cuterebrid flies in Asia (as also of *Dermatobia* itself in North America), we must recognize it to be very unlikely either that the flies accompanied the human invasion of the American continent or that they had time to both species-switch away from an unrelated mammal host and to evolve their fantastic egg-laying strategem as they now apply it to *Homo*, all after *Homo* arrived. Thus even the most jaunty of Vulgar Darwinists must admit to difficulty about the shortness of time available (this even if the Chilean site of Monte Verde has long vanquished the Clovis site of North America as the earliest proven Amerindian settlement in the New World, with the consequent approximate doubling of available millennia).

But there are signs coming that there will be a way out of this difficulty, too, albeit we will still need some other hosts and some more jigsaw pieces to be hunted for on the floor. Ever since I heard that *Dermatobia* is not really just *hominis* but a generalist warble maker who affects many hosts, and, also, that another monkey in South America suffers severely from a related warble fly, I have had confidence that the gaps in the puzzle will finally close.

That other monkey is the howler *Alouatta* and it here stands ready to re-enter our story; finally, like in that of the Musicians of Bremen, there can come back into it the crowing cockerel too. The howler monkey suffers terribly from a related cuterebrid warble fly, *Alouattamyia baeri*.[28] From this parasite being specific there is a clear hint of a long history of persecution. The maggots of *Allouattamyia* particularly specialize on the neck region, perhaps because there they are very difficult for the monkey to notice and extirpate in the early stages and because the loose skin gives plenty of room for development. A badly infested howler's neck is a terrible sight and is certainly the diametrical opposite of what any human could describe as beauty. Given this as the preferred site of infection, however, and the severity and loss of activity and condition that results,[28] it has come to me as a strong likelihood that his appearance at close range is not the main thing that the male howler is concerned about in his mating affairs. Like the cardinal, and in the interest, probably, likewise of his EPCs, he wants to project machismo to as great a distance as he possibly can and that is what that amazing noise, which beside the Tocantins River in 1964 made me imagine two embattled jaguars, is really for. But once I had heard of the howler warble flies I realized it could have another function too; remembering the sharp pains that just one warble had caused me in my arm, or later two in the relatively immobile skin of my back, I realized that to howl like a sergeant major across the forest canyons of leaves must be a quite excruciating task if one's neck supported six or so boil-like warbles such as I had winced just to look at in some of the photographs. Even if the monkey itself could stand the pain during the howling, how could it keep its deep notes steady when huge larvae were for their part turning in the walls of the inflated sound sack to make themselves more comfortable under the novel strains? Howling, then, could be a signal evolved with a particular purpose: listen all ye distant hearers, it was to say, how I am free from—how I am dominating—this special curse of our kind! Come to me all ye discontented; here are the genes that you need! Now, remembering 'gapeworms', *Syngamus trachea*, and the chicken yard at my childhood home (Chapter 6), I suggest to you that you can understand the cockerel and his neck-stretching and far-carrying call as well: the message, in fact, is exactly the same and it surely is no coincidence that *Syngamus*, like *Alouattamyia*, shows signs of being an ancient enemy of the group of its host (which are the pheasants in the case of the domestic fowl).

It might be more exciting and in a way more human if the signals used in sexual selection were arbitrary and purely symbolic; we can accept from Thorstein Veblen and Amotz Zahavi that they must usually each involve genuine cost and effort,[29] without necessarily committing ourselves to the view that every example evolves out of a need to demonstrate mastery over some particular health affliction and, for sure, a degree of the arbitrary must be in many. R. A. Fisher's process of exaggeration through correlation of female choice almost guarantees it, but it is worth remembering that Fisher himself acknowledged the importance of a utilitarian beginning and in his great book[30] chose the cockerel to illustrate this—and that he personally kept chickens during his period at Rothamsted Experimental Station in Hertfordshire so as to check his ideas about them by direct observation.[31] In general, both Darwin and Fisher would certainly agree that sexual selection is launched from natural selection, indeed *is* natural selection with certain extra knobs turned on. Therefore, it is reasonable to expect a utilitarian history underlying many of the features that it exaggerates.

My theme in the paper of this chapter, as also in Chapter 6, is, of course, that history can have a lot to do with parasite affliction. I believe that this theme has advanced considerably since I wrote the paper and I could now cite various further examples that are actually more firmly founded than my chickens and howlers—which latter, however, I still favour as among the most entertaining examples of Vulgar Darwinism that have yet come my way and which I have, for this reason, if for no other, included here. Most of the examples of specific indicators that have come to light since 1982 are already covered in my introduction to Chapter 6: almost equally entertaining examples of exaggerated under tail-covert feathers tending to be specially exaggerated and downy and ornamented in species prone to 'evacuative' bowel parasites have also been mentioned. What is worth adding here, however, is that theory concerning how an animal's choice process should effectively attend to multiple signals when a receiver gives different weightings to various traits has shown considerable strides since I wrote that introduction,[32] and various cases have been discussed in which either probably or certainly animals may so choose.[33] Special idiosynchratic preferences for different genotypes should probably also be summarized here[34] because these special preferences are usually only a part of the choice story and preferences for generalized ideals of sexual selection overlap with them. Notable amongst the idiosynchratic preferences, of course, are those for MHC variants, which I mentioned in earlier chapters.[35]

Perhaps in practice assessing multiple signals is quite difficult; it can be done, given time, and is done in intimate situations and when long-term bonding is in prospect; in contrast when a quick appreciation of 'quality' is needed, sexual selection concentrates its effort, on a single character that best averages a variety of important qualities. This was exactly the theme of Chapter 6 and is still more the theme of the present chapter as I summarized it in the word 'far' in the title of the paper. Over the past decade and half a host of broad indicator characters have been increasingly researched: among them may be mentioned carotenoid pigments,[36] ultra-violet reflectances,[37] song[38] (which may, however, also transmit varied messages within single performances[39]), persistence in display and in lek attendance,[6,40] counts of numbers of stones collected,[41] and exaggerated and easily visible plumes[42] (which we now have evidence has probably been a general indicator factor for birds ever since they were the still-toothed but feathered versions among small dinosaurs[43]). Finally I mention developmental symmetry, a general indicator to which it seems almost all animals attend, including us, and to which a lot of study has been recently devoted, often indicating connections with infectious disease.[44]

Taking all these studies together, the swing in favour of regarding female choice as being mainly in pursuit of 'good genes' rather than arbitrary, chance-initiated characters[45] has been dramatic and even is referred to as a 'paradigm shift'.[46] But, of course, not all who are now happy that mate choosers are indeed choosing 'good genes'[47] would concur further that they are choosing genes for health, and even amongst those that accept that they are, many I think would still say, following lines of the Deleterious (Deterministic) Mutation Hypothesis of sex (DMH: Chapter 16), that the 'health' has more to do with being un- (or at least less) mutated rather than of being resistant to infection. In other words, they would take the new view as a recognition that sexual selection is unavoidably harnessed to the fight against mutation.[48] And other dissenting voices continue, too,[49] and are certainly partly right: prominent, for example, is the 'direct-benefit' school (which, perhaps, I rather too summarily dismiss in the paper) that maintains that choosers seek a mate who will be competent right now to help contribute to the next generation as territory and resource protector, home builder, auxilliary or sole feeder, and so on; while a variant of the direct-benefit view emphasizes the infectious danger of the close collaboration, and especially of the mating act itself, in passing on disease to the chooser him- or herself and to the offspring.[12,50] This is the contagion idea of sexual selection, which was first championed by William Freeland:[51] in it the parasite theme joins the 'direct-benefits' view but leaves aside all subtlety of inheritance.

That contagion is a factor in mate choice is easily proved by direct introspection and it is probably true also that amongst all the weird '-philias' into which human sexual frustrations can become channelled, even necrophilia will be found at a higher frequency rating than what I might call morbophilia—a desire to have sex with the manifestly diseased. But here I can appeal again to Marlene's and my original study: for reasons already given (notably generalism, sporadic acuteness of the diseases, and intermittency of manifestation in the blood) those we studied were not ideal for assessing the evolutionary influence of chronic disease on sexual selection, but one thing about them that could be almost guaranteed with all was that they were not transmitted directly, either from mate to mate or from parent to offspring. So the fact that we found a correlation where we had hardly dared to hope for it, and one which, though weak (as expected), has stood up to extremely careful reanalysis,[52] seems a strong refutation—at least as to the sufficiency of the 'contagion' hypothesis. To catch the diseases Marlene and I studied a biting insect just has to intervene and the nature of the parasite's cycle in the insect makes it especially unlikely that even repeated bites by a given mosquito, delivered at or near to a bird-frequented site (say a nest or roosting place), would be able to transfer the parasite in the manner the hypothesis requires.[53] We know that human malaria can be transferred within an isolated family home but this is not its normal mode of travel: indeed, were malaria usually transferred so locally, like leprosy is, modern theory, with much evidence to support it, would predict malaria not to be the highly virulent and acute disease that it is.[54]

Other lines of evidence also speak out strongly against the contagion hypothesis—for example, the whole long story of the EPCs. Why is a female mating around; why, if contagion fears rule, risk dalliance with a male who is certainly not going to help you and at the same time risk loss of support by a male who is helping?

In the 'good genes' perspective (and especially, I would say, within that, from the sosigonic perspective) the reason for infidelity is plain enough. The evidence that the males who are preferred to established mates are of superior quality, and that the offspring sired by them are likewise of special vigour, is steadily accumulating.[55]

There can be little doubt that the same incentive applies to extrapair sexual relations as they are seen in humans. Here again the evidence about preferences has been accumulating fast. Open discussion of the issues has also increased dramatically,[56] but I am not yet aware of evidence bearing on health, vigour, and success in the EPC offspring as compared to the rest in the way that I am for the animal examples. To my mind there can be little doubt that they are often very successful; otherwise the activity would, over the generations, die away. Celebrated and successful careers by known bastards are commonplace in history.[57] Considering how such origins tend normally to be concealed where possible, we can safely assume that legions of successful people exist, the truth of whose irregular siring has never come to light. Personally my confidence of this point is such that I conjecture that the strong stream of bastardy that flows continually—seemingly in all civilized societies (averaging perhaps 1 in 10 in lower and middle levels of our 'Western' society)—may even be enough to resolve the puzzle of non-decline of national levels of intelligences, despite what seem quite well-founded predictions based on careful testing, heritabilities, and family sizes.[58]

Of course both EPCs and IPCs (the latter the intrapair copulations) are less solely about offspring survival and quality in humans than they are in animals, although probably more often that we realize such eventual successes may still dangle, invisible to ordinary consciousness, as our ultimate rewards. In other words what may seem almost idle entertainment, or else merely a prostitute's fee, may end as an item contributing to, in another place and far away in time, a healthy baby with an assured cultural endowment to follow. Money, some of it working via doctor's fees and for medicines but mostly assuring abundant food and hygiene that can make the doctor less necessary, can offset reductions of intrinsic health. Moving to higher social levels traditionally brought with it better survival for both sexes and their offspring[59] as well as, more obviously, better opportunities for males to sire EP offspring. The additional opportunities occur especially under socially approved polygyny, of course, but also—smile or frown at this as you will—outside marriage. Starting with a lag of just one generation, large parts of the same benefit can also accrue to high-class women and among them, as it turns out, even the diluting factor of sex ratio may assume an adaptively less diluent slant. A bias in sex ratio—more boys born to more dominant women—seems to arise specifically in those women who are assertive and set to climb,[60] and similar effects have also been discovered in animals.[61] Even though we must not forget that pair-bonding and even possible 'recreational' sex may occur outside the human world (and at that even amongst 'vegetarian' dung beetles[62]), still, comparing ourselves to animals, we still very easily see numerous sources of confusion that are sure to cloud any adaptive interpretation of all copulation and siring activities as they occur in humans—and this is true quite apart from trivial decorations upon 'spandrels', 'squinches', and the like that must also abound.[63] Such factors loom, for example, in wealth and money, in social status, in our still-evolving intelligence and language abilities; finally they loom in the very fact that for almost none of the foregoing has there been time in the fully human scheme of things for patterns to become stably adapted.

In all this the ancient factor of health, however, is obviously by no means missing for humans and it has been so constant[64] that its effects are probably more accurately interpretable than most. It is probably true of sexual preference, as of most other characters in human development, that we 'climb up our evolutionary tree'. If so we would expect to see health playing an especially strong part in attractiveness to the other sex just after puberty. This may be the case. For girls at that age, for example, male basic good looks (which, it is turning out, largely means symmetry plus averageness plus slight above-average ratios for a few key attributes such as chin and head hair) seem especially important; but along with these soon come proven abilities at sport and dance. Music-making both by voice and instrument is evidently also very attractive and the fine coordination needed for these performances, as already argued in the context of bird song, is probably only possible in states of good health.

The attraction of 'pop music stars' and of 'sport stars' for young people is, of course, well known and anyone who doubts that such attractions sometimes lead to offspring should read about the paternity cases that surround American star basketball players.[65] Virtually every player in the national team has at least one acknowledged illegitimate offspring and one has seven. Money, however, as an alternative incentive to health again rears its head because some points in the stories suggest mothers who are more interested in chasing an easy source of child-support payments that can help with the rest of a single-mother family than in obtaining a love child *per se* or in rationally seeking 'athlete genes'. But how do you distinguish? Would these women equally have sought and offered themselves to unhealthy millionaires or to decrepit but still wealthy scions of the fading families of Victorian railway magnates? Possibly, I suppose, if they could find them and could attract them and I am sure some manage this: but the baseball star is by far more visible and it is much easier to find the hotel where to go for sex. All the same, that health and physical achievement still are factors in attraction I think few will deny. Moreover, cultures certainly exist today in which virtually a human lek system of mating still operates, at least for a few days of each year—which, of course, is enough to allow a lot to happen given motivation; in these days wrestling or dancing, not displays of wealth, are the keys of success.[66] The story of one of these systems included 'married' husbands besides the 'lek' partners who are the victors in the wrestling, and for whom the women were said virtually to line up after the wrestling ceremony. The normal husbands are stated to have no call to object. Moreover, the reasons given by the women for accepting the victors is explicitly sosigonic: procuring the healthiest offspring for the tribe is the justification they give. Thinking of the European Bacchanalian orgies of the classical period, as well as of carnival and its entailments as they occur in many countries even today, we can see ourselves still in progress, metaphorically, out from those old South African caves in which were found the scattered australopithecine bones telling of two sexes of ancestors as different in size and physique as are the polygynous gorillas today—so different that the bones once were mistaken by palaeontologists as belonging to different species.

And what are we moving to out of that cave? Sunlight? Until recently I'd call it more a Midas light simply of money—the bride price, the dowry, generally the arranged marriage. Now it's the light of medicines, but still much of the money remains mixed in. 'Do not question my value; pass me on', printed an ancient eastern king on his failing coins: he was asking you not to notice that you would get better value by abandoning the convenience of money and going back to barter. Maybe we are failing to notice something rather similar in respect of people, real people: so might we also save in National Health Service taxes by going back, in matters of mating, to the old and tried recipes of love and health, and, after that, to letting nature take its course with defective neonates at around the time of birth.

'Do not question my value; buy me and believe your doctor' is stamped in effect on most pills. The ancient king didn't say what he was mixing with his gold; the modern drug companies and the medical freemasonry don't show either. Perhaps deliberately they avoid imagining the future they are leading us too. It's the future in which metastable people will be saying: 'Don't question my beauty, marry me. Let the Hospital manage whatever may come of the things I haven't told you. Aren't I fixed well enough for your taste—hasn't every one had their teeth straightened, their insulin fixed? Isn't every one the same?' But I must avoid starting to ride that horse again and rather think of ways to close this topic and move to the next and last chapter for which it is time.

One or two other points about the paper that follows deserve comment before I end. One is that this was yet another paper, which, like that of Chapter 15, was written long after the effort and early versions of the paper in Chapter 16 had been completed but still a considerable time before that paper got to be published: this explains the repeated references in the original version to that paper as 'unpublished'. I note also evidence in this one that I was already deep into and getting results from my next set of models, which, unlike HAMAX, had deterministic genetics and selection. The Red Queen chase was applied with fewer loci (up to about six, depending on the capacity of your machine) but the results still strongly reinforced the multilocus conclusions of HAMAX. Eventually this work was orally presented as another 'symposium' paper and printed in a symposium issue of the *Journal of Heredity* in 1993.[67] Once again the present paper may show the frustration I was feeling about the rejection of the HAMAX model. And clearly I was not lacking in confidence that I was on the right lines in that model because in the 'Conclusion' section I say that I regard 'the parasite theory of sex itself as very nearly established'.

The reader who has come this far will know that the phrase 'sex itself' is my code for all the 'maintenance of sex' side of the general problem of sex, and most notably that any solution to be acceptable must include an ability to defeat twofold more efficient parthenogenetic clones whenever (and in however optimized starting forms) they appear. This is a challenge that I think no one except me and my colleagues have solved realistically. In the present paper, however, I am more concerned with slower cycles—cycles still multitudinous but dictating positive heritability of fitness. As for evidence, at least as it regards sexually preferred characters independent of any implications about parasites,[68] and sometimes with respect to parasites specifically,[69] this becomes ever better attested. Related kinds of positive evidence also accumulate.[70] My general confidence about Parasitic Red Queen at the time of the meeting must have come from the combination of the HAMAX program having shown that, given the multiple shortish cycles, clones could be swiftly beaten back even by low-fecundity hosts, and my feeling that I had sufficiently surveyed the varied ecology and expression of sex throughout life to make my case convincingly stronger than the case of its main rival, which already, of course, in my eyes was DMH. But many will not agree even today that a victory of PRQ is so obvious and doubtless will say that, at that time especially, I was far in advance of my facts.

Well, again, let us wait and see. Meanwhile, before leaving this chapter, let me admit here to at least one on-going puzzle that I raise in my paper and still feel defeated by theoretically; and let me refer also to just one other exciting topic that I discussed but on which neither I nor any one else has proceeded further.

Let us take this second topic first. Restriction to poor diets necessitating assistance of microbial symbionts for animals to digest and survive upon, seems to correlate with far-gone sexual selection in the hosts in many cases. This correlation still seems to me striking and I am surprised that during the present enthusiasm for comparative studies[71] nothing has been done to test it more critically and find out what it means. A notion that apparent mutualist or neutral symbionts can be facultatively pathogenic comes not only out of study of the current immune deficiency diseases but is reinforced by such cases as the emergence of recent highly pathogenic strains of *E. coli*. The idea that former mutualists in the right conditions can rapidly evolve to become dangers has also received recent

evidence from more than one side.[72] The counterexamples I mention in
the paper come in two groups: there are the geese that have large guts to
digest poor grassy diets (but perhaps no worse than diets of horses?) and
don't use symbionts much nor go for high sexual selection (neither, of
course, do horses); and there are the pigs with relatively rich diets and
again with nothing obvious in the way of symbionts and yet tending to
carry their sexual selection—as seen in dimorphism, tusk development,
male shields, and actual fighting—quite far. But still for the most part,
even without using phylogenetic regression, the evolutionary convergences
seem to be too numerous to be a matter of chance. They cover such
unexpected examples as kangaroos, both body lice and feather mites of
birds (the former showing some of their highest developments, remarkably,
on pheasants, which, of course, are themselves highly sexually selected),
dung beetles (remarkably again, tending to live on the excreta of horned
ruminants), hoatzins, many groups to be found connected with diets of
dead wood and woody fungi, and so on. The claim of such a connection
from the comparative evidence seems particularly open to further
investigation via certain close generic pairs, such as the flour beetles in the
genus *Tribolium* (long-time favourites of lab ecologists and with a lot even
of their genetics known), and contrasting these with the closely allied
species *Gnatoceras cornutus*, which has a horned and fighting male and
polygynous habits and, in its natural habitat under bark, a coarser diet than
the seeds favoured by *Tribolium*. Similar close pairs may be available among
the mentioned body lice and feather mites of birds but these are presumably
more difficult to breed. Others pairs probably exist among cisid beetles of
dry fungi, which would be easy to breed but the group is little known.

Almost all dung beetles use gut symbionts. Most of them are dull but
among those horned for fighting the small proportion that are colourful
rises and some of the most horned are resplendent and metallic in greens,
blues, and even shimmering red. Take *Proagoderus rangifer*, for example,
whose two long, swept-back hear horns seem, from the name, to have
reminded Linnaeus of a reindeer. We may ask, why is this wonderfully
horned beetle so colourful to boot? It is hard enough to imagine that the
horns are used in any realistic male combat although experience with other
equally weird beetle armatures assures me that they will prove so. But surely
it is impossible that the beetle can fight just by its colour: it might use it on
the other hand to display quality and thus its potential to fight; here the

human perception is that such 'metallic' structural colours as the green of *P. rangifer* are rather standard and seem to give little scope for displaying differences. Dung beetles sometimes concentrate the stenches of their food supply into their own bodies, seemingly to deter predators. Pinned in a collection, their smells can still be unpleasant after several years. Then, might the colours be warnings of powerful and nauseating chemicals, like the advertisement of the dramatic stripes of a skunk? I leave this aside for a moment and just comment that, if it be so, I wish they could make a stench nauseating enough to deter humans from endeavoring to obliterate these beetles' habitats because that, far more than predators (and vastly more than collecting), is driving the reduction of some of the strangest and loveliest species.

A wonderful congener of African *P. rangifer*, *Proagoderus imperator* of India, is distinguished by having a strikingly and differently horned female—an almost unique combination—and this in addition to a weirdly horned male. This species may once have been quite common in the Western Ghats of India. Both sexes of this beetle are also colourful. But, so far as I can see, no specimens have entered either British or major Indian insect collections for a very long time, and my own efforts to find the species by searching dung in one of the recorded sites were in vain while another site seen from afar looked hopelessly deforested and had probably lost many of its large mammalian dung producers. This species is quite possibly extinct; we may never know the story underlying those very different styles of horns, and especially we will not know what I would most keenly wish to—what the female fought for with hers.

But now let's turn to the other case where to my mind a great puzzle has continued since the paper of this chapter came out, and where deforestation again poses the risk that we may lose for ever yet another key to the social and sexual mysteries of the world.

In the section 'Ability to work for the brood' (p. 713), and stating my expectation that the male pied flycatcher would look much more drab and be more similar to his female, if working for the brood was all that (in a sense of pure natural selection) the choosing female was aiming for in her mate choice, I might have added in parentheses within the sentence: '. . . (and still more so the male quetzal) . . .'. I was certainly aware at the time that the male quetzal was claimed to work for his brood but possibly didn't yet fully believe it and wanted not to be caught out in a wistful error. After a recent and beautiful article on the quetzal in *National Geographic* magazine[73] I simply have to believe it: the male's astonishing metre-long tail plumes really do wave in the breeze outside the tree hole in which the male sits on the eggs or warms the brood. The great feathers are bent forward over his back. Anything more like a fluorescent sign to inform a predator winging or walking down his cloud forest 'Sunset Strip' that 'FOOD IS HERE' is hard to imagine, though maybe a colour like bright red might be more conspicuous than the bright metallic green actually used. Apart from this even the feathers' motion in the breeze and their iridescent barbules acting as arrows directed to the hole are touches that hardly a Las Vegas gambling saloon could better. One wonders, then, why the authors of the article didn't add the unfortunate 'tell-tail' feature as combined with the male's monogamous devotion in the list of possible reasons for the species' decline. As we have seen in many ways and places in both of my volumes of *Narrow Roads of Gene Land*, the vast tail in other contexts may indeed be optimal advertising of a male's worth and this unfortunately may be outweighing the disadvantage it creates for the brood. For if the male can sire eggs by extrapair copulations these are created outside of his care and that is a great advantage; moreover, he may even expect that the unfortunate male whom he cuckolds may actually have fewer banner-like tail feathers to endanger the extra offspring, so he scores twice: were that other male not a bit inferior that other male's mate probably wouldn't have come to our resplendent and nest-endangering quetzal king for an EPC in the first place. Dilemmas of biology as revealed in these grim channels of evolutionary analysis, may appear almost too contorted to be believed, but they are undoubtedly real and potentially such trends of advertisement as I discuss here can indeed be dangerous to a whole species.

Before noticing the case of the quetzal I used to rationalize a somewhat similar though less-extreme situation with the North American rose-breasted grosbeak, whose showy male also works at the nest, by supposing his glowing sunset breast must betray the nesting enterprise little because, first, the book told that these birds nested in dense thickets, and, second, because everyone—in a list extending from Wallace and Darwin in the nineteenth century to, most recently, in the 1990s, Johnson and John (whose findings I discuss in the Appendix to Chapter 6)—had found agreement (and in the case of the latter names nigh to proof), that colourful breasts bring less danger from predators than do colourful backs.

Whatever hues the 'rose-breast' male evolves to adorn his front, he carries a very sober mix of 'thicket grey' and 'charcoal black', as menswear shop assistants would call them, on his back. I hate to say this but in contrast to this circumscribed and reasonable beauty, the quetzal seems to have flown right over the top, to have made itself a cheeky challenge to extinction to catch it: like the Californian condor or Irish elk, it seems determined to get everything wrong. Some novelties of these animals' lives—such as the advent of Los Angeles, power lines, or the coming of forests to catch your horns in for the elk or the vanishing of them for the quetzal—are not the species' faults. But it is difficult to say the same of some evolved attributes: in the case of the condor, of having so much head devoted to bulbous wattle, so little to brains; or in the case of the quetzal, of having so much of a tail devoted to advertisement, so little to fatherly good sense. Darwin's ghost may shrug, Wallace's sigh, as they murmur to one another: 'Didn't we predict it?', when finally the last quetzel nest gets zapped by an observant jay and species-level selection sounds another faint click of its ratchet.

I retained just one hope, however, just one more possible rationalization that might save the bird if not from its extinction at least from the ignominy of being said to have deserved it, instead of we evolutionists being mistaken. In particular, this idea excuses the quetzal from being an evolutionary anomaly, a bird that ought not to have arisen. Quetzals, along with many other conspicuous birds, might be examples, again, of warning coloration,[74] rather as I suggested for the brilliant dung beetles. I read of a Chilean pigeon that is at most times esteemed by human hunters as food but in a particular season is spared because a wild avocado tree is fruiting. The bird's flesh, it is said, becomes intensely bitter with secondary compounds from the fruit. Perhaps quetzals, who are also avocado-addicts, as am I, are similarly bad-tasting most of the time. Perhaps what the feather sign is really saying is 'DON'T EVEN THINK OF IT—AND OUR BABIES ARE TERRIBLE TOO', and perhaps the very sight of long plumes waving from a hollow makes hawks and coatimundis clutch wings or paws to their stomachs and, in wattles or lips, turn green as the quetzals themselves.

References and notes

1. Lord Byron, *The Poetical Works of Lord Byron* (Murray, London, 1905).
2. Anon. in E. Jesse (ed.), *Gleanings in Natural History*, 2nd series (Murray, London, 1834).

3. T. R. Birkhead and A. P. Møller, *Sperm Competition in Birds: Evolutionary Causes and Consequences* (Academic Press, London, 1992).

4. A. P. Møller, P. Christe, and E. Lux, Parasitism, host immune function and sexual selection: a meta-analysis of parasite-mediated sexual selection, *Quarterly Review of Biology* **74**, 3–20 (1999).

5. G. Hausfater, H. C. Gerhardt, and G. M. Klump, Parasites and mate choice in gray treefrogs, *Hyla versicolor*, *American Zoologist* **30**, 299–311 (1990).

6. N. Hillgarth, Parasites and female choice in the ring-necked pheasant, *American Zoologist* **30**, 227–33 (1990); H. McMinn, Effects of the nematode parasite *Camallanus cotti* on sexual and non sexual behaviors in the guppy (*Poecilia reticulata*), *American Zoologist* **30**, 245–9 (1990).

7. M. Zuk, R. Thornhill, and J. D. Ligon, Parasites and mate choice in red jungle fowl, *American Zoologist* **30**, 235–44 (1990).

8. M. Zuk, T. S. Johnsen, and T. Maclarty, Endocrine-immune interactions, ornaments and mate choice in red jungle fowl, *Proceedings of the Royal Society of London B* **260**, 1659–63 (1995).

9. As general guides I cite again here A. P. Møller, Immune defence, extra-pair paternity, and sexual selection in birds, *Proceedings of the Royal Society of London B* **264**, 561–6 (1997); Møller *et al.* (1999), note 4; J. L. John, Haematozoan parasites, mating systems and colorful plumages in songbirds, *Oikos* **72**, 395–401 (1995); J. L. John, The Hamilton–Zuk theory and initial test: a review of some parasitological criticisms, *Parasitology Today* **27**, 1269–88 (1997); J. L. John, Seven comments on the theory of sosigonic selection, *Journal of Theoretical Biology* **187**, 333–49 (1997).

10. F. C. Rohwer and S. Freeman, The distribution of conspecific nest parasitism in birds, *Canadian Zoologist* **67**, 239–62 (1989); M. Petrie and A. P. Møller, Laying in others' nests: intra-specific brood parasitism in birds, *Trends in Ecology and Evolution* **6**, 315–20 (1991); A. Yamauchi, Theory of evolution of nest parasitism in birds, *American Naturalist* **145**, 434–56 (1995); B. E. Lyon, Optimal clutch size and conspecific parasitism, *Nature* **392**, 380–3 (1998).

11. Later I found that Marion Petrie had suggested this type of intrusion before me (Petrie and Møller (1991) in note 10) and see also T. R. Birkhead, T. Burke, R. Zarn, F. M. Hunter, and A. P. Krupa, Extra-pair paternity and intraspecific brood parasitism in wild zebra finches, *Taeniopygia guttata*, revealed by DNA fingerprinting, *Behavioral Ecology and Sociobiology* **27**, 315–24 (1990).

12. B. C. Sheldon, Sexually transmitted disease in birds: occurrence and evolutionary significance, *Philosophical Transactions of the Royal Society of London B* **339**, 491–7 (1993).

13. For example, J. S. Ash, A study of the Mallophaga of birds with particular reference to their ecology, *Ibis* **102**, 93–110 (1960).

14. L. L. Wolf, 'Prostitution' behavior in a tropical hummingbird, *Condor* **77**, 140–4 (1975).

15. F. M. Hunter and L. S. Davis, Female Adelie Penguins acquire nest material from extrapair males after engaging in extrapair copulations, *The Auk* **115**, 526–8 (1998).

16. B. Low, Marriage systems and pathogen stress in human societies, *American Zoologist* **30**, 325–39 (1990).

17. Via the Baldwin effect, as is well explained by Daniel Dennett in *Consciousness Explained* (Little, Brown, Boston, 1991), the plasticity implicit in learning and culture has potential to be important for speeding physical evolution too.

18. A. F. Read, Passerine polygyny: a role for parasites, *American Naturalist* **138**, 434–59 (1991); for a start on the non-human cases see Volume 1 of *Narrow Roads of Gene Land*, Chapter 4, note 37.

19. D. M. Buss and D. P. Schmitt, Sexual strategies theory—an evolutionary perspective on human mating, *Psychological Review* **100**, 204–32 (1993); S. W. Gangestad and D. M. Buss, Pathogen prevalence and human mate preferences, *Ethology and Sociobiology* **14**, 89–96 (1993).

20. E. D. Shields and R. W. Mann, Does a parasite have a better chance of survival if an Inuit or a Mayan spits on it?, *Journal of Craniofacial Genetics and Developmental Biology* **18**, 171–81 (1996).

21. D. Singh and P. M. Bronstad, Sex differences in the anatomical locations of human body scarification and tattooing as a function of pathogen prevalence, *Evolution and Human Behavior* **18**, 403–16 (1997).

22. R. Boyd and P. J. Richerson, *Culture and Evolutionary Process* (University of Chicago Press, Chicago, 1985).

23. T. K. Shackleford and R. J. Larsen, Facial attractiveness and physical health, *Evolution and Human Behaviour* **20**, 71–6 (1999).

24. T. F. Brewer, M. E. Wilson, E. Gonzalez, and D. Felsenstein, Bacon therapy and furuncular myiasis. *JAMA—Journal of the American Medical Association* **270**, 2087–8 (1993).

    The bacon in this remedy seems not substitutable by Brazilian process cheese as I discovered during my first encounter with *Dermatobia hominis* in 1968. The Royal Geographical/Royal Society Expedition to Mato Grosso had refused to release any rasher from its small stock brought with us from Britain—it was reserved for feast occasions and special visitors; however, I was given freedom of several slabs of the standard Brazilian cheese, which had limited life and which no one liked much anyway.

The idea in the bacon therapy is that an impervious layer applied to the skin cuts off the larva's air supply, which is normally obtained through the hole in the host's skin that it keeps open by movements of the narrow trunk-like posterior of its body. Thus when a sufficiently thick slab of bacon is held over the larva's hole with sticking paster, the maggot in its efforts to re-establish contact with the air (probably accomplished with proteolytic enzymes as well as by movement, backs from its burrow at least briefly into the bacon; the observant sufferer monitors the sharp sensations of its efforts (made especially noticeable, doubtless, by the rings of backward directed 'thorns' worn around the wider frontward segments of the larva's body) and waits for them to reach a climax, which is just when the hole will be seen opening; suddenly he pulls away the bacon and the larva comes with it.

The available cheese proved too easy for the larva to bore through and too friable—a good fresh and tough Gouda might have given me a better chance. When I came to look, not having noticed much of the expected movements, I found the hole had been bored but seemingly only by the animal's extensible thin tail and breathing tube: the larva's body was as completely within its bur-row as before. In fact the cheese had cracked several ways, including right along a line where the hole was, thus making the air supply to the larva plentiful.

It is an easy but claimed dangerous alternative to the bacon idea to suffocate the larva completely with impermeable sticking plaster. The danger is of sepsis from the decaying but still unextractable maggot. Nevertheless, after my failure with cheese and on the advice of an experienced Brazilian collector with the expedition, I next tried this method and it was successful. We killed the larva overnight and next morning its corpse was squeezed out by my advisor's thumbs, its final exit resembling that of a finger-squeezed lemon pip—which, apart from the thin tail and small black holdfast thorns, the larva quite resembled in size, shape, and colour. It flew about a metre: we gathered it from the dust, washed it, and under the binocular microscope that I had always ready for my wasp studies, I marvelled at its neat form and clever array of 'contra hominem' devices. Later I presented this specimen, along with a second that I extracted in the same way by myself, to the Zoology Department of Imperial College London for teaching purposes, and for all I know the two are still there.

25. R. Dawkins, *The Blind Watchmaker* (Longman, Harlow, 1986).

26. R. Dudley and K. Milton, Parasite deterrence and the energetic costs of slapping in howler monkeys, *Alouatta palliata, Journal of Mammalogy* **71**, 463–5 (1990).

27. P. Buchner, *Endosymbiosis of Animals with Plant-Like Micro-organisms* (Wiley Interscience, New York, 1961).

28. K. Milton, Effects of bot fly (*Alouattamyia baeri*) parasitism on a free-ranging howler monkey (*Alouatta palliata*) population in Panama, *Journal of Zoology* **239**, 39–63 (1996).

29. T. Veblen, *The Theory of the Leisure Class* (MacMillan, New York, 1899); A. Zahavi and A. Zahavi, *The Handicap Principle: A Missing Piece of Darwin's Puzzle* (Oxford University Press, Oxford, 1997).

30. R. A. Fisher, *The Genetical Theory of Natural Selection*, 2nd edn (Dover, New York, 1958).

31. J. F. Box, *R. A. Fisher: The Life of a Scientist* (Wiley, New York, 1978).

32. R. A. Johnstone, Honest advertisement of multiple qualities using multiple signals, *Journal of Theoretical Biology* **177**, 87–94 (1995); R. A. Johnstone, Multiple displays in animal communication: backup signals and multiple messages, *Philosophical Transactions of the Royal Society of London B* **351**, 329–38 (1996).

33. C. Wedekind, Detailed information about parasites revealed by sexual ornamentation, *Proceedings of the Royal Society of London B* **247**, 169–74 (1992); C. Wedekind, Lek-like spawning behaviour and different female mate preferences in roach (*Rutilus rutilus*), *Behaviour* **133**, 681–95 (1996); K. L. Buchanan and C. K. Catchpole, Female choice in the sedge warbler, *Acrocephalus schoenobaenus*: multiple clues from song and territory quality, *Proceedings of the Royal Society of London B* **264**, 521–6 (1997); A. P. Møller, N. Saino, G. Taramino, P. Galeotti, and S. Ferrario, Paternity and multiple signaling: effects of a secondary sexual character and song on paternity in the barn swallow, *American Naturalist* **151**, 236–42 (1998).

34. N. Burley, Mate choice by multiple criteria in a monogamous species, *American Naturalist* **117**, 515–28 (1981); J. Goodall, *The Chimpanzees of Gombe: Patterns of Behavior* (Harvard University Press, Cambridge, MA, 1986); P. W. Trail and E. Adams, Active mate choice at cock-of-the-rock leks: tactics of sampling and comparison, *Behavioral Ecology and Sociobiology* **25**, 283–92 (1989).

35. J. Bonner, Major histocompatibility complex influences reproductive efficiency: evolutionary implications, *Journal of Craniofacial Genetics and Developmental Biology* **52**, 5–11 (1986); R. Ferstl, F. Eggert, E. Westphal, N. Zavazava, and W. Müller-Ruchholtz, MHC-related odors in humans, in R. L. Doty and D. Müller-Schwarze (ed.), *Chemical Signals in Vertebrates*, pp. 205–11 (Plenum Press, 1991); W. K. Potts, C. J. Manning, and E. K. Wakeland,

Mating patterns in semi-natural populations of mice influenced by MHC genotype, *Nature* **352**, 619–21 (1991); W. K. Potts, C. J. Manning, and E. K. Wakeland, The role of infectious disease, inbreeding and mating preferences in maintaining MHC diversity: an experimental test, *Philosophical Transactions of the Royal Society of London B* **346**, 369–78 (1994); C. Wedekind, T. Seebeck, F. Bettens, and A. J. Paepke, MHC-dependent mate preferences in humans, *Proceedings of the Royal Society of London B* **260**, 245–9 (1995); T. von Schantz, H. Witzell, G. Göransson, and M. Grahn, and K. Persson, MHC genotype and male ornamentation: genetic evidence for the Hamilton–Zuk model, *Proceedings of the Royal Society of London B* **263**, 265–71 (1996): C. Wedekind, M. Chapuisat, E. Macas, and T. Rülicke, Non-random fertilisation in mice correlates with the MHC and something else, *Heredity* **77**, 400–9 (1996); T. von Schantz, H. Wittzell, G. Göransson, and M. Grahn, Mate choice, male condition-dependent ornamentation and MHC in the pheasant, *Hereditas* **127**, 133–40 (1997); C. Wedekind and S. Füri, Body odour preferences in men and women: do they aim for specific MHC combination or simply heterozygosity?, *Proceedings of the Royal Society of London B* **264**, 1471–9 (1997); N. Zavazava and F. Eggert, MHC and behaviour, *Immunology Today* **18**, 8–10 (1997); S. V. Edwards and P. W. Hedrick, Evolution and ecology of MHC molecules: from genomics to sexual selection, *Trends in Ecology and Evolution* **13**, 305–11 (1998); D. Penn and D. Potts, Chemical signals and parasite mediated sexual selection, *Trends in Ecology and Systematics* **13**, 391–6 (1998); T. Rülicke, M. Chapuisat, F. R. Homberger, E. Macas, and C. Wedekind, MHC-genotype of progeny influenced by parental infection, *Proceedings of the Royal Society of London B* **265**, 711–16 (1998).

36. D. A. Gray, Carotenoids and sexual dichromatism in North American passerine birds, *American Naturalist* **148**, 435–80 (1996); G. E. Hill, Redness as a measure of the production cost of ornamental coloration, *Ethology, Ecology and Evolution* **8**, 157–75 (1996); F. Skarstein and I. Folstad, Sexual dichromatism and the immunocompetence handicap—an observational approach using Arctic Charr, *Oikos* **76**, 359–67 (1996); C. W. Thompson, N. Hillgarth, M. Leu, and H. E. McClure, High parasite load in House Finches (*Carpadocus mexicanus*) is correlated with reduced expression of a sexually selected trait, *American Naturalist* **149**, 270–94 (1997); G. E. Hill, Plumage redness and pigment symmetry in the House Finch, *Journal of Avian Biology* **29**, 86–92 (1998). But see G. Seutin, Plumage redness in redpoll finches does not reflect hemoparasitic infection, *Oikos* **70**, 280–6 (1994). For other references see note 1 of the Appendix to Chapter 6.

37. J. Radwan, Are dull birds still dull in UV?, *Acta Ornithologica* **27**, 125–30

(1993); A. T. D. Bennett, I. C. Cuthill, J. C. Partridge, and K. Lunau, Ultraviolet plumage colors predict mate preference in starlings, *Proceedings of the National Academy of Sciences, USA* **94**, 8618–21 (1997); S. Andersson, J. Ornborg, and M. Andersson, Ultraviolet sexual dimorphism and assortative mating in blue tits, *Proceedings of the Royal Society of London B* **265**, 445–50 (1998); S. Hunt, A. T. D. Bennett, I. C. Cuthill, and R. Griffiths, Blue tits are ultraviolet tits, *Proceedings of the Royal Society of London B* **265**, 451–5 (1998); A. Johnsen, S. Andersson, J. Ornborg, and J. T. Lifjeld, Ultraviolet plumage ornamentation affects social mate choice and sperm competition in bluethroats (Aves: *Luscinia s. svecica*): Field experiment, *Proceedings of the Royal Society of London B* **265**, 1313–18 (1998).

38. D. J. Mountjoy and R. E. Lemon, Female choice for complex song in the European Starling—a field experiment, *Behavioral Ecology and Sociobiology* **38**, 65–71 (1996); Buchanan and Catchpole (1997) in note 33; D. J. Mountjoy and R. E. Lemon, Male song complexity and parental care in the European starling, *Behaviour* **134**, 661–75 (1997).

39. Buchanan and Catchpole (1997) in note 33; R. A. Suthers, Peripheral control and lateralization of birdsong, *Journal of Neurobiology* **33**, 632–52 (1997).

40. M. S. Boyce, The Red Queen visits sage grouse leks, *American Zoologist* **30**, 263–70 (1990); N. Saino, P. Galeotti, R. Sacchi, and A. P. Møller, Song and immunological condition in male barn swallows (*Hirundo rustica*), *Behavioral Ecology* **8**, 364–71 (1997).

41. M. Soler, J. J. Soler, A. P. Møller, J. Moreno, and M. Linden. The functional significance of sexual display: stone carrying in the black Wheatear, *Animal Behaviour* **51**, 247–54 (1996).

42. R. V. Alatalo, J. Hoglund, and A. Lundberg, Patterns of variation in tail ornaments of birds, *Biological Journal of the Linnean Society* **34**, 363–74 (1988); J. Hoglund, R. V. Alatalo, and A. Lundberg, The effects of parasites on male ornaments and female choice in the lek-breeding Black Grouse (*Tetrao tetrix*), *Behavioral Ecology and Sociobiology* **30**, 71–6 (1992); Møller *et al.* (1998) in note 33; A. P. Møller, A. Barbosa, J. J. Cuervo, E. de Lope, S. Merino, and N. Saina, Sexual selection and tail streamers in the barn swallow, *Proceedings of the Royal Society of London B* **265**, 409–14 (1998).

43. J. Ackerman and O. L. Mazzatenta, Dinosaurs take wing, *National Geographic* **194**, 75–99 (1998).

44. A. P. Møller, Parasites differentially increase the degree of fluctuating asymmetry in secondary sexual characters, *Journal of Evolutionary Biology* **5**, 691–9 (1992); A. P. Møller, R. T. Kimball, and J. Erritzoe, Sexual ornamentation, condition, and immune defense in the house sparrow *Passer*

domesticus, Behavioral Ecology and Sociobiology **39**, 317–22 (1996); N. Saino and A. P. Møller, Sexual ornamentation and immunocompetence in the barn swallow, *Behavioral Ecology* **7**, 227–32 (1996); P. Agnew and J. C. Koella, Virulence, parasite mode of transmission and fluctuating asymmetry, *Proceedings of the Royal Society of London B* **264**, 9–15 (1997); E. Markusson and I. Folstad, Reindeer antlers: visual indicators of individual quality?, *Oecologia* **110**, 501–7 (1997); R. Thornhill and A. P. Møller, Developmental stability, disease and medicine, *Biological Reviews of the Cambridge Philosophical Society* **72**, 497–548 (1997); N. Saino, A. M. Bolzern, and A. P. Møller, Immunocompetence, ornamentation, and viability of male barn swallows (*Hirundo rustica*), *Proceedings of the National Academy of Sciences, USA* **94**, 549–52 (1997); Hill (1998) in note 36; A. P. Møller and R. Thornhill, Bilateral symmetry and sexual selection: a meta-analysis, *American Naturalist* **151**, 174–92 (1998). But see also B. Leung and M. R. Forbes. Fluctuating asymmetry in relation to indices of quality acid fitness in the damselfly, *Enallagma ebrium* (Hagen), *Oecologia* **110**, 472–6 (1997); and K. W. Dufour and P. J. Weatherhead, Bilateral symmetry as an indicator of male quality in red-winged blackbirds: associations with measures of health, viability, and parental effort, *Behavioral Ecology* **9**, 220–31 (1998).

45. W. D. Hamilton, Methods in March-hare madness [review of *Mate Choice*, ed. P. Bateson], *Nature* **304**, 563–4 (1983).

46. R. V. Alatalo, J. Mappes, and M. A. Elgar, Heritabilities and paradigm shifts, *Nature* **385**, 402–3 (1997); T. Tregenza and N. Wedell, Natural selection bias, *Nature* **386**, 234 only (1997).

47. J. D. Reynolds and M. R. Gross, Female mate preference enhances offspring growth and reproduction in a fish, *Poecilia reticulata*, *Proceedings of the Royal Society of London B* **250**, 57–62 (1992); A. P. Møller, Male ornament size as a reliable cue to enhanced offspring viability in the barn swallow, *Proceedings of the National Academy of Sciences USA* **91**, 6929–32 (1994); A. J. Moore, Genetic evidence for the 'good genes' process of sexual selection, *Behavioral Ecology and Sociobiology* **35**, 235–41 (1994); M. Petrie, Improved growth and survival of offspring of peacocks with more elaborate trains, *Nature* **371**, 598–9 (1994); D. Hasselquist, S. Bensch, and T. von Schantz, Correlation between male song repertoire, extra-pair paternity and offspring survival in the great reed warbler, *Nature* **381**, 229–32 (1996); Mountjoy and Lemon (1996) in note 38; K. Otter and L. Ratcliffe, Female initiated divorce in a monogamous songbird: abandoning mates for males of higher quality, *Proceedings of the Royal Society of London B* **263**, 351–4 (1996); B. Kempenaers, G. R. Verheyren, and A. A. Dhondt, Extrapair paternity in the blue tit (*Parus*

caeruleus): female choice, male characteristics, and offspring quality, *Behavioral Ecology* **8**, 481–92 (1997); S. Perreault, R. E. Lemon, and U. Kuhnlein, Patterns and correlates of extrapair paternity in American redstarts (*Setophaga ruticilla*), *Behavioral Ecology* **8**, 612–21 (1997); B. C. Sheldon, J. Merilä, A. Qvarnström, L. Gustafsson, and H. Ellegren, Paternal genetic contribution to offspring condition predicted by size of male secondary sexual character, *Proceedings of the Royal Society of London B* **264**, 297–302 (1997); M. Petrie and B. Kempanaers, Extra-pair paternity in birds: explaining the variation between species and populations, *Trends in Ecology and Evolution* **13**, 52–8 (1998).

48. H. Kokko and J. Lindstrom, Evolution of female preference for old mates, *Proceedings of the Royal Society of London B* **263**, 1533–8 (1996).

49. M. Enquist, R. H. Rosenberg, and H. Temrin, The logic of ménage à trois, *Proceedings of the Royal Society of London B* **265**, 609–13 (1998).

50. C. Loehle, The pathogen transmission avoidance theory of sexual selection, *Ecological Modelling* **103**, 231–50 (1997).

51. W. J. Freeland, Pathogens and the evolution of primate sociality, *Biotropica* **8**, 12–24 (1976). In addition to his major discussion of contagion, Freeland mentioned in this paper the possibility of 'good genes' mate choice for health being also involved.

52. John (1995) in note 9.

53. John (1997) in note 9.

54. P. W. Ewald, Host–parasite relations, vectors and the evolution of disease severity, *Annual Reviews of Ecology and Systematics* **14**, 465–85 (1983).

55. K. Norris, Heritable variation in a plumage indicator of viability in male great tits, *Parus major*, *Nature* **362**, 536–9 (1993); and the following references in note 47: Møller (1994), Moore (1994), Petrie (1994), Hasselquist *et al.* (1996), Sheldon *et al.* (1997), and Petrie and Kempanaers (1998).

56. R. Thornhill and N. W. Thornhill, Human rape: an evolutionary analysis, *Ethology and Sociobiology* **4**, 137–73 (1983); R. L. Smith, Human sperm competition, Chapter 19 in R. L. Smith (ed.), *Sperm Competition and the Evolution of Animal Mating Systems*, pp. 601–59 (Academic Press, Orlando, 1984); D. M. Buss, The strategies of human mating, *American Scientist* **82**, 238–49 (1994); D. Buss, The evolution of mating—Reply, *American Scientist* **82**, 303–4 (1994); R. R. Baker and M. A. Bellis, *Human Sperm Competition* (Chapman and Hall, London, 1995); R. R. Baker, *Sperm Wars* (Fourth Estate, London, 1996).

57. P. Laslett, K. Oosterveen, and R. M. Smith, *Bastardy and its Comparative History* (Edward Arnold, London, 1980).

58. R. Lynn, *Dysgenics: Genetic Deterioration in Modern Populations* (Praeger, Westport, CT, 1996).

59. B. Low, Reproductive life in nineteenth century Sweden: an evolutionary perspective on demographic phenomena, *Ethology and Sociobiology* **12**, 411–48 (1991); B. Low, *Why Sex Matters* (Princeton University Press, Princeton, NJ, (1999).

60. U. Mueller, Social status and sex, *Nature*, **363**, 490 only (1993); V. J. Grant, *Maternal Personality, Evolution and the Sex Ratio* (Routledge, London, 1998).

61. N. Burley, Sex ratio manipulation and selection for attractiveness, *Science* **211**, 721–2 (1981); H. Ellegren, L. Gustafsson, and B. C. Sheldon, Sex ratio adjustment in relation to paternal attractiveness in a wild bird population, *Proceedings of the National Academy of Sciences, USA* **93**, 11723–8 (1996); E. Svensson and J. Nilsson, Mate quality affects offspring sex ratio in blue tits, *Proceedings of the Royal Society of London B* **263**, 357–61 (1996).

62. G. H. Montieth and R. J. Storey, The biology of *Cephalodesmius*, a genus of dung beetles which synthesises 'dung' from plant material (Coleoptera: Scarabaeidae: Scarabaeinae), *Memoirs of the Queensland Museum* **20**, 253–71 (1981).

63. See Chapter 2 and its note 6.

64. C. L. Greenblatt (ed.), *Digging for Pathogens: Ancient Emerging Diseases—Their Evolutionary, Anthropological and Archaeological Context* (Balaban, Rehovot, Israel, 1998).

65. G. Wahl and L. J. Wertheim, Paternity ward, *Sports Illustrated*, 62–71 (4 May 1998).

66. L. Riefenstahl, *The People of Kau* (Collins, London, 1976); L. Manso, Into Africa, *The Adventurers* **1**, 65–71, 88 (1988).

67. W. D. Hamilton, Haploid dynamical polymorphism in a host with matching parasites: effects of mutation/subdivision, linkage and patterns of selection, *Journal of Heredity* **84**, 328–38 (1993).

68. D. Houle, D. K. Hoffmaster, S. Assimacopoulus, and B. Charlesworth, The genomic mutation rate for fitness in *Drosophila melanogaster*, *Nature* **359**, 58–60 (1992); M. Andersson, *Sexual Selection* (Princeton University Press, Princeton, NJ, 1994); M. Charalambous, R. K. Buttin, and G. M. Hewitt, Genetic variation in male song and female preference in the grasshopper *Chorthippus brunneus* (Orthoptera: Acrididae), *Animal Behaviour* **47**, 399–41 (1994); also from note 47: Reynolds and Gross (1992), Moore (1994), Petrie (1994), Hasselquist *et al.* (1996), Kempenaers *et al.* (1997), and Sheldon *et al.* (1997); and from note 55: Norris (1993).

69. Møller (1994) in note 47; Saino *et al.* (1997) in note 44; G. Sorci, A. P. Møller, and T. Boulinier, Genetics of host–parasite interactions, *Trends in Ecology and Evolution* **12**, 196–200 (1997).

70. Møller (1994) in note 47; A. P. Møller and J. Erritzoe, Parasite virulence and host immune defense: host immune response is related to nest reuse in birds, *Evolution* **50**, 2066–72 (1996); A. P. Møller and J. T. Nielsen, Differential predation cost of a secondary sexual character: sparrowhawk predation on barn swallows, *Animal Behaviour* **54**, 1545–51 (1997); Saino *et al.* (1994) in note 44; M. Petrie, C. Doums, and A. P. Møller, The degree of extra-pair paternity increases with genetic variability, *Proceedings of the National Academy of Sciences USA* **95**, 9390–5 (1998).

71. P. H. Harvey and M. D. Pagel, *The Comparative Method* (Oxford University Press, Oxford, 1991).

72. T. F. Thingstad, M. Heldal, G. Bratbak, and I. Dundas, Are viruses important partners in pelagic food webs?, *Trends in Ecology and Evolution* **8**, 209–13 (1993).

73. S. Winter, The elusive quetzal, *National Geographic* **193**, 34–45 (1998).

74. H. B. Cott and C. W. Benson, The palatability of birds, mainly based on observations of a tasting panel in Zambia, *Ostrich, Supplement* **8**, 357–84 (1970); F. Götmark, Are bright birds distasteful? A reanalysis of H. B. Cott's data on the edibility of birds, *Journal of Avian Biology* **25**, 184–97 (1994). Gotmark confirmed most of Cott's finding using more valid statistical procedures; however, weakening my quetzal case, he found significance only for the conspicuousness-bad taste correlation of females. Hence, weakening a bit my idea for the quetzal, the evidence remains that sexual selection, not distastefulness, overwhelmingly drives the excesses common in males. Also notable in this context is that the *Pitohui* birds of Papua New Guinea, which put toad-like poison into their skin and feathers (J. P. Dumbacher, B. M. Bechler, T. F. Spande, H. M. Garaffo, and J. W. Daly, Homobatrachotoxin in the genus *Pitohui*: chemical defence in birds, *Science* **258**, 799–801 (1992)), are not particularly conspicuous, at least to our eyes.

MATE CHOICE NEAR OR FAR[†]

W. D. HAMILTON

When strong positive heritability of fitness arises due to host–parasite coevolution, consequent sosigonic mate preference undermines monogamy through tendencies to extra-pair copulation. 'Low' females bonded to 'low' males try to parasitize their partnership by obtaining fertilization, surreptitiously if possible, from 'high' males: correspondingly, in the case of birds, 'low' males may parasitize by encouraging egg-dumping in their nests by 'high' females who have allowed copulation. It follows that nests of birds of low status should sometimes show evidence of both types of parasitism while nests of high status should show faithful monogamy. Rather differently from the argument in an earlier paper (see Chapter 6), showiness in monogamous species is more likely to be related to such extra-pair objectives than to pair-bonding for nesting.

Venereal disease makes males cautious about copulating with any female. Although venereal disease is prevented by true monogamy, under partial monogamy it may become the incentive for increasing female sexual advertisement. In extreme cases, it may combine with other ecological factors to initiate sex-role reversal, in which the female becomes the non-parenting sex of the species.

As regards the source of the heritability that backs the sosigonic selection assumed in such speculations, reasons are given for preferring a coevolutionary cycling of ancient, preserved, parasite-defence alleles to the alternatives of an abundant stream of good new defence mutations, or a process of elimination of purely deleterious mutations.

Since Marlene Zuk and I first put forward the idea that parasites might be important in sexual selection (see Chapter 6), my view of the situation has changed in several ways. This paper will discuss four. The topics might be classified as relatively sociobiological, parasitological, pragmatical, and genetical, although this is unimportant. At the end, I will summarize what I see as the present position of the idea as a whole.

MATING OUTSIDE THE PAIR BOND

In monogamous birds courtship is often a fairly lengthy affair. For example, for temperate birds, preliminaries may begin during bursts of territoriality,

---

[†]*American Zoologist* **30**, 341–52 (1990). First presented at the symposium on 'Parasites and Sexual Selection' during the Annual Meeting of the American Society of Zoologists, 27–30 December 1988, at San Francisco, California.

pairing, and even mock nesting in late months of the previous year. Then or at other times, if health is the issue, there are opportunities for observing stamina in flight or calling contests, inciting dominance confrontations with conspecifics, noting daring towards predators. Altogether, it has come to seem to me unlikely that a bird needs to rely on looking for bright, tidy plumage or listening to momentary expressions of energy and co-ordination made in complex melodic song, although it should certainly check those features also if available. It is hard to imagine that a greylag goose, for example (a bird whose courtship has been exceptionally thoroughly and sympathetically recorded[1]), makes many mistakes about the health of prospective partners just because they happen to be all dull in colouring. Geese move and nest and even migrate in family groups, and many prospective mates will have been known as individuals all their lives. But if all this casts doubt on showiness for monogamous choice it leaves us with a dilemma: if not for monogamy, and yet still used for sexual display, what else is showiness for?

Consider any typical bird species. Imagine that an individual has bonded with a mate, but imagine also that that mate is not all that might have been hoped for. If the species has an ecology that means biparental brood-raising is crucial to fledging success, and if 'good genes' are relatively unimportant, we would expect the pair bond to be very stable: one or other bird might sigh but it would probably get on with the job. An extra-pair copulation by a female with her most desired male in the local population would achieve little for her and, if it could result in her mate abandoning or reducing care,[2,3] it would risk a great deal. If, on the other hand, having 'good genes' is important to nestling and fledgling survival for reasons of health, a condition of selection to be referred to as *sosigonic*, a female, while probably hiding all signs of her doubt of her partner's quality most of the time, might actually take the same risks to her relationship to get the genes.

In my paper coauthored with Marlene Zuk (Chapter 6) it was a part of our argument that sometimes the 'good genes' imperative becomes so strong for reasons of changing parasite ecology, and that this causes infidelity to become so frequent, that monogamy breaks down altogether. The idea is that it might be a combination of parasite ecology and permissive states of 'ordinary' ecology that starts to create promiscuous manakins, icterids, or birds of paradise out of their monogamous passeriform forebears. Of course, the 'permission' from 'ordinary' ecology in such cases would mean essentially that the female is able to rear a brood by herself, though perhaps having to reduce to a smaller clutch relative to her taxonomic norm, especially if young are altricial. Thus a correlation of ecological and clutch-size variables with polygyny is to be expected and does not exclude a part played by parasites. I will take this view for granted and below will speculate further. Before leaving

the present topic, however, it is worth pointing out that it is going to be difficult to assign relative importance to permissive ecology and to parasites. For
example, maybe birds that live by foraging in damp litter on the ground have
an inevitable exposure to the parasite propagules of the whole local community of birds, this arising through having to search for food among rem-
nants of many kinds of fallen faeces. Perhaps this creates extra chances to
acquire new parasites. The parasites, through the coevolution they initiate,
may help to explain why so many birds feeding on the ground show high sexual selection (as in lyre birds, pittas, bustards, pheasants, etc.). However, it is
almost certainly also true that quickly self-supporting young, such as are
more easily evolved when nesting on the ground, facilitate polygyny and
consequent sexual selection more directly. Interest therefore may have to
focus on birds unusually earth-bound for their groups that seem to be 'going
pheasant' without abandoning sexual similarity and apparent monogamy;
likewise we need to look at galliforms like Numididae and some partridges
and quails where monogamy is also retained or where there may even be a
bias towards male care of the young (e.g. ref. 4).

Usually, in birds, brood contributions by the male (or at the very least
brood defence, which often continues even when polygyny is advanced) is
very important. This suggests the possible success of strategies where females
try to obtain both a nurturing mate and the best genes, the latter coming
from surreptitious extra-pair copulation. There is evidence for a growing list
of species that such strategies are sometimes followed.[5–7] It now seems to me
that it may be much more in regard to these events than in regard to seasonal
(or lifetime) pair bonding that bright coloration of males plays its part. The
female who is dissatisfied with the seeming genetic quality of her mate probably has had no chance to perform any detailed assessment of other males in
her vicinity; but she may well be able to appreciate the relative qualities of
advertisements that are being broadcast through brilliant plumage and song
from males at their territory posts on nearby trees. If this general interpretation is correct it will turn out that as more and more species are searched for
evidence, it will be those supposed monogamists where brightest males are
contrasted to dull females that reveal the most 'infidelity' by females and
variable negligence of maternal care by males. Already we know of possible
examples of the latter (e.g. refs 5, 8 and 9) and of suggestive cases of tendencies
to polygyny—for example, the indigo bunting[10] and the pied flycatcher[11]—
but also know contrary cases—for example, the dunnock,[12] which must be
almost Britain's dullest bird though with a good song. There are many known
cases of appropriate territory intrusions and extra-pairbond copulations in
bright-male birds,[5,13] but only sometimes does the female seem to solicit
them. The species where the female has been seen to solicit most, the
fulmar,[14] is equally bright in both sexes. Somewhat contrary also seems the
species that is becoming most notorious for its extra-pair paternity and egg-

dumping, the starling.[10,15] Marlene Zuk and I found it hard to make up our minds how to score this species, both about how dimorphic it is and how showy. With the thoughts now added it becomes almost a prediction of the theory that as appreciated by other starlings the male starling should be showy. Perhaps it would be profitable to look at starlings using ultraviolet light. (The same search might be applied to the rather similarly dark and iridescent corvids, which, like the starling, stood out in our data with anomalously high parasite loads: the black corvids as well as the magpie, have shown signs of fertilization strategies outside of the pair[5].)

But if the male fulmar and the male starling are showy, their females are equally so or very nearly. There are many cases where both sexes are alike and brilliant—birds like macaws and crowned cranes. Showiness of females gives difficulty for the view just stated, and with the abundance of this condition among the so-called monogamous birds the difficulty is grave. It would clearly be unsatisfactory to have an explanation of the brightness of the male chaffinch or American goldfinch and to have to give some completely different kind of explanation for the brightness of both sexes in the closely related European goldfinch. The usual view is that a copulation is an inexpensive act for a male; hence he should never refuse any female who offers herself to him whether she seems of high genetic quality or low; thus a female should never need to advertise.

Two factors may be suggested to resolve the difficulty. The second of them leads on to the next subheading and will be dealt with there. The first probably sounds a baseless anthropomorphism but in view of the ever-increasing parallels being found between bird behaviour and the thought-conditioned behaviour of humans, whatever the way in which these parallels are brought about, a conjecture of the kind given does not seem unreasonable.

The conjecture is that the female might act as follows. First, she would use her brightness to gain attention of other potentially high-quality males from afar; second, she would make sure that her signal emphasizes her quality to them; and third, by other signs, probably not by approach, she would induce a distant fine male to make a first move towards her, so exonerating herself in the eyes of her mate if the subsequent encounter, which may culminate in a simulated forced copulation, is detected. Underlying these interpretations there is, of course, a biological interpretation of similar events and styles in human life. Why did Beau Brummell in Regency England dress up as he did? Was it to find a wife, or to find an 'affair'? And, why did the women who admitted themselves attracted to such men dress themselves in equal finery? Many were married; but we may note a possibility that in a time when marriages were often arranged for financial advantage, their mates may have been more rich than attractive or healthy. There is, of course, a very great deal of complexity in such a human situation that birds could never have. Particularly there is the transferable wealth, which to my mind overshadows even the role of rational thought. The cases have to be very different in detail, but, even so, some accounts from long and careful observations of bird sexual and social relations suggest a degree of complexity of motivation and deceit that makes the comparison worthwhile—in both directions.[16–20]

In the background of this view of bird behaviour there is an assumption of a rank ordering of the population from the most desirable to the most undesirable. Individuals are predicted to be pursuing different goals according to their position in the order. Searching for an island in the morass of possibilities that opens, one might at first seize on the top pair as inevitably stable: surely the female of that pair must have the male that she wants. But if 'good genes' have begun to rule, even this is not so. While we easily imagine that the top male is distracted from guarding his nest by invitations from other females, what is it we expect his mate, the top female, to be doing? Is she working hard on her own to support her mate's brood? Surely not: the top female may be having plenty of invitations too. These come from the lowly males. They are sending hints to her that she is welcome to play cuckoo at their nests—with the proviso, of course, that she allows a copulation first. Their signals may be imagined to convey something like this: 'Look how dull I am. How can you possibly doubt—I am your true working father. I and my mate (but quiet about her) will look after your eggs better than you can.'

In the light of this thought it becomes not even clear that a top male will want the absolutely top female to be his mate, or vice versa. The situation calls for a game-theory type of analysis which cannot be attempted here. Could pair infidelity really exist on the scale this implies? How could this be decided? While some suggestive features could probably be made out by patient observations (crowded seabird colonies where dozens of pairs can be under observation at once would seem promising), and thus could verify or negate the general possibility, it is the new techniques offered by electrophoresis and DNA fingerprinting that most obviously open up the new issues of fidelity—regarding the actions of both sexes and for all 'monogamous' animals. Egg-dumping, like extrapair copulation, is a growing theme in bird studies[6,21,22] and already there are hints of situations plentifully complex.

VENEREAL RISK AND POLYANDRY

Now consider the other factor that could make females need to advertise 'health at a glance,' rather as males advertise it and directed to achieve the same desirable and concealed extra-pair fertilizations that they are seeking. This is a long associated presence of serious venereal diseases in a species or group. Overt signs of disease certainly deter humans. With infections present in the population, the female may need to advertise that she is uninfected or that anyway she offers so good a chance to produce a healthy offspring that the risk is worthwhile. As with the factor already discussed, decisions must still be made fast, for the male, like the female, has an interest to conceal the act—that is, he, too, wants to keep the female's mate ignorant so that the mate's care for the nest and future brood continues.

All this says that neither partner in a hasty act can insist on careful tests and observations on the other's health. However, a quick look at important points may be possible. I suspect that the issue of venereal infection, for example, may stand at the origin of the precopulatory cloacal pecking in the dunnocks. This does not exclude a subsequent role of the behaviour in causing extrusion of the previous ejaculate.[12] The cloaca is, of course, the region where the birds will make contact and the male's action may amount to seeking a better look. Needless to say similar suspicions can be raised about the genital swellings of primates, although, unfortunately for the present idea, there seem to be no suitable known diseases in those animals one can point to. (But—would we know, for example, about chimpanzee chlamydias? Do we yet know enough about the HIV-like retroviruses of primates? The fact that we knew nothing about them 20 years ago surely tells the level of our ignorance about most animals.) Back to birds and with another kind of parasite, it is already well known that some Mallophaga cluster in the cloacal region. This is just as would be expected from the imperative of dispersal,[23] and there is accumulating evidence that lice and mites do indeed transfer during the act (N. Hillgarth, personal communication). As noted below, bird lice, like Anoplura and ticks, can transmit microbial diseases.

Since venereal infection can occur both ways, it is interesting, if peripheral to the present theme, that many male displays expose the cloacal region to the female's view. The same line offers an explanation of certain characteristic postcopulatory acts also, such as the vigorous shaking and wing beating that occurs in many birds (e.g. galliforms and anatids, with washing of water over the plumage added in the latter[24]).

The immediate contagion factor in social behaviour, in which the above mating problem is involved as a special case, was first emphasized and reviewed by Freeland.[25] As regards our present theme of showiness, we need to know more about venereal diseases and contagion generally before the factor can be assessed. The distinction of venereal and non-venereal from the present point of view may not always be clear, as shown in the discovery that a rather innocent-seeming louse of swans transmits a fatal heartworm.[26] The heartworm does not seem remotely likely to be venereal at first, but it turns out not only that it might be so but that it might expose faint hints of why, with the louse dark, swans should be white, and even why their neck feathers, which is where the lice mainly live, should be so opened in their display.

As humans are currently being pressured by events to understand, true monogamy is one way to prevent venereal diseases from flourishing. But imagine a monogamous species with a venereal disease lurking nearby— perhaps currently mainly infecting a closely related species. If the general rates of disease and parasite attacks of the right kind go up in this host, so that it comes under the selection already described—that is, increased intensity of search for good genes and correspondingly decreased emphasis on offspring nurturance—then, with the venereal disease occasionally taking toll, it is not so clear as usual that it is the male who is set to become what might be called the 'sanimeter' sex of the species. Once males are less eager for extra copulations because of the risks to their own lives, it is relatively more likely that the principal form of infidelity might become male-abetted egg-dumping. Low males might now be initiators, encouraging favoured healthy paramour females to lay eggs in their nests. Just as the end result of too much female infidelity combined with male eagerness and low cost is explicit polygyny (with the female left with the burden of raising the brood), so the end result of excessive egg-dumping has to be the condition of so-called 'sex reversal'. Most females would be too involved in advertising and with the possibility of laying a further clutch, to remain faithful to a particular nest and male.[27,28] Most males therefore become unhelped raisers of brood. All this suggests that it might be worth looking for particular problems with venereal diseases in birds and other animals that have evolved sex-role reversal. As with the case of polygyny, the ecological factors offered to explain polyandry[29,30] have never seemed at all convincing as the sole determinants,[31] and some hidden factor like disease transmission might make the divergences more understandable. Thus, for example, some wetlands of the arctic have ruffs, swans, and phalaropes all breeding within a space of a few miles: their foods admittedly vary—but what else would contribute to their profound differences in styles of sex?

Since venereal diseases are expected to be rare in monogamous species while opportunities for cheap, safe copulations are by the same token abundant, it is easy to see why it is much more common for the male to become the sanimeter sex. The initiation of an opposite trend seems to need something like the parasite overflowing from another species, this parasite being transmitted venereally. Thus if an already polygynous bird like the ruff evolved to suffer venereally transmitted ectoparasites, encouraged by its promiscuity, and if some of these parasites transmit diseases which at least are dangerous for the other waders around when they reach them (they could be supposed generalist enough at least to bite), only then do the conditions seem set for evolution to take off along the line that leads to a bird like a phalarope. For all these conditions to be met must be quite rare, but it might be worth searching for a partial conjunction of them in charadriforms, where trends to polyandry seem more common than in most other groups. We know that there are abundant sets of ectoparasites providing the potential venereal factor,[32–34] and perhaps such parasites may be particularly prevalent in phalaropes.[35] But it is as yet hard to say what serious diseases, if any, transmit. The typical blood parasites are low in charadriforms.[36,37] Tinamous also show frequent sex role reversal and have varied and abundant Mallophaga.[38] However, the same association is not maintained in the polyandrous ratites. These have few Mallophaga, whereas in the generally polygynous galliforms they are again abundant,[38] so the indications from this potentially 'venereal' group are rather confusing and can be given little weight.

ABILITY TO WORK FOR THE BROOD

There is yet another reason why a monogamous or quasi-monogamous animal should be concerned about the health of its potential mate. It is rather obvious and unexciting. Moreover, although almost certainly it is playing its part in mating choices, it does not seem to me relevant to the particular phenomenon we want to explain. I will therefore treat it briefly.

If both sexes normally work for the brood, both will want partners who actually succeed in working—and work hard—right up to the end of the season. Hence it could be availability of this utilitarian aspect that they advertise health back and forth via their displays and colours. Several writers seem to hint that this idea is the main competitor with the idea of 'good genes'. Indeed, when showiness is accepted as being an index of health at all, it appears that most ornithologists might bet on health being advertised in token of immediate nurturance, and on this being the reason for both sexes being showy.

If the idea is valid, it will lead to a problem of distinguishing two contributions—the sosigonic factor and this one of immediate nurture. Although an answer to this is needed (and does not seem easy to give) I intend to avoid discussing the issue on the grounds that the idea is implausible at the root. If there is a nurturance factor in showiness, I suspect it is small. Taking another image from British history, the idea seems to me like assuming that a Royalist style of dress can derive from a Puritan ideology. In other words, the idea that showiness beyond mere cleanliness serves utilitarian decisions runs counter to intuition as well as counter to the arguments given at the beginning of the essay which stressed that there is plenty of opportunity to assess health in a potential bonded partner in more penetrating ways. Long observations of others close up and under hardship seem a far better test than bright plumage and bursts of song. Once at the nest, moreover, bright plumage is a hazard for everyone. The most royal Royalist, Charles II, had already put on woodman's clothing when he hid in the famous oak tree. A male pied flycatcher might be expected to have evolved likewise and to resemble the female if it is survival and work that the female most needs (as implied recently by Lifjeld and Slagsvold[39]). Hence I have come to much prefer the idea that bright colours and song are concerned with signalling health to relatively distant, relatively unknown receivers for the possible purposes already outlined.

CONSTANT HERITABILITY: MUTATION VERSUS HOST–PARASITE DYNAMIC

In all the above sections except the last, the idea that there can be a role for sosigonic mate choice depends on there being a plentiful supply of unfixed genetical variation undergoing selection. Accumulating evidence for sexually selected species shows that heritable variation exists,[40] although how much of it is sosigonic is not yet investigated—except, as a beginning, in pheasants.[41] In theory, whether related to health or other matters, there are roughly three ways to get the flux that is needed.[42] One way invokes sets of relatively ancient variant alleles that acquire ever-changing values under selection: the alleles hardly ever go globally extinct but do not stabilize either—in short, they cycle. The second way involves a rapid pace of incorporation of good new alleles into the population (this accompanied by, of course, extinction of some of the old alleles). The third way involves a rapid pace of occurrence of bad mutations of all kinds. Bad mutations are never incorporated but their stream can be selecting what might be called 'bad genes avoidance' by potential mates and this in turn, can be fuelling the excesses of sexual selection.

The cycling alternative is the one that Marlene Zuk and I emphasized in our paper and it is the one that I still most favour. Against it, it must be admitted that there are not yet any real examples in nature—even population dynamical true cycles have been hard to prove, and population genetical cycling is still only a gleam in a theorist's eye. The nearest thing we have is a phytopathology resistance cycle in a crop recorded over two periods where a human, anticipating a fungus, plays the selective role of the latter.[43] Almost the most we can say is that the potential for genotypic cycling exists in the very nature of host–parasite relations and resistances,[44,45] and it is hard to see what would prevent cycles happening. Meanwhile circumstantial evidence grows stronger. Such evidence is the recent argument for ancient constancy of some elements of defence polymorphisms,[46,47] plus that for molecular mimicry in the same system,[48–50] opening the further possibility of dynamical pursuits. On the theoretical side, recent simulations I have conducted with others (see ref 51 and Chapter 16) confirm how easily cycling occurs in antagonistic coevolution, and show that the conditions for the protection of polymorphism and constant heritability are easier to obtain than I once thought. Such protection is found, for example, in multilocus models under weak selection where the loci are superficially unitary and independent in their conferred resistances. The insistent dynamical behaviour of such models seems often to have been regarded as an embarrassment, as if it was an outcome of unreasonable assumptions and needed to be explained away (ref. 44 and see also Chapter 15).[44] Needless to say, to my own view the dynamism is needed; permanent dynamical behaviour tells of hope for sex and, in certain forms, gives additional hope for unravelling sexual selection.

At first such models, admittedly, often seem dynamical to an excessive degree. Keeping alleles from being lost altogether appears a problem. But it turns out that simple and reasonable assumptions about how selection could be modelled in *K*-selected climax-type species—that is, those species we know to have the most insistent emphasis on sex and, often, the most extreme signs of sexual selection as well—easily provide conditions for a restless kind of preservation. In fact it has proved that weak coevolutionary cycling for many loci based on a naturally arising 'coalpit' pattern of fitness (most genotypes of much the same value but a few deadly), is very suitable to provide a uniform force for good-genes mate choice. This fits well with the idea that numerous, chronic parasites with moderate detrimental effects can be the ones involved. 'Chronic', of course, tends to imply that they are still present to be assessed during mate choice, but the model also works for parasites that act earlier and die out, provided that the course of their pathology leaves a permanent record in host development, saying how well they were resisted.

For the view adopted here, it is necessary for our view that host deaths are indeed caused by the parasites. This requirement seems to give trouble to parasitologists because it conflicts with their claim that the specific, long-associated parasites of a host species hardly do any harm at all. However, parasites would not be parasites unless they did some harm. They would instead be mutualists and thereby outside the scope of the theory. Although the harm of a particular parasite species may be small, the small harms can add up and may in the end cause death. It needs to be made clear here that the host does not have to be shown at the time of death to be suffering from parasite-induced lesions that are sufficient in themselves to have caused the death. It is enough if we can show that the chance of death from any of a set of proximate causes, such as predation, increases substantially as the parasite load goes up. There have been minor studies and many opinions to the effect that such increased susceptibility to predators does occur, but thorough work has been lacking. This gap has now been filled in a survey of small mammals caught by a tame hawk, contrasting postmortem findings on these with others from a sample of animals shot in the wild, for which type of attack physical fitness would play much less part in their ability to escape.[52] It was clearly demonstrated that the animals the hawk caught were much more parasitized than the controls, especially if the species was one that the hawk found hard to catch. This elegant study combined with previous ample demonstration of selectability of resistance by many hosts to many classes of parasites (see Chapter 5) indicates a fate that parents can be avoiding for their offspring when they choose maximal health in their mates.

It may seem at first true that a mutualist cannot be a parasite in the sense of the theory, because it does not induce coevolutionary cycling. However, there are several ways in which microbial mutualistic symbionts could create problems for their hosts. It may well be, for example, that symbionts continually mutate in ways that tend towards parasitism. A deficiency mutation may simply block a pathway that leads to the benefit on the basis of which the relationship evolved—so creating a need for hosts to be both physiologically (within generations) and genetically (over generations) 'alert' to discriminate. It may also be that a lot of microbial symbionts exist on a borderline between exploitation and mutualism, and exist in variants that are only under proper control when paired with corresponding variants of their hosts. Finally, it may be that housing microbial symbionts in itself involves relaxing certain kinds of routine antimicrobial measures normally used. This may open the way to habitation by unequivocally harmful parasites. Focusing on structures created for gut symbionts, the rumens of ungulates and the caecae of birds do indeed house harmful parasites that are not found elsewhere. Such points taken together make it not surprising that animals using microbial symbionts to aid in digestion and synthesis have often gone in for high sexual selection as well. An example is a striking series in the galliform birds culminating in the Tetraonidae with their huge caecae, decorative plumage and skin, and, often, their lek mating. The artiodactyl ruminating ungulates make another major vertebrate series, with horns of various types here being the outstanding morphological manifestation of the sexual selection.

Looking on and under the ground where the ungulates graze it is fascinating to find the food material already processed by them and dropped as dung being worked over again by tribes of strikingly horned scarabaeid beetles, once again using gut caecae and microbial symbiont flora. The dung for the beetles, of course, does not have to come from artiodactyls, but the poorness of the medium expected to result from the efficiency of digestion in the ruminants is perhaps in favour of a need for further symbiont effort within the secondary processors. Related dynastine beetles living on equally poor media in rotting wood also use symbionts and are also dramatically horned as adults. Along with these are other wood feeders such as tipulids that show the strong sexual selection over again, doing so with colour and other morphology rather than horns. However, other xylophagous beetles that also employ symbionts show no striking adult sexual characters.[53] Looking to non-gut symbionts, several other lekking and/or strikingly ornamented groups of insects have them—for example, cicadas, membracids, various heteropterans, and picture-winged trypetid fruit flies.[53,54]

This theme of symbionts and sexual selection is very tentative, and much needs to be clarified. Why do omnivorous non-ruminating pigs retain fairly high sexual exaggeration, as manifested in dimorphism, tusks, etc? Why do some scarabaeids retain horns even though feeding on rich diets?[55] Geese, grazers living on poor diets have big guts but seemingly make no important use of symbionts in them and have no polygyny—in fact, rather the opposite. At the same time they seem to have no lack of parasites. But hoatzins, also grazers, have even heavier guts than geese and begin to look grouse-like . . . Obviously there is a need for much more study of the various cases before a real theme becomes clear.

Could there be a constant stream of beneficial new mutations being selected into a population such that indicator traits of benefits currently experienced could be evolved and used in mate choice? On the whole I am less enthusiastic about this than I am about protected cycling, but I also think that if this version does work at all it is almost certain that the genes concerned will prove to be ones involved in defence against parasites and diseases. One good new feature is that all resistance mutations can now count: we do not have to worry under this scheme about the paradox of very short cycles (two or three generations long) from powerful pathogens that might induce 'worse genes' mate choice! When parasite selection is slower, however, the system grades into the one already discussed though with the protection lacking: we can think of it as being a limit cycle that keeps hitting sticky boundaries of its space (i.e. fixation). If after travelling part way along a boundary hyperplane a mutation enables the system take off and increases its dimensionality once again, it does not matter on the sexual selection side whether the allele that 'comes back' is the same as the one that was lost or is something completely different. All that matters is that it can contribute to the almost steady heritability of fitness. As regards selection for sex itself, the system brings us back to a Weismann–Muller–Fisher type of theory but is more specific than that one was about the stream of good genes that the theory is postulating; it is the abundance of the probable stream that makes it more plausible that the model can really work. As regards sexual selection, we are back to something like a Lande model but, again, with a more specific idea about the source of the constant heritability that is being assumed: it is not now a matter of mutation of genes that provide the blueprint of the epigamic character but rather of genes that simply affect health as a whole, and thereby determine what surplus of materials and energy the individual can afford to divert to the implementation of the blueprint.[56] There is reason to believe that characters concerned with defence against parasites may involve unusual amounts of gene substitution and diversity due to the rapidity of coevolving change,[57,58] so the stream may indeed be abundant enough to account for sexual selection alone, unhelped by a more proper cycling of ancient and conserved alleles. From the point of view of the character of the sexual selection engendered and the correlation of this with parasitism, the distinction probably does not matter.

The last mechanism for generating a high-enough constant heritability differs from the above ideas in that it invokes all bad mutations taken together, whether they have to do with parasites or not. Thus the special correlations of sexual selection with parasite ecology, the varying afflictions of individuals, and so on, are only expected in so far as a large fraction of all the functional

DNA that lies open to deleterious mutation is actually concerned with parasite defence. The fraction is probably surprisingly high: this not just because the total tissue of the immune system forms a larger part of body mass than one might think (immune system tissues may even outweigh the nervous system), but also because almost any tissue or body protein can potentially become involved, via pathogen mimicry, in a disease-defence story (see Chapter 5 and ref. 50). We are once again talking about health, but it is now health of a different kind and only incidentally connected with coevolving biotic agents. In so far as the idea works it is again overall a Lande-type situation, but it needs to be emphasized, as I don't think Lande ever emphasized, that while the epigamic character can undoubtedly be brought in fast by the Fisher positive feedback, the arrival has now become a past event for most traits, or at most has been transformed into an ongoing slow change that cannot be regarded as either cause or effect of the variation in the epigamic character today. Most of that variation is still, as under the cycling alternative, a reflection of the state of well-being. What I here add is that the variation, while it includes some chance favouring by this or that generation's environment, or by this or that individual's particular circumstances within it, most importantly is reflecting the summed effects of all the deleterious mutations each individual is carrying. The epigamic character then may contribute to species-wide health in so far as it serves as an amplifier of differences[58] and in so far as females are declining to mate those individuals whose poor ornaments show that they carry exceptional loads of mutations. Chance evolution of preference by the Fisher process (or Kirkpatrick process, he having emphasized more than Fisher the arbitrariness of direction) may equally carry a character towards large size or towards disappearance. But the trends to enlargement, if sex-limited to a relatively throwaway sanimeter sex, are the ones that can serve to improve species health because size tests ability to produce and absence does not. Thus enlargement characters may give some sexual species an extra buoyancy to ride on the flowing tide of bad mutation. I use 'may' here because a more proper theoretical treatment is needed before we can say that they can, leave alone that they do.

On the whole I feel that the connections of sexual selection with parasites already shown and discussed at the symposium at which this paper was first presented, vague and sometimes negative as the connections still are, are enough to indicate that this last 'avoidance' or 'pure negative eugenics' view of sexual selection looks at only a minor part of the whole. If the line were the whole truth, all sexual animals should manifest about the same amount of sexual selection because all, presumably, have about the same amounts of functional DNA mutating badly at approximately the same rates. This does not seem to be the case. In particular, the avoidance explanation will be almost rejected if it turns out, for example, that the set of birds with wattle-like developments suffers more parasitaemia (easily revealed by wattles) than the set without wattles, if the set of birds with exaggerated showy peri-cloacal feathers suffer more diarrhoeal diseases, and if other similar special advertisements appropriate to special parasite and health problems can be shown. In general, the apparatus of sex and sexual selection looks to me more commensurate with problems created by active enemies than with problems created by an organism's failure to buffer and correct its own internal processes, the sort of failure that leads to bad mutation: to remedy that other more efficient means would serve.

CONCLUSION

In the light of ever-increasing knowledge of the complexity of the anti-parasite defence system of higher organisms, of the subtlety of weapons that parasites bring against them and how rapidly these can change, of yet other kinds of circumstantial evidence,[60,61] and of the convincing success of models concerning how the interplay of these factors with recombinant multilocus genetics could work (Chapter 16), I regard the parasite theory of sex itself as very nearly established. The main challenge, the idea of sex serving for the efficient elimination of harmful mutations,[62] clearly identifies another important function of sex, but, when offered as a primary force, does not seem able to explain major features of the geography, ecology, and actual practice of sex and parthenogenesis. Nor does it even attempt to explain a major feature of genetics, the much greater DNA copying fidelity and/or mutation control that is achieved by prokaryotes without any seeming cost to their extreme efficiency of growth.

The situation for the sexual-selection extension of the parasite theory is more dubious. The case for 'good genes' sexual selection is becoming strong, but this only implies heritability and not the source. My own feeling is that it is hard to see any other source than parasites, the only really viable and different contender being again the one last mentioned—high rates of damaging mutation. But this view does not seem to be widely shared: there is a vague but perhaps reasonable supposition of 'something else out there in all that complexity' that is capable of ensuring high heritability.

Such doubt notwithstanding, the number of studies indicating correlation of parasites consistent with the proposed connection as opposed to indicating hostile or uninformative correlations is probably already approaching significance on a sign test. But most of the results show correlations: much more needed are studies that put parasites, genetics, and natural history convincingly together either for a group or a focal species.

Acknowledgements
I thank N. Hillgarth for comments and J. B. Hainsworth for the suggestion of sosigonic.

References

1. K. Z. Lorenz, *Studies in Animal and Human Behaviour*, Vol. 1 (Methuen, London, 1970).

2. D. E. Gladstone, Promiscuity in monogamous colonial birds, *American Naturalist* **114**, 545–57 (1979).

3. P. R. K. Richardson and M. Coatzee, Mate desertion in response to female promiscuity in the socially monogamous aardwolf *Proteles cristatus*, *South African Journal of Zoology* **23**, 306–8 (1988).

4. S. Spano and D. Csermely, Male brooding in the red-legged partridge *Alectoris rufa*, *Bolletino di Zoologia* **52**, 367–9 (1985).

5. F. McKinney, K. M. Cheng, and D. J. Bruggers, Sperm competition in apparently monogamous birds, in R. L. Smith (ed.), *Sperm Competition and the Evolution of Animal Mating Systems*, pp. 523–45 (Academic Press, New York, 1984).

6. A. P. Møller, Intruders and defenders on avian breeding territories: the effect of sperm competition, *Oikos* **48**, 47–54 (1987).

7. P. W. Sherman and M. L. Morton, Extra-pair fertilisations in mountain white-crowned sparrows, *Behavioral Ecology and Sociobiology* **22**, 413–20 (1988).

8. N. Burley, The differential allocation hypothesis: an experimental test, *American Naturalist* **132**, 611–28 (1988).

9. M. V. Studd and R. J. Robertson, Differential allocation of reproductive effort to territorial establishment and maintenance by male yellow warblers (*Dendroica petechia*), *Behavioral Ecology and Sociobiology* **23**, 199–210 (1988).

10. D. F. Westneat, Extra-pair fertilisations in a predominantly monogamous bird: observations of behaviour, *Animal Behaviour* **35**, 865–76 (1987).

11. T. Jarvi, E. Roskaft, M. Bakken, and B. Zamsteg, Evolution of variation in male secondary characteristics: a test of eight hypotheses applied to pied flycatchers, *Behavioral Ecology and Sociobiology* **20**, 161–9 (1987).

12. N. B. Davies, Polyandry, cloaca-pecking and sperm competition in dunnocks, *Nature* **302**, 334–6 (1983).

13. A. P. Møller, Spatial and temporal distribution of song in the Yellowhammer *Emeriza citrinella*, *Ethology* **78**, 321–31 (1988).

14. S. A. Hatch, Copulation and mate guarding in the northern fulmar, *Auk* **104**, 405–61 (1987).

15. H. W. Power, E. Litovich, and P. Lombardo, Male starlings delay incubation to avoid being cuckolded, *Auk* **98**, 386–9 (1981).

16. J. Verwey, Paarungsbiologie des Fischreihers, *Zoologisches Jahrbücher* **48**, 1–120 (1930.

17. J. M. Dewar, Ménage à trois in the mute swan, *British Birds* **30**, 178–9 (1936).

18. T. R. Birkhead, Mate guarding in the magpie *Pica pica*, *Animal Behaviour* **27**, 866–74 (1979).

19. G. Stenmark, T. Slagsvold, and J. T. Lifjeld, Polygyny in the pied flycatcher *Fidecula hypoleuca*, a test of the deception hypothesis, *Animal Behaviour* **36**, 1646–57 (1988).

20. A. P. Møller, House sparrow, *Passer domesticus*, communal display, *Animal Behaviour* **35**, 203–10 (1987).

21. Y. Yom-Tov, Intraspecific nest parasitism in birds, *Biological Reviews* **55**, 93–108 (1980).

22. A. P. Møller, Intraspecific nest parasitism and anti-parasite behaviour in swallows, *Hirundo rustica*, *Animal Behaviour* **35**, 247–54 (1987).

23. W. D. Hamilton and R. M. May, Dispersal in stable habitats, *Nature* **269**, 578–81 (1977) [reprinted in *Narrow Roads of Gene Land*, Vol. 1, pp. 377–85].

24. A. P. Johnsgard, *Water Fowl: Their Biology and Natural History* (University of Nebraska Press, Lincoln, 1968).

25. W. J. Freeland, Pathogens and the evolution of primate sociality, *Biotropica* **8**, 12–24 (1976).

26. W. S. Seegar, E. L. Schiller, W. J. L. Sladen, and M. Trpis, A mallophaga, *Trinoton anserinum*, as cyclodevelopmental vector for a heartworm parasite of waterfowl, *Science* **194**, 739–41 (1976).

27. J. D. Reynolds, Mating system and nesting biology of the red-necked phalarope *Phalaropus lobatus*: what constrains polyandry?, *Ibis* **129**, 225–42 (1987).

28. D. Schamel and D. Tracy, Polyandry, replacement clutches, and site tenacity in the red phalarope (*Phalaropus fulicarius*) at Barrow, Alaska, *Bird Banding* **48**, 314–24 (1977).

29. F. A. Pitelka, R. T. Thomas, and S. F. MacLean, Jr, Ecology and evolution of social organisation in arctic sandpipers, *American Zoologist* **14**, 185–204 (1972).

30. W. D. Graul, Adaptive aspects of the mountain plover social system, *Living Bird* **12**, 69–94 (1974).

31. J. Maynard Smith, Parental investment: a prospective analysis, *Animal Behaviour* **25**, 1–9 (1977).

32. H. S. Peters, A list of external parasites from birds of the eastern part of the United States, *Bird Banding* **7**, 9–27 (1936).

33. R. Meinertzhagen and T. Clay, List of mallophaga from birds brought to the Society's Prosectorium, *Proceedings of the Zoological Society of London* **117**, 675–9 (1948).

34. R. O. Malcolmson, Mallophaga from birds of North America, *Wilson Bulletin* **72**, 182–97 (1960).

35. D. M. Yanez and A. G. Canaris, Metazoan parasite community composition and structure of migrating Wilson's phalarope, *Steganopus tricolor* Viellot, 1819 (Aves) from El Paso County, Texax, *Journal of Parasitology* **74**, 754–62 (1988).

36. E. C. Greiner, G. F. Bennett, E. M. White, and R. F. Coombs, Distribution of the avian haematozoa of North America, *Canadian Journal of Zoology* **53**, 1762–87 (1975).

37. M. A. Pierce, Distribution and host-parasite check-list of the Haematozoa of birds of Western Europe, *Journal of Natural History* **15**, 459–62 (1981).

38. W. Eichler, *Mallophaga* (Akademische Verlagsgesellschaft Geest & Portig K.-G., Leipzig, 1963).

39. J. T. Lifjeld and T. Slagsvold, Mate fidelity of renesting pied flycatchers *Fidecula*

hypoleuca in relation to characteristics of the pair mates, *Behavioral Ecology and Sociobiology* **22**, 117–23 (1988).

40. A. V. Hedrick, Female choice and the heritability of attractive male traits: an empirical study, *American Naturalist* **132**, 267–76 (1988).

41. N. Hillgarth, Parasites and female choice in the ring-necked pheasant, *American Zoologist* **30**, 227–33 (1990).

42. A. N. Pomiankowski, The evolution of female mate preferences for male genetic quality, *Oxford Surveys in Evolutionary Biology* **5**, 136–84 (1988).

43. J. A. Barrett, Frequency-dependent selection in plant–fungal interactions, *Philosophical Transactions of the Royal Society B* **319**, 473–83 (1988).

44. J. W. Lewis, On the coevolution of pathogen and host. Parts I and II, *Journal of Theoretical Biology* **93**, 927–85 (1981).

45. I. Eshel and E. Akin, Evolutionary instability of mixed Nash solutions, *Journal of Mathematical Biology* **18**, 123–33 (1983).

46. D. A. Lawlor, F. E. Ward, P. D. Ennis, A. P. Jackson, and P. Parham, HLA-A and B polymorphisms predate the divergence of humans and chimpanzees, *Nature* **335**, 268–71 (1988).

47. F. Figueroa, E. Günther, and J. Klein, MHC polymorphism predating speciation, *Nature* **335**, 265–7 (1988).

48. J.-G. Guillet, M.-Z. Lai, T. J. Briner, S. Buus, A. Settle, H. M. Grey, J. A. Smith, and M. L. Gefter, Immunological self, nonself recognition, *Science* **235**, 865–70 (1987).

49. I. R. Cohen, The self, the world and autoimmunity, *Scientific American* **258** (4), 34–42 (1988).

50. D. Vidovic and P. Matzinger, Unresponsiveness to a foreign antigen can be caused by self tolerance, *Nature* **336**, 222–5 (1988).

51. W. D. Hamilton, Haploid dynamic polymorphism in a host with matching parasites: effects of mutation/subdivision, linkage, and patterns of selection, *Journal of Heredity* **84**, 328–38 (1993).

52. S. A. Temple, Do predators always capture substandard individuals disproportionately from prey populations?, *Ecology* **68**, 669–74 (1987).

53. P. Buchner, *Endosymbiosis of Animals with Plant Microorganisms* (Interscience, New York, 1965).

54. D. J. Howard, G. L. Bush, and J. A. Breznak, The evolutionary significance of the bacteria associated with *Rhagoletis*, *Evolution* **39**, 405–17 (1985).

55. G. Halffter and W. D. Edmonds, *The Nesting Behavior of Dung Beetles (Scarabeinae): An Ecological and Evolutionary Approach* (Instituto de Ecologia, Mexico, DF, 1982).

56. W. Dominey, Sexual selection, additive genetic variance and the 'phenotypic handicap', *Journal of Theoretical Biology* **101**, 495–502 (1983).

57. R. E. Hill and N. D. Hastie, Accelerated evolution in the reactive centre regions of serine protease inhibitors, *Nature* **326**, 96–9 (1987).

58. M. Laskowski Jr, I. Kato, W. Ardelt, J. Cook, A. Denton, M. W. Empire *et al.*, Ovomucoid third domains from 100 avian species: isolation, sequences and hypervariability of enzyme-inhibitor residues, *Biochemistry* **26**, 202–21 (1987).

59. O. Hasson, The role of amplifiers in sexual selection: an integration of the amplifying and the Fisherian display, *Evolutionary Ecology* **4**, 277–89 (1990).

60. A. Burt and G. Bell, Mammalian chiasma frequencies as a test of two theories of recombination, *Nature* **326**, 803–5 (1987).

61. C. M. Lively, Evidence from a New Zealand snail for the maintenance of sex by parasitism, *Nature* **328**, 519–21 (1987).

62. A. S. Kondrashov, Deleterious mutations and the evolution of sexual reproduction, *Nature* **336**, 435–40 (1988).

HEALTH AND HORSEMEN

The Seething Genetics of Health and the Evolution of Sex

Now new-coming nations
That island did rule,
Who on outlying headlands
Abode ere the fight.
ICELANDIC[1]

SECURELY under the seat in front of me stayed my backpack and in it was the paper I had meant to revise and the talk I intended to give: I was looking out of the window. It was late March in 1990 and I in the air travelling from London to Japan, by way of (or near to) the disc of the Arctic Circle. A sector of darkness had passed me—night, I supposed, but from the point of view of time for sleep it had seemed little more than an eclipse. Now over East Asia all of the down slope of the world lay beneath me clear of cloud and I watched the mountains and plains of Siberia slide by below. Three papers that I considered important for my topic of sex were either recently in print or soon to become so, and I was happy, almost exultant. At this meeting I would have new work to show and already still other models were in progress—testing, for example, dynamics of deterministic versions of those individualistic simulations I have shown in Chapter 16. I felt increasingly confident I was on a broad path with the potential to explain all the manifestations of sexuality.

I was, of course, delighted to be called on and to be paid to fly such a distance and, again, delighted at this moment to be in the air over that vast land, Siberia, so thin with people, so full with natural marvels, where I had never set foot. In a parallel reality of my imagination, somewhere below me moved a life-long hero, Anton Chekhov, in ant-like three-month progress towards the east. Already seriously tubercular, he threaded doubtless these same thin lines and corridors of cultivation that I now saw from time to time crossing the wilderness. He was travelling to see and report on conditions of the convicts of Sakhalin Island; elsewhere and at quite other times (and boasting oppositely to Chekhov their overflow of health and population) Mongol armies and their baggage trains moved west. Was it on the same lines, in the same corridors? That really I didn't know, nor indeed quite where I was; by their way of life the Mongols would seek grass I supposed. From here I watched both kinds of history through inward eyes that neither Genghis Khan nor gentle Chekhov, doctor as he was, could have imagined. Almost nothing living, it seemed to me, could be unaffected by the new ideas if they were right: plants, animals, and even humans. Indeed, for the last, many of those mysteries and fooleries of passion Chekhov had worked into his human tales—far back at least, all of it must be subject to the new interpretation of life that I carried. Now necessarily would all features be transformed but all had to have a new light and a new angle, for anyone who wanted seriously to understand. As ever when I am in the air, such a flotsam of a relaxed period, even a euphoria, drifted through my mind.

What made all the brownness of tundra and moor below me, for example? Was it sex or lack of it, I wondered. I had read (I have never set foot in any expanse of lowland Arctic tundra) how *Polygonum viviparum* was one of the most abundant plants of Arctic moors, and this was *viviparum* for Linnaeus because literally viviparous. Close in its leaves and stems to any of those pink-flowered bistorts or persicarias you may have creeping on your rockery or in or beside your pond, *Pee-vivip* itself is much less showy in its flowers because it replaces almost all of them on its finger-high vertical 'flower' spike with small vegetative bulbils. As the spike matures these simply fall off. After a short trip, dodging this way and that, perhaps, down a tiny rill of summer melt water, they grow up as clonal offspring—that is, as baby plants that are identical genetically to their mother. Only at the very tip of the flower spike several small and faintly pink proper flowers still open but, although competent to be pollinated, these flowers only rarely are so and they seldom set seed. Race-wise or just by developmental forms, it seems that the higher on the mountain or farther into the Arctic, the fewer the flowers.[2] *Polygonum viviparum* has long been for me a prime example of what I think of as the 'arctic–alpine' syndrome of parthenogenesis: it is an extremely successful plant, one that I could find on any high Scottish moor, and, equally, I could see on a 10 000-foot summit in the American Rockies; and now again, from what I had read, it must be abundant below. Yet if indeed abundant enough to be making much of that giant purply-brown of the land I looked down on, and if my theory was right, how could this be so? How could there be so much of parthenogenesis in one place? Wouldn't it be a monoculture, a giant feast for any adapted herbivore or parasite? How did abundance and consequent extra potential for transmission of parasites, trade off against the harshness of the habitat? So if indeed so abundant, shouldn't this plant be as sexual as the reindeer trampling and eating it? The high state of sexuality of reindeer, despite their arctic and alpine home, I had long ago rationalized (as you will have seen in previous chapters) through their high sociability and the fact that their parasites were already known to be diverse and serious. Why wouldn't the same hold for the plant? I knew it had some fungal parasites in the Arctic,[3] but these didn't sound to be terribly common or limiting . . .

There were many such big holes in my theory waiting to be filled and were I not the jet-setting parasite that at that moment I was, I supposed I'd be on the ground somewhere, perhaps right here below though more likely in the Scottish Highlands, working it all out. I also knew well, and with some nervousness, that it wasn't for these ideas at all that I was being called across the globe to Kyoto; those who had thought of me would probably much rather I talked about sex ratio or sociality. I had no reason to think that anyone in Japan regarded my theories on sex any more favourably than did my fellow evolutionists in the West. Never mind; somehow this did not kill the euphoria. For me the ideas were increasingly proved right or, at least, the evidence for them was rapidly broadening— and practically the whole of this volume has been my attempt to convince you. Especially I have tried to bring matters roughly up to date (by about 10 years more of data) in the last chapter. Like waves of all sizes from ripples to tsunamis, surely the disease problems must come and go in the Far East just as they did in the West. As to humans, the diseases my mother had feared for her parents as a young doctor—TB, scarlet fever, the great (but not minor) swings of influenza[4]—had faded and others, as surely everyone of my age had noticed, had come instead; and all of this was on the required scale of two or a few generations. In the field of infectious disease nothing, just nothing, seemed to be static. This was known history in China perhaps still better than it was history in Europe.[5] Meanwhile the tricks of defence against parasites were turning out to be as universally abundant and diverse as the tricks and the uneven abundances of the enemies necessitating them. Moreover, as is ideal for the theory, many, perhaps most, of such tricks were also turning out to be what I will call 'particular' as opposed to 'additive' or 'quantitative' in other ways[6]—that is, they prove to be matters of the fairly precise interlock of molecules: some key either turns or it doesn't. Such few facts, and especially this visible restlessness of the disease factors affecting all life combined with the *particularity* of the defence, were enough.

A 'Red Queen' foundation for sex doesn't absolutely require the particularity but it certainly helps, and it was already clear that this quality of defence was abundant and that its genetic base in hosts is widely scattered, as is also required, across chromosomes—clustered perhaps, but seemingly, for any organism's entire defence system, usually spanning most or all of them.[7] Therefore I would talk at the coming conference yet again about my sex and parasites. From the point of view of what perhaps my inviters hoped for, I would be filling, after all, what genuinely seemed to me the biggest hole of the sociality theory I had been engaged on before—the limits, as illustrated in any example of insistence on outbreeding or risk taken for unrelated partner, to the calculus of kinship. Therefore, in effect, I would indeed be talking about that. Scientists of both East and West attending the meeting would have to listen. If the Easterners chose, let them also dismiss my new sex models as 'too vague' along with my older ones; let them neglect to remember their own vast history with cholera and all the other epidemics; . . . and again never mind, and never mind.

I looked down. These edges of Arctic deserts where roamed the reindeer and where grew the polygonum, sharpened the contrasts I needed for tests I had in mind, I would look down at those. At times vague colours far off to the right made me think I might be seeing the Gobi Desert, another real edge, in drought and insolation, to the scene of life; but, later, looking at the map in the airline magazine, I decided it was too far away. Asia indeed isn't Britain where on a clear day from a high plane you might see the whole of our southern and eastern coast—all East Anglia, say—and the Gobi was not Thetford Heath. Anyway, I thought, straining to look back, out that way somewhere there would surely be examples of desert parthenogenesis just as I knew in the Sonoran of Arizona, all paralleling those of the cold deserts of the North. At the Gobi extreme, then, what are the parallels to the reindeer, what mighty-horned caprids, what equivalents to ibex or big-horn sheep, in their contradictions to the principle of more harshness, less sex? Would there be actual ibex still, as there were, say, in Anatolia and in Israel? I thought not.

In any case in any of these extreme places my ideas could eventually be tested, proved right either from the excess or the diminished sides of sexuality if one only paid enough attention to the parasites. Else let the ideas lie, die, and dry there—and nobly if possible, like the kit of an abandoned expedition, like a Franklin's or a Burke's. Long flights to Japan, to the Far East—many ways to come here, long hours for me of this cricked neck trying to look down: Hong Kong–Singapore, Singapore–Brunei, Singapore–Narita . . . In this extreme opposite to the tundra sometimes the jungles of Southeast Asia were spread beneath me where grew those huge trees I had seen so doubtfully claimed to be parthenogenetic[8] and also probably those smaller flood-forest wild 'mangosteens' that (like the cultivated fruit tree) probably really were so.[9] That flood-forest tree, plus— with less difficulty for the theory—a few small trees and shrubs from disturbed South American savannahs[10] and also on Swiss mountainsides[11] were the exceptions and they made me a challenge exactly opposite in their implication to the supersexed reindeer and ibex. So there, too, in lofty forests, my Expedition to the Source of Sex could also languish and die. More likely, though, just now my expedition was not so much dying as stalling under the attacks of its immediate enemies, rather like those Japanese soldiers who lived on where they had become lost or had been told to wait in the forests of Oceania, stumbling out of them at last many years later and imagining World War II still being fought . . .

I thought again about the coming symposium and of how, aside from this drear image of fading and dying in jungle or tundra, I should try to fall right there in Kyoto and more spectacularly if that were possible, because historically and in the long run what better place has there been to die for some cause than in Japan, provided you can be fighting to the end? Japan is where traditionally, if enemies catch you, you end decapitated by sword and head-displayed on some castle parapet (while behind its wall all of your family may lie decapitated dead, too, if the old feudal spirit of Japanese thoroughness could get to them). Yet Japan is where in the long term even this disaster and ignominy can have its reward, finding a secure path into the heart of a people where you are revered simply because you died fighting.[12] Indeed aiming humorously to touch this very theme, I'd packed among my slides one I'd made from a print that showed the Kusonoki brothers as last survivors defying a hail of arrows at the battle of Minato River. I meant to tell (and did) that this was how things had been recently for me in Oxford. Who was it who called Oxford the City of Lost Causes— Cardinal Newman? And who, in the Japanese counterpart, in my Battle of Sex at Oxford, was to be my twice-turncoat, my Takauji? In the contempt/reverence that, simultaneously, our 'City of Lost Causes' conjures for failure in Britain, I like to think that there is a similarity to Japan, and perhaps especially a parallel to Kyoto where I was to speak, its line being drawn through all those tragic stories and including the city's final lost prime status, which it had to yield first to Kamakura and then to Edo (Tokyo) and in which loss the battle of Minato River seems to have been the key turning point.

King Charles lost his civil war at Oxford; poet Emperor Go-Daigo of Japan, a man seemingly of rather similar personal character to Charles, lost his at Kyoto and Minato River was the key fight. These men led my thought, oddly enough, to another *Polygonum* plant, one with fine flowers and, in particular, led to a tiny Aladdin tree dangling its jewel flowers as foreground for a white swan-goose: this was in a painting by artist Emperor Hui Tsung of China in the twelfth century. Perhaps that was *scabra*, or perhaps it was nothing more than our old British *lapathifolia*. Anyway that emperor, if it was really by his hand, knew how to catch a bistort's very soul as he had also caught that of the goose, and it was over the Manchurian land of that emperor's northern enemies and the place of his sad death in captivity that, just possible, I might now be flying. He yet again, of course, and probably because so good an artist, had ended a loser. An addiction that had made him so capable in capturing plant and goose on paper was perhaps that which made him so incapable with administration and military affairs, causing him to end both his dynasty and that whole marvellous Northern Sung culture of China.

Jostling with this driftwood of history came the other 'polygonums' of this region I had known—the climbing *sachaliensis* of Sakhalin, the tall-standing *japonica* of Japan, plants that now make their super-vigorous subinvasions all over Britain: a success that, again, I believed had been a matter of parasites all left behind—that is, in the cases of those just mentioned, left right here below me. Meanwhile their sexuality in Britain, as judging from their flowers, seemed, just as my theory predicted, already half-lost because of lost parasites. Finally I thought of *acuminatum*, that large and truly floating *Polygonum* I knew from still farther down the long slope but on the other side; indeed it was as far down under the sun as that slope goes, a plant adorning the flooded margins, and sometimes (marooned there—neophyte, awkward climber) the green walls all along the side lanes of the Amazon. I had no reason to think either that plant or Emperor Hui Tsung's polygonum not to be fully sexual but the same problems applied to them: why did some plants and not others become inbreeding or asexual? As another example there'd been *Symmeria*, a truly tree-like 'polygonum' or at least member of its family, puffing up in dark green and laurel-like explosion out of the swampy Amazon watersides. Why should that, so very opposite to *P. vivip.*, be separate-sexed as if it was a yew tree or a human? Could the plunging and deeply inundating Amazon watersides be 'deserts' of their kind, too, for these plants and this tree another 'reindeer' or an 'ibex', supersexed like them, for the Amazon? Might it be heavily parasitized because it became, by its very success in water roots and anoxic muds, so abundant locally and by this and by its confinement to the watersides very easy for its pests to find (as I discussed in Chapter 8)? I had noted the tree (along with Amazionia's one willow—likewise separate-sexed and often cheek-by-jowl with the *Symmeria*) was indeed much and long parasitized.

Really, like this, from on high in a modern aeroplane, the world seems a ball and one can think down all its sides. How utterly different from that which Marco Polo travelled, reaching in years destinations I overflew in a half night and a day. Floods of the Yellow River, of the Amazon—for me it was easy to conjure them both, visit them both if I tried. Apart from ephemeral mudbank weeds, virtually all the rooted water plants along the Amazon I had tested, had had, like the *Symmeria*, obvious and deep sexuality. This was also true with the floaters, but usually these in addition had bulbils and fragmentation for their increase, these modes fairly obviously set to take advantage of the enormous flood-season increase in their space. Many animal plankton in the water with them—water fleas, ostracods, and the like—were similar (or for the theory worse) because my colleague Peter Henderson who studied them could discover no males at all. Yet how could I really liken any of these dwellers beneath what was perhaps, in general, the world's most benign climate, to inhabitants of mountain and arctic pools where so much of life again was asexual? To heighten my puzzle there were also the exceptions the other way. I have already mentioned (and roughly excused) the reindeer; but there were also the bright flowers on the tundra—the mountain avens, diapensia, arctic bramble, and so forth. As with the reindeer, one could, as I said, invoke their abundance and the consequent ease of transmission of all their diseases,[4] but why, then, that viviparous polygonum or others like an arctic *Bouteloua*, a wholly asexual grass? Was this penal servitude—a sentence enough for the asexuality to happen (which I suppose means senile and harmless in Stalin's terms), but not (yet) enough time for parasites to catch up? Holes and solidity in my theory, more holes it seemed than solidity, shifted back and forth, but one thing stayed sure: my opponents, promoters of the Deleterious Mutation Hypothesis, generals in Takauji's army, weren't even thinking, not even when accorded the euphoria of 30 000 feet, of these geographical and ecological puzzles. How could they neglect them?

The brown below me changed to a pale green swathing a hilly and finally mountainous country. Looking hard at the hillsides I decided the green to be recent bud-burst in a vast forest of larches. If correct, confidently I'd predict for these trees again that they would have the same high sexuality as now is well proven for the huge similar natural stands, little mixed with other species such as the ponderosa pine forests on the plateaus of upland and western North America. If this parallel holds, the diversity and numbers of the parasites on these larches would also be guaranteed. Were it possible for me to see the effects of pests from where I flew, it would be a fine mosaic all across that apparent uniformity. When low enough over the Rockies—in a flight out and still low from Fort Collins, say—I had indeed sometimes visually picked out the brown spots of actual groups and individuals of pines dying under the attacks of mountain pine beetles, this being one of their worst enemies and a case consequently for which the complexity of the resistance picture in the forests has been partly worked out.[13] With the right imaging applied to satellite pictures possibly now a more general discrimination of the different pest and disease attacks could already be done and the differences revealed in either larches or ponderosas down to the scale of the individual trees. But that picture would be also partly reflecting the non-biotic factors of ecology; much farther in the future must wait the films and lenses that might differentiate what I would most like to see of all, the parallel mosaic ('out-of-step' always with the pests, as I believed it) that would lie beneath cuticles of needles or leaves, within cells beneath the bark. In short, what I was yearning for, of course, was to see coloured below me the actual defence polymorphisms imparted by the tree genes themselves.

The lines on the maps in the airline magazine showed our route arch far to the north; well, it might be drawn so, but could I trust that these were the true great circles that we were flying and that they would enable me to work out my course past this river and that? As for my imagined travellers, clearly they had been constrained to creep more along the lines of latitudes. I looked at the lines on the map and then tried to imagine them out of the window, stretching away to my right and half behind; I thought of the people now below there, the languages, and then again of all that trickle of back and forth across Asia, and then again of Sherwood Washburn's objections to my comments about pastoralists and their gifts to history.[14] How wrong I still believed him to have been! Aleksander Pushkin, who, with Nikolai Gogol, started what we call the great literature of Russia, was the grandchild of that curious alien serf who is usually named 'Peter the Great's Negro'; Alexander Dumas senior (and therefore his son too—also a successful writer) had African (via West Indian) ancestry; and for France, again, I have even heard of a sub-Saharan touch suggested for Napoleon.

Those real or conjectured African ancestors have no particular reasons that I know of to have been African pastoralists in particular, the theme I had been on about in that paper, and I would expect they weren't. The points instead for me, here high in the air, were that, first, these gene incursions into Europe from the south definitely did their recipients no disservice—in fact, as to originality they seem very beneficial; and, second, in these cases of African ancestry, the presence of African genes, or at least of the presumably rather unimportant accompanying visible markers, remained faintly recognizable in the portraits. As to genes from the Asiatic pastoralist warlike incursions into Europe, and any special settlements from these into individuals, I knew virtually nothing. Perhaps some Russian historian may fill in or deny what I strongly suspect.[15] Meanwhile I'll challenge Washburn's ire again by repeating my heresy as follows. It is very hard to imagine a language being transfered onto an alien culture (such as in the cases of Finnish and Estonian in northern Europe or English in South Africa) without some genes going with it. Genes seemed to me fundamentally more sociable as invaders and travellers down the ages than words are, and they are still more so, as linguists inform us, than are the structures that words make in sentences . . . By the ancient rules of Mendelian 'marriage', which rules almost define, with few exceptions,[16] sex's inner self, genes are always outbreeders, always interested to bed with strangers. Whatever the details and causes of this oddity of life may be, it is obviously some part of their function in the Mendelian system to be of this inclination. On the contrary, languages, the linguists seem to insist, evolve like parthenotes: only branching is allowed. When languages are in major collision, we are told, the structure of one ends practising a genocide against the other; they may exchange words but the structures never mix. In short, languages admit no 'recombination'; geneticists instead, even out of the worst histories of enslavements and the like, always expect to see mingling.

　　Well, maybe Washburn was right, I thought, regarding any massive immediate impacts of, say, the Hunnish and later Mongol incursions on European gene pools and what they left;[17] however, he was almost certainly not right as regards such miniature examples as of that steppic Naryskin family, who provided for Peter the Great not merely a black serf to play with but his very mother—and a half of his own genes. Nor was Washburn right about various other incursions that stayed. The first equestrian possibilities, of course, are the early Indo-European speakers who are just possibly the steppic and sub-montane 'Kurgans' recorded in archaeology.[17] Joining in prehistoric times (or themselves changing to become) settled farmers, these 'Indo-Europeans' certainly did change the gene pool and also the language of all of Europe. Then after a delay of some thousands of years, which were probably not without unknown minor events of the same kind, came those increasingly historically remembered invasions by Turks and other kindred peoples into both Europe and the Middle East and India.

　　The type of genetic interchange I discussed in the paper that I had with me on the plane—basically the essay that follows—was on a less grand scale than that of the Golden Hordes or, in terms of time, of any years' (or still less of what amounted to generations) of march in the corridors of Central Asia. But surely these differences were in type only and in the processes they seemed to me still quite similar: all of them were evolutionary and subject to selection and all, ultimately, whether they were changes of genes or artefacts, were part of a race of races to own land for their breeding, and a race, too, of lines within cultures, to own wealth and thus dominance, and all this again ultimately applied to unconscious purposes of breeding. Genghis Khan had more wives even than a cock capercaillie if the histories are true and took many with him on his campaigns. Doubtless and again in his own way, too, he looked after his multitudinous offspring's upbringing better than a cock capercaillie ever does. High leaders are like that[18]—polygamous. We may see the ancient imperative reflected even in the modern cases where contraception and other substitute acts ensure that no offspring result: Mao Tse-tung, for example, praising peasantry and proclaiming opposition of communism to lords and khans of all kinds, still certainly garnered female consorts to himself in a way no peasant could afford to.

My 'ghost of a polgynous lord'—was this already in my paper as it came with me, that mention you will see in its penultimate paragraph? Or did the phrase 'happen' during the meeting, or even during the flight I am just describing? In the dream of the short night that turned to day over the tundra, well might I have imagined, besides warriors of various accoutrement trotting or on foot below me, some lord carried by palanquin to inspect the Wall of China and beyond that wall imagined the folded yurts of the many wives of a great khan out on the plain. Well might I have imagined too, earlier in the flight, much later in the history, as we started out north-east from Britain over Europe, that Count Karnéev of far European Russia whom again Chekhov painted as a disgusting degenerate in his story *The Shooting Party*, deeming him, evidently, exactly descendant to those who strove to keep alive the *jus primae noctis* of feudal Europe, even when already professing to a monogamous religion. Or (who knows, please tell me) might that 'Karnéev' itself, the name, instead have derived out of 'khan'? Either way, every item of Chekhov's characterization described a man down-bound in a fast lane of a 'seething' highway, as I name the matter of social-class change in my paper.

If a theme is true it should be true everywhere—a truth of history should be encapsulated within the farthest atom of the Universe, as one might put it. Thus grandly, at least, things may seem to an imagination dizzied by night and day in northern Asia at 30 000 feet. If good literature is the most true literature, I conjectured, one should read almost as much of such truth in fiction as in social histories, and perhaps in some ways more—as much, for example, in Jane Austen as in Thomas Hardy; as much in Chekhov (even apart from his Sakhalin Report) as in Gibbon or, more directly, in other histories of the Golden Horde and its legacy and its break up. White Sheep Turks, White Hordes lay tangled in my mind with the Black Sheepers and the Golden Hordes; Volga Bulgars and Bulgar Volgas tangled with Khazars and Patzinaks, and Ghuzz, all surging from place to place on the steppes, spreading and sucking under, like a swarm of eddies under a bridge.[19] But after what is admittedly a long struggle in one's approach to this kind of history, one learns a bit to put aside the distortions of a period or of a historian—their Romanticism, for example, or Nationalism—as passing ephemera.

With Chekhov I hardly needed even to make effort to do that because he allowed virtually no distortions, or only those necessitated for the words, but not the hearts, of his characters. His seemed to me the absolute honesty of a scientist and this mixed with just the slightest of heightening and concentration by which the artist in him, needing to be focused and popular, keeps our attention. If, indeed, I did have that phrase of 'polygynous lord' in my paper already when I travelled, it would probably have been partly inspired by some combination of the characters drawn from the early story by Chekhov, near to his first, *The Shooting Party* (1885), and those from his very last play, in which parvenu Lopahin ends buying and axing the orchard amidst whose trees his forebears had worked as serfs; these characters along with many others out of Chekhov undoubtedly became elements in my imagining of those 'rising lines of plebeians' that I mention. Sex and marriage, I thought, were especially to be valued in the down-bound lines; it was their last chance, without it they were vanishing. Chekhov knew all this from his sympathy and his observation of human nature: he might have been interested in, but would have no need to follow, any of my arguments to see this point (instead he would probably simply have related those arguments mockingly somewhere, as the opinions of some pompous idiot). For him it's like the politician telling the sociobiologist he needs none of that stuff to win his elections, our so-called 'science' simply regurgitates to him what he well knows from the marketplace.

Either way, by reason or by experience, outmarriages were to be these lines' salvation. Economically via direct interest their benefit was obvious but also, reluctantly from one standpoint (because I suppose the aristocrats would always like to believe both in their purity and in a supposed continuance of the brains or whatever that originated their distinction), they were to be embraced via love and procreation— evolution's admission of typological inadequacy. Maxim Gorky more explicitly than Chekhov had described my 'seething' of family lines as supposedly observed by him in Russia although, it had seemed to me in reading, he had done so with a less-believable realism than had Tolstoy or Chekhov in their more casual and seemingly slanted style. Nevertheless it was certainly partly through such Russian authors that the idea of the theme of this paper took root in my mind even in my teens and early twenties, long before I had arrived at any interest in Galtonian 'regressions to mean' of family lines, or interpretation of such regression by Ronald Fisher and genetics: instead all this came when I started to buy the Garnett translations of Chekhov and also *The Artamanovs' Business* and other novels by Gorky, translated and put out by the Foreign Languages Publishing House in Moscow as propaganda, and thus coming dead cheap and within my range as a student.

Eventually the communist book shop (Collett's) I frequented in Charing Cross Road in London to buy those Moscow translations sold me a copy of a meeting proceedings entitled *The Situation in Biological Science*, which I mentioned in an earlier introduction.[20] This book was the end for me of any interest in communism: it finished, I think, all further purchases from the shop, even of the pre-revolutionary works of Gorky that I had no reason to consider particularly tainted with doctrinaire nonsense. If the System for which that shop had made itself window could only, in the end, come out with such monstrosities of philosophy as I saw in that 'Proceedings', every person of the era who submitted to, or who handled, such nonsense had to be tainted as if by a disease. I still bought, however, as I remember, the Chekhov that the Moscow unit republished because he had been wholly a pre-revolutionary and they issued some translations I had not found in the Garnett versions; moreover, I doubted that a mere act of translation could contaminate such crystal writing as his. But from then on I shunned the rest. What an inconceivable notion it was, that a verbatim report of the horrors of that Lysenko meeting could win over converts from the non-communist world—or at least that it could win anyone, who, in any other bookshop, had ever bought just one dog-eared copy of, say, the dialogues of Plato, or win over anyone who had ever understood anything of science in the least degree![21]

I wander, however, rather far from those thoughts that I can really remember from that plane flight: some I recall well, such as the brown moors and the larch-green hills and the thin roads and the political and parasitical 'wanderings' they gave me; but did I actually think of Gorky and his limitations? I doubt this a bit; so let me here cease the pretence, and indeed leave the aircraft for the last hours of dive towards the ecliptic and the Pacific—that is, its passage over the Sea of Japan and its slow descent to where it is due to land in Tokyo.

In October 1998, I read in *New Scientist* magazine for the first time hints that people knowing themselves handicapped genetically may sometimes choose not to outbreed—that is, not to try to prevent the same manifestation they have themselves from recurring in their children. Instead they rather seek to fix the manifestation to make a distinctive line. Presumably they almost aim to create a caste or a microspecies. From interviews and questionnaires that she applied during a meeting of The Deaf Nation in Britain, the student who reported this also reported some of her subjects telling even that they might consider identifying for abortion fetuses that lacked genes for deafness, if this was possible and legal![22] I read this amazed, but then thought how it might be just at such meetings as that of a Deaf Nation society that optimism might run so high as to deny that the handicap was substantial. Second, I thought how gentle welfare states are at present to those with such handicaps so that, in so far as their alleviations can be expected to continue, the traits actually aren't handicaps any more. When my mind drifts amongst such crystallizing and 'anti-seething' tendencies of human breeding today never far from my thoughts is the instance of the caste system of India. Surely even if its only success had been as a massive experiment for the world's information, India deserves all of our thanks, as well as, perhaps, our condolence for having, by the same experiment, denied herself the opportunity ever to become a true nation in a normal sense at all. So far as I have found no massive Indian encyclopaedia covers the matter of caste in the way one does find one to cover the complexities of, say, Judaism, and this fact by itself no doubt illustrates my point. Nevertheless one may still learn much about the system through lucid snatches.[23]

What such an experiment as that of India appears to demonstrate is the non-necessity of seething for maintaining either health or any of the other prerequisites of advanced civilization (though some small 'castes', through their exogamous sublineages or *gotra*, seem to show their awareness of an inbreeding problem[17]). But how well do the experiments convince us? Indeed, have they convinced the experimenters themselves? If supreme health (or physical specialization that ought to be an unusual possibility of the system) is reflected in winning Olympic medals, India is not, in the light of its huge population, doing very well. Many will say this is due to the poverty of so large a part of the population: provide the labouring castes of India with nutrition and the wealth that gives opportunities for training, and maybe the picture will become very different. For sure, more excellence would emerge if the conditions could be changed, but whether, if still of 'pure caste' origin, the newly found athletes would be of world class I remain doubtful. Meanwhile signs of the 'seething' scenario of this chapter seem latent in, for example, the permanent potential for outmarriage revealed in intercaste sexual attraction[24] and also in the tendency of local castes themselves to rise and to fall to a limited degree as wholes.[25]

Could there have come about, somewhere in all that marvellous diversity in India (which I understand to comprise tens of thousands of groups) a caste of blind or of deaf persons? The former have possibility to exist as beggars, I suppose, but I doubt such a prospect for the latter. India doesn't have enough social security; associations of such people would be surely too un-useful and therefore too down-bound to persist economically. But wait—consider for a moment these deaf again. For sure, there the kindly Planetary Hospital will have space for them but maybe there will be important jobs too. Can deaf people perhaps become *extra* literate in a kind of compensation? Once past a barrier to understanding what words are for, then, by writing everything, can they carry prose to more logical heights than the rest of us? Are we hearers and listeners held back through our everyday use of a more slovenly mode, and could the deaf, on lines of such reasoning, become a special part of our symbiosis of aptitudes, almost in the way of the polymorphism of colourblindness (colour vision and its lack) has been suggested to help, through its specialisms on fruit or leaf colours, or, contrarily, on the breaking of crypsis, troupes of squirrel monkeys and marmosets in their search for varied food?[26] If any of this is true, rooms will be set aside in the Planetary Hospital, or set even outside of it in, for example, computer software houses—perhaps there's even a new caste opportunity in India that seems so good at this. Demosthenes of the Internet doesn't need to fill his cheeks with stones to check his stammer any more, and if not both blind and deaf nowadays he doesn't need to talk to librarians or aides either; nothing impedes him to search the Internet for all he needs and he can put back there his great 'speeches'.

In short, the Indian caste system seems to me still to be posing some of the deepest questions about the pan-global civilization we are headed to make. Thoughts and questions about this system return to me again and again, much more than return thoughts about that other world system, communism, that I told you that I rejected. Is some version of a caste system, perhaps not particularly close to that of India, unavoidable in our future? Maybe unique sole stability of some such system is as surely proven as a mathematical lemma, this applying to both the modular multicellularity invented in biology and to a similar organization bound to hold in a Planetary Hospital of humans. Because India currently, as a state, officially denies its own caste system and (still officially following Gandhi) expresses devotion to an ideal of equal opportunities, the answer seems to be no: instead India is supposed to no longer have an intact system at all and therefore nothing to export. But most know that this is not the India that exists. Many Indians tell me that if anything the caste system is re-trenching, at least rurally, after that brief though real flurry that threatened its dissolution during and immediately after the British Raj: the flurry has given an opportunity for some readjustments of status and for some new castes altogether (deaf programmers?), but that is all. Gandhi's major objective was not first of all social reform but rather the ousting of a supercaste system imposed by the British, in which he succeeded.

In other parts of the Middle East, similar to India in being among civilization's most ancient homes, castes seem also to be solidifying along what amount to racial lines—for example, in Moslem countries and in Israel. Likewise in China, yet another ancient home, it is not clear what is going to happen after the fervour of the recent Maoist revolution, so like the enthusiasm engendered by Gandhi in India, finally drains away. As a sign of what is to come there, the anti-democracy of the present self-styled communist regime seems, as usual, striking. Even Japan keeps strong traces of its former feudal class structure, including a deep underclass in the Eta. As for those countries where I happen to have lived—mainly Britain and the USA—fortunately (as I see it) there remains a very free interclass migration in progress, especially in America and affecting the structures that that continent inherited from Europe; but we in the 'West' are all very young cultures and, as everywhere, the assortative tendencies are present. While 'equal opportunity' remains a passionate belief for most in Europe and America, belief in 'equality' in the deepest sense (equal abilities), though still strongly pretended to by some intellectuals, seems to be fading, and to judge by the family actions such as the schools chosen for children, I see this even amongst the intellectuals who say they believe. The strongest mixing process I personally have witnessed in places where I have spent time, is in Brazil and perhaps it is no coincidence that this is also exactly the country from which I heard of that *Que saude!* exclamation ('What health!') being a high accolade that young men give for feminine beauty. For a contrast, think instead of versions of this one might hear in Britain. Nearest, perhaps, would be 'Cor, wadda peach!' of East London, but in other places it might be 'What elegance!' or even still 'Oh, look, a real English rose!'.

The conference in Kyoto I was flying to was entitled 'Evolution of Life' and it aimed to cover every kind of evolution from origins of life to culture. I remember the complex of spectacular modern buildings that contained the meeting standing a little out from Kyoto city and, not far across the road, the more orthogonal but equally spectacular modern hotel where we stayed. Between the two I noticed, aiming no doubt to catch the adventurous from both buildings, a sign of an entrepreneur advertising 'Lental Bicycles'. Amongst the speakers, Allan Wilson of the University of California at Berkeley stands out in my memory for reasons of his talk, and Susumu Ohno for the surprises of both his talk and his personality. I think I met Susumu at this meeting for the first time. Taking Wilson first, I remember he arrived only shortly before his talk and disappeared immediately after it: as with the brief appearances of angels in the Bible, this seems a good step towards being remembered. I also recall his vehement refusal to be photographed and that he seemed to expect it to be known, even in camera-mad Japan, that he didn't allow it, and I remember my wondering whether a certain unhealthy paleness might not be the reason. Yet with all this he stood a tall, well-built man and gave a fine talk, and the combination cast my mind back in memory to my sole sighting of J. B. S. Haldane at that lecture in London around 1962. In less than a year

Haldane died; and I proved right about Wilson's paleness for in less than a year after the Kyoto meeting, I think, he also had died and I learned later that he had already known by that time he was in Kyoto that he had cancer. Hence it was that, although I heard him talk about that vast and twiggy tree of the mitochondria he had constructed for us humans, I have had to wait in vain for revelations he claimed to be working on at the time about why he thought one new mitochondrial mutant might have started such a victorious spread in Africa and then out of it a mere 200 000 years ago: I stayed with only his mysterious question-time hint that the spread might have had something to do with language. Again, as with Haldane, one could always expect something surprising and new from Allan Wilson; one might agree or not but his new things were never boring. Indeed he was one of the few molecular geneticists that even before this I was always eager to read or hear, a molecular tannoy of major evolutionary ideas. Having missed anything direct about what he meant, I have spent some time trying to guess and this has led on to what I call my 'Chatty Woman Hypothesis', all of it based on his mitochondrial Eve and that hint in the lecture.

In the antique wisdom of my high-school cellular physiology, which is virtually all I have on this subject, mitochondria provide simply an energy supply and it was hard to see what could be particularly 'hominizing' about any particular mutant occurring in them. If the chimpanzees that had become newly bipedal on the savannah had 'gone cheetah', say— specializing in sprinting instead of in brains—I might have understood, but the idea of becoming more sapient helped on by a successful new mitochondrion was an idea that had never before crossed my mind. But Wilson's clue to what he was thinking of had been 'language' and another was that 200 000 years ago could be about the right time for this, being about the time when we, the *sapiens*-oriented branch, were already distinct and therefore reproductively separate from the Neanderthals. Mental processes seem to be extraordinary consumptive of physical power. I knew that an immobile thinking person's thoughts were said to take up to 20 per cent of his energy budget, so Wilson's remark sparked the thought that a new mutant mitochondrion might deliver more power in the special way that extra brain activity might require, rather than in the way that, previously, it had been best delivered for the purpose of muscular exertion—the cheetah's run. So the pre-construction of fluent speech within compact masses of nervous tissue evolved for the purpose could have been this great new demand or opportunity.

Having acquired a new mutant mitochondrion that produced the new possibility, and having evolved the beginnings of its use, imagine next how the new speaker and communicator might start to spread his genes by its aid within Africa and outside. The carrier we may suppose to have been, through his brain, a little more expert at planning and at teaching in every way, but how actually is his advantage expressed? Well, of course, I have deliberately slipped here, my pronoun is definitely wrong, and actually it doesn't matter at all what advantage a male carrier had from a new mitochondrion because he does not (commonly) pass on any mitochondrion to anyone. Instead it has to have been *she* who was more expert at planning *her* household, teaching *her* children, and so on: *he* affects the spread of the mitochondrion he carries almost only through principles of kin selection—that is, insofar as he benefits his sisters who come from the same mother. Perhaps the mitochondrion's DNA does have code promoting a love for sisters. But my major point is it is just in this domain of nuclear kin selection that the old mitochrondrion was overruled by the sweet persuasive chat of an incomer. In short, grunting or silent pre-sapient *Homo* proved a sucker for the New Woman's gift of speech. Sweet words, flattery, appeals, chatter cheerful and unlimited—how else but thus by soft talking, does the bearer of an ambitious mitochondrion (transmitted it must always be remembered, in humans at least, only in the female line) make its way in the mere tens of thousands of years the fossil record allows us, across the face of a planet that is already populated with reasonably successful and adapted races of humans? Outside Africa the species was scarce, admittedly, but it was holding its own. Moreover, it was holding, probably, quite fiercely against competing strangers, this in many climes and biomes to which the old races have already become adapted. Since long overlaps of *Homo sapiens* with the older *erectus* and the parallel cold-adapted *neanderthalensis* seem not to have been common (although each has an example—a longish overlap with *erectus* in Java and again on Mount Carmel in the Levant with *neanderthalensis*, although for the latter case alternation is probably more the word), it seems to me the main alternatives are as obvious as they are stark.

Verbally persuasive and ultimately 'capturable' women (for which we have a prototypical pattern less the talking in chimpanzees, as well as plenty of suggestive parallels in the forceful-capture theme in the marriage customs of a great variety of extant human cultures, as shown us by anthropology) these seem to be a major way the spread could have gone; and either it is such women passing ever onwards, generation to generation, tribe to tribe, or else it virtually has to be genocidal aggression and/or competition (for which we admittedly also have good prototypical behaviour observed in chimpanzees[27]). The only alternative for the apparent pure extinctions of the 'other' *Homo* species have to be extraordinary coincidences. Evidence will eventually decide and certainly I shouldn't throw out the second alternative simply because I don't like the sound of it; however, it is my suspicion that it was the former course that Allan Wilson was thinking of when, tall and pale, a fleeting angel at the meeting and with both his own genes and his mitochondria nearing the ends of their tasks, he passed out those remarks about mitochondria and language that I heard in Kyoto.

Since thinking about this, it has always surprised me that this rather obvious course of reconciliation of the so-called 'Out of Africa' and the 'Multiregional' hypotheses of *sapiens* origin is not more widely discussed.[28] I even vaguely imagined I might win Brownie points from the feminist movement for my Wilsonian 'Chatty Woman Hypothesis', but no luck so far on this and it was probably a foolish thought: I fancy that the word 'chatty' has the wrong academic ring or something and I should choose another—or else, perhaps, I should expound the whole matter under a feminine pseudonym.

To be blunt from another quarter, the evidence hasn't been swimming my way particularly well recently either. It seems that the tree of the human Y chromosome 'converges' to about the same date in Africa as did the mitochondrion. Of course as regards its mode of inheritance, the Y chromosome is the mitochondrion's reflection onto the male side: it goes like a surname, descending only through the males, which it causes to exist. Therefore if a particular Y did spread out into us all at the same rate as the mitochondrion it must mean that we males have been travelling as far and as fast as our women, which roughly means, apart from possibilities that again involve strange and repeated coincidences, that we probably travelled together. That leaves practically only warfare or competition to explain the disappearance of the erstwhile subhumans in all the lands that the new *sapiens* form invaded. Rumour, however, tells me that all is not settled as to the dates on either side of the story yet, and in any case they have always had very wide confidence limits, so perhaps there is still hope for my so-likeable 'speech power' lady (let's try this alternative), who, from Somalia to Australia, while most of her menfolk skulked and stayed behind, not wishing to face what they saw, carried forward a flag of love and humanhood against, as I was seeing it, a towering wall of rednecks of her time—all the Leonardoesque caricatures, the heavy-browed, the prognathous, the ultra-macho if male, all the walls of subhumans who barred her way.

Like Allan Wilson, Susumu Ohno is a geneticist of very molecular bent, a man originating and reared in Japan but who for most of his life has worked in California. He and I have some overlap in our interest in the sex ratio, I coming to it more from the selectionist and he more from the developmental point of view. At the 'Evolution of Life' meeting we talked a lot and I found him not only the most unusual character I encountered there but in an odd way one of the most compatible. Aristocratic himself but not at all caring about it, a loose cannon and even a cretin like me for any dictates of 'correctness'—by which I mean simply a person who for at least a quarter of his time is saying exactly what he thinks. At that very meeting, having been told by him about this background and deeper, I ended thinking of him almost as an ideal illustration of my remarks about 'pastoralists' that irritated Sherwood Washburn. I have written too much about this above but, speaking of pastoralism reminds of the saying that you might as well be hung for a sheep as for a lamb. Therefore here I will write still more but with immediate caveat that in Ohno's case the key concept should be 'horseman' rather than pastoralist: this is partly because I have difficulty imagining Susumu personally driving sheep.

Susumu's father's ancestors, however, were from Mongolia. His family entered the Japanese aristocracy in the twentieth century at a point when an emperor (I am afraid I have forgotten which, as also I have forgotten even the exact era, though he told me) decided Japan needed cavalry in its army. He therefore invited an officer from Mongolia to bring horses and staff for training them and soldiers to Japan to initiate Japanese arts of military equitation—an entire enterprise that even if it happened a century ago would seem to me to have been oddly anachronistic. Because the ancestor was, I think, his grandfather, and his line has since gathered Japanese elements, Susumu seems by both present attributes—notably unconventionality—and by his origin almost exactly the type to illustrate what I have written above and wrote in the 1970s. (This leads me to the aside that I would like to know, by the way, more about the uxorial sides that produced such men as Ulugh Beg, or Babur of India, and, as a more general comment, how much less in history one is usually told of the mothers' lines for all exceptional people—whichever side of the nature–nurture debate you fall, this is a truly foolish inequity.)

Again surprising to me at the time, but according exactly with my archetypal offending passage of 1975, Susumu proved to have a hearty and outspoken contempt for, of all people, Motoo Kimura, then the doyen of Japanese genetics and another speaker at our meeting. He described Motoo to me as having the 'spirit of a shopkeeper'. I told him (and have said in my introduction to Chapter 11 in these memoirs) that I had tremendous admiration for Kimura's mathematics and methods and that I believed he had laid the scene for much of the modern population genetics of evolution theory. Susumu replied to this with some species of tchah! followed by: but then why doesn't he *use it* for something himself, instead of all his silliness of neutral evolution? The fact is he's a coward, he takes a soft course, he criticizes. I confess this attack on Kimura was rather stunning for me, and especially so as coming from a man I took at the time to be purely Japanese. Now I look back on it, that fine moustache might have made me doubt it; but, had he been purely Japanese, a deft silence concerning a last-mentioned character would have been the much more usual form of disparagement. Still, I knew immediately what Susumu meant and remembered my own disappointment after the early years in which I had been so much inspired by the Kimura papers of the 1950s and grateful for the extra entrance they had given me to population genetics, following Fisher and Haldane. Kimura has been a critical and an analytical guiding star, I think, for all of us, a baseline for us to fall back to and a whetstone on which to sharpen wits . . . But what else?

On top of this surprise I was further stunned by Susumu's expectation that I would know a lot about horses, know about the tastes and operations of British aristocracy in this line; and also was a bit by his assumption that I'd be happy to spend the evening drinking—and at his expense if I didn't fight to avoid it—while discussing all this. There seemed to be an idea that we were to rewrite the history of genetics and evolution and laugh at many a jolly demotion and demolition as we went along. If

you wish to contrast Ohno's style with that of Kimura, go to the whole
volume in which the paper of this chapter appears[29] and contrast first the
titles of the papers: 'Neutral evolution' for Kimura and 'The grammatical
rule of DNA language: messages in palindromic verse' for Ohno. Then take
the concluding sentences of summaries. Kimura gives us: 'I conclude that
since the origin of life on Earth, neutral evolutionary changes have
predominated over Darwinian changes, at least in number' (well, yes, given
that 'in number' it's hard to disagree). Ohno instead gives us—well, oh
gosh, it's complicated, but it all refers to a long palindrome that is in the
summary itself as an example, and the like of which I have never seen in
any summary of any paper. The palindrome says: 'Doom! Sad named was
dog DNA, devil's deeds lived, and God saw demand as mood.' I am not
enough a molecular geneticist to comment on Ohno's theory in the paper
or on its status today, but, I ask any population geneticist, try to imagine
that introduction and palindrome being written by Kimura. Actually I have
little more ability to comment on Kimura's own conclusion either except
to say that to me it has always been surpassingly obvious, ever since reading
Darwin, that organisms in all important ways are shaped by natural
selection. Little about them that matters has anything to do with
neutrality. What happens to variation that is not of selective significance
may indeed contribute measuring rods—scales for the shop as Susumu
might put it. But if I wanted an exciting idea to follow, and was prepared to
get myself some 'genetic-code' background (that side of evolution which I
always neglected) I know which of the above two theories I would choose
to look into first.

Well, Susumu and I started our demolition and as we talked and I
learned more about his roots, my mind hunted in the background, far away
from Kyoto, for possible related antecedents for the other 'unconventional'
Japanese I had met; again it seemed to me they were disproportionately
peripherals, those not drawn from successful, mainstream, urban-Honshu
Japan. One who came to my mind was from the far north; another from the
southern big island, Kyushu, where the general character already seems
somewhat different in a way parallel with the differences in Europe; a third,
the most striking (in more ways than one) from a smaller island that is
noted, I had been told, for its general output of dunces. Farther than that, I
also recalled wild-looking men I had met in the University of Naha,
Okinawa, men who one felt came with beach sand gritting their thin bright

clothes and coral seaweeds stirring in longer hair. They brought me
thoughts of an infinity of islands somewhere to the back and smiles of
Margaret Mead's laughing and hoaxing informants were touching their lips
and eyes and even reminding me a little of pictures in my mother's books
about the Maoris—in short, they were did-I-want-to-go-swimming-or-else-
fishing-or-dancing kinds of people. Now here I sat and drank with one who
was from the opposite 'end' to those last and yet in many ways the same:
had there been horses at hand instead of just the 'lental' bicycles I have
little doubt it would have proved fair to call Susumu a 'did-I-want-to-go-
riding-and-hunting' type of man.

Another adventure vast as the Pacific seemed to me lurking behind
him too, except that this time all of it solid land, all of eastern and
northern Asia, a terrestrial Pacific, a whole other side to the planet.
Between such vastnesses Japan lay sandwiched, a row of islands and a
people all riding on an unstable and smoking wave of colliding plates.
Surely, I thought, listening to Susumu Ohno it had been quite wrong what
Dan Freedman, another brave spirit,[30] had once told me in Chicago after I
had put it to him that Japan should be like Britain, namely a refuge that
had benefited genetically from all rebellious, free-thinking persons—all
tending to be bastards, I suppose, in various senses, which its nearby
mainland had thrown to it, as well as from all the self-reliant sailors who
simply arrived. I thought it was from tradition Freedman's dictum had
come, from Japanese 'in-laws' and so had believed it. Dan, in fact, had been
maintaining (I had stayed in his home) that Japan's case had been very
different from Britain's: the Japanese had historically been very exclusive
and they appeared to ethnographers an unusually pure race of shrouded
origins, but probably ancient Korean and, it seemed, just one main
colonization. Certainly in the historical period no outside conquest ever
succeeded until World War II; previous attempts had included even an
'Armada' from China like the 'English' version, and it also had been
scattered and destroyed by providential storms. No new settlers by conquest
then; but here Susumu seemed to me, through his own story, to provide an
almost exact counterexample to Dan's first statements; and even before
hearing Dan's opinion I had learned of similar exceptions. Susumu's
forebears had come exactly where Kublai Khan's armada had failed—but in
peace and invited as alien teachers.

Esmond de Beer, historian and wealthy Jewish friend to my mother's family, in Reading, England, and then also later their friend in New Zealand, even in my teens once told me while we looked over fine Japanese ceramics in his home, how the complex crackelature on some pieces had made them despised rejects in China but this same 'imperfection' had founded a style due to become much prized in Japan, and he told that in fact these ceramicists entered Japan directly, invited, with the Japanese objective to improve their standards in ceramic crafts, just as I was learning now had come much later Susumu's equestrian ancestor. And all this, it seemed to me, was not at all unlike how Dutch dykes and land drainage had come to swampy British East Anglia via the engineers England explicitly invited from Holland—a country that was, like China for Japan, at most times England's enemy. Well known also for Britain is how Gothic cathedrals came to us in styles already pioneered and perfected in another still-closer neighbour and rival nation, so that when you visit them, any pamphlet you pick tells of their French-named architects, who had been in some cases explicitly invited from France to design and erect the building. If British cathedrals rose to less wondrous heights than those of France it was either that we were less rich in the ordinary sense or else perhaps more rich in a spiritual sense in having hearkened to the anti-religious free-thinkers amongst the other imports. No doubt civilized cultures always, and without any necessary intention, draw to themselves the arts and artisans that they need, the missing elements in the symbiosis of aptitudes to which all civilizations inevitably tend. But when a substantial sea happens to impede more random and casual arrival, perhaps a culture has to be more explicit and there is positive activity to get them to come; in these ways a higher quality is obtained.

But who else, besides these privileged invited, might have drifted or wrecked on the coast of Japan and been spared the legally enjoined execution normally meted to strangers, because of their captors' curiosity over some artefact discovered in a cast-up hull, something that the locals thought—probably wrongly—the sailors would teach them?

Propping up a bar, drinking whisky alongside this determined, passionate, soldierly man, artist as well as scientist as I was to learn later when listening to his setting to music of the genetic code at another meeting (this was in Yokohama, later in 1990), it appeared to me that I could indeed trace in Susumu those heroic, rejuvenating virtues I had postulated might accrue to a nation-sized unit; and especially accrue to those standing as islands offcoast to a civilized and ostensibly culturally superior mainland. I felt sure that Japan, like Britain, had had advantage from this. More exclusive than we were the Japanese might have been, as Dan Freedman had suggested; but perhaps also they were more unified in knowing what they wanted; perhaps also their broader Sea of Japan had been a harsher filter for the parasitical and simply incompetent that Korea and China would also willingly have thrown off.

I do not like greatly the paper I contributed to the Kyoto meeting. My prime aim was to put in words the mechanisms that underlay the success of my recent models, but I think I failed. Partly the problem was that I myself didn't yet understand how my processes worked in any detail, I just knew the general idea did work and that the models showing it were realistic. Partly the problem might be that even for those parts I did understand, I was trying for the impossible—that is, to give three-dimensional visualizations to the reader for processes that need many more dimensions for proper description. For example, early in the paper I think I explain well the nature of that inevitable intrinsic instability and cyclical dynamic that onsets out of what may seem at first sight natural states of 'balance', 'compromise', 'stand off'—however it is normally put—between hosts and their parasites; but then, having shown how situations will start to move in what must become inevitably somewhat repetitive ways, I fail rather badly to explain why they will not proceed to such escalations of change that gene variety is lost. I like, however, my explanation of why truncation selection is bound to destabilize any attempted point of balance (on p. 767); yet now find on the following page my explanation of why oscillations do not lead on to allele extinction quite unconvincing and feel that the emphasis implying almost a necessity for 'pitty' patterns of selection but to support sex under frequency-dependent selection rather overdone. ('Pitty' selection to remind you, is the pattern in which most genotypes have a roughly OK fitness with small differences between them but a few genotypes are extremely bad—see the paper for more explanation of this and also of its opposite, 'peaky' pattern, which has most genotypes again similar but a few outstandingly good.[31]) Indeed even in the study I was at the time engaged in, it soon appeared that for three-locus pursuit situations, the combination of pitty-selected parasites with pitty-selected hosts resulted in such an extremely 'loose' frequency-dependent responsiveness—that is, the system allowed such long overshoots—that it was the most likely of all the combinations to lose allele variation. But I won't bore the reader with these issues now—perhaps more of that will come in a third volume, if ever.[32]

What I do like in the paper, and still strongly believe, are, early on, my reasons why *K*-selected species (big, philopatric, and slow breeding . . .) are likely to be much more steadily sexual than are *r*-selected species (the opportunists: small, fast, far-dispersing . . .), and then, later, my reasons why the dynamics of sex-supporting Red Queen selection processes are likely to be cyclical only in a rough statistical sense while in detail they are highly irregular and non-repetitive (see p. 774). This is how I now expect the case to be in nature; therefore it is also expected that the cycles will be cryptic, not easily to be distinguished from the natural buffeting of frequency by inanimate varying selection—and this a good reason for why biology wasn't long ago forced to see the power of the parasite-level biotic interaction, it was all hidden in the ups and downs of worms and viruses that no one knew about, or even in the ups and downs of genotypes amongst them that people knew of even less.[33] I also still stand strongly by all my remarks about the potential importance of Parasite Red Queen agitation for macroevolution and still regard the 'parasite' experiment in artificial intelligence performed by Danny Hillis to illustrate how his 'Thinking Machines' (the name of his company) adapt to think, an important pioneer in the genre of approach that is bound to bear rich fruit not only within its own field of AI but also through the repercussions this will have for the understanding of natural evolution.

As for predictions already under study, I hope that most of this volume and especially the two 'appendix' chapters (6 and 16) will be agreed by the reader to bear out my claim (on p. 775) of a correspondence between what may roughly be called the ecology of sex and sex selection and that, in effect, the Parasite Red Queen (PRQ) theory is showing as I put it 'slowly accumulating evidence'. In some fields and studies, the accumulation of the evidence has seemed to me spectacular. A half dozen especially penetrating experimentalists, each heading a group of colleagues, deserve parting special mention here as outstanding amongst the others. Compared to me—sneaker on the forested mountain ridge that I describe in the Preface, a peerer through this small hole and that in the canopy, trying to fit together bit by bit out of these separate tiny glimpses the landscape of sex that makes most sense—these other men instead take the direct approach. They hew at the trees, and each opens his own gap, a clear window onto the vista. On what I have called the 'sex-itself' side of the problem, the woodsmen I have most in mind are the Americans Curt Lively and Steven Kelley; and on the sexual-selection side Anders Møller, the Dane, has been notably the most active. All teams have extended and transformed the ideas; however, I am glad to say all three, too, seem to be confirming the general outline that, roughly, I have put forth since the first paper of this volume. From the parasite side, the factors imposed on hosts are strong and multitudinous with the parasites themselves always adjusting rapidly; on the host side, multiple polymorphisms for resistance, non-multiplicative interaction, and high heritability are becoming ever more widely proven. Only the cyclical nature of the selection, explaining the persistent high heritability, and a quantification to the point showing that the halving inefficiency of sex can indeed be overcome by advantages through these causes alone, remain to be established, and even on these points the evidence is coming.[33,34] It would be nice to have, as well, more actual genes to talk about; this, too, may be coming although unfortunately it is from another quarter and for fungal systems (I am thinking here of the details gleaned by Jeremy Burdon in Australia, as mentioned below). One collaborator with Curt Lively, Stephen Howard, has strongly made the case that deleterious mutation needs to be considered as integrating with a PRQ-type process, a view suggesting that the latter would be insufficient alone. I am not sure—that is, I'm unsure how *necessary* a factor deleterious mutation is for sexuality. For certain it must be helping in the selection that occurs, but does it have to be there? Or is it that high mutation rate rides on the back of—become permitted by—a selection by biotic forces that alone are truly unavoidable? If we could magically switch all organisms to lower genome mutation rates, would sex die out? It is my belief that if we switch them almost equally magically to a condition of no parasites, sex would indeed die out and as it did so mutation rates would start to fall.

Steven Kelley, now at Emory University, opened his front on the 'sex-itself' problem differently from the above in three ways. First, he researches plant sexuality. Second, he focuses selection by plant viruses, not the more usual and easily studied bacteria, fungi, or insects. Third, choosing his course perhaps because the immobility of plants gives the dispersal they have by seeds and vegetative spread a particular importance, he has studied sexuality in conjunction with the problem of escaping parasites in space. This is also true of the work of Jeremy Burdon, working instead with plant fungi. Unfortunately for simple interpretation, escape in space is not a synergizing factor for sex selection but an alternative; however, the spatial patterning is unavoidably there in virtually all populations and its role has to be considered. In more recent years another group in Kyushu, Japan, founded by Tetsukazu Yahara, has also been producing excellent studies of the interactions of plant breeding systems (in this case comparing populations that may be either sexual or parthenogenetic through seed) with viruses, again with striking results that are generally favourable to PRQ.[35] By focusing on viruses Kelley and Yahara are doing a great service for all naturalists, including me, who tend not to notice viruses in the field even when they drop on one's toes. Indeed viruses often seem specialized to tread lightly even on proper victims, their effects there being commonly subtle even though, in terms of fitness (especially in competitive situations), they are still very important. Unless to be purely transmitted to descendants of a present host through that host's genome (in which, while shut away, they do no harm), every virus must insert itself into a cell without killing it if it is to multiply. The necessary precision of virus attack that this entails virtually constitutes the particularity I mentioned earlier: hence viruses are certainly going to be very important supporters for PRQ theory when enough is known.

Indeed my main working notion to resolve the mystery of why sexuality is so much more uniformly supported in the sea than it is on land (or even within inconstant bodies of fresh water) is that the sea provides a more kindly medium for the direct transport of viruses, making them often extremely abundant there (see early notes in my introduction to Chapter 12). In the contrasted smaller freshwater domains, thus often dealing with alternative sex and asex in small zooplankton of water bodies ranging from puddles to small lakes, all variably subject to drying up, Dieter Ebert, now in Basle, has opened a new field in evolutionary and 'parasite' limnology. Again with numerous and varied collaborations he has done great service to the background of PRQ theory by showing that, even without descending to viruses, microbial parasites are varied, abundant, and influential even for sexual–asexual fence-sitters such as daphnia. Thus at least conjecturally, and with some positive hints from ecological distribution, they can justify these tiny animals in being sexual when they are sexual, while dispersal (in time through dormant eggs when pools dry up; in space through the help of such as far-wandering birds and blown dust) can justify them in being not sexual when they aren't. We need far more quantification, but Dieter has made the start and in doing so has clarified what virulence is in an evolutionary setting and who is normally ahead in the race.

Viruses of Kelley and Yahara lead me to viruses of the sea, and they on to Dieter Ebert. He and Basle can lead still within Switzerland to other explosions in the parasite connection of evolutionary ecology, referring to quite other extreme groups. I refer to Paul Schmid Hempel and his group at Zurich. This group's outstanding studies on parasites of social insects are now well summarized, along with a wide-ranging review, in his recent book on the subject.[36] For me the group's revelations have been important not only through the confirmation that these insects, for all their antibiotic secretions plus wonderful hygiene in their colonies, escape parasites no more than the rest, but also because they bring my life of speculation in a sense full circle to where I started on the puzzle of sex: Paul has at long last begun to exorcise with real data my earliest 'kin-selection' nightmare and demon, that infested from the moment when I was first forced to realize that the honeybee queen, riding her broomstick to a sky-borne wedding orgy whose acts lie beyond even the weirdest of medieval dreams, flouted my notion of high co-operation via high relatedness as completely as she possibly could.[37]

Danish tree-feller extraordinary, Anders Møller, now at Marie Curie University in Paris, wielding what must be the world's sharpest chainsaw, has done more to enlarge the window that looks out to the parasite–sexual selection connection that Marlene Zuk and I saw in dim outline through the canopy, than anyone else. But, working almost exclusively on birds, he has been revealing the causal connection as normally strongly modified by the filter of the vertebrate immune system. This 'immunocompetence' addendum to our idea was first outlined by Andrew Karter, Ivar Folstad, and then Claus Wedekind;[38] it is an addendum that I am afraid I tend to oversimplify in this volume to the point of leaving it out, even though I absolutely agree as to its importance for all animals that indeed have an epigenetically evolving and memorizing immune system. It leads, however, to an overall view still preserving most of the predictions Marlene and I originally proposed (Chapter 6). As for supporting his points and modifications, Møller's rate of accumulation of evidence, helped on by numerous colleagues, is extraordinary and unequalled, and, wrapped along with the modern concepts of costly and hence honest signalling,[39] the huge and still-growing theme of extra-pair copulations (EPCs) has gained, through his multifarious additions and approaches, a deep and proven involvement in sexual selection in which the parasite connection is increasingly to the fore. Again, I have covered these contributions in previous chapters and I need not enlarge here. Anders's work has also drawn out illuminating distinctions between the timescales of health to which elements of sexual display are relevant, and, along with that theme, I have been happy to see rehabilitated the sosigonic role Marlene and I suggested for birdsong.[40] Indeed the latest finding from Møller's group on song is both unexpectedly significant and far beyond anything I had imagined: birds with bigger spleens (hinting their general level of parasitism and thus their need for a strong immune system—for which the lymphocyte-producing spleen is a primary organ) are found to sing more varied songs. Even with some six potentially confusing variables that might affect song taken into account, the pattern of big spleens existing in song-rich species still dramatically stands out.[41] Prior to our present insight into the general parasite connection, who could have guessed this? It seems to me that only an impresario, long chastened in efforts to bring prima donnas and tenors to a satisfactory point of performance in his hall, may sometimes have mopped his brow and muttered about them, not so much '*Que saude!*' as 'What a spleen!'.

To those birds able to invest a little more in display while not simultaneously damning their immune system to failure, there come the rewards of the extra copulations and the extra offspring; but the above chain still implies, of course: the correlation of first and last across the species that Marlene and I claimed and found, and which Jeremy John later, with better statistical methodology and controls, confirmed for us. We mentioned but did not much highlight the EPCs, a topic then far less known. We also mentioned only briefly the matter of Zahavian handicap, which also at the time was little understood.[39] Instead we emphasized more explicit polygynous tendencies in birds as well as assuming a potential two-way mate attraction. Possibly we overemphasized roles of both female display and of male song but, to me, this is not yet clear. Explicit polygyny turned out to have its special complications,[42] which, as discussed in the Appendix to Chapter 6, aren't yet fully explained. Certainly, however, I don't accept yet that either female display or song have to be excluded

from the ever-expanding sosigonic selection scheme. As usual I accept
Møller's points about all this—and who, anyway, will try to climb against
his customary Niagara of evidence? Whatever: song has always seemed to
me eventually to regain its place in the parasite–sexual-selection scheme
and the signs of this happening seem already to be set. Surely, as I have said
to many people at many times, it was only necessary to think of that prima
donna, whether the splenetic one or not, asked to perform her most
troublesome piece when she has influenza, or even just a touch of fever, to
see how, in a general way, its place has to be there. Generally I see a steady
causal convergence to the theme that, at least by the time of the
conference in San Francisco (Chapter 17—and excepting our omission to
discuss the mediating role of immunocompetence) I long ago came to
support. Møller builds a fortress where Marlene and I set up a shepherd's
tent: nevertheless, whether by luck or strength of our tent pegs, our wind-
torn canopy still flutters beside his keep.

   Anders Møller has my admiration on all of three other separate
counts too. Two concern his first careful evidence accumulated for matters
I guessed at only vaguely in my papers and in some cases have mentioned
again in these introductions. One is that high sexual selection is
accompanied by high speciation rate[43] and, second, that hole- and colony-
nesting birds, through their nest places being also parasite exchange
centres, become particularly ornamented by sosigonic sexual selection.[44]
The third matter is yet more 'first careful evidence', in this case supporting
the idea of a higher level of parasite-mediated sexual selection, which I had
likewise assumed probable for the tropics.[45] Early on my initial enthusiasm
for this idea had been dampened by Marlene's correlation for tropical
South American birds, which, though still significant and positive, was
weaker than that which we had found for North American birds. These
birds and results were, one might say, one of those 'glimpses through a
canopy hole'—that of the already recorded tropical blood parasites. Shortly
I added to these positive though dilute results the misgivings I have
mentioned above about the obvious high sexuality of some large arctic and
desert animals and plants. Finally, I added yet others from my own
experience with certain groups along the equatorial Amazon. But Møller's
study has tended to put the generality about latitude back more or less as I
had first guessed at it.

   I would still, however, avoid projecting the tropical–arctic contrast too far; as shown in my high-altitude musings of some pages back, there are at present just too many exceptions and ambiguities and the whole subject needs a much more careful and 'J-Johnian' kind of analysis. In the penultimate section of the present paper, my concern with latitude was given explicit attention but this, of course, was without benefit of the recent studies that I now cite. Although it is true that the so well-known birds of paradise and the bower birds are tropical, and manakins and pennant-winged nightjars too, I find, as I said, the tropics to be very far from embracing all the world's most bright, exaggerated, and lekking species, even though this is a trend that our own near relatives among monkeys[46] and some human cultures themselves[47] might encourage us to believe. Through my quetzal-tinted PRQ spectacles I still see what may be a prevailing rule; but, at least, out in the animal world, I also only too easily imagine that *'Que saude!'* from tropical Brazil, taught me by Carlos Fonseca, being uttered by shivering and lovesick lesser walruses in the sea and by adoring doe reindeer on icy land as particularly warm and lovely specimens pass them by. While Low's dramatic results[48] show that tropic-originated *H. sapiens* join with elephants,[49] birds of paradise,[50] and guinea fowl[51] (the last exaggerating, just like we do, their ornaments in both their sexes), and thus emphasize tropical sexual exuberance, grouse, ruffs, reindeer, walrus, narwhals, and even phalaropes line up in colder climates as their counter theme.

Whatever, wherever, tropic or arctic, far-spaced or socially grouped, you can't escape it: everyone has parasites, everyone needs choice but some, according to correlations of the successive (not spatial) environments, need it more. But now imagine those potentials for seductive cajolement by language that I mentioned only a little back going beyond the precision and stamina in 'chat' through the selection of nuclear genes improving the information content and the intelligence of language. The genes can be on any chromosome but the Y is unlikely; and it seems that, in the reality, for reasons not yet fully understood, the X may have an unfair share of coding both intelligence and language.[52] An initial advantage having started the ball in motion, now sexual selection in both sexes pushes it on. By such a route, of course, we arrive at a brain not so much preadapting to some 'useful' life-history duty—as would powerful wings, say, that had started as ornaments by serving for actual dispersal by the female as well as the male, as in my previous speculation—but instead deputizing for more indirect

virtues like those of antlers or the scent brushes of a male monarch
butterfly, or the sex pheromones and sexual skin in a female primate. In
short, it plays a part in the more normal scenario of sexual selection. In this
way, as hinted in Chapter 10 in Volume 1, we find an alternative—or an
adjunct?—to my suggestion about the 'Eve' mitochondrion, and this now
applying to the male or to both sexes. Whichever, whoever—let me call
these possibilities variously the 'Suave', 'Boasting', or 'Insulting Person
Hypothesis'. You can pick as you please but keep hold, if you will, in each
case, of its background in animal caricature: the peacock's fake 'watching
crowd', and bull frog's deep croak, the stag's antlers, the female pheasant-
tailed jacana's—well what?—so pheasant-like and easily muddled tail. All
such accelerations of sexual selection, including, of course, my own dear
('posthumous-Wilson'?) chatty woman, may have contributed to what we
have become—that is, an animal with potential to be boring in a totally
new way.

Possibly even that is an understatement: doubtless we, or even an
animal, can manage to be pretty boring without use of words, but I doubt
that in all evolution any such skull-crushing boredom has been possible for
or from animals as can be evoked by listening to top human performers—
let's say to those with BQs (boringness quotients) of 150 and above, the
Mensa community of bores. I claim boredom and boringness to be, indeed,
a truly human emergent condition. A trifle more serious than this is to
suggest that perhaps some other great gifts of evolution have come to life
more generally in roughly the same way as boringness came to us—insect
flight is a possible example, with the well-known (and, of course, 'boring')
blue-bottle buzz, giving us a taste for the precursor in this instance. It has
been conjectured that sexual selection based on insect wavable stubs,
'paranota' as technically named (flat excrescences of the insect's mid-
section) may have started amidst the younger coal forests or thereabouts,
for signalling between sexes and having, probably, for the usual sexual
reasons, males take the lead.[53] By one or more of various alternative
intermediate steps, such as wind-scudding on ponds like toy boats,[54] this
enterprise may have set insects on the road towards buzzing and then to
flying; and thus, finally, through their flight, to become the most speciose
group of our planet,

In our case, our play first with phonemes and then with words was
the preamble 'buzz' still far from take off—and its purpose: to impress the

opposite sex, to wear down an opponent by talking more and faster than he or she could. Perhaps in its early stage the contest was quite exciting to attend, like listening to a reed bed of warblers warbling in which mostly all are ignoring any incipient symbolism the song may have—liken it all to simultaneous disk jockeys gabbling from rostrums set up in a teenage crowd. Then, at last, a special use for the contest was found and words set us on the road to think. At this point the old style, those rippling streams of platitudes and opinions, became boringness—liked by some as better (more social) than the condition of silence, but that's about all—but thence onward and, as I hope, upward, even towards this book.

References and notes

1. From the 'supernatural' poem *The Woof of War* about celtic King Brian's defeat by Icelandic and Orkney Vikings in Ireland, transl. G. W. Dascent, *The Story of Burnt Njal* (Dent, London).

2. A. J. Richards, *Plant Breeding Systems* (Allen & Unwin, London, 1986).

3. D. B. O. Savile and J. A. Parmelee, Parasitic fungi of the Queen Elizabeth Islands, *Canadian Journal of Botany* **46**, 211–18 (1964); D. B. O. Savile, Arctic adaptations in plants, *Canada Department of Agriculture, Research Branch Monographs* **6**, 1–81 (1972).

4. W. D. Hamilton, Recurrent viruses and theories of sex, *Trends in Ecology and Evolution* **7**, 277–8 (1992).

5. W. H. McNeill, *Plagues and Peoples* (Anchor Press/Doubleday, New York, 1976).

6. S. A. Frank, Recognition and polymorphism in host–parasite genetics, *Philosophical Transactions of the Royal Society of London B* **346**, 283–93 (1994).

7. S. H. Hulbert and R. W. Michelmore, Linkage analysis of genes for resistance to downy mildew (*Bremia lactucae*) in lettuce (*Lactuca sativa*), *Theoretical and Applied Genetics* **70**, 520–8 (1985); B. N. Kunkel, A useful weed to put to work: genetic analysis of disease resistance in *Arabidopsis thaliana*, *Trends in Genetics* **12**, 63–9 (1996); for other examples see Chapter 16 and its Appendix; in the paper of that chapter especially notice O'Brien and Evermann (1988), note 46.

8. A. Kaur, C. O. Ha, K. Jong, V. E. Sands, H. Chan, E. Soepadmo, *et al.*, Apomixis may be widespread among trees of the climax rain forest, *Nature* **271**, 440–1 (1978); A. Kaur, K. Jong, V. E. Sands, and E. Soepadmo, Cytoembryology of some Malaysian dipterocarps, with some evidence of apomixis, *Botanical Journal of the Linnean Society* **92**, 75–88 (1986).

9. S. C. Thomas, Geographic parthenogenesis in a tropical forest tree, *American Journal of Botany* **84**, 1012–15 (1997). Similar certainty of apomixis applies to at least two small trees of the same family as *Garcinia* in the Caribbean region (B. Maguire, Apomixis in the genus *Clusia* (Clusiaceae)—a preliminary report, *Taxon* **25**, 241–4 (1976)).

10. P. E. Berry, H. Tobe, and J. A. Gomez, Agamospermy and the loss of distyly in *Erythroxylum undulatum* (Erythroxylaceae) from Northern Venezuela, *American Journal of Botany* **78**, 595–600 (1991); P. E. Oliveira, P. E. Gibbs, A. A. Barbosa, and S. Talavera, Contrasting breeding systems in 2 *Eriotheca* (Bombacaceae) species of the Brazilian cerrados, *Plant Systematics and Evolution* **179**, 207–19 (1992); L. C. Saraiva, O. Cesar, and R. Monteiro, Breeding systems of shrubs and trees of a Brazilian savanna, *Arquivos de Biologia e Tecnologia* **39**, 751–63 (1996).

11. A. Rudow and G. Aas, *Sorbus latifolia* sl in Central Northern Switzerland: distribution, site and population biology, *Botanica Helvetica* **107**, 51–73 (1997).

12. I. Morris, *The Nobility of Failure* (Holt, Rinehart and Winston, New York, 1975).

13. K. B. Sturgeon, Monoterpene variation in ponderosa pine xylem resin related to western pine beetle predation, *Evolution* **33**, 803–14 (1979).

14. See *Narrow Roads of Gene Land*, Vol. 1, Chapter 9, p. 317.

15. The Naryskin family, which provided the mother for Peter the Great, came from the steppes. As Tsaritza, Natalia Naryshkina was a mother eager for change and far more willing for her royal son to grow up a European than was pleasing to the established aristocrats of Russia of the time, and courtly rebellions arose because of this. If my interpretation is right, this particular 'steppic' genetic incursion would reflect a common pattern: little impact of pastoralists on the gene pool of Russia as a whole (or on other affected peoples) (L. L. Cavalli-Sforza, P. Menozzi, and A. Piazza, *The History and Geography of Human Genes* (Princeton University Press, Princeton, NJ, 1994)) but much more upon upper classes and the rich, with which, as ruling elites, if only for brief periods, new pastoral lords found themselves associated.

   There are also, of course, the examples of the intellectual descendants of Timur Lenk (Tamerlane) whom I have mentioned already. Grandson Ulugh Beg in Samarkand was to establish the best stellar observatory and accumulate and publish the best star tables of his time. Shortly Babur, that so vigorous as well as intellectual hybrid who counted both Timur Lenk and Gengkis Khan in his ancestry and who became the first Moghul in India, became a similar case; again I would like to think that there might be significant uxorial hybridity in these families going beyond the two pastoral groups and ancestors mentioned, but so far records I have seen have told me almost nothing of the mothers—a situation all too common with brief histories, unfortunately. As showing the style of the background of Babur and its setting for a potential hybridity, however, consider the following quotation concerning Babur in his youth:

> True to the time, he found beneath the walls of Samarkand two of his cousins already engaged in the same pursuit [intention to conquer], though it turned out that one of them had only come to try to carry off a girl in the city with whom he was in love. They pooled their resources, but winter came with Samarkand intact and they had to retire. Only the lover had attained his end (B. Gascoigne, *The Great Moghuls* (Jonathan Cape, London, 1971)).

16. W. M. Shields, *Philopatry, Inbreeding, and the Evolution of Sex* (State University of New York Press, Albany, 1982).

17. Cavalli-Sforza *et al.* (1994) in note 15.

18. L. Betzig, *Depotism and Differential Reproduction: A Darwinian View of History* (Aldine de Gruyter, New York, 1986).

19. C. McEvedy, *The Penguin Atlas of Medieval History* (Penguin, London, 1961).

20. See Chapter 9, note 41.

21. A. Cromer, *Uncommon Sense: The Heretical Nature of Science* (Oxford University Press, Oxford, 1993).

22. P. Cohen, Not disabled, just different, *New Scientist* **160**, 18 only (24 October 1998).

23. M. Gadgil and R. Guha, *This Fissure Land: An Ecological History of India* (Oxford University Press, Oxford, 1992).

24. A. Roy, *The God of Small Things* (Flamingo, London, 1997); see also the works of R. P. Jhabvala.

25. The last process even can happen in genetic isolation, as when a group does win a higher status and privileges. It may be interesting to compare this to my comments on numeric changes in clonal mixtures in the Appendix to Chapter 16.

26. M. J. Morgan, A. Adam, and J. D. Mollon, Dichromates detect colour-camouflaged objects that are not detected by trichromates, *Proceedings of the Royal Society of London B* **248**, 291–5 (1992); M. J. Tovee, J. K. Bowmaker, and J. D. Mollon. The relationship between cone pigments and behavioral sensitivity in a new-world monkey (*Callithrix jacchus jacchus*), *Vision Research* **32**, 867–78 (1992); J. D. Mollon, J. K. Bowmaker, and G. H. Jacobs, Variations of color-vision in a New World primate can be explained by

polymorphism of retinal photopigments, *Proceedings of the Royal Society of London B* **222**, 373–99 (1984).

27. R. Wrangham and D. Peterson, *Demonic Males: Apes and the Origins of Human Violence* (Houghton Mifflin, Boston, 1996).

28. But for a related line on different rates of spread of genomic and mitochondrial genes see M. Treisman, The multiregional and single origin hypotheses of the evolution of modern man: a reconciliation, *Journal of Theoretical Biology* **172**, 23–9 (1995).

29. S. Osawa and T. Hojo (ed.), *Evolution and Life: Fossils, Molecules, and Culture* (Springer, Berlin, 1991).

30. He had published his book *Human Sociobiology* (Collier Macmillan, London), enclosing many provocative opinions on heredity, in 1979 when it was still far from cool to join the two words.

31. H. Suzuki. The optimum recombination rate that realizes the fastest evolution of a novel functional combination of many genes, *Theoretical Population Biology* **51**, 185–200 (1997).

32. Others have assumed that what I call 'peaky' patterns of host fitness under parasite attack are more likely (S. Gandon, D. Ebert, I. Olivieri, and Y. Machalakis, Differential adaptation in spatially heterogeneous environments and host-parasite evolution, in S. Mopper and S. Y. Strauss (ed.), *Genetic Structure and Local Adaptation in Natural Insect Populations*, pp. 325–42 (Chapman & Hall, New York, 1998); S. A. Frank, The evolution of host parasite diversity, *Evolution* **47**, 1721–32 (1993)). This a different choice as to likelihood from mine, which was that a 'pitty' fitness pattern for hosts and a 'peaky' one for parasites would be most common: it appears to me that even the typical single-locus diploid 'gene-to-gene' pattern supports this and that it becomes even more probable when interactions involve several loci, as arises when many different mild parasites are acting in combination. I have found the combined pattern of pitty hosts with peaky parasites most conducive to permanent dynamic polymorphism, and also potentially very conducive to the support of sex, as in the HAMAX model of Chapter 16. But it is not true that other combinations of patterns cannot protect polymorphism (W. D. Hamilton, Haploid dynamical polymorphism in a host with matching parasites: effects of mutation/subdivision, linkage and patterns of selection, *Journal of Heredity* **84**, 328–38 (1993)) and unpublished runs of HAMAX showed that they could sometimes also support sex.

33. M. F. Dybdahl and C. M. Lively, Host–parasite coevolution: evidence for a rare advantage and time lagged selection in a natural population, *Evolution* **52**, 1057–66 (1998).

34. S. E. Kelley, Viral pathogens and the advantage of sex in the perennial grass *Anthoxanthum odoratum*, *Philosophical Transactions of the Royal Society of London B* **346**, 295–302 (1994).

35. T. Yahara and K. Oyama, Effects of virus-infection on demographic traits of an agamospermous population of *Eupatorium chinense* (Asteraceae), *Oecologia* **96**, 310–15 (1993); T. Yahara, K. Ooi, S. Oshita, I. Ishii, and M. Ikegami, Molecular evolution of a host-range gene in geminiviruses infecting asexual populations of *Eupatorium makinoi*, *Genes & Genetic Systems* **73**, 137–41 (1998); K. Ooi and T. Yahara, Genetic variation of geminiviruses: comparison between sexual and asexual host plant populations, *Molecular Ecology* **8**, 89–97 (1999).

36. P. Schmid-Hempel, *Parasites in Social Insects* (Princeton University Press, Princeton, NJ, 1998).

37. All this is further discussed in the introduction to Chapter 2 and especially note 10: for more on the social evolutionary implications of multiple mating of the queen see also Chapters 2 and 8 in *Narrow Roads of Gene Land*, Vol. 1.

38. I. Folstad, and A. J. Karter, Parasites, bright males and the immunocompetance handicap, *American Naturalist* **139**, 603–22 (1992); C. Wedekind and I. Folstad, Adaptive or non-adaptive immunosuppression by sex hormones?, *American Naturalist* **143**, 936–8 (1994); A. P. Møller, P. Christe, and E. Lux, Parasitism, host immune function and sexual selection, *Quarterly Review of Biology* **74**, 3–20 (1999).

39. A. Grafen, Biological signals as handicaps, *Journal of Theoretical Biology* **144**, 517–46 (1990).

40. P. Galeotti, N. Saino, R. Sacchi, and A. P. Møller, Song correlates with social context, testosterone and body condition in male barn swallows, *Animal Behaviour* **53**, 687–700 (1997); N. Saino, P. Galeotti, R. Sacchi, and A. P. Møller, Song and immunological condition in male barn swallows (*Hirundo rustica*), *Behavioral Ecology* **8**, 364–71 (1997); A. P. Møller, N. Saino, G. Taramino, P. Galeotti, and S. Ferrario, Paternity and multiple signaling: effects of a secondary sexual character and song on paternity in the barn swallow, *American Naturalist* **151**, 236–42 (1998).

41. A. P. Møller, P.-Y. Henry, and J. Erritzoe, the evolution of song repertoires and immune defense in birds, *Proceedings of the Royal Society of London* **267**, 165–9 (2000). In detail, the study has found, across a set of species, after controlling for six potentially confounding variables affecting the birds and their songs, that variation in spleen size explained about half the variation of the extent of song repertoire. See also the Appendix to Chapter 6.

42. A. F. Read, Passerine polygyny: a role for parasites, *American Naturalist* **138**, 434–59 (1991).

43. A. P. Møller and J. J. Cuervo, Speciation and feather ornamentation in birds, *Evolution* **52**, 859–69 (1998).

44. A. P. Møller and J. Erritzoe, Parasite virulence and host immune defense: host immune response is related to nest reuse in birds, *Evolution* **50**, 2066–72 (1996).

45. A. P. Møller, Evidence of larger impact of parasites on hosts in the tropics: investment in immune function within and outside the tropics, *Oikos* **82**, 265–70 (1998),.

46. J. Burne, Love in a cold climate dents monkey's macho image, *New Scientist* **145**, 15 only (22 February 1995).

47. S. W. Gangestad and D. M. Buss, Pathogen prevalence and human mate preferences, *Ethology and Sociobiology* **14**, 89–96 (1993).

48. See Chapter 17 and also note 47 above.

49. P. Bagla, Longer tusks are healthy signs, *Science* **276**, 1972 only (1997).

50. H. G. Plimmer [Various reports on deaths of animals, especially birds, in the London Zoo], *Proceedings of the Zoological Society of London* (1912–1934); S. G. Pruett-Jones, H. L. Pruett-Jones, and H. L. Jones, Parasites and sexual selection in birds of paradise, *American Zoologist* **30**, 287–98 (1990).

51. J. H. Oosthizen and M. B. Markus, The haematozoa of South African birds. II. Blood parasites of some Rhodesian birds, *Journal of the South African Veterinary Association* **38**, 438–40 (1967).

52. G. Turner, Intelligence and the X chromosome, *Lancet* **347**, 1814–15 (1996); G. Turner, Intelligence and the X chromosome—Reply, *Lancet* **348**, 826 only (1996); E. B. Hook, Intelligence and the X chromosome, *Lancet* **348**, 826 only (1996); K. Rheinhold, Sex linkage among genes controlling sexually selected traits, *Zoology. Analysis of Complex Systems* **101**, Suppl. 1, 33 only (1998).

53. Either sex in many insect groups can retain its wings when the other sex evolves to lose them but the retaining sex is much more likely to be the male: this theme, with other reasoning than ancient tendency and constraint, is much discussed and exemplified in Chapters 4, 12, and 13 of Volume 1.

54. J. H. Marden and M. G. Kramer, Locomotor performance of insects with rudimentary wings, *Nature* **377**, 332–4 (1995).

THE SEETHING GENETICS
OF HEALTH AND THE
EVOLUTION OF SEX[†]

W. D. HAMILTON

—

It is well known that the adaptations that fit a species to be a colonist of newly opened habitats are not the same as those that fit one to persist in a mature community. In a transition from the one successful lifestyle to the other, fast growth and maximal production of uncared-for offspring, which best increase descendance in uncrowded habitats, have to give place to slower growth and setting aside of material and energy for competition. This implies a range of social and antisocial activities which are mainly concerned with conspecifics. Some of the activities appropriate to a colonist are reduced or abandoned. For example, adaptations for dispersal, although still needed,[1] are usually focused nearer at hand and changed in character.[2] Other activities have to be increased. Reproduction is through fewer, larger, better cared-for offspring destined to contest for places in their environment just as their parents contested for them. The comparison of adaptation under the two regimes described—colonization, versus persistence in a developed community—summarize a field of evolutionary ecology that is often called 'r- and K-selection', r referring to the growth rate of population in ideal empty habitats and K to the limiting level of population that can be reached in crowded ones. Obviously, however, a much more complex picture is being projected in the changes instanced above than could be drawn directly from the logistic growth equation that was the original source of the r and K parameters referred to. The picture is indeed more complex even than is suggested in the formal treatments that later more properly integrated r and K into genetical selection theory (e.g. ref. 3). Nevertheless, reference to 'r and K' has become for the time being a common usage, and, with reservations to be discussed later, its ideas still categorize fairly well a major trend of covariation that has been detected in various unrelated sets of species.[4-6] I will continue use of the r and K symbols and the ideas in most of this chapter but will mention some reservations at the end.

[†]In S. Osawa and T. Honjo (ed.), *Evolution of Life: Fossils, Molecules, and Culture*, pp. 229–51 (Springer-Verlag, Tokyo, 1991).

GENETIC LANDSCAPES AND ECOLOGY

Some of the extra complexity must be faced at once because it introduces my main theme. I will focus on what I believe is a change in the very landscape of selection as a transition of the type outlined above. It seems to me that there is a typical change whose far-reaching consequences have not had due attention. Fortunately, at least in so far as an 'adaptive landscape' idea is justified at all (for doubts see ref. 7), the change is easy to characterize: generally a peaky landscape of the colonist species (a plain with emergent hills) changes to a karst (a plateau dotted with deep pits). In the following paragraphs I will first justify this idea of characteristically 'peaky' and 'pitty' landscapes, and second will argue that the change has important consequences for the nature of genetic variation of species. Third, I will argue that when numerous co-evolving specific parasites add unceasing change to the landscape, which, however, is still expected to preserve a typical land form (i.e. to remain a karst), we have compounded the system of selection which has provided the main support for sexual reproduction and perhaps its original cause.

The process of going from a known genotype to an expressed fitness, whether it is an individual's actual reproduction or an average of a type, can be thought of in two main stages. The first is the determination of what may here be called intrinsic fitness. This is akin to the everyday and non-Darwinian notion of physical fitness. Intrinsic fitness is dependent on factors of genotype, local environment, favourable, and unfavourable accidents. The second stage starts with intrinsic fitness and determines from it the success in reproduction: realized fitness. In the case of the r-species the determination or 'mapping' is generally simple and nearly proportional: other than sudden chance annihilation (such as an encounter with a grazing goat or a falling rock), there is nothing to stop the individual eventually transforming whatever intrinsic fitness it has into reproduction. In such cases for a formal account of the selection process the second stage is hardly needed: the intrinsic fitnesses are by themselves a sufficient basis. Selection defined through such a proportional mapping is known as 'hard' selection.[8] In the case of the K-species, however, the second stage becomes all important. Now it is assumed that the individual must win contests and obtain a place in the community before it can reproduce.[9] The 'mapping' from intrinsic to expressed fitness in this case, therefore, while still presumably monotonic, is unlikely to be anything like proportional and instead probably approximates to a step function. The position of the step up to success on the scale of intrinsic fitness is not fixed but depends on local conditions, which normally also means on social conditions. Selection of this kind, again following Wallace who defined the terms,[8] is called 'soft' selection. It is important to realize that 'peaky' and 'pitty' patterns of intrinsic fitness do not in the second-stage evaluation have to transform into corresponding peaky or pitty expressed patterns; a peaky pattern could become pitty and vice versa. But the likelihood for the *K*-species is that both intrinsic patterns, after a social struggle, would end, expressing in the pitty or karst pattern although there will be likely differences in whether relatively few or many genotypes form a pit—that is, in whether pits are in this sense narrow or broad.

A plateau with pits implies that variation in success among those that succeed in becoming reproducers is small but this does not mean that overall fitness variance is also small. Premature death may be common. The overall variance is constrained only by the difference between maximum fecundity and zero. Low performers—the dead and non-reproductive living—are in practice those that, in any fairly dense and stable occupation of space, fail to obtain territories or nest sites or places high enough in a hierarchy. For example, if they are trees they are those that do not attain crown space in a forest canopy; if barnacles, they are those similarly overgrown and unable to filter feed to maturity on a rock; and, if territorial birds, they are the floaters. The distribution of expressed fitness under such a scheme tends to bimodality with the highest mode at zero and with an overall positive skew. At the opposite extreme of an *r*-habitat, consider a sparsely weedy fallow field: here most plants that have at least a few leaves are able to produce some seeds while those that are lucky in their position and in escaping being eaten by herbivores or attacked by parasites (their luck can include their innate ability to resist the enemies) are able to produce immense numbers.[10] Between such extremes the distribution of achievement in the field is smooth, at least in principle, and overall it is probably negatively skewed. Performance depends partly on chance, partly on genotype, partly on advent of enemies, but usually (again with some exceptions mentioned below) the social interactions of such species are few. What social interactions they have occur mostly other than with peers. They occur not with like-age conspecifics, usually not even with individuals of different species having similar life form. This picture, of course, excludes frequency-dependent effects both of population growth and differential selection. As has often been pointed out this cannot be entirely the truth because if it were, the *r*-species would be either taking over the world or going extinct. There have to be stages and places where adverse effects of high density of population are felt. Nevertheless, the *r*-species is by definition relatively free from frequency-dependent forces compared with the *K*-species.

For the *K*-species, on the contrary, frequency effects are all important and once a situation has more than one species or even more than one genotype such effects imply a likely constant change in the landscape. The characteristic pits are likely to be constantly moving around, perhaps continuously, perhaps by infilling and then reforming elsewhere. The typical selection resulting from this changing landscape for *K*-species, surprisingly, can be described as both destabilizing and stabilizing.

Consider the destabilization aspect first. Imagine a haploid population of four genotypes of two loci, *AB, Ab, aB,* and *ab*, under a gentle frequency-dependent hard selection. Let the selection be such that the four genotypes are stable when at equal frequencies (see Chapter 2) 'Hard' selection, it will be recalled, refers to proportional or smoothly monotonic mapping from intrinsic to actual fitness. Imagine that starting at this equilibrium the pattern of selection now changes to 'soft': that is, instead of some individuals reproducing more and some less on a graded scale, a fixed small quota of deaths now occurs every generation with frequency dependence such as to ensure that the deaths are always of the most common genotype. The change inevitably destabilizes because the subtraction of a fixed number from a single genotype cannot be consistent with unchanging ratios. A little thought shows that the end state has to be some cycle, regular or otherwise. Simulation reveals two alternative cycles, depending on the starting condition. In view of the special nature of the exactly central equilibrium being discussed, and also of the unlikelihood that real ecology would accurately detect which was a merely slightly more or less common genotype, the existence of just two types of cycle is not very exciting or realistic except perhaps as a hint of complexity to come when more loci are added. What is more important to note is that if there is a delay in 'recognizing' the newly most common genotype there is more rather than less certainty of cycling and, due to over-swings, the expected amplitude is greater.

Now consider the stabilizing or, more accurately, protective effect[11] that the same type of selection regime introduces once the system is far from its potential equilibrium. Lags of response and environmental perturbation can optionally be present. With the most common and therefore worst genotype in part counterselected, and with others uniformly successful, the truncating system tends to preserve all alleles.[11,12] Any run of selection against a particular combination genotype loses effectiveness rapidly as the disfavoured genotype becomes rare. The process is a multilocus analogue to that which makes recessive deleterious genes slow to decline once they are rare. Copies of currently 'bad' alleles in a multilocus karst are eliminated only after recombinant segregation and gamete union brings them together with enough other bad alleles to put their bearers into the pit. In the same way, gamete union after single-locus segregation has to bring bad recessives together for them to be counterselected. In both cases the factor helping to preserve the bad alleles is their increasing relative frequency of occurrence, as they become more rare, in combinations where their reproduction is the same as that of the companion genes that are currently optimal. Going from the suggested two-locus model to multiple loci, we see that once individuals of a population are spread out over a plateau of possibilities, they will tend to remain so and will only slowly concentrate towards regions that are distant from pits or else, more definitely, congregate towards any slight local rise.

The corresponding behaviour under an equal amount of variance on a peaky landscape is very different. Then over the generations the population moves rapidly towards the peak that is nearest to the centre of gravity of its current members. In a relatively short time it goes to monomorphy at this peak, full uniformity being prevented only if the peak depends on either heterozygote advantage, on special patterns of epistasis, or, in haploids, on even more special patterns and frequency dependences.[13,14] In summary, for the K-species frequency-based truncation transforms a destabilization near to a potential equilibrium to allele protection once far from the equilibrium, an overall situation that contrasts strongly to the expected trends in an r-species.

Independently of whether soft truncation is present, frequency dependence provides a strong force preserving of variation.[15,16] A direct favouring of rare alleles is most effective; however, except occasionally as in mating incompatibility systems, direct favouring is unlikely in nature. Selection against abundant genotypes, on the other hand, is probably common, and is especially so when species are subject to harmful specialist parasites as discussed below. At the same time, it is not obvious that such genotype selection is always protective because a rare allele restricted to an otherwise common gene combination may have its own fitness contribution, perhaps neutral or good, outweighed by the disadvantage associated with the rest of the common genotype and so be carried to extinction. More theory or simulation is needed to clarify this.

The potential of frequency-dependent selection to explain allele variability and to resolve the problem of the genetic load has long been understood (e.g.refs 12 and 17). Frequency-dependent selection seems, however, to have been little discussed in the context of K-selection and still less in that of likely K fitness landscapes. This means it has not been put in the context of the 'soft' and 'truncation' selection forms that I have argued to be correlated with K. The disconnection of themes may be from two causes. The first is the intrinsic difficulty of analytical approaches to soft selection, causing it to be neglected generally in theoretical population genetics compared to hard selection in spite of its obvious importance in the wild.[9] The second is that although Wright's best-known diagram illustrating his influential metaphor of landscape showed pits and although his writing always manifested awareness of the changeability of landscape, including summits that sink, his discussion on the whole has much more emphasized valley crossing and ascent of hills than avoidance of pits. So has everyone else's. This is not surprising because in the evolutionist's world of reversed gravitation caused by the tendency of fitness to increase under natural selection, it seems rather obvious that activity must move away from pits and towards hills. If, however, pits are forming very fast all the time wherever crowded aspiring individuals accumulate, the neglect of selection near to pits could be unfortunate.

PARASITES

Given the presence of parasites, just such a consequence to the landscape is expected. Parasites are especially abundant for K-selected species and an appropriate image of a K-landscape may be one in which, wherever population temporarily accumulates, parasites multiply so that in the space of genotypes the crowded region drops. In this situation segregation from doubly and multiply heterozygous genotypes close to any new-formed pit throws some offspring into it but others away; the general movement over several generations is away but not far or fast because the plateau offers nowhere much better to go. Overall in such a system individuals stay widely dispersed (i.e. remain highly variable). There seems to be nothing in our knowledge of genetic polymorphism that is inconsistent with this view of a changing dynamic.[12,18] It does partly conflict with the claim that within-generation heterozygote advantage is the most important factor maintaining polymorphism (ref. 19, but see also 16); the evidence supporting this claim, however, seems sometimes open to reinterpretation.[20]

By this point it may reasonably be felt that within the setting of ecology and parasitism the whole metaphor of an adaptive landscape has become strained even if it was not so before. Whether such a landscape is imagined to have peaks, pits, or both together, the genetic and mathematical realities it attempts to depict simply do not resemble any imaginable landscape or possible behaviour on one, and these defects merely add to another long recognized which is that nothing, certainly not fitness or inclusive fitness, has yet been found to maximize consistently when selection affects two or more loci at once. It thus seems we would do well to look for other metaphors if we must use them. In the landscape notion, for the purpose of a picture, numerous dimensions of genotype space were compacted into two, a procedure that obviously has to give misleading impressions. Really we ought to visualize, if it were possible, fitness ascribed to genotypes that are at the vertices of a very high-dimensional polytope.[21] As such a structure is built up each new locus opens a new simplex with dimensionality equal to its number of alleles: simplexes are then dragged from one position to another, trailing as it were spiderlines from their vertices as they go. The spiderlines form 'edges' as demanded by pairs or sets of alleles at other loci; the new combination polytopes are then dragged again for the next locus; and so on. The end result is something that has to be imagined, even in its three-dimensional shadow, as densely web-like and carrying potentially an immense number of genotype vertices.

In such a mental exercise of constructing the polytope, which simply continues what is so easily begun when there are merely two alleles at each locus and few loci (line segment, square, cube, . . .), we are quickly forced to realize that distance on the landscape is different from what it appears to be in the classic two-dimensional metaphor of Sewall Wright's diagrams.[22] The generally useful idea of distance is 'Manhattan' rather than euclidean, 'Manhattan' meaning in this case the number of gene substitutions necessary to get from one point to another. But important sets also associate with points in the genetic model when these are taken in pairs, and these sets are not respectful even of the Manhattan distances. Each comprises the vertices that the pair as parents are jointly capable of producing in their offspring through recombination; these are the vertices of the sub-polytope of which the parents form opposite corners. With free recombination all vertices are equally likely for offspring; hence a seemingly distant point can be 'close by' to a prospective parent if intelligent choice of a mate is allowed. If there is linkage, however, the polytope has anisotropies, which implies that some of the possible offspring genotypes may be relatively likely whether others needing multiple unlikely crossover combinations have virtually no chance of appearing. In all this, one modification to the original picture stands out: everywhere is actually connected more closely to everywhere else than is suggested in the 'landscape'. If we think of the two-dimensional paper of the contour-type plot of the landscape (e.g. see fig. 13.1 in ref. 19) being folded or crumpled into a tight ball and then having holes punched through the ball to establish the multiple connections we are nearer to the truth. But even after we have so recreated a ground plan for the landscape, there remain the hills and pits still needing to be shown—the actual fitnesses. These have to be imagined projecting up or down in some yet further dimension. With native space so overloaded already it hardly makes things worse if we imagine the extra fitness dimension being 'outward', making an extra prickly shell surrounding the shadow of the polytope that has been projected in the familiar three: a jagged skin for an object such as a litchi or pandanus fruit. But—the point to be reiterated—we must take care not to be biased by familiar, easily imagined images, and remember that a caricatured ripe straw-berry with its seeds sunken out of sight into pits or, better still, a simple apple with just one deep pit at the stalk is at least as suitable. As already pointed out it is forms like a hyperstrawberry or a hyperapple that are most likely to keep large numbers of alleles extant and that can therefore utilize the ex-tensive recombination sets of parent pairs in continually placing offspring almost everywhere on the surface. A hyperapple pitted at the stalk in par-ticular can be emphasized over the strawberry because it is not clear that the truncation dynamics we are discussing ever creates pits in multiple unless there is spatial heterogeneity of habitat as well. This is a point to which I will return.

In suggesting such new possible images, apart from my emphasizing pits over peaks, I do not think I have departed much from Wright's original concept of adaptive landscape. Although Wright's prose is often difficult, I think the various quotations of Provine's book[7] make clear that Wright had in mind fitnesses raised over a framework of no greater a dimensionality than the number of alleles at all loci minus one for each locus, just as is the case for my polytope. He was never thinking of the space of all possible genotypes in the way suggested by Provine and much more nearly was thinking of the polytope model discussed by Haldane in 1931,[21] albeit his own development was quite independent, as Provine rightly points out. More specifically countering Provine's criticism, I do not see strong evidence for Wright's supposed inconsistency and in particular do not follow Provine on p. 310 that Wright's 'use of one fitness value assigned to each genotype' implies that he assigned 'as many dimensions to the surface as there were gene combinations, plus one for adaptive value'. One fitness value can be assigned much more reasonably to each vertex on the hypercube or polytope. If it is accepted that Wright assigned only as many dimensions as there were non-wild-type alleles, the weight of Provine's criticism falls upon Wright's power of explanation as the cause of misunderstanding, and not on any changing of his own imagined model. The problem of how to represent sometimes individuals and sometimes subpopulations means in the setting of one type of diagram is also resolved: mean genotypes become centroid points within the polytope and mean fitnesses can be raised over these in the yet extra dimension with no more difficulty than they are raised over individual fitnesses.

COEVOLUTION AND AN ALTERNATIVE DEPICTION

In repeatedly running a model of frequency-dependent co-evolution of hosts with short-lived parasites while increasing the number of loci interacting with the parasites, various kinds of regular oscillation may be watched giving place to the irregular but still allele-conservative patterns whose nature has been outlined (see Chapter 16). The patterns of change very quickly become so complex that they would be called chaotic except that whether they meet criteria recently associated with the word in dynamical theory remains uncertain. As already indicated, even when there are only two resistance alleles at each of two loci, several modes of oscillation are already possible. To the two cycles of period two described by me (Chapter 2) but decided to be practically infeasible by others (ref. 23 and Chapter 2), there may be added two cycles of period four, one resembling a long-winged butterfly perched at the centre and the other a rotation around the centre of a fixed pattern of frequencies of the four genotypes. The last, which for each genotype has three steps up in frequency followed by one down, is the cycle that seems to arise most naturally when actual parasite numbers are simulated in the model and it is also the easiest to understand. Like the 'alternate diagonal' mode of oscillation I first described in 1980 (Chapter 2), this rotation is very supportive of sex against competing clones. The reason is easy to see. Recombination always reduces linkage disequilibrium set up by a previous selection. In the present case the class currently raised to highest frequency by selection is exactly the class destined to be the next pit. So recombination moves offspring, on average, away from a coming pit.

When there are three loci, even when each has only two alleles so that the polytope is a cube, it is easily seen to be impossible for all of the eight genotypes to be treated quite similarly as all four were for the rotating square defined by two loci, as just mentioned. As more loci are added more 'diagonal' (allele abundances changed equally at two or more loci at once) have to occur. Erratic sequences also appear more inevitable. A generic feature expected and also found in such motions is that the newly most disfavoured genotype—the current pit— moves away from a penultimate most disfavoured genotype and towards the general region of those vertices that have been longest not disfavoured, this region, of course, having had longest to become populous and now taking turn to be plagued by parasites adapting to it. But the awkwardness of dynamical description couched all in such negative terms hints that it is time that even our crumpled, pierced paper ball, or our dented apple, should go the way that may have seemed appropriate for them from the first—to a rubbish bin—and that some image more custom-made for pure biotic coevolution should replace them. Although for a visualization there is never any way to get around the problem of multi-dimensionality, I have recently, with others (Chapter 16), suggested the following three-dimensional picture. It should appeal to notions of dynamics familiar to anyone who has ever played with an air-filled rubber ball in the bath and yet it still represents the genetical process surprisingly well.

A light ball inscribed with the outline of a cube is imagined to float on water. There are only three loci and the vertices are the genotypes *ABC*, *ABc*, etc., as above described. The part of the cube or ball under the water is undergoing the truncation imposed by the *K*-environment: under the water all the currently worst genotypes are dying and falling off. This makes the wetted part of the ball lighter. Weight, which is a metaphor in the model for badness of general health, is imagined to be accumulating on corners out of the water all the time in proportion to successful attack by parasites. Because parasites adapt most to common genotypes, weight follows abundance but is not directly caused by it. In effect and slightly vaguely, a side of the ball that has had most genotypes 'out of the water' for longest and therefore has exerted the greatest past evolutionary conditioning on parasites, gathers weight, rolls down, and sends the next genotypes under the water. For more loci than three a hyperball is needed, and here the usual difficulty comes in; nevertheless, although once more vaguely, an extension of the imagined process helps to make understandable the stertorous process of change that we actually witness in our simulation runs of coevolution. Our experiments have had up to 15 defence loci in the host species with these matched by corresponding offence loci in chromosomes of up to 12 parasite species.

Obviously the one-to-one matching of loci does not do justice to the complex interplay of traits that mediate the struggle between hosts and their parasites. Nevertheless, such a simple choice may reasonably be hoped to give an essence, and at least in its multiplicity of both loci and parasite species, our model is a more realistic genetical plank to reality than any other yet offered.

From what has been said about low-level truncation in the two and three loci cases it is fairly obvious that the process is permanently unstable, and also that it will have great difficulty finding any regular sequence or cycle; thus, for example, a turning sequence of the floating hyperball involving more than a minute amount of submergence of its surface cannot always satisfy a rule of next swamping the vertex that has been longest dry. On the basis of this last difficulty, plus the empirical observation of seemingly chaotic changes of gene frequency in all our high-order simulations, I believe that permanently irregular sequences of disfavour must be the rule in multilocus antagonistic coevolution, and I take this as a cue to suggest a new term, 'coevolutionary seething', for the population genetical process that is implied. The term draws on the parallel between sequences of fitnesses observed in the model and the sequences of position of particles suspended in a turbulent, simmering liquid. In the liquid, the regions currently coolest are those due to move soonest to the hot bottom but at the same time, because of turbulence, no particle is allowed to follow a regular cycle: all this is very similar to the expectations for host genotypes and their loadings of parasites.

SEXUALITY

Let us now turn to the question of whether the model can help us to understand how a mysterious and seemingly wasteful process of organic reproduction can be successful. Without sexuality and its associated recombination the polytope of genotypes that has been described still exists but is much less effectively interconnected: genotypes on the polytope are now close to one another only by virtue of sequences that can interchange by mutation: Manhattan distances rule now absolutely. The same restricted kind of closeness applies to the ease of pursuit by asexual parasites that also only use mutation. Because pursuit in these terms is as easy as fleeing and because parasites have generally many more generations in which mutations can occur and be selected, asexual hosts are at a disadvantage. In a sexless world they are liable in the long run to be very much exploited by parasites (see also ref. 24). In fact if 'swampings' on the floating hypercube correspond to zero fitness, asexual lines are killed off quite rapidly and leave sexuals, with all their inefficiency, victorious by default, an inevitable consequence which Treisman[25] was the first to point out. Even if reversible mutation brings some genotypes back it is far less efficient at dispersing them into varied combinations than it could be if aided by recombination in the presence of sex.

In the light of the image suggested above, which is a fairly faithful analogue of the expected process, it is predicted—and is observed in the simulations—that the combination of coevolving parasites and truncation selection gives an extremely robust selection for sex. Parthenogenesis, even when of twofold or sometimes even greater relative efficiency, cannot replace it. Provided parasite types and parasite resistance loci are numerous in the model, the above result holds even when selection differentials per locus and per generation are small. In such cases the changes in gene frequency and linkage disequilibrium associated in the selection process are likewise small. Moreover, random variation in phenotype—intrinsic fitness increased or decreased by a supposedly erratic environment as if the ball is being floated upon a choppy sea—is found only moderately to weaken the advantage of sex. Thus in runs with 12–15 resistance loci even when the random environmental variance was four times the variance occasioned by the parasites sexuality could still keep parthenogenesis to a small minority.[23] Such tests were often run over hundreds of host generations and still no resistance alleles were lost: in short, the protection in the model provides for them is powerful.

In my opinion this model for the maintenance of sex is not only theoretically robust but is the model best favoured by the evidence. Its nearest rival is the mutation clearance theory which has been especially championed by Kondrashov.[26,27] Unfortunately for discrimination, mutation clearance works most strongly in K-environments for similar 'landscaping' reasons. Mutation clearance has the advantage that there can be much less doubt (at least qualitatively) about raw material because bad mutations are undoubtedly ubiquitous, whereas alleles with fluctuating favourability occurring on a wide scale in nature (as in my coevolution model) are yet to be demonstrated. However, the mutation theory has difficulties also. For example, it may be asked if there are enough bad mutations for Kondrashov's scheme to work,[27] and, if there are, then what would preclude better buffering or correction of the genome as a more efficient alternative: is it credible that the whole labyrinthine edifice of sexuality would be invented and maintained just to get bad mutations packaged (in the form of unhatching eggs, abortuses, runts, etc.) ready for elimination? Most important, the mutation theory offers no explanation of the ecological distribution of sexuality other than what has been already mentioned—that sex should be safest in K-environments because of superadditive reduction of fitness under multilocus loads. Thus the mutation theory can explain rather well why K-species are more sexual than r-species even though curiously this has not yet been emphasized by its proponents. Its corresponding great weakness, however, is to be unable to explain why a mutagenic environment and a susceptible life form do not provide an even stronger incentive than the soft selection factor for a species to remain sexual. Nor has the theory yet suggested an explanation for all the striking variations in kinds and degrees of sex, as, for example, in the differing forms of sexual selection; these forms, given some additional factors to create different periods of cycling (Chapter 6, but see also ref. 24), can be explained naturally, both in principle and with slowly accumulating evidence, by the parasite coevolution theory.[28]

There is no question from our perspective but that mutation clearance is indeed a function of sex. The important question is of relative magnitude and whether mutation alone could have brought about the initial stabilization and perfection of such a gross inefficiency as maleness when a simple improvement of buffering and the correction of errors in replication seem an easily available option. Mutation rates can be reduced by selection,[29] and prokaryotes, with genomes far more exposed to destructive outside influences, seem to have set their rates much lower than eukaryotes, with protected genomes (this means also lower than the mutation theory can easily explain). In contrast to mutations, parasites impose pressures for recombination on alleles at all frequencies, including in the mid-range where the effects of recombination are greatest. Doubtless parasites could begin their influence when their hosts were still unicells. As hosts became larger and longer-lived, parasites for their part became more diverse, specialized, and numerous, and so must have posed ever greater problems.

Once again, however, the common features need to be emphasized. Both mutation clearance and parasite coevolution theories feature burdened families and both have the possibility of the production of 'clean' offspring through recombination. In peaky landscapes of r-selection, sex aids the rapid predominance of a few genotypes: this leads to loss of alleles and renders recombination useless, so offering parthenogenesis is efficient and, in the long run, self-destructive victory.[30,31] In other words, by speeding progress to peaks, inbreeding and asex under r work towards eventual species extinction.[31] In a K-selection landscape, on the contrary, recombination helps to prevent uniformity of resistance loci. In mutation terms, in the r-landscape every mutation must eventually take its toll; there is no extra efficiency arising through outbreeding and packaging of them together. Both through maintained resistance and efficient elimination of truly bad mutation, effective and outbred recombination raises the population to steadier higher average levels of health.

OTHER THEORIES

Can any other theories of sex revive or gain new force under the present view? I see two possibilities and take the older one first.

In so far as due emphasis on parasites implies that good mutations may often affect health with respect to infection, the new idea partly revitalizes the old orthodoxy, the Weismann–Fisher–Muller theory of sex.[32] The speed of coevolution of parasites and the unusually varied options that exist with respect to biochemical defence make it at last more conceivable that good mutations occur fast enough for the demands of the old theory to be met.[33,34] The process revived, however, is now based on a Red Queen coevolution and not directly on an 'onward and upward' macroevolution as the authors of the old theory saw it. Whether sex might still be indirectly beneficial to macroevolution I return to shortly.

Second, regarding more recent ideas, the parasite view seems to overlap and potentially complement certain models that have used sibling competition to make sex stable (ref. 35, and see also Chapter 8). In those models, whether the lifestyle is K or not, offspring can be imagined to be part of a local set that includes many siblings. The selection is soft: in the 'lottery' model of Williams, for example, it is specifically truncating but differs from mine in that the major part of the distribution is killed leaving only the extreme of the better tail.[36] The scheme above, at least in any particular year, truncates only a small part of the worse tail. Thus the ideas at first sight seem quite different. A key piece of reality left out of my theory so far, however, may draw them much closer. Spatial variation is explicit in Williams's model and so far not invoked in mine. In the present situation, if our 'ball' sinks lower in the water there comes a time where entire populations of alleles disappear in single rounds of selection. This, of course, is very bad for the process because soon there is not enough variability left for any worthwhile recombination to utilize. In the lottery model the same losses occur all the time to extreme degree but occur merely locally: the lottery in effect has a multitude of balls floating in independence. Each ball is a local coevolution, and, provided the dynamics of the different localities do not synchronize (and there are signs that they may be expected not to—ref. 37 and Chapter 8), then migration between localities may continually reintroduce swamped-out alleles. The Williams–Maynard Smith models lacked explicit rationale for the maintenance of variability but asynchronous cycling plus dispersal of both hosts and parasites may provide this. Although much remains conjectural and needs modelling the most vital point brought into focus here is that we should not expect all colonizing r-species to reduce emphasis on sex equally. Those with the most panmictic dispersal of offspring and other traits that reduce sibling competition should decrease it the most, as perhaps is occurring in the apomictic pappus-borne composite plants where many genera are largely asexual (e.g. *Taraxacum, Antennaria, Heiracium,* and *Eupatorium*) and in those psychid and other moths blown by winds when they are tiny long-haired or silk-lofted larvae. Otherwise, similar plants and insects that are more feebly dispersed but lay batched eggs, or form clumps or colonies of offspring, may retain it strongly. Such a prediction follows (1) because clumps give easy conditions to a parasite once it has entered a population and (2) because whether through parasites or not, the 'best man'[32] or 'lottery'[36] process favouring sex is likely to occur within the clumps.

MACROEVOLUTION

Let us now turn to how seething may affect our understanding of phenomena that are not directly genetical or connected with the parasite source. I will first consider how seething may affect macroevolution, and then in the following section will discuss effects on local demographic reactions and on microevolution. Under the latter two headings come two interesting and probably controversial implications for humans that seem best left to the end.

If Wright was correct that genetic drift in metapopulations made up of small almost isolated demes is important for macroevolution because it makes easy the crossing of adaptive valleys and hence the attainment of new peaks, we may expect the dynamical changes of adaptive landscapes induced by parasites to play a similar role. The effect no longer has to come from subdivision into demes although, as just pointed out, subdivision may still help. As might be expected there is little empirical evidence. A suggestive scrap comes not from biology but from the techniques of artificial intelligence. Hillis[38] set a 'genetic algorithm' (GA, a computer procedure adapting by natural selection; for its origin see ref. 39) to find the highest peak among an extreme multiplicity of foothills. He used contrived functional landscapes and also semiserious problem ones. The latter demanded the discovery of fastest fixed-exchange sort algorithms. It turned out that his search succeeded best when 'parasites' were added to the simulation. The achievements whether parasites were present or not were considerable: the GA eventually finished with a world second-equal fastest sort for any initial arrangement of 16 elements (up to 1969 the GA's discovery would have been the fastest sort algorithm for the task known). Hillis took as the 'parasites' of his procedure simply the lists presented to it to be sorted; these were rewarded and multiplied—or coevolved—in such a way as to acquire resistance to being sorted, a procedure closely equivalent to that being applied to the sort algorithms themselves. Because the lists had the same generation time as their hosts the step was perhaps more analogous to creating evolving prey for a predator; however, the idea of the presence of a coevolving suite of resistances is the same. In effect, like parasites, the lists evolved fastest to be resistant to these sorts that through success had become common. Thus their evolution must have repeatedly driven down the fitness of discovered minor peaks. Presumably this helped populations to spill from one hilltop catchment to another.[40] The end result should be that the whole landscape is searched more extensively for its higher ground, although it should be pointed out that even without coevolution a conspicuous strength of the GA approach to artificial intelligence is its ability not to be held up on the nearest minor peaks.[39]

The computer used by Hillis was the Connection Machine in which massive parallelism permitted handling of a very natural and fast 'stepping stone' spatial structure of the population: individuals were arrayed in a square grid toroidally wrapped to avoid edge effects, and each individual was mated only to near neighbours as determined by a low-variance Gaussian distribution of distance. Although not essential for the effect we are discussing, such a spatial structure probably makes progress even faster on the lines of Wright's original 'shifting balance' idea.[41,42] It is likely that recombination of locally discovered 'partial good ideas' about sorting contributed to the striking over-all success.

The models of Hillis had 'parasites' keying to exactly the same alleles that in the right combinations caused the fixed macroevolutionary peaks of various altitude. Whether this scenario of identical alleles serving for defence and for macroevolution is at all reasonable for organic evolution as well as for evolution, as in this software, is unknown. It is possible that every spreading good mutation that contributes to macroevolutionary change not only changes morphogenesis but, through the primary proteins affected, changes vulnerability to parasites as well. If this is the case then, as built into the Hillis models, the locus is indeed one for resistance even if its most important macroevolutionary effect is elsewhere.

Even if such abundance of pleiotropy is unrealistic, genes primarily concerned with resistance are probably so widespread in the genome[43] that any new mutation of macroevolutionary importance is likely to be closely enough linked to a resistance allele for the combinations at the two loci never to have time to reach linkage equilibrium. Then linked pairs are treated by the selection almost as if the effects are from pleiotropy. It is moot theoretically whether the disadvantage when a macroevolutionarily 'good' mutation occurs linked to a temporarily 'bad' and therefore declining health gene, and hence has increased chance to be lost, is outweighed by the advantage of a doubly rapid rise in frequency of the mutation when it occurs linked to a 'good' and currently increasing allele. In other words it is uncertain whether the seething in such models helps or hinders passage of unequivocally good mutations to fixation. Intuition suggests that the advantageous phase should outweigh the disadvantages because even without the 'bad' companion, the mutation in the first case has a fairly high chance in its early uncertain days to be lost,[44] whereas in the second case the early rapid rise gets the allele to a numerical representation where even rare recombinations can be expected, and once it is in company with both 'good' and 'bad' health alleles the further upward progress on its own merits is assured: it certainly cannot do worse than the 'bad' companion does alone and this, as I have already argued, is preserved. All this suggests that seething may speed macroevolution even where multipeakedness of landscape is not a problem and even in cases where resistance alleles do not pleiotropically affect macroevolutionary fitness.

At the same time contributary benefit comes always from mutation clearance. It follows from what was said earlier that sex, especially in K-environments, can tolerate more mutation of all kinds. Out of the stream sex sifts the good while limiting the burden of the rest. In this respect also sex increases the fixation rate of beneficial mutations, showing yet another point on which the parasite and mutation clearance theories make a similar prediction.

If the ideas above are right, species much subject to parasites, if surviving at all, should be more innovative in their evolutionary history than the average. In other words K-species, due to a combination of their more numerous parasites, stronger expected emphasis on outbreeding, and higher mutation rates, should be more innovative than r-species. I know of little evidence that directly bears on all this, but the idea hints at some possible resolutions for old puzzles. One is the way, always mysterious to me although surprisingly little discussed by botanists, that root ancestors of angiosperm taxa are projected not only to be perennials rather than annuals but often to be woody and arborescent;[45–47] in other words, to be large and long-lived, quite contrary to the obvious expectation of achievable rates based on generation time. Main stems of angiosperm trees are woody: stems that persist do not lie among life forms where the turnover is fastest as might be expected but among those where it is an order of magnitude or more slower. It seems to be a race where a tortoise is winning, but perhaps the explanation is also rather as in the fable: irked onward by its parasites, the tree or bush plods and never sleeps whereas the annual (or the herb turning annual) sprints fast in one regard (having mutations available) but slow in another (sorting good from bad) and, worse still, once gone on the diversion to inbreeding or asexuality, it bogs down—misses the peak shifts and good combinations. Finally, when disease, competition, or other hardship press hard on it, it drops completely from the race. Because large organisms suffer more from more kinds of parasites than small organisms, although with a large overlap, we find here a possible subtle advantage to being large and a direct compensation to be set against slow turnover. As longer generations and larger size in a host make its parasites more diverse and numerous it seems that the same traits also may paradoxically increase the species' chance to be original and to take macroevolutionary strides. It follows from these thoughts that the direction from r to K used to illustrate the first paragraphs of this chapter may not have been well chosen. Regarding evolution on remote islands some likely examples could be shown (e.g. some arborescent daisies and others), but more generally at least on continents the r–K transition seems rare and it is likely that r species generally share some of the irreversible differentiation and final ephemerality that everyone agrees to be the fate of clones.

If valid for plants the same idea should hold for animals. This may make it

more understandable that key fossils believed to be on or close to the vertebrate stem leading to mammals (and thus humans) seem often to be quite
large, as, for example, in the transitional forms in the fish–amphibian–reptile
sequence. Against the background of other primates humans are fairly large
and are exceptionally long-lived; and again their acquisition of size and
longevity seems to have been accompanied by a speeding rather than a
slowing of the rate of evolution.

LATITUDE

It may seem that the above argument, if correct, would confirm the tropics to
be the region of fastest evolution of all kinds because the tropics have the
most parasites. Certainly the tropics have most as judged from the set afflict-
ing humans. I think, however, that we should be cautious of this general conclusion for two reasons. One is that the great diversity of hosts of all kinds in
tropical biomes spaces them and impedes parasite transmission. A case has
been made for this being the reason why rather passive dispersing parasites
(e.g. aphids) are actually rare in the tropics,[48] and a similar factor may abet a
self-escalation of diversity generally.[49] Generalists, coming where the specialists
cannot live, less entrained to fluctuations by particular hosts, and imposing
pressures on each one perhaps more as would an abiotic force such as the
weather, may not have the same genetically enlivening effect. Lives and
deaths of generalist parasites on any one host species make only a very minor
impression on the total parasite gene pool (see Chapter 1). As with so many
of the speculative issues raised in this chapter we are again mostly in the dark
here, even theoretically having little idea how the dynamics of the suggested
system could work. But until the dynamics of generalist parasites are better
understood it seems prudent not to jump to the conclusion that the temperate hosts, with parasites less diverse but at least as prevalent and at least
as tightly keyed to coevolving host genotypes, are less 'seethed' by them
genetically than are tropical counterparts. I easily imagine, in fact, and see
some evidence, that ancient footprints of geographic phylogeny point both
ways across latitudes, some changing and multiplying more towards the
equator and others away from it.

Let me now turn to different smaller-scale issues among the potential
effects of seething while still attending to the role of latitude. Consider
marriage. This may seem at first quite outside of biology. Marriage is cultural,
unparalleled by anything in other animals; however, marriage also (at least
most of the time) reflects a proclivity to bond. This proclivity is almost
certainly innate and has numerous parallels in fish and in birds as well as in
other mammals. Concern and grievance over fidelity within pairs are insepar-
able from bonding and are paralleled in all the mentioned groups as well as
being found in humans.[50,51] It seems not unreasonable to assume that the

range of patterns stretching from promiscuity or harem polygyny on the one hand to strong, faithful monogamy on the other, as occurring both between and within animal species, arises from divergent forces of the same kind as cause the similar range found within *Homo*. There is growing evidence that diversity and prevalence of parasites in birds, fish, and lizards affects the evolution of their mating patterns.[29] It is of great interest, therefore, to find a rather similar suggestive pattern connecting parasitism and human marriage.

Low[52,53] found that human societies in which polygyny is sanctioned occur significantly most often in the areas most affected by human macroparasites. A little thought shows macroparasites to be exactly the kinds for which recent health in a prospective mate could most reliably guide to innate resistance. Although Low's study showed both parasitism and polygyny rising towards the equator the association remained significant even when latitude was controlled for. An alternative interpretation of polygyny emphasizes that actual polygynist males are rich and powerful, and these are indeed more obvious factors than health in determining which males get extra wives. Being powerful in a tribal environment, however, often depends not only on an individual's own efforts, for which health is needed, but also on inheritance and the numerical strength of kin.[54] These in turn probably reflect the average health of an extended family and this from the mate chooser's point of view may be a still better guide. Women who join polygynous groups may often not seem to do so by choice; nevertheless their protest usually appears less vigorous than would be within reasonable non-self-injuring effort. While it may be argued for some cases and perhaps for most that they actually get more resources from the polygynous mate because of his wealth and power (e.g. ref. 55), still, on the basis that paternal support is often conspicuously absent in polygynous animals and this is tolerated by females, I predict that wealth and power will not always prove to be crucial with humans either. English literature often shows women choosing glamorous and healthy rakes over competent honest adorers. Examples are in Thomas Hardy's *Far from the Madding Crowd* and Charlotte Bronte's *Jane Eyre*. The characters concerned are far from being artistic figments. Thus the latter book is said to be the most-loved classic of British women and it is hard to believe a story would be so absorbing unless Charlotte's hero, Mr Rochester, was taken to be reasonably realistic and touched deep chords. I believe that women ultimately and on average prefer marriage patterns in which they have best options for their inclusive fitness, and that this remains true even when their choice appears to be restricted by male-imposed rules. It seems to me that the usual ultimate deciders in human mate choice, used by both human sexes and well adjusted to evolutionary priorities, are wealth, health, and (partly reflecting still health but partly independent) an elusive aura of sexiness *per se*.[56]

Low found that latitude does not correlate either with parasites or with polygyny as strongly as these correlate with each other in humans and this raises the point that the *r* and *K* theme may be to some extent a factor for sex orthogonal to the factor of parasites. Wherever populations become dense there is opportunity for specialist parasites to flourish, and wherever enough do flourish and make health of their hosts uncertain and highly heritable, it may become a mate-selector's best longterm interest to pick the healthiest as its mate rather than the individual most willing to provide parental care (Chapter 17). On these lines it seems less surprising that some dramatic examples of both polygyny and polyandry are found in the Arctic in spite of the commonly *r*-selected nature of Arctic habitats. *Tarandus rangifer,* the reindeer or caribou, is an example of extreme polygyny. The reindeer has larger horns in proportion to body size than any other deer, a trend continued even in the most northerly and isolated population on Svalbard. Reindeer are sociable, common, warm-blooded herbivores in their habitats and doubtless connected with this they do indeed support a large variety of parasites, some of which are known to be very damaging to their health[57] and may commonly be the cause for particular animals to fall to wolves. The various Arctic eiders and the ruff are birds with approximate and outright polygynous systems. In contrast to these, equally Arctic phalaropes show the opposite, sex-reversed pattern. They are seconded by several near polyandrous and male care-biased cases among other waders. Monogamous species are, however, also common in the Arctic, as with geese and swans; and in better accord with the expected *r* to *K* correlation, arctic ptarmigan, willow grouse, and hazel hens are less polygynous than their more temperate blackcock and capercaillie relatives. Unfortunately species lists remain almost the sum of knowledge of the parasites and diseases of these various groups.

It has been suggested that all the branches of extant *Homo* might descend from an arctic or periglacial ancestor.[58] The accepted picture, however, has a complex but relatively unreversing expansion out of Africa in all the possible directions.[59] But if the periglacial suggestion were true and all of what was due to become *H. sapiens* went through at least a brief northerly phase it might make more understandable an otherwise odd claim by Rushton[60] that the seemingly most thoroughly tropic-adapted humans, Africans, show most signs of apparent *r*-selection, while what we believe to be the most Arctic-adapted, Mongoloids, show the extreme of opposite and apparent *K* characters. This is contrary to what would be expected, at least from the limited viewpoint of *r* and *K*, if Africans had never left the hominid native land because that would predict that the far-flung Mongoloids ought to be most endowed with colonist characters. To confuse the picture still more, Mongoloids turn out to be low in both parasites and polygyny (at least in the Old World) while Africans are high in both. If, however, *K*-ness and parasites could be regarded as nearly orthogonal factors contributing importance to sexuality, none of the above facts would seriously contradict the present theme but instead would merely point to the need for more discriminating study. It can be suggested that Rushton's impressively covariant sets for the major races may turn out to be more connected with backgrounds of disease in the various habitats than with his claims for r and K. If health is an all-important source of mortality and if disease shows little response to extra nurturance given to offspring, as seems to be true, for example, of such diseases as malaria and cholera, and if on the other hand there is always the possibility of acquiring exceptional resistance by choosing the right mate, selection will favour numerous offspring rather than a few that are carefully nurtured.

In similar vein, until the advent of microscopes and scientific medicine an almost random search by humans for herbal remedies, aided perhaps by growing taxonomic insight, was probably the most that intellect could do towards solving the very difficult problem of disease. Even in Europe long after applied science was well established in other fields it was only approximately at the time of Louis Pasteur that infectious disease began to respond to treatment. This may be contrasted to the ease of at least choosing appropriate directions for effort in combating prevalent causes of death in boreal habitats with their inescapable demand for weatherproof homes, clothing, and food storage. Here would seem to lie possible explanations of the differences in some types of cognitive ability reported in Rushton's review.[60] However, there are two points that need to be emphasized before leaving Rushton's interesting claims. First, the ethnic overlap of distributions in all traits is broad, just as we would expect from the fact that most challenges faced by *H. sapiens* in whatever climate are similar. Second, an ecological point affecting the claimed connection with 'r and K' is that species with high potential r certainly do not always have higher fecundities, as Rushton seems to assume. Thus aphids, which are among the most extreme colonist species during the temperate spring and which in optimal conditions show some of the highest rates of growth of biomass known, do not have very high fecundities. Moreover, aphids are viviparous and nurture their embryos internally through a structure resembling a placenta. Such careful nurturance again seems little in accord with Rushton's claims. Similar remarks apply to many other colonizing small arthropods (e.g. *Heteropeza, Ptinella*, certain Scolytidae, and *Triops*); high r in all these cases is not achieved by exceptional fecundity but by very rapid maturation combined with iteroparity and short birth intervals. It would seem that this type of infecund r-species and the possibly more typical type with higher fecundity are little different when it comes to the chance of abandoning sex. Examples of parthenogens are easily found in both groups, as in aphids, *Daphnia*, and ptiliid beetles[61] on the one hand and in ticks[62] and *Stylops*[63] on the other. At the K extreme similar exceptions to any simplistic scheme appear. Thus some scarab beetles have lower total fecundity than humans. On the social side scarabs are excellent nurturing parents for the most part but are opportunistically thieves as well.[64,65] Some birds senesce less fast than humans and yet in a successful lifetime are more fecund, a point that again questions Rushton's assumed r and K correlation. As other ecologists have cautioned,[66,67] it is naive to expect a single r and K axis to explain very much. This, however, criticizes my own makeshift use above as much as Rushton's: *Primo medici cura teipsum.*

SEETHING OF FAMILY LINES

With such a caution in mind and more in a spirit of wistfulness for what may sometime be understood than in confidence in what is, let us now consider the predictions that genetic seething may have for human sociology. As promised earlier we take up demographic patterns and the descendances of individuals. There comes to be an almost moralistic turn. 'Blessed are the meek, for they shall inherit the earth.' If this Biblical saying of a psalmist and of Jesus meant the meek to be those failing to be dominant because of weak health, we might glimpse of a more material meaning than the one usually favoured, a real future redemption to be achieved through descendants. A genotype lacking in current intrinsic fitness that nevertheless succeeds in reproducing (or in terms of the floating-ball image, a vertex skimming near the waterline) may expect, some generations hence, to find its genetic descendance revive and grow. This would be because such a type had carried through uncommon, currently disfavoured alleles that at length had become due for new periods of favour. Through the same timespan, looking on the other side, the ghost of a polygynous lord might be doomed to watch his extremely numerous genetical descendance of the first few generations dwindle as the line goes on and to see its survivors becoming outnumbered and humbled by rising lines of plebeians. On the mutation-clearance theory a positive correlation of fitness to lineal descendants is strong for offspring and from there drops down exponentially towards zero, more than halving each generation, as determined by the deleterious mutation rate. On the theory of cycling parasites such a correlation will not necessarily even be falling all of the time. Significant non-linearity of fall on a log scale actually may provide one possible way of discriminating the mutation from the parasite factor. If major cycles in fitness contribution from alleles are few and have similar periods there could be brief reversals of fall, or the coefficient might return from negative to positive. However, to detect fluctuating correlations, supposing they occur, it would almost certainly be necessary to look farther down than grandchildren, and even for grandchildren the correlations would already be so weak as to be difficult to measure. Pressing the same subjective and speculative thought further into sociology, the parasite theory leads us to expect, in parallel to such correlations, more rises and falls of family fortunes if seething really occurs than under a pure mutation-clearance process.

Fluctuations in prevalent parasites are expected to affect whole related families in parallel. This might be difficult to pick out but where groups are endogamous it might manifest in the changes in group dominance. Possibly changes observed in rhesus macaques on Cayo Santiago[68-70] or Yanomami Indians in Venezuela[71] could have their foundation in averages of health. On this view, in the monkeys and human parallels, newly dominant groups would be analogous to up-bound portions of the liquid in convectional turbulence or, in the other image, to an up-bound 'patch' of connected vertices on the floating hypercube.

Turning from demography to the mating that in a sexual species necessarily underlies it, let us look further at humans and macaques. Given some evolved sense of the cues of disease resistance, members of a lowly but very healthy caste might be expected to be sought after as mates. This could be the origin of the strong tendency to hypergyny based on the mysterious perception of 'beauty' that is found in so many broad human cultures including European. Such an interpretation would receive support if the ideal of beauty were to switch from time to time for no obvious reason from one race, or caste, or type of physique to another with constancy of preference after the switch continuing for at least several human generations. Fairly obviously the change in taste that is being suggested need not be innate, and rather arbitrary fashions—for example, in styles of clothing or in approval of male showiness—might well ride along. As a corollary we would expect groups that became closely endogamous, perhaps to help preserve property or some special civilized aptitude, would pay eventually for their elitism by becoming progressively less healthy. To test the last idea it would be necessary to show that the loss of health was not due to loss of heterosis or exposure of deleterious recessives but simply to the retention in the group of too many currently bad alleles. The postulated decline would be most marked in susceptibility to infections but should also show up in abundance and severity of autoimmune reactions in so far as the latter are misguided responses to pathogen challenge. It is difficult to suggest even anecdotal evidence for any of this at present. Possibly the popular image of the ectomorphic (and, not long ago, tubercular) intellectual is one hint. Reading in late nineteenth century novelists such as Dickens, Zola, Dostoevsky, or Gorky, of the lives of poor people who were exposed through malnutrition and poor hygiene to the full force of the infections of their time, one gains an impression of an immense, unexplained variation of healthiness. And if it could be assumed that male lines in typical patrilineal aristocracies rise to power partly through innate health, as could be true of British aristocratic families initiated in the Norman Conquest (distinguished contribution during any campaign probably needs very good health), then the extinction of such lines a few hundred years later at greater rates than predicted from demography of the population at large might be looked on as possible evidence. But, again, more conventional effects of inbreeding in aristocracies as well as various other processes[72] seem at present confounded.

References

1. W. D. Hamilton and R. M. May, Dispersal in stable habitats, *Nature* **269**, 578–81 (1977) [reprinted in *Narrow Roads of Gene Land*, Vol. 1, pp. 377–95].

2. D. A. Roff, The genetic basis of wing dimorphism in the sand cricket, *Gryllus firmus* and its relevance to the evolution of wing dimorphisms in insects, *Journal of Heredity* **57**, 221–31 (1986).

3. J. Roughgarden, Density-dependent natural selection, *Ecology* **52**, 453–68 (1971).

4. D. W. Tinkle, H. M. Wilbur, and S. G. Tilley, Evolutionary strategies in lizard reproduction, *Evolution* **24**, 55–74 (1970).

5. M. L. Cody Ecological aspects of avian reproduction, in D. S. Farner, J. R. King (ed.), *Avian Biology*, Vol. 1, pp 463–503 (Academic, London, 1971).

6. M. Gadgil and O. T. Solbrig, The concept of r- and K-selection: evidence from wild flowers and some theoretical considerations, *American Naturalist* **106**, 14–31 (1972).

7. W. B. Provine, *Sewall Wright and Evolutionary Biology* (University of Chicago Press, Chicago, 1986).

8. B. Wallace, Hard and soft selection revisited, *Evolution* **29**, 465–73 (1975).

9. A. Lomnicki, *Population Ecology of Individuals* (Princeton University Press, Princeton, NJ, 1988).

10. E. J. Salisbury, *The Reproductive Capacity of Plants* (Bell, London, 1942).

11. S. Karlin and R. B. Campbell, The existence of a protected polymorphism under conditions of soft as opposed to hard selection in a multi-deme population system, *American Naturalist* **117**, 262–75 (1981).

12. C. Wills, Rank order selection is capable of maintaining all genetic polymorphisms, *Genetics* **89**, 403–17 (1978).

13. C. Gliddon and C. Strobeck, Necessary and sufficient conditions for multiple-niche polymorphism in haploids, *American Naturalist* **109**, 233–5 (1975).

14. M. W. Feldman, I. R. Franklin, and G. Thomson, Selection in complex genetic systems I: the symmetric equilibria of the three locus symmetric viability model, *Genetics* **76**, 135–62 (1973).

15. S. Wright, Adaptation and selection, in G. L. Jepson, G. G. Simpson, and E. Mayr (ed.), *Genetics, Palaeontology and Evolution*, pp. 365–89 (Princeton University Press, Princeton, NJ, 1949).

16. B. C. Clarke and P. O'Donald, Frequency dependant selection, *Journal of Heredity* **19**, 201–6 (1964).

17. B. C. Clarke, The evolution of genetic diversity, *Proceedings of the Royal Society of London*, **B205**, 453–74 (1979).

18. J. H. Gillespie and M. Turelli, Genotype-environment interactions and the maintenance of polygenic variation, *Genetics* **121**, 129–38 (1989).

19. J. B. Mitton, Theory and data pertinent to the relationship between heterozygosity and fitness, in N. W. Thornhill (ed.), *The Natural History of Inbreeding and*

Outbreeding: Theoretical and Empirical Perspectives, pp. 17–41 (University of Chicago Press, Chicago, 1993).

20. W. D. Hamilton, Inbreeding in Egypt and in this book, in N. W. Thornhill (ed.), *The Natural History of Inbreeding and Outbreeding: Theoretical and Empirical Perspectives*, pp. 429–50 (University of Chicago Press, Chicago, 1993).

21. J. B. S. Haldane, A mathematical theory of natural selection, part VIII: metastable populations, *Proceedings of the Cambridge Philosophical Society* **27**, 137 –42 (1931).

22. S. Wright, *Evolution and the Genetics of Populations*, Vol. 3, p. 452 (University of Chicago Press, Chicago, 1977).

23. R. M. May and R. M. Anderson, Epidemiology and genetics in the correlation of parasites and hosts, *Proceedings of the Royal Society of London* **B219**, 281–313 (1983).

24. S. Nee, Antagonistic coevolution and the evolution of genotypic randomisation, *Journal of Theoretical Biology* **140**, 499–518 (1989).

25. M. Treisman, The evolution of sexual reproduction reproduction: a model which assumes individual selection, *Journal of Theoretical Biology* **60**, 421–31 (1976).

26. J. T. Manning, Males and the advantage of sex, *Journal of Theoretical Biology* **108**, 215–20 (1984).

27. A. S. Kondrashov, Deleterious mutations and the evolution of sexual reproduction, *Nature* **336**, 435–40 (1988).

28. G. Hausfater and R. Thornhill (ed.), Symposium: Parasites and sexual selection, *American Zoologist* **30**, 225–352 (1990).

29. H. Nothel, Adaptation of *Drosophila melanogaster* populations to high mutation pressure: evolutionary adjustment of mutation rates. *Proceedings of the National Academy of Sciences* USA **84**, 1045–9 (1987).

30. L. Nunney, The maintenance of sex by group selection *Evolution* **43**, 245–57 (1989).

31. V. Thompson, Does sex accelerate evolution?, *Evolutionary Theory* **1**, 131–56 (1976).

32. G. Bell, *The Masterpiece of Nature: The Evolution and Genetics of Sexuality* (University of California Press, Berkeley, 1982).

33. R. E. Hill and N. D. Hastie, Accelerated evolution in the reactive centre regions of serine protease inhibitors, *Nature* **326**, 96–9 (1987).

34. M. Laskowski Jr, I. Kato, W. Ardelt, J. Cook, A. Denton, M. W. Empie *et al.* Ovomucoid third domains from 100 avian species: isolation, sequences, and hypervariability of enzyme-inhibitor contact residues, *Biochemistry* **26**, 202–21 (1987).

35. J. Maynard Smith, *The Evolution of Sex* (Cambridge University Press, Cambridge, 1978).

36. G. C. Williams, *Sex in Evolution* (Princeton University Press, Princeton, NJ, 1975).

37. S. A. Frank, Spatial variation in coevolutionary dynamics, *Evolutionary Ecology* **51**, 193–217 (1991).

38. W. D. Hillis, Co-evolving parasites improve simulated evolution in an optimisation procedure, *Physica D* **42**, 228–34 (1990).

39. J. H. Holland, *Adaptation in Natural and Artificial systems* (University of Michigan Press, Ann Arbor, 1975).

40. S. Wright, The roles of mutation, inbreeding, crossbreeding, and selection in evolution, *Proceedings of the Sixth International Congress on Genetics*, **1**, 356–66 (1932).

41. R. Tanese, Parallel genetic algorithms for a hypercube, in J. J. Grefenstette (ed.), *Genetic Algorithms and their Applications: Proceedings of the Second International Congress on Genetic Algorithms, July 28–31, 1987*, pp. 177–83 (Lawrence Erlbaum, Hove, 1987).

42. B. H. Sumida, A. E. Houston, J. M. McNamara, and W. D. Hamilton, Genetic algorithms and learning, *Journal of Theoretical Biology* **147**, 59–84 (1990).

43. S. J. O'Brien and J. F. Evermann, Interactive influence of infectious disease and genetic diversity in natural populations, *TREE* **3**, 254–9 (1988).

44. J. F. Crow and M. Kimura, *An Introduction to Population Genetics Theory* (Harper and Row, New York, 1970).

45. A. H. Church, *Thalassiophyta and the Subaerial Transmigration*, Oxford Botanical Memoirs 3 (Clarendon, Oxford, 1919).

46. E. J. H. Corner, *The Life of Plants* (Weidenfeld & Nicolson, London, 1964).

47. S. Carlquist, Wood anatomy of Compositae: a summary with comments on factors controlling wood evolution, *Aliso* **6**, 25–44 (1966).

48. A. F. G. Dixon, J. Holman, P. Kindlemann, and J. Lips, Why are there so few species of aphids, especially in the tropics?, *American Naturalist* **129**, 580–92 (1985).

49. P. Becker, L. W. Lee, E. D. Rothman, and W. D. Hamilton, Seed predation and the coexistence of tree species: Hubbell's models revisited, *Oikos* **44**, 382–90 (1985).

50. M. Daly and M. Wilson, *Sex, Evolution, and Behaviour*, 2nd edn (Willard Grant, Boston, 1983).

51. R. L. Smith, *Sperm Competition and the Evolution of Animal Mating Systems* (Academic, London, 1984).

52. B. Low, Pathogen stress and polygyny in humans, in L. L. Betzig, M. Borgerhof Mulder, and P. W. Turke (ed.), *Human Reproductive Behaviour: A Darwinian Perspective*, pp. 115–27 (Cambridge University Press, Cambridge, 1987).

53. B. Low, Parasite virulence in relation to mating system structure in humans, *American Zoologist* **30**, 325–39 (1990).

54. A. L. Hughes, *Evolution and Human Kinship* (Oxford University Press, Oxford, 1988).

55. M. Borgerhof Mulder, Marital status and reproductive performance in Kipsigis women: re-evaluating the polygyny–fertility hypothesis, *Population Studies* **43**, 285–304 (1989).

56. J. W. Curtsinger and I. L. Heisler, On the consistency of sexy son models: a reply to Kirkpatrick, *American Naturalist* **134**, 979–81 (1990).

57. O. Halvorsen, Epidemiology of reindeer parasites, *Parasitology Today* **2**, 334–9 (1986).

58. V. Geist, *Life Strategies, Human Evolution, Environmental Design: Towards a Biological Theory of Health* (Springer, New York, 1978).

59. L. L. Cavalli-Sforza, A. Piazza, P. Menozzi, and J. Mountain, Reconstruction of human evolution: bringing together genetic, archaeological, and linguistic data, *Proceedings of the National Academy of Sciences, USA* **85**, 6002–6 (1988).

60. J. P. Rushton, Toward a theory of human multiple birthing: sociobiology and r/K reproductive strategies, *Acta Genetica Medica* Gemellol (Roma) **36** 289–96 (1987).

61. V. A. Taylor, Coexistence of two species of *Ptinella* Motulsky (Coleoptera: Ptiliidae) and the significance of their adaptation to different temperature ranges. *Ecological Entomology* **5**, 397–411 (1980).

62. J. H. Oliver Jr, Parthenogenesis in mites and ticks (Arachnida: Acari), *American Zoologist* **11**, 283–99 (1971).

63. R. Kinzelbach, Strepsiptera (Facherfluger), *Handbuch der Zoologie* **4** (2), 1–68 (1971).

64. G. Halffeter and E. G. Matthews, The natural history of dung beetles of the sub-family Scarabaeinae (Coleoptera: Scarabaeidae), *Folia Entomologica Mexicana* **12–14**, 1–312 (1966).

65. G. Halffeter and E. G. Edmonds, *The Nesting Behavior of Dung Beetles (Scarabaeinae): An Ecological and Evolutive Approach* (Instituto de Ecologia, Mexico, 1982).

66. S. C. Stearns, The evolution of life history traits, *Annual Review of Ecology and Systematics* **8**, 145–71 (1977).

67. J. P. Grime, *Plant Strategies and Vegetation Processes* (Wiley, Chichester, 1979).

68. D. S. Sade, A longitudinal study of rhesus monkeys, in R. Tuttle (ed.), *The functional and Evolutionary Biology of Primates* (Aldine-Atherton, Chicago, 1972).

69. B. D. Chepko-Sade and D. S. Sade, Patterns of group splitting within matrilineal kinship groups, *Behavioral Ecology and Sociobiology* **5**, 167–86 (1979).

70. T. J. Olivier, C. Ober, J. Buettner-Janusch, and D. S. Sade, Genetic differentiation amond matrilines in social groups of rhesus monkeys, *Behavioral Ecology and Sociobiology* **8**, 279–85 (1981).

71. N. A. Chagnon, Genealogy, solidarity, and relatedness: limits to local group size and patterns of fissioning in an expanding population, *Yearbook of Physical Anthropology* **19**, 95–110 (1975).

72. R. A. Fisher, *The Genetical Theory of Natural Selection* (Clarendon, Oxford, 1930).

APPENDICES

OUR PAPER THEN AND NOW
Appendix to Chapter Six

*The sere, comb and wattles become much paler and the blood
vessels beneath the wing also look pale. The head appendages
gradually become more and more pale as the disease progresses, and
finally acquire a peculiar bluish tinge. This tint is also shown by the
eyelids and ears and the legs are affected although to a less extent.
The feathers on the head tend to fall off so that the forepart of the
head and the region round the bill become almost bald, and the bird
presents a very peculiar appearance, owing to the bluish coloration.*

*The plumage of the infected birds is affected in regions other than
the head, and the quills are less rigid than in normal birds. The
feathering of the legs is ragged, and the sheen on the neck and tail
coverts is not so well developed.*

H. B. FANTHAM[1]

AFTER it was published, the paper by Marlene Zuk and myself
('H&Z') experienced rapid changes in its scientific acceptance. In the first
year or so objections that had been based largely on misunderstandings died
away, giving place to a lot of enthusiasm and a surprising amount of
apparent confirmation of our predictions. Confirmation came especially
from tests that compared individuals within species but some followed the
path of our own evidence and compared species.[2]

Within populations, our prediction had been that the showier
individuals would be less afflicted than the duller ones. Some studies on
this point found no effect and some were contrary or strongly equivocal,[3]
but on the whole the results were positive, as reviewed by Marlene some
10 years later.[4]

As to the cross-species prediction, retests of this side of our idea were
more mixed and changeable. The prediction had been that those species
most affected by chronic parasitism would have evolved to make states of

health more visible to prospective partners: long-term states of health mattered more in such species, we argued, and the partners would consequently be evolved to be picky and to watch for signs of resistance and susceptibility. Characters naturally apt to reveal health would be seized on by preference and become exaggerated. Thus the advertising sex would evolve to be 'showier' via bright feathers, horns, or whatever, and would display excesses that only a healthy animal could afford to display. Choosers in species having less trouble with chronic parasites would be less interested in such health-revealing displays ('sanimetric' displays as I called them in a later paper (Chapter 17) so that fewer costly and revealing characters would evolve. Costliness would make the displayed health difficult to bluff and the ecology and genetics of cycling would ensure that health was substantially heritable.

Looking for the predicted association across species Marlene and I had found and published a weak but significant correlation. Perhaps mainly due to an independently mounting interest in controlling for 'correlations due to phylogeny' in comparative studies, a suspicion soon arose, however, that ours might after all have arisen by chance and a drive commenced to do everything again with improvements of statistical technique and also with supposedly more 'objective' estimates of degrees of showiness. In the outcome it was claimed that our approach had indeed produced spurious results and that with better technique our effect went away.[5]

All the new studies as they accumulated have been reviewed several times[4] and it is not my intention to go over them in detail. Marlene rebutted the first adverse report;[6] then she and I responded again in print to further criticism later in the same year.[7] One critic had read our paper so superficially that the point he made in his letter, which he supposed to give the *coup de grâce* to our hypothesis, instead strengthened it by a very plausible explanation of a point that had puzzled us,[8] as will be shown below. The current situation is best summarized in recent papers by John,[9] who seems to have been almost alone with Møller[10] in keeping in sight what Marlene and I originally wrote. The present situation looks very bright for both the across-species comparisons and for effects within species, the latter being especially supported, with numerous extensions to the ideas, in a veritable swarm on the parasite–sexual selection connections emanating from Anders Møller.

What is worthwhile to do here is to try to summarize the approach that underlay our work and to clarify more fully than was possible in our notes in *Nature* some points on which we have been misunderstood; nevertheless, the most-thorough countercritique will remain the papers of John. Our publication covered all the bird blood-parasite data we had found for eastern North America by mid-1982. The parasite recording had been done in various years and places in the past but basically all by the same methods.

Marlene and I had been aware of the statistical problem of the non-independence due to phylogeny of the species units we were regressing. We had tried to reduce it a little by using only the passerine birds from the haematozoa tabulations in the literature. From early on our indications were that whether we restricted our set like this or included all bird species, and whether we looked at birds in the West or East of the USA or in Europe, we always found the positive regression slopes predicted by our hypothesis. With the slant yet again appearing in a later tropical study by Marlene,[11] it began to seem true that anywhere in the world if you mist-net a bird, smear a drop of its blood (this can be harmlessly taken I should add) onto a slide, and then search for blood parasites, you are more likely to find the blood infected, and to be infected with more diseases, if the bird is brightly coloured or has other signs of high sexual selection. The extra chance is not great but is demonstrable. I believe even our critics don't doubt the above statement; their attack has rather been on our interpretation of the implications.

Why didn't we ourselves split the data still further taxonomically and why didn't we also try to control for more, possibly confounding ecological factors?[12] Providing categorization and measurements are possible without too extreme an effort, we certainly believe in principle in trying to control for any variables or underlying structure that may be obscuring a target relationship in a statistical study; but all along two problems had to be recognized.

First, the best method for control isn't always obvious. In Oxford, even 7 years after the paper, at the time of Read and Harvey's first critical report,[13] there was no general method for controlling over-multibranching phylogenies. Alan Grafen was working on such a method[14] but it was incomplete and neither Andrew Read nor Paul Harvey,[13] who were to become prominent among our critics, had any better idea for coping with the phylogenetic problem than to use an extension of that which Marlene and I had used in a minor way when we restricted our survey to passerines —that is, to look at each taxon separately.

Second, regarding all the other possibly confounding factors, even those for which the method is clear in principle, there are so many that could possibly be treated that an attempt to quantify them all would ensure that the test would be delayed for years. An example will illustrate this. Imagine that you have carried out an agricultural experiment that uses clonally produced plants laid out in blocks in a 'latin-square' design. The latin square, which ensures that as far as possible irregularities in the field don't favour any particular batch of plants, is generally considered an admirable course for yielding unambiguous results. But even if given clones used such a layout, there are further sources of variation that in principle could be controlled for. The order of planting, for example: I hope it was pre-arranged to have a random insertion order over the whole field and that nothing was done row by row. Even if this precaution was taken, still the planting order actually used ought to be controlled for just in case (maybe the plants were increasingly wilted as you put them in: it is true that the wiltedness is being randomized in the blocks but by chance there may be more initially wilted ones in some block and, as I now come to, viruses or fungi may be more taking their hold in the more wilted plants even as you plant them). In the same spirit and still more important, did you control for exact heights of the plantlets at the time of insertion? Heights are easily measured and we certainly must not assume that these are a case of random variation. Short plants might have a virus or fungus that could spread to neighbours and create patches of reduced yield, immediately plunging us into the same mess of Type I Errors as if you had somehow allowed, in another context, God forefend, the 'patches' of a taxonomic artefact! Of course, all this is not saying that the 'phylogenetic' statistical controls are not desirable, rather that the idealist who tries to think of controls for absolutely everything is likely to be still thinking about his study whilst others by crude methods have obtained strong hints.

Returning to the birds, if and when phylogenetic statistical controls are applied to the whole world's aggregated haematozoa data and there turns out to be no significant connection of haematozoa with showiness, then, of course, our result, as we presented it, will have to be abandoned although, even then, an interesting questions will remain about why in Europe and the Americas so many slopes of regression in studies by us and others had the same sign. Such an inclusive test, however, hasn't been done, and in fact the most careful work yet, using phylogenetic regression on roughly the aggregate set we had tackled, indicates that in the NE American birds the showiness–parasitism association is still significant at about the level we originally claimed.[15] This most recent study controls a much wider range of the potentially confusing factors than any previous study had done. It vindicates our decision that the original test was worth publishing in spite of our doubts about problems of phylogeny and other matters. Other careful studies have also confirmed the general validity of our 'cross-species' approach. Most recently the inexhaustible Møller and his colleagues have further shown how much stronger and more significant correlations of our type can be obtained[16] when measures of the capability (and activity) of the immune system are used instead of attempting to quantify the parasites directly, so that the test will reflect the whole spectrum of parasite attack as the immune system perceives it (as opposed to just looking at very imperfect direct estimates of five or so species, some of them believed fairly harmless). In short, verifications for both our intraspecies and interspecies correlations[17] seem to be accumulating steadily. As pointed out by Møller in 1998 and 1999 other possible explanations, some of them suggested or emphasized since our paper, continue as possible explanations; however, at the very least we can claim our paper began the revelation of an important connection between parasitism and sexual selection that had not been noticed previously.

Bird colours and songs are obviously very labile in their degrees of development within genera and consequently are not characters favoured by taxonomists except perhaps at the lowest (subspecific) level. In the absence of obvious rationales from ecology one has to wonder why one bird species has a high level of showy characters when another has a low one. Some factors are fairly obvious as Darwin and Wallace already knew,[18] and these can now be confirmed statistically.[19] But they leave much unexplained. Phylogenetic correlations must, of course, be admitted; without them phylogenies couldn't be constructed. But it is not obvious that we should calmly accept, say, that colourful genera (e.g. of finches) are usually large while dull genera (e.g. of dunnocks) are small, or that by chance the ancestors of large genera just happened to be more prone to parasitism.[11] Are cryptic species of dunnocks perhaps being overlooked by bird taxonomists just because dunnocks are all dull whereas obvious colourful characters have already easily differentiated another group, as in our redpoll, twite, and linnet? I doubt it. A better guess seems that finches, made evolutionarily more 'buoyant' in the choppy sea of their parasites, and therefore buoyant also in their ecology (because of their high sexual selection), might be speciating faster.[20] In short, even if it had been true (which is wasn't) that our finding couldn't survive a phylogenetic regression analysis, I would still claim that our simplistic interspecies correlation pointed to real and interesting phenomena. And given the studies done since its publication, I still in fact claim that the phenomenon that showier birds have more parasites has never been better or more plausibly explained than by our hypothesis.

A more misguided criticism of our methods than that concerning the phylogenetic background, and one seeming prominent in the temporary rejection of our result, was directed at our method of assessing showiness. What Marlene and I agreed was that anything contributing to an impression of a high level of sexual selection in a species, with the displays informing about health, would gain marks for the species. Because even Wallace and Darwin had discussed the idea that there might be a trade-off across bird species in the use of visual showiness versus song in displays, we decided to assess male song separately. We also assessed visual showiness of the two sexes separately. My vague expectation underlying both decisions was that each *species* might evolve, depending on its level of parasitism, towards characteristic norms of investment in some kind of display of health, some display sufficiently expensive for unhealthy individuals to have difficulty in imitating it. The display could be song, colour, or some energetic performance; it could be expressed in male or female or partly in each. If the general idea was right, the indications were that the males would normally be the 'sanimeter' (or 'health-index' sex), but that in supposedly monogamous species, with equal sexual contributions to brood raising, a situation of the sexes assessing each other equally for quality seemed very likely. As for reversals, in which the females became 'sanimeter' (as seemingly she is in phalaropes, for example), there happened to be no instances of these among our passerines (not surprisingly because they are very rare). Nevertheless that case needed in principle to be provided for in our approach.

Scientific objectivity in approach and repeatability in results being rightly so venerated, it may seem surprising in all this that I have not yet mentioned any misgiving about the validity of our subjective assessments of showiness scores in the three categories above. In fact, I at least have no misgiving; we are unapologetic, even a bit aggressive. We think (at least I do) that our measures are even now probably better than anything we could have produced out of months of classification of feather tracts and summations of subscores. All that seemed to us clearly essential was that it should be Marlene, or someone else who also hadn't looked at any of the parasite prevalence lists, who should do the assessments.

I still find objections to our procedure in this regard strange and believe our method to be probably (given the present state of understanding of how a bird's perception works and what focuses its interest) superior to more 'objective' ones. Independent replication of our scoring by others, as demanded by Read and Harvey,[21] seemed to us mildly desirable but not very important. The big question is whether the hypothesis we were led to launch stands up to investigation on all its fronts as these develop in the future. It seems to be standing up; meanwhile let others choose their own preferred ways to check the various aspects that are arising. If replication of our study is insisted on, I note again that the most careful study yet, that of John, did use the scores 'objectively' derived by Johnson (though not pursued by Johnson to test our point) and yet John came back ultimately with correlations very similar to our own.[15]

In the case of sexual selection, in short, we doubt the superiority of 'objective' methods because of doubt whether the information content of combinations of measured items can distinguish nuances that the human senses almost automatically detect and find interesting. In my case faith in this is based on a belief that Darwin was right about a deep parallel existing in the aesthetic sense in animals and that of humans; if such a parallel did not exist we wouldn't find bird coloration and their displays beautiful to us and birdwatching wouldn't be the world-dominant naturalists' pastime that it is. To focus on the songs first, unless a lot of preliminary work has established the nuances that are appreciated by birds in their songs, and unless it is certain that these are being noticed and measured by suitably sophisticated sonograms, I will continue to bet on a more meaningful assessment by a human scorer who has been told to treat bird songs as if they were operatic arias, and to assess them in terms of a combination of their difficulty for a singer and their intrinsic beauty. Such a course seems to me to have a much better chance to mesh with the birds' own

appreciation than does an 'objective' summary statistic from crude
sonograms, such as those that appear in field guides. Those are fairly
attractive to look at but, even just glancing at them, we can see that they
fall far short of rendering the detail and complexity of the songs we hear in
the field. In the nineteenth century art books were mainly illustrated in
black and white but the loss of colour hardly ever had the effect of mixing
mediocre painters with masters. If all colour can be foregone and we can
still recognize the master, why worry too much about subjective
impressions of birds, at least in first approaches? If you want to go very
deep, that may be another matter.

When Marlene was studying the mate choice of field crickets for her
thesis in Michigan she discovered that greater age of a male was the most
significantly attractive character to females listening to the songs at a
distance but she could find no clue whatever in the sonograms as to what
was telling her females the singer's age. These were sonograms that the best
resources of the Museum of Zoology in Ann Arbor had enabled her to
record—and the museum was no novice at sonograms, having used them in
evolutionary and taxonomic study of cicadas and birds as well as various
crickets. Yet again, considering human acoustic discrimination, is a simple
sonogram of the type usually published for birds able to distinguish a first
class from an average violin? I doubt it. Is it unreasonable to suppose that a
female cricket engaged in a choice extremely important to her in creating
the best offspring she can—as taught to her by aeons of natural selection—
is capable of a discrimination of the same degree? All this also applies to
birds. I was therefore disappointed, but not very surprised, to learn that
Read and Weary, using scores they derived from the sonograms in field
guides, were unable to obtain a significant association of song with parasite
prevalence across species, whereas we had.[22] Our statistically significant
result had used scores provided by a field ornithologist, who had always
been extremely appreciative of bird song, and who was familiar with almost
all of the specific songs (he referred to recordings when he was not
familiar). General evidence that exceptional ability in song reveals health
and vigour, and that females take acute notice of song in mate choice,
continues to accumulate rapidly[23] and sometimes parasites are known to be,
or can be presumed to be, involved in the expression. But although this
accumulation adds strongly to conviction as to a song's importance in mate
choice, from our point of view it adds only a circumstantial probability that

because we were right about the correlation of plumage style (which reflects health) with parasite pressure, we are likely to be right also in our claimed similar correlation for song. It does not directly address the shortcoming of our comparative study.[24]

No doubt sometime it will be desirable to have more objective and quantitative measures and these may give us more detail about mate choice, but generally that is a long way ahead. As an example of the potential complexity of bird song, it is known that a starling (like various other birds) sings two songs at once, one with its right syrinx and one with its left, and the songs are quite different in style and content.[25] Ultimately for sexual selection studies we may need, for such birds, measures of the complexity, cost, and specific function of each 'half' as well as of how (and why!) they are combined. Meanwhile, however, a human's intuitive assessment via even vague expressions of appreciation—such as 'startling', 'varied', 'a lot of surprises', 'not very melodious', or even 'seems almost singing two songs at once'—followed by some subjective numerical estimate comparing this performance with other birds probably do as well to assess one starling's view of another as anything that can be derived by callipers from sonograms or that can emerge via parameters estimated by time-series analysis. In summary, because human appreciation probably intuively offsets much irrelevant variation and is designed to pick out structured complexity wherever it occurs, including in black-and-white renderings of colourful master painters, a correlation under our hypothesis may well emerge more readily, at least in the early stages, from 'subjective' than from 'objective' assessments.

If, as was the case with both our 'song' and our 'brightness' scores, the scorer is truly ignorant of the parasite data, statistical significance of such correlation cannot be due to subjective adjustments of scores to suit the hypothesis. Indeed if the doubters are right and the above claim of human intuitive assessment being better is wrong, then the same doubters should be even more impressed than we were when our method finds a correlation. If, of course, the modes of human sensual assessment are very different from those normal for the animals, it is quite possible that our phenomenon could exist but fail to be detected by human appreciation. This might well have happened with fish where a result similar to ours was reported shortly after our paper came out but then later was found insignificant (and seemingly this was not contested) when the phylogenetic analysis replaced straightforward regression:[26] I had been pleased, of course, to see the positive result but wasn't greatly surprised at its later disappearance. The world of fishes, with their usual heavy dependence on smell, is so different from ours that it is not at all surprising that visual assessment alone doesn't give the result, especially when no attempt was made to exclude the fishes of turbid and deep waters in which smell and perhaps touch and sound, or even electric discharges, are likely to overrule sight almost completely. No attempt seems to have been made to look at that correlation again while restricting the survey to surface and clear-water fish, and applying more ecological controls. Because in clear waters fish are colourful and obviously do look at one another, I strongly suspect that such a restricted survey would show our correlation.

   Critics are, of course, justified to be concerned about how strict we had been in ensuring no feedback from me, who had seen at least the first few parasite tables—enough of them to make me enthusiastic to try the test—while Marlene was scoring. We were starting in a tentative way so we certainly weren't perfect in the advance planning. I think there were perhaps three or four birds where Marlene expressed doubts to me about a score assignment. How was she to assess visual showiness in, again, our dear old starling, for example—what about all that iridescence, the regular speckling that you see if you are close enough? This was one case where there certainly was some back and forth. Possibly the starling mattered more to Marlene personally than most others because she liked starlings and had kept one as a pet: almost as I write I see her starling-imitating arms as she complains to me about the difficulty. I gave her some opinion, I forget what, and I forget whether she made any change. As to the colour we knew that birds generally see further into the ultraviolet than we do and faint iridescences on dark birds at the violet end of the spectrum are suggestive of more to come in that direction—perhaps there was even a brilliance to starling eyes as of a morpho butterfly to ours if the ultraviolet colours are perceptible (some other species of starling are largely glossy blue or green even in the visible range). But what might be there in the ultraviolet and might be seen was at the time pure speculation.[27] If Marlene did make a change after our discussion of iridescence it may have affected her scoring of the grackles, the crow, and one or two other 'glossy-black' birds. In giving my opinion, I would have tried to divorce from my mind anything I knew of the starling's, crow's, or whatever's parasites but I may not have been completely unbiased. Anyway, there cannot have been more than half a dozen in the 109 species we surveyed where such a factor could have had any effect, and even for these the discussion would not have changed Marlene's first estimate by more than one point in our scale of six.

I also remember some discussion, after we had started, about non-utilitarian oddities of morphology that contribute to appearance and displays, such as long tails, crests, and wattles, and also behavioural enhancements of displays. We unfortunately omitted to mention in the paper that we had decided to include these in making our scores; our excuse can only be that it is very hard to mention all that you may initially intend to when writing a *Science* paper: I suspect, in fact, that it was there initially and at some stage had to be cut. I was in favour of including such characters in the analysis and persuaded Marlene; I think she initially objected on grounds of the difficulty of definition—for example, there were too many 'almost' crests in our birds. Intermediate or not I thought such details must be seen by the bird and potentially could convey an important point, so that including them might bring us closer to the display's true cost and the viewers' interest; if a bird's 'hairstyle' was noticeable to us, we should attend to it just as we would to an unusual hairstyle in the street— our notice was acknowledgement, whether we liked the style or not, that someone had made an effort and had forced us to look at him or her. This directive concerning morphology and behaviour was agreed upon, as I say, after we had started the work but not after we had passed beyond our first suggestive but non-significant data set. The decision may have caused small though widespread changes, mostly one score point, throughout Marlene's table. The yellow chat was one bird that had particularly glowing accounts devoted to its behavioural display: I vaguely recall this may have raised it by two points but, if it did, it was probably the only case; though had it been in the set I would probably have also given a similar two extra points for the amazing song flight of the European skylark.

I took it as an extremely positive starting factor that Marlene knew most of the birds she was assessing from her personal experience (she admitted not knowing, however, a few birds—including, for example, the yellow chat). Her knowledge came from her own birdwatching and help she had sometimes given others with netting and ringing. When the doubts over our results began at Oxford, Andrew Read and Paul Harvey persuaded six members of the Department of Zoology to rescore the relevant North American passerines in our fashion on a scale from one to six. None of the six new scorers, I was told later, had more than a fleeting acquaintance with North American birds and all of their scoring was done from the pictures of one field guide. I don't criticize this if better experts were lacking, but the unfamiliarity of these scorers with the birds as they appear in the field and non-use of their displays easily accounts, I think, for the result that, while all scorers obtained the same sign of regression as we had, only one of their regressions was significant (as it happens just 1 per cent more significant than ours). It had also turned out that the re-run of our own regression of parasites on the male brightness score (using our original parasite data set, which was a little less comprehensive than the later one Read and Harvey had discovered), provided a stronger correlation than any of the new ones by a small difference. This fact puzzled and worried me a little at first but when, at my request, Andrew Read kindly showed me the lists of new scorings (minus the names of scorers) I felt relieved. Even from my own knowledge of the birds in eastern North America, it seemed clear to me that Marlene's assessments of showiness agreed better with the birds as I knew them than did any of the new score sets.

To mention just two cases, the boat-tailed grackle was scored by most new scorers as less showy than the common grackle and by none of them as more showy. This was on the basis apparently of the latter being portrayed with bronzy purple rather than with blue-green iridescence in the field guide. How to give marks to faint iridescences is tricky, as I mentioned above, but for now I will simply admit that to me, too, blue-green does seem slightly less 'interesting' a colour than bronzy purple does though I am not sure why—perhaps because it is more unusual. The recent result of Bennett and his colleagues[28] shows that starlings may have the same opinion as we do, seemingly showing a preference for what might be called a 'UV' extension of 'bronzy purple' (see their figure 3). The major point here, however, is that in attending only to the colour difference, the six new scorers had been neglecting what was to us a more striking feature— the male boat-tailed grackle's very distinctive, long, and boat-like tail. We had rated this grackle a point higher than the common grackle for that. Likewise the mockingbird was scored uniformly lower by the others than by us for its brightness. This difference might be expected for assessors who had seen a grey thrush-like bird standing on a page of their field guide but would be less expectable in an observer who had seen the mockingbird's white wing patches being flaunted in flight.

But I sense that any more of such point-by-point defence of Marlene's scores will irritate readers and dispose them against us: it's better to offer a prediction. Marlene doesn't esteem herself highly as a bird person although she is clearly more experienced in American birds than I am. Assuming expert American birders are still available and willing, and who still don't know about the parasite situation or about all the fuss our result aroused, I predict that if the scoring is done yet again by American birders (another set of six perhaps) and they are instructed to include in their assessment every kind of visual showiness, just as each person appreciates it in recognizing the bird and assessing its beauty, then a significant average correlation not less than ours will be found once again. If they are indeed more experienced birders than Marlene quite probably it will be higher.

In this line it was reassuring that Read and Harvey's new scorers obtained a statistical result at about the same level as ours when testing the European birds with which they were much more familiar. That significance was retained in a sign test on the regressions within taxa but, surprisingly at first, it diminished when they cut out species that seemed to have been inadequately sampled. It was here that the experienced London parasitologist Frank Cox came to our aid with a point about the quality of attention that different species received that we had already toyed with ourselves but hadn't dared to suggest. He wrote 'The explanation for this [the better correlation of "brightness and parasites" among rarely than among commonly trapped birds] is simple—and here I plead guilty—for when examining blood smears, one takes more care if they come from rarer birds because these are more likely to produce observations justifying publication.'[8] The picture from Read and Harvey's results combined with Cox's point suggest that for the European birds where the six scorers were scoring birds that they knew well and the parasitologists were examining blood smears with maximum care, our prediction stood up to all the extra attention very well.

Of course it was disappointing for us that all the North Temperate regressions, based on whatever the scorers had done, didn't show the right slant (as virtually all did) nor also always a statistical significance to that slant. But we had always expected the correlation would be low. The magnitude that we found before applying any controls had surprised us;

there had been good reason to expect the predicted correlation to be almost undetectable if we studied just one small group of parasites. If it had been high it would have meant that in the bird's blood, even though it is just one of the very numerous niches of a bird's body that were inhabited by parasites of the right general type, we had discovered the parasite group that was the major motivator of all passerine sexual selection. With haematozoa never having seemed ideal chronic parasites for our theory anyway (this mainly because of their dubious lags, their usual though not universal low virulence,[29] and, above all, their high generalism) and with no sign of a connection ever having been noticed before either by parasitologists or ornithologists, indeed rather the opposite,[30] to have made such a discovery would have been astonishing.

Weak for any single species but strong for many: even under this difficult presentation, why wasn't the connection noticed earlier? One curious historical fact suggests that it may, actually, have been very near to being recognized 70 or so years before I started looking for haematozoa tables in the bird library in the University of Michigan Museum and that only a unique historical accident may have prevented the step being taken then. In June 1910, a tiny steamship, still fitted with masts for sails when needed, ploughed down the Atlantic from England bound for Cape Town. Each day before dawn, with no space in his cabin to work seated, and therefore standing to write (having, moreover, already even before this spent some hours heaving coal around in the ship's coal bunker to keep himself fit), one of the passengers, Dr Edward A. Wilson, was hurriedly putting together a paper on his work of the previous year concerning the effects of a caecum-infesting nematode on the health of red grouse in Scotland. In his paper he noted the effects of the parasite on parts of the bird very distant from the main seat of infection, the caecae. He claimed that a severely infected bird was easy to recognize by the reduced feathering of its legs (such feathering, by the way, is a principal family-level character of grouse within the wider group of chicken-like birds) and also by an unusual proportion of broken and malformed feathers all over the bird. He gave his belief that the worm infestations are important not only in red grouse biology but for their ecology, and in particular that they may even be responsible for the mysterious cycles in abundance that affect them on the grouse moors. This in fact had been the job starting him on this topic: engaged by a federation of British grouse-moor owners he had been set to find out for them what was causing the dramatic cycles in grouse abundance and how they could be prevented. Finally arrived at Cape Town, Wilson put his manuscript in the post to the editor of the *Proceedings of the Zoological Society of London*. He reboarded and steamed onward with his ship, now with an easier pre-breakfast routine though doubtless he was still heaving the regulation piles of coal to keep fit . . .

Who made the strange exercise regulations on the ship and where was it bound? Robert Falcon Scott made them: Wilson, Scott, and others of his team were on their way to Antarctica hoping to become the first to stand at the South Pole. After many adventures, including a side expedition organized by Wilson himself to gather materials for another theory (in this case wrong) about the primitivity of penguins, which involved him in enormous dangers, five of the Scott team did reach their ultimate goal and found, as is well known, that they had been preceded there by Roald Amundsen and his fellow Norwegians by a few weeks. Also well known is the story of the return and how, as the last survivors, Scott, Wilson, and Lieutenant Bowers died in a final encampment only 11 miles from the key depot where they could have found supplies and shelter.

Accompanying Wilson's paper in the Zoological Society's *Proceedings*[31] were some unexplained watercolour illustrations of ptarmigan. The paper gave no mention of these or of ptarmigan at all. The paper had been edited and published during Wilson's absence and at a time when his fate in Antarctica was still unknown—he was indeed at the time still alive. The fine watercolour plates he had provided seemed intended to emphasize the fleshy parts on the head (which, though less specific than the feathered legs, are again very characteristic of the grouse family). It is these illustrations that suggest to me that Wilson may have had plans to continue his work on the effects of parasites on the health and ecology of grouse and possibly to add further notes on their effects on display characters such as wattles. At least, it seems he had already painted illustrations highlighting what were in effect other easily visible signs whose loss of brilliance indicated infections. Possibly he had intended to mention these in the paper that was accepted but had no time for it due to his difficult voyage on Scott's ship. He may have sent off from Cape Town, to editor Ogilvie-Grant in London (or may simply have left not too well explained with his wife in Britain) a set of both documents and paintings relevant to his work.

Another strange coincidence involving the journal leads me to speculate a further connection between Wilson's plates and another paper on a related issue that came out in the same year, 1910—in the same journal, although what the nature of the connection is I can hardly guess. In this paper, H. B. Fantham,[32] a zoologist of whom I managed to find out nothing further than I give here, reported his studies of the effects of coccidiosis—a protozoan disease—on grouse, chickens, and pigeons. As can be seen in the quotation from his paper that I have set at the head of this chapter (where he writes specifically about chickens), Fantham, like Wilson, was emphasizing the effects of the parasite on visible yet 'remote' characters. In a kind of counterpart to the feather signs that Wilson emphasized, Fantham gave more attention to effects on fleshy parts. As with Wilson's (but not so far as I know for any reason of its author's death) this paper seems also a very isolated piece of work, and after the two came out in that one volume, a strange neglect began that was to last 60 years, right up until when, in the 1970s, poor sales of chicken carcasses with pale legs in the newly invented help-yourself supermarkets stimulated some direct biochemical and physiological studies on carotenoid metabolism in gallinaceous birds, comparing healthy birds to others that were chronically parasitized.[33]

Here is a glimpse of British history through worm-lovers' spectacles: our best horn-rimmed 'distance' pair shows us grouse moors and their 'damned bad' seasons, the huff and puff of tall men in hunting hats and 'plus-fours', all of it long ago—an almost Sherlock Holmes kind of scene; next, through our multifocals of the 1970s, we peer instead into the new freezer bins of a supermarket and we make our choice amongst the bare, ready-plucked chicken bodies. With luck, however, if we glance up from this task, we may still see grouse moors in the background: the men out there now are all in Barbour jackets, wearing 'green wellies' instead of the plus-fours, and they are fussing and confederating more about 'right-to-roamers' than about bad seasons. But still, quietly in the background out there, Edward Wilson's start on the practical issue of cycles is at last being brought to its proof.[34]

Back into the older period, however, and re-examining those two papers that came out in the same journal in 1910, I repeat that I cannot avoid a sensation of a near miss of a discovery of parasites as a major general influence in sexual selection in the early 1900s. It is one I conjecture might have come fully that whole three-quarter century earlier had Wilson—artist, naturalist, ecologist, doctor, and, most unusually of all, during his last years of life, parasitologist and expert skier—not received (twice) the fateful invitation to accompany Robert Scott to the Antarctic.[35] Perhaps it was the skiing that particularly recommended him; it came from Norwegian experience because he went there, of all places, to fight down a chronic disease of his own, tuberculosis.

How inevitably, how tragically, this man shaped up as Britain's ideal naturalist for Scott's two expeditions! Looking back on his story I cannot escape a thought of how easily Charles Darwin, too, might have been lost in some similar way—finally endangered perhaps more in his case by his own bravado on the Andes or in war-torn Argentina, less through the mistakes of his leader. But such seeds in the wind, one way or another, seem to have been so many exploratory Britons, sailing, enquiring, hoisting flags, literally everywhere in the world[36] in those days. Even as I continue to reserve my judgement about the importance of accidents for evolution's main achievements on our planet, such historical cases (especially the thought of those last dozen miles that Scott and Wilson might have but did not attempt to struggle to reach their storage depot, and then the thought of what might have happened to the theme of 'parasites and sexual selection' if they had reached it) incline me to allow quite a substantial role for accidents in the evolution of evolution history.

Wilson had a fair way to go before a general idea and its focus on display characters would be apparent, but the coincidence (was it?) with Fantham was a start. What if they had put their heads together? By themselves, Wilson's caecal worms and the mostly caecal coccidia that Fantham studied were, as far as the story went, either single-species or single-genus systems: as with Marlene's and my start on a small set, the haematozoa, the cards were at that time, as now, stacked against any broad support for a sexual-selection idea. What was really needed to make the matter plain earlier on, I like to fantasize, was something more magical.

Imagine that Röntgen besides his X-rays invented binoculars for ornithologists, using some other peculiar ray—let us call it the P-ray. Through the binoculars the ornithologist would see every kind of parasite but no flesh nor bone nor feather to surround it. Even viruses, inside their animals, would be visible. In that case for sure, I fantasize, the discovery of the connection of animal showiness with proneness to parasites would have

come much sooner. Through the 'P-ray' binos the birds as birds would still
be visible in a shadowy outline. A fuzz of bacteria and viruses would to
some degree fill out all fleshy parts but would add a particularly sharp
outline to all that coiled 'internal exterior' of an animal that makes its gut.
Other interior parts would also be enlivened here and there in the P-ray
view with heavy and forceful lines, these being the sinuous or leaf-like
shapes of the worms and flukes lying embedded in various tissues and
organs. Finally, around the level of the skin, an exterior speckling would
again be evident due to all the fleas, lice, and mites that are crawling about
or plugged in on this true exterior. Mites yet again might also add faint
outline not only to another 'internal exterior', the lungs, but also to
feathers; and amidst the latter their shadows would mingle with serried
ranks of yet other elongate hyphens of parasitic life—some of the more
specialized lice that secure themselves most of the time between the barbs
of the larger feathers. All these external types together sometimes would
well outline the bird's plumage. The main point here, however, is not the
Rembrandtesque sketch of the essentials of a bird etched by all its parasites,
rather the two points emerging from comparisons of different kinds of male
birds' loads that a birder could now easily examine using the wonderful
instrument. Through the binoculars he would quickly notice how the
second-rate, shabby peacock (as seen by ordinary eyes), for example, would
be showing dark with his load while by comparison, on the biggest lawn, a
first-rate cock, the rock-star squawking darling of the whole mansion's
garden and bearer of the most immense and multi-eyed train, would show
up (at least early in the breeding season, before virile efforts wore him
down) rather uninterestingly faint—a much less well-stocked zoo of small
troubles. A second point that our birder would notice would come as he
swung his gaze from one species to another. However sparsely the body of
that first-class peacock was occupied, it would still (etched as if by many
kinds of artist's tools, these reflecting the many differing kinds of parasites
within) be sketched more richly than any of the males of related, dull-
plumaged, and un-habitually polygygnous species: male chachalacas,
partridges, and quails, for example; while still less varied in comparison
even to these would come your dull dunnock and your lack-lustre lark.

　　Such binoculars, whose magical potential effect on bird studies was
suggested to me by the diagram in Miriam Rothschild and Teresa Clay's
book *Fleas, Flukes and Cuckoos*,[37] don't exist. Alas for that, but something

like them does—the energy and ingenuity of Anders Møller and his allies. In recent years they have been achieving something quite similar to the effects of the binoculars. In particular, it seems to be coming out that, as better and better indices of total parasitism are built up and take in ever-more comprehensive biochemical and physiological indices of activity of immune and detoxification systems, correlations of both the within-species and the between-species kinds (that Marlene and I predicted) appear to be quite dramatically improved.[10]

While this has been going on, another front has opened, improving control for all the other factors known to affect sexual selection. As with the inclusion of more parasites, it would be very disappointing to us if, with these more statistical controls added, we didn't again see a strengthening of the correlations. As already mentioned, a start in this direction by Steven Johnson in Kansas had sharply identified factors that obviously needed to be controlled. But having found them, surprisingly, Johnson didn't use them to retest our hypothesis. When his lead was extended to such control by Jeremy John in Oxford, who also applied his own extensive additional set of controls, not only were Johnson's findings confirmed (plus confirmation of another interesting point due to Andrew Read—an intriguing though still mysterious 'anti-H&Z' correlation applying in cases of overt polygyny; see below) but John also showed that, if these effects were taken into account, the regression with parasites, dismissed by Johnson as apparently too weak to deserve further attention (as by Read also previously), came strongly back into view.[38]

Countering yet another criticism that our work has received[39] during those tough middle-distance days of public opinion against us, a point worth stressing for the present context is that any observation through the magic binoculars that every peacock and peahen without exception had at least some of a given parasite (species X say) living upon it is definitely *not* an indication that species X cannot be in significant coevolutionary cycling with its host and, through this, contributing to the sosigonic selection pressure on its host. Even an observation that most individuals have about the same absolute numbers of species X would be no such indication. Take our own relation with our most-abundant and best-known gut bacterium as an example. Our observations of each other through the binoculars will certainly reveal (amongst many other perhaps unexpected and unattractive details—such study without a viewee's permission won't make one popular) that every single human is carrying enormous numbers of *Escherichia coli*. It is a good example of a symbiont with '100 per cent prevalence' as parasitologists style the situation. Every human in the world carries *E. coli*; indeed, every woman (or man) carries more cells of this species than she carries cells of *H. sapiens*, though fortunately for her appearance the bacterial cells are, of course, individually much smaller than the host cells. Although in most of us *E. coli* seems fairly stable and harmless in its normal human habitat (the lower gut), when we 'catch' a new strain from someone else we usually suffer at least brief enteric upheavals.[40] This suggests that the apparent 'stability' of the normal infestation may be really a dynamic affair involving control exerted by the host. This possibility is only too strongly confirmed when the immune system is somehow weakened: *E. coli* then often joins the other so-called 'opportunist' symbionts in becoming invasive, it enters then into what one could call the 'internal internal' habitats, which even occasionally include the brain. It is then very obviously harmful. In addition, some strains of *E. coli* strains seem able to be dangerous to a considerable proportion of the population all the time: as I write, *E. coli* strain 0157:H7 is appearing as a very dangerous pathogen around the world, the form seemingly reaching us from its ultimate niches in cattle[41] and/or possibly in seabirds.[42] The bacterium as we generally know it, therefore, probably is always under positive control by the immune system in all of us. Hosts with fortunate genetic constitutions with resect to *E. coli* may control all strains at little cost; other hosts may have to spend effort via their immune system to control it, reducing the resources that the system can apply to other challenges.

Suggestively supporting a dynamic situation for *E. coli* in our species, the situation just described definitely exists in pigs.[43] In these a gene-for-gene polymorphism between the host and the parasite had been identified. This means that piglets having the wrong genotype for the *E. coli* strain that happens to infest them suffer diarrhoea, fail to thrive, and may die.[44] The variation is a matter of cell adhesion mediated by the whiskery pili of the bacterium's surface, features whose molecular details depend in turn on plasmids (those small loops of DNA often carried in bacteria and inherited like the main chromosome but also occasionally lost or reacquired like infectious particles). But similar plasmid involvement is probably true for the slighter effects seen in *E. coli* infections in human hosts. In any case all this hardly affects the essential point that in both humans and pigs the bacterium is furthering a genetic polymorphism that is probably in constant change, from one generation to the next, due to selection. Left to ourselves with no medical intervention, some time in the future *E. coli* strain 0157:H7, for example, may come to matter to us no more than most of the strains now endemic in our guts. In short, it certainly cannot yet be said that resistance to *E. coli* is not a factor, albeit possibly a small one, affecting human health. If this is true, *E. coli* and genes against *E. coli* must affect our sexual selection too. Similar arguments can be applied to *Candida*, *Staphylococcus*, the common cold virus, and no doubt to many more parasites, including many out of the hundreds of species that live in us and on us that are not at all well known.[45] All the common species are easily seen to give some people more trouble than others—*Staphylococcus aureus*, for example, is my special and cyclical enemy.

Two other critical points that have come up several times about 'H&Z' are intertwined and can be dealt with simultaneously. One is the idea that, if our hypothesis is valid and sexual selection works against, say, haematozoa, then perhaps it might be so successful that we could expect to find fewer haematozoan infections, not more, in the bird species with the highest signs of sexual selection. Read and Harvey[5] believed this was as reasonable an expectation as the prediction we made and hence said we should have used two-tailed significance tests. In a bizarre twist of the re-assessments of that middle-distance period, we had used two-tailed tests, probably not having given the matter much thought.[46] But this seems to have been forgotten by us as well as others and we responded as if we needed to justify the use of one-tailed tests—tests, which, if we had used them, would have given us higher significances!

The main point against the postulated successful reduction of parasites by sexual selection is that it seems to be very rare for a host's defence to evolve to be so strong as to throw off a parasite enemy completely. Many cards in the game, especially that of the short lifespan enabling rapid counteradaptation, seem stacked in favour of the parasite. The 'life-dinner principle'[47] also plays an important part. Changing ecology—moving into a desert, for example, where intermediate hosts are absent or parasite transmission may be harder to achieve—is far more likely to eliminate a parasite than is any change in resistance connected with increased choosiness for mates or different coevolutionary cycling. But if sosigonic choice in an afflicted species leads to a relative advantage compared with no choice, then this is still enough for choice to evolve. Thus although it is a very interesting possibility that some effect of the kind indicated may occur,[48] I do not believe we can expect a species as a whole to reduce very much its average parasite prevalence (the proportion of species infected) through more intense sexual selection. I would be happier to believe it might somewhat reduce the impact of that prevalence or, adopting another parasitologists' term, reduce the intensity of the infections. But here again the results of Møller suggest that even this may not happen.

If sosigonic effects more powerful than we imagined do reduce the intensity of parasitism enough to reverse our prediction, that, of course, will be exciting, but because such a possibility hadn't even crossed our minds when we started, our use of a one-tailed test (if we had used it) should have caused little surprise to readers. (The reason for our two-tailed test was, I guess, just general scepticism that our idea would work.) It is tempting to believe that a high efficacy of sosigonic control might explain the seeming absence of correlation that Read and Harvey found in the common, and therefore much sampled, European species of birds. This finding was strongly contrasted with the highly significant positive correlation they found in the rare species (see above). Frank Cox's point about detection could be heeded here with the common, therefore 'boring', bright species hypothesized to have their parasitaemia coevolved to be so mild as to be commonly missed, making them apparently similar to the common dull species (which, according to us, are likely to be simply not encountering the parasite). It is also tempting to believe that a similar effect on intensity might bear on Read's surprising finding of a negative correlation of showiness with parasites in explicitly polygynous species; but as yet I don't see either of these interpretations as very likely.

The other point has to do with the fact that ours was always a theory invoking multiple parasite pressures. We made this quite clear in the publication and it has been surprising to see several critical comments and even a counter model based on single-locus coevolution. We explicitly supposed *polygenic* cycling with the polygenic alleles changing their effects from plus to minus and then back as various cycles of various periods swing back and forth, and we were trying to look at such a postulated multi-cycling, multiperiodic system through the window of the particular set of diseases for which we had data. From this point of view, as stated, we were not expecting that characters supposedly expressive of the overall sexual selection would show high correlation with the particular small subset that was seen through our window; nor would we expect a high correlation even if the preceding points about parasite dominance in the arms race were invalid. If it becomes possible eventually to identify a particular facet of sexual selection that is solely relevant to blood infections, then there may be a better chance to detect effects that sosigonic selection has on intensities within the disease set. This idea of multiple signals informing different aspects of health has already received some attention, as I discuss in more detail in the introduction to Chapter 17. Thus I have an idea that looking at bare skin and fleshy excrescences, which seem almost as if designed to reveal blood problems, may help in this direction, but, again, I am not very hopeful.

After saying all this I have to admit that there was one way in which we did imagine a regression of opposite slope might arise within our general scenario, and we even indicated this possibility in our paper. This point again has seemingly not been noticed by our critics. I have mentioned several times the fact that blood parasites never seemed ideal to test sosigonic sexual selection: they weren't very host-specific and had other reasons to be not very suitable to generate coevolutionary cycles; for example, many sources stated them to be of trivial pathogenicity. Suppose, then, that they actually didn't cause coevolutionary cycles and that there was really nothing to be gained with respect to these particular diseases from picking the mate who seemed most resistant (because there was no heritability): how could any regressions still arise? A reasonable possibility is that the blood diseases might be non-cycling themselves but might still be stressors for the animals they infected. As I argued for *E. coli*, a generalist parasite of humans, they might necessitate activity of the immune system to keep them in check and thus might reduce resources of that system for facultative responses to host-specific and genetically cycling diseases—that is, to those diseases for which mate choice did matter. Thus a species more prone to blood parasitism due to its environment of biting insects could be the more thrown back onto its innate defences against the challenges of the cycle-prone chronic disease. The positive correlation that we found in our paper would then be paralleling another stronger one applying to certain cycling but as yet unknown diseases.[49]

This idea would move us away from the positive correlation; but could a negative correlation ever arise? Cases of the opposite regression—that is, lower prevalence associated with higher showiness—might be explicable by the theory for very trivial generalist diseases (and, it seems to me, only for these) in the following way. Prevalence of the trivial agents could correlate negatively with prevalences of either the serious generalist stressors or/and the serious and cycling specific diseases. The reason would be that raising the level of immune activity in response to either kind of serious disease might have the side-effect of a nearly absolute cross-resistance to the trivial disease. This resistance would be especially likely if the trivial disease had characteristics in common with the more serious ones, making it susceptible to similar defence. This is actually what we suggest in a footnote in our paper for the 'opposite' regression of *Toxoplasma* in the birds we studied: contrary to the pattern of all the other diseases, the brightest species were less likely to have *Toxoplasma*.

The protozoan *Toxoplasma gondii* is one of the most extreme generalists known among all vertebrate parasites: it has been found in blood and tissues in a wide range of homeotherms. It is, however, rarely implicated in manifest disease except in individuals with an impaired immune system (as in AIDS victims) and also in its definitive hosts, which seem to be cats. Thus although I am generally reluctant to believe any parasite to be completely harmless—indeed it shouldn't be called a parasite if it is—it seems unlikely to be coincidence that out of the six diseases that our surveys covered, the one usually correlating most in our favour was *Plasmodium*, bird malaria, a fairly common disease and the one of the set we studied that is by far most often stated to be virulent in birds,[29,50] while at the other extreme the one correlating most 'against us', *Toxoplasma*, was the rarest, the least specialized as to hosts, and the one always treated by authors as the most trivial. Possibly something like the contrast of the *Toxoplasma* and malaria could be showing up again in another puzzling case where two Scandinavians working on arctic charr in lakes found most parasites rose in prevalence in the most brightly coloured fish and with several significant positive correlations.[3] This happening within a population is obviously against our prediction. Most of the parasites concerned here, however, are ones that are believed very mild in their effect on the host: the one species that is notoriously detrimental was the one that did show a significant negative correlation, at least in the female charr. Thus, as if the resistance/display system had been concentrating on this parasite alone (which actually I don't believe), this one fulfilled our prediction.

My last counter-comment is one that again has been made much more thoroughly by John[51] and it is directed to critics who have treated of our argument as if we had excluded combat as a factor in sexual selection. Both in the text and the footnotes of our paper it is made so clear that we didn't exclude it, so that it is again hard to see how this idea arose. Although we referred to 'choice' throughout much of the paper, and in some cases used examples that can only be interpreted in terms of females directly inspecting and choosing males, we clearly mentioned cases of female 'choice' mediated via male competition. If we didn't greatly emphasize the combat factor this was probably because it seemed too obvious that picking victors of fights would be a good way of finding healthy males.

The interesting point about choice via male competition is that in waiting to be mated by victors instead of choosing males directly, the 'run away' part of Fisherian sexual selection is eliminated. Fisher himself pointed this out,[52] a fact we also mentioned in our paper. It may be that the very highest exaggerations of sexual characters disproportionately occur in species that have recently evolved from displaying in leks in which all the males see each other and in which fighting is important, to displaying in 'exploded' leks where fighting is reduced and females inspect males individually and sequentially. Peacocks are an example and bower birds another: these situations have a perhaps unique potential to compound Fisherian and Zahavian sexual selection.

The Zahavian element ensures cost; this cost in turn ensures that such mixed cases stand within our scenario. If parasites prove as important as preliminary data suggest they may be for such species,[53] then they may mediate a kind of selection, which, with my usual dead hand for neologisms, I might try to call 'zufishamian selection'. Made out of Fisher, Hamilton, Zahavi, and Zuk (and perhaps sounding with repetition better than it looks) I present the word not in serious hope to see it adopted but instead hoping it may scare evolutionists into accepting a lesser evil— 'sosigonic selection', my simpler and slightly differently applied adjective, which I'm sad to admit, for all its careful and classical (New College) pedigree, I haven't even persuaded Marlene to like as yet.

References and notes

1. H. B. Fantham, Experimental studies in avian coccidiosis, especially in relation to young grouse, fowls, and pigeons, *Proceedings of the Zoological Society of London*, 708–22 (1910). He is describing the effects of coccidiosis on the appearance of chickens and says they are similar for grouse. Paling appendages and plumage is in the first place due to loss of carotenoid pigmentation. This and its connection with sosigonic selection is by now detailed and discussed in many other studies. For example, T. von Schantz, S. Bensch, M. Grahn, D. Hasselquist, and H. Wittzell, Good genes, oxidative stress and condition-dependent sexual signals, *Proceedings of the Royal Society of London B* **266**, 1–12 (1999); G. A. Lozano, Carotenoids, parasites, and sexual selection, *Oikos* **70**, 309–11 (1994); B. Kouwenhoven and C. J. G. Van der Horst, Disturbed intestinal absorption of vitamin A and carotenes and the effect of a low pH during *Eimeria acervulina* infection in the domestic fowl (*Gallus domesticus*), *Zeitschrift für Parasitenkunde* **38**, 152–61 (1972); C. W. Thompson, N. Hillgarth, M. Leu, and H. E. McClure, High parasite load in House Finches (*Carpadocus mexicanus*) is correlated with reduced expression of a sexually selected trait, *American Naturalist* **149**, 270–94 (1997); M. D. Ruff, W. Reid, and M. J. K. Johnson, Lowered blood carotenoid levels in chickens infected with coccidia, *Poultry Science* **53**, 1801–9 (1974); G. E. Hill, Plumage redness and pigment symmetry in the House Finch, *Journal of Avian Biology* **29**, 86–92 (1998); F. Skarstein and I. Folstad, Sexual dichromatism and the immunocompetence handicap—an observational approach using Arctic Charr, *Oikos* **76**, 359–67 (1996); P. Yvoré and P. Maingut, Influence de la coccídiose duodénale sur la tenseur en carotenoides du serum chez le poulet, *Annales de Recherches Vetérinaire* **3**, 381–7 (1972); D. A. Gray, Carotenoids and sexual dichromatism in North American passerine birds, *American Naturalist* **148**, 435–80 (1996); G. E. Hill, Redness as a measure of the production cost of ornamental coloration, *Ethology, Ecology, and Evolution* **8**, 157–75 (1996); V. A. Olson and I. P. F. Owens, Costly sexual signals: are carotenoids rare, risky or required?, *Trends in Ecology and Evolution* **13**, 510–14 (1998); I. P. F. Owens and K. Wilson, Immunocompetence: a neglected life history trait or conspicuous red herring?, *Trends in Ecology and Evolution* **14**, 170–2 (1999). See also note 36 in the introduction to Chapter 17.

2. A. F. Read, Comparative evidence supports the Hamilton–Zuk hypothesis on parasites and sexual selection, *Nature* **328**, 68–70 (1987); P. I. Ward, Sexual dichromatism and parasitism in British and Irish fresh-water fish, *Animal Behaviour* **36**, 1210–15 (1988).

3. F. Skarstein and I. Folstad, Sexual dichromatism and the immunocompetence

handicap—an observational approach using Arctic Char, *Oikos* **76**, 359–67 (1996).

4. M. Zuk, The role of parasites in sexual selection: current evidence and future directions, *Advances in the Study of Behavior* **21**, 39–68 (1992).

5. A. F. Read and P. H. Harvey, Reassessment of comparative evidence for the Hamilton and Zuk theory on the evolution of secondary sexual characters, *Nature* **339**, 618–20 (1989); A. F. Read, Passerine polygyny: a role for parasites, *American Naturalist* **138**, 434–59 (1991); S. G. Johnson, Effects of predation, parasites and phylogeny on the evolution of bright coloration in North American passerines, *Evolutionary Ecology* **5**, 52–62 (1991).

6. M. Zuk, Validity of sexual selection in birds, *Nature* **340**, 104–5 (1989).

7. W. D. Hamilton and M. Zuk, Parasites and sexual selection: Hamilton and Zuk reply, *Nature* **341**, 289–90 (1989).

8. F. E. G. Cox, Parasites and sexual selection, *Nature* **341**, 289 only (1989).

9. J. L. John, Haematozoan parasites, mating systems and colorful plumages in songbirds, *Oikos* **72**, 395–401 (1995); J. L. John, The Hamilton–Zuk theory and initial test: a review of some parasitological criticisms, *Parasitology Today* **27**, 1269–88 (1997); J. L. John, Seven comments on the theory of sosigonic selection, *Journal of Theoretical Biology* **187**, 333–49 (1997).

10. A. P. Møller, R. Dufva, and J. Erritzoe, Host immune function and sexual selection in birds, *Journal of Evolutionary Biology* **11**, 703–19 (1998); A. P. Møller, P. Christe, and E. Lux, Parasitism, host immune function and sexual selection: a meta-analysis of parasite-mediated sexual selection, *Quarterly Review of Biology* **74**, 3–20 (1999).

11. M. Zuk, Parasites and bright birds: new data and a new prediction, in J. E. Loye and M. Zuk (ed.), *Bird–Parasite Interactions*, pp. 179–204 (Oxford University Press, Oxford, 1991).

12. We did in effect control for one variable for which we are not often given credit—individual study. This implies a rather weak and mixed control for locality, date, and authorship of the survey. None of the subsequent repeat studies of the North American birds touched on any of this.

13. Read and Harvey (1989) in note 5.

14. A Grafen, The phylogenetic regression, *Philosophical Transactions of the Royal Society of London B* **326**, 119–57 (1989).

15. John (1995) in note 9.

16. A. P. Møller, Immune defence, extra-pair paternity, and sexual selection in birds, *Proceedings of the Royal Society of London B* **264**, 561–6 (1997); Møller *et al.* (1998) and Møller *et al.* (1999) in note 10.

17. S. M. Yezerinac and P. J. Weatherhead, Extra-pair mating, male plumage coloration and sexual selection in yellow warblers (*Dendroica petechia*), *Proceedings of the Royal Society of London B* **264**, 527–32 (1997); Møller *et al.* 1998 in note 10.

18. H. Cronin, *The Ant and the Peacock* (Cambridge University Press, Cambridge, 1992).

19. Johnson (1991) in note 5; John (1995) in note 9.

20. We mention this idea in our joint reply to F. E. G. Cox (see Hamilton and Zuk (1989) in note 7) while a more general complaint about controlling for phylogeny, roughly in the line of 'throwing out the baby with bath water', has been made by Westoby and others (M. Westoby, M. R. Leishman, and J. M. Lord, On misinterpreting the phylogenetic correction, *Journal of Ecology* **83**, 531–4 (1995); M. Westoby, M. Leishman, and J. Lord, Further remarks on phylogenetic correction, *Journal of Ecology* **83**, 727–9 (1995)). As other kinds of interpretation of the cross-species evidence as detailed in our paper, see T. G. Barraclough, P. H. Harvey, and S. Nee, Sexual selection and taxonomic diversity in passerine birds, *Proceedings of the Royal Society of London B* **259**, 211–15 (1995); G. Sorci and A. P. Møller, Comparative evidence for a positive correlation between haematozoan prevalence and mortality in waterfowl, *Journal of Evolutionary Biology* **10**, 731–41 (1997); R. V. Alatalo, L. Gustafsson, and A. Lundberg, Male coloration and species recognition in sympatric flycatchers, *Proceedings of the Royal Society of London B* **256**, 113–18 (1994).

21. Read and Harvey (1989) in note 5.

22. A. F. Read and D. M. Weary, Sexual selection and the evolution of song: a test of the Hamilton and Zuk hypothesis, *Behavioral Ecology and Sociobiology* **26**, 47–56 (1990); P. H. Harvey, A. F. Read, J. L. John, R. D. Gregory, and A. E. Keymer, An evolutionary perspective, in C. A. Toft and A. Aeschlimann (ed.), *Parasitism, Coexistence or Conflict: Ecological, Physiological and Immunological Aspects*, pp. 344–55 (Oxford University Press, Oxford, 1991).

23. D. J. Mountjoy and R. E. Lemon, Female choice for complex song in the European Starling—a field experiment, *Behavioral Ecology and Sociobiology* **38**, 65–71 (1996); D. Hasselquist, S. Bensch, and T. von Schantz, Correlation between male song repertoire, extra-pair paternity and offspring survival in the great reed warbler, *Nature* **381**, 229–32 (1996); K. L. Buchanan and C. K. Catchpole, Female choice in the sedge warbler, *Acrocephalus schoenobaenus*: multiple clues from song and territory quality, *Proceedings of the Royal Society of London B* **264**, 521–6 (1997); D. J. Mountjoy and R. E. Lemon, Male song complexity and parental care in the European starling, *Behaviour* **134**, 661–75

(1997); N. Saino, P. Galeotti, R. Sacchi, and A. P. Møller, Song and immunological condition in male barn swallows (*Hirundo rustica*), *Behavioral Ecology* 8, 364–71 (1997); P. Galeotti, N. Saino, R. Sacchi, and A. P. Møller, Song correlates with social context, testosterone and body condition in male barn swallows, *Animal Behaviour* 53, 687–700 (1997); B. Kempenaers, G. R. Verheyren, and A. A. Dhondt, Extrapair paternity in the blue tit (*Parus caeruleus*): female choice, male characteristics, and offspring quality, *Behavioral Ecology* 8, 481–92 (1997); A. P. Møller, N. Saino, G. Taramino, P. Galeotti, and S. Ferrario, Paternity and multiple signaling: effects of a secondary sexual character and song on paternity in the barn swallow, *American Naturalist* 151, 236–42 (1998).

24. A paper that does address the comparative aspect and uses a more inclusive measure (spleen weight) for assessing species involvement with parasites than we used (haematozoa prevalence) has been seen by me in manuscript. The result gives excellent support for our belief in the importance of song as a sosigonic character in our original paper and is, as so often, due to the inexhaustible industry of the group of Anders Møller. For more detail see the remention of song in the introduction to Chapter 18 and its note 41.

25. L. F. Klinger, S. A. Elias, V. M. Behan Pelletier, and N. E. Williams, The bog climax hypothesis: fossil arthropod and stratigraphic evidence in peat sections from South East Alaska, USA, *Holarctic Ecology* 13, 1–9 (1990).

26. A. F. Read, Parasites and the evolution of host sexual behaviour, in C. J. Barnard and J. M. Behnke (ed.), *Parasitism and Host Behaviour*, pp. 117–57 (Taylor and Francis, London, 1990).

27. Later this was investigated in a series of neat studies and we now know not only that starlings have special reflectant patterns in the UV, they also notice them and choose adversely when they are poorly present, just as with other colours (J. Radwan, Are dull birds still dull in UV?, *Acta Ornithologica* 27, 125–30 (1993); A. T. D. Bennett, I. C. Cuthill, J. C. Partridge, and K. Lunau, Ultraviolet plumage colors predict mate preference in starlings, *Proceedings of the National Academy of Sciences USA* 94, 8618–21 (1997). In revealing sexual dimorphism in cases where not apparent to the unaided human eye, this is an exciting development and has led on to demonstrations that these invisible colours are indeed used for sexual selection (M. Andersson, J. Ornborg, and S. Andersson, Ultraviolet plumage ornamentation affects social mate choice and sperm competition in bluethroats (Aves: *Luscinia s. svecica*): a field experiment, *Proceedings of the Royal Society of London B* 265, 1313–18 (1998). In the blue tit, for example, in which the sexes appear similar in colour to the human eye, the male has a UV reflectant patch covering the erectible feathers

of his crest (S. Andersson, J. Ornberg, and M. Andersson, Ultraviolet sexual dimorphism and assortative mating in blue tits, *Proceedings of the Royal Society of London B* **265**, 445–50 (1998); S. Hunt, A. T. D. Bennett, I. C. Cuthill, and R. Griffiths, Blue tit heads, *Proceedings of the Royal Society of London B* **265**, 451–5 (1998)).

28. Bennett *et al.* (1997) in note 27.

29. G. Valkiunas, Ecological implications of hematozoa in birds, *Bulletin of Scandinavian Parasitology* **6**, 103–13 (1996).

30. R. W. Ashford, Blood parasites and migratory fat at Lake Chad, *Ibis* **113**, 100–1 (1971).

31. E. A. Wilson, The changes of plumage in the red grouse (*Lagopus scoticus* Lath.) in health and disease, *Proceedings of the Zoological Society of London* 1000–33 (1910).

32. Fantham (1910) in note 1.

33. See Ruff *et al.* (1974) and also Kouwenhoven and Van der Horst (1972) and P. Yvoré and Maingut (1972) in note 1.

34. P. J. Hudson, A. P. Dobson, and D. Newborn, Prevention of population cycles by parasite removal, *Science* **282**, 2256–8 (1999); see also K.-A. Rorvik and J. B. Steen, The genetic structure of Scandinavian Willow Ptarmigan (*Lagopus lagopus lagopus*), *Hereditas* **110**, 139–44 (1989).

35. Peter Scott (PS), the son of Robert Scott (RS), the Antarctic explorer, became a well-known wildlife artist with a speciality in painting waterfowl while at the same time he advanced conservation and taxonomy of the group. I have wondered whether his mother—Kathleen, Lady Scott, the sculptor—may have specifically encouraged in PS the talents, artistic and scientific, she knew to have been lost to the world in Wilson, her husband's much admired companion in death. If so, she perhaps has succeeded indirectly in fostering additions to the theory that may have died with him (which I have no reason to think she guessed at) in the following ways.

    Not long after PS's death, Nigella Hillgarth worked as parasitologist's assistant at the sanctuary and worldwide waterfowl aviary that PS set up at Slimbridge in Gloucestershire, England. There she gained expertise that shortly (a) attracted her to the 'H&Z hypothesis' when she heard of it as a mature student at Oxford (which was about the same time as I came there in 1984; see Chapter 9), (b) suggested to her to become my and Alan Grafen's graduate student for her DPhil degree at Oxford, and (c) helped her to help me to confirm that susceptibility to coccidia indeed affects mate choice in pheasants, with a slant appropriate to the 'H&Z' prediction. In effect, via PS's creation of Slimbridge, her work synthesized Wilson, Fantham, and 'H&Z'. When I first met Nigella she knew from her personal experience most of the two men's earlier knowledge concerning visible and remote parasite effects. She told me, for example, how the colour of a duck's bill could change almost overnight if it became sick, and that she and her superior at Slimbridge, Dr John Beer, were often able to spot looming problems in the aviary through such signs.

I like to think that the work of Nigella Hillgarth—parasitologist, 'sexual selectionist', and now aviarist as well in Salt Lake City—became one way (doubtless among many) in which Edward A. Wilson and his unusual start in a line of science that combined natural history and medicine/parasitology did not die.

A convergent Slimbridge thread runs through RS's granddaughter Dafila, who, together with her husband zoologist Tim Clutton-Brock, tested the species-comparative aspect of the 'H&Z' notion as applied again in the father's favourite group, waterfowl (D. K. Scott and T. H. Clutton-Brock, Mating systems, parasites and plumage dimorphism in waterfowl, *Behavioral Ecology and Sociobiology* **26** 261–73 (1990)); however, these two concluded (rather as did Steven Johnson later—see under note 5 and also the introduction to Chapter 17), and much more negatively than Nigella, that the effect of our correlation, if present at all, was too weak to be seen amidst other more powerful influences—these of the Darwin–Wallace type—that acted upon plumage evolution. Well, par for the course from an evolutionist's point of view, I suppose—that is to say, so said in a field where if you win three and lose two you can still fancy you are advancing. The 'Scott&C' study could perhaps profit by being repeated when the data are better, when an extra careful analyst like Jeremy John stands by to help, and when spleen considerations such as have been now highlighted by both John and Anders Møller are available to serve also as overall measures of parasite severity (for references see notes 9 and 10). For what it is worth I mention here that Gene Mesher and I in unpublished work on similar data to that used by Scott and Clutton-Brock previously found clear signs that one prediction of the theory—that more colourful duck species would prove subject to more *specialist* parasite species—was upheld; but then we were at a stage of applying only crude correctives for any phylogenetic correlations.

36. R. Kipling, *Rudyard Kiplings's Verse* (Doubleday, New York, 1939); see 'The English Flag', pp. 221–3.

37. M. Rothschild and T. Clay, *Fleas, Flukes and Cuckoos* (Collins, London, 1953).

38. John (1995) in note 9.

39. D. H. Clayton, S. G. Pruett-Jones, and R. Lande, Reappraisal of the interspecific prediction of parasite-mediated sexual selection: opportunity knocks, *Journal of Theoretical Biology* **157**, 95–108 (1992); R. Poulin and W. L. Vickery, Parasite distribution and virulence: implications for parasite-mediated sexual selection, *Behavioral Ecology and Sociobiology* **33**, 429–36 (1993).

40. D. A. Caugant, B. R. Levin, and R. K. Selander, Distribution of multilocus genotypes of *Escherichia coli* within and between host families, *Journal of Hygiene* **92**, 377–84 (1983).

41. A. Coghlan, Killer strain raises urgent questions, *New Scientist* **153**, 7 only (25 January 1997).

42. P. Brown, Seabirds' dirty diet spreads disease, *New Scientist* **153**, 11 only (22 March 1997).

43. D. R. Baker, L. O. Billey, and D. R. Francis, Distribution of K88 *Escherichia coli* adhesive and non-adhesive phenotypes among pigs of four breeds, *Veterinary Microbiology* **54**, 123–32 (1997).

44. J. M. Rutter, M. R. Burrows, R. Sellwood, and R. A. Gibbons, A. genetic basis for resistance to enteric disease caused by *E. coli*, *Nature* **257**, 135–6 (1975); I. G. W. Bijlsma, A. Denijs, C. Vandermeer, and J. F. Frick, Different pig phenotypes affect adherence of *Escherichia coli* to jejunal brush-borders by K88AB-antigen, K88Ac-antigen or K88AD-antigen, *Infection and Immunity* **37**, 891–4 (1982).

45. H. M. Sommers, The indigenous microbiota of the human host, in G. P. Youmans, P. Y. Paterson, and H. M. Sommers (ed.), *Biologic and Clinical Basis of Infectious Diseases*, pp. 83–94 (W. B. Saunders, Philadelphia, 1990).

46. Our paper seems not to state explicitly that we used two-tailed tests but that we did so can be deduced from the spacing of standard error bars in relation to means in our Fig. 1 and relating these to the levels of significance given.

47. R. Dawkins and J. R. Krebs, Arms races between and within species, *Proceedings of the Royal Society of London B* **205**, 489–511 (1979).

48. As I and Marlene Zuk suggested in our paper; see also John (1995) in note 9.

49. Blood parasites seem to be more host-specific in some groups of birds than in others. Galliforms (pheasants, grouse, quails, etc.), for example, seem to have more specific parasites (G. Valkiunas, *Bird Haemosporida* (Acta Zoologica Lituanica, Vilnius, 1997)). I have already hinted in the text that bare skin patches and wattles might be particularly appropriate sexual 'ornaments' for birds with blood diseases in which the destruction of the cells often causes

changes of blood colour. This suggests an explanation of the abundance of wattles and bare patches in galliform birds and makes the prediction that unusual prevalence and variety of such diseases will be found in other unrelated but similarly ornamented birds. Birds with colourful legs and ceres on bills deserve similar attention.

50. I. van Riper, C. S. G. van Riper, M. L. Goff, and M. Laird, The epizootiology and ecological significance of malaria in Hawaiian land birds, *Ecological Monographs* **56**, 327–44 (1986); G. Valkiunas, Pathogenic influence of hemosporidians on wild birds in the field conditions: facts and hypotheses, *Ekologija* No. 1, 47–60 (1993).

51. John (1997) in note 9.

52. R. A. Fisher, *The Genetical Theory of Natural Selection* (Dover, New York, 2nd edn, 1958).

53. G. Borgia and K. Collis, Parasites and bright male plumage in the satin bowerbird (*Ptilonorhynchus violaceus*), *American Zoologist* **30**, 279–85 (1990); S. G. Pruett-Jones, H. L. Pruett-Jones, and H. L. Jones, Parasites and sexual selection in birds of paradise, *American Zoologist* **30**, 287–98 (1990).

FURTHER EVIDENCE

Appendix to Chapter Sixteen

> *For while the tired waves vainly breaking,*
> *Seem here no painful inch to gain,*
> *Far back, through creeks and inlets making,*
> *Comes silent, flooding in, the main.*
> A. H. CLOUGH[1]

AS PROMISED, I now summarize how the theme of the paper in Chapter 16 appears to me to have progressed since I, Bob Alexrod, and Reiko Tanese wrote it. Although a section for negative findings is included, I have far more to add that is positive. I compiled the list, so this is not surprising. But because such a bias seems inevitable it is a pity to be unable to challenge individually the particular scientists who were so negative on the early submissions of the paper to add their own (still?) negative lists to contend with or amplify mine—but perhaps at least someone will be stimulated by my bias, if it is there, to write the needed negative review. It would also, of course, be interesting to see positive lists of those people supporting whatever alternative 'Queen of Nature', if not the Deterministic Mutation Hypothesis (DMH), they prefer; however, my general impression of many of the critics whose reports we received was rather one of opposition to theorizing about such a subject generally.

FINDINGS FOR PARASITIC RED QUEEN SINCE 1990 (AND SOME EARLIER)

Power and ubiquity of parasite selection

Three ecological followed by six genetical lines of evidence followed by one mixed are currently strengthening the setting for PRQ and are as follows:

1. Parasites are admitted to deeply influence ecology and extinction.[2,3]

2. Episodic and multiple simultaneous parasite pressures for various host species accumulate further evidence. Heritabilities of resistance in wild plants are now demonstrated, citing only well-worked examples, in goldenrods,[4] willows,[5] forest peppers,[6] parsnips,[7] and aquatic snails (see below).

3. Virulence correlates with vagility of parasites and hosts (see the papers in Chapters 4 and 5), a factor that accentuates PRQ pressure for sex in well-mixed species and populations—the situations where, in fact, sex is very constantly maintained.[8] (But a caution from theoretical studies is that supervagility of parasites of immobile hosts may lead to asexuality, probably because asynchronous local cycles are prevented.[9])

4. Instances of genetic and variable resistances in plants and animals accumulate rapidly,[10] including new whole classes of variable defence molecules—for example, defensins.[11] A search on keyword combinations like '[domestic animal/plant name] + resistance + gene + map' in *Science Citation Index* from 1990 on will confirm an almost explosive increase.

5. Resistances to parasites are shown (a) to be costly to fitness when parasites are absent[12] and/or (b) to involve trade-offs of effective resistance for different parasites.[13]

6. Ever more effective methods[14] are showing resistance-related genes to be widely scattered in the genome, although often accompanied by local clustering. As one example, after only 7 years of study, *Arabidopsis* resistances now span all the plant's chromosomes and overtake other known gene systems both in the dispersion of their loci and in numbers of alleles.[15] Other well-studied hosts include potato, tomato, lettuce,[16] flax,[17] cereals,[18] chickens,[19] mice,[20] cattle,[21] pigs,[22] humans,[23] and yeast.[24] In yeast, more than 50 gene loci associated with 'managing' viruses underlying the 'killer' trait are known. Apart from the sometimes expressed advantage to infecteds in competition these are generally disadvantageous to hosts (with which they have obviously been long associated). Genes for the 'complement' defence system in vertebrates are found scattered over several, perhaps many loci.[25] Some components have shown interchromosomal shifts (as between mice and humans[26]).

7. Lists of parasite anti-host loci and plasmids grow in parallel with host anti-parasite lists although per species they are less long—again as

expected.[27] Rapidly changing frequencies are in evidence but usually due to scientific interference as by pesticides, antibiotics, or vaccinations[28] than by purely natural selection.

8. Molecular mimicry of host molecules by pathogens is proven abundant and can have several styles,[29] evidence for which is continually increasing.[30] Because all can be set back by the host by changes in its own molecules, all have potential to generate dynamical polymorphisms.[31] In a common form, parasites expose molecules resembling those exposed on cells, thus driving human immune systems continually to the edge of self-attack. Many degenerative and autoimmune diseases of mammals are now conjectured to involve 'misdirected' attacks, the prime targets of which are or were pathogens.[32] Publication on this topic has accelerated in almost every year from 1982 to 1997, with major increases in 1991 and 1997:

Year	1981	82	83	84	85	86	87	88	89	90	91	92	93	94	95	96	97
Papers	0	0	2	2	7	7	11	7	16	14	57	62	82	72	87	93	127

(Data from Institute of Scientific Information databases.)

As the first to see the ubiquity of this source of host variation, and its potential to cause dynamic polymorphism, it appears that Raymond Damian's time for recognition as a prophet has come. Some probable examples in polymorphisms that he cited in his defining paper of 1964 have been little followed up.[29]

So far such papers only occasionally mention or discuss the response polymorphism in target molecular systems but attention here is also increasing.[33] Fewer still discuss the necessarily dynamical aspects of the polymorphism but that topic, too, may be expanding.[34] On the negative side, a warning against false cases based solely on sequence similarities also needs to be heeded.[35]

9. Ancient (transpecific) polymorphisms are dated and increasingly shown linked to disease. The trends include examples in plants.[36]

10. New human polymorphic resistances are discovered frequently[37] and generally, as is predicted by PRQ, fail to show heterozygote advantage (see my paper in Chapter 15). Examples are malaria-resistance variation within HLA with heterozygote intermediate[38] and AIDS-resisting genes.[39] More generally, heterozygote advantage appears to me

to have lost ground as the pervasive explanation of polymorphism unless heterozygote advantage in 'geometric' averages formed over time are allowed as examples—and these, of course, could derive from the disease-caused fluctuations this volume is about. (For various items pro and con common heterozygote advantage since 1990 the reader can turn to notes 40 and then against 41.)

11. Ecological clustering of like genotypes leads to heavier parasitism.[42] Sexuality, including use of polyandry, is shown able to counter this effect.[43] Low genotypic variability in hosts may lead to asexuality in parasites.[44]

The following four published topics illustrate the closeness of host–parasite coevolution, a primary assumption of PRQ:

12. Direct evidence confirmatory on almost all aspects of PRQ predictions accumulates for particular cases, notably New Zealand and other aquatic snails,[45,46] barn swallows,[47] cladoceran waterfleas,[48] sweet vernal grass[49] and *Eupatorium* herbs[50] (but on the snails see also note 51).

13. Increasing ecological or experimental exposure to parasites increases facultative expression of sexuality in hosts.[52] Increasing the changeable resistance mounted by hosts, either as innate[53] or facultative variation,[54] increases sexuality in parasites.

14. Inbred and asexual races have to stagnate genetically[55] with the consequence that abundant and/or old clones become more subject to disease.[56] (But, against, see note 57: are these cases of young clones starting with high resistance?)

15. Parasites are shown generally to lead in local arms races, so intensifying the local incentive for hosts to escape via dormant protected stages, distant dispersal, or/and sex.[58]

Modelling and more data

Parallel and further feasibility study and still more evidence for PRQ have been found as follows:

16. No other model besides HAMAX yet promotes sex against a twofold cost under realistic assumptions and panmixia. A spatially structured population, however, has been shown competent against the cost[59] and this model again depends on the assumption of parasites.

17. The robustness of permanent dynamic polymorphism (PDP) arising from pursuit situations generates increasing theoretical interest[60] and especially, it seems, in Haifa.[61] Unfortunately these studies rather seldom and only recently focus selection produced in any observed multilocus dynamic on modifiers of linkage or on sex versus asex. Strain mixtures of asexuals in process of changes,[62] especially if seeming to 'take turns' for bursts of abundance, as shown by Dybdahl for the aquatic snail *Potamopyrgus*, as referred to on p. 639 (last reference),[63] are particularly favourable to PRQ with only the drawback that such mixtures may be a major alternative to sex if only a few loci are interacting, as mentioned in the introduction to Chapter 16. Dynamical polymorphism is increasingly documented for barley as it manoeuvres under human encouragement against its *Erysiphe* mildew but (a) the resistance is mainly single locus in the host and (b) the system is obviously mainly controlled by the human schemes of breeding and planting[64] rather than naturally. All the same, cycles occur and reusability of 'old' resistance alleles after decline of their specific 'virulent' strain of parasite has been observed:[65] we might invoke Darwin in this context and point out how much earlier, in parallel to the present argument, he instanced human breeding practices to illustrate through a more systematic version the process he believed to be occurring in nature. As suggested of opportunities for useful recombination, resistance alleles out of linkage equilibria with other loci are sometimes observed.[66] Allele cycling associated with population cycling is also documented for an island population of an ungulate.[67] But, despite implications of parasites in polymorphism in a previous paper on the same wild herd by the same team,[68] the second does not mention parasites even as partial causative agents.

18. Even asexual strain mixtures that are not yet shown to show sharp selective changes remain very hard to rationalize on DMH if the strain divergences are deep and such that closest-other strains are scattered in other subpopulations rather than found locally. A set of references to such situations is given in the paragraphs on multiclonal populations in the main part of this chapter (p. 623).

19. PDP is shown to result in model populations allowed to react demographically as well genetically to their parasites.[69]

20. Metapopulation models of PDP confirm two points:
    (a) Typical desynchrony of cycling over a whole connected but partially
        subdivided range. Sometimes this takes the form of systematic spiral
        waves; other parameter conditions give chaotic, patchy fluctuations:
        through their systematic desynchrony both cases greatly strengthen
        the basis for allele protection for the whole. Possibilities of similar
        processes generating mobile and fine-grain patterns also apply in
        wider ecological settings to the matter of species protection[70] and, in
        fact, the most intensive theoretical studies so far done on these
        patterning phenomena of metapopulations have been ecologically
        oriented.[71] But genetical parallels, including the spirals and chaotic
        patches, are obviously at least as likely to turn up in contexts of
        genetic antagonisms in species pairs. Recently they were indeed
        shown by A. Sasaki in metapopulation model schemes. These follow
        Judson, Ladle, and Johnstone (these authors themselves using the
        model structure of HAMAX), and also follow Frank.[72]
        Simultaneous sampling of various demes in such a metapopulation
        potentially show all phases of cycling as in the HAMAX-like
        original situation.[73]
    (b) High disparity of rates of migration in hosts and parasites creates
        conditions for adaptive asexual metapopulations with the overall
        arena developing a 'chequerboard' of different mixtures of highly
        disparate clones (A. Sasaki, unpublished). Such a chequerboard can
        apply to both hosts and parasites, and asex in such a model can be
        stable even when parasites are important.[74] The favour to asex being
        most common under a high disparity in the mobility hints at an
        explanation for many cases of long-standing parthenogenesis: such
        organisms appear disproportionately often either very mobile or very
        immobile compared to the average for their wider groups.
21. No organism is too small or too primitive to experience important in-
    fluences from parasites.[75] Nor probably is any too small to have sex.[76]
    Models of artificial life such as Tom Ray's *Tierra* program show parasite
    'species' evolving right from the time of origin of the simplest self-
    sufficient organisms. In cases where parasitism in large organisms is
    non-obvious, viruses may justifiably be suspected:[77] these may exert
    important but subtle and, to humans, non-obvious effects, which
    nevertheless stress the plants and greatly affect competition.

22. Further study and discussion, against earlier criticism,[78] confirms that the cost of sexuality is real and perhaps sometimes *under*estimated as twofold.[79]

Evidence from sexual selection
(see also Chapters 6 and 17)

23. There has been dramatic strengthening of evidence for heritable fitness and 'good genes'[80] and it is shown that the more genetic variability the more a species 'invests' in sexual selection.[81]

24. Evidence accumulates that parasites affect behavioural dominance and attractiveness even in cases where contagiousness and parental care are not at issue.[82]

25. Favour to offspring (as manifested by retention *in utero*, non-abortion of seeds, etc.) and mate choice are both increasingly shown to be connected with genetic polymorphisms having known relevance to disease.[83]

AGAINST DMH THEORY

Problems faced by DMH and PRQ have increased since 1990 as follows:

1. (a) Mutation rates[84] and (b) synergistic mutation damage[85] have so far been marginal or insufficient for this theory to provide a sole cause of sexuality. Of these (a) is the more important, while against taking too seriously the studies cited under (b) it needs to be noted that the findings concern small fast-breeding *r*-selected organisms that are not typically persistently sexual; in cases of large and highly sexual organisms, DMH can always fall back (as also does PRQ in the paper of Chapter 17) on rank-order truncation selection to provide the required pattern.

2. DMH has so far produced hardly any explanation for the ecological/geographic distribution of sex. I present just one example. Why should sexuality be relaxed and even often lost in probably more mutagenic mountain and arctic environments.[86] Many arctic and high alpine species are dark or black in contrast to more southerly or lowland relatives, and this trend is presumed due to the high and damaging levels of UVB light imposed during the long summer daylight.[87] If this protective adaptation is enough needed to have occurred convergently

in animals as remote as the tiny (normally transparant) water fleas and moderate-sized insects, then surely in parallel, on DMH reasoning, enhanced rather than reduced sexuality would be expected.

3. DMH cannot account for large almost immediate differences in fitness of offspring according to whether sexually produced or not, such as were found by Kelley and coworkers in sweet vernal grass.[88] (On the contrary swift acquisition of disease by an unchanged genotype could well explain them.[89]) More experiments like this one are much needed.

AGAINST PRQ THEORY

1. Gene-for-gene modelling in PRQ has been criticized: real world resistance is claimed to be different.[90] The importance of the gene-for-gene resistance generally has also been questioned but in a context, seemingly, only of familiarity with crop problems and of strong advocacy of 'horizontal resistance' (roughly, polygenic and quantitative—similar also to my so-called 'static' resistance in my Dahlem paper of Chapter 5) in the crop context.[91] These systems cycle quite readily too (Chapter 15) but support sex less strongly (Chapter 16). The importance of heritable resistance for human health has also been questioned in a review emphasizing the facultative aspect of the immune system and the imprecision of the genetic or innate side;[92] my impression, as shown by the citations sampled above under 4 and 6–8 of my positive points (pp. 828–9), is, however, that this view, once so common in medical circles, is losing ground.

2. Examples of complementary resistances remain relatively few. Parker is correct on this but see 5 above (p. 828).

3. More is known of highly sexual organisms inhabiting extreme environments where parasites might be supposed few; for example, reindeer and ibex, and, among plants, note *Dryas* and *Diapensia*. High rates of high sexuality are highlighted on certain islands.[93] But reindeer and ibex certainly don't escape parasites[94] (and see Chapter 6), nor are arctic environments necessarily sparse in pathogens.[95] Much more needs to be researched about the interrelations of diversity, density, and harshness of environment as these relate to the prevalence of infections.

4. Little is yet recorded to show either cycling with parasites or temporal changes of interlocus gene association.[96]

5. There is conflicting evidence on parasite loads and fitness decline in clones.[62,97] But information on the ages of clones and their histories of parasite escape and re-encounter is generally lacking so that the implication is uncertain. For example, in one case a sexual gecko is shown invading Pacific islands and ousting an asexual gecko with fewer parasites—the re-invasion is good for PRQ but the parasite situation is bad, or at least currently inexplicable.[98]

6. Some findings seem to support biotic RQ interpretations of biodiversity at both species and genotype-within-species levels without involving parasites. For example, Taylor and Aarssen[99] found complex and sometimes intransitive relations at the genotype level in experiments that competed clonal shoots of three meadow herb species under greenhouse conditions. The conditions would, seemingly, exclude pathogens from having a major influence and it was concluded that the complex competitive interactions could, by themselves, promote mosaics of coexistence. But the possibility that the field-sampled individuals multiplied for the experiment were already carrying viruses and endophytes that could affect competitive results seems not to have been examined.

References and notes

1. A. H. Clough, Say not the struggle nought availeth, in M. Van Doren (ed.), *An Anthology of World Poetry*, p. 1109 (Cassell, London, 1929).

2. J. B. Gillet, Pest pressure, an underestimated factor in evolution, in D. Nichols (ed.), *Taxonomy and Geography, A Symposium*, pp. 37–46 (London Systematics Association, London, 1962); J. M. Behnke, C. J. Barnard, and D. Wakelin, Understanding chronic nematode infections—evolutionary considerations, current hypotheses and the way forward, *International Journal for Parasitology* **22**, 861–907 (2992); R. S. Fritz and E. S. Simms (ed.), *Plant Resistance to Herbivores and Pathogens* (Chicago University Press, Chicago, 1992); R. B. Root, Herbivore pressure on goldenrods (*Solidago altissima*)—its variation and cumulative effects, *Ecology* **77**, 1074–87 (1996); P. A. Henderson, W. D. Hamilton, and W. G. R. Crampton, Evolution and diversity in Amazonian floodplain communities, in D. M. Newberry, N. Brown, and H. H. T. Prins (ed.), *Dynamics of Tropical Communities, 37th Symposium of the British Ecological Society, Cambridge University*, pp. 385–419 (Blackwell Science, Oxford, 1998); P. A. Henderson, On the variation in dab *Limanda limanda* recruitment: a zoogeographic study, *Journal of Sea Research* **40**, 131–42 (1998).

3. D. W. Macdonald, Dangerous liaisons and disease, *Nature* **379**, 400–1 (1995).

4. G. D. Maddox and R. B. H. Root, Structure of the encounter between goldenrod (*Solidago altissima*) and its diverse fauna, *Ecology* **71**, 2115–24 (1990); Root (1996) in note 2; W. G. Abrahamson and A. E. Weis, *Evolutionary Ecology across Three Tropic Levels: Goldenrods, Gallmakers, and Natural Enemies* (Princeton University Press, Princeton, NJ, 1997); B. Schmid, Effects of genetic diversity in experimental stands of *Solidago altissima*: evidence for the potential role of pathogens as selective agents in plant-populations, *Journal of Ecology* **82**, 165–75 (1994).

5. R. S. Fritz, Community structure and species interactions of phytophagous insects on resistant and susceptible host plants, in R. S. Fritz and E. L. Simms (ed.), *Plant Resistance to Herbivores and Pathogens*, pp. 240–77 (Chicago University Press, Chicago, 1992).

6. R. J. Marquis, Genotypic variation in leaf damage in *Piper arieanum* (Piperaceae) by a multispecies assemblage of herbivores, *Evolution* **44**, 104–20 (1990).

7. M. R. Berenbaum and A. R. Zangerl, Quantification of chemical coevolution, in R. S. Fritz and E. S. Simms (ed.), *Plant Resistance to Herbivores and Pathogens*, pp. 69–87 (Chicago University Press, Chicago, 1992).

8. B. Low, Marriage systems and pathogen stress in human societies, *American Zoologist* **30**, 325–39 (1990); P. W. Ewald, Waterborne transmission and the evolution of virulence among gastrointestinal bacteria, *Epidemiology and Infection* **106**, 83–119 (1991); S. A. Frank, Kin selection and virulence in the evolution of protocells and parasites, *Proceedings of the Royal Society of London B* **258**, 153–61 (1994); K. L. Mangin, M. Lipsitch, and D. Ebert, Virulence and transmission modes of two microsporidia in *Daphnia magna, Parasitology* **111**, 133–42 (1995); D. Ebert and W. D. Hamilton, Sex against virulence: the coevolution of parasitic diseases, *Trends in Ecology and Evolution* **11**, 79–82 (1986); D. Ebert and K. L. Mangin, The influence of host demography on the evolution of virulence of a microsporidian gut parasite, *Evolution* **51**, 1828–37 (1996); D. Ebert and E. A. Herre, The evolution of parasitic diseases, *Parasitology Today* **12**, 96–100 (1996); S. A. Frank, Models of parasite virulence, *Quarterly Review of Biology* **71**, 37–78 (1996); P. Agnew and J. C. Koella, Virulence, parasite mode of transmission, and fluctuating asymmetry, *Proceedings of the Royal Society of London B* **264**, 9–15 (1997).

9. R. J. Ladle, R. A. Johnstone, and O. P. Judson, Coevolutionary dynamics of sex in a metapopulation: escaping the Red Queen, *Proceedings of the Royal Society of London B* **253**, 155–60 (1993); O. Judson, Preserving genes: a model of the maintenance of genetic variation in a metapopulation under frequency dependent selection, *Genetical Research, Cambridge* **65**, 175–91 (1995).

10. L. Gedda, G. Rajani, G. Brenci, M. T. Lun, C. Talone, and G. Oddi, Heredity and infectious diseases—a twin study, *Acta Geneticae Medicae et Gemellologiae* **33**, 497–500 (1984); J. B. Owen and R. F. E. Axford (ed.), *Breeding for Disease Resistance in Farm Animals* (CAB International, Wallingford, UK, 1991) (within this book see especially N. Bumstead, B. M. Millard, P. Barrow, and J. K. A. Cook, Genetic basis of disease resistance in chickens (pp. 10–23)); E. D. Grosholz, The effects of host genotype and spatial-distribution on trematode parasitism in a bivalve population, *Evolution* **48**, 1514–24 (1994); J. K. M. Brown, Chance and selection in the evolution of barley mildew, *Trends in Microbiology* **2**, 470–5 (1994); P. L. Wilcox, H. V. Amerson, E. G. Kuhlman, B. H. Liu, D. M. Omalley, and R. R. Sederoff, Detection of a major gene for resistance to fusiform rust disease in loblolly pine by genomic mapping, *Proceedings of the National Academy of Sciences, USA* **93**, 3859–64 (1996); A. G. Clark and L. Wang, Molecular population genetics of *Drosophila* immune system genes, *Genetics* **147** 713–24 (1997); A. Date, Y. Satta, N. Takahata, and S. I. Chigusa, Evolutionary history and mechanism of the *Drosophila* cecropin gene family, *Immunogenetics* **47**, 417–29 (1998).

11. T. Dork and M. Stuhrmann, Polymorphisms of the human beta-defensin-1 gene, *Molecular and Cellular Probes* **12**, 171–3 (1998).

12. P. B. Siegel, W. B. Gross, and J. A. Cherry, Correlated responses of chickens to selection for production of antibody to sheep erythrocytes, *Animal Blood Groups and Biochemical Genetics* **13**, 291–6 (1983); R. R. Dietert, R. L. Taylor, and M. E. Dietert, The chicken major histocompatibility complex: structure and impact on immune function, disease resistance and productivity, *Monographs in Animal Immunology* **1**, 7–26 (1990); Behnke *et al.* (1992) in note 2; A. Kloosterman, H. K. Permentier, and H. W. Ploezer, Breeding cattle for resistance to gastro-intestinal nematodes, *Parasitology Today* **8**, 330–5 (1992); B. C. Sheldon and S. Verhulst, Ecological immunity: costly parasite defences and trade-offs in evolutionary ecology, *Trends in Ecology and Evolution* **11**, 317–21 (1996); J. Bergelson, C. B. Purrington, C. J. Palm, and J. C. Lopez-Gutierrez, Costs of resistence: a test using transgenic *Arabidopsis thaliana*, *Proceedings of the Royal Society of London B* **263**, 1659–63 (1996); E. A. Stahl, G. Dwyer, R. Mauricio, M. Kreitman, and J. Bergelson, Dynamics of disease resistance polymorphism at the *Rpm1* locus of *Arabidopsis*, *Nature* **400**, 667–71 (1999); A. W. Gemmill and A. F. Read, Counting the cost of disease resistance, *Trends in Ecology and Evolution* **13**, 18–19 (1998); A. R. Kraaijeveld and H. C. J. Godfray, Trade-off between parasitoid resistance and larval competitive ability in *Drosophila melanogaster*, *Nature* **389**, 278–80 (1997); M. D. E. Fellowes, A. R. Kraaijeveld, and H. C. J. Godfray, Trade-off

associated with selection for increased ability to resist parasitoid attack in *Drosophila melanogaster*, *Proceedings of the Royal Society of London B* **265**, 1553–8 (1998).

13. As noted by M. A. Parker (Pathogens and sex in plants, *Evolutionary Ecology* **8**, 560–84 (1994)), examples of what I have termed 'complementary resistances' (see the paper in Chapter 5) are not commonly recorded; far more common is the pattern where a resistance can be presumed balanced by what in the same work I referred to as a 'static cost'. Parker doubts that even these can be very common but seems to have no explanation for how without costs to reduce the frequencies of resistance (and virulence) genes when they are not needed, the immense variability of genetic resistance exists, for if costless, of course, every allele conferring resistance (or virulence), even if by only occasional episodes of selection, should be brought to fixation (S. A. Frank, Problems of inferring the specificity of plant-pathogen genetics (a reply to Parker), *Evolutionary Ecology* **10**, 323–5 (1996)).

In any case examples of complementarity may be difficult to detect because the opposite biotic selection force has not yet been noticed. Some examples where complementarity has been detected recently are: Bumstead *et al.* (1991) in note 10; S. Via, The genetic structure of host plant adaptation in a spatial patchwork: demographic variability among reciprocally transplanted pea aphid clones, *Evolution* **45**, 827–52 (1991); Berenbaum and Zangerl (1992), note 7; Fritz (1992), note 5; E. L. Simms and M. D. Rausher, Patterns of selection on phytophagous resistance in *Ipomaea purpurea*, *Evolution* **47**, 970–6 (1993); J. A. Lane, D. V. Child, G. C. Reiss, V. Entcheva, and J. A. Bailey, Crop resistance to parasitic plants, in I. R. Crute, E. B. Holub, and J. J. Burdon (ed.), *The Gene-for-Gene Relationship in Plant-Parasite Interactions*, pp. 81–97 (CAB International, Wallingford, UK, 1997).

Older references, missed in my previous papers mentioning this topic, include: M. J. Way and G. Murdie, An example of varietal variations of resistance of Brussels Sprouts, *Annals of Applied Biology* **56**, 326–8 (1965); G. G. Kennedy and M. F. Abou-Ghadir, Bionomics of the turnip aphid on two turnip cultivars, *Journal of Economic Entomology* **72**, 754–62 (1979); T. L. Carpenter, W. W. Neel, and P. A. Hedin, A review of host plant resistance of Pecan, *Carya illinoensis*, to insecta and acarina, *Bulletin of the American Entomological Society* (1979); N. Blakley, Biotic unpredictability and sexual reproduction: do aphid genotype–host genotype interactions favor aphid sexuality?, *Oecologia* **52**, 396–9 (1982); M. R. Berenbaum, A. R. Zangerl, and J. K. Nitao, Constraints on chemical evolution: wild parsnips and the parsnip webworm, *Evolution* **42**, 958–68 (1986).

In any case, 'static' cost-opposed resistances and virulences, the seemingly more typical genetic pattern, cycle readily (S. D. Jayakar, A mathematical model for interaction of gene frequencies in a parasite and its host, *Theoretical Population Biology* **1**, 140–64 (1970)) and under reasonable assumptions almost always yield dynamic multilocus polymorphisms in theoretical models (S. A. Frank, Coevolutionary genetics of plants and pathogens, *Evolutionary Ecology* **7**, 45–75 (1993*a*)). When involving many loci (and especially if each locus is selected slightly differently to others so that different period tendencies from different loci are to be expected), upward selection on recombination, and consequent contributions from these systems to support for sex, seem almost inevitable.

14. V. Lefebvre and A. M. Chevre, Tools for marking plant-disease and pest resistance genes—a review, *Agronomie* **15**, 3–19 (1995); L. G. Adams and J. W. Templeton, Genetic resistance to bacterial diseases of animals, *Revue scientifique et technique de l'Office International des Epizooties* **17**, 200–19 (1998).

15. E. B. Holub, Organisation of resistance genes in *Arabidopsis*, in I. R. Crute, E. B. Holub, and J. J. Burdon (ed.), *The Gene-for-Gene Relationship in Plant-Parasite Interactions*, pp. 5–43 (CAB International, Wallingford, UK, 1997).

16. I. R. Crute, E. B. Holub, and J. J. Burdon (ed.), *The Gene-for-Gene Relationship in Plant-Parasite Interactions* (CAB International, Wallingford, UK, 1997); A. H. Paterson, T. H. Lan, K. P. Reischmann, C. Chang, Y. R. Lin, S. C. Liu, *et al.*, Toward a unified genetic map of higher plants, transcending the monocot–dicot divergence, *Nature Genetics* **14**, 80–2 (1997).

17. M. R. Islam and K. W. Shepherd, Present status of genetics of rust resistance in flax, *Euphytica* **55**, 255–67 (1991).

18. J. K. M. Brown, Chance and selection in the evolution of barley mildew, *Trends in Microbiology* **2**, 470–5 (1994); J. H. Jorgensen, Genetics of powdery mildew resistance in barley, *Critical Reviews in Plant Sciences* **13**, 97–119 (1994); Lefebvre and Chevre (1995) in note 14; T. Miedaner, Breeding wheat and rye for resistance to *Fusarium* diseases, *Plant Breeding* **116**, 201–20 (1997).

19. Bumstead *et al.* (1991) in note 10; J. Hu, N. Bumstead, D. Burke, F. A. P. Deleon, E. Skamene, P. Gros, *et al.* Genetic and physical mapping of the natural resistance-associated macrophage protein-1 (Nramp1) in chicken, *Mammalian Genome* **6**, 809–15 (1995); N. Bumstead, Genomic mapping of resistance to Marek's disease, *Avian Pathology* **27**, S78–S81 (1998).

20. W. F. Dietrich, D. M. Damron, R. R. Isberg, E. S. Lander, and M. S. Swanson, Lgn1, a gene that determines susceptibility to *Legionella pneumophila*, maps to mouse chromosome-13, *Genomics* **26**, 443–50 (1995); N. Urosevic,

J. P. Mansfield, J. S. Mackenzie, and G. R. Shellam, Low-resolution mapping around the flavivirus resistance locus (Flv) on mouse Chromosome-5, *Mammalian Genome* **6**, 454–8 (1995); S. Vidal, P. Gros, and E. Skamene, Natural resistance to infection with intracellular parasites: molecular genetics identifies Nramp1 as the Bcg/Ity/Lsh locus, *Journal of Leukocyte Biology* **58**, 382–90 (1995); S. J. Kemp, A. Darvasi, M. Soller, and A. J. Teale, Genetic control of resistance to trypanosomiasis, *Veterinary Immunology and Immunopathology* **54**, 239–43 (1996); M. C. Beckers, E. Ernst, E. Diez, C. Morissette, F. Gervais, K. Hunter, *et al.*, High-resolution linkage map of mouse chromosome 13 in the vicinity of the host resistance Locus Lgn1, *Genomics* **39**, 254–63 (1997); L. M. M. Araujo, O. G. Ribeiro, M. Siqueira, M. DeFranco, N. Starobinas, S. Massa, *et al.*, Innate resistance to infection by intracellular bacterial pathogens differs in mice selected for maximal or minimal acute inflammatory response, *European Journal of Immunology* **28**, 2913–20 (1998); M. A. Brinton, I. Kurane, A. Mathew, L. L. Zeng, P. Y. Shi, A. Rothman, *et al.*, Immune mediated and inherited defences against flaviviruses, *Clinical and Diagnostic Virology* **10**, 129–39 (1998); J. F. Bureau, K. M. Drescher, L. R. Pease, T. Vikoren, M. Delcroiz, L. Zoecklein, *et al.*, Chromosome 14 contains determinants that regulate susceptibility to Theiler's virus-induced demyelination in the mouse, *Genetics* **148**, 1941–9 (1998); R. F. Paulson and A. Bernstein, A genetic linkage map of the mouse Chromosome 9 region encompassing the Friend virus susceptibility gene 2 (Fv2), *Mammalian Genome* **9**, 381–4 (1998).

21. M. Amills, V. Ramiya, J. Norimine, and H. A. Lewin, The major histocompatibility complex of ruminants, *Revue scientifique et technique de l'Office International des Epizooties* **17**, 108–20 (1998); S. Sharif, B. A. Mallard, B. N. Wilkie, J. M. Sargeant, H. M. Scott, J. C. M. Dekkers, *et al.*, Associations of the bovine major histocompatibility complex DRB3 (BoLA-DRB3) alleles with occurrence of disease and milk somatic cell score in Canadian dairy cattle, *Animal Genetics* **29**, 185–93 (1998).

22. P. Vogeli, H. U. Bertschinger, M. Stamm, C. Stricker, C. Hagger, R. Fries, *et al.*, Genes specifying receptors for F18 fimbriated *Escherichia coli* causing oedema disease and postweaning diarrhoea in pigs, map to chromosome 6, *Animal Genetics* **27**, 321–8 (1996).

23. K. F. Lindahl, Minor histocompatibility antigens, *Trends in Genetics* **7**, 219–24 (1991); J. M. Blackwell, C. H. Barton, J. K. White, S. Searle, A. M. Baker, H. Williams, *et al.*, Genomic organization and sequence of the human Nramp gene: identification and mapping of a promoter region polymorphism, *Molecular Medicine* **1**, 194–205 (1995); L. Abel, F. O. Sanchez, J. Oberti,

N. V. Thuc, L. VanHoa, V. D. Lap, *et al.*, Susceptibility to leprosy is linked to the human NRAMP1 gene, *Journal of Infectious Diseases* **177**, 133–45 (1998).

24. Y. Ohtake and R. B. Wickner, Yeast virus propagation depends critically on free 60s ribosomal-subunit concentration, *Molecular and Cellular Biology* **15**, 2772–81 (1995); W. Magliani, S. Conti, M. Gerloni, D. Bertolotti, and L. Polonelli, Yeast killer systems, *Clinical Microbiology Reviews* **10**, 369 only (1997).

25. M. J. Hobart, Phenotypic genetics of complement components, *Philosophical Transactions of the Royal Society of London B* **306**, 325–31 (1984).

26. M. J. Hobart, M. J. Walport, and P. J. Lachmann, Complement polymorphism and disease, *Clinics in Immunology and Allergy* **4**, 647–64 (1984).

27. O. Soderlind, B. Thafvelin, and R. Mölby, Virulence factors in *Escherichia coli* strains isolated from Swedish piglets with diarrhoea, *Journal of Clinical Microbiology* **26**, 879–84 (1988); S. Gupta and A. V. S. Hill, Dynamic interactions in malaria: host heterogeneity meets parasite polymorphism, *Proceedings of the Royal Society of London B* **261**, 271–7 (1995); A. A. Escalante, A. A. Lal, and F. J. Ayala, Genetic polymorphism and natural selection in the malaria parasite *Plasmodium falciparum*, *Genetics* **149**, 189–202 (1998).

28. Soderlind *et al.* (1988) in note 27.

29. R. T. Damian, Molecular mimicry: antigen sharing by parasite and host and its consequences, *American Naturalist*, **98**, 129–49 (1964); R. T. Damian, Molecular mimicry in biological adaptation, in B. B. Nickol (ed.), *Host–Parasite Interfaces*, pp. 103–26 (Academic Press, New York, 1979); R. T. Damian, Parasite immune evasion and exploitation: reflections and projections, *Parasitology* **115**, S169–S175 (1997).

30. Å. Lernmark, T. Dyrberg, L. Terenius, and B. Hökfelt (ed.), *Molecular Mimicry in Health and Disease* (Excerpta Medica, Amsterdam, 1988); G. Myers and G. N. Pavlakis, Evolutionary potential of complex retroviruses, in J. Levy (ed.), *The Retroviridae*, pp. 51–105 (Plenum, New York, 1991); P. W. Dempsey, M. E. Allison, S. Akkaraju, and C. C. Goodnow, C3d of complement as a molecular adjuvant: bridging innate and acquired immunity, *Science* **271**, 348–50 (1996); Z. S. Zhao, F. Granucci, L. Yeh, P. A. Schaffer, and H. Cantor, Molecular mimicry by Herpes Simplex Virus—Type 1: autoimmune disease after viral infection, *Science* **279**, 1344–7 (1998).

31. V. Apanius, D. Penn, P. R. Slev, L. R. Ruff, and W. K. Potts, The nature of selection on the major histocompatibility complex, *Critical Reviews in Immunology* **17**, 129–224 (1997); D. Penn and W. K. Potts, The evolution of mating preferences and MHC genes, *American Naturalist* **153**, 145–64 (1999).

32. Myers and Pavlakis (1991) in note 30; G. Strobel and S. Dickman, Prime suspects lined up in MS mystery, *New Scientist*, **146**, 16 only (1 April 1995); H. Baum, H. Davies, and M. Peakman, Molecular mimicry in the MHC: hidden clues to autoimmunity?, *Immunology Today* **17**, 64–70 (1996); M. Day, Undercover agents, *New Scientist* **155**, 6 (9 August 1997); P. Brown, Over and over and over . . ., *New Scientist* **155**, 27–31 (2 August 1997).

33. R. Grubb, Immunogenetic markers as proves or polymorphism, gene-regulation and gene-transfer in Man: the Gm system in perspective, *Apmis* **99**, 199–209 (1991); A. Husainchishti and P. Ruff, Malaria and ovalocytosis: molecular mimicry, *Biochimica et Biophysica Acta* **1096**, 263–4 (1991); M. Baines and A. Ebringer, HLA and disease, *Molecular Aspects of Medicine* **13**, 263–378 (1992); Apanius *et al.* (1997) in note 31; G. R. Vreugdenhil, A. Geluk, T. H. M. Ottenhoff, W. J. G. Melchers, B. O. Roep, and J. M. D. Galama, Molecular mimicry in diabetes mellitus: the homologous domain in coxsackie B virus protein 2C and islet autoantigen GAD(65) is highly conserved in the coxsackie B-like enteroviruses and binds to the diabetes associated HLA-DR3 molecule, *Diabetologia* **41**, 40–6 (1998).

34. Penn and Potts (1999) in note 31.

35. C. Roudier, I. Auger, and J. Roudier, Molecular mimicry reflected through database screening: serendipity or survival strategy?, *Immunology Today* **17**, 357–8 (1996).

36. Anon., Diverse genes bring humans and chimps closer, *New Scientist* **119**, 35 only (22 September 1988); J. Klein and N. Takahata, The major histocompatibility complex and the quest for origins, *Immunological Reviews* **113**, 5–25 (1990); U. Brandle, O. Hideki, V. Vincek, D. Klein, M. Golubic, B. Grahovac, *et al.*, Trans-species evolution of MHC-DRB haplotype polymorphism in primates: organisation of DRB genes in the chimpanzee, *Immunogenetics* **36**, 39–48 (1992); M. K. Uyenoyama, A generalised least-squares estimate for the origin of sporophytic self-incompatibility, *Genetics* **139**, 975–92 (1995); A. D. Richman, M. K. Uyenoyama, and J. R. Kohn, Allelic diversity and gene genealogy at the self-incompatibility locus in the Solanaceae, *Science* **273**, 1212–16 (1996); M. Kusaba, T. Nishio, Y. Satta, K. Hinata, and D. Ockendon, Striking similarity in inter- and intra-specific comparisons of class I *SLG* alleles from *Brassica oleracea* and *B. campestris*: implications for the evolution and recognition mechanism, *Proceedings of the National Academy of Sciences, USA* **94**, 7673–8 (1997).

37. See, for example, note 11.

38. A. V. S. Hill, C. E. M. Allsopp, D. Kwiatkowski, N. M. Anstey, P. Twumasi,

P. A. Rowe, *et al.*, Common West African HLA antigens are connected with protection from severe malaria, *Nature* **352**, 595–600 (1991).

39. R. A. Kaslow, M. Carrington, R. Apple, L. Park, A. Munoz, A. J. Saah, *et al.*, Influence of combinations of human major histocompatibility complex genes on the course of HIV-1 infection, *Nature Medicine* **2**, 405–11 (1996); W. A. Paxton, S. R. Martin, D. Tse, T. R. Obrien, J. Skurnick, N. L. VanDevanter, *et al.*, Relative resistance to HIV-1 infection of CD4 lymphocytes from persons who remain uninfected despite multiple high-risk sexual exposures, *Nature Medicine* **2**, 412–17 (1996); C. Thompson, The genes that keep AIDS at bay, *New Scientist* **150**, 16 only (6 April 1996).

40. M. M. Ferguson and L. R. Drahushchak, Disease resistance and enzyme heterozygosity in rainbow trout, *Heredity* **64**, 413–17 (1990); D. E. Jelinski, Associations between environmental heterogeneity, heterozygosity, and growth-rates of *Populus tremuloides* in a Cordilleran landscape, *Arctic and Alpine Research* **25**, 183–8 (1993); J. B. Mitton, Theory and data pertinent to the relationship between heterozygosity and fitness, in N. W. Thornhill (ed.), *The Natural History of Inbreeding and Outbreeding*, pp. 17–41 (University of Chicago Press, Chicago, 1993).

41. B. J. McAndrew, R. D. Ward, and J. A. Beardmore, Lack of relationship between morphological variance and enzyme heterozygosity in the plaice, *Pleuronectes platessa*, *Heredity* **48**, 117–25 (1982); J. M. Pemberton, S. D. Albon, F. E. Guinness, T. H. Clutton-Brock, and R. J. Berry, Genetic variation and juvenile survival in red deer, *Evolution* **42**, 921–34 (1988); A. R. Beaumont, Genetic studies of laboratory reared mussels, *Mytilus edulis*: heterozygote deficiencies, heterozygosity and growth, *Biological Journal of the Linnean Society* **44**, 273–85 (1991); D. B. Goldstein, Heterozygote advantage and the evolution of dominant diploid phase, *Genetics* **132**, 1195–8 (1992); J. E. Fairbrother and A. R. Beaumont, Heterozygote deficiencies in a cohort of newly settled *Mytilus edulis* spat, *Journal of the Marine Biological Association of the United Kingdom* **73**, 647–53 (1993); E. Zouros, Associative overdominance—evaluating the effects of inbreeding and linkage disequilibrium, *Genetica* **89**, 35–46 (1993); R. Demelo and P. D. N. Hebert, Allozymic variation and species-diversity in North-American Bosminidae, *Canadian Journal of Fisheries and Aquatic Sciences* **51**, 873–80 (1994); A. R. Beaumont, J. E. Fairbrother, and K. Hoare, Multilocus heterozygosity and size—a test of hypotheses using triploid *Mytilus edulis*, *Heredity* **75**, 256–66 (1995); H. B. Britten, Meta-analyses of the association between multilocus heterozygosity and fitness, *Evolution* **50**, 2158–64 (1996).

42. C. Mortiz, H. McCallum, S. Donnellan, and J. D. Roberts, Parasite loads in

parthenogenetic and sexual lizards (*Heteronotia binoei*): support for the Red
Queen hypothesis, *Proceedings of the Royal Society of London B* **244**, 145–9
(1991); A. C. Newton, Cultivar mixtures in intensive agriculture, in I. R. Crute,
E. B. Holub, and J. J. Burdon (ed.), *The Gene-for-Gene Relationship in Plant-
Parasite Interactions*, pp. 65–80 (CAB International, Wallingford, UK, 1997);
M. F. Dybdahl and C. M. Lively, Host–parasite coevolution: evidence for a
rate advantage and time lagged selection in a natural population, *Evolution* **52**,
1057–66 (1998).

43. J. A. Shykoff and P. Schmid-Hempel, Parasites and the advantage of genetic-
    variability within social insect colonies, *Proceedings of the Royal Society of London
    B* **243**, 55–8 (1991); P. Schmid-Hempel, Infection and colony variability in
    social insects, *Philosophical Transactions of the Royal Society of London B* **346**,
    313–21 (1994); S. Liersch and P. Schmid-Hempel, Genetic variation within
    social insect colonies reduces parasite load, *Proceedings of the Royal Society of
    London B* **265**, 221–5 (1998).

44. N. Jones and J. S. Madriewiez, Naturally occurring trioploidy and
    parthenogenesis in *Atractolytocestus huronensis* Anthony (Cestoidea:
    Caryophylloidea) from *Cyprinus carpio* L. in North America, *Journal of
    Parasitology* **55**, 1105–118 (1969); R. M. Cable, Parthenogenesis in parasitic
    helminths, *American Zoologist* **11**, 267–72 (1971).

45. S. G. Johnson, Parasitism, reproductive assurance and the evolution of
    reproductive mode in a fresh-water snail, *Proceedings of the Royal Society of
    London B* **255**, 209–13 (1994); M. F. Dybdahl and C. M. Lively, Host–parasite
    interactions: infection of common clones in natural populations of a fresh-
    water snail (*Potamopyrgus antipodarum*), *Proceedings of the Royal Society of
    London B* **260**, 99–103 (1995); M. F. Dybdahl and C. M. Lively, Diverse,
    endemic and polyphyletic clones in mixed populations of a fresh-water snail
    (*Potamopyrgus antipodarum*), *Journal of Evolutionary Biology* **8**, 385–98 (1995);
    S. G. Johnson, C. M. Lively, and S. J. Schrag, Evolution and ecological
    correlates of uniparental reproduction in fresh-water snails, *Experientia* **51**,
    498–509 (1995); C. M. Lively and V. Apanius, Genetic diversity in
    host–parasite interactions, in B. T. Grenfell and A. P. Dobson (ed.), *Ecology of
    Infectious Diseases in Natural Populations*, pp. 421–49 (Cambridge University
    Press, Cambridge, 1995); M. F. Dybdahl and C. M. Lively, The geography of
    coevolution: comparative population structures for a snail and its trematode
    parasite, *Evolution* **50**, 2264–75 (1996); C. M. Lively, S. G. Johnson, L. F.
    Delph, and K. Clay, Thinning reduces the effect of rust infection in jewelweed
    (*Impatiens capensis*) *Ecology* **76**, 1851–4 (1996); C. M. Lively and J. Jokela,
    Clinal variation for local adaptation in a host–parasite interaction, *Proceedings*

of the Royal Society of London B **263**, 891–7 (1996); J. Jokela, C. M. Lively, M. F. Dybdahl, and J. A. Fox, Evidence for a cost of sex in the freshwater snail *Potamopyrgus antipodarum*, *Ecology* **78**, 452–62 (1997*a*); J. Jokela, C. M. Lively, J. A. Fox, and M. F. Dybdahl, Flat reaction norms and 'frozen' phenotypic variation in clonal snails (*Potamopyrgus antipodarum*), *Evolution* **51**, 1120–9 (1997*b*).

46. Dybdahl and Lively (1998) in note 42.

47. A. P. Møller and T. R. Birkhead, The evolution of plumage brightness in birds is related to extra-pair paternity, *Evolution* **48**, 1089–100 (1994); A. P. Møller, Immune defence, extra-pair paternity, and sexual selection in birds, *Proceedings of the Royal Society of London B* **264**, 561–6 (1997); N. Saino, A. M. Bolzern, and A. P. Møller, Immunocompetence, ornamentation, and viability of male barn swallows (*Hirundo rustica*), *Proceedings of the National Academy of Sciences, USA* **94**, 549–52 (1997); A. P. Møller, P. Christe, J. Erritzoe, and A. P. Meller, Condition, disease and immune defence, *Oikos* **83**, 301–6 (1998); A. P. Møller, R. Dufva, and J. Erritzoe, Host immune function and sexual selection in birds, *Journal of Evolutionary Biology* **11**, 703–19 (1998); A. P. Møller, G. Sorci, and J. Erritzoe, Sexual dimorphism in immune defense, *American Naturalist* **152**, 605–19 (1998); A. P. Møller, P. Christe, and E. Lux, Parasitism, host immune function and sexual selection, *Quarterly Review of Biology* **74**, 3–20 (1999).

48. D. Ebert, Virulence and local adaptation of a horizontally transmitted parasite, *Science* **265**, 1084–6 (1994*a*); D. Ebert, Genetic differences in the interactions of a microsporidian parasite and four clones of its cyclically parthenogenetic host, *Parasitology* **108**, 11–16 (1994*b*); D. Ebert, The ecological interactions between a microsporidian parasite and its host *Daphnia magna*, *Journal of Animal Ecology* **64**, 361–9 (1995); K. L. Mangin, M. Lipsitch, and D. Ebert, Virulence and transmission modes of two microsporidia in *Daphnia magna*, *Parasitology* **111**, 133–42 (1995); T. J. Little and D. Ebert, Associations between parasitism and host genotype in natural populations of *Daphnia* (Crustacea: Cladocera), *Journal of Animal Ecology* **68**, 134–49 (1999).

49. S. E. Kelley, Viral pathogens and the advantage of sex in the perennial grass *Anthoxanthum odoratum*, *Philosophical Transactions of the Royal Society of London B* **346**, 295–302 (1994); G. Park and S. E. Kelley, The consequences of genetically homogeneous and heterogeneous host passage for virulence in a viral pathogen. I. Effects on population size and virulence [ref. untraced—Ed.]; S. E. Kelley and A. F. Kirkley, Factors affecting the incidence of natural virus infection in the grass, *Anthoxanthum odoratum*: spatial patterns of offspring dispersal and aphid preference [ref. untraced—Ed.].

50. T. Yahara and K. Oyama, Effects of virus-infection on demographic traits of an agamospermous population of *Eupatorium chinense* (Asteraceae), *Oecologia* **96**, 310–15 (1993); K. Ooi and T. Yahara, Genetic variation of geminiviruses: comparison between sexual and asexual host plant populations, *Molecular Ecology* **8**, 89–97 (1999).

51. R. S. Howard and C. M. Lively, The maintenance of sex by parasitism and mutation accumulation under epistatic fitness functions, *Evolution* **52**, 603–10 (1998).

52. S. J. Schrag, A. O. Mooers, G. T. Ndifon, and A. F. Read, Ecological correlates of male outcrossing ability in a simultaneous hermaphrodite snail, *American Naturalist* **143**, 636–55 (1994).

53. H. J. Schouten, Preservation of avirulence genes of potato cyst nematodes through environmental sex determination—a model involving complete, monogenic resistance, *Phytopathology* **84**, 771–3 (1994).

54. A. W. Gemmill, M. E. Viney, and A. F. Read, Host immune status determines sexuality in a parasitic nematode, *Evolution* **51**, 393–401 (1997).

55. M. A. Sanjayan and K. Crooks, Skin grafts and cheetahs, *Nature* **381**, 566 only (1996); B.-E. Saether, Environmental stochasticity and population dynamics of large herbivores: a search for mechanisms, *Trends in Ecology and Evolution* **12**, 143–9 (1997).

56. Mortiz *et al.* (1991) in note 42; Johnson (1994) in note 45; P. Chaboudez and J. J. Burdon, Frequency-dependent selection in a wild plant-pathogen system, *Oecologia* **102**, 490–3 (1995).

57. D. M. Weary, K. J. Norris, and J. B. Falls, Song features birds use to identify individuals, *Auk* **107**, 623–5 (1990); K. A. Hanley, D. M. Vollmer, and T. J. Case, The distribution and prevalence of helminths, coccidia and blood parasites in two competing species of gecko: implications for apparent competition, *Oecologia* **102**, 220–9 (1995a); K. A. Hanley, A. N. Fisher, and T. J. Case, Lower mite infestations in an asexual gecko compared with its sexual ancestors, *Evolution* **49**, 418–26 (1995b). See also S. G. Brown, S. Kwan, and S. Shero, The parasitic theory of sexual reproduction—parasitism in unisexual and bisexual geckos, *Proceedings of the Royal Society of London B* **260**, 317–20 (1995); S. C. Weeks, A reevaluation of the Red-Queen model for the maintenance of sex in a clonal-sexual fish complex (Poeciliidae, *Poeciliopsis*), *Canadian Journal of Fisheries and Aquatic Sciences* **53**, 1157–64 (1996).

58. R. Karban, Fine-scale adaptation of herbivorous thrips to individual host plants, *Nature* **340**, 60–1 (1989); Ebert (1994a,b) in note 48; Ebert and

Hamilton (1996) in note 8; Jokela *et al.* (1997*b*) in note 45; Kelley and Kirkley and Park and Kelley in note 49.

59. M. J. Keeling and D. A. Rand, A spatial mechanism for the evolution and maintenance of sexual reproduction, *Oikos* **74**, 414–24 (1995).

60. S. A. Levin, L. A. Segel, and F. R. Adler, Diffuse coevolution in plant–herbivore communities, *Theoretical Population Biology* **13**, 171–91 (1990); Frank (1993*a*) in note 13; S. A. Frank, Evolution of host–parasite diversity, *Evolution* **47**, 1721–32 (1993*b*); G. D. Ruxton, Low levels of immigration between chaotic populations can reduce system extinctions by inducing asynchronous regular cycles, *Proceedings of the Royal Society of London B* **256**, 189–93 (1994); A. B. Korol, I. A. Preygel, and S. I. Preygel, *Recombination, Variability and Evolution* (Chapman and Hall, London, 1994); V. Andreasen and F. B. Christiansen, Slow coevolution of a viral pathogen and its diploid host, *Philosophical Transactions of the Royal Society of London B* **348**, 341–54 (1995); U. Dieckmann, P. Marrow, and R. Law, Evolutionary cycling in predator–prey interactions: population dynamics and the Red Queen, *Journal of Theoretical Biology* **176**, 91–102 (1995); J. D. Van de Laan and P. Hogeweg, Predator–prey coevolution: interactions across different timescales, *Proceedings of the Royal Society of London B* **259**, 35–42 (1995); P. Marrow, U. Dieckmann, and R. Law, Evolutionary dynamics in a predator–prey system: an ecological perspective, *Journal of Mathematical Biology* **34**, 556–78 (1996); S. Gavrilets, Coevolutionary chase in exploiter–victim systems with polygenic characters, *Journal of Theoretical Biology* **186**, 527–34 (1997); A. I. Khibnik and A. S. Kondrashov, Three mechanisms of Red Queen dynamics, *Proceedings of the Royal Society of London B* **264**, 1049–56 (1997); S. Gandon, D. Ebert, I. Olivieri, and Y. Michalakis, Differential adaptation in spatially heterogeneous environments and host–parasite evolution, in S. Mopper and S. Y. Strauss (ed.), *Genetic Structure and Local Adaptations in Natural Insect Populations*, pp. 325–42 (New York, Chapman and Hall, 1998); K. McCann, A. Hastings, and G. R. Huxel, Weak trophic interactions and the balance of nature, *Nature* **395**, 794–8 (1998); G. A. Polis, Stability is woven by complex webs, *Nature* **395**, 744–5 (1998).

61. See other references to Korol and the Haifa team in the main chapter (16).

62. J. G. Vernon, B. Okamura, C. S. Jones, and L. R. Noble, Temporal patterns of clonality and parasitism in a population of fresh-water bryozoans, *Proceedings of the Royal Society of London B* **263**, 1313–18 (1996).

63. See also Dybdahl and Lively (1996) in note 45.

64. Brown (1994) in note 18.

65. M. S. Wolfe, Trying to understand and control powdery mildew, *Plant Pathology* **33**, 451–66 (1984).

66. M. A. Parker, Disequilibrium between disease-resistance variants and allozyme loci in an annual legume, *Evolution* **42**, 239–47 (1988); F. Libert, P. Cochaux, G. Beckman, M. Samson, M. Aksenova, A. Cao, *et al.*, The Delta CCR5 mutation conferring protection against HIV-1 in Caucasian populations has a single and recent origin in Northeastern Europe, *Human Molecular Genetics* **7**, 399–406 (1998).

67. P. R. Moorcroft, S. D. Albon, J. M. Pemberton, I. R. Stevenson, and T. H. Clutton-Brock, Density-dependent selection in a fluctuating ungulate population, *Proceedings of the Royal Society of London B* **263**, 31–8 (1996).

68. F. M. D. Gulland, S. D. Albon, J. M. Pemberton, P. R. Moorcroft, and T. H. Clutton-Brock, Parasite-associated polymorphism in a cyclic ungulate population, *Proceedings of the Royal Society of London B* **254**, 7–13 (1993).

69. S. A. Frank, Spatial variation in coevolutionary dynamics, *Evolutionary Ecology* **5**, 193–217 (1991); S. A. Frank, Models of plant–pathogen coevolution, *Trends in genetics* **8**, 213–19 (1992); Frank (1996) in note 8.

70. See Chapter 8, also I. Hanski, T. Pakkala, M. Kuusaari, and G. C. Lei, Metapopulation persistence of an endangered butterfly in a fragmented landscape, *Oikos* **72**, 21–8 (1995); E. Ranta and V. Kaitala, Travelling waves in vole population dynamics, *Nature* **390**, 456 only (1997).

71. M. Boerlijst, M. E. Lamers, and P. Hogeweg, Evolutionary consequences of spiral waves in a host-parasitoid system, *Proceedings of the Royal Society of London B* **153**, 15–18 (1993); A. White, M. Begon, and R. G. Bowers, Host–pathogen systems in a spatially patchy environment, *Proceedings of the Royal Society of London B* **263**, 325–32 (1996); S. Nee, R. M. May, and M. P. Hassell. Two species metapopulation models, in I. Hanski and M. E. Gilpin (ed.), *Metapopulation Biology: Ecology, Genetics and Evolution*, pp. 123–47 (Academic Press, London, 1997).

72. Frank (1991, 1992) in note 69; Frank (1993*a*) in note 13; Frank (1993*b*) in note 60; and Ladle *et al.* (1993) and Judson (1995) in note 9.

73. For example, Maddox and Root (1990) in note 4.

74. Ladle *et al.* (1993) and Judson (1995) in note 9; O. P. Judson, A model of asexuality and clonal diversity: cloning the Red Queen, *Journal of Theoretical Biology* **186**, 33–40 (1997).

75. T. F. Thingstad, M. Heldal, G. Bratbak, and I. Dundas, Are viruses important partners in pelagic food webs?, *Trends in Ecology and Evolution* **8**, 209–13 (1993); G. Turner, Intelligence and the X chromosome, *Lancet* **347**, 1814–15 (1996).

76. P. McGrath, Lethal hybrid decimates harvest, *New Scientist* **155**, 8 only (30 August 1997). This paper presents evidence that two species of virus that attack manioc in Africa can hybridize and thereby acquire extra virulence. It may even be that, if sex is defined as symmetrical exchange, retroviruses actually have 'better' sex by eukaryote standards than do bacteria (H. M. Temin, Sex and Recombination in retroviruses, *Trends in Genetics* **7**, 71–4 (1991); R. J. Redfield, M. R. Schrag, and A. M. Dean, The evolution of bacterial transformation: sex with poor relations, *Genetics* **146**, 27–38 (1997)). In the group including the AIDS virus (SIVs and of HIVs), for example, recombinants generated in doubly parasitized hosts appear to be playing a part in long-term evolution (F. Gao, E. Bailes, D. L. Robertson, Y. L. Chen, C. M. Rodenburg, S. F. Michael, *et al.*, Origin of HIV-1 in the chimpanzee *Pan troglodytes troglodytes*, *Nature* **397**, 436–41 (1999)). As victims we tend to think of this sexuality of viruses as mainly important to virus adaptation in large hosts but virology has also found a range of types of parasitic, incomplete viruses (P. E. Turner and L. Chao, Prisoner's dilemma in an RNA virus, *Nature* **398**, 441–3 (1999)), besides an assortment of RNA and DNA 'bits', and against all these a complete virus may itself need recombining defences.

77. S. E. Kelley, Viruses and the advantages of sex in *Anthoxanthum odoratum*: a review, *Plant Species Biology* **8**, 217–23 (1993); S. Balachandran, V. M. Hurry, S. E. Kelley, C. B. Osmond, S. A. Robinson, J. Rohozinski, *et al.*, Concepts of plant biotic stress: some insights into the stress physiology of virus-infected plants, from the perspective of photosynthesis, *Physiologia Plantarum* **100**, 203–13 (1997).

78. R. Y. Lamb and R. B. Willey, Are parthenogenetic and related bisexual insects equal in fertility?, **313**, 774–5 (1979).

79. P. Bierzychudek, Pollinators increase the cost of sex by avoiding female flowers, **68**, 444–7 (1987); Kelley (1994) in note 49; R. J. Redfield, Male mutation rates and the cost of sex for females, *Nature* **369**, 145–6 (1994); Jokela *et al.* (1997*a*) in note 45.

80. B. Kempenaers, G. R. Verheyren, and A. A. Dhondt, Extrapair paternity in the blue tit (*Parus caeruleus*): female choice, male characteristics, and offspring quality, *Behavioral Ecology* **8**, 481–92 (1997); and see Chapter 17.

81. M. Petrie, C. Doums, and A. P. Møller, The degree of extra-pair paternity increases with genetic variability, *Proceedings of the National Academy of Sciences, USA* **95**, 9390–5 (1998).

82. H. Ellegren, L. Gustafsson, and B. C. Sheldon, Sex ratio adjustment in relation to paternal attractiveness in a wild bird population, *Proceedings of the National Academy of Sciences, USA* **93**, 11723–8 (1996); K. Lagesan and

I. Folstad, Antler asymmetry and immunity in reindeer, *Behavioral Ecology and Sociobiology* **44**, 135–42 (1998); D. L. Marshall, Pollen donor performance can be consistent across maternal plants in wild radish (*Raphanus sativus*, Brassicaceae): a necessary condition for the action of sexual selection, *American Journal of Botany* **85**, 1389–97 (1998); D. Penn and D. Potts, Chemical signals and parasite mediated sexual selection, *Trends in Ecology and Systematics* **13**, 391–6 (1998).

83. W. K. Potts, C. J. Manning, and E. K. Wakeland, Mating patterns in semi-natural populations of mice influenced by MHC genotype, *Nature* **352**, 619–21 (1991); W. K. Potts, C. J. Manning, and E. K. Wakeland, The role of infectious disease, inbreeding and mating preferences in maintaining MHC diversity: an experimental test, *Philosophical Transactions of the Royal Society of London B* **346**, 369–78 (1994); C. Wedekind, T. Seebeck, F. Bettens, and A. J. Paepke, MHC-dependent mate preferences in humans, *Proceedings of the Royal Society of London B* **260**, 245–9 (1995); C. Wedekind, M. Chapuisat, E. Macas, and T. Rülicke, Non-random fertilisation in mice correlates with the MHC and something else, *Heredity* **77**, 400–9 (1996); C. Wedekind and S. Füri, Body odour preferences in men and women: do they aim for specific MHC combination or simply heterozygosity?, *Proceedings of the Royal Society of London B* **264**, 1471–9 (1997).

84. B. Charlesworth, Mutation-selection balance and the evolutionary advantage of sex and recombination, *Genetical Research, Cambridge* **55**, 199–221 (1990); D. Charlesworth, E. E. Lyons and L. B. Litchfield, Inbreeding depression in two highly inbred populations of *Leavenworthia*, *Proceedings of the Royal Society of London B* **258**, 209–14 (1994); P. D. Keightley, The distribution of mutation effects on viability in *Drosophila*, *Genetics* **138**, 1315–22 (1994); M. O. Johnston and D. J. Schoen, Mutation rates and dominance levels of genes affecting total fitness in 2 angiosperm species, *Science* **267**, 226–9 (1995); P. D. Keightley, Nature of deleterious mutation load in *Drosophila*, *Genetics* **144**, 1993–9 (1996); H.-W. Deng and M. Lynch, Inbreeding depression and inferred deleterious-mutation parameters in *Daphnia*, *Genetics* **147**, 147–55 (1997); P. D. Keightley and A. Caballero, Genomic mutation rates for lifetime reproductive output and lifespan in *Caenorhabditis elegans*, *Proceedings of the National Academy of Sciences, USA* **94**, 3823–7 (1997); J. W. Drake, B. Charlesworth, D. Charlesworth, and J. F. Crow, Rates of spontaneous mutation, *Genetics* **148**, 1667–86 (1998).

85. J. A. G. M. De Visser, Deleterious mutations and the evolution of sex, thesis, Wageningen (CIP-Data Koninklijke Biblioheek, Den Haag, 1996); J. A. G. M. De Visser, R. F. Hoekstra, and H. van den Ende, The effect of sex and

deleterious mutations on fitness in *Chlamydomonas*, *Proceedings of the Royal Society of London B* **263**, 193–200 (1996);S. F. Elena and R. E. Lenski, Test of synergistic interactions among deleterious mutations in bacteria, *Nature* **390**, 395–8 (1997); D. Greig, R. H. Bots, and E. J. Louis, The effect of sex on adaptation to high temperature in heterozygous and homozygous yeast, *Proceedings of the Royal Society of London B* **265**, 1017–23.

86. P. Bierzychudek, Patterns in plant parthenogenesis, *Experientia* **41**, 1255–64 (1985); but see also R. A. Bingham, Efficient pollination of alpine plants, *Nature* **391**, 238–9 (1998); D. B. O. Savile, Arctic adaptations in plants, *Canada Department of Agriculture, Research Branch Monographs* **6**, 1–81 (1972). For mosses see P. Convey and R. I. L. Smith, Investment in sexual reproduction by Antarctic mosses, *Oikos* **68**, 293–302 (1993). For arthropods (crustaceans, insects, and mites) see J. A. Downes, What is an arctic insect?, *Canadian Entomologist* **94**, 143–62 (1964*a*); J. A. Downes, Arctic insects and their environment, *Canadian Entomologist* **96**, 279–307 (1964*b*),

87. Downes (1964*a,b*) in note 86; L. J. Weider and P. D. N. Hebert, Microgeographic genetic heterogeneity of melanic *Daphnia pulex* at a ow-arctic site, *Heredity* **58**, 391–9 (1987).

88. Kelley (1994) in note 49.

89. Kelley (1993) in note 77.

90. M. A. Parker, Pathogens and sex in plants, *Evolutionary Ecology* **8**, 560–84 (1994); but see S. A. Frank, Problems of inferring the specificity of plant-pathogen genetics (a reply to Parker), *Evolutionary Ecology* **10**, 323–5 (1996).

91. N. W. Simmonds, Genetics of horizontal resistance to diseases of crops, *Biological Reviews* **66**, 189–241 (1991).

92. C. Svanborg-Eden and B. R. Levin, Infectious disease and natural selection in human populations: a critical reexamination, in A. C. Swedlund and G. J. Armelagos (ed.), *Disease in Populations in Transition*, pp. 31–41 (Bergin and Garvey, New York, 1991).

93. S. C. H. Barrett, The reproductive biology and genetics of island plants, in P. R. Grant (ed.), *Evolution on Islands*, pp. 18–34 (Oxford University Press, Oxford, 1998).

94. O. Halvorsen, Epidemiology of reindeer parasites, *Parasitology Today* **2**, 334–9 (1986).

95. R. L. Kepner, R. A. Wharton, and C. A. Suttle, Viruses in Antarctic lakes, *Limnology and Oceanography* **43**, 1754–61 (1998).

96. But see Halvorsen (1986), note 94; J. A. Barrett, Frequency-dependent selection in plant–fungal interactions, *Philosophical Transactions of the Royal*

Society of London B **319**, 473–83 (1988); P. J. Hudson, A. P. Dobson, and D. Newborn, Prevention of population cycles by parasite removal, *Science* **282**, 2256–8 (1999). See also K.-A. Rorvik and J. B. Steen, The genetic structure of Scandinavian Willow Ptarmigan (*Lagopus lagopus lagopus*), *Hereditas* **110**, 139–44 (1989).

97. Hanley *et al.* (1995a,b) and Weeks (1996) in note 57.

98. Hanley *et al.* (1995 a,b) in note 57; R. R. Radtkey, S. C. Donnellan, R. N. Fisher, C. Mortiz, K. A. Hanley, and T. J. Case, When species collide: the origin and spread of an asexual species of gecko, *Proceedings of the Royal Society of London B* **259**, 145–52 (1995).

99. D. R. Taylor and L. W. Aarrssen, Complex competitive relationships among genotypes of three perennial grasses: implications for species coexistence, *American Naturalist* **136**, 305–27 (1990).

Credits

—

We thank the publishers for permission to reproduce the copyright material listed below:

CHAPTER 1
'Fluctuation of environment and coevolved antagonist polymorphism as factors in the
maintenance of sex', Ch. 22 in R. D. Alexander and D. W. Tinkle (ed.), *Natural Selection and
Social Behavior*, pp. 363–81 (Chiron Press, New York, 1981).
© 1981 Chiron Press.

CHAPTER 2
'Sex versus non-sex versus parasite', *Oikos*, Vol. 35, pp. 282–90 (1980).
© 1980 Oikos.

CHAPTER 3
'Coefficients of relatedness in sociobiology', *Nature*, Vol. 288, 694–7 (1981).
© 1980 Macmillan Journals Ltd.

CHAPTER 4
'The evolution of co-operation', *Science*, Vol. 211, pp. 1390–6 (1981).
© 1981 American Association for the Advancement of Science.

CHAPTER 5
'Pathogens as causes of genetic diversity in their host populations', Ch. 5 in
R. M. Anderson and R. M. May (ed.), *Population Biology of Infectious Diseases*, Dahlem
Konferencen, 1982, pp. 269–96 (Springer-Verlag, Berlin, 1981).
© 1981 Springer-Verlag Berlin.

CHAPTER 6
'Heritable true fitness and bright birds: a role for parasites', *Science*, Vol. 218,
pp. 384–7 (1982).
© 1982 American Association for the Advancement of Science.

CHAPTER 7
'Parent–offspring correlation in fitness under fluctuating selection', *Proceedings of the
Royal Society of London B*, Vol. 222, pp. 1–14 (1984).
© 1984 The Royal Society.

CHAPTER 8
'Instability and cycling of two competing hosts with two parasites', in S. Karlin and
E. Nevo (ed.), *Evolutionary Processes and Theory*, pp. 645–68 (Academic Press, New York, 1986).
© 1986 Academic Press, Inc.

CHAPTER 9
'Discriminating nepotism: expectable, common, overlooked', Ch. 13 in D. J. C. Fletcher
and C. D. Michener (ed.), *Kin Recognition in Animals*, pp. 417–37
(Wiley, New York, 1987).
© 1987 John Wiley & Sons Ltd.

CHAPTER 10

'Kinship, recognition, disease, and intelligence: constraints of social evolution', Ch. 6
in Y. Itô, J. L. Brown, and J. Kikkawa (ed.), *Animal Societies: Theories and Facts*, pp. 81–102
(Japan Scientific Societies Press, Tokyo, 1987).
© 1987 Japan Scientific Societies Press.

CHAPTER 11

'Parasites and sex', Ch. 11 in R. E. Michod and B. R. Levin (ed.), *The Evolution of Sex:
An Examination of Current Ideas*, pp. 176–93 (Sinauer Associates, Sunderland, Mass., 1988).
© 1988 Sinauer Associates, Inc.

CHAPTER 12

'Sex and disease', Ch. 4 in G. Stevens and R. Bellig (ed.), *Nobel Conference XXIII:
The Evolution of Sex*, pp. 65–99 (Harper and Row, San Francisco, 1988).
© 1988 Harper and Row, Inc.

CHAPTER 13

'This week's citation classic', *Current Contents*, No. 40, p. 16 (1988).
© 1988 ISI® Current Contents®.

CHAPTER 14

'Selfishness re-examined: no man is an island', *Behavioral and Brain Sciences*,
Vol. 12, pp. 699–700 (1989).
© 1989 Cambridge University Press.

CHAPTER 15

'Memes of Haldane and Jayakar in a theory of sex', *Journal of Genetics*,
Vol. 69, pp. 17–32 (1990).
© Printed in India.

CHAPTER 16

'Sexual reproduction as an adaptation to resist parasites (a review)', *Proceedings of the
National Academy of Sciences, USA*, Vol. 87, pp. 3566–73 (1990).
© 1990 National Academy of Sciences.

CHAPTER 17

'Mate choice near or far', *American Zoologist*, Vol. 30, pp. 341–52 (1990).
© 1990 Society for Integrative and Comparative Biology.

CHAPTER 18

'The seething genetics of health and the evolution of sex', Ch. 3.3 in S. Osawa and
T. Honjo (ed.), *Evolution of Life: Fossils, Molecules, and Culture*, pp. 229–51
(Springer-Verlag, Tokyo, 1991).
© 1991 Springer-Verlag Tokyo.

NAME INDEX

SUBJECT INDEX